REGULATION OF SMOOTH MUSCLE CONTRACTION

ADVANCES IN EXPERIMENTAL MEDICINE AND BIOLOGY

Recent Volumes in this Series

Volume 300
MECHANISMS AND SPECIFICITY OF HIV ENTRY INTO HOST CELLS
Edited by Nejat Düzgüneş

Volume 301
MECHANISMS OF ANESTHETIC ACTION IN SKELETAL, CARDIAC, AND SMOOTH MUSCLE
Edited by Thomas J. J. Blanck and David M. Wheeler

Volume 302
WATER RELATIONSHIPS IN FOODS: Advances in the 1980s and Trends for the 1990s
Edited by Harry Levine and Louise Slade

Volume 303
IMMUNOBIOLOGY OF PROTEINS AND PEPTIDES VI:
Human Immunodeficiency Virus, Antibody Immunoconjugates, Bacterial Vaccines, and Immunomodulators
Edited by M. Zouhair Atassi

Volume 304
REGULATION OF SMOOTH MUSCLE CONTRACTION
Edited by Robert S. Moreland

Volume 305
CHEMOTACTIC CYTOKINES: Biology of the Inflammatory Peptide Supergene Family
Edited by J. Westwick, I. J. D. Lindley, and S. L. Kunkel

Volume 306
STRUCTURE AND FUNCTION OF THE ASPARTIC PROTEINASES: Genetics, Structures, and Mechanisms
Edited by Ben M. Dunn

Volume 307
RED BLOOD CELL AGING
Edited by Mauro Magnani and Antonio De Flora

REGULATION OF SMOOTH MUSCLE CONTRACTION

Edited by

Robert S. Moreland

Bockus Research Institute
The Graduate Hospital
and
The University of Pennsylvania School of Medicine
Philadelphia, Pennsylvania

PLENUM PRESS • NEW YORK AND LONDON

Library of Congress Cataloging in Publication Data

Research Symposium on Regulation of Smooth Muscle: Progress in Solving the Puzzle (1990: Philadelphia, Pa.)
Regulation of smooth muscle contraction / edited by Robert S. Moreland.
p. cm.—(Advances in experimental medicine and biology; v. 304)
"Proceedings of the Sixth Annual Graduate Hospital Research Symposium: Regulation of Smooth Muscle: Progress in Solving the Puzzle, held September 24-26, 1990, in Philadelphia, Pennsylvania"—T.p. verso.
Includes bibliographical references and index.
ISBN 0-306-44041-5
1. Smooth muscle—Congresses. 2. Muscle contraction—Regulation—Congresses. I. Moreland, Robert S. II. Title. III. Series.
[DNLM: 1. Calcium—physiology—congresses. 2. Muscle Contraction—physiology—congresses. 3. Muscles, Smooth—physiology—congresses. W1 AD559 v. 304 / WE 500 R4325r 1990]
QP321.R376 1990
591.1′852—dc20
DNLM/DLC 91-29113
for Library of Congress CIP

Proceedings of the Sixth Annual Graduate Hospital Research Symposium:
Regulation of Smooth Muscle—Progress in Solving the Puzzle,
held September 24-26, 1990, in Philadelphia, Pennsylvania

ISBN 0-306-44041-5

A Division of Plenum Publishing Corporation
233 Spring Street, New York, N.Y. 10013

Printed in the United States of America

Preface

Sixth Annual Graduate Hospital Research Symposium

REGULATION OF SMOOTH MUSCLE

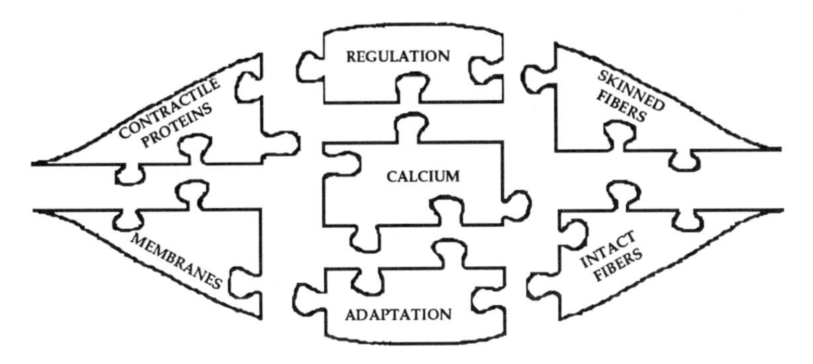

PROGRESS IN SOLVING THE PUZZLE

Every so often a scientific conference comes at a time when everyone has new and exciting information, when old "dogmas" do not seem to be as well established, and when speakers and participants alike are ready to challenge interpretations of old and new experimental data. This was such a conference.

What turns on a smooth muscle cell? The precise answer to this question has eluded scientists for much longer than I have been involved in the field. We know that an increase in cytosolic calcium is necessary and we know that phosphorylation of the 20 kDa myosin light chain is an important step in the process. We do not know if other processes are necessary for the initiation and/or maintenance of a smooth muscle contraction nor do we know if other processes modulate the regulation of contraction. The goal of the symposium on which this volume is based was to explore the most current hypotheses for the answers to these questions. I believe that after reading the chapters included in this volume, you will agree that this goal was achieved.

The importance of calcium and calmodulin dependent myosin light chain phosphorylation in the regulation of smooth muscle contraction was reinforced by many presentations. However, the status of myosin light chain phosphorylation as a simple calcium dependent switch came under serious suspicion. One underlying theme that could be discerned from this conference, and is reflected in the chapters in this volume, is that the response of smooth muscle to a stimulus can be dramatically altered by intracellular processes. This theme is exemplified by the strength of the arguments suggesting (demonstrating?) a role for thin filament regulation, and results clearly demonstrating the physiological alteration in myofilament calcium sensitivity. There are also new data adding credence to a role for protein kinase C in contraction and the intriguing possibility of cooperativity in crossbridge activation.

The complex dynamics of calcium handling were also critically addressed at this conference. New information concerning the mechanisms involved in pharmacomechanical and electromechanical coupling were presented. Studies are included in this volume which describe the specific characteristics of both voltage-dependent and receptor-operated calcium channels, the molecular basis of calcium release from the sarcoplasmic reticulum, the possible role of calcium oscillations in cellular signalling, and the effects of the cellular calcium "history" on calcium metabolism. This information clearly demonstrates that calcium movements involve the integration of numerous interacting mechanisms.

One must not forget that the smooth muscle cell possesses the ability to adapt to a variety of changes in its environment. This adaptive ability was addressed by several presentations. Adaptations occur in the membranes, in the control of cytosolic calcium, and in both the function and structure of the contractile proteins. Thus, the level of complexity involved in the adaptive ability of this cell equals if not exceeds that we know to be involved in the normal functioning of this important muscle.

The mechanisms responsible for determining the level of smooth muscle tone are integrated in such a way that the level of precision achieved is remarkable. My final thoughts as I was leaving this conference and during the time I was editing this volume were of the tremendous amount of knowledge we have concerning the regulation of smooth muscle and yet how little we really know. Our charge over the next few (hopefully not many) years is to use this knowledge to develop models and hypotheses that encompass all of the available information no matter how seemingly insignificant or aberrant. This is a difficult task and I look forward with great anticipation to the new insights that will be obtained concerning this fascinating tissue.

In closing, I would like to express my sincere thanks to Dr. Suzanne Moreland, Ms. Jacqueline Cilea, and Ms. Laura A. Trinkle for their significant contributions in the planning and organization of this symposium and for helping me maintain a semblance of sanity during the process. I would also like to acknowledge the session chairmen, Drs. David J. Hartshorne, James T. Stull, Andrew P. Somlyo, and Robert H. Cox for helping to develop a strong program. Lastly, I would like to recognize the most important contributors to the symposium, the participants. Without the excitement and interest of the audience, the symposium would not have succeeded and this volume would not have surfaced.

Robert S. Moreland, Ph.D.

Support of the following patrons is gratefully acknowledged:

Corporate Sponsors

Hässle/Astra Cardiovascular

Merck Sharp & Dohme

SmithKline Beecham Pharmaceuticals

Sponsors

American Cyanamid Company

Berlex Laboratories, Inc.

Boehringer Ingelheim Pharmaceuticals, Inc.

Bristol-Myers Squibb Pharmaceutical Research Institute

Burroughs Wellcome Company

CIBA-GEIGY Corporation

Glaxo Research Laboratories

ICI Pharmaceutical Group

Lilly Research Laboratories

Merck Sharp & Dohme Research Laboratories

Merrell Dow Research Institute

Miles Inc./Bayer AG

Monsanto Company

Optical Apparatus

Rhône-Poulenc Rorer Central Research

R.W. Johnson Pharmaceutical Research Institute

Schering-Plough Research

SmithKline Beecham Research and Development

Sterling Drug Inc.

Wyeth-Ayerst Research

Contents

INTACT FIBER STUDIES

EXCITATION-CONTRACTION COUPLING

ADAPTATION

POSTER PRESENTATIONS

TRIBUTE TO PROFESSOR EDITH BÜLBRING

Hirosi Kuriyama

Edith Bülbring, FRS, Emeritus Professor of Pharmacology at Oxford University passed away on July 5th, 1990 at the age of 86. During her distinguished career, she received many honors for her work on the fundamental properties of smooth muscle including: Schmiedeberg-Plakette der Deutschen Pharmakologischen Gesellschaft; honorary memberships in the Italian Pharmaceutical Society, British Physiological Society, and German Physiological Society; honorary degrees from the Universities of Gröningen (Netherlands), Leuven (Belgium), and Homburg Saar (Germany); and the Wellcome Gold Medal in Pharmacology.

Edith was born into a scholarly family in Nürnberg and was educated in Germany, moving to England only when Hitler came to power. She made England her home and lived in Oxford for many years. In 1960 she was made a Professorial Fellow of Lady Margaret Hall and worked in the Department of Pharmacology until her "retirement" in 1970.

The pioneering work on smooth muscle physiology which she did during her latter years in Oxford involved many discoveries on the fundamental properties of smooth muscle. Scientific collaborators were drawn to her from the four corners of the world, from Asia, Oceania, Europe, and America. Those who worked with her then became international authorities in this field and leaders in their own countries, and they in turn each generated scientific "grandchildren" for this remarkable woman.

Her early life as the youngest of four children was spent in Germany. It was in some ways a rather sheltered childhood. All four children of the Professor of English at Bonn were gifted artistically and linguistically. Edith spoke German, Dutch, English, and some Italian, and became a very proficient pianist. However, in spite of these talents and to her family's surprise, she took to the study of physiology and medicine. During this period of her life, her father died and her eldest brother was killed in action in the First World War.

At Munich University her medical education began in earnest. She continued her scientific career at Freiburg and Berlin, and considered at times a career as a pediatrician at Jena. But then Hitler came to power, and because her mother was Jewish, she was dismissed from her hospital post in Berlin. She went briefly to Holland where many of her family lived, but in 1933 went to England where she found a position under Professor J. H. Burn in the laboratories of the Pharmaceutical Society of Great Britain.

Thus began the second phase of her life, developing scientifically under Burn's guidance and moving with him to Oxford when he was appointed to head the Pharmacology Department. During the war she played a major part in keeping the department running and rose to Reader in 1960 and Professor in 1967, although she was disappointed not to be ultimately appointed head of the department.

In this period, her great zest for life showed in her passion for experimental science but also in her ardor in playing piano duets. She loved social occasions and meeting her friends, but had no love for the minutiae of administration.

By the age of 48, Edith had acquired a wide experience of life and science. Her determination and singlemindedness were legendary, but there was a warmth and humanity which drew those around her.

At the Tuesday evening sherry parties held in her garden or sitting room in North Oxford, her scientific collaborators would relax and enjoy discussing their ideas. Edith was in her element leading the discussions and "drawing out" shy newcomers. She enjoyed kindred spirits who were meticulous, rigorous, and uncompromising.

By then she was famous for uncovering the fundamental workings of the smooth muscles which surround and encase the hollow viscera. Her particular contributions centered on the electrical events which control and regulate contraction and relaxation, the influence of nerve activity on these electrical events, and the mechanisms underlying the events. For this work she was elected to a Fellowship of the Royal Society in 1958.

Edith never married. Her many collaborators became her family, and when they in turn grew more senior, their collaborators became her "grandchildren". They all knew and respected her, and as she grew older, she loved to hear news of their progress and large and small triumphs. Her scientific family is spread world-wide and now has a second generation. We have lost a great scientific pioneer and leader.

(Taken, in part, from "Independent" and "The Guardian" written by Professor T. B. Bolton)

STRUCTURE-FUNCTION RELATIONSHIPS IN SMOOTH MUSCLE MYOSIN LIGHT CHAIN KINASE

Masaaki Ito,[2] Vince Guerriero, Jr.,[1] and David J. Hartshorne[1]

[1]Muscle Biology Group
Department of Animal Sciences
The University of Arizona
Tucson, AZ 87521

[2]The First Medical Clinic
Mie University Hospital
Tsu-City, Mie, Japan

INTRODUCTION

Regulation of the contractile apparatus in vertebrate smooth muscle is thought to involve the phosphorylation-dephosphorylation of the two 20,000-dalton light chains of myosin (Hartshorne, 1987). It has been known for many years that contraction in smooth muscle, as in skeletal muscle, is determined by the concentration of intracellular Ca^{2+}. Contractile activity is coupled to the Ca^{2+} transients via the formation of the Ca^{2+}-calmodulin (CaM) complex and subsequent activation of the CaM-dependent enzyme, myosin light chain kinase (MLCK). Phosphorylation of the myosin light chain increases the actin-activated ATPase activity of myosin and this event is thought to be reflected, under physiological conditions, by an increased rate of cross-bridge cycling and the development of tension. As long as the intracellular Ca^{2+} concentration remains above the threshold determined by the affinity of Ca^{2+} binding to the Ca^{2+}-CaM-MLCK complex (i.e. ~ 0.5 μM) myosin remains phosphorylated and contraction persists. Reduction in the Ca^{2+} level leads to a commensurate decrease in kinase activity and varying extents of myosin dephosphorylation. It is assumed (in the absence of conflicting data) that the myosin phosphatase is unregulated and dephosphorylation occurs at the same rate both in the presence and absence of Ca^{2+}. However, the identity of the phosphatase involved in myosin dephosphorylation has not been established and one of the intriguing questions to be answered is whether phosphatase activity is regulated *in vivo*? Thus, by decreasing the MLCK activity the balance of phosphorylation to dephosphorylation is altered. In the simplest scenario the role of myosin phosphorylation is to initiate contraction and dephosphorylation leads to relaxation. While this simple relationship is not obvious under physiological conditions, and several papers in this volume attest to the complexity of the situation, it is clear that two key regulatory enzymes in smooth muscle are MLCK and myosin phosphatase.

MYOSIN LIGHT CHAIN KINASE

Molecular Plan

The influence of myosin phosphorylation on actin-activated ATPase activity was observed first with platelet myosin. Subsequently MLCK was isolated from both platelets and skeletal muscle. These studies preceded its discovery in smooth muscle (see Hartshorne, 1987 for references). MLCK is now known to exist in a wide variety of eucaryotic, non-muscle cells including fibroblasts, macrophages, lymphocytes, intestinal epithelial brush border cells, teleost retinal cones, thyroid, etc. In addition, it has been isolated from the invertebrate skeletal muscle of *Limulus* (see Hartshorne, 1987). It is reasonable to assume that the role of MLCK is coupled to phosphorylation (activation) of myosin and the induction of some mechanochemical process involving the interaction of myosin with actin, for example secretion or motility. In addition, it has been suggested (Suzuki et al., 1978; Scholey et al., 1980) that phosphorylation of myosin alters its state of aggregation, and this may be particularly important in non-muscle systems where myosin filaments are not seen.

The CaM dependence of MLCK was established with both the smooth (Dabrowska et al., 1978) and skeletal muscle forms (Yazawa et al., 1978) and it is now established that the stoichiometry of CaM to MLCK is 1. It is generally accepted that each of the four Ca^{2+}-binding sites of CaM need to be occupied in order to generate the conformation required for activation of MLCK.

Many early studies using limited proteolysis, were focused on elucidating the domain structure of MLCK, and several models were proposed. In general, these depicted an N-terminal segment of variable length, a central active site and a CaM-binding site on the C-terminal side of the active site. The identification of specific residues involved with these areas was facilitated by sequences derived from a cloned partial cDNA (Guerriero et al., 1986) and recently the complete cDNA (Olson et al., 1990) for the gizzard MLCK. Complete sequences are also available for the skeletal muscle MLCK's from rabbit (Takio et al., 1986) and rat (Roush et al., 1988).

For the gizzard MLCK, the active site is thought to span residues D^{517} to R^{762}, with the consensus ATP binding sequence (G-X-G-X-X-G) beginning at its N-terminal end at G^{526}. Based on studies with isolated and synthetic peptides the CaM-binding site was determined as A^{796} to S^{815} (Lukas et al., 1986). The two sites for phosphorylation by the cAMP-dependent protein kinase are S^{815} and S^{828}. The function of the N-terminal part of MLCK is not known, but it has been suggested for one of the early models (Mayr et al., 1983) that the N-terminal asymmetric tail may be involved in binding of MLCK to either actin or myosin. It is known that MLCK under *in vitro* conditions binds to both F-actin and myosin, although its intracellular location is not established. The N-terminal side of the active site is rich in proline residues, and the extended structure that this generates may be important in the interaction with the actin filament.

An unusual feature of the MLCK molecule is that the C-terminal part is synthesized as an independent protein in smooth muscle. The initiation site is not established but is probably M^{818}. Assuming this, the protein would be composed of 153 residues of approximate molecular weight 17,600. Since this protein contains the "tail" (Greek, *telos*) of the kinase molecule it has been termed telokin (Ito et al., 1989). In the earlier studies of Guerriero et al. (1986) a puzzling feature was that the 2.1-kb DNA (to MLCK) hybridized to two sizes of RNA: a 5.5-kb RNA, thought to be the mRNA for the complete molecule, and a smaller 2.7-kb RNA. The latter was determined to be related to the C-terminal part of the molecule, but its translation product was not detected by a polyclonal antibody to MLCK on Western blots. This was particularly surprising since the smaller mRNA was present at higher concentrations then the 5.5-kb RNA. Subsequently it was realized that if different conditions were used during the transfer to nitrocellulose a second antigen could be detected, and this was telokin.

The evidence to indicate that telokin indeed is synthesized independently of the rest of the molecule (rather than a degradation product of MLCK) is: 1) its N-terminal amino acid is blocked; 2) the concentration of telokin is at least 15 μM in gizzard (compared to about 4 μM MLCK). This is consistent with the relatively-abundant 2.7-kb RNA; and 3) no fragments corresponding to the central and N-terminal regions of MLCK are detected. These would be expected if telokin was generated via proteolysis. Thus, although it is not conclusive, it appears likely that the MLCK gene is under the control of two promoters. It is possible that regulation of each is independent since the message for telokin, not MLCK, is hormonally regulated in the oviduct (Russo et al., 1987).

To date the function of telokin is not known. Because of its acidic nature (pI ~ 4.5) and metachromatic staining with stains-all (Ito et al., 1989) it was suspected to bind cations. As shown in Figure 1, it does not bind Ca^{2+} with high affinity and clearly the Ca^{2+}-binding is competitive with Mg^{2+}. It is possible that its interaction with Mg^{2+} may be related to function but this remains to be determined. A Scatchard plot of the binding data in the absence

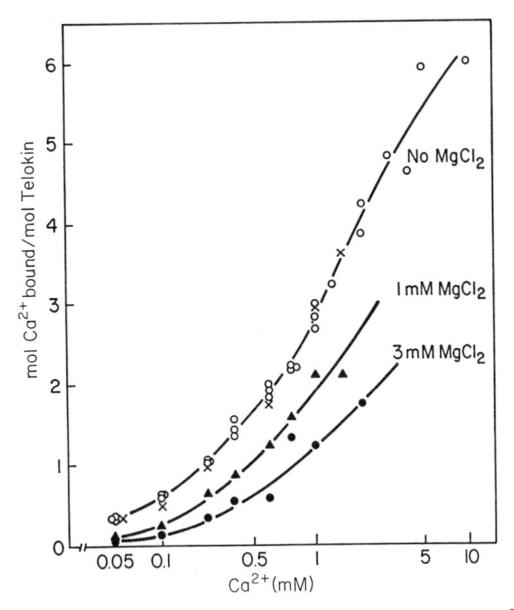

Figure 1. Ca^{2+} binding by telokin at different concentrations of Mg^{2+}. Conditions: 30 mM imidazole (pH 7.0), 20 mM KCl, 1 mg/ml telokin, $^{45}CaCl_2$ as indicated, $MgCl_2$ as indicated. Bound and free Ca^{2+} determined following dialysis overnight at 4° C. Dephosphorylated telokin used except (X) which was telokin phosphorylated to 0.8 mol P_i/mol telokin by cAMP-dependent protein kinase.

of Mg^{2+} showed two classes of binding sites of different affinity. The higher affinity class had a K_D of 540 μM and an n value of 4 mol Ca^{2+} per mol telokin. From the data shown in Figure 1 it is also clear that phosphorylation of telokin by cAMP-dependent protein kinase has no effect on the Ca^{2+}-binding properties of telokin.

Autoinhibitory Domain

CaM-dependent enzymes are usually inactive in the absence of Ca^{2+}-CaM and this raises the interesting question of how the apoenzyme is maintained in the inactive state. For the skeletal muscle MLCK the inactivity of the apoenzyme was suggested to reflect inhibition via an inhibitory domain (Edelman et al., 1985). For the gizzard MLCK this situation may also occur, but it has been suggested that the recognition by the active site of the inhibitory domain is due to the latter's similarity to the light chain substrate, leading to the term pseudosubstrate domain (Kemp et al., 1987). In the apoenzyme the active site would be occupied by the pseudosubstrate domain and the binding of Ca^{2+}-CaM would release this interaction. This hypothesis is based on similarities in juxtaposition of the basic residues in both the MLCK sequence and the N-terminal sequence of the 20,000-dalton light chain, where H^{805} of MLCK is aligned with S^{19} of the light chain. This concept is illustrated in Figure 2. The pseudosubstrate domain would therefore extend over 19 residues from S^{787} to H^{805}. Strong support for this idea comes from the observation that the synthetic peptide 783-804 inhibits MLCK activity even at saturating CaM concentrations (Kemp et al., 1987). This peptide also is a potent inhibitor of the constitutively active CaM-independent kinase fragment (Ikebe et al., 1987).

The concept of autoinhibition of a kinase via a sequence similar to its substrate is not unique to MLCK and in fact was proposed initially for cAMP-dependent protein kinase (for review see Taylor, 1989) and has subsequently been proposed for other protein kinases, i.e. protein kinase C (House and Kemp, 1987) and CaM-dependent protein kinase II (Hanley et al., 1987; Payne et al., 1988). Attractive though this concept is, it is not universally accepted for MLCK and Ikebe et al. (1989) have suggested that the inhibitory domain of gizzard MLCK is contained within the sequence Asp^{777} to Lys^{793}, as shown in Figure 2. Obviously this sequence is close to the pseudosubstrate sequence, and

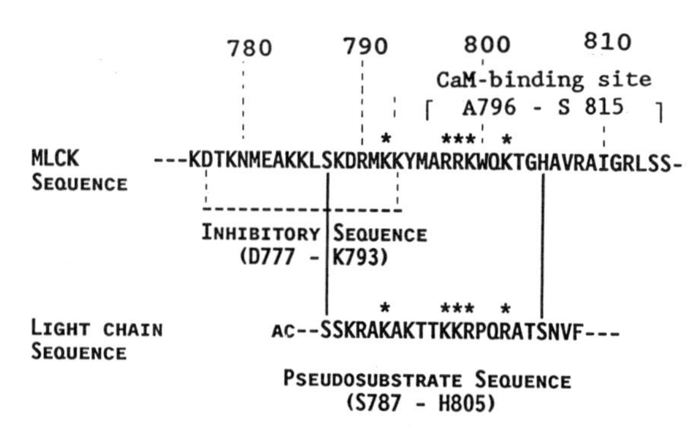

Figure 2. Diagram indicating various sequences in MLCK. Shown are the CaM-binding site, the inhibitory sequence of Ikebe et al. (1989), and the pseudosubstrate sequence. For the latter, the MLCK sequence is aligned to the pertinent segment of the 20,000-dalton light chain sequence. Similarities in positions of basic residues indicated by asterisks.

in fact overlaps for several residues, but it is not based on the pseudosubstrate concept, and as such poses a challenge to the latter.

Studies on the limited proteolysis of MLCK (Ikebe et al., 1987) were important to our understanding of the autoinhibitory mechanism. It was shown that on tryptic hydrolysis, the CaM-dependent activity of MLCK decreased progressively to yield an inactive kinase fragment (IKF). Further hydrolysis by trypsin caused an increase in activity but this was CaM-independent. During this phase the constitutively-active kinase fragment (CAKF) was formed. Neither IKF nor CAKF bound CaM. These results were interpreted as follows: 1) initially the CaM-binding site was removed by proteolysis but the inhibitory sequence was retained. This generates an unregulated kinase fragment frozen in the inhibited state. The activity of IKF was determined to be about 0.1% of that of the native enzyme (Ikebe et al., 1987). Theoretically this value should and does approach that observed for the apoenzyme; 2) on further proteolysis the inhibitory site is removed and this generates an unregulated kinase fragment frozen in the active state. Theoretically the activity of CAKF should equal that of the holoenzyme. In practice, the activity of CAKF is slightly less than that of the native enzyme plus Ca^{2+}-CaM. In addition, it was found that neither fragment bound to actin and therefore additional cleavage sites would be predicted towards the N-terminal half of the molecule. The molecular weight of IKF and CAKF were estimated as 64- and 61-kDa, respectively, and the inhibitory region must be contained with a 3-kDa span of the enzyme. Obviously this segment is large enough to accommodate both the inhibitory sequence and the pseudosubstrate sequence (see Fig. 2).

Attempts to further define the autoinhibitory region using other proteases were not eminently successful, although these studies did provide some useful information. It was shown (Ito et al., 1989) that thermolysin generated only IKF and this proved useful for the preparation of this fragment. Endoproteinase Lys-C generated both IKF and CAKF. (Other proteases were tried but did not offer any obvious advantages). The objective of these studies was to determine the N- and C-terminal boundaries of each fragment and in this way to define limits of the autoinhibitory sequence. It was found (Olson et al., 1990) that R^{808} is the C-terminal residue of the tryptic IKF and K^{779} the C-terminal residue of CAKF (Ito et al., 1989). For the other IKF fragments the C-terminal residues were A^{806} and K^{802} for thermolysin and endoproteinase Lys-C, respectively. Analysis of CAKF fragments from thermolysis and endoproteinase Lys-C, gave C-terminal residues that were to the N-terminal side of the tryptic cleavage site and therefore were not useful. From these results it can be concluded that the conversion of IKF to CAKF is due to proteolysis entirely on the C-terminal end of MLCK and the inhibitory region is contained within the sequence N^{780} (tryptic CAKF) and K^{802} (endoproteinase Lys-C IKF). This stretch of 23 residues is too long to discriminate between the inhibitory sequence of Ikebe et al. (1989) and the pseudosubstrate hypothesis.

Mutagenesis of MLCK

One method to define more precisely those residues involved in autoinhibition is by application of site-directed mutagenesis. In these preliminary studies only truncated mutants were expressed. The area of the molecule that was focused on was that indicated from the proteolysis experiments, namely to the C-terminal side of K^{779}. This region includes the autoinhibitory sequence, the CaM-binding site, and telokin.

The expression vectors that were used were mostly pRIT2T (Pharmacia LKB Biotechnology) and to a lesser extent pET-3a (Studier et al., 1990). The advantage of the former is that this vector contains a fragment coding for Protein A, that has IgG binding ability, followed by the pUC9 polylinker region for the insertion of cDNAs. A cDNA inserted in the correct reading frame and orientation will be expressed as a fusion protein containing the segment of Protein A. This facilitates purification of the mutants using a monoclonal IgG_1 affinity column. The purpose of the second expression vector was to determine whether the inclusion of a fragment of Protein A altered markedly the biological properties of the mutants. Where possible the expressed proteins were purified by both CaM-affinity and IgG_1-affinity chromatography and where mutants lacked either the CaM-binding site or the Protein A fragment only one affinity column was used. The majority of the expressed protein was incorporated into inclusion bodies and was not soluble. Attempts to solubilize and renature this protein have failed. Consequently the amount of available protein was limited and therefore only enzymatic parameters for the mutants were determined.

Each mutant was initiated at L^{447} and therefore the mutants are designated according to their C-terminal amino acid. These were: pMK.Glu^{972}; pMK.Trp^{800}, pMK.Lys^{793}, pMK.Thr^{778}, and pET.Glu^{972}.

The two "complete" mutants, pMK.Glu^{972} and pET.Glu^{972} contained autoinhibitory and CaM-binding domains and were predicted to be CaM-dependent. This was observed and there was no detectable difference between the pMK- and pET-systems, indicating that the Protein A fragment did not influence kinase activity. The specific activity of the two mutants in the presence of Ca^{2+} and CaM was between 2.2 to 2.6 μmol P_i transferred/min.mg^{-1} (based on the percentage of the pertinent band on SDS-electrophoresis). The three truncated mutants each possessed lower kinase activity, with pMK.Trp^{800} being considerably lower than the others. For each of the five mutants K_m values for ATP and light chain were constant and were similar to values obtained with the native enzyme. The important distinction between these expressed proteins was that the three truncated mutants were constitutively-active and were not dependent on Ca^{2+}-CaM.

Unfortunately, however, the finite level of kinase activity obtained with the pMK.Trp^{800} mutant was too high to allow an unequivocal decision concerning the presence or absence of the autoinhibitory sequence. Theoretically, based on either of the two proposals concerning the inhibitory domain, pMK.Trp^{800} should have been inactive. The failure to realize this is not understood but may be due to either the presence of an incomplete inhibitory sequence (i.e. residues C-terminal to W^{800} may be involved), incorrect folding of the expressed protein, or a limited extent of proteolysis occurring during isolation. Whatever the reason, a second procedure was required to evaluate the mutants for the presence of the inhibitory domain. To achieve this we resorted to the use of limited proteolysis by trypsin, since there was ample precedence to document that the removal of the inhibitory domain by trypsin resulted in activation of kinase activity. When this procedure was applied to the three truncated mutants, activation was observed in only one instance, namely that of pMK.Trp^{800}. The two shorter mutants, pMK.Lys^{793} and pMK.Thr^{778}, were not markedly affected by tryptic hydrolysis. Thus, the conclusion from both the specific activities of the expressed proteins and the effects of tryptic hydrolysis is that an important component of the inhibitory sequence is present only in pMK.Trp^{800}. Thus the sequence, Tyr^{794} Met Ala Arg Arg Lys Trp^{800} must be regarded as critical to the mechanism of autoinhibition. The last four residues of this sequence are probably important to the inhibitory effect, but this must be validated using additional mutants.

Assignment of the critical region of the inhibitory domain to this sequence would argue strongly in favor of the pseudosubstrate hypothesis and would be inconsistent with the proposal of Ikebe et al. (1989). The CaM-binding site extends from A^{796} to S^{815} and would therefore overlap with the inhibitory sequence by five residues. With this arrangement it is not improbable that the binding of Ca^{2+}-CaM to the CaM-binding site would release the constraints on the apoenzyme imposed by the sequence Y^{794} to W^{800}.

ACKNOWLEDGMENTS

Authors supported by grants HL-43651 (to V.G.) and HL-23615 (to D.J.H.) from the National Institutes of Health.

REFERENCES

Dabrowska, R., Sherry, J. M. F., Aromatorio, D. K., and Hartshorne, D. J., 1978, Modulator protein as a component of the myosin light chain kinase from chicken gizzard, *Biochemistry,* 17: 253.

Edelman, A. M., Takio, K., Blumenthal, D. K., Hansen, R. S., Walsh, K. A., Titani, K., and Krebs, E. G., 1985, Characterization of the calmodulin-binding and catalytic domains in skeletal muscle myosin light chain kinase, *J. Biol. Chem.,* 260: 11275.

Guerriero Jr., V., Russo, M. A., Olson, N. J., Putkey, J. A., and Means, A. R., 1986, Domain organization of chicken gizzard myosin light chain kinase deduced from a cloned cDNA, *Biochemistry,* 25: 372.

Hanley, R. M., Means, A. R., Ono, T., Kemp, B. E., Burgin, K. E., Waxham, N., and Kelly, P. T., 1987, Functional analysis of a complementary DNA for the 50-kilodalton subunit of calmodulin kinase II, *Science,* 237: 293.

Hartshorne, D. J., 1987, Biochemistry of the contractile process in smooth muscle, *in:* "Physiology of the Gastrointestinal tract", L. R. Johnson, ed., Raven Press, New York, p. 423.

House, C. and Kemp, B. E., 1987, Protein kinase C contains a pseudosubstrate prototype in its regulatory domain, *Science,* 238: 1726.

Ikebe, M., Maruta, S., and Reardon, S., 1989, Location of the inhibitory region of smooth muscle myosin light chain kinase, *J. Biol. Chem.,* 264: 6967.

Ikebe, M., Stepinska, M., Kemp, B. E., Means, A. R., and Hartshorne, D. J., 1987, Proteolysis of smooth muscle myosin light chain kinase. Formation of inactive and calmodulin-independent fragments, *J. Biol. Chem.,* 260: 13828.

Ito, M., Dabrowska, R., Guerriero Jr., V., and Hartshorne, D. J., 1989, Identification in turkey gizzard of an acidic protein related to the C-terminal portion of smooth muscle myosin light chain kinase, *J. Biol. Chem.,* 264: 13971.

Ito, M., Hartshorne, D. J., Pearson, R., and Kemp, B. E., 1989, Proteolysis of smooth muscle myosin light chain kinase, *Biophys. J.,* 55: 494a.

Kemp, B. E., Pearson, R. B., Guerriero Jr., V., Bagchi, I. C., and Means, A. R., 1987, The calmodulin binding domain of chicken smooth muscle myosin light chain kinase contains a pseudosubstrate sequence, *J. Biol. Chem.*, 262: 2542.

Lukas, T. J., Burgess, W. H., Prendergast, F. G., Lau, W., and Watterson, D. M., 1986, Calmodulin binding domains: Characterization of a phosphorylation and calmodulin binding site from myosin light chain kinase, *Biochemistry,* 25: 1458.

Mayr, G. W. and Heilmeyer Jr., L. M. G., 1983, Skeletal muscle myosin light chain kinase. a refined structural model, *FEBS Lett.*, 157: 225.

Olson, N. J., Pearson, R. B., Needleman, D. S., Hurwitz, M. Y., Kemp, B. E., and Means, A. R., 1990, Regulatory and structural motifs of chicken gizzard myosin light chain kinase, *Proc. Nat'l. Acad. Sci. U.S.A.*, 87: 2284.

Payne, M. E., Fong, Y.-L., Ono, T., Colbran, R.-J., Kemp, B. E., Soderling, T. R., and Means, A. R., 1988, Calcium/calmodulin-dependent protein kinase II. Characterization of distinct calmodulin binding and inhibitory domains, *J. Biol. Chem.*, 263: 7190.

Roush, C. L., Kennelly, P. J., Glaccum, M. B., Helfman, D. M., Scott, J. D., and Krebs, E. G., 1988, Isolation of the cDNA encoding rat skeletal muscle myosin light chain kinase. sequence and tissue distribution, *J. Biol. Chem.*, 263: 10510.

Russo, M. A., Guerriero Jr., V., and Means, A. R., 1987, Hormonal regulation of a 2-7 kb mRNA that shares a common domain with myosin light chain kinase in the chicken oviduct, *Mol. Endocrinol.*, 1: 60.

Scholey, J. M., Taylor, K. A., and Kendrick-Jones, J., 1980, Regulation of non-muscle myosin assembly by calmodulin-dependent light chain kinase, *Nature,* 287: 233.

Studier, W. S., Rosenberg, A. H., and Dunn, J. J., 1990, Use of T7 RNA polymerase to direct the expression of cloned genes, *Methods Enzymol.*, 185: 60.

Suzuki, H., Onishi, H., Takahashi, K., and Watanabe, S., 1978, Structure and function of chicken gizzard myosin, *J. Biochem.*, 84: 1529.

Takio, K., Blumenthal, D. K., Walsh, K. A., Titani, K., and Krebs, E. G., 1986, Amino acid sequence of rabbit skeletal muscle myosin light chain kinase, *Biochemistry,* 25: 8049.

Taylor, S. S., 1989, cAMP-dependent protein kinase. Model for an enzyme family, *J. Biol. Chem.*, 264: 8443.

Yazawa, M., Kuwayama, H., and Yagi, K., 1978, Modulator protein as a Ca^{2+}-dependent activator of rabbit skeletal myosin light chain kinase, *J. Biochem.*, 84: 1253.

REGULATION OF SMOOTH MUSCLE MYOSIN LIGHT CHAIN KINASE BY CALMODULIN

Anthony R. Means,[1] Indrani C. Bagchi,[1]
Mark F. A. VanBerkum,[1] and Bruce E. Kemp[2]

[1]Department of Cell Biology
Baylor College of Medicine
Houston, TX, 77030

[2]St. Vincent's Institute of Medical Research
Melbourne, Australia

INTRODUCTION

Whereas Ca^{2+} serves as the primary intracellular messenger for regulation of contraction in all types of muscle, the underlying molecular mechanisms are different. Striated muscles utilize troponin C as the Ca^{2+} receptor and contain an array of cell specific proteins involved in maintenance of Ca^{2+} homeostasis. Phosphorylation reactions are important in regulation of contraction but phosphorylation of myosin light chains does not play a major role in force generation (Sweeney and Stull, 1990). Smooth muscles utilize calmodulin as the Ca^{2+} receptor and phosphorylation of myosin light chains is the rate-limiting reaction in contractility (Kamm and Stull, 1985). Since myosin light chain kinase is the primary enzyme that catalyzes light chain phosphorylation, analysis of this complex activation process is crucial to the understanding of how smooth muscles move.

One approach to an appreciation of smooth muscle contractility is to dissect the molecular mechanics by which calmodulin binds to and activates myosin light chain kinase (MLCK). Our laboratory has initiated this approach by cloning and sequencing full-length cDNAs for calmodulin and smooth muscle myosin light chain kinase (Putkey et al., 1983; Guerriero, 1986; Olson et al., 1990), ligating each nucleic acid into a bacterial expression vector (Putkey et al., 1985; Bagchi et al., 1989), and employing site-specific mutagenesis to evaluate the role of various amino acids. These studies, whereas not complete, have resulted in an appreciation of the complexity of this seemingly simple enzyme activation reaction. The MLCK is inhibited in the resting state due to a portion of the enzyme interacting with its active site in a manner that prevents substrate binding (Kemp et al., 1987). Calmodulin interacts with an overlapping region in the primary sequence of MLCK (Lukas et al., 1986). This interaction can result in relief of the autoinhibition, allowing substrate access to the active site and productive myosin light chain phosphorylation (Pearson

Regulation of Smooth Muscle Contraction
Edited by R.S. Moreland, Plenum Press, New York, 1991

et al., 1988). However, introduction of only minimal amino acid substitutions into calmodulin can produce a protein capable of binding MLCK with high affinity without a resultant enzyme activation step (George et al., 1990).

Calmodulin is a remarkably conserved protein. The primary amino acid sequence is identical in all vertebrate species examined with the exception of a single conservative substitution present in the electric eel (Lagace et al., 1983). Even when non-vertebrate sequences are included in the comparison, greater than 80% homology is generally observed. The most different calmodulin reported to date is in the yeast *Saccharomyces cerevisiae* which is only 60% identical to its vertebrate counterpart (Davis et al., 1986). Although multiple genes encode calmodulin in vertebrate species such as rat and man, the proteins produced are identical at the primary sequence level. In only one instance, that of the sea urchin, *Arbacia punctulata*, have two genes been shown to encode different proteins (Hardy et al., 1988). Even in this case, the functional significance, if any, is unknown.

THE MYOSIN LIGHT CHAIN KINASES

Much less information is available concerning the primary sequence of MLCK. The first three published sequences were from rabbit skeletal muscle (Takio et al., 1986), chicken smooth muscle (Guerriero et al., 1986; Olson et al., 1990) and the slime mold *Dictyostelium discoidium* (Tan and Spudich, 1990). The three enzymes differ in size from 34 kDa to 110 kDa and, even within the kinase homology domain, are only about 50% identical. They also differ in structural organization and functional domains. The simplest enzyme is that from *Dictyostelium* which is not regulated by calmodulin and consists of little other sequence than the 270 amino acids that constitute the kinase homology. The rabbit skeletal muscle enzyme has 202 amino acids NH_2-terminal to the kinase region and ends in a 35 amino acid calmodulin binding/autoregulatory domain. The most complicated structural organization is found in the chicken smooth muscle enzyme. This protein contains 4 copies of two different 100 amino acid repeats that are not present in skeletal or *Dictyostelium* MLCK but are common to the *unc*-22 gene product of the nematode *Caenorhabiditis elegans* (Benian et al., 1989) and the vertebrate striated muscle protein called titin (Labeit et al., 1990). The organization of repeats around the kinase regulatory domain is very similar in smooth muscle MLCK and the *unc*-22 gene product, twitchin, which is a component of the myosin-containing A-band of the striated body wall musculature. The major difference in this region is the apparent absence of a calmodulin binding prototope in twitchin. The similar organization, coupled with the fact that deletions of the twitchin gene in this kinase-like area results in alteration of muscle contraction, leads to the suggestion that this protein may be the MLCK of *C. elegans* striated muscle. If so, this enzyme would probably not be regulated by calmodulin. Twitchin is a large protein (600 kDa) and contains 31 and 26 copies of the 100 amino acid repeats I and II respectively. Titin may be the largest polypeptide yet described (3000 kDa) and also consists of many copies of the two repeated elements but does not appear to contain a kinase domain. Since titin is a structural protein that seems to span between M and Z lines in the muscle sarcomere, Benian, et al. (1989) have suggested that twitchin might be a chimera of titin and MLCK and serve both functions in the *C. elegans* body wall muscle. By analogy, the 4 repeat elements in gizzard MLCK could indicate a structural role for this molecule as well. Labeit et al. (1990) have suggested that one or both repeat elements may bind to myosin. If true, this could be the feature common to all three repeat-containing proteins.

Shoemaker et al. (1990), have cloned a cDNA representing the chicken non-muscle form of MLCK. Whereas the overlapping portions of this enzyme and smooth muscle MLCK are virtually identical, the non-muscle form contains an additional 285 amino acids at its NH_2-terminus. Whether these enzymes are derived from the same transcription unit by differential exon usage or represent products of different genes remains to be determined. However, from preliminary studies on a truncated version of a bacterially expressed nonmuscle enzyme it is highly likely that the kinetic properties and substrate specificity of the non-muscle and smooth muscle enzymes will be similar if not identical.

The rabbit uterine smooth muscle cDNA has now been cloned and sequenced in the Stull laboratory. It is larger than the chicken smooth muscle homolog due to the presence in the NH_2-terminal half of the protein of an insertion that consists of multiple short amino acid repeats. However, the NH_2-terminus, twitchin-like repeat elements, kinase region, autoregulatory domain, and COOH-terminus are 85% identical to the homologous regions of the chicken smooth muscle MLCK. Another conserved feature between the rabbit and chicken smooth muscle MLCK genes is the presence of another mRNA that encodes a protein identical in primary sequence to the COOH-terminal region of MLCK immediately distal to the calmodulin binding domain (Guerriero et al., 1986). In the chicken gene, the initiator Met residue is at position 816 and a protein is produced that has been called telokin by Ito et al. (1989). This mRNA is generated by utilization of an alternative promoter that is present in the 3' portion of an intron of the MLCK gene and is regulated by steroid hormones (Russo et al., 1987). Whereas the 5' nontranslated region of the mRNA is unique, the amino acid sequence of telokin is identical to the COOH-terminal 156 amino acids of MLCK. The only obvious functional motif in this protein is one 100 amino acid repeat element common to MLCK. Whereas the function of the protein has not yet been resolved, it seems to exist in both non-muscle and smooth muscle cells of the chicken and the rabbit and be derived from the MLCK gene by a common molecular mechanism. Together the available data suggest that the non-muscle and smooth muscle forms of MLCK will be very similar within and between vertebrate species. The interspecies similarities are considerably greater than intraspecies similarities between non-muscle/smooth muscle and skeletal muscle isoforms of MLCK. The degree of similarity between the rabbit, rat and chicken skeletal muscle MLCKs can now be compared (Roush et al., 1988; Leachman et al., 1989). The skeletal enzymes are much more similar to each other than to either smooth muscle enzyme. Thus interspecies similarities seem to hold for the skeletal muscle MLCKs as well.

REGULATION OF MYOSIN LIGHT CHAIN KINASES

A number of enzymes that are regulated by calmodulin can be activated by limited proteolysis. Both skeletal muscle and smooth muscle MLCK isoforms are in this category. Such experiments led to the suggestion that in the absence of calmodulin, MLCK might be regulated by an intramolecular mechanism (Kemp et al., 1987; Kennelly et al., 1987). Calmodulin binding would relieve this autoinhibition and allow a catalytically active conformation to be favored. The calmodulin binding fragments of rabbit skeletal muscle and chicken smooth muscle MLCK were identified by Blumenthal et al. (1985) and Lukas et al. (1986) respectively. The amino acid sequence of the rabbit skeletal muscle enzyme had also been determined (Takio et al., 1986) and the equivalent information about the chicken smooth muscle MLCK was shortly deduced from analysis of the cDNA sequence (Guerriero et al., 1986). Examination of the MLCK sequences surrounding the calmodulin binding residues led Kemp et al. (1987) to appreciate the similarity to the amino acid sequence in the light chain substrate proximal to the phosphorylated Ser residue. In the smooth muscle MLCK, this region began at Ser^{787} and ended with Val^{807} whereas the equivalent region of rabbit skeletal muscle MLCK was from Ser^{567} - Val^{587}. The arrangement of basic residues in this region of the enzymes was similar to the pattern of basic amino acids previously identified as substrate specificity determinants in the myosin light chains (see Table 1). Taken together these observations led to the concept that a sequence within the calmodulin binding region could be responsible for maintaining the enzyme in the inactive form by mimicking a protein substrate.

Confirmation of the autoinhibitory nature of the calmodulin binding region of smooth muscle MLCK came from the demonstration that synthetic peptide analogs of this region were potent substrate antagonists. Results from synthetic peptide studies showed that the calmodulin/autoinhibitory region extended from Ala^{783} to Leu^{813} and could be roughly divided into thirds. The NH_2-terminal 2/3 were both necessary and sufficient for substrate inhibition whereas the COOH-terminal 2/3 were necessary and sufficient for calmodulin binding. Ikebe et al. (1987) used limited proteolysis to produce an inactive 64 kDa fragment of gizzard MLCK that neither bound calmodulin nor exhibited catalytic activity. Additional proteolysis resulted in generation of a 61 kDa peptide that demonstrated about 60% of the activity of the native enzyme. Sequence analysis revealed that the 64 and 61 kDa proteins had an identical NH_2-terminus beginning with Thr^{283} (Pearson et al., 1988; Ikebe et al., 1989). The 64 kDa peptide ended with Arg^{808} and thus contained all of the proposed substrate inhibitory region. In addition, this result revealed that the removal of only 5 amino acids (Ala^{809}-Leu^{813}) completely prevented calmodulin binding, a result that has been confirmed using synthetic peptides (Foster et al., 1990). The COOH-terminal amino acid of the active 61 kDa peptide was Lys^{779} (Olson et al., 1990). Therefore, activity was restored by deletion of the suggested substrate mimicking sequence. A synthetic peptide from Leu^{774} to Ser^{787} did not inhibit MLCK activity so the autoinhibitory region was confined to the sequence between Asn^{780} and Arg^{808} (Olson et al., 1990).

Edelman et al. (1985) have characterized a proteolytically-derived active fragment of rabbit skeletal muscle MLCK that extends from residues 256 - 584.

Table 1

smMLCK (787-813)	S	K D R M K K Y M A R R K W Q K T G H A	V	R A I G R L
skMLCK (567-593)	S	Q R L L K K Y L M K R R W K K N F I A	V	S A A N R F
smMLC (1-23)	S	S K R A K A K T T K K R P Q R A T S N	V	F A

However, this fragment is only about 5% as active as the native enzyme. Studies with the synthetic peptides revealed the essential determinants of high affinity calmodulin binding corresponded to a 17 amino acid region of the enzyme from 577-593. Since the active fragment does not bind calmodulin then amino acids mandatory for calmodulin binding must exist from 585-593. The sequence of these amino acids is Ile^{585}-Ala-Val-Ser-Ala-Ala-Asn-Arg-Phe^{593}. Whereas most authors have acknowledged the similarities between the regulation of smooth and skeletal muscle MLCKs, the results described above suggest some substantial differences. In order to better appreciate these apparent inconsistencies the regulatory regions of both kinases as well as the corresponding region of smooth muscle myosin light chain are detailed in Table 1. The substrate inhibitory region of the smooth muscle enzyme would be predicted from sequence alignment to extend from Ser^{787} to Val^{807}. Proteolysis and synthetic peptide studies are entirely consistent with the prediction and define the region to exist between Asn^{780} and Arg^{808}. The 64 kDa inactive fragment ends in Arg^{808} and additional fragments ending with Thr^{803} and Lys^{799} behave similarly. In all three cases, activity can be restored by proteolysis. In the case of the 64 kDa fragment, the active 61 kDa peptide terminates with Lys^{779}; COOH-termini of the other two re-activated enzyme fragments have not been defined.

Based on sequence alignment the substrate inhibitory fragment of rabbit skeletal muscle MLCK should reside between Ser^{567} and Val^{587}. However, the proteolytically active fragment described by Edelman et al. (1985) terminates at Phe^{584} and therefore still contains the NH_2-terminal 18 amino acids of the putative inhibitory region (Takio et al., 1985). It was reported by Blumenthal et al. (1985; 1987) that a synthetic peptide from Lys^{577}-Met^{603} bound calmodulin with high affinity and could be truncated from the COOH-end to Phe^{593} without demonstrable loss of binding affinity. Removal of Phe^{593} resulted in a 50-fold loss of calmodulin binding. Synthetic peptides from Lys^{577} to Phe^{593} not only bind calmodulin but also inhibit the activity of the proteolytically-derived fragment (Kennelly et al., 1987). This inhibition is competitive with respect to a peptide substrate exhibiting a K_i of $\sim 3\ \mu M$. If the proteolytically active fragment terminates with Phe^{584} but removal of the Phe^{593} residue from the end of synthetic peptides decreases calmodulin binding by 50-fold, then one must question the compatibility of these results with the pseudosubstrate hypothesis. A simple explanation would be that the poor activity of the proteolytic fragment is not stable and looses additional residues from its COOH end. In such a case, the active protein could have a considerably different COOH-terminus than the initially derived proteolytic fragment. On the other hand, if the active enzyme fragment does terminate in Phe^{584} then the substrate inhibitory hypothesis cannot be correct and must be distinct from the regulatory process controlling the smooth muscle MLCK.

A similar treatment of the calmodulin binding data to that described above for the substrate inhibitory region, results in the conclusion that calmodulin binding determinants might be quite similar between the smooth and skeletal muscle forms of MLCK. Calmodulin binding to smooth muscle MLCK absolutely requires Arg^{812} since alteration of this basic residue to Ile abolishes calmodulin binding. Similarly deletion of Gly^{811}-Arg^{812}-Leu^{813} abolishes calmodulin binding. Therefore, all of these alterations result in an inactive enzyme whose activity can be restored by limited proteolysis. The amino acids at the equivalent positions in skeletal muscle MLCK are also required for calmodulin binding since peptides ending with Phe^{593} retain full calmodulin interacting properties. As mentioned above, removal of this single Phe residue decreases calmodulin binding affinity by 50 fold. These apparent similarities and differences point to the requirement for further analysis of the properties of the isoforms of MLCK.

ANALYSIS OF MYOSIN LIGHT CHAIN KINASE BY MUTAGENESIS

It is possible that results obtained with proteolytically derived enzyme fragments and synthetic peptides might not be directly relevant to studies on the native form of the enzyme. As a first step towards resolving this question, Bagchi et al. (1989) introduced a cDNA encoding an appropriately regulated fragment of smooth muscle MLCK (smMLCK) into a bacterial expression vector. The resulting 50 kDa peptide began with Asp^{450} and terminated with Glu^{913}. This molecule exhibited very similar calmodulin binding, calmodulin regulation, and kinetic properties to the native 107,500 M_r enzyme isolated from chicken gizzard. Furthermore removal of 98 amino acids from the COOH-end did not change the properties. The shorter version terminated two residues COOH-terminal to the end of the calmodulin binding site at Ser^{815}. As mentioned earlier, truncation experiments were used to confirm results from the proteolysis experiments. That is, termination at Arg^{808} produced an inactive, non-calmodulin binding fragment that could be activated by limited proteolysis similar to the 64 kDa peptide. Removal of the autoinhibitory sequences did not activate the enzyme as was the case for the 61 kDa peptide. Whereas the reasons for this inconsistency remain unknown, they are likely to be related to problems in protein folding and/or stability. By preparing a fusion protein with an immunoglobulin fragment added to the NH_2-terminus, we have now been able to confirm that a peptide that terminates with Lys^{779} does contain enzyme activity similar to the proteolytically derived 61 kDa peptide.

The next concern was to define the contribution of individual amino acids in the autoregulatory region of smMLCK. Site-specific mutagenesis of the 40 kDa bacterially synthesized peptide has been utilized to evaluate amino acids involved in calmodulin binding and substrate inhibition. Both oligonucleotide-directed and polymerase chain reaction technology have been utilized to create mutations. In every case, authenticity of the mutation was assured by direct DNA sequencing. Mutant polypeptides were partially purified and analyzed by gel electrophoresis. The gels were: a) stained for protein; b) incubated with ^{125}I-CaM in the absence or presence of saturating Ca^{2+}; or c) transferred to membrane filters and reacted with affinity purified antibodies to the chicken gizzard MLCK. Enzyme activity was assayed using the synthetic light chain peptide as substrate in the absence and presence of Ca^{2+}/CaM.

The one feature characteristic of all CaM binding peptides is that they can be modeled as amphipathic α-helices (Malencik et al, 1982; DeGrado, et al, 1987). When the residues are incorporated on to a helical wheel, one face has several positive charges, while the other face contains numerous hydrophobic residues. However, both globular halves of CaM and both ends of a CaM binding peptide interact in solution (O'Neill and DeGrado, 1989). The NH_2-terminal half of the peptide binds the COOH-terminal domain of CaM and vice-versa. Both enthalpic and entropic interactions are involved. Finally, it

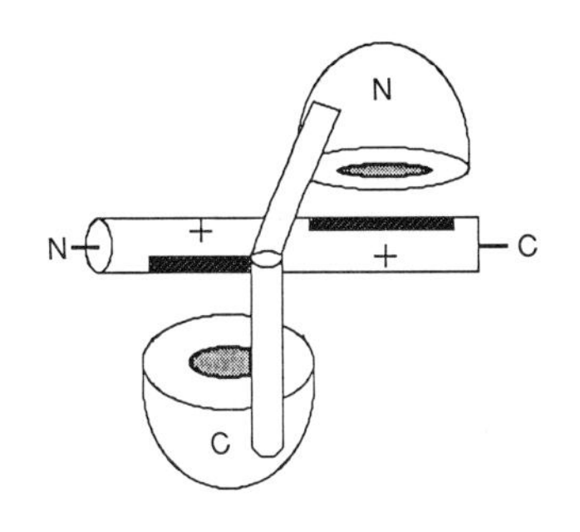

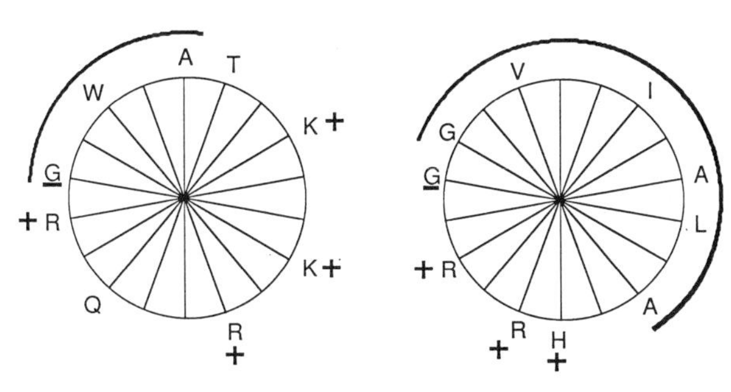

Figure 1. A model for calmodulin binding to an amphipathic helix. A distribution of hydrophobic and basic amino acids along each half of this amphipathic helix is depicted in B. (See text for details)

has been suggested by Persechini and Kretsinger (1988) that at least for the MLCKs, the central region of CaM acts as a flexible tether. This model suggests that a bend in this region occurs to allow each half of CaM to appropriately associate with the MLCK CaM-binding region.

The interaction of CaM with a region of smMLCK is schematically depicted in Figure 1A. In this configuration, the hydrophobic pocket in each half of CaM is positioned to interact with the corresponding hydrophobic region on the CaM binding area of MLCK. Such a configuration requires flexibility of the central region of CaM and also allows interaction of oppositely charged residues in the two molecules. When viewed in this manner, the smMLCK CaM binding region can be considered as 2 half-helices rather than one longer one composed of 18 amino acids. We have depicted how the individual residues would be distributed on two wheels in Figure 1B. A single helical wheel can be reconstituted by simply placing the left wheel on top of the right wheel and aligning the common Gly residue (the underlined G at the left center of each wheel which corresponds to G^{804}; Table 1).

The left wheel represents the NH_2-terminal half of the smMLCK CaM binding region. The hydrophobic face is only about 25% of the total and would contain a prominent W as well as an A. The charged face would contain 4 positively charged amino acids. Thus the charged region is considerably larger than the hydrophobic region. The COOH-terminal wheel (right part of Fig. 1B) is arranged in an opposite manner. In this case, the hydrophobic region is 3/4 of the wheel whereas the charged portion consists of only 2 (or 3 if one counts the H) closely spaced basic amino acids. Mutation of the single W residue to A in the left wheel completely abolishes CaM binding which results in an inactive enzyme. On the other hand, mutation of R^{797}, R^{798}, K^{799} has considerably less effect. Changing one of these basic residues to A decreases CaM dependent activity by only 5%, 2 alterations to A causes a 35% reduction in activity, and even when all 3 basic residues are neutralized, 35% enzyme activity is retained. Mutation of the fourth basic residue ($K^{802} \rightarrow$ A) alone leads to a 60% reduction in CaM dependent kinase activity. From these data we conclude that the primary hydrophobic residue is more important for CaM binding than are any of the 4 basic amino acids alone or in combination.

The results obtained with mutation of the basic residues in the NH_2-terminal half of the MLCK CaM binding region are considerably different from those obtained with the COOH-terminal half mutants. In the latter case mutation of L^{813} to A, R^{812} to I, and G^{811} to E completely abolish CaM binding and enzyme activity. Alteration of G^{811} to A or R^{808} to A decreases CaM-dependent activity by 60-70%. Therefore R^{812} and L^{813} are absolutely required for CaM binding whereas G^{811} is important but not mandatory. It is likely that mutation of G^{811} to E abolishes CaM binding due to the introduction of a negative charge or due to salt bridge formation with R^{808} thus extinguishing the positive charge at this site. Interestingly, the G at position 811 must be important in its own right since the A mutant is considerably less active.

Results obtained by mutating G^{804} were unexpected. Changing this amino acid to A causes a 97% reduction in CaM dependent activation due to a large decrease in CaM binding. These data suggest that the CaM binding region of MLCK may also be flexible and could "bend" around the central G residue. If true, then the flexible tether hypothesis (Persechini and Kretsinger, 1988) might be considerably more complicated even for CaM activation of MLCK. Taken together, all results regarding CaM binding suggest that the COOH-terminal few residues of the CaM binding domain of smMLCK may be the most important for high affinity CaM binding and subsequent enzyme activation. To summarize these observations: 1) truncation of the CaM binding region of the enzyme by removing the terminal GRL residues abolishes CaM binding; 2) synthetic peptides that terminate before the GRL sequence do not bind CaM; 3) mutation of R^{812} to I, or G^{811} to A, or L^{813} to A abolishes CaM binding; and 4) limited proteolysis of native MLCK results in an inactive 64 kDa peptide. This peptide ends in R^{808} and does not bind CaM. These results are also compatible with the first 2/3 of the regulatory domain being more important for auto-inhibition than CaM binding.

Site specific mutations were also carried out within the autoinhibitory region in order to identify the amino acid residues involved in substrate inhibition. Several basic amino acids within this region which closely resemble the structural positions of the basic residues within the phosphorylation site of the MLC substrate (see Table 1) were targeted for mutation because of their potential involvement in substrate inhibition. The residues K^{788}, R^{790}, K^{792}, K^{793} and R^{797}, R^{798}, K^{799}, were mutated to either Ala or Glu. Mutation of these amino acids to Ala retained the CaM dependent kinase activity of the enzyme. However, mutation of either K^{788} R^{790}, to E^{788} E^{790} or R^{797}, R^{798}, K^{799}, to E^{797}, E^{798}, E^{799}, resulted in an enzyme which exhibited partial (10-20%) constitutive activity. Consistent with the peptide studies, our results thus suggest an important role of these basic amino acids in substrate inhibition.

Initial studies suggested that both entropic and enthalpic interactions occur during CaM binding to target enzymes (O'Neil and DeGrado, 1985). The above mutagenesis experiments on the regulatory domain of MLCK support these previous studies. However, it should also be possible to create a CaM which binds but does not activate MLCK. The initial experimental evidence of this nature was obtained by George et al. (1990) who created a series of chimeric Ca^{2+} binding proteins by ligating together various portions of CaM and the cardiac (slow skeletal muscle) isoform of troponin C (cTnC). One of these proteins called TaM consisted of the first 58 amino acids of cTnC ligated to the last 99 amino acids of CaM. This protein, which contains an inactive first Ca^{2+} binding site was a poor agonist of smMLCK. Restoration of Ca^{2+} binding to this site by mutagenesis abolished agonist activity. Instead this chimeric molecule (TaM-BM1) was a potent competitive inhibitor of smMLCK activation by CaM (K_i = 66 nM). These data support the concept that it is possible to dissociate binding of CaM to a target enzyme from activation.

The first 58 amino acids of TaM-BM1 differ from the first 49 amino acids of CaM due to an 8 amino acid NH_2-terminal extension and 11 amino acid substitutions. In order to evaluate the relative importance of the numerous changes, individual amino acid substitutions present in the first domain of TaM-BM1 were introduced into CaM. Extension of the CaM molecule by the 8 amino acids normally present in cTnC did not alter the ability of this protein to activate CaM. Even when multiple copies of the extension were added, the activation kinetics of smMLCK could not be distinguished from native CaM. Therefore, the changes in TaM-BM1 that convert the molecule from agonist to antagonist must reside in some of the 11 amino acid substitutions that exist in the 49 amino acids that are homologous between cTnC and CaM.

The 11 amino acid differences between cTnC and CaM are listed in Table 2. The NH_2-terminal residue of CaM is A whereas the corresponding V in cTnC is actually amino acid 9. Since TnC substitutions were introduced into CaM, the CaM numbering system will be used in the discussion. We initially concentrated on residues 9, 14, 17 and 34 since each substitution resulted in an alteration in the charge of CaM and each of these residues resides on the external surface of CaM. Site-specific mutagenesis was utilized to introduce each substitution individually into CaM. The I^9K, $E^{14}A$, and $S^{17}E$ substitutions did not have major effects on K_{act} or percent maximal activation of smMLCK. However, the single substitution of $T^{34}K$ resulted in a 72 fold increase in K_{act} and only a 30% maximal activation. The next step in the analysis was to introduce combinations of the 4 amino acid substitutions that result in charge changes. The most effective combination was the double mutant $E^{14}A/T^{34}K$. This protein showed a 100 fold increase in K_{act} and only 11% maximal activation relative to CaM. The quadruple mutant ($I^9K/E^{14}A/S^{17}E/T^{34}K$) was no worse an agonist than the double mutant. However, in comparison to the parental TaM-BM1 which completely fails to activate MLCK but binds tightly, all of these charge mutants have an elevated K_{act} and 10-15% activity. Therefore, other alterations in addition to the charge changes had to be important in establishing the phenotype of TaM-BM1. Of the remaining 5 changes introduced in TaM-BM1, the next least conservative alteration was residue 38 which is normally an S in CaM but an M in cTnC. Methionine is particularly abundant in CaM and has been suggested to have an important role in CaM binding to target enzymes by increasing the flexibility of hydrophobic interactions (O'Neil and Degrado, 1985). Altering S^{38} of CaM to M increased the K_{act} by only 10 fold and also decreased maximal activation of smMLCK to 30%. The double mutant ($T^{34}K/S^{38}M$) had 14% maximal activity and if the mutation $E^{14}A$ was also included less than 4% maximal activity is retained.

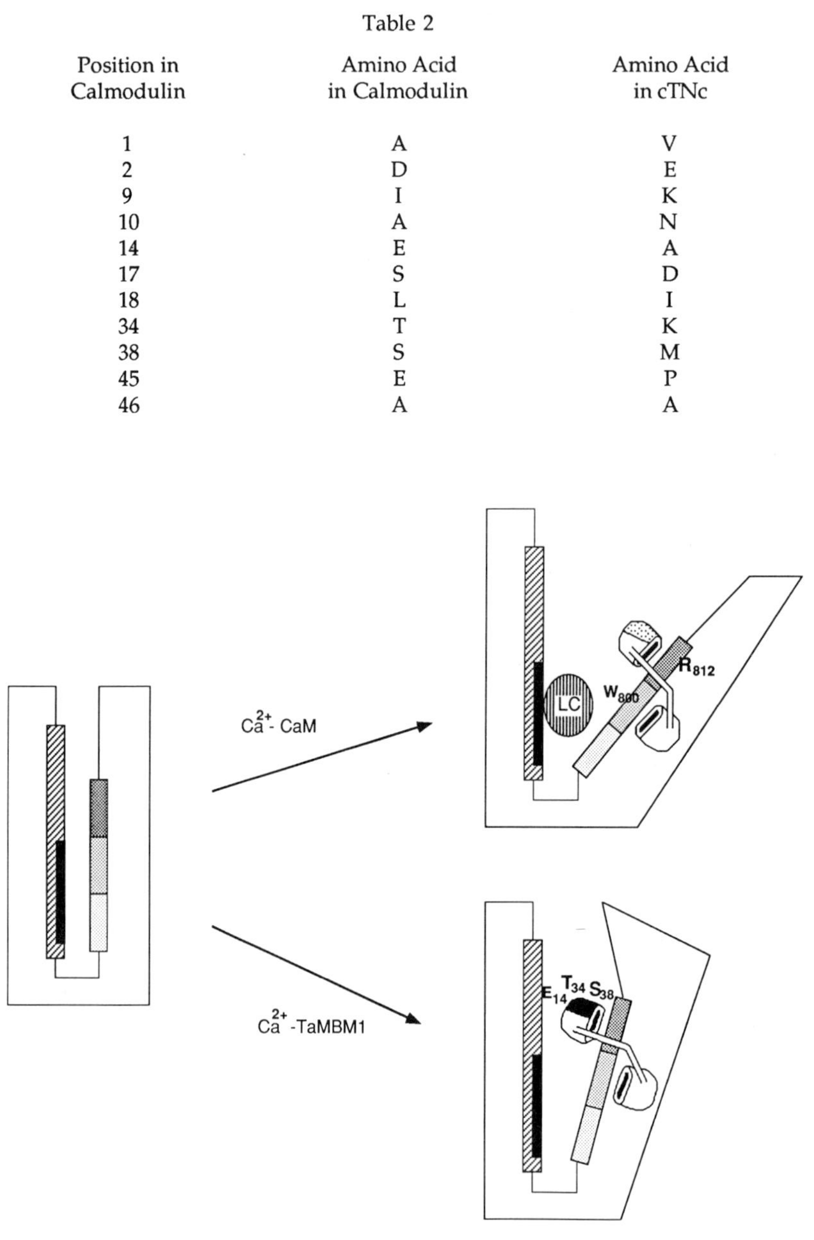

Table 2

Position in Calmodulin	Amino Acid in Calmodulin	Amino Acid in cTNc
1	A	V
2	D	E
9	I	K
10	A	N
14	E	A
17	S	D
18	L	I
34	T	K
38	S	M
45	E	P
46	A	A

Figure 2. Summary of MLCK activation by CaM (upper right) and TaM-BM1 (lower right) showing important residues of MLCK and CaM. (See text for details)

Therefore, the activation of smooth muscle MLCK by CaM requires a series of complex electrostatic interactions between residues on the surface of CaM, particularly in the first calcium binding domain, and unknown residues on MLCK. The two residues on the B helix of CaM (T^{34} and S^{38}) appear to be especially crucial for the interaction of CaM and MLCK. The triple mutant ($E^{14}A/T^{34}K/S^{38}M$) seems to have completely lost the ability to activate MLCK. Mutants with 4 and 5 of these substitutions are being created at present to confirm this possibility.

SUMMARY

The mutagenesis work described in this paper has been instrumental in furthering our understanding of how CaM binds to and activates MLCK. Figure 2 schematically represents this interaction. The inactive MLCK appears to have a catalytic domain that is repressed by a substrate inhibitory domain that overlaps with the CaM binding domain, a basic amphipathic helix. In the presence of Ca^{2+}, CaM undergoes a conformational change that exposes two hydrophobic pockets, one in each globular lobe, that are important for binding to MLCK. Upon binding CaM, MLCK undergoes a conformational change that derepresses the catalytic site, allows substrate access and light chain phosphorylation (Figure 2, upper right). Calmodulin antagonist drugs intercalate within these hydrophobic pockets to interfere with target enzyme binding. The total loss of activity if W^{800} is altered to A illustrates the importance of these hydrophobic interactions within the enzyme. The basic residues are also important; most of the basic residues in the binding domain of MLCK appear to aid in CaM binding but are not in themselves crucial, this includes the RRK triad. However, a specific electrostatic interaction between R^{812} of MLCK and CaM is suggested by the complete failure in MLCK activation if this residue is changed to an A. Electrostatic interactions between MLCK and CaM are also indicated by the TaM-BM1 mutant. This mutant can bind to but not activate MLCK. It is hypothesized that TaM-BM1 will bind to the basic amphipathic helix of MLCK but that the alterations in the surface charges (especially E^{14} and T^{34}) and/or hydrophobicity (S^{38}) prevent the proper conformational change in MLCK necessary for light chain phosphorylation (Figure 2, lower right). The resulting MLCK-CaM complex is therefore, inactive but can bind TaM-BM1. The exact interaction of these amino acids in CaM with MLCK will have to await the elucidation of a CaM-MLCK co-crystal.

REFERENCES

Bagchi, I. C., Kemp, B. E., and Means, A. R., 1989, Myosin light chain kinase structure function analysis using bacterial expression, *J. Biol. Chem.*, 264: 15843.

Benian, G. M., Kiff, J. E., Neckelmann, N., Moerman, D. G., and Watterson, R.H., 1989, Sequence of an unusually large protein implicated in regulation of myosin activity in C. elegans, *Science*, 342: 45.

Blumenthal, D. K. and Krebs, E. G., 1987, Preparation and properties of the calmodulin-binding domain of skeletal muscle myosin light chain kinase, *Methods Enzymol.*, 139: 115.

Blumenthal, D. K., Takio, D., Edelman, A. M., Charbonneau, H., Titani, K., Walsh, K. A., and Krebs, E. G., 1985, Identification of the calmodulin binding domain of skeletal muscle myosin light chain kinase, *Proc. Nat'l. Acad. Sci. U.S.A.*, 82: 3187.

Davis, T. N., Urdea, M. S., Masiarz, F. R., and Thorner, J., 1986, Isolation of the yeast calmodulin gene: Calmodulin is an essential protein, *Cell*, 47: 423.

DeGrado, W. F., Erickson-Viitanen, S., Wolfe, H. R. and O'Neil, K. T., 1987, Predicted calmodulin binding sequence in the γ-subunit of phosphorylase b kinase, *Proteins*, 2: 20.

Edelman, A. M., Takio, K., Blumenthal, D. K., Hansen, R. S., Walsh, K. A., Titani, K., and Krebs, E. G., 1985, Characterization of the calmodulin-binding and catalytic domains in skeletal muscle myosin light chain kinase, *J. Biol. Chem.*, 260: 11275.

Foster, C. J., Johnston, S. A., Sunday, B., and Gaeta, F. C. A., 1990, Potent peptide inhibitors of smooth muscle myosin light chain kinase: Mapping of the pseudosubstrate and calmodulin binding domains, *Arch. Biochem. Biophys.*, 280: 397.

George, S. E., VanBerkum, M. F. A., Ono, T., Cook, R. G., Hanley, R. M., Putkey, J. A., and Means, A. R., 1990, Chimeric calmodulin-cardiac troponin C proteins differentially activate calmodulin target enzymes, *J. Biol. Chem.*, 265: 9228.

Guerriero, V., Russo, M. A., Olson, N. J., Putkey, J. A., Means, A. R., 1986, Location of functional domains in a cDNA for chicken gizzard myosin light chain kinase, *Biochemistry*, 25: 8372.

Hardy, D. O., Bender, P. K., and Kretsinger, R. H., 1988, Two calmodulin genes are expressed in Arbacia punctulata. An ancient gene duplication is indicated., *J. Mol. Biol.*, 199: 223.

Ikebe, M., Malgorzata, S., Kemp, B. E., Means, A. R., and Hartshorne, D. J., 1987, Proteolysis of smooth muscle myosin light chain kinase: Formation of inactive and calmodulin-independent fragments, *J. Biol. Chem.*, 262: 13828.

Ikebe, M., Maruta, S., and Reardon, S., 1989, Location of the inhibitory region of smooth muscle myosin light chain kinase, *J. Biol. Chem.*, 264: 6764.

Ito, M., Dabrowska, R., Guerriero, V. J., and Hartshorne, D. J., 1989, Identification in turkey gizzard of an acidic protein related to the C-terminal portion of smooth muscle myosin light chain kinase., J. *Biol. Chem.*, 264: 13971.

Kamm, K. E. and Stull, J. T., 1985, The function of myosin and myosin light chain kinase phosphorylation in smooth muscle., *Ann. Rev. Pharmacol. Toxicol.*, 25: 593.

Kemp, B. E., Pearson, R. B., Guerriero, V., Bagchi, I. C., and Means, A. R., 1987, The calmodulin binding domain of chicken smooth muscle myosin light chain kinase contains a pseudosubstrate sequence, *J. Biol. Chem.*, 262: 2542.

Kennelly, P. J., Edelman, A. M., Blumenthal, D. K., and Krebs, E. G., 1987, Rabbit skeletal muscle myosin light chain kinase: The calmodulin binding domain as a potential active site-directed inhibitory domain., *J. Biol. Chem.*, 262: 11958.

Labeit, S., Barlow, D. P., Gautel, M., Gibson, T., Holt, J., Hsieh, C. L., Francke, U., Leonard, K., Wardale, J., Whiting, A., and Trinick, J., 1990, A regular pattern of two types of 100-residue motif in the sequence of titin, *Nature*, 345: 273.

Lagace, L., Chandra, T., Woo, S. L. C., and Means, A. R., 1983, Identification of multiple species of calmodulin messenger RNA using a full length complementary DNA, *J. Biol. Chem.*, 258: 1684.

Leachman, S. A., Herring, B. P., Gallagher, P. J., and Stull, J.T., 1989, Isolation of cDNA clone encoding chicken skeletal muscle myosin light chain kinase, *J. Cell Biol.*, 107: 678a.

Lukas, T. J., Burgess, W. H., Prendergast, F. G., Lau, W., and Watterson, D. M., 1986, Calmodulin binding domains: Characterization of phosphorylation and calmodulin binding site from myosin light chain kinase, *Biochemistry*, 25: 1458.

Malencik, D. A., Anderson, S. R., Bahrert, J. L., and Shalitin, Y., 1982, Functional interactions between smooth muscle myosin light chain kinase and calmodulin, *Biochemistry*, 21: 4031.

Olson, N. J., Pearson, R. B., Needleman, D., Hurwitz, M. Y., Kemp, B. E., and Means, A.R., 1990, Regulatory and structural motifs of chicken gizzard myosin light chain kinase, *Proc. Nat'l. Acad. Sci. U.S.A.*, 87: 2284.

O'Neil, K. T. and Degrado, W. F., 1985, A predicted structure of calmodulin suggests an electrostatic basis for its function, *Proc. Nat'l. Acad. Sci. U.S.A.*, 82: 4954.

O'Neil, K. T. and Degrado, W. F., 1989, The interaction of calmodulin with fluorescent and photoreactive model peptides: Evidence for a short interdomain separation, *Proteins*, 6: 284.

Pearson, R. B., Wettenhall, R. E. H., Means, A. R., Hartshorne, D. J., and Kemp, B. E., 1988, Autoregulation of enzymes by pseudosubstrate prototypes: Myosin light chain kinase, *Science*, 241: 970.

Persechini, A. and Kretsinger, R. H., 1988, The control helix of calmodulin functions as a flexible tether, *J. Biol Chem.*, 263: 12175.

Putkey, J. A., Slaughter, G. R., and Means, A. R., 1985, Bacterial expression and characterization of proteins encoded by the chicken calmodulin gene and a calmodulin processed gene, *J. Biol. Chem.*, 260: 4704.

Putkey, J. A., Ts'ui, K. F., Tanaka, T., Lagace, L., Stein, J. P., Lai, E. C., and Means, A. R., 1983, Chicken calmodulin genes: A species comparison of cDNA sequences and isolation of a genomic clone, *J. Biol. Chem.*, 258: 11864.

Roush, C. L., Kennelly, P. J., Glaccum, M. B., Helfman, D. M., Scott, J. D., and Krebs, E. G., 1988, Isolation of the cDNA encoding rat skeletal muscle myosin light chain kinase, *J. Biol. Chem.*, 263: 10510.

Russo, M. A., Guerriero, V., and Means, A. R., 1987, Hormonal regulation of a chicken oviduct mRNA that shares a common domain with gizzard myosin light chain kinase, *Mol. Endocrinol.*, 1: 60.

Shoemaker, M. O., Lau, W., Shattuck, R. L., Kwiatkowski, A. P., Matrisian, P. E., Guerra-Santos, L., Lukas, T. J., Van Eldik, L. J., and Watterson, D. M., 1990, Use of DNA sequence and mutant analyses and the antisense oligodeoxynucleotides to examine the molecular basis of nonmuscle myosin light chain kinase autoinhibition, calmodulin recognition, and activity, *J. Cell Biol.*, 111: 1107.

Sweeney, H. L. and Stull, J. T., 1990, Alteration of cross-bridge kinetics by myosin light chain phosphorylation in rabbit skeletal muscle: Implications for regulation of actin-myosin interaction, *Proc. Nat'l. Acad. Sci. U.S.A.*, 87: 414.

Takio, K., Blumenthal, D. K., Edelman, A. M., Walsh, K. A., Krebs, E. G., and Titani, K., 1985, Amino acid sequence of an active fragment of rabbit skeletal muscle myosin light chain kinase, *Biochemistry*, 24: 6028.

Takio, K., Blumenthal, D. K., Walsh, K. A., Titani, K., and Krebs, E. G., 1986, Amino acid sequence of rabbit skeletal muscle myosin light chain kinase, *Biochemistry*, 25: 8049.

Tan, J. L. and Spudich, J. A., 1990, Dictyostelium myosin light chain kinase, *J. Biol. Chem.*, 265: 13818.

REGULATION OF SMOOTH MUSCLE ACTOMYOSIN FUNCTION

Mitsuo Ikebe, Toshiaki Mitsui, and Shinsaku Maruta

Department of Physiology and Biophysics
Case Western Reserve University School of Medicine
Cleveland, OH 44106

INTRODUCTION

It is widely accepted that the smooth muscle contractile apparatus is primarily regulated by reversible phosphorylation and dephosphorylation of the 20 kDa light chain of myosin (Hartshorne, 1987). The enzyme which is responsible for the phosphorylation is the Ca^{2+}/calmodulin dependent myosin light chain kinase (MLCK) and this is the key enzyme to confer Ca^{2+} sensitivity to the smooth muscle contractile apparatus (Hartshorne, 1987). Another critical component of the phosphorylation hypothesis is the dephosphorylation of the myosin light chain which is catalyzed by a protein phosphatase. Several protein phosphatases have been purified from the soluble fraction of smooth muscle cells. Pato and collaborators (Pato and Adelstein, 1983a, 1983b; Pato and Kerc, 1985) reported four protein phosphatases from gizzard smooth muscle soluble fraction and three of which (SMP I, II and IV) have been purified. Among them, SMP III and IV can dephosphorylate intact myosin. The purified SMP-IV has a molecular weight of 150 kDa and is composed of two subunits of MW 58 kDa and 40 kDa. SMP-IV preferentially dephosphorylates the α-subunit of phosphorylase kinase (a known substrate of Type 1 phosphatases) but is insensitive to Inhibitor-2 which is known to inhibit Type 1 phosphatases specifically. Phosphatases have also been purified from the soluble fraction of aortic smooth muscle (Werth et al., 1982; DiSalvo and Gifford 1983; Erdodi et al., 1989). Erdodi et al. (1989) recently purified two types of phosphatases from dog aortic smooth muscle which dephosphorylate native actomyosin. A 260 kDa phosphatase was similar in its properties to the Type 2A phosphatases. On the other hand, a 150 kDa phosphatase dephosphorylated the α-subunit of phosphorylase kinase although it was not inhibited by Inhibitor 1 or Inhibitor 2, which are known to be specific inhibitors of Type 1 protein phosphatase. These reports suggest that smooth muscle phosphatases purified from the soluble fraction, which dephosphorylate myosin, may be different from both Type 1 and Type 2 phosphatases. However, it is still obscure whether or not these phosphatases are responsible for the regulation of contraction *in vivo*. Furthermore, it is not known if phosphatase activity is physiologically regulated, and this is one of the most important areas to be elucidated.

Regulation of Smooth Muscle Contraction
Edited by R.S. Moreland, Plenum Press, New York, 1991

Another important problem to be solved is how the phosphorylation of the 20 kDa light chain activates actomyosin ATPase activity. It should be pointed out that the site of phosphorylation that activates actomyosin ATPase is very specific. It is known that the 20 kDa light chain of smooth muscle myosin can be phosphorylated by several different sites by two protein kinases. Myosin light chain kinase phosphorylates preferentially serine 19 (Jakes et al., 1976) and subsequently threonine 18 (Ikebe et al., 1985). Phosphorylation of serine 19 activates actomyosin ATPase and initiates tension development (Hartshorne, 1987). The phosphorylation of threonine 18 further activates actomyosin ATPase activity *in vitro* (Ikebe and Hartshorne, 1985a), but does not effect the extent of tension development (Haeberle et al., 1988). On the other hand, protein kinase C phosphorylates serine 1, serine 2, and threonine 9 (Bengur et al., 1987; Ikebe et al., 1987). However, none of these phosphorylations activate actomyosin ATPase activity and instead decrease the ATPase activity of myosin which has been prephosphorylated at serine 19 (Nishikawa, 1984; Ikebe et al., 1987). It is also interesting that the phosphorylation of skeletal muscle light chain at a similar serine site by myosin light chain kinase does not alter actomyosin ATPase activity (Cooke and Stull, 1981). These results suggest the requirement of a particular structure to express the phosphorylation mediated regulation.

The molecular mechanism by which the phosphorylation of the 20 kDa light chain at serine 19 regulates ATPase activity is still obscure, however, several results suggest the involvement of the structure at the head-rod junction for the regulatory mechanism. Initially, it was found by Dr. Watanabe's group (Onishi et al., 1978; Suzuki et al., 1978) that smooth muscle myosin filaments are disassembled on addition of Mg^{2+}-ATP at 0.15 M KCl and formed a 10S component. This component was thought to be a dimer of myosin (Onishi et al., 1978; Suzuki et al., 1978) because conventional myosin monomer shows a sedimentation coefficient of 6S. Subsequently, it was found (Onishi and Wakabayashi, 1982; Trybus et al., 1982; Suzuki et al., 1982; Craig et al., 1983) that smooth muscle myosin can form two distinct conformations, folded and extended, and that a folded myosin shows a sedimentation coefficient of 10S. An important point is that these conformations are characterized by distinct enzymatic properties (Ikebe et al., 1983) and although several parameters can affect the conformational transition, the phosphorylation of the 20 kDa light chain favors the 6S conformation (Trybus et al., 1982; Craig et al., 1983; Ikebe et al., 1983; Onishi et al., 1983). According to these findings we proposed the hypothesis that the 10S-6S transition or some associated conformational change may be a component of the regulatory mechanism in smooth muscle contraction. That the conformational change occurs at the head-rod junction is suggested by several results: 1) The resistance to proteolysis at the S-1/S-2 junction is markedly altered with 10S-6S transition (Onishi and Watanabe, 1984; Ikebe and Hartshorne, 1984). Furthermore, the change in the resistance to proteolysis is observed for HMM which does not form a looped structure (Ikebe and Hartshorne, 1984). 2) The position of the heads is distinct between phosphorylated and dephosphorylated HMM at low ionic strength (Suzuki et al., 1985; Hartshorne and Ikebe, 1987). 3) Using fluorescently labelled light chain or the myosin $\varepsilon ADP \cdot V_i$ complex as probes, it was found that the rotational relaxation time of 10S myosin is markedly larger than S-1 whereas the relaxation time of 6S myosin is similar to that of S-1 (Morita et al., manuscript submitted; Ikebe and Takashi, 1990).

Several findings suggest that the change in conformation at the head-rod junction is important in determining the actomyosin ATPase activity. 1)

Acto-S-1 ATPase activity is not regulated by phosphorylation although the 20 kDa light chain is intact (Ikebe and Hartshorne, 1985c). 2) A monoclonal antibody which inhibits the 10S formation markedly activates actin activated ATPase activity of dephosphorylated myosin and HMM (Higashihara et al., 1989). 3) A monoclonal antibody which recognizes S-2 also alters the ATPase activity (Ito et al., 1989; Higashihara and Ikebe, 1990). In addition to the change in the conformation at the head-rod junction, it was suggested (Ikebe and Hartshorne, 1986) that the actin binding site is hindered by 10S formation. Since phosphorylation markedly activates ATPase activity, one may assume that the active site conformation may be affected by phosphorylation, however, to date, there is no such information available.

In the first part of this chapter, we describe some preliminary results that may provide information toward understanding the molecular mechanism of regulation of smooth muscle myosin. In the second part, we report the purification and characterization of smooth muscle myosin associated phosphatase which is likely to be a physiologically important myosin light chain phosphatase.

PHOTOAFFINITY LABELLING OF SMOOTH MUSCLE MYOSIN ACTIVE SITE

To detect the possible change in the active site conformation due to 10S-6S transition or 20 kDa light chain phosphorylation, we synthesized Methylanthraniloy1-8-azido ATP (Mant-8-N_3-ATP). First, we tested to see if Mant-8-N_3-ATP also induces 10S conformation at low ionic strength. The conformational transition was monitored by viscosity measurements (Ikebe et al., 1983) and we found that the 10S conformation was induced by Mant-N_3-ATP as well as ATP below 0.2 M KCl. As was shown previously for ATP (Ikebe et al., 1983), the hydrolysis of Mant-8-N_3-ATP in the presence of Mg^{2+} was also

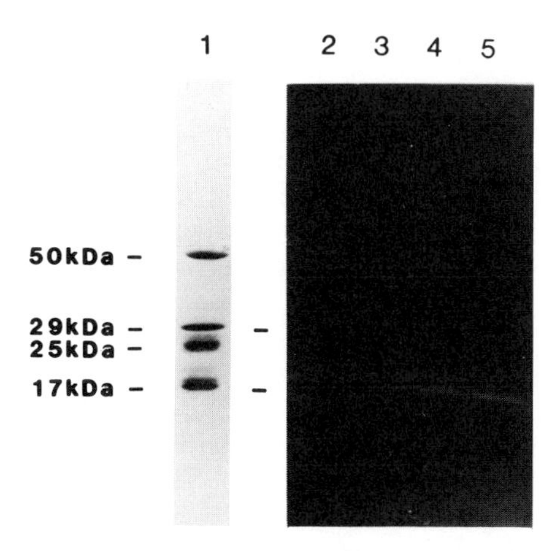

Figure 1. SDS Page of tryptic fragments of S-1 labelled with Mant-8-N_3-ATP. Myosin (15.4 mg/ml) was labelled with 80 μM Mant-8-N_3-ATP in 5 mM $MgCl_2$, 30 mM Tris-HCl pH 7.8, and 180 mM KCl (Lanes 2,4; 10S condition) or 500 mM KCl (Lanes 3, 5; 6S condition) in the absence (Lanes 4, 5) or presence (Lanes 2, 3) of 2 mM ATP. Photolabelling was performed by irradiating myosin with Mant-ATP at 366 nm using UVL-56, Ultra-Violet Product for 3 min at a distance of 2 cm above the surface of a 0.5 ml stirred solution. Fluorescence in the gel was detected under a 366 nm UV lamp and photographed. Lane 1: Coomassie Brilliant Blue staining of tryptic S-1; Lanes 2 - 5: Fluorescence of the labelled tryptic S-1.

markedly decreased below 0.3 M KCl where the myosin conformation changes from 6S to 10S. These results indicate that smooth muscle myosin forms 10S and 6S conformations in the presence of Mant-8-N_3-ATP and that the hydrolysis of Mant-8-N_3-ATP is depressed by the formation of 10S myosin.

To study the binding site of Mant-8-N_3-ATP, myosin was photoreacted with Mant-8-N_3-ATP in the 6S and 10S conformations. Fluorescence of the Mant group was observed in both myosin heavy chain and in the 17 kDa light chain but not in the 20 kDa light chain (data not shown). The photoaffinity labelled myosin was then digested by papain to obtain S-l. No fluorescence was observed in the rod fraction. The papain S-l was further digested by trypsin to analyze the labelled fragments. As shown in Figure 1, the fluorescence of Mant-8-N_3-ATP was predominantly found at the 17 kDa light chain and 29 kDa N-terminal fragment of the heavy chain. The intensity of the fluorescence incorporated in the 29 kDa fragment was the same for 6S and 10S myosins and was almost completely abolished when the labelling was carried out in the presence of excess ATP. The labelled 29 kDa fragment was isolated and subjected to complete lysyl-endopeptidase digestion. The obtained peptides were analyzed by C-8 reverse phase chromatography to find out the possible difference in the labelling sites between 10S and 6S myosins. However, the labelled 6S and 10S myosin 29 kDa fragments showed the same elution profile monitored by fluorescence (data not shown). On the other hand, the fluorescence incorporated in the 17 kDa light chain was reduced to a small extent in the presence of excess ATP for 6S myosin whereas it was significantly reduced for 10S myosin in the presence of excess ATP. This suggests that ATP specific labelling on the 17 kDa light chain occurs for 10S myosin but not for 6S myosin. Previously, it was reported that the ATP analog NANTP labels both the 17 kDa light chain and the heavy chain of smooth muscle S-l in the presence of vanadate (Okamoto et al., 1986) and therefore suggested the possible involvement of the 17 kDa light chain in active site structure. The present results further suggest that the 17 kDa light chain may contribute to the difference in active site structure between 10S and 6S myosins. Since significant amounts of non-specific labelling occur on the 17 kDa light chain, further study is required to evaluate the significance of 17 kDa light chain labelling by ATP analogs. The involvement of the 17 kDa light chain on the regulation of ATPase activity has also been suggested by Higashihara et al. (1989) who showed that a monoclonal antibody against the 17 kDa light chain activates actomyosin ATPase activity without light chain phosphorylation. The role of the 17 kDa light chain on the regulatory mechanism of smooth muscle myosin should be further elucidated.

FORMATION AND PROPERTIES OF SMOOTH MUSCLE MYOSIN 20 kDa LIGHT CHAIN - SKELETAL MUSCLE MYOSIN HYBRID

It has been shown that skeletal muscle myosin is also phosphorylated at the equivalent site of the regulatory light chain by MLCK. However, in the case of skeletal muscle myosin, no equivalent change in actin activated ATPase activity or myosin shape has been detected as a consequence of light chain phosphorylation as that seen with smooth muscle myosin. It is known that smooth muscle and skeletal muscle regulatory light chains are different in the sense that the former can confer the Ca^{2+} sensitive regulation on the molluscan myosin while the latter cannot (Chantler and Szent-Gyorgyi, 1980). Recently, Trybus and Lowey (1988) used the skeletal muscle regulatory light

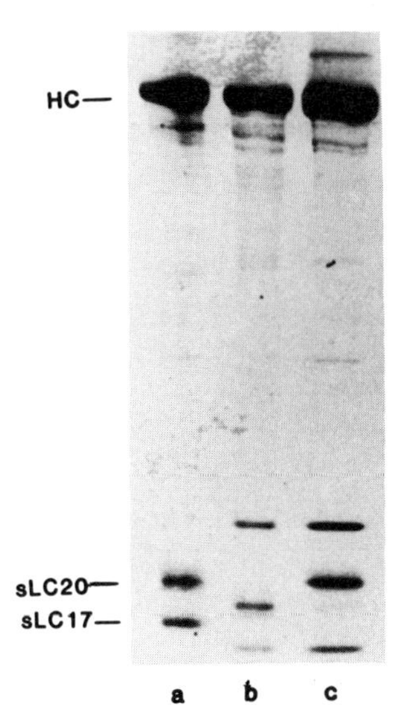

Figure 2. SDS-PAGE analysis of skeletal muscle myosin - smooth muscle myosin 20 kDa light chain hybrid. Lane a: smooth muscle myosin; Lane b: skeletal muscle myosin; Lane c: skeletal muscle myosin - smooth muscle myosin 20 kDa light chain hybrid.

Table 1. Actin-activated Mg^{2+}-ATPase activity of myosin hybrids

Form of myosin	Rate of ATP hydrolysis (sec^{-1}) -EGTA	+EGTA
Myosin control	1.1	1.28
Myosin/DTNB LC	2.1	--
Myosin/s20 kDa LC	1.8	1.7
Myosin/s20 kDa LC-P	1.4	1.8
Myosin/Al LC	1.5	1.7
Myosin/A2 LC	1.4	1.5

Actin activated Mg^{2+}-ATPase activity was assayed in 37.5 mM KCl, 50 mM imidazole, 5 mM $MgCl_2$, 3 mM ATP, pH 7.0 at 25° C at a myosin and actin concentration of 0.098 mg/ml and 0.5 mg/ml respectively. DTNB LC, skeletal muscle myosin regulatory light chain; s20 kDa LC, smooth muscle myosin 20 kDa light chain; LC-P, phosphorylated light chain; Al LC, alkali light chain 1; A2 LC, alkali light chain 2.

chain - smooth muscle myosin hybrid and determined the effect of phosphorylation on the shape of this hybrid myosin. Dephosphorylation of skeletal muscle regulatory light chain did not lead to a complete folding of the hybrid myosin. These results suggest that regulatory light chain properties are important in the regulation of myosin.

In the present study, we made a smooth muscle 20 kDa light chain skeletal muscle myosin hybrid and studied the effect of phosphorylation on its properties. The regulatory light chain exchange was carried out in the presence of EDTA and ATP. A relatively high temperature (37° C) was required for the effective exchange of light chain, the extent of exchange at 25° C was only about 50 - 60% of that seen at 37° C. ATP was required to protect myosin from thermal denaturation, therefore, we used NaCl instead of KCl in the solvent to reduce ATP hydrolysis by myosin (Schliselfeld and Bárány, 1968). The hybrid myosin thus obtained was separated from excess light chain by gel filtration and the subunit composition of the purified myosin hybrid was examined by gel electrophoresis (Fig. 2). The regulatory light chain was completely exchanged with the 20 kDa smooth muscle light chain. This suggests that the binding interactions and affinities of these two regulatory light chains to skeletal muscle myosin heavy chain are quite similar. This implies that the active site on heavy chain for light chain attachment in skeletal myosin cannot discriminate between the two light chains. In contrast to the case of the exchange of smooth muscle 20 kDa light chain into skeletal myosin, smooth muscle myosin showed an extremely low extent of exchange with skeletal regulatory light chain. This implies that the smooth muscle myosin heavy chain has significantly lower affinity for skeletal regulatory light chain despite the high degree of sequence homology between the two light chains. Therefore, it appears that the regulatory light chain attachment site of smooth myosin heavy chain is significantly more specific than the analogous site in its skeletal myosin counterpart.

Mg^{2+}-ATPase activities of the hybrid myosin were investigated to determine whether the exchange of the 20 kDa light chain for the skeletal muscle regulatory light chain had affected the active site of the protein. We were also interested in determining whether the reconstituted protein now bearing the regulatory light chain of the smooth muscle system would exhibit any phosphorylation dependent regulation of its ATPase activity. The rates of actin activated Mg^{2+}-ATPase activity are summarized in Table 1. The activity of the hybrid myosin was similar to that of skeletal muscle myosin and was not affected by light chain phosphorylation.

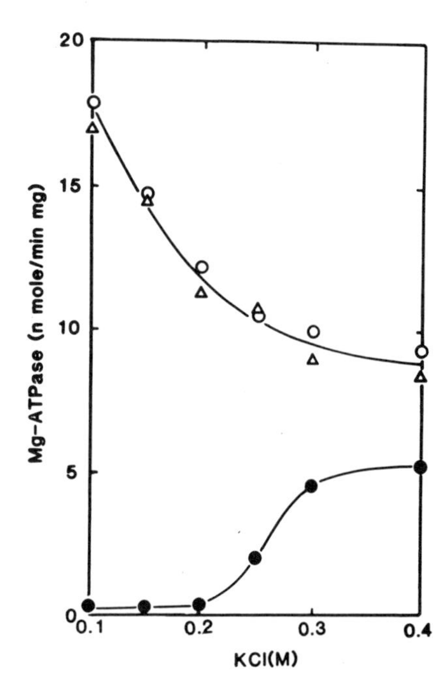

Figure 3. KCl concentration dependence of Mg^{2+}-ATPase activity of hybrid myosin. ○, skeletal muscle myosin; ●, smooth muscle myosin; Δ, skeletal muscle myosin-smooth muscle myosin 20 kDa light chain hybrid.

We next examined the conformational state of the hybrid myosin to determine if any change could be linked to the presence of the smooth muscle regulatory 20 kDa light chain. This was tested by the method of gel filtration which separates 10S and 6S myosins (Ikebe et al., 1983). Gel filtration of dephosphorylated myosin hybrid using TSK 4000 SW column was carried out at 0.2 M and 0.5 M KCl, salt concentrations at which smooth muscle myosin forms 10S and 6S myosin, respectively. However, a 10S hybrid myosin was not seen with either elution condition (data not shown). In addition, the KCl concentration dependence of Mg^{2+}-ATPase activity of the hybrid myosin in the dephosphorylated state was found to be quite similar to that exhibited by skeletal muscle myosin (Fig. 3). It has been shown previously (Ikebe et al., 1983) that Mg^{2+}-ATPase activity of dephosphorylated smooth muscle myosin decreases below 0.3 M KCl due to the transition to the 10S conformation. Therefore, the above results suggest that the hybrid myosin did not form a 10S conformation and that the type of heavy and/or "essential" light chain present in the myosin plays a role in determining whether conformational changes in the head/rod junction can occur. The hybrid myosin also did not exhibit phosphorylation mediated regulation of actin activated ATPase activity. This finding is consistent with our shape-activity hypothesis that the regulation of ATPase activity occurs through the alteration of the conformation of myosin rather than by a direct effect of phosphorylation itself (Ikebe et al., 1983; Ikebe and Hartshorne, 1984, 1985; Hartshorne and Ikebe, 1987).

Table 2. Actin activated ATPase activity of smooth muscle HMM containing various 20 kDa light chain fragments.

	Actin activated ATPase activity (nmol/min/mg)	
	Dephosphorylated	Phosphorylated
SAP-HMM	17.5	769
SAP-HMM (LYS-$C_{(1)}$)	34.0	769
SAP-HMM (LYS-$C_{(2)}$)	38.0	523
SAP-HMM (Trypsin)	55.2	77
SAP-HMM (Exchanged Trypsin-LC_{20})	59.3	63
SAP-HMM (LYS-$C_{(2)}$) (Exchanged LC_{20})	40.5	690
SAP-HMM (Trypsin) (Exchanged LC_{20})	50.2	635

Actin activated ATPase activity was assayed in: 2 mg/ml actin, 20 mM KCl, and 1 mM $MgCl_2$, pH 7.5 at 25° C. LYS-$C_{(1)}$: 20 kDa light chain cleaved at the C-terminus of lysine-6; LYS-$C_{(2)}$: 20 kDa light chain cleaved at the C-terminus of lysine-12; Trypsin: 20 kDa light chain cleaved at the C-terminus of arginine-16.

AMINO ACID STRUCTURE OF THE 20 kDa LIGHT CHAIN REQUIRED FOR THE REGULATION OF ACTOMYOSIN ATPase

As is described in the introduction, it has been suggested that the location of the phosphorylation site of the 20 kDa light chain must be very specific in order to activate actomyosin ATPase. It has also been reported (Chantler and Szent-Gyorgyi, 1980; Trybus and Lowey, 1988) that skeletal muscle regulatory light chains are different from smooth muscle regulatory light chain in terms of their regulatory activity even though both have the same MLCK catalyzed phosphorylation site. These facts suggest the requirement of a specific light chain structure necessary to confer phosphorylation dependent regulation on actomyosin.

We studied the relationship between the structure of the 20 kDa light chain and phosphorylation dependent regulation of actin activated ATPase activity by using the subunit exchange technique. The 20 kDa light chain was proteolyzed by lysylendopeptidase and trypsin. Lysylendoprotease cleaved the light chain initially at the C-terminus of lysine-6 and subsequently at the C-terminus of lysine-12. On the other hand, trypsin cleaved at the C-terminus of arginine-16. HMM prepared by *S. aureus* protease digestion (Ikebe and Hartshorne, 1985c) was hybridized with the tryptic 20 kDa light chain and thus HMM containing the tryptic 20 kDa light chain was prepared. HMM was also digested by either trypsin or lysylendoprotease and the digested HMM was purified by sephacryl S-300 gel filtration. The HMMs which retained the truncated light chains were exchanged with the exogenous intact 20 kDa light chain. Actin activated ATPase activities of these HMMs were measured and the effect of light chain phosphorylation on the activity was examined (Table 2). HMM, whose 20 kDa light chain was cleaved at the C-terminus of lysine 6, showed the same phosphorylation dependent actin activated ATPase activity as that of undigested HMM. The actin activated ATPase activity of HMM was still phosphorylation dependent when the light chain was further cleaved at the C-terminus of lysine-12, although the activity of phosphorylated HMM was slightly lower than the undigested one. In contrast, the HMM digested by trypsin or the HMM whose light chain was exchanged with the tryptic 20 kDa light chain did not show phosphorylation dependent activity (Table 2). The abolition of phosphorylation dependent activation of the ATPase is due to cleavage of the regulatory light chain because the exchange of the cleaved light chain of HMM with the exogenous intact light chain restored the phosphorylation dependence (Table 2). These results clearly show that the amino acid sequence from arginine-13 to arginine-16 is essential to express phosphorylation dependent regulation of smooth muscle myosin. Consistent with this finding, the primary structure of the 20 kDa light chain from proline-14 to arginine-16 is quite different from that of skeletal and cardiac muscle myosin regulatory light chains in which phosphorylation does not regulate actin activated ATPase activity. Smooth muscle 20 kDa light chain contains no acidic residues, but rather contains a basic residue (arginine-16) in contrast to striated muscle regulatory light chains. Possibly this change in the charge may affect phosphorylation dependent regulation.

PURIFICATION AND CHARACTERIZATION OF MYOSIN ASSOCIATED PHOSPHATASE

We have noticed that purified smooth muscle myosin contains significant phosphatase activity. The phosphatase activity was tightly associated with myosin as washing with buffer containing a physiological salt concentration did not separate phosphatase activity from myosin. This suggests that this phosphatase is likely to be a physiologically important enzyme for myosin dephosphorylation *in vivo.* Therefore, we attempted to purify the myosin associated phosphatase. Myofibrils were washed several times with Triton X-100 (3%) containing buffer and 100 mM KCl containing buffer to eliminate the phosphatases in the soluble fraction (Table 3). Myosin was then extracted and purified according to the method described previously

(Ikebe and Hartshorne, 1985b). As shown in Table 3, purified myosin contained significant phosphatase activity. The myosin associated phosphatase activity was dissociated from myosin by washing with 100 mM $MgCl_2$. The phosphatase binding site of myosin was studied by affinity chromatography of various myosin fragments. Myosin rod and HMM affinity columns absorbed 100% of the phosphatase activity while the phosphatase hardly bound to an S-l affinity column (Table 4). This result suggests that the myosin associated phosphatase binds at the S-2 portion of the myosin molecule. The molecular weight was estimated by Sephacryl S-300 gel filtration to be 170 kDa for the myosin associated phosphatase and 90 kDa for the soluble phosphatase (SMP IV) The molecular weight of SMP IV was previously estimated to be 150 kDa (Pato and Kerc, 1985). We do not, at this time, understand this apparent discrepancy.

Table 3. Distribution of the myosin light chain phosphatase activity in triton X- 100 extracts, 100 mM KCl extracts, and myosin fraction from chicken gizzard

Fraction	MLCK activity $[nmol \cdot min^{-1} \cdot (g\ tissue)^{-1}]$	
Triton X-l00 extract	1st	3.12
	2nd	2.34
	3rd	1.17
	4th	not detected
	5th	not detected
100 mM KCl extract	1st	0.9
	2nd	0.2
	3rd	not detected
	4th	not detected
Myosin fraction		4.17

Phosphatase activity was assayed in the presence of 25 mM HEPES pH 7.5, 4 mM EDTA, 2 mM EGTA, and 30 mM KCl at 25° C using ^{32}P-labelled 20 kDa light chain as a substrate.

The phosphatase was purified by DEAE sephacel column, myosin rod affinity column, thiophosphorylated 20 kDa light chain affinity column, and DEAE 5 PW HPLC column. The purified enzyme fraction showed two polypeptide bands, 60 kDa and 34 kDa, as judged from SDS gel electrophoresis. The activity of the purified phosphatase was dependent on neither Mg^{2+} nor Ca^{2+}/calmodulin. The activity was inhibited by okadaic acid (half-maximal inhibition at 20 nM) at 10 times less than the concentration required for the inhibition of SMP IV. The myosin associated phosphatase was also inhibited by skeletal muscle phosphatase inhibitor 2, although 10 times higher concentration of the inhibitor was required as compared to that for the inhibition of skeletal muscle type I phosphatase. These results suggest that the myosin associated phosphatase is similar to type I phosphatase although the properties are a little different from skeletal muscle type I phosphatase. Recently, Ishihara et al. (1989) reported that the phosphatase activity in crude actomyosin (myosin B) preparation is similar to type I phosphatase. Our results of myosin associated phosphatase are consistent with their results.

Table 4. Binding of myosin-associated MLCP to proteolytic fragments of myosin

Myosin fragments	% bound
Intact myosin	96
Rod	100
HMM	100
S1	21

Binding of phosphatase was examined using myosin fragment affinity column (1 cm x 15 cm). The phosphatase was applied to the column equilibrated with 25 mM Tris-HCl pH 7.5, 2 mM EGTA, and 0.5 mM DTT and the bound phosphatase was eluted by 50 mM NaCl and then 0.3 M NaCl and 100 mM $MgCl_2$. Almost all phosphatase activity was recovered in the flow through fraction for S-1 column. Practically no phosphatase was eluted by 50 mM NaCl and bound phosphatase was eluted by 0.3 M NaCl and 100 mM $MgCl_2$.

The regulation of myosin associated phosphatase is still obscure. In skeletal muscle, it has been suggested that type-I phosphatase activity can be regulated by inhibitor proteins (Cohen and Cohen, 1989). It has also been suggested that glycogen binding phosphatase I activity may be regulated via the phosphorylation of glycogen binding subunit (Cohen and Cohen, 1989). However, it is not known whether or not similar mechanisms are operating for the smooth muscle system and this needs further study to clarify the regulatory mechanism for myosin dephosphorylation. The involvement of a myosin binding subunit and putative phosphatase inhibitor in smooth muscle on the regulation of this enzyme are currently being studied.

REFERENCES

Bengur, A. R., Robinson, E. A., Apella, E., and Sellers, J. R., 1987, Sequence of the sites phosphorylated by protein kinase C in the smooth muscle light chains, *J. Biol. Chem.*, 262: 7613.

Cohen, P. and Cohen, P. T. W., 1989, Protein phosphatases come of age, *J. Biol. Chem.*, 264: 21435.

Cooke, R. and Stull, J. T., 1981, Myosin phosphorylation: A biochemical mechanism for regulating contractility, *in:* "Cell and Muscle Motility, Vol. 1", J. W. Shay, and R. M. Dowben, eds., Plenum Press, New York, p. 99.

Craig, R., Smith, R., and Kendrick-Jones, J., 1983, Light-chain phosphorylation controls the conformation of vertebrate non-muscle and smooth muscle myosin molecules, *Nature,* 302: 436.

DiSalvo, J. and Gifford, D., 1983, Spontaneously active and ATPMg-dependent protein phosphatase activity in vascular smooth muscle, *Biochem. Biophys. Res. Commun.*, 111: 912.

Erdodi, F., Rokolya, A., Bárány, M., and Bárány, K., 1989, Dephosphorylation of distinct sites in myosin light chain by two types of phosphatase in aortic smooth muscle, *Biochim. Biophys. Acta,* 1011: 67.

Haeberle, J. R., Sutton, T. A., and Trockman, B. A., 1988, Phosphorylation of two sites on smooth muscle myosin, effects on contraction of glycerinated vascular smooth muscle, *J. Biol. Chem.,* 263: 4424.

Hartshorne, D. J.. 1987, Biochemistry of the contractile process in smooth muscle, *in:* "Physiology of the Gastrointestinal Tract", L. R. Johnson, ed., Raven Press, New York, p. 423.

Hartshorne, D. J. and Ikebe, M., 1987, Phosphorylation of myosin, *in:* "Platelet Activation", H. Yamazaki, and J. F. Mustard, eds., Academic Press, Tokyo, p. 3.

Higashihara, M., Young-Frado, L.-L., and Craig, R., 1989, Inhibition of conformational change in smooth muscle myosin by a monoclonal antibody against the 17-kDa light chain, *J. Biol. Chem.*, 264: 5218.

Higashihara, M. and Ikebe, M., 1990, Alteration of the enzymatic properties of smooth muscle myosin by a monoclonal antibody against subfragment 2, *FEBS Lett.*, 263: 241.

Ikebe, M. and Hartshorne, D. J., 1984, Conformation-dependent proteolysis of smooth muscle, *J. Biol. Chem.*, 259: 11639.

Ikebe, M. and Hartshorne, D. J., 1985a, Phosphorylation of smooth muscle myosin at two distinct sites by myosin light chain kinase, *J. Biol. Chem.*, 260: 10027.

Ikebe, M. and Hartshorne, D. J., 1985b, Effects of Ca^{2+} on the conformation and enzymatic activity of smooth muscle myosin, *J. Biol. Chem.*, 260: 13146.

Ikebe, M. and Hartshorne, D. J., 1985c, Proteolysis of smooth muscle myosin by Staphylococcus aureus protease: Preparation of heavy meromyosin and subfragment 1 with intact 20,000-dalton light chains, *Biochemistry*, 24: 2380.

Ikebe, M. and Hartshorne, D. J., 1986, Proteolysis and actin-binding properties of 10S and 6S smooth muscle myosin: Identification of a site protected from proteolysis in the 10S conformation, *Biochemistry*, 25: 6177.

Ikebe, M. and Takashi, R., 1990, Change in the flexibility of the head-rod junction of smooth muscle myosin by phosphorylation, *Biophys. J.*, 57: 331a.

Ikebe, M., Hinkins, S., and Hartshorne, D. J., 1983, Correlation of enzymatic properties and conformation of smooth muscle myosin, *Biochemistry*, 22: 4580.

Ikebe, M., Hartshorne, D. J., and Elzinga, M., 1986, Identification, phosphorylation, and dephosphorylation of a second site for myosin light chain kinase on the 20,000-dalton light chain of smooth muscle myosin, *J. Biol. Chem.*, 261: 36.

Ikebe, M., Hartshorne, D. J., and Elzinga, M., 1987, Phosphorylation of the 20,000-dalton light chain of smooth muscle myosin by the calcium-activated, phospholipid-dependent protein kinase, *J. Biol. Chem.*, 262: 9569.

Ishihara, H., Martin, B. L., Brautigan, D. L., Karaki, H., Ozaki, H., Kato, Y., Fusetani, N., Watabe, S., Hashimoto, K., Uemura, D., and Hartshorne, D. J., 1989, Calyculin A and okadaic acid: Inhibitors of protein phosphatase activity, *Biochem. Biophys. Res. Commun.*, 159: 871.

Ito, M., Pierce, P. R., Allen, R. E., and Hartshorne, D. J., 1989, Effect of monoclonal antibodies on the properties of smooth muscle myosin, *Biochemistry*, 28: 5567.

Jakes, R., Northrop, F., and Kendrick-Jones, J., 1976, Calcium binding regions of myosin 'regulatory' light chains, *FEBS Lett.*, 70: 229.

Morita, J., Takashi, R., and Ikebe, M., 1991, Exchange of the fluorescence labelled 20,000 dalton light chain of smooth muscle myosin, *Biochemistry*, in press.

Nishikawa, M., Sellers, J. R., Adelstein, R. S., and Hidaka, H., 1984, Protein kinase C modulates in vitro phosphorylation of the smooth muscle heavy meromyosin by myosin light chain kinase, *J. Biol. Chem.*, 259: 8808.

Okamoto, Y., Sekine, T., Grammer, J., and Yount, R. G., 1986, The essential light chains constitute part of the active site of smooth muscle myosin, *Nature*, 324: 78.

Onishi, H. and Wakabayashi, T., 1982, Electron microscopic studies of myosin molecules from chicken gizzard muscle I: The formation of the intramolecular loop in the myosin tail, *J. Biochem.*, 92: 871.

Onishi, H. and Watanabe, S., 1984, Correlation between the papain digestibility and the conformation of 10S-myosin from chicken gizzard, *J. Biochem.*, 95: 899.

Onishi, H., Suzuki, H., Nakamura, K., Takahashi, K., and Watanabe, S., 1978, Adenosine triphosphatase activity and "thick filament" formation of chicken gizzard myosin in low salt media, *J. Biochem.*, 83: 835.

Onishi, H., Wakabayashi, T., Kamata, T., and Watanabe, S., 1983, Electron microscopic studies of myosin molecules from chicken gizzard muscle. II: The effect of thiophosphorylation of the 20K-dalton light chain on the ATP-induced change in the conformation of myosin monomers, *J. Biochem.*, 94: 1147.

Pato, M. D. and Adelstein, R. S., 1983a, Purification and characterization of a multisubunit phosphatase from turkey gizzard smooth muscle. The effect of calmodulin binding to myosin light chain, *J. Biol. Chem.*, 258: 7047.

Pato, M. D. and Adelstein, R. S., 1983b, Characterization of a Mg^{2+}-dependent phosphatase from turkey gizzard smooth muscle, *J. Biol. Chem.*, 258: 7055.

Pato, M. D. and Kerc, E., 1985, Limited proteolytic digestion and dissociation of smooth muscle phosphatase-I modifies its substrate specificity. Preparation and properties of different forms of smooth muscle phosphatase-I, *J. Biol. Chem.*, 260: 12359.

Schliselfeld, L. H., and Bárány, M., 1968, The binding of adenosine triphosphate to myosin, *Biochemistry*, 7: 3206.

Sellers, J. R., Chantler, P. D., and Szent-Gyorgyi, A. G., 1980, Hybrid formation between scallop myofibrils and foreign regulatory light chains, *J. Mol. Biol.*, 144: 223.

Suzuki, H., Onishi, H., Takahashi, K., and Watanabe, S., 1978, Structure and function of chicken gizzard myosin, *J. Biochem.*, 84: 1529.

Suzuki, H., Kamata, T., Onishi, H., and Watanabe, S., 1982, Adenosine triphosphate-induced reversible change in the conformation of chicken gizzard myosin and heavy meromyosin, *J. Biochem.*, 91: 1699.

Suzuki, H. Stafford III, W. F., Slayter, H. S., and Seidel, J. C., 1985, A conformational transition in gizzard heavy meromyosin involving the head-tail junction, resulting in changes in sedimentation coefficient, ATPase activity, and orientation of heads, *J. Biol. Chem.*, 260: 14810.

Trybus, K. M., Huiatt, T. W., and Lowey, S., 1982, A bent monomeric conformation of myosin from smooth muscle, *Proc. Nat'l. Acad. Sci. U.S.A.*, 79: 6151.

Trybus, K. M. and Lowey, S., 1988, The regulatory light chain is required for folding of smooth muscle myosin, *J. Biol. Chem.*, 263: 16485.

Werth, D. K., Haeberle, J. R., and Hathaway, D. R., 1982, Purification of a myosin phosphatase from bovine aortic smooth muscle, *J. Biol. Chem.*, *257: 7306*.

BIOCHEMICAL AND FUNCTIONAL CHARACTERIZATION OF SMOOTH MUSCLE CALPONIN

Steven J. Winder, Cindy Sutherland, and Michael P. Walsh

Department of Medical Biochemistry
Faculty of Medicine
University of Calgary
3330 Hospital Drive N.W.
Calgary, Alberta T2N 4N1, Canada

INTRODUCTION

Calponin (calcium- and calmodulin-binding troponin T-like protein) was first described by Takahashi et al. (1986) and isolated from chicken gizzard smooth muscle taking advantage of its heat-stability. The isolated protein was shown to bind calmodulin (by affinity chromatography) in a Ca^{2+}-dependent manner, and F-actin or F-actin-tropomyosin (by analytical ultracentrifugation) in a Ca^{2+}-independent manner. Its tissue content was estimated to be equimolar to tropomyosin. These properties suggested that calponin may be a thin filament-associated protein involved in regulating the contractile state of smooth muscle. This notion was supported by our analysis of thin filament preparations from chicken gizzard which were designed to retain actin-binding protein components (Ngai et al., 1987). Such thin filament preparations (cf. Marston and Lehman, 1985) contain, in addition to actin and tropomyosin, caldesmon (140 kDa) and a 32 kDa protein later shown to be identical to calponin (Fig. 1 right-hand panel). This figure also shows the calponin in gizzard actomyosin preparations (left-hand panel).

Our intention here is to review the biochemical characteristics of calponin and its functional effects and regulation in an *in vitro* "contractile" system in which the purified contractile and regulatory proteins of smooth muscle are reconstituted.

BIOCHEMICAL CHARACTERIZATION

Purified calponin migrates as a protein of 34 kDa on denaturing polyacrylamide gel electrophoresis. Gel filtration under non-denaturing conditions indicates the protein is a monomer (Takahashi et al., 1986). The possibility that calponin exists as two isoforms in chicken gizzard was suggested by the chromatofocusing results of Takahashi et al. (1986) and confirmed by us using SDS-PAGE and isoelectric focusing (Winder and Walsh, 1990a). Scatchard

Regulation of Smooth Muscle Contraction
Edited by R.S. Moreland, Plenum Press, New York, 1991

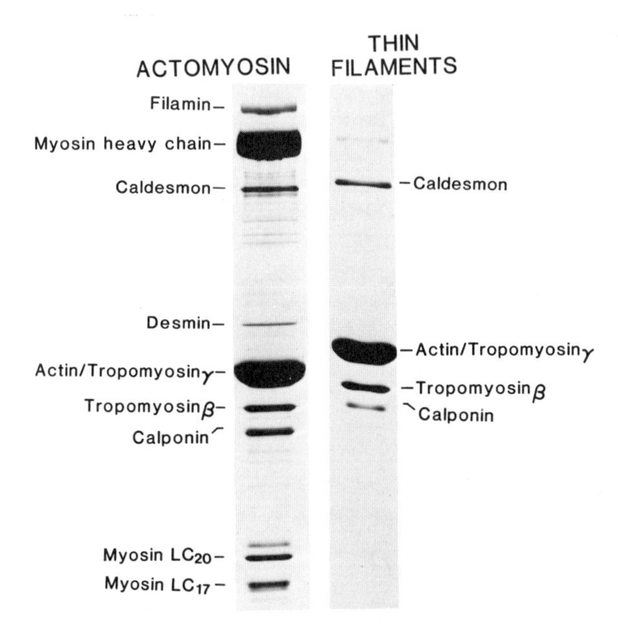

Figure 1. Calponin in smooth muscle actomyosin and native thin filament preparations. Chicken gizzard actomyosin was prepared according to Sobieszek and Bremel (1975) and native thin filaments according to Marston and Lehman (1985). Protein compositions were analyzed by SDS-PAGE with Coomassie Blue staining. Protein components were identified by molecular weight, co-electrophoresis with authentic purified proteins, and immunoblotting with specific polyclonal antibodies raised in rabbits to the isolated chicken gizzard proteins.

analysis of F-actin-calponin and F-actin-tropomyosin-calponin sedimentation data gave the binding constants and stoichiometries indicated in Table 1. Tropomyosin has no significant effect on the affinity of actin for calponin. Binding to skeletal muscle actin or actin-tropomyosin is significantly weaker (7- to 8-fold) than to the corresponding smooth muscle proteins. In all cases, the maximum stoichiometry of binding is ≈ 3 actin monomers/calponin.

TISSUE AND CELLULAR DISTRIBUTION

Using polyclonal antibodies raised in rabbits against calponin purified from chicken gizzard, Takahashi et al. (1987a) detected immunoreactive proteins in the molecular weight range 33-35 kDa in all bovine smooth muscle tissues examined (aorta, esophagus, stomach, trachea, and uterus). In addition, we have observed immunoreactive proteins in toad stomach, rat uterus, and *Aplysia* gut (Winder, Kang, and Walsh, unpublished results). No immunoreactive species were detected in bovine intestinal membrane vesicles, cardiac atrium and ventricle, or brain cortex, but a 36 kDa cross-reactive protein was identified in bovine adrenal medulla and cortex. The bovine aortic protein was subsequently purified (Takahashi et al., 1988a) and shown to resemble chicken gizzard calponin in many respects. Both proteins exhibited some, albeit weak, immunological cross-reactivity with striated muscle troponins T, and bound troponin C of rabbit skeletal muscle in a Ca^{2+}-dependent manner. The common antigenic site was localized to the C-terminus of troponin T, known to be the site of binding of tropomyosin. Furthermore, bovine aortic and chicken gizzard calponins bound to immobilized smooth muscle tropomyosin. At this time, Takahashi et al. (1988a) coined the term "calponin" for this protein. Electron microscopy of calponin bound to tropomyosin paracrystals revealed 40 nm periodicity and indicated the calponin binding site is located ≈17 nm from the C-terminus of tropomyosin (Takahashi et al., 1988b). This is the same site at which troponin T binds to tropomyosin. Immunofluorescence microscopy of cultured aortic smooth muscle cells localized calponin to the microfilament bundles (stress fibers) (Takahashi et al., 1988c).

Table 1. Binding parameters of calponin to actin and actin/tropomyosin of smooth and skeletal muscles

	K_d		n	
Actin (and tropomyosin)	-Tm	+Tm	-Tm	+Tm
Smooth muscle	4.6×10^{-8} M	4.3×10^{-8} M	3.3	2.9
Skeletal muscle	3.6×10^{-7} M	3.2×10^{-7} M	3.1	3.1

The binding parameters were obtained by Scatchard analysis of sedimentation data obtained as described by Winder and Walsh (1990b). Tm = tropomyosin; n = maximum binding stoichiometry (actin monomer/calponin).

FUNCTIONAL PROPERTIES

Several features of calponin suggest that it may play a role in Ca^{2+}-mediated regulation of actin-myosin interaction and therefore the contractile state of smooth muscles: (1) it is a thin filament protein, binding to both actin and tropomyosin; (2) it binds calmodulin in a Ca^{2+}-dependent manner and may also bind Ca^{2+} directly; (3) it is present in smooth muscle at the same molar concentration as tropomyosin; and (4) it shows some similarities to the troponin complex of striated muscles. For these reasons, we have examined the effects of purified calponin on an *in vitro* system reconstituted from the purified contractile and regulatory proteins of smooth muscle: actin, myosin, tropomyosin, calmodulin, and myosin light chain kinase (Winder and Walsh, 1990b,c). Figure 2 shows that calponin inhibits the actin-activated MgATPase activity of myosin in a dose-dependent manner, but has no effect on the Ca^{2+}/calmodulin-dependent phosphorylation of myosin catalyzed by myosin light chain kinase. We consistently observe maximal inhibition of ≈80% of the ATPase rate.

Calponin also inhibited superprecipitation of actomyosin in the reconstituted system (Fig. 3). Inhibition was shown to be due to calponin since it was lost following immunoprecipitation with specific polyclonal antibodies raised in rabbits against the chicken gizzard protein. Furthermore, inhibition was not a nonspecific effect due to the basic nature of calponin since two other basic proteins, chymotrypsinogen (pI = 9.5) and ribonuclease A (pI = 9.6), even at concentrations as high as 10 μM, had no effect on the actin-activated myosin MgATPase. Inhibition by calponin of the actomyosin ATPase was not dependent on the presence of tropomyosin. It is likely that an inhibitor of skeletal muscle actomyosin MgATPase partially purified from frozen chicken gizzard by Makioka and Hirabayashi (1978) is identical to calponin.

Since it is unlikely that the tissue would exhibit a constitutively active actomyosin ATPase inhibitor, we considered the possibility that calponin's inhibitory function may be regulated. Takahashi et al. (1987b) demonstrated by UV absorption difference spectroscopy that calponin can bind Ca^{2+} directly, albeit with relatively low affinity ($K_d \approx 7$ μM). We confirmed this using a $^{45}Ca^{2+}$ transblot overlay method. Therefore we investigated the possibility that calponin-mediated inhibition of the actin-activated myosin MgATPase may be Ca^{2+}-dependent by observing the effect of calponin in the presence and absence of Ca^{2+} on the ATPase activity of acto-phosphorylated myosin (Fig. 4) or acto-thiophosphorylated myosin (Table 2). In the case of acto-phosphorylated myosin in the absence of calponin (Fig. 4), ATPase rates of 114.7 and 91.4 nmol P_i/mg myosin.min were determined in the presence and absence, respectively, of Ca^{2+}. Inhibition by calponin was observed both in the presence (to 40.4 nmol P_i/mg myosin.min) and absence (to 25.5 nmol P_i/mg myosin.min) of Ca^{2+}, i.e. 65% and 78% inhibition, respectively. The lower ATPase rates observed in the absence of Ca^{2+} in each case were due to a low level of dephosphorylation of

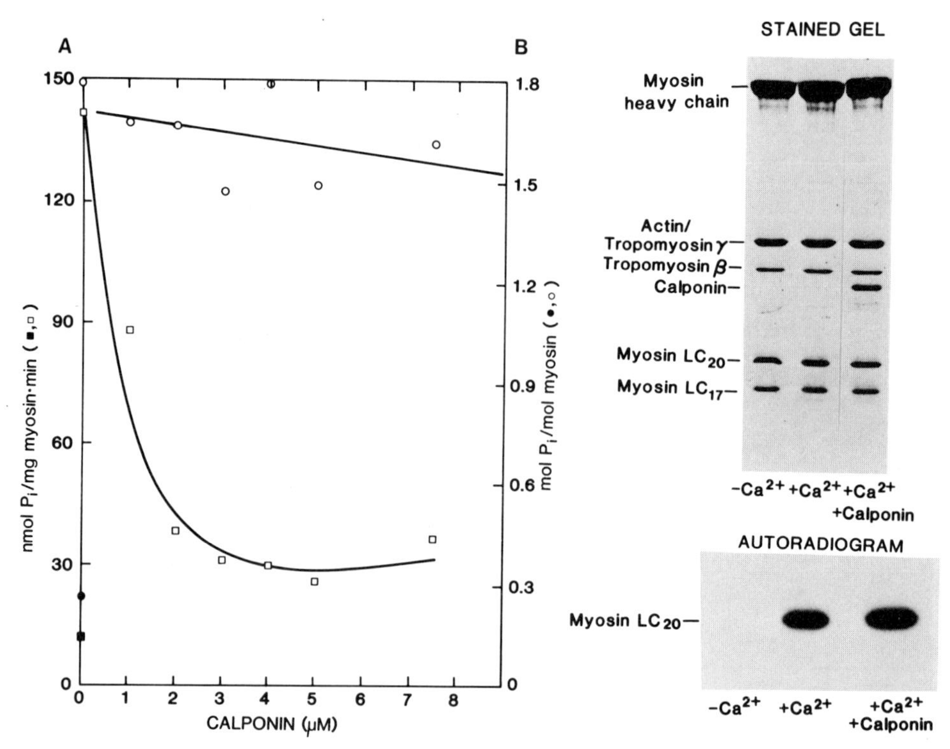

Figure 2. Calponin-mediated inhibition of the actin-activated MgATPase activity of smooth muscle myosin. A: actomyosin ATPase rates (□, ■) and myosin phosphory-lation levels (○, ●) in a reconstituted contractile system of: 1 μM myosin, 6 μM actin, 2 μM tropomyosin, 1 μM calmodulin, and 74 nM myosin light chain kinase in 25 mM Tris-HCl (pH 7.5), 10 mM $MgCl_2$, 60 mM KCl, 1 mM [γ-^{32}P]-ATP (≈10 cpm/pmol) were quantified at the indicated calponin concentrations as described by Winder and Walsh (1990b) in the presence (○, □) or absence (●, ■) of Ca^{2+}. B: selected reaction mixtures from A were analyzed by SDS-PAGE and autoradiography. These represent the fully reconstituted system incubated with Mg[γ-^{32}P]-ATP in the absence of Ca^{2+} (-Ca^{2+}), in the presence of Ca^{2+} (+Ca^{2+}), and in the presence of Ca^{2+} and 3 μM calponin (+Ca^{2+} + calponin).

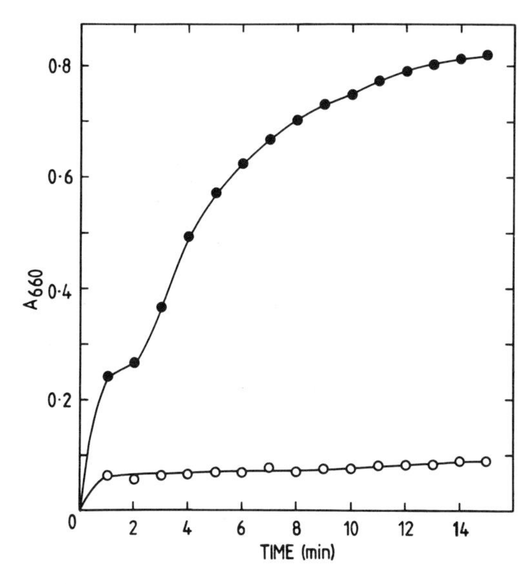

Figure 3. Inhibition of actomyosin superprecipitation by calponin. Super-precipitation of actomyosin was measured by an increase in A_{660nm} in the fully reconstituted system (see legend to Fig. 2) in the presence of Ca^{2+} and in the presence (○) and absence (●) of 5 μM calponin. Non-radiolabeled ATP (1 mM) was added to the reaction mixtures at time zero.

Table 2. Calponin-mediated inhibition of acto-thiophosphorylated myosin MgATPase is Ca^{2+}-independent

Additions at time zero			ATPase rate	
Actin	Calponin	Ca^{2+}	(nmol P_i/mg myosin.min)	% inhibition
+	-	+	66.5	-
+	-	-	62.3	-
+	+	+	24.2	63.6
+	+	-	23.0	63.1

myosin by Ca^{2+}-independent phosphatase activity once Ca^{2+} was removed from the assay system (Table 3). In separate experiments the myosin preparations used were shown to be contaminated with low levels of myosin phosphatase activity. Due to this problem of partial dephosphorylation of myosin, we repeated these experiments using thiophosphorylated rather than phosphorylated myosin, since the thiophosphorylated protein is resistant to the action of myosin phosphatase (Sherry et al., 1978). Experimental conditions were exactly as described in Figure 4 except that ATPγS replaced ATP in the prephosphorylation stage. ATP hydrolysis was then quantified following addition of radiolabeled ATP and actin in the presence and absence of calponin and in the presence and absence of Ca^{2+}. The data in Table 2 indicate that calponin inhibited the acto-thiophosphorylated myosin MgATPase rate by 63.6% in the presence of Ca^{2+} and by 63.1% in the absence of Ca^{2+}. From these experiments, we concluded that the inhibitory effect of calponin on the actin-activated myosin MgATPase is not regulated by direct binding of Ca^{2+} to calponin.

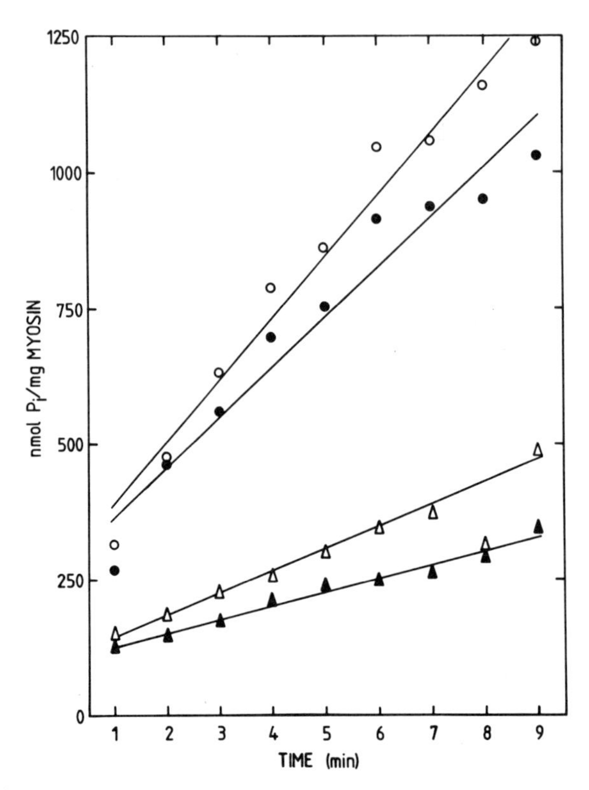

Figure 4. Calponin-mediated inhibition of acto-phosphorylated myosin MgATPase is Ca^{2+}-independent. Myosin was phosphorylated to 1.7 mol P_i/mol by incubation for 8 min at 30° C as described in the legend to Figure 2, but omitting actin. At time zero in this figure, actin (6 μM) was added with (Δ, ▲) or without (O, ●) calponin (5 μM) in the presence of 0.1 mM $CaCl_2$ (O, Δ) or 1mM EGTA (●, ▲). Samples were withdrawn at the indicated times following actin addition for quantification of ATP hydrolysis by the method of Ikebe and Hartshorne (1985).

We proceeded to investigate the possibility that calponin may be regulated by phosphorylation. Five purified protein kinases were examined for possible phosphorylation of calponin. As shown in Figure 5, Ca^{2+}/calmodulin-dependent protein kinase II and the Ca^{2+}- and phospholipid-dependent protein kinase (protein kinase C) phosphorylated purified calponin. On the other hand, cAMP- and cGMP-dependent protein kinases and myosin light chain kinase did not phosphorylate calponin. Both Ca^{2+}-dependent kinases exhibited a high degree of phosphorylation site specificity. Tryptic digestion of calponin phosphorylated by protein kinase C followed by two-dimensional thin-layer electrophoresis/chromatography revealed 3 phosphopeptides (Winder and Walsh, 1990b). These phosphopeptides were purified by reversed-phase HPLC and sequenced (Table 4). Phosphoamino acid analysis by thin-layer electrophoresis of acid hydrolysates of the purified phosphopeptides revealed that the phosphorylated residue in each case was phosphoserine. The phosphorylation sites are indicated in Table 4. Since these phosphopeptides were generated by tryptic digestion, the preceding amino acid in each case must be lysine or arginine. These phosphorylation sites are therefore consistent with the established site specificity of protein kinase C (House et al., 1987).

Calponin phosphorylated by either protein kinase C or Ca^{2+}/calmodulin-dependent protein kinase II had no significant inhibitory effect on the actin-activated myosin MgATPase (Fig. 6). Calponin inhibition of the ATPase can therefore be prevented by phosphorylation by either of these Ca^{2+}-dependent kinases.

Table 3. Myosin light chain dephosphorylation in the absence of Ca^{2+}

Additions at time zero			mol P_i/mol myosin	
Actin	Calponin	Ca^{2+}	t = 1 min	t = 9 min
+	-	+	1.71	1.64
+	-	-	1.86	1.23
+	+	+	1.89	1.78
+	+	-	1.72	1.32

Myosin phosphorylation levels were measured at times 1 and 9 min on the samples used for ATPase assays in Figure 4 as described by Walsh et al. (1983).

Table 4. Amino-terminal sequences of tryptic phosphopeptides derived from calponin phosphorylated by protein kinase C.

Peptide #	Amino-terminal sequence
1	FAS*QQGMTAYGTR
2	GAS*QQGMTVYGLP
3	NHS*GHVQ

Peptide # as described by Winder and Walsh (1990b). The asterisks in the amino terminal sequences indicate the phosphorylated serine residue in each case.

With a view to defining the mechanism of calponin inhibition of the actomyosin ATPase, we compared the binding properties of phosphorylated and unphosphorylated calponins. Both species bound to immobilized tropomyosin or Ca^{2+}/calmodulin. However, unlike unphosphorylated calponin, the phosphorylated protein did not bind to F-actin (Fig. 7A). Similar results were obtained with actin/tropomyosin (Winder and Walsh, 1990b). Similarly, in the fully reconstituted actomyosin system, phosphorylated calponin did not sediment with actin and the other components of the system (Fig. 7B). We concluded from these binding studies that it is the interaction of calponin with actin that is responsible for inhibition of the actomyosin ATPase.

PHYSIOLOGICAL SIGNIFICANCE

Figure 8 illustrates a model which may explain the physiological role of calponin in regulating smooth muscle contraction. The *in vitro* results described above suggest that, in the resting smooth muscle cell in which the sarcoplasmic free [Ca^{2+}] is low (≈130 nM) (Williams and Fay, 1986), calponin would be bound to actin. Stimulation of the cell leading to an elevation of

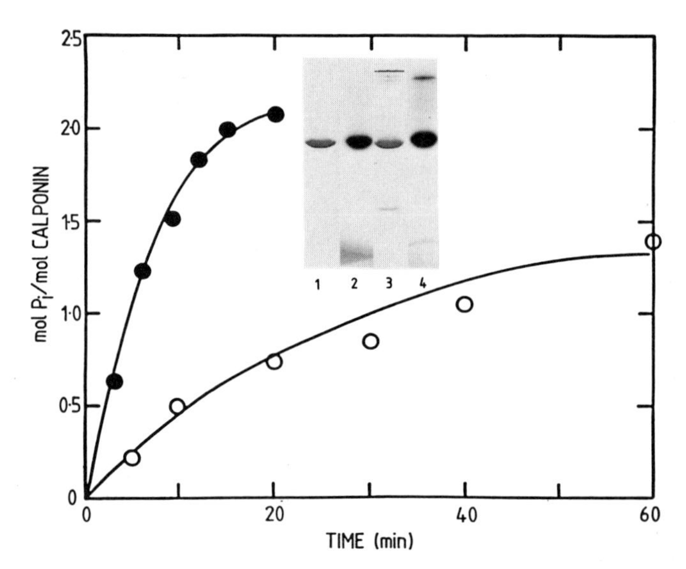

Figure 5. Phosphorylation of calponin. Calponin phosphorylation by protein kinase C (○) and Ca^{2+}/calmodulin-dependent protein kinase II (●) was quantified as described by Walsh et al. (1983). Key to inset: lanes 1 and 3, Coomassie Blue-stained SDS gels of calponin phosphorylated with protein kinase C and Ca^{2+}/calmodulin-dependent protein kinase II, respectively; lanes 2 and 4, corresponding autoradiograms.

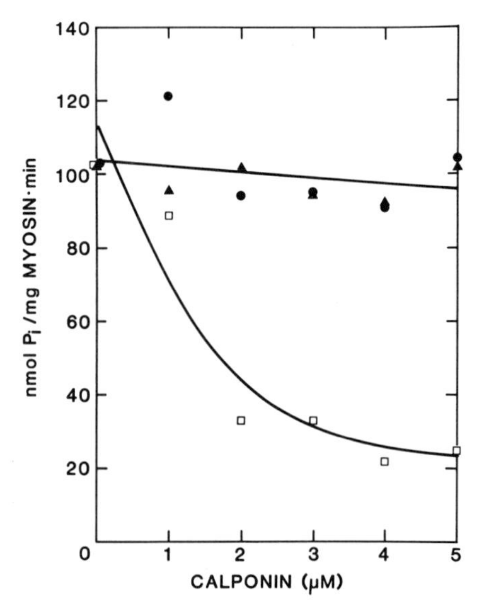

Figure 6. Phosphorylation blocks calponin-mediated inhibition of the actin-activated myosin MgATPase. ATPase rates were measured as described in the legend to Figure 2 in the presence of the indicated concentrations of unphosphorylated calponin (□), calponin previously phosphorylated with protein kinase C (▲), or Ca^{2+}/calmodulin-dependent protein kinase II (●). Reproduced from Winder and Walsh (1990b) with permission.

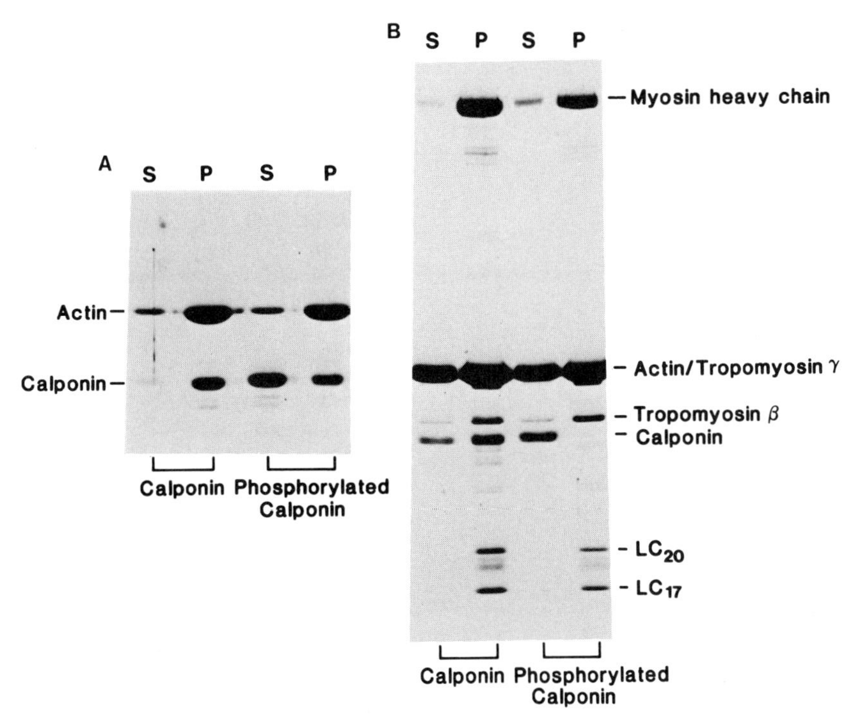

Figure 7. Phosphorylation blocks the calponin-actin interaction. A: actin (11μM) and calponin (phosphorylated or unphosphorylated; 4 μM) were incubated at 25° C for 30 min in 20 mM Tris-HCl (pH 7.5), 100 mM KCl, 2 mM $MgCl_2$, 1 mM ATP, 1 mM dithiothreitol, 0.1 mM $CaCl_2$ prior to centrifugation at 100,000 x g for 1h at 2° C to sediment actin and actin-bound calponin. Supernatants (S) and pellets (P) were subjected to SDS-PAGE. Lanes 1 and 2, actin + unphosphorylated calponin; lanes 3 and 4, actin + phosphorylated calponin. When centrifuged alone, calponin was recovered exclusively in the supernatant. B: sedimentation assays were carried out with the fully reconstituted contractile system in the presence of phosphorylated or unphosphorylated calponin (2 μM) following a 7 min incubation at 30° C under ATPase assay conditions. Lanes 1 and 2, with unphosphorylated calponin; lanes 3 and 4, with phosphorylated calponin.

sarcoplasmic free [Ca^{2+}] (to 500 - 700 nM) (Williams and Fay, 1986; Williams et al., 1987) would activate Ca^{2+}/calmodulin-dependent protein kinase II, and possibly protein kinase C, resulting in calponin phosphorylation and its dissociation from the thin filament. In this condition, myosin would be phosphorylated by Ca^{2+}/calmodulin-dependent myosin light chain kinase and crossbridge cycling would proceed at a maximal rate. Since myosin and calponin are phosphorylated by distinct Ca^{2+}-dependent kinases which have different Ca^{2+} sensitivities, it is possible that, at intermediate [Ca^{2+}], myosin may be fully or partially phosphorylated while calponin is completely or predominantly dephosphorylated and therefore reassociates with the thin filament. In this condition, phosphorylated myosin heads would have limited access to actin and therefore the number of attached crossbridges would be low. The force developed could be very finely tuned by alterations in the relative proportions of phosphorylated and dephosphorylated calponin and myosin.

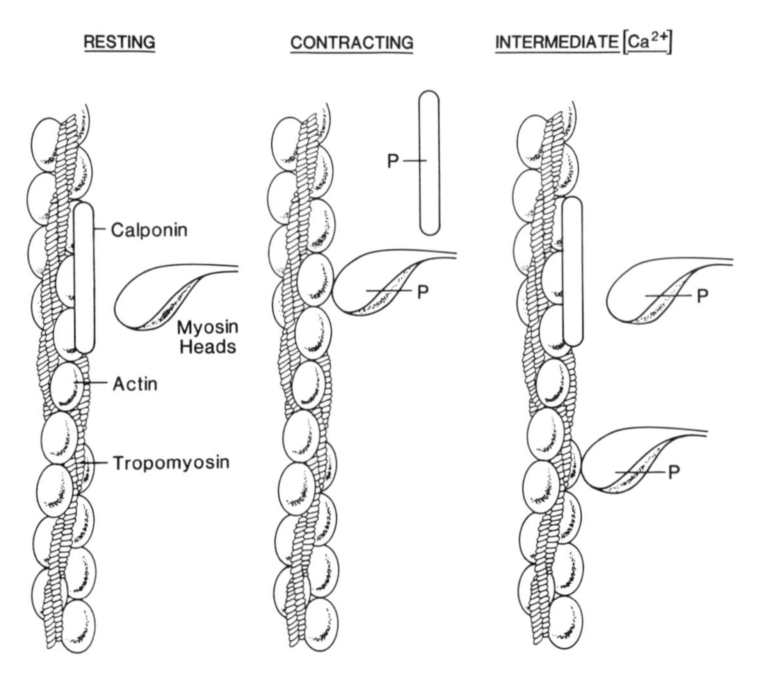

Figure 8. A model depicting the possible physiological role of calponin in regulating smooth muscle actin-myosin interaction. Calponin is shown spanning three actin monomers simply because this is the maximal binding stoichiometry achieved *in vitro*. The calponin content *in situ* is estimated at 1 mol/7 actin monomers. Only the S-1 regions of myosin are shown for simplicity. P denotes phosphorylation.

FUTURE PERSPECTIVES

The model presented in Figure 8 is based exclusively on *in vitro* experimental observations. It will be necessary to test its validity using skinned and intact fiber preparations. For example, it may be possible to remove calponin selectively from skinned fibers, as has been done for troponin C of skeletal muscle (Kerrick et al., 1985), and observe the consequences for the contractile properties. It may additionally be possible to add back purified calponin or phosphorylated calponin and evaluate their effects. With regards to phosphorylation, it will be necessary to determine whether or not calponin is phosphorylated in skinned fibers in response to Ca^{2+}/calmodulin-dependent protein kinase II or protein kinase C, and in intact fibers in response to stimuli which cause an elevation of sarcoplasmic free [Ca^{2+}], e.g. K^+ depolarization, and/or polyphosphoinositide turnover, e.g. phenylephrine. In such studies, it would be most desirable to identify the phosphorylation sites in calponin in the skinned and intact fiber preparations for comparison with the sites phosphorylated *in vitro* by the purified kinases, as has been done for the 20 kDa light chain of myosin (Colburn et al., 1988).

If calponin phosphorylation is to be of physiological significance, there must also be a phosphatase in smooth muscle which is capable of dephosphorylating calponin. The data in Figure 9 indicate that such a phosphatase does exist in chicken gizzard smooth muscle. In this experiment, which was done for another purpose, i.e. to investigate calponin phosphorylation in thin filaments rather than with the isolated protein, thin filaments were

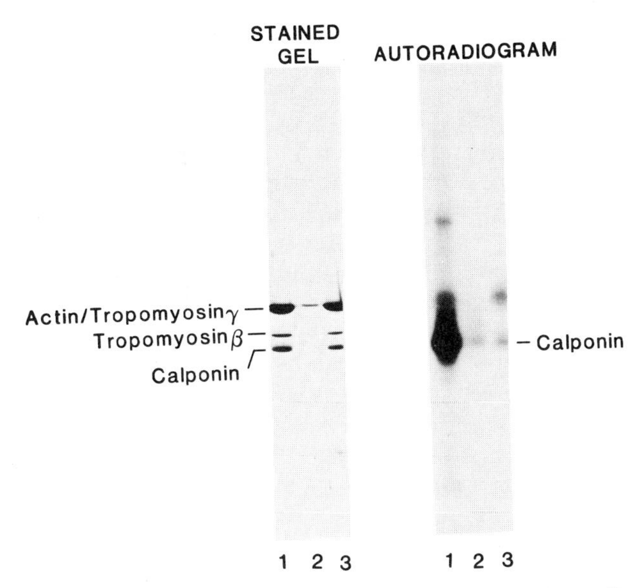

Figure 9. Chicken gizzard contains a calponin phosphatase. Actin (21 μM), tropomyosin (3 μM), and calponin (3 μM) were incubated for 30 min at 25° C in 20 mM Tris-HCl (pH 7.5), 100 mM KCl, 2 mM $MgCl_2$, 1 mM dithiothreitol, and 1 mM ATP. Reconstituted thin filaments were obtained by centrifugation at 2° C for 1h at 100,000 x g. The resultant pellet was resuspended in 20 mM Tris-HCl (pH 7.5), 100 mM KCl, 2 mM $MgCl_2$, and 1 mM dithiothreitol. Protein kinase C, phosphatidylserine and diolein were added (Kikkawa et al., 1982) and the reaction started by addition of [γ-^{32}P]-ATP (0.1 mM). After incubation for 60 min at 30° C, a sample was withdrawn for SDS-PAGE and autoradiography (lane 1). The remainder of the reaction mixture was centrifuged at 2° C at 100,000 x g for 1h prior to SDS-PAGE and autoradiographic analysis of the supernatant (lane 2) and pellet (lane 3).

reconstituted by combining gizzard actin, tropomyosin and calponin in 7:1:1 molar ratio. The sedimented thin filaments were resuspended and phosphorylated with protein kinase C prior to SDS-PAGE and autoradiography (Fig. 9, lane 1). Alternatively, the phosphorylated thin filaments were centrifuged for 1h at 100,000 x g prior to SDS-PAGE and autoradiography of the resolved supernatant (lane 2) and pellet (lane 3). The phosphorylation of calponin is clearly evident in lane 1 but, following centrifugation of the reaction mixture, most of the calponin was recovered in the actin/tropomyosin pellet and was clearly dephosphorylated (compare lanes 2 and 3). The reconstituted thin filaments therefore must contain a protein phosphatase which is presumably bound to some component of the thin filaments. During the centrifugation step protein kinase C and the phosphatase probably separate, partitioning to the supernatant and pellet, respectively. It will be of interest to isolate this phosphatase and compare its properties with known phosphatases classified according to Cohen (1989).

Another consideration for the future is the functional relationship between calponin and caldesmon (Walsh, 1990). These two proteins have a number of similar properties: (i) they inhibit the actin-activated MgATPase activity of smooth and skeletal muscle myosins; (ii) they both bind F-actin and tropomyosin in a Ca^{2+}-independent manner and calmodulin in a Ca^{2+}-dependent manner; (iii) both are substrates of Ca^{2+}/calmodulin-dependent protein kinase II and protein kinase C; and (iv) both lose their ability to inhibit the actomyosin ATPase upon phosphorylation. It is clear, however, that caldesmon and calponin are quite distinct proteins and calponin is not derived from caldesmon (M_r 140 kDa) by proteolysis since antibodies against the two proteins recognize only the antigen to which they were raised. Caldesmon has been shown to bind to myosin, the binding site on caldesmon being located near the amino-terminus, i.e. at the opposite end of the molecule from the actin-, tropomyosin- and calmodulin-binding sites (Sutherland and Walsh, 1989). On the other hand, calponin does not interact with myosin (Winder and Walsh, 1990b). The tissue concentrations of caldesmon and calponin differ substantially: 11 μM and 150 μM, respectively, suggesting that they probably do not function together as a complex analogous to the troponin complex of striated muscles. We have initiated a series of experiments designed to elucidate the functional relationship between caldesmon and calponin. In the

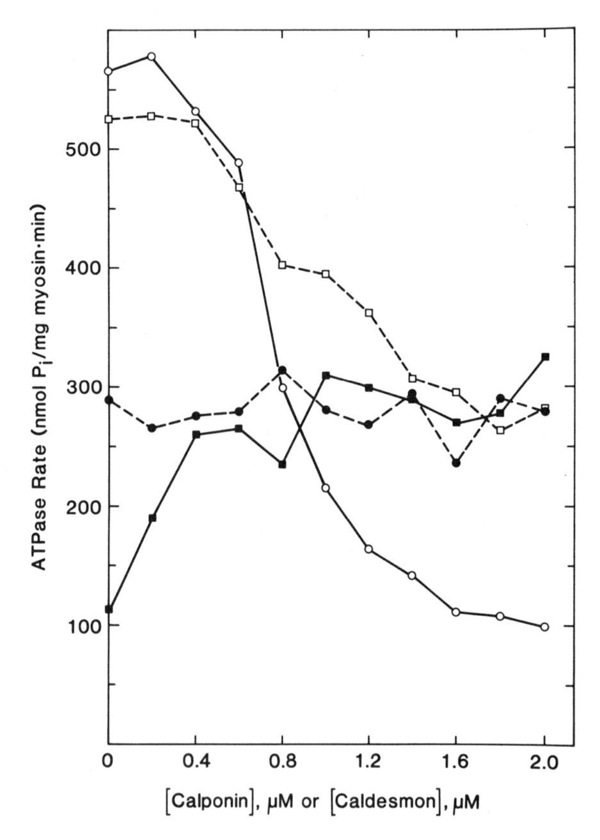

Figure 10. Combined effects of calponin and caldesmon on skeletal muscle actin-activated myosin MgATPase. Reaction conditions: 3.6 μM skeletal actin, 0.57 μM skeletal myosin, 25 mM Tris-HCl (pH 7.5), 50 mM KCl, 1 mM dithiothreitol, 3.5 mM $MgCl_2$, 0.2 mM EGTA, 1 mM [γ-^{32}P]-ATP, 30° C. Reactions were started by addition of radiolabeled ATP. Samples were withdrawn at 1, 2, 3, 4, and 5 min for quantification of ATP hydrolysis (Ikebe and Hartshorne, 1985). ATPase rates were calculated by linear regression analysis of the time-course data. Key to symbols: ○, varying calponin, no caldesmon; ●, varying calponin, 2 μM caldesmon; □, varying caldesmon, no calponin; ■, varying caldesmon, 2 μM calponin.

initial stages of this work, we are examining the effects of caldesmon and calponin on the skeletal muscle actomyosin MgATPase since this unregulated system has no requirement for myosin phosphorylation. In preliminary experiments (Fig. 10), we observed that calponin was more potent than caldesmon in inhibiting this ATPase. When both proteins were included together in the assay system, the level of inhibition achieved was equivalent to that of caldesmon alone. Most interestingly, addition of increasing concentrations of caldesmon to the assay system containing a concentration of calponin giving maximal inhibition of the ATPase caused a reduction of inhibition to the level seen with caldesmon alone. The effects of caldesmon on the binding of calponin to actin, and the effects of calponin on the binding of caldesmon to actin must be examined in detail.

Takahashi et al. (1990) recently reported, in abstract form, the cloning and sequencing of a calponin cDNA. This will pave the way for site-directed mutagenesis studies of the structure-function relations of calponin, enabling us to elucidate the relative importance of the many properties of smooth muscle calponin in regulation of smooth muscle contraction.

ACKNOWLEDGMENTS

The authors gratefully acknowledge the financial support of the Alberta Heart and Stroke Foundation, the Alberta Heritage Foundation for Medical Research, and the Medical Research Council of Canada, and the secretarial assistance of Gerry Garnett.

REFERENCES

Cohen, P., 1989, The structure and regulation of protein phosphatases, *Ann. Rev. Biochem.*, 58: 453.

Colburn, J. C., Michnoff, C. H., Hsu, L.- C., Slaughter, C. A., Kamm, K. E., and Stull, J. T., 1988, Sites phosphorylated in myosin light chain in contracting smooth muscle, *J. Biol. Chem.*, 263: 19166.

House, C., Wettenhall, R. E. H., and Kemp, B. E., 1987, The influence of basic residues on the substrate specificity of protein kinase C, *J. Biol. Chem.*, 262: 772.

Ikebe, M. and Hartshorne, D. J., 1985, Proteolysis of smooth muscle myosin by Staphylococcus aureus protease: Preparation of heavy meromyosin and subfragment 1 with intact 20,000-dalton light chains, *Biochemistry*, 24: 2380.

Kikkawa, U., Takai, Y., Minakuchi, R., Inohara, S., and Nishizuka, Y., 1982, Calcium-activated, phospholipid-dependent protein kinase from rat brain: Subcellular distribution, purification and properties, *J. Biol. Chem.*, 257: 13341.

Kerrick, W. G. L., Zot, H. G., Hoar, P. E., and Potter, J. D., 1985, Evidence that the Sr^{2+} activation properties of cardiac troponin C are altered when substituted into skinned skeletal muscle fibers, *J.Biol. Chem.*, 260: 15687.

Makioka, A. and Hirabayashi, T., 1978, Isolation and localization from chicken gizzard of an inhibitory protein for Mg^{2+}-activated skeletal muscle actomyosin ATPase, *J. Biochem.*, 84: 947.

Marston, S. B., and Lehman, W., 1985, Caldesmon is a Ca^{2+} regulatory component of native smooth muscle thin filaments, *Biochem. J.*, 231: 517.

Ngai, P. K., Scott-Woo, G. C., Lim, M. S., Sutherland, C., and Walsh, M. P., 1987, Activation of smooth muscle myosin Mg^{2+}-ATPase by native thin filaments and actin/tropomyosin, *J. Biol. Chem.*, 262: 5352.

Sherry, J. M. F., Gorecka, A., Aksoy, M. O., Dabrowska, R., and Hartshorne, D.J., 1978, Roles of calcium and phosphorylation in the regulation of the activity of gizzard myosin, *Biochemistry*, 17: 4411.

Sobieszek, A. and Bremel, R. D., 1975, Preparation and properties of vertebrate smooth-muscle myofibrils and actomyosin, *Eur. J. Biochem.*, 55: 49.

Sutherland, C. and Walsh, M. P., 1989, Phosphorylation of caldesmon prevents its interaction with smooth muscle myosin, *J. Biol. Chem.*, 264: 578.

Takahashi, K., Hiwada, K., and Kokubu, T., 1986, Isolation and characterization of a 34000-dalton calmodulin- and F-actin-binding protein from chicken gizzard smooth muscle, *Biochem. Biophys. Res. Commun.*, 141: 20.

Takahashi, K., Hiwada, K., and Kokubu, T., 1987a, Occurrence of anti- gizzard P34K antibody cross-reactive components in bovine smooth muscles and non-smooth muscle tissues, *Life Sci.*, 41: 291.

Takahashi, K., Hiwada, K., and Kokubu, T., 1987b, Vascular smooth muscle calponin; A novel calcium- and calmodulin-binding troponin T-like protein, *Hypertension*, 10: 360 (abstr).

Takahashi, K., Hiwada, K., and Kokubu, T., 1988a, Vascular smooth muscle calponin. A novel troponin T-like protein, *Hypertension*, 11: 620.

Takahashi, K., Abe, M., Hiwada, K., and Kokubu, T., 1988b, A novel troponin T-like protein (calponin) in vascular smooth muscle: interaction with tropomyosin paracrystals, *J. Hypertension*, 6: S40.

Takahashi, K., Abe, M., Hiwada, K., and Kokubu, T., 1988c, Smooth muscle calponin: a novel troponin T-like protein in thin filaments, *in:* "Proceedings Int. Symp. on Calcium-Binding Proteins in Health and Disease", 6: 69 (abstr.).

Takahashi, K., Abe, M., Nishida, W., Hiwada, K., and Nadal-Ginard, B., 1990, Primary structure, tropomyosin-binding and regulatory function of smooth muscle calponin: A new homolog of troponin T, *J. Muscle Res. Cell Motility*, 11: 435 (abstr.).

Walsh, M. P., 1990, Smooth muscle caldesmon, *in:* "Frontiers in Smooth Muscle Research", N. Sperelakis and J. D. Wood, eds., Wiley-Liss, New York, p. 127.

Walsh, M. P., Hinkins, S., Dabrowska, R., and Hartshorne, D. J., 1983, Purification of smooth muscle myosin light chain kinase, *Methods Enzymol.*, 99: 279.

Williams, D. A. and Fay, F. S., 1986, Calcium transients and resting levels in isolated smooth muscle cells as monitored with quin-2, *Am. J. Physiol.*, 250: C779.

Williams, D. A., Becker, P. L., and Fay, F. S., 1987, Regional changes in calcium underlying contraction of single smooth muscle cells, *Science*, 235: 1644.

Winder, S. J. and Walsh, M. P., 1990a, Structural and functional characterization of calponin fragments, *Biochem. Int.*, 22: 335.

Winder, S. J., and Walsh, M. P., 1990b, Smooth muscle calponin. Inhibition of actomyosin MgATPase and regulation by phosphorylation, *J. Biol.Chem.*, 265: 10148.

Winder, S. and Walsh, M. P., 1990c, Inhibition of the actomyosin MgATPase by chicken gizzard calponin, *in:* "Frontiers in Smooth Muscle Research", N. Sperelakis and J. D. Wood, eds., Wiley-Liss, New York, p. 141.

IN VITRO EVIDENCE FOR SMOOTH MUSCLE CROSSBRIDGE MECHANICAL INTERACTIONS

David Warshaw[1] and Kathleen Trybus[2]

[1]University of Vermont
Department of Physiology & Biophysics
Burlington, VT 05405

[2]Brandeis University
Rosenstiel Research Center
Waltham, MA 02254

INTRODUCTION

Force production and shortening in smooth muscle result from the cyclic interaction between myosin crossbridges and actin filaments. Although the crossbridge mechanism in smooth muscle is qualitatively similar to that in skeletal muscle, smooth muscle has the unique ability of sustaining prolonged contractions with very little energy expenditure (Butler and Siegman, 1983). This economy of energy consumption may reflect a modulation of crossbridge cycling rate within the time course of a contraction. Investigators have postulated that this modulation is related to the regulatory processes that govern the crossbridge's interaction with actin.

In the relaxed muscle, crossbridges are detached or weakly bound to actin resulting in little resting tension. However, upon stimulation, intracellular calcium rises initiating a biochemical cascade resulting in phosphorylation of the 20 kDa myosin light chain (Kamm and Stull, 1985). Once phosphorylated, the crossbridge cycles rapidly, generating force and muscle shortening. This regulatory scheme suggests that phosphorylation acts merely as a switch to turn on crossbridge cycling. However, during the period of force maintenance, Murphy and coworkers observed that both the extent of light chain phosphorylation and velocity of shortening were reduced and correlated in time (Dillon et al., 1981). If shortening velocity is directly related to crossbridge cycling rate, then in addition to acting as a switch, the extent of phosphorylation may also modulate the rate of crossbridge cycling.

To account for the reduced shortening velocity, Murphy and coworkers proposed that crossbridges that were dephosphorylated while attached (i.e. latchbridges) would cycle slowly and thus act as an internal load to the remaining rapidly cycling phosphorylated crossbridges (Dillon et al., 1981). An alternative explanation is that cooperativity exists between crossbridges within a myosin filament such that the cycling rate of the entire crossbridge

Regulation of Smooth Muscle Contraction
Edited by R.S. Moreland, Plenum Press, New York, 1991

population is modulated and dependent on the extent of phosphorylated crossbridges within the filament.

We have used an *in vitro* motility assay to determine whether either of these two mechanisms are operative at the level of the myosin filament (Warshaw et al., 1990). In this assay the motion of single fluorescently labeled actin filaments are observed as they move over myosin filaments, containing known proportions of dephosphorylated and thiophosphorylated heads (Fig. 1B).

METHODS

The isolation and preparation of contractile proteins for the motility assay have been described in detail (Warshaw et al., 1990). In brief, smooth muscle myosin was purified from turkey gizzards and then desired proportions of dephosphorylated and thiophosphorylated monomers were mixed at high ionic strength (300 mM KCl). Once mixed, the ionic strength was lowered (100 mM KCl), forming myosin filaments (250 µg/ml) having varying proportions of dephosphorylated and thiophosphorylated heads. To stabilize these filaments against the depolymerizing effects of MgATP (see Fig. 1D), a monoclonal antibody to the smooth muscle myosin rod region (LMM.1) was used at a three-fold molar excess (Trybus and Henry, 1989). The presence of antibody had no effect on either actin-activated myosin ATPases or actin filament motility. In experiments requiring monomeric myosin mixtures, dephosphorylated and thiophosphorylated monomers were simply mixed and diluted to 250 µg/ml in a solution containing 300 mM KCl. Unregulated actin was isolated from chicken pectoralis and fluorescently labeled with rhodamine phalloidin.

A small flow through chamber (30 µl) was created by placing a nitrocellulose coated coverslip on a glass microscope slide (Fig. 1A). Either

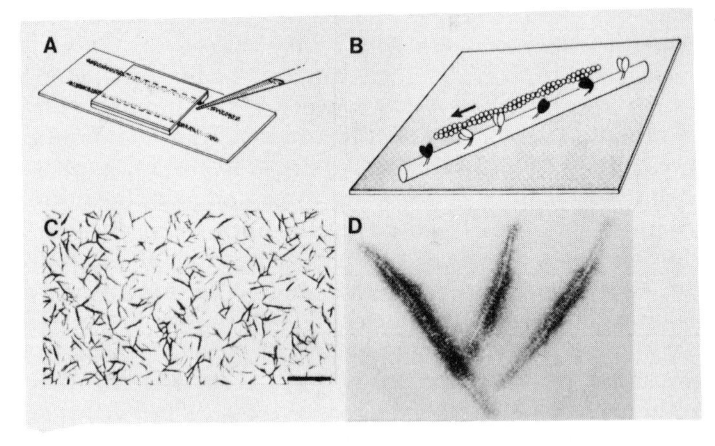

Figure 1. A) Flow cell illustration. B) Motility assay illustration with myosin filament containing dephosphorylated (black) and phosphorylated (white) crossbridge heads. Myosin filament is adhered to a nitrocellulose coated coverslip. C) Electron micrograph of negatively stained smooth muscle myosin filaments on nitrocellulose. Calibration bar = 1 µm. D) Electron micrograph of negatively stained antibody stabilized dephosphorylated myosin filaments in the presence of MgATP.

filamentous or monomeric myosin was introduced into the chamber and allowed to adhere to the coverslip. Myosin filaments were approximately 0.5 to 1.0 μm in length and randomly dispersed on the coverslip (see Fig. 1C). Once fluorescently labeled actin was perfused into the flow cell in the presence of millimolar MgATP, actin filaments were observed moving in a directed manner for distances greater than 10 μm. In control experiments considerably higher concentrations of monomeric myosin (1 mg/ml) were required for actin filament motility in contrast to filamentous myosin. However, in the presence of 0.5% methylcellulose, greater actin binding and motility were observed at concentrations (<250 μg/ml) similar to that used with filamentous myosin. Actin motility was observed in assay buffer containing either 25 or 80 mM KCl with 25 mM imidazole, 4 mM $MgCl_2$, 1 mM EGTA, 1 mM DTT at pH 7.4. Experiments were performed at 30° C.

The flow cell was placed on the stage of an inverted microscope equipped with epifluorescence. Actin images were recorded on video tape through an image intensified video camera. At a later time, video images were digitized by computer and analyzed for velocity of motion (Warshaw et al., 1990).

RESULTS AND DISCUSSION

Figure 2 show the results of varying the percentage of thiophosphorylated (lower horizontal axis) and dephosphorylated (upper horizontal axis) crossbridge heads in either filaments (open symbols) or monomeric mixtures (closed symbols) on actin filament velocity at low ionic strength (i.e. 25 mM KCl). Notice that the observed dependence of actin

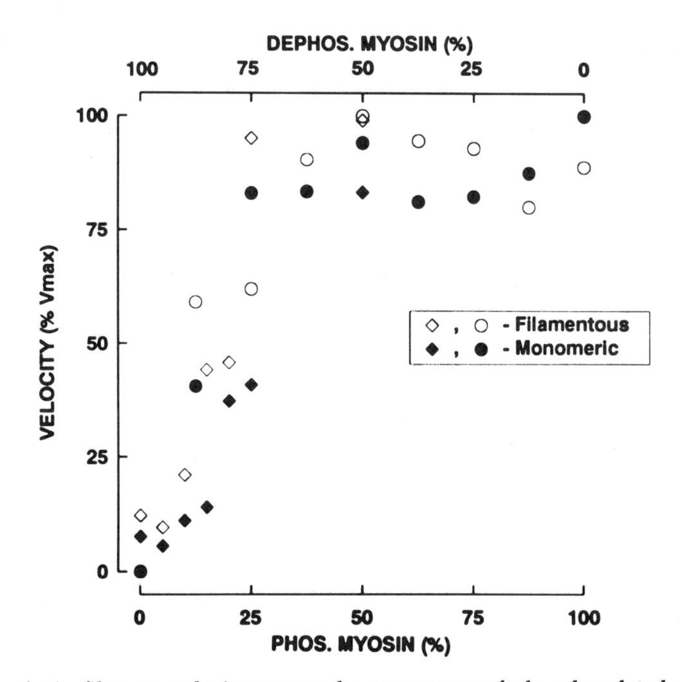

Figure 2. Actin filament velocity versus the percentage of phosphorylated myosin in both filamentous and monomeric forms. Experiments performed in 25 mM KCl assay buffer.

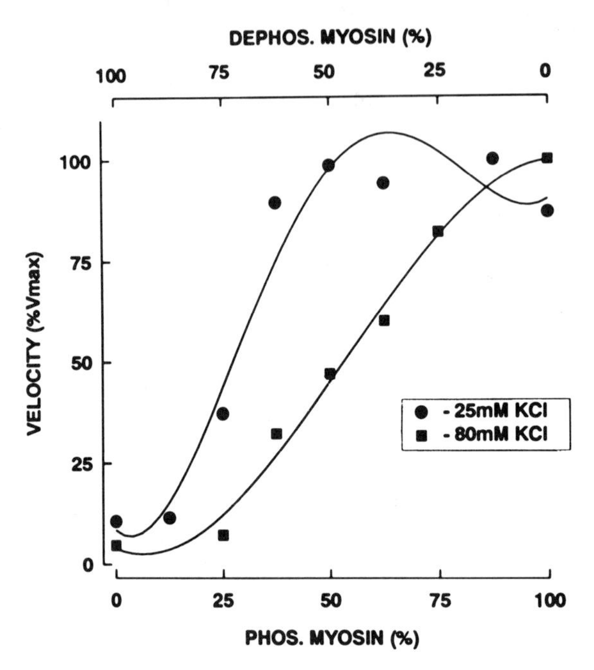

Figure 3. Actin filament velocity versus the percentage of monomeric phosphorylated myosin. Experiments were performed in both 25 and 80 mM KCl assay buffer.

filament velocity on extent of thiophosphorylation is similar for both filamentous and monomeric myosins. When 100% dephosphorylated myosin filaments or monomers were used, actin filaments were bound tightly to the myosin but no motion was detected. However, when the myosin was 100% thiophosphorylated, actin filaments moved at 0.6 μm/s. Thus thiophosphorylation acts as a switch to turn on rapid crossbridge cycling. As the percentage of dephosphorylated heads increased, actin filament velocity remained constant until the percentage of dephosphorylated heads increased to 75%, above which actin filament velocity slowed. A similar relationship was observed by Sellers et al. (1985) using the *Nitella* based movement assay. Thus in addition to acting as a switch, it appears that the extent of thiophosphorylation can also modulate crossbridge cycling rate.

The apparent modulation of actin filament velocity could be due to cooperativity between crossbridge heads within the myosin filament, where the cycling rate of the entire crossbridge population would be determined by the number of thiophosphorylated heads within the filament. An alternative explanation is that the presence of dephosphorylated heads which cycle slowly or not at all impose a load that retards actin movement by rapidly cycling thiophosphorylated heads. These two possible mechanisms can be distinguished by comparing the results of experiments in which filamentous or monomeric myosins were used. If cooperativity between crossbridge heads exists within a copolymer, then information about the thiophosphorylation state of a given crossbridge head could be passed to neighboring heads through the intramolecular interactions that exist in the filament. If these intramolecular interactions are essential for modulation of velocity, then the use of monomers in the motility assay should alter the relationship between actin filament velocity and myosin light chain thiophosphorylation. Given no obvious difference in this relationship, it appears that cooperativity between crossbridge heads that is transmitted through the myosin filament backbone cannot account for the apparent modulation of actin filament velocity. Cooperativity between crossbridges could exist via information transfer through the actin filament, but we do not believe this is the case given that cooperative effects were not observed by solution measurements of the actin-activated myosin ATPase (Warshaw et al., 1990). Therefore we propose that the slow or noncycling dephosphorylated heads mechanically interact and impose a load on the more rapidly cycling thiophosphorylated heads.

One concern is that these experiments were performed at low ionic strength where it is known that the binding of myosin to actin is several fold higher than at more physiologic ionic strengths (Greene et al., 1983). Therefore the internal load caused by the dephosphorylated myosin may be a consequence of the tighter binding of myosin to actin at low ionic strength. To examine this possibility, experiments were done at a higher ionic strength (i.e. 80 mM KCl). Figure 3 shows the dependencies of actin filament velocity on the percentage of dephosphorylated and thiophosphorylated monomeric myosin at both 25 (circles) and 80 mM KCl (squares). Note that with increasing ionic strength less dephosphorylated myosin is needed to impede the thiophosphorylated heads. Therefore we believe that this internal load could exist within the muscle fiber.

The ability of dephosphorylated crossbridges to impose a load is surprising given our present understanding of the kinetics of crossbridge cycling. The hydrolysis of ATP by myosin occurs through a series of biochemical intermediates which are identical in both smooth and skeletal muscle. This kinetic scheme is characterized by myosin with bound nucleotide or the products of ATP hydrolysis (ADP, P_i) attaching first to actin in a weakly bound state. Weakly bound states are defined as intermediates that rapidly attach to and detach from actin, with an equilibrium constant of $\approx 10^4\ M^{-1}$. Following the release of P_i, myosin undergoes a conformational change to a state which is strongly bound to actin. Sellers (1985) has reported that the rate limiting step for ATP hydrolysis by dephosphorylated myosin is P_i release. If so, dephosphorylated crossbridges should exist predominantly in a weakly bound state to actin. How could these weakly bound crossbridges impede actin filament motion if they rapidly attach and detach from actin? This question really arises given the data of Brenner et al. (1982) which suggest that weakly bound bridges at physiologic ionic strength offer little resistance to stretch in a skinned skeletal muscle fiber. Recent experiments in our laboratory have confirmed that chemically modified skeletal muscle myosin (using pPDM), that no longer hydrolyzes ATP but is biochemically defined as weakly bound to actin, does impede faster cycling phosphorylated crossbridges. In addition, we have successfully modeled the modulation of actin filament velocity with the extent of light chain thiophosphorylation by a mechanical interaction between crossbridges having distinctly different cycling rates and affinities for actin (Warshaw et al., 1990).

The motility assay provides evidence that crossbridges within the same filament do interact mechanically, which could account for the apparent modulatory role of light chain phosphorylation on smooth muscle velocity. In addition, the motility data suggest that the dephosphorylated crossbridge's ability to bind actin and thus impose an internal load is an inherent property of dephosphorylated myosin. One need not invoke that the load bearing capacity of the dephosphorylated head appears only after an attached phosphorylated head is dephosphorylated. If dephosphorylated myosin can bind with relatively high affinity to actin, how is the relaxed state in smooth muscle ever attained? Evidence for actin linked regulatory systems (e.g. caldesmon, calponin) exists; these proteins could prevent the dephosphorylated bridges from attaching to actin in the relaxed state (Hemric and Chalovich, 1988; Winder and Walsh, 1990). Since the motility data described here were obtained with unregulated skeletal muscle actin, the ability to prevent dephosphorylated crossbridge attachment may have been lost. Reconstitution of actin-linked regulatory systems in the motility assay should provide a means of addressing these concerns.

ACKNOWLEDGMENTS

We would like to thank Janet Desrosiers for her technical assistance, Steven Work for his development of the computer based filament tracking system, and Trish Warshaw for the illustrations. This work was supported by grants from the National Institutes of Health HL35684 and HL45161 to D. Warshaw and HL38113 to K. Trybus. D. Warshaw is an Established Investigator of the American Heart Association.

REFERENCES

Brenner, B., Schoenberg, M., Chalovich, J. M., Greene, L. E., and Eisenberg, E., 1982, Evidence for cross-bridge attachment in relaxed muscle at low ionic strength, *Proc. Nat'l. Acad. Sci. U.S.A.*, 79: 7288.

Butler, T. M. and Siegman, M. J., 1983, Chemical energy usage and myosin light chain phosphorylation in mammalian smooth muscle, *Fed. Proc.*, 42: 57.

Dillon, P. F., Aksoy, M. O., Driska, S. P., and Murphy, R. A., 1981, Myosin phosphorylation and the crossbridge cycle in arterial smooth muscle, *Science*, 211: 495.

Greene, L. E., Sellers, J. R., Eisenberg, E., and Adelstein, R. S., 1983, Binding of gizzard smooth muscle myosin subfragment 1 to actin in the presence and absence of adenosine 5'-triphosphate, *Biochemistry*, 22: 530.

Hemric, M. E, and Chalovich, J. M., 1988, Effect of caldesmon on the ATPase activity and the binding of smooth and skeletal myosin subfragments to actin, *J. Biol. Chem.*, 263: 1878.

Kamm, K. E. and Stull, J. T., 1985, The function of myosin and myosin light chain kinase phosphorylation in smooth muscle, *Ann. Rev. Pharmacol. Toxicol.*, 25:593.

Sellers, J. R., 1985, Mechanism of the phosphorylation-dependent regulation of smooth muscle heavy meromyosin, *J. Biol. Chem.*, 260: 15815.

Sellers, J. R., Spudich, J. A., and Sheetz, M. P., 1985, Light chain phosphorylation regulates the movement of smooth muscle myosin on actin filaments, *J. Cell Biol.*, 101: 1897.

Trybus, K. M. and Henry, L., 1989, Monoclonal antibodies detect and stabilize conformational states of smooth muscle myosin, *J. Cell Biol.*, 109: 2879.

Warshaw, D. M., Desrosiers, J. M., Work, S. S., and Trybus, K. M., 1990, Smooth muscle myosin cross-bridge interactions modulate actin filament sliding velocity in vitro, *J. Cell Biol.*, 111: 453.

Winder, S. J. and Walsh, M. P., 1990, Smooth muscle calponin. Inhibition of actomyosin MgATPase and the regulation by phosphorylation, *J. Biol. Chem.*, 265: 10148.

REGULATION OF A SMOOTH MUSCLE CONTRACTION: A HYPOTHESIS BASED ON SKINNED FIBER STUDIES

Robert S. Moreland,[1] Jan Willem R. Pott,[1] Jacqueline Cilea,[1] and Suzanne Moreland[1,2]

[1]Bockus Research Institute
Graduate Hospital
Philadelphia, PA, 19146

[2]Department of Pharmacology
Bristol-Myers Squibb Pharmaceutical Research Institute
Princeton, NJ 08543

INTRODUCTION

Although it is generally believed that smooth muscle will contract in response to an increase in cytosolic free calcium ion concentration, there is still considerable controversy concerning the explicit mechanism(s) coupling calcium to contraction. Bremel (1974), using filament displacement studies, showed that the Ca^{2+} dependence of vertebrate smooth muscle contraction is associated primarily with the thick filament. A few years later, Aksoy et al. (1976) and Sobieszek (1977) demonstrated that the Ca^{2+} sensitivity of actomyosin ATPase activity was associated with phosphorylation of the 20 kDa myosin light chain (MLC) which was subsequently shown to result from activation of MLC kinase, a Ca^{2+} and calmodulin dependent enzyme (for reviews see Kamm and Stull, 1985; Hartshorne, 1987). Correlations have been shown between MLC phosphorylation and both Ca^{2+} dependent actin-activated myosin ATPase activity (Dabrowska et al., 1978; DiSalvo et al., 1978) and force development in either skinned (Kerrick et al., 1980; Chatterjee and Murphy, 1983) or intact (Barron et al., 1980; Driska et al., 1981) muscle fibers. These findings have brought about the widespread belief that this system is the primary regulator of smooth muscle contraction.

It is becoming increasingly clear that the regulation of a smooth muscle contraction is significantly more complicated than the model in which MLC phosphorylation acts as a simple switch. This additional complexity is defined and described in numerous chapters in this volume suggesting the involvement of thin filament regulatory proteins (Winder et al.), alterations in the activity of the MLC kinase (Stull et al.), alterations in the Ca^{2+} sensitivity of the contractile apparatus by either inhibition of the MLC phosphatase (Kitazawa and Somlyo) or activation of protein kinase C (Nishimura et al.), and coopera-

Regulation of Smooth Muscle Contraction
Edited by R.S. Moreland, Plenum Press, New York, 1991

tivity between myosin heads (Siegman et al.; Somlyo et al.) as integral components of the overall regulation of contraction.

The goal of this chapter is to briefly discuss (we apologize that numerous important studies will be necessarily omitted due to space limitations) the historical importance of permeabilized, or "skinned", fibers in the development of our current understanding of the regulation of a smooth muscle contraction. In addition, this chapter will outline studies using detergent skinned fibers that have led us to postulate that two Ca^{2+} dependent pathways, acting in parallel, regulate smooth muscle contractile activity.

SKINNED FIBERS AND SMOOTH MUSCLE CONTRACTION

Hasselbach and Ledermair (1958) and Briggs (1963) were, to our knowledge, the first to publish studies using permeabilized smooth muscle fibers. Both of these studies, using glycerinated fibers, clearly demonstrated the dependence of a contraction on ATP. However, although a large dependence on Mg^{2+} was demonstrated for force development, only a small dependence on Ca^{2+} was reported. Filo et al. (1965) were the first to use permeabilized smooth muscle fibers to demonstrate that force development was dependent on the free calcium ion concentration ($[Ca^{2+}]$). Interestingly, these investigators, in apparent agreement with the findings of Hasselbach and Ledermair (1958) and Briggs (1963), demonstrated a striking dependence of contraction on high (≥ 1 mM) levels of Mg^{2+} relative to that found for glycerinated striated muscle fibers. Subsequently, important studies emerged demonstrating the usefulness of detergent-permeabilized fibers in the study of smooth muscle regulation. Gordon (1978) using Triton X-100 and Saida and Nonomura (1978) using saponin to permeabilize smooth muscle fibers described preparations that not only exhibited Ca^{2+} sensitivity but also produced near pre-skinning levels of force development. These studies added another intriguing piece of information to the puzzle in that force development in both preparations could be elicited in the absence of Ca^{2+} by millimolar concentrations of Mg^{2+}.

During the time that the skinned smooth muscle fiber was being established as a model for contractile regulation, biochemical studies on purified proteins were demonstrating the importance of MLC phosphorylation in the initiation of actomyosin ATPase activity. It was therefore a natural extension of the purified protein studies to examine the role of MLC phosphorylation in skinned fibers. Three important pieces of information were gained from these studies resulting in some of the strongest evidence in support of the MLC phosphorylation hypothesis. First was the initial demonstration that the level of MLC phosphorylation correlated with both the level of developed force and the $[Ca^{2+}]$ (Hoar et al., 1979). Second was the demonstration that irreversible MLC phosphorylation with the ATP analog ATPγS (which produces a MLC phosphatase resistant thiophosphorylated MLC) resulted in a Ca^{2+} independent but ATP dependent contraction (Cassidy et al., 1979). The third finding was that the introduction of a Ca^{2+} and calmodulin independent tryptic fragment of MLC kinase to the skinned smooth muscle fiber resulted in the Ca^{2+} independent development of force (Walsh et al., 1982).

Although the early studies with isolated proteins and skinned fibers illuminated the importance of Ca^{2+} dependent MLC phosphorylation in the activation of smooth muscle, subsequent experiments hinted at a significantly greater level of complexity. For example, there appeared to be a temporal component in the regulation of contraction. The maximal shortening velocity of rat portal vein was found to be greater during the development of a twitch than during the time of peak twitch force (Hellstrand and Johansson, 1975). In rabbit taenia coli, energy utilization was reported to increase rapidly during force development, then fall during force maintenance (Siegman et al., 1980). The discovery of this temporal component was foreseen by Bozler (1930) who found that smooth muscle maintained high levels of force with low energy expenditure. Time-dependent changes like these were explained in the "latch hypothesis" (Dillon et al., 1981) which suggested the increase in myoplasmic Ca^{2+} upon stimulation brought about an increase in MLC phosphorylation which resulted in force development by rapidly cycling crossbridges. With continued stimulation, force could be maintained at high levels even though MLC phosphorylation levels and isotonic shortening velocity fell to suprabasal levels. Thus, force maintenance was suggested to be supported by latchbridges, defined originally as slowly cycling, dephosphorylated crossbridges.

In the original latch hypothesis, Ca^{2+} dependent MLC phosphorylation was associated with the development of force (Dillon et al., 1981; Driska et al., 1981). During force maintenance, the dephosphorylation of the rapidly cycling crossbridges led to slowly cycling latchbridges. Therefore, the level of MLC phosphorylation determined the rate of crossbridge cycling (Aksoy et al., 1982; Dillon and Murphy, 1982). The latchbridges were shown to be regulated by a second Ca^{2+} dependent system with a higher Ca^{2+} sensitivity than that for MLC phosphorylation (Dillon et al., 1981; Aksoy et al., 1983; Moreland and Murphy, 1986). The concept of a latch state is generally well-accepted, although the suggestion that crossbridge cycling rate is regulated by the degree of MLC phosphorylation has not been supported under all experimental conditions (Siegman et al., 1984; Haeberle et al., 1985; Moreland and Moreland, 1987; Moreland et al., 1987). The latch hypothesis can be used to explain much, but not all, of the energetic (Siegman et al., 1980; Paul, 1983), mechanical (Dillon and Murphy, 1982; Aksoy et al., 1983; Kamm and Stull, 1985), and biochemical results (Aksoy et al., 1983; Kamm and Stull, 1985; Moreland and Murphy, 1986) obtained from numerous stimulation regimens in various smooth muscle tissues.

As experiments to test the latch hypothesis were conducted, the original model was revised (Hai and Murphy, 1988a; 1988b). The new model again provided a simple and testable paradigm for the Ca^{2+} dependent regulation of smooth muscle. This model predicted:

- A hyperbolic relationship between force and MLC phosphorylation.
- A linear relationship between maximal shortening velocity and MLC phosphorylation as demonstrated under a variety of conditions (Aksoy et al., 1982; Rembold and Murphy, 1988).
- A requirement of only 0.3 mol P_i/mol MLC for maximal force development.
- A second Ca^{2+} dependent regulatory system responsible for latchbridge attachment was not necessary.
- Significant ATP consumption ($\approx$ 85%) due to the pseudo-ATPase as compared to only $\approx$ 15% ATP consumed by crossbridge cycling (but see Hai et al., this volume, for a more recent calculation suggesting 50% of the steady state ATP utilization is the result of crossbridge cycling).

The new model could predict the time-dependent changes in shortening velocity, MLC phosphorylation, and force in isolated strips of swine carotid media (Singer and Murphy, 1987), but was less successful in accounting for other aspects of contraction. Some of the primary pieces of information that the Hai and Murphy model does not take into account are:

- Thiophosphorylation of MLC in skinned taenia coli fibers depressed ATP consumption approximately 15% in the absence and 25% in the presence of Ca^{2+} (Hellstrand and Arner, 1985). The Hai and Murphy model predicts a 50 - 85% decrease in ATP consumption due to blockade of the pseudo-ATPase.
- Inhibition of MLC phosphatase activity in permeabilized portal vein by okadaic acid did not affect the curvilinear relationship between force and MLC phosphorylation (Siegman et al., 1989, see Fig. 1 in Siegman et al., this volume). The Hai and Murphy model predicts a linear relationship following inhibition of the MLC phosphatase.
- Altering smooth muscle fiber length was associated with a tightly coupled decline in both O_2 consumption and force (Paul and Peterson, 1975; Glück and Paul, 1977). The Hai and Murphy model cannot predict a linear dependence of energy utilization on force without postulating a length dependence of the pseudo-ATPase.
- Alterations in fiber length in either intact (Aksoy et al., 1983) or skinned fibers (Moreland et al., 1988) had no effect on the magnitude, temporal profile, or Ca^{2+} sensitivity of net MLC phosphorylation levels. Therefore, there is no length dependence of the pseudo-ATPase.

The latch hypothesis proposed by Hai and Murphy (1988a; 1988b) is exquisite in its simplicity and ability to predict several parameters of a smooth muscle contraction. It has certainly proven to be a useful, testable model for studying crossbridge regulation. However, it is precisely its simplicity that may account for its inability to predict numerous fundamental aspects of the regulation of a contraction. Therefore, other hypotheses have been advanced to explain the biochemical regulation of smooth muscle contraction.

AN ALTERNATE HYPOTHESIS FOR CONTRACTILE REGULATION

Shortly after the initial publication of the latchbridge hypothesis (Dillon et al., 1981), Chatterjee and Murphy (1983) demonstrated that force could be maintained with disproportionately low levels of MLC phosphorylation in the Triton X-100 detergent skinned swine carotid medial fiber. These investigators used a protocol which intentionally mimicked the transient increase in intracellular [Ca^{2+}] observed in the intact smooth muscle cell stimulated with various agonists (Morgan and Morgan, 1984; Himpens et al., 1988). As the [Ca^{2+}] was increased, force and MLC phosphorylation concomitantly increased. In contrast, if the fibers were first contracted with a high [Ca^{2+}] and then exposed to a lower [Ca^{2+}], MLC phosphorylation levels fell in proportion to the [Ca^{2+}] but force was maintained at significantly elevated levels. This "force hysteresis" which resulted in an increased Ca^{2+} sensitivity of force following the initial exposure to high [Ca^{2+}] was suggested to be evidence of the latch state in skinned smooth muscle fibers (Chatterjee and Murphy, 1983). Thus, similar to sustained contractions in the intact swine carotid fiber, this protocol

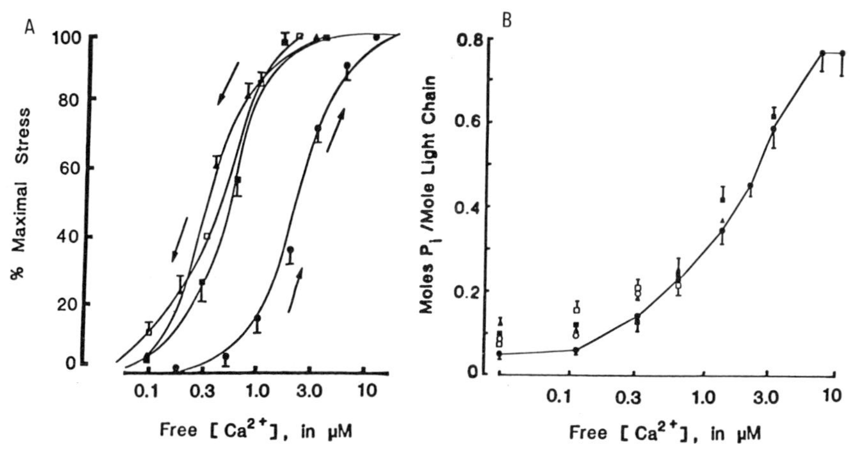

Figure 1. A. Ca^{2+} sensitivity for force development and force maintenance in skinned swine carotid media. The direction of change in $[Ca^{2+}]$ is indicated by the direction of the arrows. The curve for increasing Ca^{2+} (●) is on the right. Data for decreasing Ca^{2+} are presented as percent of maximal response to 7.2 (■), 2.1 (▲), or 1.3 (□) μM Ca^{2+}. B. Ca^{2+} dependence of MLC phosphorylation during force development and force maintenance. MLC phosphorylation levels were determined in all fibers shown in A. Symbols as described for A. (Reprinted from Moreland and Murphy, 1986; by permission).

resulted in a state of high maintained force with reduced levels of MLC phosphorylation.

Following the work of Chatterjee and Murphy (1983), we used a similar protocol to determine if the initial level of activation influenced the formation of the latch state (Moreland and Murphy, 1986). We demonstrated that if the detergent skinned fibers were stimulated with varying $[Ca^{2+}]$ and then exposed to lower $[Ca^{2+}]$, the magnitude of maintained force, when normalized to the initial maximum, was similar regardless of the initial level of contraction. As shown in Figure 1, the dependence of force on Ca^{2+} was described by two distinct curves, one with an apparent $K_m \approx 1.3$ μM for the Ca^{2+} sensitivity of force development and MLC phosphorylation and a second with an apparent K_m of ≈ 0.6 μM for the Ca^{2+} sensitivity of the latch state. The finding of two distinct curves rather than a family of curves dependent on the initial activation conditions was taken as evidence for two distinct Ca^{2+} dependent regulatory systems (Moreland and Murphy, 1986; but see Chatterjee et al., 1987).

In an attempt to further define the characteristics of these two putative Ca^{2+} dependent mechanisms, we performed a series of experiments examining the effect of fiber length on the phenomenon of force hysteresis (Moreland et al., 1988). Aksoy et al. (1983) had previously demonstrated that net MLC phosphorylation levels were not affected, although the magnitude of force was altered by changing the length of intact swine carotid medial fibers in a manner consistent with classical length-tension properties of muscle. We performed similar studies using the Triton X-100 detergent skinned swine carotid fiber and determined the effect of changing fiber length from 0.7 to 1.4 L_o on the Ca^{2+} sensitivity of force development, force maintenance, and MLC phosphorylation. As shown in Figure 2, decreasing fiber length from 1.0 to 0.7 L_o was not associated with significant effects on the Ca^{2+} sensitivity of any parameter, although as expected the absolute magnitude of developed force was decreased. In contrast, increasing fiber length to 1.4 L_o significantly decreased the Ca^{2+} sensitivity of force development and apparently prevented force hysteresis. Of particular interest is that although the Ca^{2+} sensitivity of force development was decreased, no change in the Ca^{2+} sensitivity of MLC phosphorylation occurred. The original study of Chatterjee and Murphy (1983) and our extension of those experiments (Moreland and Murphy, 1986) suggested that if two Ca^{2+} dependent regulatory systems did indeed exist, the Ca^{2+} dependence of force maintenance could only be expressed after a cycle of MLC phosphorylation-dephosphorylation. Therefore the two systems operated in series. However, the findings described above suggested that prior MLC phosphorylation may not be a prerequisite for expression of the second Ca^{2+} regulatory system. Moreover, we suggested the possibility that the Ca^{2+} sensitivity and magnitude of force developed at 1.0 L_o was the sum of the two regulatory systems acting in parallel. This was based, in part, on the following: 1. The Ca^{2+} sensitivity of force development could be altered without a concomitant change in the sensitivity of the Ca^{2+}-calmodulin-MLC kinase system; 2. The same perturbation that altered the Ca^{2+} sensitivity of force development abolished force hysteresis suggesting the possibility that it was the loss of a putative latchbridge population that produced the decrease in sensitivity; 3. Increasing fiber length did not result in irreversible damage to the fibers; and 4. The absolute magnitude of attained force did not determine the presence or absence of force hysteresis as the order of force developed was $1.0 > 1.4 > 0.7\ L_o$.

Several of the original studies using skinned smooth muscle fibers (Filo et al., 1965; Gordon, 1978; Saida and Nonomura, 1978) as well as more recent investigations (Arner, 1983) clearly demonstrated a significant influence of Mg^{2+} on the Ca^{2+} dependent development of force. In fact several investigators have shown that high levels of Mg^{2+} contract skinned smooth muscle fibers in the absence of Ca^{2+} (Gordon, 1978; Saida and Nonomura, 1978)

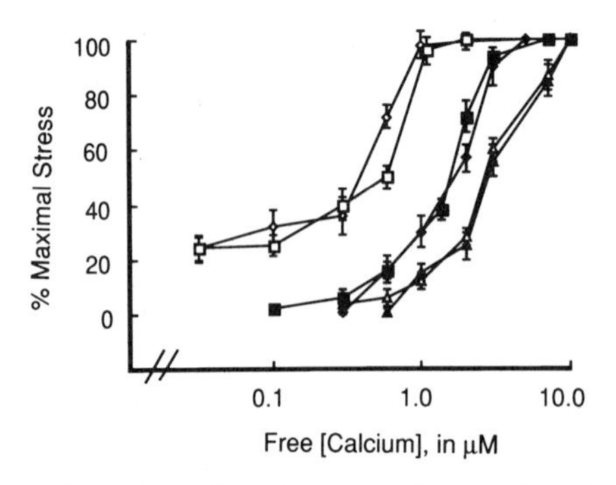

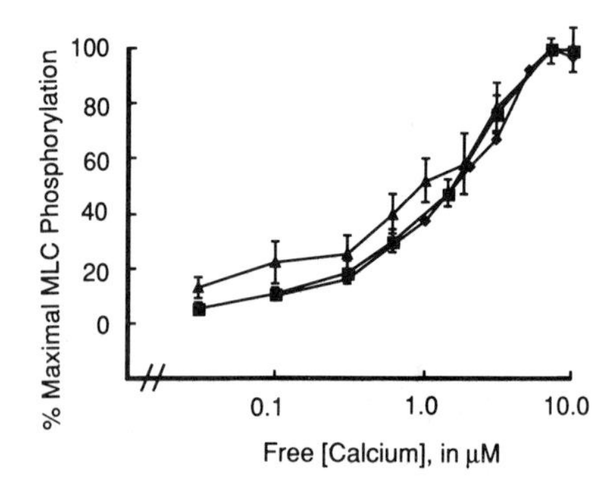

Figure 2. A. Force development and force hysteresis at different fiber lengths in skinned swine carotid media. Fibers were either contracted by exposure to increasing [Ca^{2+}] (filled symbols) or first maximally contracted and then exposed to a lower [Ca^{2+}] (open symbols). Fibers were studied at 0.7 (◆,◊), 1.0 (■,□), and 1.4 (▲,Δ) L_o. Values shown are normalized as a percent of the maximal force developed at each fiber length. B. Ca^{2+} sensitivity of MLC phosphorylation during force development and force maintenance experiments shown in A. Symbols correspond to those described in A. For clarity only the increasing [Ca^{2+}] curves are shown as there were no differences between increasing and decreasing Ca^{2+} values. (Reprinted from Moreland et al., 1988; by permission).

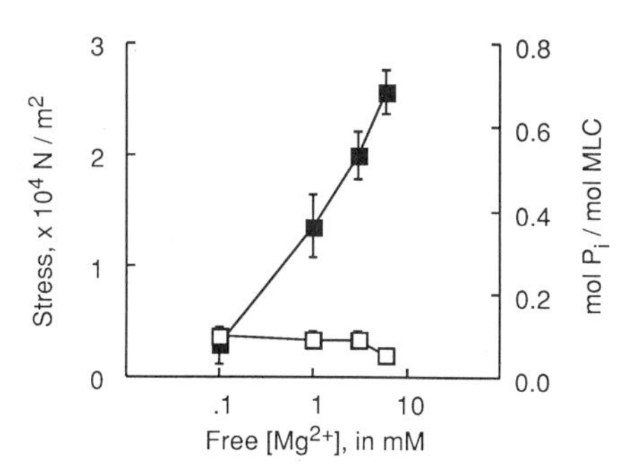

Figure 3. Effect of Mg^{2+} on maintained stress in the detergent skinned swine carotid media. Skinned fibers were contracted with 7 μM Ca^{2+} in the presence of 6 mM Mg^{2+}, then the [Ca^{2+}] was decreased to < 0.01 μM. After force stabilized, the fibers were exposed to lower [Mg^{2+}]. A decrease in the [Mg^{2+}] (denoted by the direction of the arrow) decreased the level of maintained stress (■) although MLC phosphorylation levels (□) were basal at every concentration.

and without increasing MLC phosphorylation (Ikebe et al., 1984). Mg^{2+} also modulates the magnitude of maintained force. Increasing the free [Mg^{2+}] from 0.1 to 6 mM in a step-wise fashion increases the level of force hysteresis at every [Ca^{2+}]. This is best illustrated by the data shown in Figure 3. In these experiments, detergent skinned swine carotid fibers were contracted with 7 μM Ca^{2+} in the presence of 5 mM MgATP and 6 mM free Mg^{2+}. The fibers were then exposed to a solution containing 5 mM MgATP and 6 mM Mg^{2+} but the [Ca^{2+}] was < 0.01 μM. MLC phosphorylation levels rapidly fell to basal values but force was maintained. The [Mg^{2+}] was then decreased while all other ionic conditions were held constant. As can be seen from this figure, the level of maintained force (shown as stress = force/cross-sectional area) decreased as the [Mg^{2+}] was decreased. Reducing the [Mg^{2+}] had no effect on MLC phosphorylation levels. It is important to note that this is not the result of a simple reduction in cellular [MgADP] as quantitatively similar results were obtained in the presence and absence of an ATP regenerating system. Increasing the [Mg^{2+}] decreased the magnitude and Ca^{2+} sensitivity of the MLC phosphorylation dependent force development. Therefore, Mg^{2+} appears to have differential and in fact opposite effects on the development (rapidly cycling, phosphorylated crossbridges) and maintenance (unphosphorylated latchbridges) of force. These divergent effects of Mg^{2+} support the hypothesis that two different systems regulate force.

Ikebe et al. (1984) first demonstrated that Mg^{2+}-induced contractions were independent of MLC phosphorylation. These investigators suggested that high [Mg^{2+}] induced a conformational change in myosin similar to that induced by MLC phosphorylation and therefore was able to "bypass" the phosphorylation step. Although we agree that this may account, in part, for the Mg^{2+} stimulated force development, we also believe that Mg^{2+} may directly activate the postulated second Ca^{2+} dependent but MLC phosphorylation independent development of force. We have shown, in the detergent skinned swine carotid fiber, that Mg^{2+} stimulates MLC phosphorylation independent contractions at 1.0 L_o, but at 1.4 L_o, Mg^{2+} does not cause the development of force (Moreland and Moreland, 1991). Force hysteresis, our definition of the latch state in the skinned fiber, is also abolished at these long fiber lengths. Although the fibers do not contract in response to Mg^{2+} at 1.4 L_o, they still respond to the addition of Ca^{2+} demonstrating that fiber damage does not account for this finding. Therefore, it appears that a condition in which force can be developed in the absence of MLC phosphorylation (high $[Mg^{2+}]$) can be specifically induced and this condition as well as a state of high maintained force with disproportionately low levels of MLC phosphorylation can be specifically abolished (1.4 L_o).

The results from experiments using high $[Mg^{2+}]$ and changes in fiber length as experimental tools are compelling but circumstantial evidence suggesting a novel Ca^{2+} dependence of contraction, in addition to that involving MLC phosphorylation. More direct evidence has been provided by Wagner and Rüegg (1986) and by our laboratories (Moreland and Moreland, 1991). A Ca^{2+} and calmodulin dependent contraction can be elicited in the total absence of an increase in MLC phosphorylation in freshly skinned chicken gizzard smooth muscle (Wagner and Rüegg, 1986). However, if the glycerol treated fibers are stored for a significant period of time, as is the usual treatment for this preparation, only a Ca^{2+} and calmodulin and MLC phosphorylation dependent contraction can be elicited. One possible explanation for these findings is that an integral but unknown factor is lost or degraded during storage. This may explain some of the differences in results obtained by our laboratory as compared to others because our studies are always performed on the same day the intact tissues are mounted and permeabilized for experimentation. Using this approach, we have shown that MLC phosphorylation independent development of force can be supported by 5 mM CTP in the presence of either 20 mM Mg^{2+} or 5 μM Ca^{2+} (Moreland and Moreland, 1991). CTP is a substrate for the actin-activated myosin ATPase but not for the MLC kinase. Moreover, a reasonably physiological $[Ca^{2+}]$ (5 μM), although greater than that shown by fura-2 studies (Himpens et al., 1988), elicited the MLC phosphorylation independent contraction. Although we do not know at this time whether the role played by this Ca^{2+} dependent mechanism is important in the normal physiological function of the vascular smooth muscle cell (we have shown that norepinephrine stimulation of vascular smooth muscle *in situ* causes an increase in MLC phosphorylation levels (Moreland et al., 1990)), we believe these results clearly demonstrate that actin and myosin interactions can be initiated without prior MLC phosphorylation. In terms of the widely discussed but controversial Hai and Murphy model (see Hai et al., this volume), these results would require the further modification of the model to allow the formation of a latchbridge by a Ca^{2+} dependent process without prior MLC phosphorylation, i.e. the inclusion of K8 in the model.

If Ca^{2+} dependent latchbridges can form without prior MLC phosphorylation, what are the precise biochemical steps involved? We have performed experiments demonstrating that calmodulin or a calmodulin-like protein may be important. We have previously shown (Moreland et al., 1987) that a contraction developed in the presence of Ca^{2+} and MgATP (and, therefore, associated with increased levels of MLC phosphorylation) can be supported by Ca^{2+} and MgCTP while MLC phosphorylation falls to basal levels in agreement with Hoar et al. (1985). The Ca^{2+} dependent, but MLC phosphorylation independent, maintenance of force is relaxed in a concentration dependent manner by the addition of either W-7 or trifluoperazine, both calmodulin antagonists. In addition, we have recently demonstrated that the Mg^{2+} induced contractions as well as the contraction in response to Ca^{2+} and MgCTP without prior MLC phosphorylation can be inhibited or relaxed by the addition of calmodulin antagonists. We believe that both of these protocols may directly activate a latchbridge population. The effects of these calmodulin antagonists are not non-specific actions on the crossbridges as the inhibition is reversible and the antagonists have no effect on force developed in fibers maximally thiophosphorylated by ATPγS (Moreland and Moreland, 1991). Therefore, it appears plausible that a calmodulin-like regulatory protein is involved in the Ca^{2+} dependence of latchbridge attachment (K8).

Further evidence that a smooth muscle contraction is not regulated by a simple MLC phosphorylation "switch" can be obtained from measurement of ATPase activity in skinned fibers. Butler et al. (1990) measured fiber ATPase activity during a contraction in response to Ca^{2+} in permeabilized rabbit portal vein. They demonstrated that ATPase activity, presumed to be solely or at least primarily due to actin-activated myosin ATPase activity, increased rapidly upon the addition of Ca^{2+} then fell to a slightly lower level although the $[Ca^{2+}]$ remained constant and force continued to rise. Kühn et al. (1990), using skinned guinea pig taenia coli, demonstrated that MLC phosphorylation levels are constant during the apparent transient in ATPase activity. We have recently initiated a series of studies designed to investigate the relationship between Ca^{2+}, MLC phosphorylation, and ATPase activity in the detergent skinned fiber. Figure 4 shows preliminary results from these studies. ATPase activity was measured as described by Butler et al. (1990) and Kühn et al. (1990) and stiffness was estimated by imposing a 500 Hz sine wave length change (0.5% L_o) and measuring the resultant change in force. In agreement with the results discussed above, upon stimulation of the fiber with Ca^{2+}, ATPase activity rapidly, but transiently, increased during the phase in which force was still developing. Stiffness increased in proportion to the level of developed force, suggesting that although ATPase activity was falling, the number of

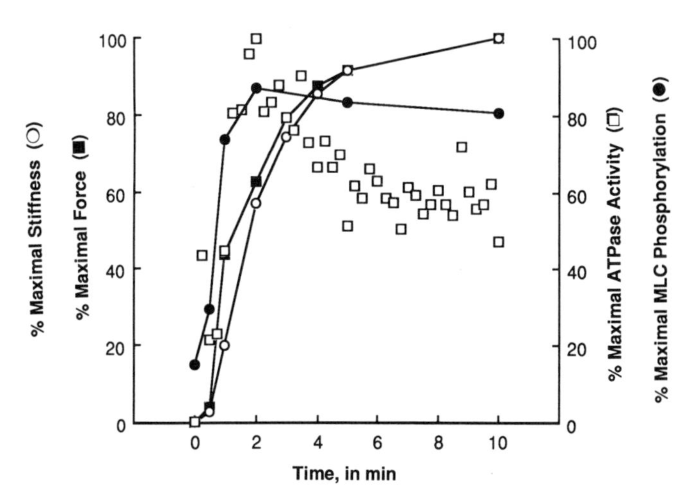

Figure 4. Time course of force, stiffness, MLC phosphorylation, and fiber ATPase activity in the detergent skinned swine carotid media. Fibers were stimulated with 7 μM Ca^{2+} in the presence of 5 mM MgATP and 1 mM Mg^{2+}. See text for details of experimental design. Note that fiber ATPase activity (□) transiently increases to high levels before attaining a stable steady state, although force (■) and stiffness (○) continue to rise and MLC phosphorylation (●) levels remain constant.

attached crossbridges was rising. Most importantly, MLC phosphorylation levels were constant during much of the change in ATPase activity. Assuming shortening velocity is an estimate of actin-activated myosin ATPase activity (Bárány, 1967), these results strongly suggest that MLC phosphorylation is not the sole determinant of crossbridge cycling rate. This information is consistent with the idea of two Ca^{2+} dependent processes occurring in parallel with the proposed "second site" responsible for the slowing of ATPase activity with time. It could also be used to support the alternate hypothesis suggesting a role for thin filament regulatory proteins (Winder and Walsh, 1990). Regardless of what process is ultimately found to control latchbridge attachment, the regulation of force development in smooth muscle certainly involves more than the simple Ca^{2+} and calmodulin dependent activation of the MLC kinase with the resultant phosphorylation of the MLC.

SUMMARY

It seems clear that a simple Ca^{2+} dependent switch (MLC phosphorylation) cannot completely explain all of the disparate mechanical and energetic results obtained under numerous experimental conditions in numerous laboratories. Some of the problems of the simple switch model are that: 1. Force can be developed in the complete absence of increases in MLC phosphorylation; 2. Crossbridge cycling rate, as measured by either shortening velocity or directly by ATPase activity, can be regulated independent of changes in MLC phosphorylation; and 3. Ca^{2+} can directly influence both force and crossbridge cycling rate. Thus, we believe that there are two distinct Ca^{2+} dependent regulatory systems which normally act in parallel to contract smooth muscle (Fig. 5). One of these is the Ca^{2+} dependent MLC phosphorylation-

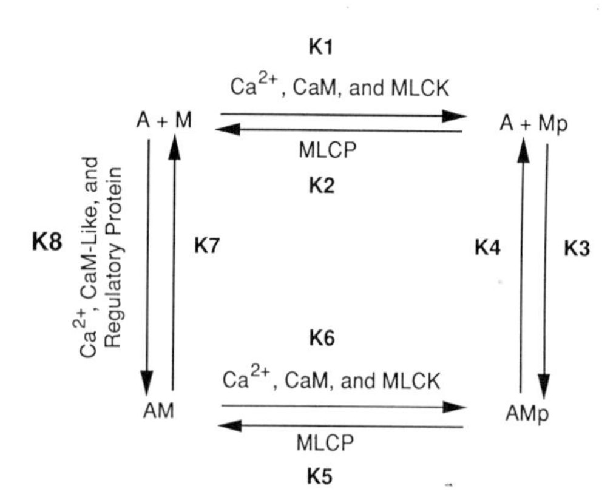

Figure 5. Proposed model for the regulation of a smooth muscle contraction, adapted from Hai and Murphy (1988a; 1988b). A = actin, M = myosin, Mp = myosin with phosphorylated MLC, AMp = actomyosin with phosphorylated MLC, AM = latchbridge. Rate constants K1 - K7 taken from Hai and Murphy (1988a). This model assumes that two Ca^{2+} dependent regulatory mechanisms acting in parallel control smooth muscle contraction. The final force developed and maintained in the cell is the sum of the total AMp and AM species. The important difference between this model and that of Hai and Murphy (1988a) is that the latchbridge (AM) can be activated directly with a rate of K8 through a Ca^{2+} dependent mechanism involving a calmodulin-like protein and a hypothesized regulatory protein.

dephosphorylation system which is likely to be responsible for the rapid development of force. The other is the hypothesized Ca^{2+} dependent system which is probably responsible for the slow development of force as well as the maintenance of previously developed force, represented in Figure 5 as K8. This second system involves a calmodulin-like protein with a higher Ca^{2+} sensitivity than that for the Ca^{2+}-calmodulin-MLC kinase system. Under most conditions, the total force attained by smooth muscle in response to stimulation is the result of the concerted activation of both of these regulatory systems.

The available information is consistent with this hypothesis of two regulatory systems functioning in parallel. In addition to the information presented in this chapter, work from a number of laboratories (Moreland and Ford, 1982; Fujiwara et al., 1989; Kitazawa et al., 1989; Somlyo et al., 1989; Kubota et al., 1990; Kitazawa and Somlyo, this volume) have suggested the possibility that a regulated MLC phosphatase may functionally alter the Ca^{2+} sensitivity of the contractile filaments. There is evidence suggesting that the sensitivity of MLC kinase to activation by Ca^{2+} and calmodulin may be regulated (Stull et al., this volume). Protein kinase C has been postulated to play an important role in the regulation of myofilament Ca^{2+} sensitivity (Nishimura et al., this volume). MgADP has been suggested to affect the kinetics of latchbridge attachment and detachment (Kerrick and Hoar, 1987; Nishimura and van Breemen, 1989). Cooperativity between crossbridges as described by Somlyo et al. (1988) and Siegman et al. (this volume) might also be an important component in the regulation of smooth muscle contraction. These findings, as well as the discovery of several thin filament proteins with potential regulatory functions (Ngai and Walsh, 1984; Takahashi et al., 1988; Winder and Walsh, this volume) must all be considered in the final solution to the complicated and intriguing puzzle of the biochemical regulation of smooth muscle contraction.

ACKNOWLEDGMENTS

This work was supported, in part, by funds from NIH HL 37956 (RSM) and by a Grant-in-Aid from The American Heart Association, Southeastern Pennsylvania Affiliate (RSM).

REFERENCES

Aksoy, M. O., Williams, D., Sharkey, E. M., and Hartshorne, D. J., 1976, A relationship between Ca^{2+} sensitivity and phosphorylation of gizzard actomyosin, *Biochem. Biophys. Res. Commun.*, 69: 35.

Aksoy, M. O., Murphy, R. A., Kamm, K. E., 1982, Role of Ca^{2+} and myosin light chain phosphorylation in regulation of smooth muscle, *Am. J. Physiol.*, 242: C109.

Aksoy, M. O., Mras, S., Kamm, K. E., and Murphy, R. A., 1983, Ca^{2+}, cAMP, and changes in myosin phosphorylation during contraction of smooth muscle, *Am. J. Physiol.*, 245: C255.

Arner, A., 1983, Force-velocity relation in chemically skinned rat portal vein. Effects of Ca^{2+} and Mg^{2+}, *Pflügers Arch.*, 397: 6.

Bárány, M., 1967, ATPase activity of myosin correlated with speed of muscle shortening, *J. Gen. Physiol.*, 50: 197.
Barron, J. T., Bárány, M., Bárány, K., and Storti, R. V., 1980, Reversible phosphorylation and dephosphorylation of the 20000 dalton light chain of myosin during the contraction-relaxation-contraction cycle of arterial smooth muscle, *J. Biol. Chem.*, 255: 6238.
Bozler, E., 1930, The heat production of smooth muscle, *J. Physiol.*, 69: 442.
Bremel, R. D., 1974, Myosin linked calcium regulation in vertebrate smooth muscle, *Nature*, 252: 405.
Briggs, A. H., 1963, Characteristics of contraction of glycerinated uterine smooth muscle, *Am. J. Physiol.*, 204: 739.
Butler, T. M., Siegman, M. J., Mooers, S. U., and Narayan, S. R., 1990, Myosin-product complex in the resting state and during relaxation of smooth muscle, *Am. J. Physiol.*, 258: C1092.
Cassidy, P., Hoar, P. E., and Kerrick, W. G. L., 1979, Irreversible thiophosphorylation and activation of tension in functionally skinned rabbit ileum strips by [^{35}S]ATPγS, *J. Biol. Chem.*, 254: 11148.
Chatterjee, M. and Murphy, R. A., 1983, Calcium dependent stress maintenance without myosin phosphorylation in skinned smooth muscle, *Science*, 221: 464.
Chatterjee, M., Hai, C-M., and Murphy, R. A., 1987, Dependence of stress and velocity on Ca^{2+} and myosin phosphorylation in the skinned swine carotid media, *in:* "Regulation and Contraction of Smooth Muscle", M. J. Siegman, A. P. Somlyo, and N. L. Stephens, eds., Alan R. Liss,., New York, p. 399.
Dabrowska, R., Sherry, J. M. F., Aromatorio, D. K., and Hartshorne, D. J., 1978, Modulator protein as a component of the myosin light chain kinase from chicken gizzard, *Biochemistry*, 17: 253.
Dillon, P. F., Aksoy, M. O., Driska, S. P., and Murphy, R. A., 1981, Myosin phosphorylation and the cross-bridge cycle in arterial smooth muscle, *Science*, 211: 495.
Dillon, P. F. and Murphy, R. A., 1982, Tonic force maintenance with reduced shortening velocity in arterial smooth muscle, *Am. J. Physiol.*, 242: C102.
DiSalvo, J., Gruenstein, E., and Silver, P., 1978, Ca^{2+}-dependent phosphorylation of bovine aortic actomyosin, *Proc. Soc. Exp. Biol. Med.*, 158: 410.
Driska, S. P., Aksoy, M. O., and Murphy, R. A., 1981, Myosin light chain phosphorylation associated with contraction in arterial smooth muscle, *Am. J. Physiol.*, 240: C222.
Filo, R. S., Bohr, D. F., and Rüegg, J. C., 1965, Glycerinated skeletal and smooth muscle: Calcium and magnesium dependence, *Science*, 147: 1581.
Fujiwara, T., Itoh, T., Kubota, Y., and Kuriyama, H., 1989, Effect of guanosine nucleotides on skinned smooth muscle tissue of the rabbit mesenteric artery, *J. Physiol.*, 408: 535.
Glück, E. and Paul, R. J., 1977, The aerobic metabolism of porcine carotid artery and its relation to force. Energy cost of isometric contraction, *Pflügers Arch.*, 370: 9.
Gordon, A. R., 1978, Contraction of detergent-treated smooth muscle, *Proc. Nat'l. Acad. Sci. U.S.A.*, 75: 3527.
Haeberle, J. R., Hott, J. W., Hathaway, D. R., 1985, Regulation of isometric force and isotonic shortening velocity by phosphorylation of the 20,000 dalton myosin light chain of rat uterine smooth muscle, *Pflügers Arch.*, 403: 215.
Hai, C-M. and Murphy, R. A., 1988a, Cross-bridge phosphorylation and regulation of latch state in smooth muscle, *Am. J. Physiol.*, 254: C99.

Hai, C-M. and Murphy, R. A., 1988b, Regulation of shortening velocity by cross-bridge phosphorylation in smooth muscle, *Am. J. Physiol.*, 255: C86.

Hartshorne, D. J., 1987, Biochemistry of the contractile process in smooth muscle, *in:* "Physiology of the Gastrointestinal Tract", L. R. Johnson, ed., Raven Press, New York, p. 423.

Hasselbach, W. and Ledermair, O., 1958, Contraction cycle of isolated contractile structures of uterine musculature and its peculiarities, *Pflügers Arch.*, 267: 532.

Hellstrand, P. and Johansson, B., 1975, The force velocity relation in phasic contraction of venous smooth muscle, *Acta Physiol. Scand.*, 93: 157.

Hellstrand, P. and Arner, A., 1985, Myosin light chain phosphorylation and the cross-bridge cycle at low substrate concentration in chemically skinned guinea pig taenia coli, *Pflügers Arch.*, 405: 323.

Himpens, B., Matthijs, G., Somlyo, A. V., Butler, T. M., and Somlyo, A. P., 1988, Cytoplasmic free calcium, myosin light chain phosphorylation, and force in phasic and tonic smooth muscle, *J. Gen. Physiol.*, 92: 713.

Hoar, P. E., Kerrick, W. G. L., and Cassidy, P. S., 1979, Chicken gizzard: Relation between calcium-activated phosphorylation and contraction, *Science*, 204: 503.

Hoar, P.E., Pato, M. D., and Kerrick, W. G. L., 1985, Myosin light chain phosphatase. Effect on the activation and relaxation of gizzard smooth muscle fibers, *J. Biol. Chem.*, 260: 8760.

Ikebe, M., Barsotti, R. J., Hinkins, S., and Hartshorne, D. J., 1984, Effects of magnesium chloride on smooth muscle actomyosin adenosine-5'-triphosphate activity, myosin conformation, and tension development in glycerinated smooth muscle fibers, *Biochemistry*, 23: 5062.

Kamm, K. E. and Stull, J. T., 1985, Myosin phosphorylation, force, and maximal shortening velocity in neurally stimulated tracheal smooth muscle, *Am. J. Physiol.*, 249: C238.

Kamm, K. E. and Stull, J. T., 1985, The function of myosin and myosin light chain kinase phosphorylation in smooth muscle, *Ann. Rev. Pharmacol. Toxicol.*, 25: 593.

Kerrick, W. G. L. and Hoar, P. E., 1987, Non-Ca^{2+}-activated contraction in smooth muscle, *in:* "Regulation and Contraction of Smooth Muscle", M. J. Siegman, A. P. Somlyo, and N. L. Stephens, eds., Alan R. Liss, New York, p. 437.

Kerrick, W. G. L., Hoar, P. E., and Cassidy, P. S., 1980, Calcium activated tension: The role of myosin light chain phosphorylation, *Fed. Proc.*, 39: 1558.

Kitazawa, T., Kobayashi, S. Horiuti, K., Somlyo, A. V., and Somlyo, A. P., 1989, Receptor-coupled, permeabilized smooth muscle. Role of the phosphatidylinositol cascade, G-proteins, and modulation of the contractile response to Ca^{2+}, *J. Biol. Chem.*, 264: 5339.

Kubota, Y., Kamm, K. E., and Stull, J. T., 1990, Mechanism of GTPγS-dependent regulation of smooth muscle contraction, *Biophys. J.*, 57: 163a.

Kühn, H., Tewes, A., Gagelmann, M., Güth, K., Arner, A., and Rüegg, J. C., 1990, Temporal relationship between force, ATPase activity, and myosin phosphorylation during a contraction/relaxation cycle in a skinned smooth muscle, *Pflügers Arch.*, 416: 512.

Moreland, R. S. and Ford, G. D., 1982, The influence of Mg^{2+} on the phosphorylation and dephosphorylation of myosin by an actomyosin preparation from vascular smooth muscle, *Biochem. Biophys. Res. Commun.*, 106: 652.

Moreland, R. S. and Moreland, S., 1991, Characterization of magnesium-induced contractions in detergent-skinned swine carotid media, *Am. J. Physiol.*, in press.

Moreland, R. S., Moreland, S., and Murphy, R. A., 1988, Effects of length, Ca^{2+}, and myosin phosphorylation on stress generation in smooth muscle, *Am. J. Physiol.*, 255: C473.

Moreland, R. S. and Murphy, R. A., 1986, Determinants of Ca^{2+}-dependent stress maintenance in skinned swine carotid media, *Am. J. Physiol.*, 251: C892.

Moreland, S., Antes, L. M., McMullen, D. M., Sleph, P. G., and Grover, G. J., 1990, Myosin light-chain phosphorylation and vascular resistance in canine anterior tibial arteries in situ, *Pflügers Arch.*, 417: 180.

Moreland, S., Little, D. K., and Moreland, R. S., 1987, Calmodulin antagonists inhibit latch bridges in detergent skinned swine carotid media, *Am. J. Physiol.*, 252: C523.

Moreland, S. and Moreland, R. S., 1987, Effects of dihydropyridines on stress, myosin phosphorylation, and V_o in smooth muscle, *Am. J. Physiol.*, 252: H1049.

Moreland, S., Moreland, R. S., and Singer, H. S., 1987, Apparent dissociation of myosin light chain phosphorylation and maximal velocity of shortening in KCl depolarized swine carotid artery: Effects of temperature and [KCl], *Pflügers Arch.*, 408: 139.

Morgan, J. P. and Morgan, K. G., 1984, Stimulus-specific patterns of intracellular calcium levels in smooth muscle of ferret portal vein, *J. Physiol.*, 351: 155.

Ngai, P. K. and Walsh, M. P., 1984, Inhibition of smooth muscle actin-activated myosin Mg^{2+}-ATPase activity by caldesmon, *J. Biol. Chem.*, 259: 13656.

Nishimura, J. and van Breemen, C., 1989, Possible involvement of actomyosin ADP complex in regulation of Ca^{2+} sensitivity in α-toxin permeabilized smooth muscle, *Biochem. Biophys. Res. Commun.*, 165: 408.

Paul, R. J. and Peterson, J. W., 1975, Relation between length, isometric force and oxygen consumption rate in vascular smooth muscle, *Am. J. Physiol.*, 228: 915.

Paul, R. J., 1983, Coordination of metabolism and contractility in vascular smooth muscle, *Fed. Proc.*, 42: 62.

Rembold, C. M. and Murphy, R. A., 1988, Myoplasmic [Ca^{2+}] determines myosin phosphorylation and isometric force in agonist-stimulated swine arterial smooth muscle, *J. Cardiovasc. Pharmacol.*, 12(Suppl 5): S38.

Saida, K. and Nonomura, Y., 1978, Characteristics of Ca^{2+}- and Mg^{2+}- induced tension development in chemically skinned smooth muscle fibers, *J. Gen. Physiol.*, 72: 1.

Siegman, M. J., Butler, T. M., Mooers, S. U., and Davies, R. E., 1980, Chemical energetics of force development, force maintenance, and relaxation in mammalian smooth muscle, *J. Gen. Physiol.*, 76: 609.

Siegman, M. J., Butler, T. M., Mooers, S. U., and Michalek, A., 1984, Ca^{2+} can affect V_{max} without changes in myosin light chain phosphorylation in smooth muscle, *Pflügers Arch.*, 401: 385.

Siegman, M. J., Butler, T. M., and Mooers, S. U., 1989, Phosphatase inhibition with okadaic acid does not alter the relationship between force and myosin light chain phosphorylation in permeabilized smooth muscle, *Biochem. Biophys. Res. Commun.*, 161: 838.

Singer, H. S. and Murphy, R. A., 1987, Maximal rates of activation in electrically stimulated swine carotid media, *Circ. Res.*, 60: 438.

Sobieszek, A., 1977, Ca^{2+}-linked phosphorylation of a light chain in vertebrate smooth muscle myosin, *Eur. J. Biochem.*, 73: 477.

Somlyo, A. V., Goldman, Y. E., Fujimori, T., Bond, M., Trentham, D. R., and Somlyo, A. P., 1988, Cross-bridge kinetics, cooperativity, and negatively strained cross-bridges in vertebrate smooth muscle, *J. Gen. Physiol.*, 91: 165.

Somlyo, A. P., Kitazawa, T., Himpens, B., Matthijs, G., Horiuti, K., Kobayashi, S., Goldman, Y. E., and Somlyo, A. V., 1989, Modulation of Ca^{2+}-sensitivity and of the time course of contraction in smooth muscle: A major role of protein phosphatases?, *in:* "Adv. Prot. Phosphatases; Vol. 5", W. Merleude and J. DiSalvo, eds., Leuven University Press, Leuven, p. 181.

Takahashi, K., Hiwada, K., and Kokubu, T., 1988, Vascular smooth muscle calponin. A novel troponin T-like protein, *Hypertension*, 11: 620.

Wagner, J. and Rüegg, J. C., 1986, Skinned smooth muscle: Calcium-calmodulin activation independent of myosin phosphorylation, *Pflügers Arch.*, 407: 569.

Walsh, M. P., Bridenbaugh, R., Hartshorne, D. J., and Kerrick, W. G. L., 1982, Phosphorylation-dependent activated tension in skinned gizzard muscle fibers in the absence of Ca^{2+}, *J. Biol. Chem.*, 257: 5987.

Winder, S. J. and Walsh, M. P., 1990, Smooth muscle calponin. Inhibition of actomyosin MgATPase and regulation by phosphorylation, *J. Biol. Chem.*, 265: 10148.

COOPERATIVE MECHANISMS IN THE REGULATION OF SMOOTH MUSCLE CONTRACTION

Marion J. Siegman, Thomas M. Butler, Tapan Vyas,
Susan U. Mooers, and Srinivasa Narayan

Department of Physiology
Jefferson Medical College
Philadelphia, PA 19107

INTRODUCTION

In 1930, Bozler reported that smooth muscle was unique in that it could maintain force with a very low energy expenditure (Bozler, 1930). More recent studies have extended this pioneering work to show that the economy of force maintenance of mammalian smooth muscle can vary to a large extent during the course of an isometric contraction (for review see Butler and Siegman, 1985). Crossbridge cycling rate varies independently of the ability to maintain maximum force, and the underlying regulatory mechanisms are not fully understood. It is known that phosphorylation of the 20 kDa light chain of myosin is central to the process of regulation, but details remain elusive. *In vitro* biochemical experiments have shown that phosphorylation increases the actin-activated myosin ATPase activity (for review see Hartshorne, 1987). To our knowledge, there has been no study showing that contraction of smooth muscle can occur without some increase in myosin light chain phosphorylation. However, high force can be generated with very low degrees of myosin light chain phosphorylation (Moreland and Moreland, 1987; Ratz and Murphy, 1987). Although there are some important exceptions, in general, the maximum velocity of shortening and energy usage are high under conditions when the degree of light chain phosphorylation is high. The major question that really distills from all of the above is what role does myosin light chain phosphorylation play in the regulation of force output and crossbridge cycling rate?

The latch-bridge hypothesis (Dillon et al., 1981) has evolved through the years into one in which the unique properties of smooth muscle result from the dephosphorylation of an attached crossbridge (Driska, 1987; Hai and Murphy, 1988). The dephosphorylation of the myosin is postulated to alter the kinetics of the completion of the crossbridge cycle. This is a very attractive model and has led to a large experimental effort which has tested its applicability. Some experimental results support the model, but in our view certain critical features are not supported. In this chapter we will present some of the data which argues against the idea that the observed high force with low phosphorylation can be accounted for by the dephosphorylation of attached myosin. We will then try to provide a framework for a model in which phosphorylation of the myosin light chain directly controls force output and crossbridge cycling rate through cooperative interactions within the contractile filaments.

The idea that high force output with low levels of myosin light chain phosphorylation could be explained on the basis of myosin light chain phosphatase activity was independently proposed by Driska (1987), and Hai and Murphy (1988). They suggested that myosin light chain phosphorylation is required for the transition of the crossbridge into the force-producing state, and that phosphorylated myosin goes through the normal crossbridge cycle in which there is attachment and detachment of the crossbridge with concomitant splitting of ATP. However, a crossbridge that is dephosphorylated while it is attached to actin and generating force has a detachment rate that is very slow. Under steady-state conditions in this model, there is a large force production from dephosphorylated crossbridges with the result that large forces can be maintained when only a small fraction of myosin is phosphorylated at any time.

The basic assumption of these models is that the activity of the phosphatase is responsible for the formation of dephosphorylated, force-producing crossbridges (latchbridges). We tested these models through the use of a phosphatase inhibitor, okadaic acid, on a skinned preparation of the portal vein of the rabbit (Siegman et al., 1989). We asked the question whether inhibition of the phosphatase activity alters the relationship between force and myosin light chain phosphorylation. The results are shown in Figure 1.

When the skinned portal vein is activated by an increase in the calcium concentration of the bathing solution, there is a curvilinear relationship between myosin light chain phosphorylation and force output. Maximum force occurs when about 50% of the myosin light chain is phosphorylated. Thus, the mechanism that causes high force output with low myosin light chain phosphorylation in the intact muscle also operates in this permeabilized muscle. According to the models of Hai and Murphy, and Driska, inhibition of

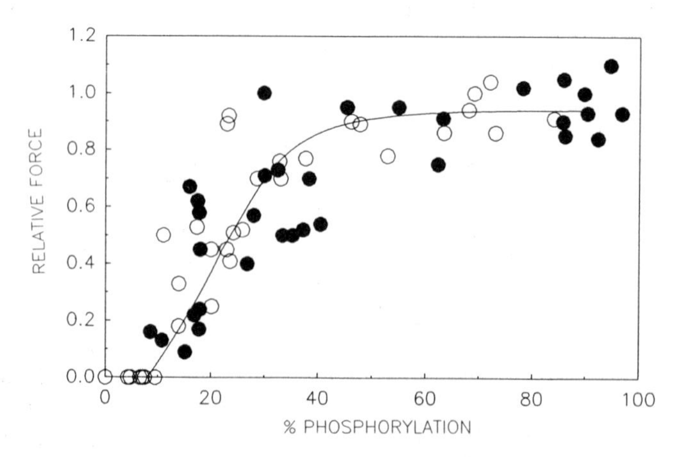

Figure 1. Relationship between force and degree of phosphorylation of the 20 kDa myosin light chain at different [Ca^{2+}] in the presence (●) and absence (○) of the phosphatase inhibitor okadaic acid (5 μM). The degree of phosphorylation represents the percent of the total light chain which has at least one phosphate. (Modified from Siegman et al., 1989).

the phosphatase should prevent the formation of latchbridges, and there should be a linear relationship between phosphorylation and force. However, the data show that inhibition of the phosphatase with okadaic acid has no effect on this relationship (see Fig. 1). Similar results were obtained when the light chain was thiophosphorylated with ATPγS, making it resistant to dephosphorylation by the phosphatase (unpublished observations). It is obvious that in this case, the mechanism responsible for high force output with low myosin light chain phosphorylation must rely on some process other than dephosphorylation of the crossbridge in the high force-generating state.

There is, however, a specific condition where there is strong evidence for dephosphorylation of myosin resulting in the formation of force-bearing crossbridges with very slow detachment rates. This occurs during relaxation of the permeabilized portal vein. Specifically, we have found that during relaxation, force declines very slowly at a time when myosin light chain phosphorylation and ATPase activity have decreased to near-resting values (Butler et al., 1990). It also appears that there is a significant excess of inorganic phosphate bound to the crossbridge under such conditions, and that the time course of release of the extra inorganic phosphate is similar to the time course of relaxation. From these data we proposed that dephosphorylation of the attached crossbridge during the onset of relaxation can result in an alteration in the kinetics of release of the products of ATP splitting, and thus lead to a condition where force is maintained by a dephosphorylated crossbridge. It should be stressed, however, that this is a very special condition, and the lack of an effect of inhibition of the phosphatase on the force vs. phosphorylation relationship (Fig. 1) shows that such a special crossbridge state is not formed to a great extent under activated conditions. If the action of the phosphatase cannot explain high force with low myosin light chain phosphorylation under activated conditions, then what can?

One possibility is that low degrees of myosin light chain phosphorylation turn on all of the crossbridges with respect to force production. Such a cooperative mechanism is shown diagrammatically in Figure 2A and could be mediated through either the thick or thin filament. Another possible mechanism for maximum force output with less than maximum myosin light chain phosphorylation is shown in Figure 2B. In this model only a very few crossbridges which are preferentially phosphorylated ever generate force. Further phosphorylation has no effect on force production. Although seemingly far-fetched because only a few of the crossbridges are ever turned on, and there is no obvious function to assign to nearly half of the myosin, this would explain the observation of maximum force with very low phosphorylation!

We attempted to test these two models and to answer the question of how many crossbridges are activated by low degrees of myosin light chain phosphorylation. We used an experimental design based on a single turnover paradigm described previously (Butler et al., 1989). Briefly, the experiments are based on the observation that myosin exists with ADP bound in both resting and activated muscles. In the resting state, the release of the bound ADP is very slow, while under activated (and fully phosphorylated) conditions ADP release is very rapid. In order to quantify the transition between the resting and activated state, we determined the relationship between phosphorylation and the fraction of crossbridges which release nucleotide diphosphate quickly.

In these experiments the permeabilized smooth muscle was incubated in formycin triphosphate (FTP), a fluorescent analog of ATP, followed by a rapid transfer to ATP. Under resting conditions there is a biphasic release of

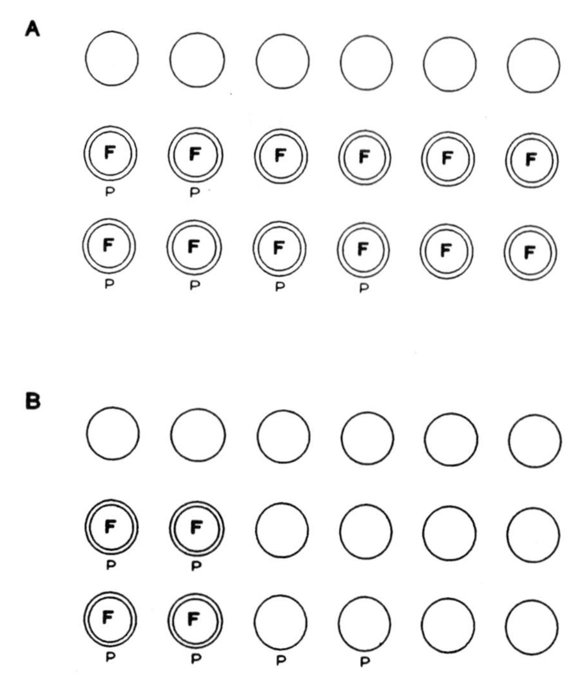

Figure 2. Two possible models for the control of force production by myosin light chain phosphorylation. Each circle represents a myosin subfragment 1. The single circles show crossbridges which do not generate force, and those with the letter **F** are crossbridges which are generating force. The letter **P** beneath the circle indicates that the myosin light chain is phosphorylated. In each model, the first row shows the resting state with none of the myosin phosphorylated, and the second and third rows show phosphorylation of one-third, and two-thirds of the myosin, respectively.

formycin diphosphate (FDP) which is complete in about 30 min. The time course is similar to results reported earlier for ADP release and most likely represents product release from myosin. Activation of the muscle by near-maximum (80%) thiophosphorylation of the myosin light chain with ATPγS results in a suprabasal FTPase activity which is similar to suprabasal ATPase activity and dramatically accelerates the rate of nucleotide diphosphate release in the single turnover experiments.

In order to determine how the extent of myosin light chain phosphorylation controls activation of the crossbridge, we determined the relationships among the fraction of FDP rapidly released, the degree of myosin light chain phosphorylation and force production. **A 20% thiophosphorylation of the myosin light chain caused the maximum shift to rapid release of FDP.** Maximum force output also occurs at less than maximal thiophosphorylation. Therefore, there is a large degree of cooperativity in the activation of myosin by light chain thiophosphorylation so that phosphorylation of a single myosin head activates the ATPase activity and force-generating capabilities of other non-phosphorylated myosin heads. This supports the model shown in Figure 2A. It is also further evidence that the idea of a latchbridge formed by dephosphorylation of attached myosin is less than satisfactory because the thiophosphorylated light chain is not a substrate for the phosphatase.

The observed high degree of cooperativity explains how high force results from low degrees of phosphorylation, but leaves open to question the function of higher degrees of myosin light chain phosphorylation. As mentioned earlier, previous studies on both intact and permeabilized muscles have generally found that higher degrees of myosin light chain phosphorylation were associated with higher ATPase activity. We have found that the suprabasal ATPase rate in the permeabilized portal vein of the rabbit is two-fold higher when the myosin light chain is thiophosphorylated to 60% as compared to 35%. This is the case even though both levels of phosphorylation are higher than is required for maximum force production and for the maximum effect on nucleotide diphosphate release in the single turnover experiments. This suggests that the degree of phosphorylation can modulate the cycling rate of an activated crossbridge. In other words, a small amount of phosphorylation activates all of the myosin and a higher degree of phosphorylation makes the myosin cycle faster.

Figure 3 shows two possible models to explain the relationships among myosin light chain phosphorylation, force, and ATPase activity of the crossbridge. Both models show the cooperative turn-on of force by low levels of myosin light chain phosphorylation. In model A, phosphorylation of some crossbridges increases the ATPase activity of all crossbridges, but those that are phosphorylated have a higher rate than those that are not. An increase in the degree of phosphorylation increases the overall ATPase activity by increasing the fraction of the crossbridges in the very high ATPase pool. In model B, all crossbridges have the same cycling rate which is determined by the overall

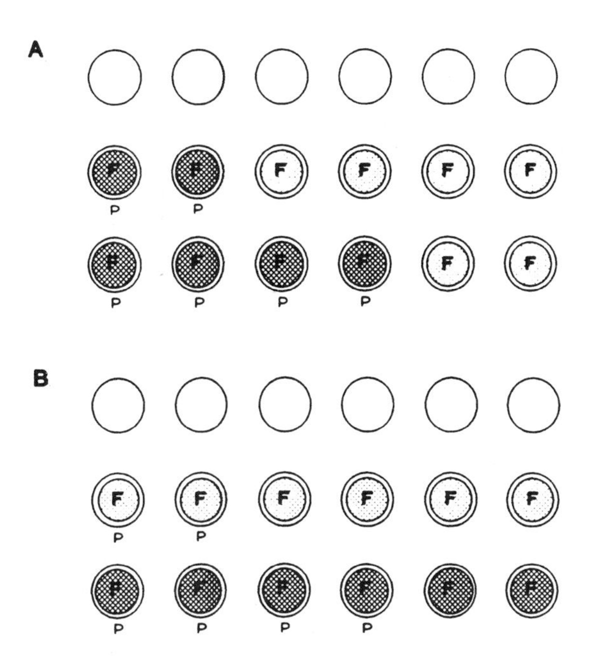

Figure 3. Two possible models for the regulation of force production and actin-activated myosin ATPase activity by myosin light chain phosphorylation. Both models are an extension of model B shown in Figure 2, with the degree of shading proportional to the ATPase activity of the myosin head. See the legend for Figure 2 and text for further details.

phosphorylation of the entire thick filament. An increase in phosphorylation increases ATPase rate because of an increase in the cycling rate of all crossbridges. Model B would require some second cooperative mechanism which results in a graded ATPase for individual myosin molecules.

It is difficult to distinguish between these two models experimentally, but some indication can be gained from the predicted effect that they would have on the energetics of work production during shortening. In model A, the pool of slowly cycling crossbridges would presumably provide an internal load against which the rapidly cycling crossbridges would have to function. This would be the simplest explanation as to why maximum velocity of shortening is graded. Under such conditions one would predict that at low levels of phosphorylation some of the energy that normally appears as work would appear as heat and that the chemical energy cost of external work production would be greater than when the levels of phosphorylation are higher. In contrast, with a single pool of crossbridges that all cycle at the same rate (model B) there would be no internal work dissipation by fast-cycling crossbridges that are impeded by slow-cycling crossbridges. Therefore, there would be no obvious prediction of a change in the chemical energy cost of work production as a function of myosin light chain phosphorylation.

In experiments performed on the intact rabbit taenia coli under conditions in which the "latch" state was induced (high force with decreased maximum velocity of shortening and chemical energy usage as well as a low degree of myosin light chain phosphorylation) there was no evidence of an increase in the chemical energy cost of active external work production (Butler et al., 1986). There is, therefore, no energetic evidence for the dissipation of work against an internal load under conditions during which most of the force would be expected to be generated by latchbridges. This suggests that force maintenance with low energy input occurs as shown in model B (Fig. 3).

Evidence for cooperativity in the regulation of smooth muscle has also come from mechanical experiments on permeabilized muscles in which the presence of some rigor crossbridges caused an increase in force output (Somlyo et al., 1988). Further, *in vitro* biochemical studies have shown that rigor crossbridges cause an increase in the actin-activated ATPase activity of phosphorylated myosin (Chacko and Eisenberg, 1990).

Our major conclusion is that phosphorylation of the myosin light chain has a dual function in driving the regulation of smooth muscle. Whereas little phosphorylation is required for activation of all of the myosin and maximum force production, more phosphorylation results in an increase in the ATPase activity of the already activated myosin. Thus, the single process of myosin light chain phosphorylation controls both force output and crossbridge cycling rate independently. Clearly, cooperative interactions through the thick or thin filament are central to this mechanism. Recent evidence from *in vitro* motility experiments by Warshaw and Trybus (this volume) suggest that such a cooperativity is not mediated through the thick filament. However, definitive experiments showing cooperativity mediated through the thin filament and/or its associated regulatory proteins are needed.

The simple model presented here explains much but not all of the available data. A specific and important observation which does not fit with such a simple model is how changes can occur in the mechanical and energetic parameters of the muscle in the face of constant levels of myosin light chain phosphorylation (Siegman et al., 1984). Such observations could be explained if other proteins such as caldesmon and/or calponin, which are known to reversibly associate with the thin filament, modulate the degree of cooperativity and give an even finer control over the relationships among myosin light chain phosphorylation, force, and ATPase rate in smooth muscle.

ACKNOWLEDGMENTS

This work was supported by grants HL15835 to the Pennsylvania Muscle Institute and DK37598 from the National Institutes of Health.

REFERENCES

Bozler, E., 1930, The heat production of smooth muscle, *J. Physiol.*, 69: 442.

Butler, T. M., Pacifico, D. S., and Siegman, M. J., 1989, ADP release from myosin in permeabilized smooth muscle, *Am. J. Physiol.*, 256: C59.

Butler, T. M. and Siegman, M. J., 1985, High-energy phosphate metabolism in vascular smooth muscle, *Ann. Rev. Physiol.*, 47: 629.

Butler, T. M., Siegman, M. J., and Mooers, S. U., 1986, Slowing of cross-bridge cycling in smooth muscle without evidence of an internal load, *Am. J. Physiol.*, 251: C945.

Butler, T. M., Siegman, M. J., Mooers, S. U., and Narayan, S. R., 1990, Myosin-product complex in the resting state and during relaxation in smooth muscle, *Am. J. Physiol.*, 258: C1092.

Chacko, S. and Eisenberg, E., 1990, Cooperativity of actin-activated ATPase of gizzard heavy meromyosin in the presence of gizzard tropomyosin, *J. Biol. Chem.*, 265: 2105.

Dillon, P. F., Aksoy, M. O., Driska, S. P., and Murphy, R. A., 1981, Myosin phosphorylation and the crossbridge cycle in arterial smooth muscle, *Science*, 211: 495.

Driska, S. P., 1987, High myosin light chain phosphatase activity in arterial smooth muscle: can it explain the latch phenomenon? *in:* "Regulation and Contraction of Smooth Muscle", M. J. Siegman, A. P. Somlyo, N. L. Stephens, eds., Alan R. Liss, New York, p. 387.

Hai, C. M. and Murphy, R. A., 1988, Cross-bridge phosphorylation and regulation of latch state in smooth muscle, *Am. J. Physiol.*, 254: C99.

Hartshorne, D. J., 1987, Biochemistry of the contractile process in smooth muscle, *in:* "Physiology of the Gastrointestinal Tract", L. R. Johnson, ed., Raven Press, New York, p. 423.

Moreland, S. and Moreland, R. S., 1987, Effects of dihydropyridines on stress, myosin phosphorylation and V_o in smooth muscle, *Am. J. Physiol.*, 252: H1049.

Ratz, P. H. and Murphy, R. A., 1987, Contributions of intracellular and extracellular Ca^{++} pools to activation of myosin phosphorylation and stress in swine carotid media, *Circ. Res.*, 60: 410.

Siegman, M. J., Butler, T. M., and Mooers, S. U., 1989, Phosphatase inhibition with okadaic acid does not alter the relationship between force and myosin light chain phosphorylation in permeabilized smooth muscle, *Biochem. Biophys. Res. Commun.*, 161: 838.

Siegman, M. J., Butler, T. M., Mooers, S. U., and Michalek, A., 1984, Ca^{++} can affect V_{max} without changes in myosin light chain phosphorylation in smooth muscle, *Pflügers Arch.*, 401: 385.

Somlyo, A. V., Goldman, Y. E., Fujimori, T., Bond, M., Trentham, D. R., and Somlyo, A. P., 1988, Cross-bridge kinetics, cooperativity, and negatively strained cross-bridges in vertebrate smooth muscle, *J. Gen. Physiol.*, 91: 165.

MECHANICS OF THE CROSSBRIDGE INTERACTION IN LIVING AND CHEMICALLY SKINNED SMOOTH MUSCLE

Per Hellstrand

Department of Physiology and Biophysics
University of Lund
Sölvegatan 19, S-223 62
Lund, Sweden

INTRODUCTION

Force development in muscle is thought to result from cyclic attachment and detachment of crossbridges extending from "thick" myosin-containing filaments to binding sites on "thin" actin-containing filaments. According to the crossbridge model formulated by Huxley (1957) and extended by Huxley and Simmons (1971), the rate constants for attachment and detachment are functions of the position of the crossbridge relative to the binding site on the actin filament, which also determines the force by the elastic extension of the crossbridge. Particularly, "negative" strain on the crossbridge, giving rise to a force tending to oppose shortening of the muscle, is associated with a high rate constant for crossbridge detachment.

The conversion of chemical to mechanical energy in the crossbridge is considered to involve a conformational change altering the strain dependence of the force on the crossbridge. Several models for this process have been proposed, based on the available biochemical and mechanical data (see e.g. Huxley and Simmons, 1971; Eisenberg and Hill, 1985; Huxley and Kress, 1985). Phenomenologically the crossbridge is found to behave as a viscoelastic body comprising an undamped elasticity in series with a damped elastic element. Relaxation of the undamped and damped elasticity after an imposed length step apparently takes place without the crossbridge going through a full cycle involving ATP breakdown, which is a much slower process than the transient responses occurring in the first milliseconds after a length step.

The investigation of crossbridge properties in smooth muscle is made difficult by the poor knowledge concerning the organization of the contractile filament system in this tissue. However, steady state mechanical and energetic properties have been characterized in a variety of smooth muscle preparations under different conditions of activation and mechanical constraints (Hellstrand and Paul, 1982). Interesting deviations from the behavior of skeletal muscle have been detected, such as a low maximal shortening velocity at nearly the same force per cross-sectional area as in striated muscle, and a high energetic "economy" of force maintenance.

Regulation of Smooth Muscle Contraction
Edited by R.S. Moreland, Plenum Press, New York, 1991

One aspect of the crossbridge mechanism in smooth muscle that has attracted considerable attention is the apparent change in the rate of crossbridge turnover during the course of a sustained contraction, resulting in lower maximal shortening velocity and higher economy of force maintenance in "tonic" vs. "phasic" contractions (Hardung and Laszt, 1966; Hellstrand and Johansson, 1975; Murphy, 1976; Hellstrand, 1977). Dillon et al. (1981) observed that a decrease in shortening velocity during tonic contraction in swine carotid artery correlated with a decreased level of phosphorylation of the 20 kDa myosin light chain and postulated that dephosphorylation of attached, tension-generating crossbridges results in a crossbridge state (termed "latch") that is characterized by a low tendency for detachment and thus could serve as an internal resistance to shortening. Under isometric conditions "latch bridges" would increase the economy of force maintenance by holding tension at low rates of ATP turnover. More detailed kinetic models of the "latch" mechanism have since been worked out (Driska, 1987; Hai and Murphy, 1988) as well as challenged (Paul, 1990; Hai et al., this volume).

As noted above, information on the kinetics of crossbridge attachment and detachment reactions can be obtained by non-steady state "transient" measurements, where the rate of change of state in the crossbridge population after a mechanical or other perturbation is observed. We have presented data from chemically skinned smooth muscle of guinea-pig taenia coli describing mechanical transient responses in the sub-millisecond time range (Hellstrand and Arheden, 1989; Arheden and Hellstrand, 1991). The present paper describes stiffness properties of chemically skinned taenia coli fibers during Ca^{2+}-activated contraction and rigor, and of living fibers activated by high K^+ solution. These experiments were performed to reveal changes in the population of attached crossbridges during a transition from isometric contraction to unloaded shortening. The study of living fibers was aimed at elucidating alterations in these reactions associated with development of the "latch" state.

METHODS

Strips with a length of 1 - 2 mm and a width of 0.1 - 0.2 mm were prepared from either fresh or chemically skinned smooth muscle of guinea-pig taenia coli. The strips were mounted between a force transducer (modified AE 801, SensoNor A/S, Horten, Norway) and a home-built electromagnetic puller capable of performing ramp-shaped length changes between two fixed positions in a time of about 300 μs. Protocols of several length steps between different positions at preset intervals could be programmed. The force transducer was modified by shortening the blade of the silicon strain gauge to a length of about 2 mm, which increased the resonant frequency in air to about 40 kHz. Length and force data were sampled at 100 kHz using a 12-bit A/D-converter.

Experiments on intact preparations were carried out at 37° C in a Krebs solution of the following composition in mM: NaCl 122, KCl 4.7, $MgCl_2$ 1.2, $CaCl_2$ 2.5, $NaHCO_3$ 15.5, KH_2PO_4 1.2, glucose 11.5, equilibrated with 96% O_2 - 4% CO_2, giving a pH of 7.3 - 7.4. The solution was continuously pumped through the temperature-controlled bath (0.5 ml) containing the muscle fiber. Activation was achieved by switching to a high K^+ solution, obtained by exchanging some of the NaCl of the Krebs solution for KCl. The concentration of KCl was usually about 30 mM and was chosen to give a maintained maximal contraction for at least 5 min. Higher concentration of KCl tended to produce contractions which faded with time. Repeated activations for 5 min with relaxed intervals of 2 min in normal Krebs solution were performed. One length step sequence consisting of a shortening step (release) with a following restretch after varying time to a longer length than the original (L_i), and then after 1 - 2 ms a release back to L_i, was elicited at the end of each 5 min contraction. These measurements using "tonic" activation were compared with a series of "phasic" contractions on the same fiber where the same protocol of length steps was applied after activation for 15 s. The variability in isometric force between the different contractions in each series, as well as between the "tonic" and "phasic" series, was only a few per cent.

Skinned preparations were obtained by treatment with 1% Triton X-100 for 3 h at 4° C, as described by Arheden and Hellstrand (1991). Skinned preparations were stored for a maximum of 4 weeks at -15° C in the skinning solution without Triton X-100 and with 50% glycerol added. Experimental solutions contained 12 mM PCr, 3.2 mM MgATP, 2 mM free Mg^{2+}, 0.5 mM NaN_3, 0.5 μM calmodulin, 15 U/ml creatine phosphokinase, and were buffered to pH 6.9 by 30 mM TES. Desired free-Ca^{2+} levels (expressed as pCa = -log[Ca^{2+}]) were obtained by mixing stock solutions of the above composition added with 4 mM of either K_2EGTA or K_2CaEGTA, using a computer program similar to that described by Fabiato and Fabiato (1979) with binding constants as given by Fabiato (1981). In solutions for rigor experiments, MgATP and PCr were replaced by an ATP trap consisting of 10 mM glucose and 20 U/ml hexokinase. All experiments on skinned preparations were performed at room temperature (21° - 23° C). In some experiments muscle fibers were first mounted as intact preparations and length step data obtained at 37° C and at room temperature, then they were skinned using 0.1% β-escin in relaxing solution (pCa 9) for 10 min (Kobayashi et al., 1989). Length steps were performed on the skinned preparations after activation at pCa 4.5

RESULTS AND DISCUSSION

Crossbridge Detachment in Shortening of Skinned Fibers

Responses of an activated skinned taenia coli fiber to a sequence of length steps consisting of a shortening and a following restretch to beyond the original length are shown in Figure 1. The upper panels show length and force records plotted against time. A shortening step (1) followed by a restretch (2) after 1 ms and a further shortening step (3) back to the initial length are shown. The right hand records show another restretch (4) at 512 ms after an initial shortening step (not shown) identical to that in the left hand records. The lower panel in Figure 1 shows length and force plotted against each other. The different steps are identified as in the upper panels. The release is large enough, about 9% of the initial length (L_i), so that the muscle goes slack and is at zero force for most of the release (the horizontal portion of the release record). Notice, however, that the muscle approaches the force baseline under an acute angle and initially slightly undershoots zero force. This suggests that the contractile system has some capacity to support negative force (c.f. Ford et al., 1977).

The shape of the restretch curve at 1 ms is similar to that during release except for a displacement upwards and towards the left. The restretch is carried beyond L_i to emphasize the linear part of the restretch record and to allow a determination of the slope of the record (indicating stiffness) in the vicinity of L_i. The slight oscillations in the record originate in the servo system. The

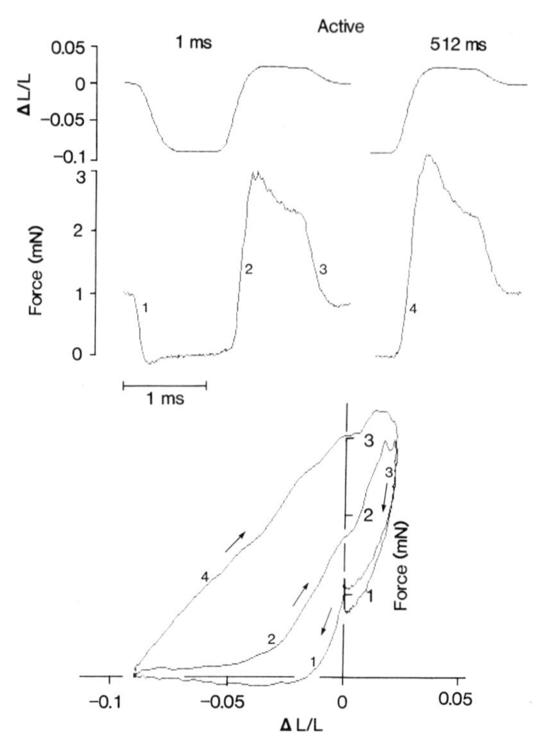

Figure 1. Force response to a sequence of length changes in an actively contracting skinned taenia coli fiber. In the upper panel the left hand records show length and force signals when the fiber was first shortened (1), then after 1 ms extended (2) to beyond L_i and finally shortened (3) back to L_i. The right hand records show a re-extension (4) performed at 512 ms after another shortening step (corresponding to step 1) during the same contraction. The lower panel shows force vs length, with the different length steps identified as in the upper records. The direction of movement is shown by arrows.

return release to L_i after 1 ms at the longer length completes the loop. The shape of the response at 1 ms represents a hysteresis loop and suggests visco-elastic behavior of the muscle, consistent with the assumed properties of the attached crossbridge system. However, close inspection of the restretch curve shows a lower slope near L_i than that in the initial release. This suggests that some net detachment of crossbridges has taken place.

With prolonged waiting time at the short length (512 ms) before restretch, the shape of the force response is changed considerably. The record now sets out from the minimal length without a horizontal segment, indicating that all slack in the muscle has been taken up. The restretch curve is fairly linear and its slope is clearly smaller than that in the original release. The amplitude of the force response to the restretch is slightly larger at 512 ms than at 1 ms (see also upper panels in Fig. 1). This is because the effective stretch distance is larger as the muscle is now intrinsically shorter, and this effect more than compensates for the decrease in stiffness (see below concerning possible detachment-attachment of crossbridges during the restretch). It is also seen that the rapid force decay after the stretch is larger at 512 ms.

Responses in Rigor

To produce rigor the skinned muscle was first contracted maximally and then transferred to an ATP-free solution (see Methods), initially at pCa 4.5 and then, after 10 - 15 min, at pCa 9. The duration of the incubation in ATP-free medium before length steps were performed was at least 30 min. During this time force slowly decreased to a plateau of about 1/3 of the active isometric force, and this force was maintained in the absence of Ca^{2+} (pCa 9). It has been shown that this state represents rigor, since force is rapidly reduced when ATP is photolytically released from inactive "caged-ATP" (Arner et al., 1987). When the muscle fiber from Figure 1 had been put into rigor and the length step protocol was repeated (Fig. 2) it is seen that the transient force responses to release and restretch at 1 ms resemble those in the activated fiber (Fig. 1) although the force before release is considerably smaller. A hysteresis loop is present which supports the conclusion that this part of the response is due to viscoelastic effects within attached crossbridges and possibly other structures in the fiber. On the other hand, inertial effects due to the fiber mass and the wave propagation time along the muscle fiber should be small in fibers of this dimension (Arheden and Hellstrand, 1991). A major deviation of the response in rigor from that in the active fiber is apparent when the restretches at 512 ms are compared (Figs. 1 and 2). In rigor there is just a small further leftward shift in the restretch response, showing little uptake of slack. However, some internal shortening of the fiber is clearly present. Huxley and Simmons (1971) ascribed the rapid (a few ms) force recovery seen after a length step in an activated frog muscle fiber to a mechanically induced damped conformational change in the attached crossbridge. Such a response might be present in rigor as well and could be one factor behind the internal shortening seen in this state

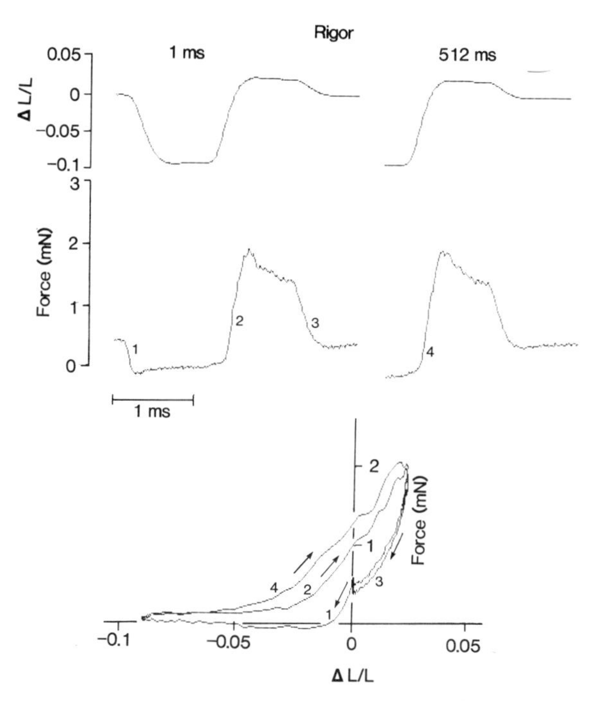

Figure 2. Mechanical transient responses in the same fiber as in Figure 1, but after induction of rigor. The sequence of length steps and their labeling are as in Figure 1.

after an unloading length step. Work on skeletal muscle has also shown evidence for transfer of crossbridges between different binding sites on the actin filament without an intervening step involving ATP-splitting (Schoenberg, 1985). While this cannot be elucidated by the present data such a mechanism may provide an explanation for the apparent (small) internal shortening after a release in rigor, since the release will induce a shift of the attached crossbridge population towards states with zero or negative strain, which might increase the otherwise low probability for detachment of rigor bridges. In contrast to this type of viscoelastic response there is no evidence for tension redevelopment as in active contraction when a muscle in rigor is released and then held at the short length.

Tension Transients in Living and Skinned Fibers

To relate the findings in the skinned fiber to events occurring in the living fiber, experiments were carried out in living taenia coli strips which were mounted in a Krebs solution at 37° C and repeatedly stimulated by high K^+ as described in Methods. When the contractions were uniform, temperature was lowered to 21° - 23° C (room temperature) and the fiber was again regularly stimulated by high K^+ solution. After a series of release-restretch

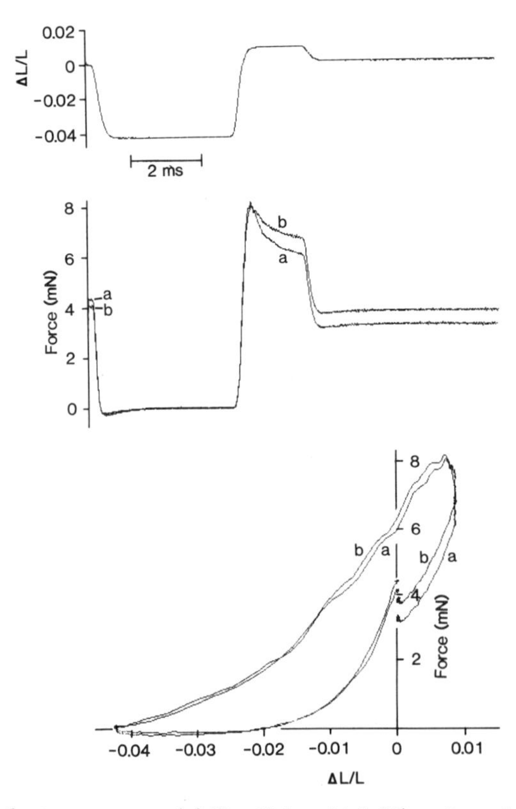

Figure 3. Length step sequence (cf. Fig. 1) in a high K^+ activated living fiber (a) and after skinning and activation at pCa 4.5 (b).

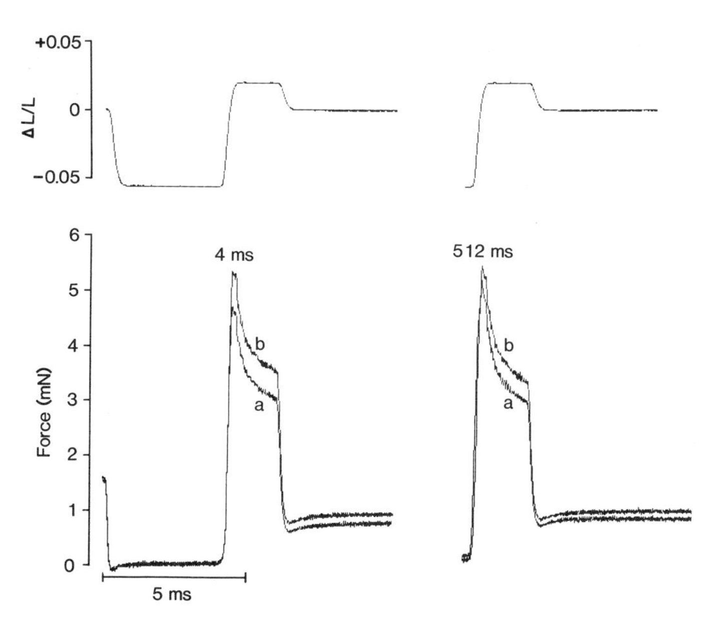

Figure 4. Length step sequences (cf. Fig. 1) in a high K^+ activated living fiber at 37° C. Records from contractions of 15 s (a) and 5 min (b) duration are shown. A restretch at 4 ms (left panel) after release is shown, as well as a restretch at 512 ms (right panel) after another release (not shown).

cycles had been performed as described above the fiber was skinned using β–escin for 10 min. Figure 3 shows responses to a 4 ms release-restretch cycle in an intact fiber activated by high K^+ solution at room temperature for 5 min (a) and in the same fiber after skinning and activation at pCa 4.5 for about 5 min (b). The isometric force before release, although marginally smaller after skinning, is close enough so that the responses can be directly compared. The initial stiffness (shape of the length-force relation during release) as well as the response to restretch is very similar. However, there is a tendency for a larger amplitude of the force response to restretch in the skinned fiber, and there is less decay of force after restretch. Return to the initial length after the 2 ms "overshoot" after restretch is associated with a drop to a lower force in the intact than in the skinned muscle. These results indicate that the response of the contractile system in the intact and skinned muscle is very similar. However, the small differences that exist may be related to alterations in the kinetic properties of the crossbridge system. The behavior during restretch suggests a somewhat smaller net crossbridge detachment after the shortening step in the skinned muscle.

Crossbridge Kinetics in Phasic and Tonic Contraction

The methodology developed by work on skinned fibers in active contraction and rigor was next applied to the question of whether the force transients would show different behavior in an intact fiber during contraction under "phasic" and "tonic" ("latch") conditions. Figure 4 shows responses in a fiber at 37° C activated by high K^+ solution for 15 s (Fig. 4a) and 5 min (Fig. 4b). Isometric force before the initial release was identical in both series of activations. This figure shows one set of release-restretch responses with restretch at 4 ms, and one set of restretch responses obtained at 512 ms (release not shown, but identical to that in the 4 ms cycle). In both restretches the amplitude of the force response is somewhat larger in the tonically activated contraction (Fig. 4b), and the force decay after restretch is smaller. This indicates a difference in the response of the contractile system to the identical length perturbation depending on the mode of stimulation. This is shown more clearly in the plots of force vs length displayed in Figure 5. The records for phasic (15 s) contractions are shown as full lines, whereas the records for tonic (5 min) contractions are shown as crosses, with each cross representing an individual sample value. The length-force record during release is virtually identical in phasic and tonic contractions, and the restretch at 4 ms also appears similar at the beginning of the restretch where force is low. As the restretch is carried beyond L_i, however, it becomes apparent that the amplitude of the force response is greater in the tonically activated contraction. The almost horizontal first part of the restretch record lies only slightly above the corresponding part of the release record and presumably represents re-extension of the fiber while it is slack, as described above. As the restretch proceeds, the length-force plot becomes fairly linear with a slope at L_i that is seen to be slightly larger in the record from the phasic contraction. Towards the end of the restretch the slope of the plot decreases, indicating a "yield" response. This is more evident in the phasic contraction. The decreasing slope is probably at least partly a consequence of the deceleration of the length change as it nears completion (cf. the records of length vs time shown on an expanded timescale in Figs. 1 and 2). The "yield" might then reflect a dependence of stiffness on rate of length change, and the greater yield in the phasic contraction indicate a greater tendency for crossbridge detachment. Note that the total length change during the stretch, even excluding the nearly horizontal part at the bottom, is larger than could reasonably be accommodated by attached cross-bridges (cf. Ford et al., 1977) and therefore an equilibrium of attachment and detachment during the stretch is likely. Different slopes of the restretch records, e.g. at the length L_i, might thus reflect different numbers of attached crossbridges remaining after the shortening step, but the quantitative relationship of the slope to the number of crossbridges attached at the start of the restretch is complicated. Nevertheless, these results indicate that during an

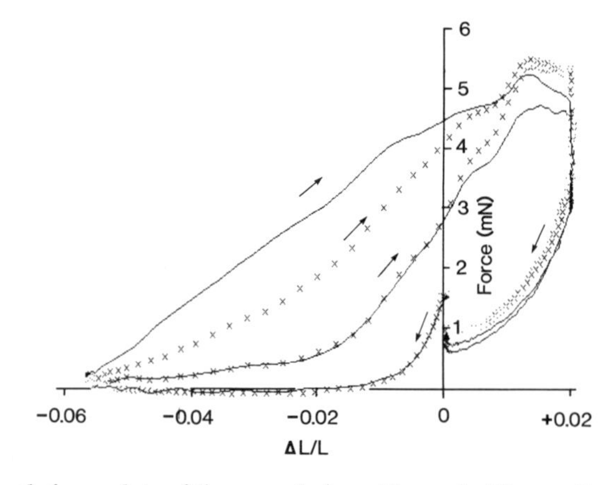

Figure 5. Length-force plots of the records from Figure 4. The continuous curves show data from the 15 s contraction (a in Fig. 4) and the crosses show data from the 5 min contraction (b in Fig. 4). Direction of movement shown by arrows. The rightmost restretches were performed at 4 ms and the leftmost restretches at 512 ms after release.

isometric contraction the number of attached crossbridges is the same in a tonic and a phasic contraction of the same force, since the release records are identical, whereas the release to zero force induces a greater decrease in the number of attached crossbridges in the phasic contraction.

The restretch records at 512 ms plotted in Figure 5 are shifted to the left by a greater amount in the phasic than in the tonic contraction. Stiffness as measured by the slope of the restretch record is clearly smaller in the phasic contraction. The release distance is large enough so that no force redevelopment has occurred before the restretch. The greater leftward shift in the phasic contraction indicates that the amount of intrinsic shortening of the muscle before restretch has been greater, consistent with a greater speed of unloaded shortening. The amplitude of the restretch response, both in the phasic and the tonic contraction, is only slightly greater in restretches at 512 ms than at 4 ms, despite a much larger "effective" stretch distance relative to the working range of the attached crossbridges. This again demonstrates that the slope stiffness during restretch represents an equilibrium between attachment and detachment of crossbridges during the stretch itself rather than just elastic extension of crossbridges remaining attached during all of the stretch. With this reservation the present analysis suggests that the unloaded shortening occurs with a smaller number of attached crossbridges in the phasic contraction.

Time Course of Change in Force and Stiffness After a Shortening Step

In the records shown in Figures 4 and 5 no force redevelopment had taken place at the time of the restretch. If the release is small enough, or the waiting time large enough, so that the muscle takes up its slack the sliding of the contractile filaments relative to each other will be arrested and isometric redevelopment of force will occur. The presence of a significant series-coupled elasticity outside the crossbridges is however likely in the present experiments since a "sarcomere-clamp" cannot be applied on the smooth muscle. This would imply a gradual transition from unloaded to isometric contraction, involving some filament sliding as force starts to develop. Klemt et al. (1981)

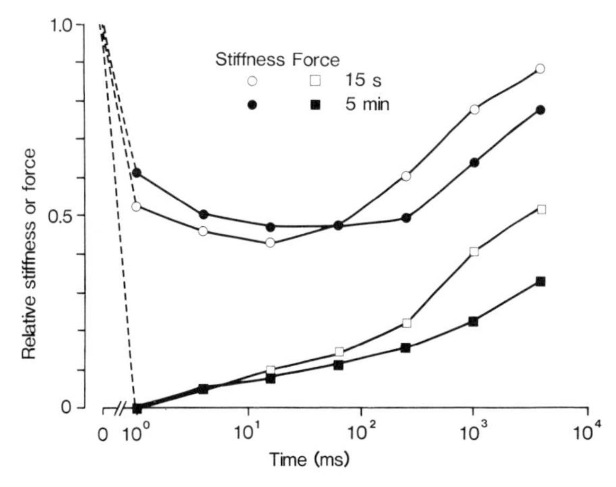

Figure 6. Relative values of force immediately before restretch and maximal stiffness during restretch at varying points in time after a shortening step in a living taenia coli fiber activated for 15 s or 5 min.

showed in electrically stimulated rat portal vein that when crossbridge attachments had been broken by large-amplitude vibration the rate of force redevelopment is faster after a contraction of short duration. Using our approach it is possible to get an indication of the relative number of attached crossbridges during unloaded shortening and isometric tension redevelopment after a shortening step. Figure 6 shows data from a series of releases by 4.1% L_i with restretches after time intervals of between 1 ms and 4096 ms (increments by a factor of 4). Force at the time of restretch is shown relative to isometric force before release, and stiffness is normalized to that measured in a stretch from the isometric state at L_i ("isometric stiffness"). The stiffness values show the greatest slope measured during restretch. Data from phasic (15 s) and tonic (5 min) contractions are shown. Note the logarithmic time scale. With the chosen amplitude of release force drops to zero, but redevelopment starts very early so that noticeable force is present already at 16 ms after release. Force redevelopment is seen to be faster in the phasically activated contractions, whereas stiffness is initially smaller. When significant force builds up, however, stiffness starts to increase faster in the phasic contraction and is back to nearly the isometric value at a time when force has recovered by only about 50%. Thus stiffness leads force during redevelopment of tension, as was also found by Kamm and Stull (1986) in electrically stimulated contractions of bovine trachealis muscle.

CONCLUSIONS

These studies of force transients in smooth muscle have shown that a rapid detachment of crossbridges occurs when the muscle is shortened and then held at slack length. This response is not seen in rigor, although a viscoelastic behavior that may reflect mechanical relaxation in attached crossbridges or crossbridge detachment-attachment is seen in this state as well. In intact smooth muscle fibers the number of attached crossbridges, as reflected in the stiffness, is identical under isometric conditions when force is equal during either phasic or tonic activation. There is evidence for a lowering of the energetic cost of force maintenance with time during contraction in smooth muscle (e.g. Paul, 1989). This implies a greater average force per crossbridge <u>cycle</u>. However, the present results indicate unchanged force per <u>attached</u> crossbridge when the turnover rate is decreased in latch. Thus, the time spent in a force-generating state in each cycle is longer. A shortening step to slack length leads to a decrease of the relative number of attached crossbridges, and this decrease is larger in a phasic than a tonic contraction. Although this effect could arise from a greater rate constant for detachment of phosphorylated vs unphosphorylated crossbridges as envisaged in the "latch" hypothesis, it cannot at present be excluded that the rate constant for attachment, or more precisely for transition to a strongly bound state, is greater in the phasic contraction. This would be suggested by the faster redevelopment of stiffness and force in this state.

ACKNOWLEDGMENTS

The study was supported by the Swedish Medical Research Council (project 14x-28), the Medical Faculty, University of Lund, and AB Hässle, Mölndal, Sweden.

REFERENCES

Arheden, H. and Hellstrand, P., 1991, Force response to rapid length change during contraction and rigor in skinned smooth muscle of guinea-pig taenia coli, *J. Physiol.*, in press.

Arner, A., Goody, R. S., Rapp, G., and Rüegg, J. C., 1987, Relaxation of chemically skinned guinea pig taenia coli smooth muscle from rigor by photolytic release of adenosine-5'-triphosphate, *J. Muscle Res. Cell Motil.*, 8: 377.

Dillon, P. F., Aksoy, M. O., Driska, S. P., and Murphy, R. A., 1981, Myosin phosphorylation and the cross-bridge cycle in arterial smooth muscle, *Science*, 211: 495.

Driska, S. P, 1987, High myosin light chain phosphatase activity in arterial smooth muscle: can it explain the latch phenomenon?, *in:* "Regulation and Contraction of Smooth Muscle", M. J. Siegman, A. P. Somlyo, and N. L. Stephens, eds., Alan R. Liss, New York, p. 387.

Eisenberg, E. and Hill, T. L., 1985, Muscle contraction and free energy transduction in biological systems, *Science*, 227: 999

Fabiato, A., 1981, Myoplasmic free calcium concentration reached during twitch of an intact isolated cardiac cell and during calcium-induced release of calcium from the sarcoplasmic reticulum of a skinned cardiac cell from the adult rat or rabbit ventricle, *J. Gen. Physiol.*, 78: 457.

Fabiato, A. and Fabiato, F., 1979, Calculator programs for computing the composition of the solutions containing multiple metals and ligands used for experiments in skinned muscle cells, *J. Physiol. (Paris)*, 75: 463.

Hai, C.-M. and Murphy, R. A., 1988, Cross-bridge phosphorylation and regulation of latch state in smooth muscle, *Am. J. Physiol.*, 254: C99.

Hardung, V. and Laszt, L., 1966, Die Beziehung zwischen Last und Verkürzungsgeschwindigkeit beim Gefässmuskel, *Angiologica*, 3: 100.

Hellstrand, P., 1977, Oxygen consumption and lactate production of the rat portal vein in relation to its contractile activity, *Acta Physiol. Scand.*, 100: 91.

Hellstrand, P. and Arheden, H., 1989, Mechanical transients in smooth muscle, *in:* "Muscle Energetics", R. J. Paul, G. Elzinga, and K. Yamada, eds., Alan R. Liss, New York, p. 347.

Hellstrand, P. and Johansson, B., 1975, The force-velocity relation in phasic contractions of venous smooth muscle, *Acta Physiol. Scand.*, 93: 157.

Hellstrand, P. and Paul, R. J., 1982, Vascular smooth muscle: relations between energy metabolism and mechanics, *in:* "Vascular Smooth Muscle: Metabolic, Ionic and Contractile Mechanisms", M. F. Crass III, ed., Academic Press, New York, p. 1.

Huxley, A. F., 1957, Muscle structure and theories of contraction, *Prog. Biophys. Biophys. Chem.*, 7: 255.

Huxley, A. F. and Simmons, R. M., 1971, Proposed mechanism of force generation in striated muscle, *Nature*, 233: 533.

Huxley, H. E. and Kress, M., 1985, Crossbridge behaviour during muscle contraction, *J. Muscle Res. Cell Motil.*, 6: 153.

Kamm, K. E. and Stull, J. T., 1986, Activation of smooth muscle contraction: Relation between myosin phosphorylation and stiffness, *Science*, 232: 80.

Klemt, P., Peiper, U., Speden, R. N., and Zilker, F., 1981, The kinetics of post-vibration tension recovery of the isolated rat portal vein, *J. Physiol.*, 312: 281.

Kobayashi, S., Kitazawa, T., Somlyo, A. V., and Somlyo, A. P., 1989, Cytosolic heparin inhibits muscarinic and α-adrenergic Ca^{2+} release in smooth muscle. Physiological role of inositol 1,4,5-trisphosphate in pharmacomechanical coupling, *J. Biol. Chem.*, 30: 17997.
Murphy, R. A., 1976, Contractile system function in mammalian smooth muscle, *Blood Vessels*, 13: 1.
Paul, R. J., 1989, Smooth muscle energetics, *Ann. Rev. Physiol.*, 51: 331.
Paul, R. J., 1990, Smooth muscle energetics and theories of cross-bridge regulation, *Am. J. Physiol.*, 258: C369.
Schoenberg, M., 1985, Equilibrium muscle cross-bridge behavior. theoretical considerations, *Biophys. J.*, 48: 467.

MODULATION OF Ca^{2+} SENSITIVITY BY AGONISTS IN SMOOTH MUSCLE

Toshio Kitazawa and Andrew P. Somlyo

Department of Physiology
University of Virginia School of Medicine
Charlottesville, VA 22908

INTRODUCTION

The fact that Ca^{2+} is the primary physiological regulator of smooth muscle contraction is no longer questioned. Cytoplasmic Ca^{2+} can be increased through electromechanical coupling and through pharmacomechanical coupling (Somlyo and Somlyo, 1968). The two components of pharmacomechanical coupling explored, until recently, are ligand-gated Ca^{2+} influx and G-protein coupled activation of the phosphatidylinositol cascade that results in inositol 1,4,5-trisphosphate ($InsP_3$) -induced Ca^{2+}-release. However, several lines of evidence suggest that cytoplasmic Ca^{2+} concentration and force are not rigidly coupled and the Ca^{2+}-sensitivity of the regulatory/contractile apparatus can be modified by physiological mechanisms. The force/Ca^{2+} ratio in intact smooth muscles studied with Ca^{2+}-indicators is higher during agonist- than during high K^+-induced contractions (Bradley and Morgan, 1987; Himpens and Casteels, 1987; Rembold and Murphy 1988; Sato et al., 1988; Himpens et al., 1990). Recently, this agonist-induced increase in Ca^{2+}-sensitivity was unequivocally demonstrated in preparations permeabilized with Staphylococcus aureus α-toxin (Nishimura et al., 1988; Kitazawa et al., 1989) or with β-escin (Kobayashi et al., 1989) that retain agonist-coupled responses. Various agonists can markedly increase the contractile response of such preparations to a fixed, submaximal level of Ca^{2+}. Therefore, we had suggested that the third excitatory component of pharmacomechanical coupling (Somlyo and Somlyo, 1968), modulation of Ca^{2+}-sensitivity, operates by increasing myosin light chain (MLC) phosphorylation at a given level of cytoplasmic Ca^{2+} (Somlyo et al., 1989). In the following, we shall summarize our recent data (Kitazawa and Somlyo, 1990; Kitazawa et al., 1991) on the mechanism of the agonist-induced increase in Ca^{2+}-sensitivity of force associated with increased phosphorylation of MLC. In addition, we will show that phasic contractile responses of ileum smooth muscle are due, at least in part, to a time and/or Ca^{2+}-dependent desensitization of MLC phosphorylation (Kitazawa and Somlyo, 1990). The other two components of pharmacomechanical coupling have been described elsewhere (see Somlyo et al. this volume).

Regulation of Smooth Muscle Contraction
Edited by R.S. Moreland, Plenum Press, New York, 1991

DESENSITIZATION OF CONTRACTILE RESPONSE OF PERMEABILIZED PHASIC SMOOTH MUSCLE TO Ca^{2+}

Depolarization of smooth muscle with high K^+ causes a relatively rapid contraction that, in phasic smooth muscles, declines within minutes to an intermediate level, whereas in tonic smooth muscles it is followed by sustained increases in force for up to 30 min (Himpens et al., 1988, 1989). However, these differences can not be explained by the time course of the cytoplasmic Ca^{2+} transients that are similar in the two types of smooth muscle; cytoplasmic Ca^{2+} does not decrease any more during the phasic component of contraction in the phasic (ileum) than in the tonic (pulmonary artery) smooth muscle (Himpens et al., 1989). These results suggested that phasic smooth muscles are desensitized to Ca^{2+} with time, an hypothesis also supported by the rapid (within 30 seconds) dephosphorylation of MLC of intact ileum smooth muscle following activation (Himpens et al., 1988). Further evidence indicating that the decline in Ca^{2+}-sensitivity was not an artifact of the fluorescent Ca^{2+}-indicator was provided in preliminary experiments (Somlyo et al., 1989) showing that a phenomenon similar to desensitization of force to Ca^{2+} could also be observed in permeabilized preparations.

A step increase in Ca^{2+} to pCa 6.3 produces different contractile responses in the different types of smooth muscle permeabilized with Staphylococcal α-toxin (Fig. 1): a rapid rise in force, followed by a spontaneous decline to a low steady state level, in the very phasic ileum smooth muscle, and a monotonic increase in force in the (tonic) pulmonary artery smooth muscle. Since Ca^{2+} was buffered with 10 mM EGTA and all sources of intracellular Ca^{2+} were removed with A23187, the phasic contraction in ileum could not be due to a transient increase in cytoplasmic Ca^{2+}. Furthermore, ryanodine (10 μM) or inositol 1,4,5-trisphosphate (100 μM), agents that release Ca^{2+} from the SR, did not affect the time course of this phasic contraction. The size of the pores

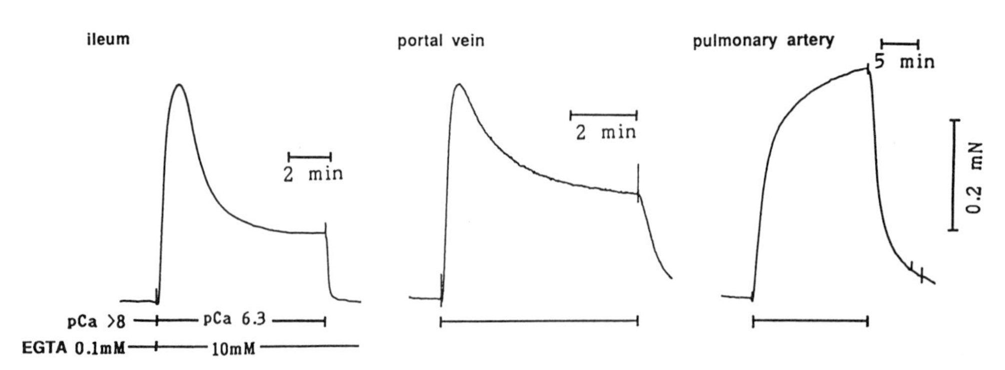

Figure 1. Phasic and tonic contractile responses of different types of α-toxin permeabilized smooth muscle to constant Ca^{2+}. Permeabilized (guinea-pig ileum and portal vein, and rabbit pulmonary artery) strips treated with A23187 had been incubated in the Ca^{2+}-free (first 10 mM and then 0.1 mM EGTA) solution for, at least, a total of 10 min before application of Ca^{2+} (pCa 6.3 buffered with 10 mM EGTA). MgATP, 4.5 mM; Mg^{2+}, 2 mM; pH 7.1. Temperature was controlled at 25°C.

created by α-toxin in the plasma membrane are so small that intrinsic, important soluble proteins, such as calmodulin, can not leak out of the cells, although Ca-EGTA penetrates into cytosol and can clamp cytoplasmic free Ca^{2+} (Bader et al., 1986; Kitazawa et al., 1989). We also confirmed the phasic responses in phasic smooth muscle permeabilized with β-escin, a saponin ester, that permeabilizes cell membranes to much higher molecular weight solutes than does α-toxin (Kitazawa and Somlyo, 1990). Preincubation of the strips in a very low concentration (0.1 mM) of EGTA, instead of 10 mM, prior to the pCa 6.3 (10 mM EGTA step), increased the peak of the phasic contraction, but not the steady state level.

When Ca^{2+} was cumulatively increased or repeatedly applied at shorter intervals, the peak of the phasic contraction was greatly reduced, but the tonic level was not significantly affected. The half time of the spontaneous decline in force at pCa 6.3 was 2 - 3 min. The decline in the phasic component of contractions induced by Ca^{2+} was not due to "run-down" of the preparations, as the decline was reversible and recovered in Ca^{2+}-free solution with a half time of about 1 min; complete recovery time was about 10 min (Kitazawa and Somlyo, 1990). Phasic contractions were seen even at pCa 4.5 in the very phasic ileum smooth muscle. In the tonic pulmonary artery smooth muscle, repeated applications of Ca^{2+} at shorter intervals or cumulative increases in Ca^{2+} resulted in little or no decrease in contractility.

PROPERTIES OF RECEPTOR/G-PROTEIN-MEDIATED ENHANCEMENT OF CONTRACTION IN PERMEABILIZED SMOOTH MUSCLES

The contractile response of α-toxin permeabilized smooth muscle to submaximal Ca^{2+} is markedly enhanced by α_1-adrenergic (Nishimura et al., 1988; Kitazawa et al., 1989, 1991) and muscarinic (Kitazawa and Somlyo, 1990) agonists, and by a thromboxane analogue, U46619 (Himpens et al., 1990). In

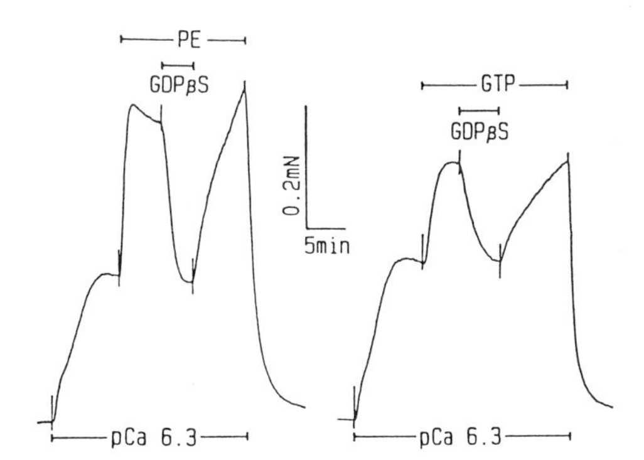

Figure 2. Potentiation of force at constant Ca^{2+} by phenylephrine (PE) or GTP and its reversible inhibition by GDPβS in α-toxin-permeabilized portal vein smooth muscle. Ca^{2+} transients were eliminated by depleting the intracellular stores of Ca^{2+} with A23187. The muscles were incubated in Ca^{2+}-free (pCa > 8), 10 mM EGTA-containing solution, followed by increase in Ca^{2+} to pCa 6.3 with a 10 mM EGTA buffered solution. The addition of PE (50 μM) or GTP (100 μM) produced additional contractile force at constant Ca^{2+}. The potentiated contraction, but not the Ca^{2+}-activated contraction, was reversibly inhibited by 1 mM GDPβS (From Kitazawa et al., 1989, by permission).

Figure 2, for example, the α_1-adrenergic agonist, phenylephrine or GTP markedly enhanced the steady state contraction activated by pCa 6.3. This agonist- and GTP-induced enhancement of contraction was not due to an increase in cytoplasmic Ca^{2+} concentration, because Ca^{2+} in the solution was clamped with 10 mM EGTA and permeabilized muscles had been treated with 10 μM A23187 to eliminate Ca^{2+}-release from the SR. Furthermore, ryanodine (10 μM) and $InsP_3$ (100 μM) had no effect on the potentiation of contraction by agonist, GTP or GTPγS. The diffuse intracellular distribution of the 10 mM Ca-EGTA buffer in the cytosol of α-toxin-permeabilized portal vein smooth muscle cells was verified with electron probe X-ray microanalysis (Kitazawa et al., 1989). In fact, we have also documented the absence of any change in free Ca^{2+} concentration, monitored with a fluorescent Ca^{2+}-indicator, during development of force induced by GTPγS (Somlyo et al., 1990). The contractions induced by agonist or GTP, in the absence of any change in Ca^{2+} were reversibly inhibited by GDPβS (Fig. 2). The latter result indicates that G-protein(s) mediate the α_1-adrenergic agonist increase in the Ca^{2+}-sensitivity of the regulatory /contractile apparatus of smooth muscle .

Figure 3 shows the different efficacies of various agonists on G-protein-mediated Ca^{2+}-sensitization in α-toxin permeabilized guinea-pig portal vein smooth muscle. Maximum concentrations (100 μM) of phenylephrine or norepinephrine markedly increased the contractile response at constant pCa 6.5. The same concentration of histamine had minimal potentiating effect on permeabilized portal vein, while in the intact muscle, both in the presence and absence of extracellular Ca^{2+}, this agonist produced a fairly large (more than 50% of a high K^+ contracture), but very transient, contraction. These results suggest that in guinea-pig portal vein, Ca^{2+}-release from the SR is the major pathway of pharmacomechanical coupling of histamine-induced contraction, whereas α_1-agonists stimulate both the Ca^{2+}-releasing (Bond et al., 1984) and Ca^{2+}-sensitizing pathways (Kitazawa et al., 1989) to produce high forces in intact muscle. U46619 had an intermediate effect in potentiating contraction in permeabilized portal vein. On the other hand, in the rabbit pulmonary artery, U46619 is a more potent potentiator than phenylephrine (Himpens et al., 1990), whereas in the guinea-pig ileum, both histamine and the muscarinic agonist, carbachol, can markedly potentiate contraction (Kitazawa and Somlyo, 1990). It

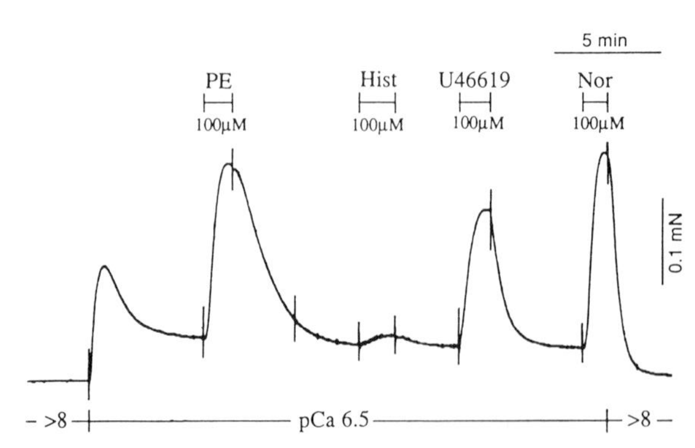

Figure 3. Different efficacies of various agonists on the G-protein-mediated Ca^{2+}-sensitization in α-toxin-permeabilized portal vein. The strips had been incubated in the solution containing 0.1 mM EGTA and no added Ca, followed by increase in Ca^{2+} to pCa 6.5 buffered with 10 mM EGTA. All solutions contained 10 μM GTP.

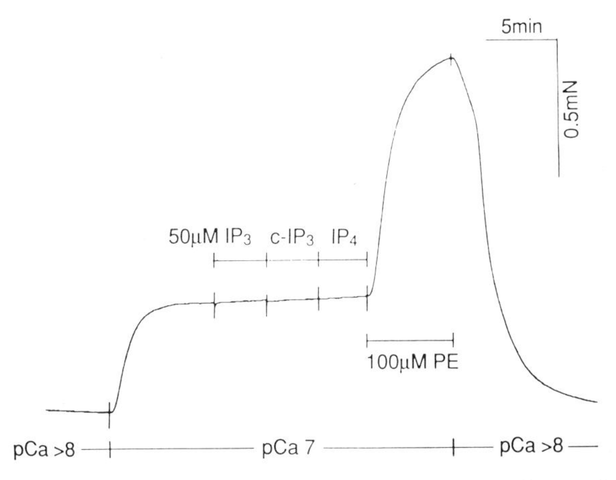

Figure 4. Inositol phosphates are not the second messengers for modulation of Ca^{2+}-sensitivity by agonists. Rabbit femoral artery was permeabilized with α-toxin and treated with A23187. All solutions contained 10 μM GTP. Inositol (1:2-cyclic, 4,5)-trisphosphate was a gift from Dr. S. Ozaki, Ehime University.

is clear that the efficacies of various agonists on G-protein-mediated Ca^{2+}-sensitization are highly variable and quantitatively very different in their action on different smooth muscles.

The agonist-induced increase in Ca^{2+}-sensitivity was inhibited by specific antagonists; 10 μM prazosin completely inhibited only phenylephrine-induced, but not U46619-induced, potentiation of contraction in portal vein (Kitazawa et al., 1991), while 10 μM atropine inhibited carbachol-induced potentiation in ileum smooth muscle. On the other hand, GDPβS inhibited the Ca^{2+}-sensitizing effect of each agonist examined. These results suggest that agonist-induced Ca^{2+}-sensitization is mediated through the receptor specific to each agonist, and is coupled by a G-protein with common properties.

The nonhydrolyzable GTP analogue, GTPγS evoked the largest contraction among these potentiators in either type of smooth muscle, pulmonary artery, femoral artery, portal vein, and ileum. Responses to maximum concentrations of GTPγS were not enhanced further by agonists, suggesting that the full extent of G-protein-mediated Ca^{2+}-sensitization can be induced by GTPγS. The concentration required for half maximum effect in rabbit femoral artery smooth muscle was 0.1 μM for GTPγS and 10 μM or higher for GTP (see Fig. 4 in Kitazawa et al., 1991), indicating that the GTP-binding protein responsible for Ca^{2+}-sensitization is a genuine G-protein with GTPase activity.

Sensitization to Ca^{2+} through G-proteins was not limited to guanine nucleotides, although GTP was most effective; the rank order of nucleotide activity was GTPγS >> GTP > ITP >> CTP = UTP (see Fig. 6 in Kitazawa et al., 1991). Non-guanine nucleotide-induced potentiation was also inhibited by GDPβS, suggesting that non-guanine nucleotides also acted on a G-protein.

Agonist-induced potentiation of contractions activated by Ca^{2+} (pCa 6.3) was observed even in the absence of added exogenous GTP (Kitazawa et al 1989), and was enhanced by the addition of GTP. However, the apparent affinity of the agonist to its receptor, as estimated through the induced potentiation, was not affected by added GTP (Kitazawa et al., 1991).

EVIDENCE AGAINST $InsP_3$ AND ITS METABOLITES OR DIACYLGLYCEROL BEING THE SECOND MESSENGER OF AGONIST-INDUCED Ca^{2+}-SENSITIZATION

As mentioned above, in smooth muscle in which Ca^{2+}-uptake by the SR is abolished by A23187 and Ca^{2+} is buffered with EGTA, $InsP_3$ neither potentiates by itself nor has any effect on agonist-induced potentiation of contraction. $InsP_3$, inositol (1;2-cyclic,4,5)-trisphosphate, and inositol (1,3,4,5)-tetrakisphosphate failed to produce any force at submaximal Ca^{2+} buffered with 10 mM EGTA in α-toxin-permeabilized rabbit femoral artery treated with A23187, in which phenylephrine could evoke a large contraction, as shown in Figure 4.

Another possible second messenger of agonist-induced sensitization to Ca^{2+} is diacylglycerol that is produced by agonist/G-protein-mediated activation of phospholipases (Berridge, 1988) and increases kinase C activity (Nishizuka, 1984). The possibility of such a mechanism is supported by the increased Ca^{2+}-sensitivity of permeabilized smooth muscle treated with phorbol esters, activators of kinase C, and by the inhibitory effect of the kinase C-inhibitor, H-7 (Chatterjee and Tejada, 1986; Nishimura et al., 1988). Therefore, we investigated the possibility that activation of kinase C is involved in the agonist/G-protein-mediated sensitization to Ca^{2+}. As shown in Figure 5A, 100 μM GTPγS evoked significant force development even at pCa > 8 in rabbit femoral artery. Increasing the GTPγS concentration to 200 μM had no additional effect, indicating that 100 μM was already a saturating concentration for force development at this pCa. Even in the presence of this saturating concentration of GTPγS, 1 μM phorbol 12,13-dibutyrate (PDBu) could further increase force. On the other hand, 1 μM PDBu alone, like GTPγS, also produced force at pCa > 8 (Fig. 5B). Increasing this concentration of PDBu caused no further increase in force, indicating that 1 μM is a saturating concentration. However, GTPγS was able to further increase force, even at this saturating concentration of PDBu, to almost the same extent as in the absence of the phorbol ester (Fig. 5B). These results suggest that the potentiation of

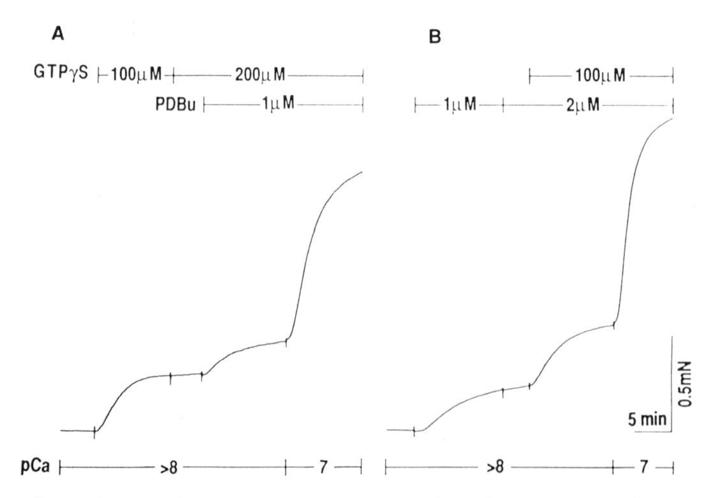

Figure 5. Potentiation of contraction induced by GTPγS is additive to that of phorbol ester. Rabbit femoral artery was permeabilized with α-toxin and treated with A23187. All solutions contained 10 mM EGTA. See text for complete details.

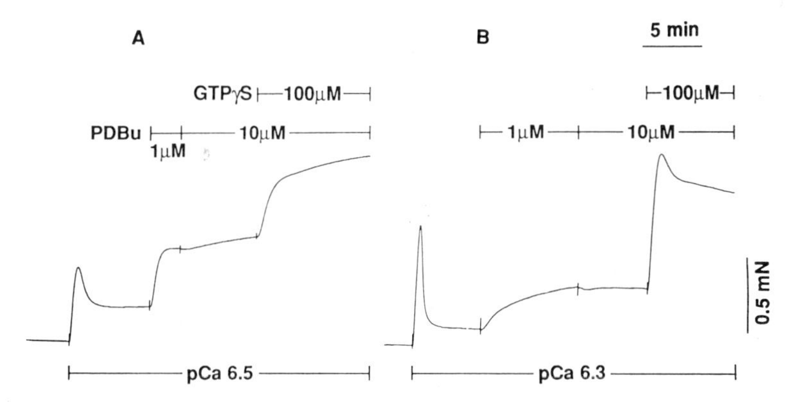

Figure 6. Persistence of potentiating effect of GTPγS on force in the presence of a saturating concentration of phorbol ester in α-toxin-permeabilized phasic, portal vein (A) and ileum (B) smooth muscle. All solutions contained 10 mM EGTA.

contraction induced by GTPγS is additive to that of phorbol ester, at least under Ca^{2+}-free (pCa > 8) conditions. This type of experiment could not be carried out on femoral artery smooth muscle in the presence of physiological concentrations of Ca^{2+} because, even at pCa 7, addition of either potentiator produced near maximum force (Fig. 5).

Therefore, we used a phasic type of smooth muscle in which either GTPγS or PDBu partially activate contractility in the presence of a physiological concentration of Ca^{2+}. In the portal vein (Fig. 6A) at pCa 6.5 and in the ileum (Fig. 6B) at pCa 6.3, GTPγS markedly enhanced force even in the presence of a saturating concentration of PDBu. All these results suggest that the potentiating effect of GTPγS is not through the activation of kinase C and that diacylglycerol is not a second messenger of G-protein-mediated sensitization of Ca^{2+}, unless both PDBu and GTPγS are incomplete activators of the smooth muscle isoform(s) of kinase C, in which case they could have additive effects mediated through a kinase C pathway.

THE EFFECT OF AGONISTS AND GDPβS ON MLC PHOSPHORYLATION

The primary mechanism regulating the contraction/relaxation cycle in smooth muscle is phosphorylation/dephosphorylation of MLC by, respectively, MLC kinase and MLC phosphatases (Hartshorne, 1987; Stull and Kamm, 1989). We have suggested that Ca^{2+}-sensitization of force by agonists is secondary to Ca^{2+}-sensitization of MLC phosphorylation (Somlyo et al., 1989). Therefore, it was necessary to determine whether or not the sensitization of force to Ca^{2+} by agonists and its inhibition by GDPβS are correlated with phosphorylation/dephosphorylation of MLC.

Tables 1 and 2 show the summary of simultaneous measurements of MLC phosphorylation and force in permeabilized, ileum and portal vein smooth muscles, respectively. At resting state (pCa > 8), 11% of total MLC in ileum (Table 1) was phosphorylated in the absence of detectable active force. A step increase in Ca^{2+}, to a submaximal concentration, produced a rapid rise in force, followed by a spontaneous decline to a low steady state level in the phasic ileum smooth muscle (Kitazawa and Somlyo, 1990). At the peak of contraction induced by pCa 6.3, MLC phosphorylation was significantly increased to 33%, and then significantly decreased to 25% of total MLC during the steady state contraction (10 min after Ca^{2+} was added), while force first increased to 36% and then decreased to 5% of the maximum force at pCa 5. The muscarinic agonist, carbachol (100 μM) significantly ($p < 0.01$) increased the amount of phosphorylated MLC further to 34% of total MLC, while force was increased from the steady state contraction (5%) to 33% of the maximum at pCa 5.

Table 1. Muscarinic agonist-induced sensitization of MLC phosphorylation and force to Ca^{2+} in permeabilized guinea-pig ileum.

Condition	Force (%)	Phosphorylation (%)	
Rest (pCa > 8; n = 13)	0	11 ± 2.0	
			$p < 0.01$
Peak (pCa 6.3; n = 7)	36 ± 2.7	33 ± 2.8	
			$p < 0.03$
Steady state (pCa 6.3; n = 14)	5 ± 0.9	25 ± 1.8	
			$p < 0.01$
Steady state + carbachol (pCa 6.3; n = 12)	33 ± 5.4	34 ± 2.5	

Phosphorylation is expressed as % of total MLC ± S.E.M. and force as % of maximum at pCa 5 ± S.E.M. All solutions contained 3 μM GTP (From Kitazawa and Somlyo, 1990, by permission).

The α-adrenergic agonist, norepinephrine (100 μM), as shown in Table 2, also significantly ($p < 0.01$) increased the amount of phosphorylated MLC at pCa 6.5, from the control value (17%) to 27% of total MLC, while force was increased from 13% to 49% of force at pCa 5 (Kitazawa et al., 1991). GDPβS (2 mM) completely inhibited both the norepinephrine-induced increase in MLC phosphorylation (to 19%) and force development (to 9%).

In conclusion, contractile sensitization to Ca^{2+} by agonists and the inhibition of this sensitization by GDPβS correlated very well with sensitization and inhibition of MLC phosphorylation, respectively. These results support our hypothesis (Somlyo et al., 1989) that the Ca^{2+}-sensitivity of force is modulated through variations in MLC phosphorylation, due to changes in the kinase/phosphatase activity ratio modulated by G-proteins.

Ca^{2+}-FORCE AND Ca^{2+}-MLC PHOSPHORYLATION RELATIONSHIP IN α-TOXIN PERMEABILIZED, TONIC AND PHASIC SMOOTH MUSCLES AND THE EFFECTS OF GTPγS

The steady state pCa-force relationship was obtained in smooth muscles permeabilized with α-toxin, by cumulatively increasing Ca^{2+}, either in the absence or in the presence of a saturating concentration (100 μM) of GTPγS.

The Ca^{2+}-sensitivity of the different types of smooth muscle was different (Fig. 7): the contractile/regulatory system of tonic muscles was about 3 times more sensitive to Ca^{2+} than that of phasic muscles (concentration of Ca^{2+} required for half maximum force was approximately 300 nM in pulmonary and femoral arteries, 800 nM in portal vein and 1 μM in ileum). The Ca^{2+}-sensitivity was similar in rabbit and guinea-pig pulmonary arteries.

Table 2. α_1-Adrenergic agonist-induced sensitization of MLC phosphorylation and force to Ca^{2+} and effect of GDPβS in guinea-pig portal vein.

Condition	Force (%)	Phosphorylation (%)	
pCa > 8; (n = 7)	0	7 ± 1.4	
			$p < 0.01$
pCa 6.5; (n = 12)	13 ± 1.8	17 ± 2.0	
			$p < 0.03$
pCa 6.5 + NE; (n = 11)	49 ± 1.5	27 ± 1.5	
			$p < 0.01$
pCa 6.5 + NE + GDPβS (n = 12)	9 ± 1.2	19 ± 1.8	

Phosphorylation is expressed as % of total MLC ± S.E.M. and force as % of maximum at pCa5 ± S.E.M. All solutions contained 3 μM GTP. Norepinephrine (NE), 100 μM; GDPbS, 2 mM. (From Kitazawa et al., 1991, by permission).

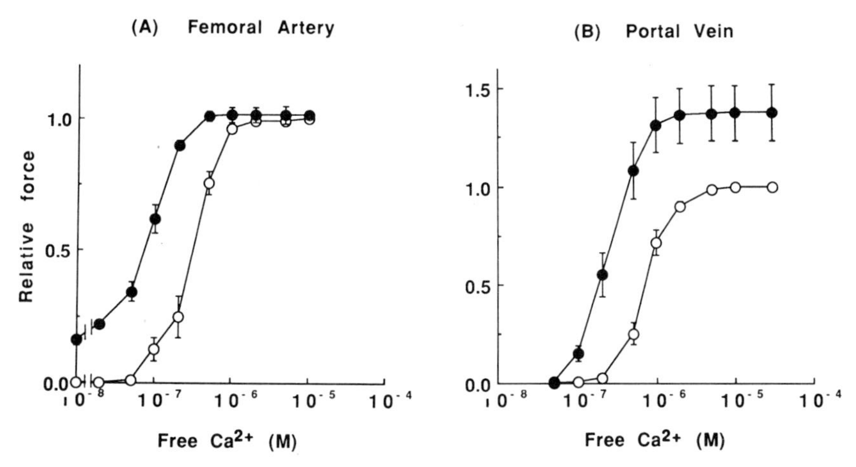

Figure 7. Steady-state Ca^{2+}-force relationship in tonic (femoral artery) and phasic (portal vein) types of α-toxin permeabilized smooth muscle with (●) and without (○) 100 μM GTPγS. (A), rabbit femoral artery: Data are mean ± S.E.M of 4 experiments for control and 5 for GTPγS. (B), guinea-pig portal vein; n=7 for control and 6 for GTPγS. Force was normalized to that produced at pCa 5. (From Kitazawa et al., 1991, by permission).

GTPγS at saturating concentration (100 μM) increased the Ca^{2+}-sensitivity of the regulatory/contractile system by about 3-fold, regardless of muscle type (Fig. 7; see also Nishimura et al., 1988; Fujiwara et al., 1989; Kitazawa et al., 1989). In the tonic smooth muscle, GTPγS produced significant force (17% of the maximum force at pCa 5) even at pCa > 8 (10 mM EGTA and no added calcium), but did not further increase the maximum force at pCa $\leq$ 5 (Fig. 7A). The maximum force (~ 2 x 10^5 N/m^2 of cross-sectional area of tissue) of permeabilized femoral artery corresponds to about 200% of the high K^+ contracture of the same muscles before permeabilization. In contrast, in the phasic muscle, very little force (1% of maximum force at pCa 5) was evoked by GTPγS in the absence of Ca^{2+}, but the steady state maximum force at pCa $\leq$ 5 was significantly increased to about 150% of the high K^+ contracture before permeabilization (Fig. 7B).

Figure 8A shows the relationship between Ca^{2+} and MLC phosphorylation and the effect of GTPγS in α-toxin permeabilized portal vein. In the control at pCa > 8, a significant amount (7% of total MLC) of MLC was phosphorylated, although no significant force was detectable; increasing Ca^{2+} to pCa 6.5 and 5 caused significant ($p < 0.01$) increases in MLC phosphorylation to 17 and 58% of total MLC, respectively, while force increased to 13 and 100% of force at pCa 5, respectively. GTPγS significantly ($p < 0.01$ in all cases) increased phosphorylation of MLC at every pCa (Fig. 8A), to 13% at pCa > 8, 41% at pCa 6.5, and 75% at pCa 5. A very small increase (1%) in force was detectable at pCa > 8, while force at pCa 6.5 and 5 was markedly increased by GTPγS to 85 and 130%, respectively, of maximum force activated by pCa 5.

Figure 8B shows the relationship between phosphorylation of MLC and force in all (steady state, 2 - 8 min following step activation) conditions. Force was normalized to that at pCa 5. Phosphorylation (P) vs. force (F) data were fitted with $F = -31.60 + 3.13 \times P - 0.0132 \times P^2$ (0.97 regression coefficient), suggesting that phosphorylation of slightly more than 10% of total MLC is required for force development in portal vein smooth muscle, and 100% of phosphorylation predictably results in 150% of the maximum force activated by saturating Ca^{2+} concentration alone.

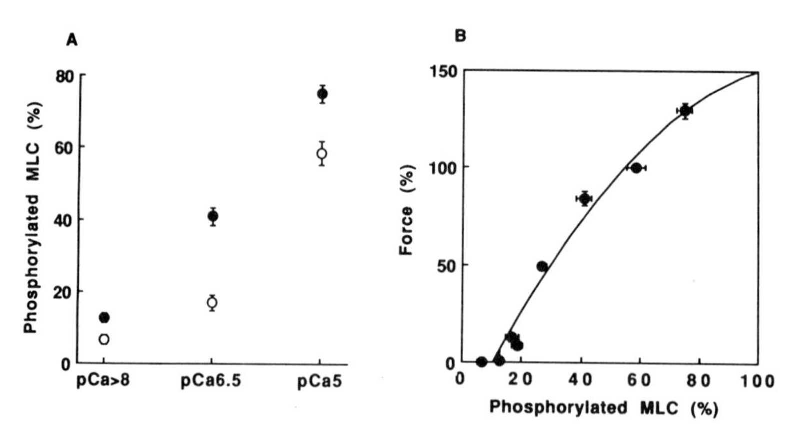

Figure 8. Effect of GTPγS on phosphorylation of MLC at various concentrations of Ca^{2+} (A), and the relationship between MLC phosphorylation and force in all experimental conditions (B) in permeabilized guinea-pig portal vein. All solutions contained 3 μM GTP. A: Open and solid symbols represent, respectively, control and + 100 μM GTPγS. B: MLC phosphorylation (P) vs. force (F) data were fitted with $F = -31.60 + 3.13 \times P - 0.0132 \times P^2$; 0.97 regression coefficient. (From Kitazawa et al., 1991, by permission).

In tonic femoral artery (Fig. 9A), the basal MLC phosphorylation levels were somewhat higher (20% of total MLC) than in phasic portal vein (7%) and ileum (11%). Increasing Ca^{2+} to pCa 7 and 5 caused significant ($p < 0.01$) increases in phosphorylation of MLC to 37 and 97% of total MLC, while force increased to 13 and 100 % of force at pCa 5, respectively.

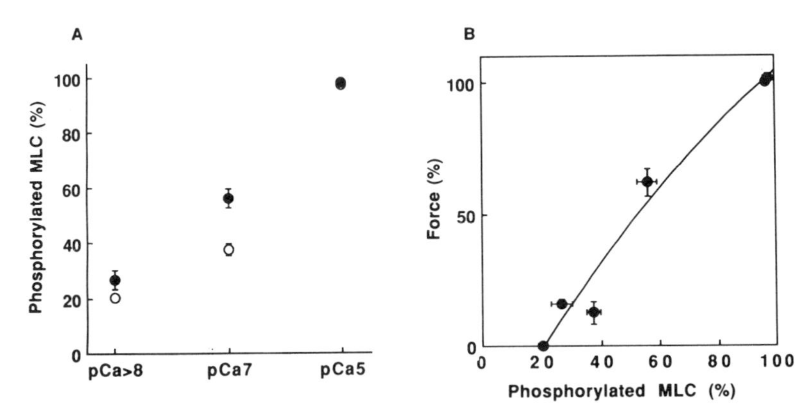

Figure 9. Effect of GTPγS on phosphorylation of MLC at various concentrations of Ca^{2+} (A), and the relationship between MLC phosphorylation and force in all experimental conditions (B) in permeabilized rabbit femoral artery. All solutions contained 3 μM GTP. A: Open and solid symbols represent, respectively, control and + 100 μM GTPγS. B: Solid fitted curve represents $F = -38.09 + 1.93 \times P - 0.00513 \times P^2$; 0.97 regression coefficient. (From Kitazawa et al., 1991, by permission).

GTPγS (100 μM) at pCa 7 significantly increased both MLC phosphorylation, to 56%, and force, to 62%; whereas at pCa 5 it did not cause a significant increase in either phosphorylation (98%) or force (102%) (Fig. 9A). At pCa > 8, force was significantly increased to 16%, but the increase in phosphorylation (to 27%) was not statistically significant ($p < 0.04$, "one-tailed" t-test). The relation between phosphorylation of MLC and force in rabbit femoral artery in all conditions is shown in Fig. 9B. The fitted curve represents $F = -38.09 + 1.93 \times P - 0.00513 \times P^2$ (0.97 regression coefficient), suggesting that force (F) development by the femoral artery requires 21% MLC phosphorylation. 100% phosphorylation is predicted to result in 104% of the maximum force developed at pCa 5 in the absence of GTPγS.

In conclusion, in both tonic and phasic smooth muscle, the contractions evoked by agonists and by GTPγS at constant Ca^{2+} ("Ca^{2+} sensitization") are associated with increases in MLC phosphorylation that can be inhibited by GDPβS. These results support the hypothesis (Somlyo et al., 1989) that the Ca^{2+}-sensitivity of force is modulated through variations in MLC phosphorylation, due to changes in the MLC kinase/phosphatase activity ratio modulated by G-proteins. The decay of force during the phasic contraction of ileum, in the presence of constant cytoplasmic Ca^{2+}, is associated with dephosphorylation (Kitazawa and Somlyo, 1990), and reflects desensitization (Himpens et al., 1988) of the contractile/regulatory apparatus to Ca^{2+}.

ACKNOWLEDGMENTS

We are grateful to Drs. A. V. Somlyo and R. A. Murphy for helpful discussions. We thank Gerald H. Denney and Dr. Bruce D. Gaylinn for technical assistance, and also Drs. E. Nishie and M. C. Gong for preparation of frozen samples. This work was supported by National Institute of Health Grant P01 HL19242-14 to the University of Virginia and HL15835 to the Pennsylvania Muscle Institute.

REFERENCES

Bader, M.-F., Thierse, D., Aunis, D., Ahnert-Hilger, G., and Gratzl, M., 1986, Characterization of hormone and protein release from α-toxin-permeabilized chromaffin cells in primary culture, *J. Biol. Chem.*, 261: 5777.

Berridge, M. J., 1988, Inositol lipids and calcium signalling, *Proc. R. Soc. Lond. B*, 234: 359.

Bond, M., Kitazawa, T., Somlyo, A. P., and Somlyo, A. V., 1984, Release and recycling of calcium by the sarcoplasmic reticulum in guinea-pig portal vein smooth muscle, *J. Physiol.*, 355: 677.

Bradley, A. B. and Morgan, K. G., 1987, Alterations in cytoplasmic calcium sensitivity during porcine coronary artery contractions as detected by aequorin, *J. Physiol.*, 385: 437.

Fujiwara, T., Itoh, T., Kubota, Y., and Kuriyama, H., 1989, Effects of guanosine nucleotides on skinned smooth muscle tissue of the rabbit mesenteric artery, *J. Physiol.*, 408: 535.

Hartshorne, D. J., 1987, Biochemistry of the contractile process in smooth muscle, *in:* "Physiology of the Gastrointestinal Tract", L. R. Johnson, ed., Raven Press, New York, p. 423.

Himpens, B. and Casteels, R., 1987, Measurement by Quin2 of changes of the intracellular calcium concentration in strips of the rabbit ear artery and of the guinea-pig ileum, *Pflügers Arch.*, 408, 32.

Himpens, B., Matthijs, G., Somlyo, A. V., Butler, T. M., and Somlyo, A. P., 1988, Cytoplasmic free calcium, myosin light chain phosphorylation, and force in phasic and tonic smooth muscle, *J. Gen. Physiol.*, 92: 713.

Himpens, B., Matthijs, G., and Somlyo, A. P., 1989, Desensitization to cytoplasmic Ca and Ca sensitivities of guinea-pig ileum and rabbit pulmonary artery smooth muscle, *J. Physiol.*, 413: 489.

Himpens, B., Kitazawa, T., and Somlyo, A. P., 1990, Agonist dependent modulation of Ca^{2+} sensitivity in rabbit pulmonary artery smooth muscle, *Pflügers Arch.*, 417: 21.

Kitazawa, T., Kobayashi, S., Horiuti, K., Somlyo, A. V., and Somlyo, A. P., 1989, Receptor-coupled, permeabilized smooth muscle. Role of the phosphatidylinositol cascade, G-proteins, and modulation of the contractile response to Ca^{2+}, *J. Biol. Chem.*, 264: 5339.

Kitazawa, T., Gaylinn, B. D., Denney, G. H., and Somlyo, A. P., 1991, G-protein mediated Ca^{2+} sensitization of smooth muscle contraction through myosin light chain phosphorylation, *J. Biol. Chem.*, 266: 1708.

Kitazawa, T. and Somlyo, A. P., 1990, Desensitization and muscarinic re-sensitization of force and myosin light chain phosphorylation to cytoplasmic Ca^{2+} in smooth muscle, *Biochem. Biophys. Res. Commun.*, 172: 1291.

Kobayashi, S., Kitazawa, T., Somlyo, A. V., and Somlyo, A. P., 1989, Cytosolic heparin inhibits muscarinic and alpha-adrenergic Ca^{2+} release in smooth muscle, *J. Biol. Chem.*, 264: 17997.

Nishimura, J., Kolber, M., and van Breemen, C., 1988, Norepinephrine and GTP-γ-S increase myofilament Ca^{2+} sensitivity in α-toxin permeabilized arterial smooth muscle, *Biochem. Biophys. Res. Commun.*, 157: 677.

Nishizuka, Y., 1984, The role of protein kinase C in cell surface signal transduction and tumour promotion, *Nature*, 308: 693.

Rembold, C. M. and Murphy, R. A., 1988, Myoplasmic $[Ca^{2+}]$ determines myosin phosphorylation in agonist-stimulated swine arterial smooth muscle, *Circ. Res.*, 63: 593.

Sato, K., Ozaki, H., and Karaki, H., 1988, Changes in cytosolic calcium level in vascular smooth muscle strip measured simultaneously with contraction using fluorescent indicator fura 2, *J. Pharmacol. Exp. Ther.*, 246: 294.

Somlyo, A. V. and Somlyo, A. P., 1968, Electromechanical and pharmacomechanical coupling in vascular smooth muscle, *J. Pharmacol. Exp. Ther.*, 159: 129.

Somlyo, A. P., Kitazawa, T., Himpens, B., Matthijs, G., Horiuti, K., Kobayashi, S., Goldman, Y. E.,, and Somlyo, A. V., 1989, Modulation of Ca^{2+}-sensitivity and of the time course of contraction in smooth muscle. A major role of protein phosphatases?, *in:* "Advances in Protein Phosphatases vol. 5", W. Merleude and J. DiSalvo, eds., Leuven Univ. Press, Leuven, p. 181.

Somlyo, A. V., Kitazawa, T., Horiuti, K., Kobayashi, S., Trentham, D., and Somlyo, A. P., 1990, Heparin-sensitive inositol trisphosphate signaling and the role of G-proteins in Ca^{2+}-release and contractile regulation in smooth muscle, *in:* "Frontiers in Smooth Muscle Research", N. Sperelakis, J. D. Wood, eds., Wiley-Liss, New York, p. 167.

Stull, J. T. and Kamm, K. E., 1989, Second messenger effectors of the smooth muscle myosin phosphorylation system, *in:* "Advances in Protein Phosphatases vol. 5", W. Merleude and J. DiSalvo, eds., Leuven Univ. Press, Leuven, p. 197.

REGULATION OF THE Ca^{2+}-FORCE RELATIONSHIP IN PERMEABILIZED ARTERIAL SMOOTH MUSCLE

Junji Nishimura,[1] Suzanne Moreland,[2,3] Robert S. Moreland,[3] and Cornelis van Breemen[1]

[1]Department of Molecular and Cellular Pharmacology
University of Miami, School of Medicine
Miami, FL 33101

[2]Department of Pharmacology
Bristol-Myers Squibb Pharmaceutical Research Institute
Princeton, NJ 08543

[3]Bockus Research Institute
Graduate Hospital,
Philadelphia, PA 19146

INTRODUCTION

It is generally accepted that the primary trigger for smooth muscle contraction is the elevation of intracellular Ca^{2+} concentration ($[Ca^{2+}]_i$) and subsequent phosphorylation of myosin light chain (MLC) by the Ca^{2+}-calmodulin dependent MLC kinase (for reviews see Kamm and Stull, 1985; Somlyo, 1985). However, simultaneous measurements of tension and $[Ca^{2+}]_i$ in intact tissues have shown that during continuous stimulation, although α-adrenergic agonist induced force is maintained at high constant levels, $[Ca^{2+}]_i$ falls close to basal concentrations (Morgan and Morgan, 1982; 1984a). It has also been demonstrated that levels of both MLC phosphorylation and shortening velocity fall to suprabasal levels during the phase of force maintenance in intact smooth muscle (Dillon et al., 1981). In order to account for these phenomena, a high Ca^{2+} sensitivity state, the latch state, was proposed to be important for the maintenance of developed force in the face of significant decreases in the $[Ca^{2+}]_i$ and levels of MLC phosphorylation (Dillon et al., 1981). Similar phenomena, increases in the Ca^{2+}-force relationship, have been demonstrated in skinned smooth muscle preparations by the addition of exogenous calmodulin (Cassidy et al., 1981; Rüegg and Paul, 1982), an initial stimulation in high $[Ca^{2+}]$ followed by exposure to a lower $[Ca^{2+}]$ (Chatterjee and Murphy, 1981; Moreland and Murphy, 1986), and stimulation by phorbol esters (Chatterjee and Tajeda, 1986; Itoh et al., 1988) or GTPγS (Fujiwara et al., 1989). However, until recently it has not been possible to increase the Ca^{2+} sensitivity of force by a physiological mode of stimulation. Recent studies utilizing the Staphylcoccal α-toxin permeabilized smooth muscle preparation have shown that receptor stimulation by a physiological agonist plus GTP can produce a significant level of additional force, at a fixed submaximal [Ca^{2+}], as compared to the force developed in response to Ca^{2+} alone (Nishimura et al., 1988; Kitazawa et al., 1989; Kobayashi et al., 1989).

In addition to the apparent effect of G protein stimulation by either GTPγS or receptor stimulation to modulate (enhance) smooth muscle myofilament Ca^{2+} sensitivity, there has been much interest in determining whether cyclic nucleotides modulate (decrease) smooth muscle contractile activity directly through an effect on the contractile elements or only indirectly through changes in transmembrane Ca^{2+} fluxes. Adelstein et al. (1978) initially proposed a model whereby cAMP dependent protein kinase catalyzed phosphorylation of the MLC kinase would inactivate this enzyme and thus result in dephosphorylation of the MLC and smooth muscle relaxation. Unfortunately, most experiments using either intact or detergent skinned smooth muscle tissues failed to substantiate a physiologic role for this proposed mechanism, leading to the suggestion that cyclic nucleotide-induced relaxation of smooth muscle operates only indirectly through their effect on membrane Ca^{2+} transport systems (Itoh et al., 1985; Kamm and Stull, 1985; 1989). However, we have recently shown, again using the α-toxin permeabilized preparation, that cyclic nucleotides may directly affect the contractile elements, producing a down regulation or decrease in the myofilament Ca^{2+} sensitivity (Nishimura and van Breemen, 1989a).

In the present report, we addressed possible mechanisms responsible for the increase in myofilament Ca^{2+} sensitivity following receptor activation and the apparent decrease in Ca^{2+} sensitivity by cyclic nucleotides. Specifically, we were interested in answering the following questions:

1. Does the myofilament Ca^{2+} sensitivity of smooth muscle change following receptor activation?
2. What kind of stimulation increases or decreases Ca^{2+} sensitivity?
3. How is the myofilament Ca^{2+} sensitivity regulated?
    a. Are G proteins involved?
    b. Is protein kinase C (PKC) involved?
    c. Does ADP or phosphate, both of which are believed to affect the actomyosin ATPase cycle, affect Ca^{2+} sensitivity?
    d. Is a change in the level of MLC phosphorylation always associated with changes in myofilament Ca^{2+} sensitivity and therefore the sole mechanism responsible for smooth muscle regulation?

MATERIALS AND METHODS

The α-toxin permeabilized arterial preparations have been described previously (Nishimura et al., 1988; Nishimura and van Breemen, 1989a). Small rings (~ 250 μm diameter x ~ 300 μm long) from the first branch of the superior mesenteric artery from male Wistar Kyoto rats or the third branch of the superior mesenteric artery from New Zealand white rabbits were prepared under a binocular microscope and two tungsten wires (40 μm diameter, California Fine Wire Co.) were passed through the lumen. One wire was fixed to the muscle chamber and the other was attached to a force transducer (U gage, Sinko Co. Ltd). Alternatively, small strips of circular muscle were dissected free and mounted by fine silk threads. Permeabilization was accomplished by

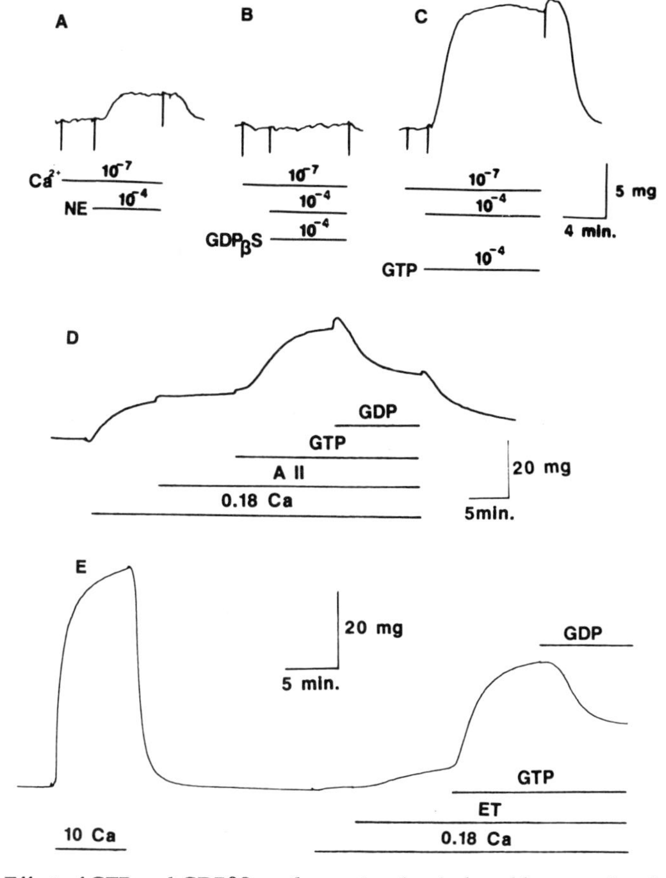

Figure 1. Effect of GTP and GDPβS on the contraction induced by norepinephrine (NE), angiotensin II (AII), and endothelin (ET) in the α-toxin permeabilized rabbit mesenteric artery at room temperature. A. NE (0.1 mM) -induced contraction in 0.1 μM Ca^{2+} solution. B. Effect of 0.1 mM GDPβS on NE-induced contraction. C. Effect of 0.1 mM GTP on NE-induced contraction. D. Effect of the addition of 30 nM angiotensin II (AII), 10 μM GTP, and 100 μM GDPβS on 0.18 μM Ca^{2+} induced contraction. The addition of AII alone did not increase the contraction. However, the addition of AII plus GTP enhanced the contraction. The GTP enhanced force was inhibited by GDPβS. E. Effect of the addition of 100 nM endothelin (ET), 10 μM GTP, and 100 μM GDPβS on 0.18 μM Ca^{2+} induced contraction. The addition of ET induced a small contraction. The addition of ET plus GTP significantly enhanced the contraction. The GTP enhanced force was inhibited by GDPβS. In A, B, and C, Ca^{2+} solutions were buffered by 2 mM EGTA; and in D and E, Ca^{2+} solutions were buffered by 10 mM EGTA.

incubating the arterial segments with Staphylococcal α-toxin (30 μg protein/ml; α-toxin was a generous gift from Dr. R. J. Hohman, NIH) for 15 minutes in a 0.1 μM Ca^{2+}-containing cytoplasmic substitution solution (CSS) or 2500 hemolytic unit/ml of α-toxin from Gibco/BRL in 2 mM EGTA containing relaxing solution for 30 - 60 min. The experimental CSS solution contained in mM: 2 or 10 EGTA, 130 K-propionate or 100 K-methanesulfonate, 5.2 $MgCl_2$, 5.1 Na_2ATP, 10 creatine phosphate (CP), 20 Tris-maleate (pH 6.8), and indicated concentrations of free Ca^{2+}. The apparent binding constant used for the Ca^{2+}-EGTA complex was 10^6 M^{-1}.

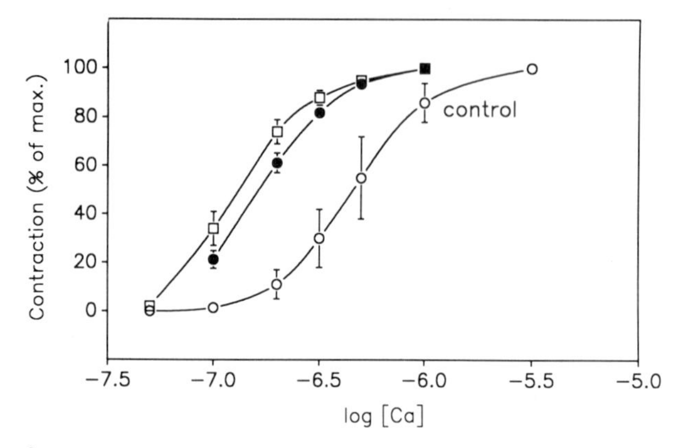

Figure 2. Ca^{2+}-force relationship in the absence (○; n = 5, illustrated as control) and presence of 10 μM NE plus 10 μM GTP (●; n = 3) or 100 nM ET plus 10 μM GTP (□; n = 3-5) in the rabbit mesenteric artery at room temperature. Force is expressed as a percent of maximum contraction. Values shown are means ± S.D. Ca^{2+} solutions were buffered by 10 mM EGTA.

Maximal Velocity of Shortening and Stiffness

For the measurement of maximal velocity of shortening (V_o) and stiffness, small helical (very close to circular) strips (2.5 - 4.5 mm long and 20 - 80 μg wet weight) were prepared from the first or second branches of the superior mesenteric artery in the rabbit. The tissues were mounted by small tweezers between a force transducer and a servolever (Güth Muscle Research Station, Heidelberg). Permeabilization was performed using 2500 hemolytic units/ml of α-toxin from Gibco BRL for 60 min as described above. During the permeabilizing period, the solution was changed every 10 - 15 min. The estimation of V_o was performed by isotonic quick releases to afterloads ranging from 0.1 to 0.5 times the force at the instant of release. The natural logarithm of the change in length between 1 and 2 sec of isotonic release was used to estimate V_o from at least five isotonic quick releases. All releases were performed at steady state, ~ 20 min after tissue activation. The servolever was interfaced to an IBM 286 compatible computer for the initiation of experimental protocols, collection of data, and data analysis. The estimation of tissue stiffness was performed by imposing a 500 Hz sine wave length change (0.05% L_o) and measuring the resultant change in force with respect to length. All stiffness measurements were normalized to that obtained during a maximal 1 μM Ca^{2+}-induced contraction.

MLC Phosphorylation Determinations

For the determination of total MLC phosphorylation levels, tissues were prepared and permeabilized in a manner similar to that described for the estimation of V_o. After twenty minutes of stimulation, the tissues were rapidly frozen in a dry ice-acetone slurry containing 6% trichloroacetic acid, allowed to slowly reach room temperature and then transferred to a chamber containing 100 μl of a buffer composed of 1% SDS, 10% glycerol, and 20 mM dithiothreitol previously frozen in liquid nitrogen. The tissue homogenization chamber (containing the tissue, frozen buffer, and a 1/4" diameter stainless steel ball)

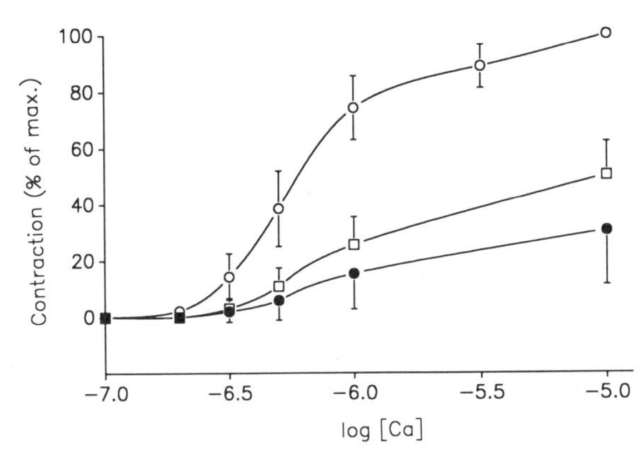

Figure 3. Ca^{2+}-force relationship in the absence (○; n = 9) and presence of 30 μM cAMP (●; n = 8) or 30 μM cGMP (□; n = 5) in the rat mesenteric artery at 37° C. Force is expressed as a percent of maximum contraction. In the case of cAMP and cGMP responses, contractions induced by 100 μM Ca^{2+} solution in the absence of the cyclic nucleotides were defined as 100 %. Values shown are means ± S.D. Ca^{2+} solutions were buffered by 2 mM EGTA.

was immersed in liquid nitrogen and then pulverized, using a dental amalgamator (Wig-a-Bug). The chamber was allowed to reach 4° C, the contents transferred to a storage tube, and placed in an ultra-low freezer. The homogenized samples were subjected to two-dimensional electrophoresis as described by Aksoy et al. (1983). Immediately following electrophoresis, the gels were subjected to high field intensity Western blotting as described by Gaylinn and Murphy (1990). Visualization of the blotted proteins was performed using an Amersham immunogold protein stain. Quantitation of the stained blots was performed by scanning densitometry of the nitrocellulose paper made transparent by immersion in decalin. Values are reported as mol P_i/mol MLC by calculating the area of the spot corresponding to the phosphorylated MLC as a percent of the total of both phosphorylated and unphosphorylated MLC.

RESULTS

Involvement of G Proteins in Receptor Stimulation

Figure 1A shows that norepinephrine (NE, 100 μM) produced a sustained contraction in 0.1 μM Ca^{2+} solution. The magnitude of this contraction varied considerably from one tissue preparation to another (34 ± 35% of the maximal Ca^{2+} induced force; mean ± S.D., n = 7). Inclusion of 100 μM guanosine-5'-O-(2-thiodiphosphate) (GDPβS) in the bathing solution almost completely abolished the NE-induced contractions (Figure 1B; 6 ± 4% of maximal force development, n = 4). On the other hand, addition of 100 μM GTP to the bathing solution enhanced the NE induced contractions to 65 ± 15% of maximum Ca^{2+}-induced force as well as significantly decreased tissue variability (Figure 1C; n = 4). Figure 1D shows a similar enhancing effect of 10 μM GTP and inhibitory effect of 100 μM GDPβS on an angiotensin II (AII)-induced contraction. AII, in the presence of 10 μM GTP, significantly increased contractile force in 0.18 μM Ca^{2+}. The addition of 100 nM endothelin (ET) to a

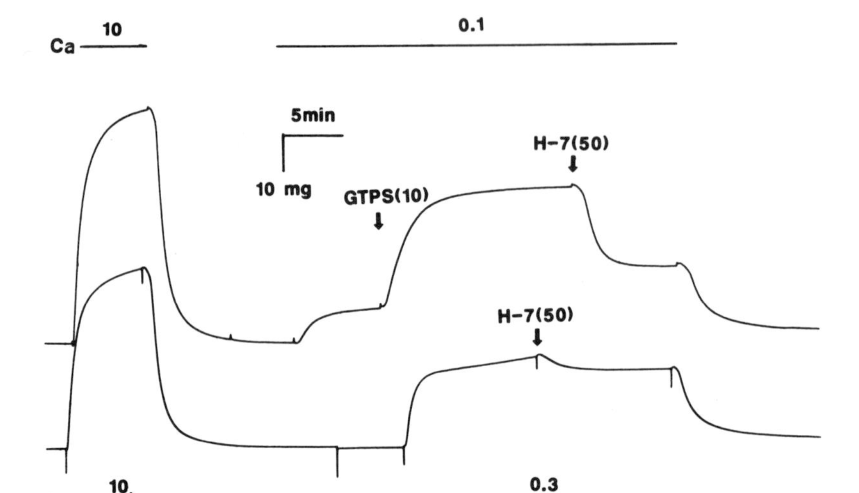

Figure 4. Effect of 50 μM H-7 on contractions induced by 0.1 μM Ca^{2+} plus 10 μM GTPγS (upper trace) or by 0.3 μM Ca^{2+} alone (lower trace). Traces shown are representative of at least three experiments. Ca^{2+} solutions were buffered by 10 mM EGTA. Initial control contractions are in response 10 μM Ca^{2+}.

0.18 μM Ca^{2+} containing solution also increased force. Similar to the effects of NE, the contractions in response to ET also varied in amplitude from one tissue to another. The addition of 10 μM GTP enhanced the level of contraction and decreased the variability and GDPβS inhibited the contraction (Fig. 1E). These results suggest that G proteins are coupled to NE, AII, and ET receptors and that stimulation of G proteins which are coupled to these receptors result in the enhancement of myofilament Ca^{2+} sensitivity.

Modulation of Myofilament Ca^{2+} Sensitivity

Figure 2 shows the effects of NE (10 μM) plus GTP (10 μM) and ET (100 nM) plus GTP (10 μM) on the pCa-tension relationships in rabbit mesenteric artery permeabilized by α-toxin. NE plus GTP and ET plus GTP significantly shifted the pCa-tension relationship to the left. The Ca^{2+} sensitivity could be decreased by the addition of either cAMP or cGMP. Figure 3 illustrates the effects of 30 μM cAMP or 30 μM cGMP on the pCa-tension relationship in rat mesenteric artery permeabilized by α-toxin at 37° C. cAMP and cGMP both decreased the Ca^{2+} sensitivity of force development.

Possible Involvement of Protein Kinase C in the Increased Ca^{2+} Sensitivity

The possible involvement of PKC in the increased myofilament Ca^{2+} sensitivity was assessed by the use of H-7 and staurosporine, relatively selective inhibitors of PKC, and phorbol 12, 13 dibutyrate, a non-physiological activator of PKC. Figure 4 shows the effect of 50 μM H-7 on a contraction in response to 0.1 μM Ca^{2+} plus 10 μM GTPγS and a contraction of comparable amplitude in response to Ca^{2+} (0.3 μM) alone. Although the amplitude of the contractions induced by these different stimulation conditions was similar, as compared

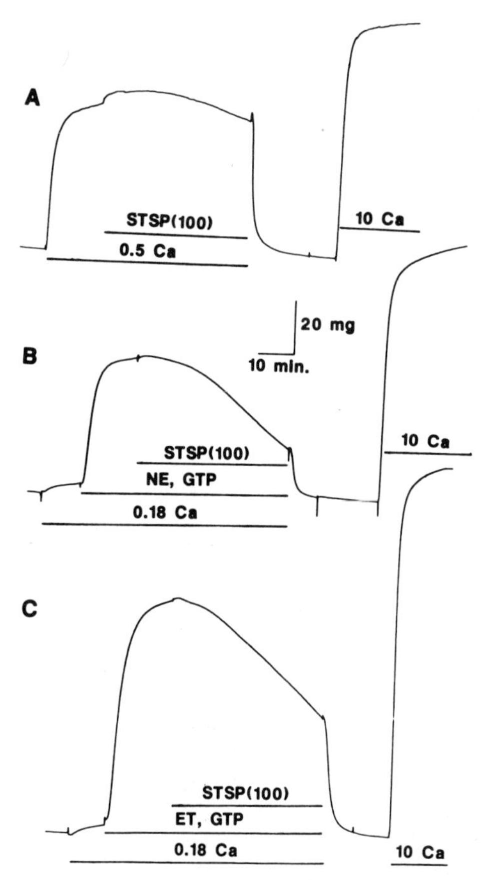

Figure 5. Effect of 100 nM staurosporine on contractions induced by 0.5 μM Ca^{2+} alone (A), by 0.18 μM Ca^{2+} plus 10 μM NE plus 10 μM GTP (B), or by 0.18 μM Ca^{2+} plus 100 nM ET plus 10 μM GTP (C) in the rabbit mesenteric artery at room temperature. Ca^{2+} concentrations expressed in μM are indicated above horizontal bars.

with a maximal Ca^{2+}-induced contraction (10 μM Ca^{2+}-induced contraction illustrated in Figure 4), the effect of H-7 was more prominent in the contraction in response to GTPγS than to Ca^{2+} alone. Staurosporine also inhibited NE- or ET-induced contractions more effectively than a contraction of similar magnitude induced by Ca^{2+} alone (Figure 5). In contrast, activation of PKC by PDBu (0.1 μM) stimulation, shifted the pCa-tension curve to the left (Figure 6). These results suggest that PKC may indeed be involved in the modulation of rabbit mesenteric arterial myofilament Ca^{2+}.

Modulation of Ca^{2+} Sensitivity by ADP and Phosphate

In addition to agonists and second messengers, myofilament Ca^{2+} sensitivity may also be manipulated by changes in substrate concentrations. We have found that removal of creatine phosphate (CP) increased the magnitude of maintained force at any given Ca^{2+} concentration. The effect of CP removal on the pCa-tension relationship is illustrated in Figure 7. Since it

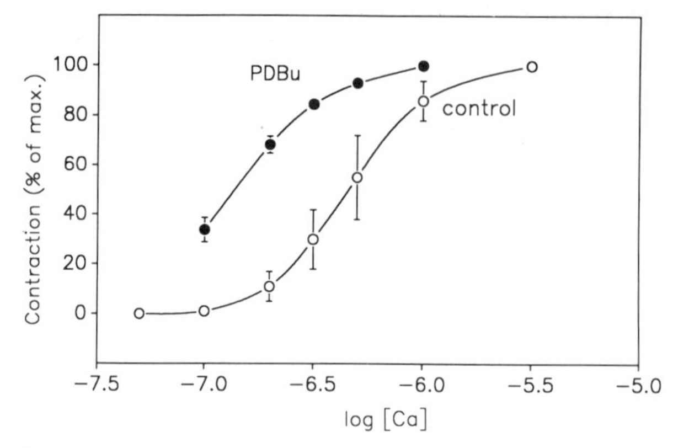

Figure 6. Ca^{2+}-force relationship in the absence (O; n = 5, illustrated as control) or presence of 0.1 μM PDBu (●; n = 3) in the rabbit mesenteric artery at room temperature. Force is expressed as a percent of maximum contraction. Values shown are means ± S.D. Ca^{2+} solutions were buffered by 10 mM EGTA.

is probable that CP functions not only in the regeneration of ATP but also to reduce ADP concentrations, we considered the possibility that this leftward shift of the pCa-tension curve shown in Figure 7 may be due to the accumulation of ADP. This hypothesis was substantiated by the data shown in Figure 8. The results in this figure demonstrate that the addition of 100 μM adenosine-5'-O-(2-thiodiphosphate) (ADPβS), a nonhydrolyzable ADP analog, while ATP and CP concentration were maintained constant, caused a similar leftward shift in the pCa-tension curve as that shown in Figure 7.

It has previously been suggested that the mechanism by which ADP increases Ca^{2+} sensitivity is by a direct affect on the actomyosin ATPase cycle (Kerrick and Hoar, 1987; and see Discussion and Figure 12). In a similar manner, excess exogenous phosphate (P_i) should decrease myofilament Ca^{2+} sensitivity. Figure 9 shows the effects of 15 mM P_i on contractions in response to several [Ca^{2+}] alone or plus NE and GTP. In all conditions, force

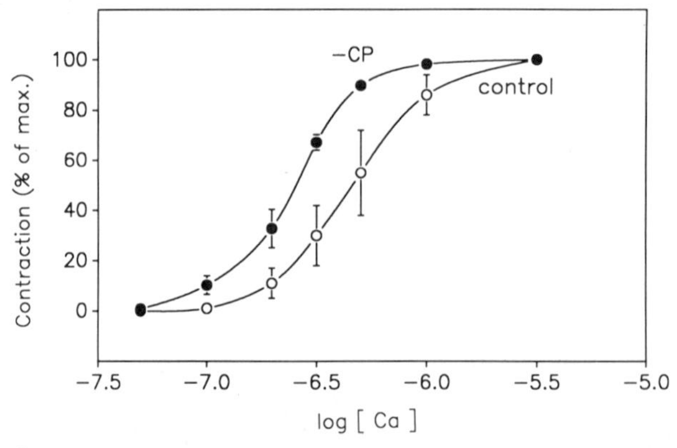

Figure 7. Ca^{2+}-force relationships in the presence (O; n = 5) or absence of 10 mM CP (●; n = 4) in the rabbit mesenteric artery at room temperature. Force is expressed as a percent of maximum contraction. Values shown are means ± S.D. Ca^{2+} solutions were buffered by 10 mM EGTA.

development was reduced by the addition of 15 mM P_i. In contractions in response to either Ca^{2+} alone or plus agonist, the relative reduction in force by exogenous P_i decreased as the magnitude of activation increased. However, if we compared contractions of similar magnitude elicited by Ca^{2+} alone and Ca^{2+} plus NE and GTP, the degree of depression by exogenous P_i is significantly greater with the agonist induced contraction (for example Fig. 9E compared to Fig. 9H; Fig 9D compared to Fig. 9G). This observation is clearly illustrated in Figure 10, in which we compared the effect of 10 mM P_i on contractions in response to 1 μM Ca^{2+} alone and 0.3 μM Ca^{2+} plus NE and GTP. Although the magnitude of force in response to 0.3 μM Ca^{2+} plus NE and GTP was similar or even larger than that in response to 1 μM Ca^{2+} alone, the depression of force by the addition of 10 mM P_i is greater in the NE-induced contraction. These results suggest that contractions in response to agonist stimulation are more sensitive to the effect of added P_i than are those in response to Ca^{2+} alone.

MLC Phosphorylation, V_o, and Stiffness

Figure 11 shows the results of experiments designed to estimate force, stiffness, V_o, and MLC phosphorylation in the α-toxin permeabilized tissue during a steady state contraction in response to 0.5 μM Ca^{2+} alone or plus NE and GTP. All data shown were obtained 20 min after the initiation of a contraction. The addition of 10 μM NE and 10 μM GTP significantly increased the force from 49.9 ± 3.3% (mean ± S.E., n = 7) to 86 ± 2.7% (n = 5; $p < 0.05$) and stiffness from 56.4 ± 7.3% (n = 7) to 78.2 ± 5.9% (n = 5; $p < 0.05$) of the maximal response to 1 μM Ca^{2+}. However, the addition of NE and GTP did not significantly increase the steady state level of total MLC phosphorylation; 0.38 ± 0.03 mol P_i/mol MLC (n = 7) for Ca^{2+} alone as compared to 0.41 ± 0.04 mol P_i/mol MLC (n = 7) for Ca^{2+} plus NE and GTP. The increased magnitude of force in response to agonist stimulation was also not associated with an increase in V_o; 0.021 ± 0.002 L_o/sec. (n = 6) for Ca^{2+} alone as compared to 0.020 ± 0.002 L_o/sec (n = 5) for Ca^{2+} plus NE and GTP.

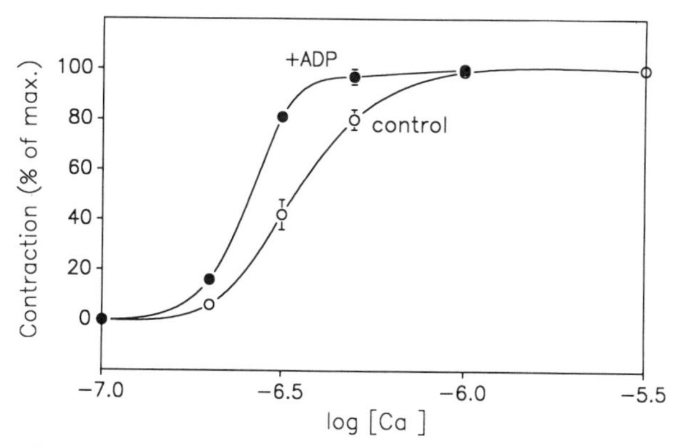

Figure 8. Ca^{2+}-force relationship in the absence (○; n = 5) or presence of 100 μM ADPβS (●; n = 3) in the rabbit mesenteric artery at 37° C. Force is expressed as percentage of maximum contraction. Values shown are means ± S.D. Ca^{2+} solutions were buffered by 2 mM EGTA.

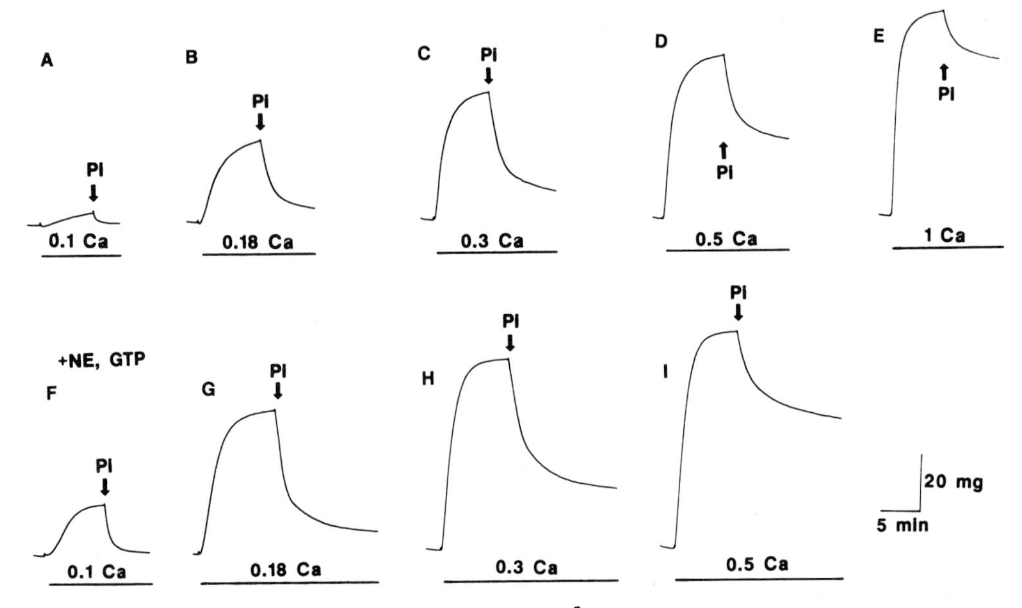

Figure 9. Effect of 15 mM phosphate (P_i) on Ca^{2+}-induced contractions (A-E) and on contractions induced by Ca^{2+} plus 10 μM NE plus 10μM GTP (F-I) in the rabbit mesenteric artery at room temperature. Ca^{2+} solutions were buffered by 10 mM EGTA. All contractions illustrated were performed using the same preparation. Traces are representative of three experiments.

DISCUSSION

In this study, we used α-toxin to prepare receptor coupled permeabilized smooth muscle fibers. This toxin (molecular weight ~33,000) is an exoprotein produced by Staphylococcus aureus which forms pores of 2-3 nm diameter upon aggregation into hexamers in the cell membrane (Hohman, 1988). The limited pore size allows equilibration of the cytoplasm with inorganic ions and small molecules, but prevents permeation by proteins. Thus, integral proteins such as calmodulin and regulatory enzymes are expected to remain in the cytoplasmic space. Furthermore, α-toxin is prevented from entering the cells and permeabilizing intracellular organelles. α-Toxin was first used to permeabilize smooth muscle tissue (Cassidy et al., 1979). However, it was not recognized until 1988 that this procedure does not disrupt receptor coupled signal transduction (Nishimura et al., 1988). We demonstrated, for the first time, that this toxin could be successfully used to permeabilize small preparations of smooth muscle tissue which retain functional receptors and receptor coupled transduction systems (Nishimura et al., 1988). We also demonstrated that calmodulin, adenylate- and guanylate-cyclases, and cAMP- and cGMP-dependent protein kinases are retained at apparently physiological levels. This enables the evaluation of the physiological role(s) of cAMP and cGMP-induced relaxing effects (Nishimura and van Breemen, 1989a). Detergent skinned smooth muscle preparation have also been of great importance in similar evaluations, however, exogenous cAMP- and cGMP- dependent protein kinases were required to observe cyclic nucleotide induced relaxations, which changed depending on the calmodulin concentration (Kerrick and Hoar, 1981; Itoh et al., 1985)

Modulation of Ca^{2+} Sensitivity

Studies in which force and $[Ca^{2+}]_i$ were simultaneously measured in intact smooth muscle tissue have suggested that the force/$[Ca^{2+}]_i$ ratio is greater in agonist-induced as compared to high K-induced contraction. Morgan and Morgan (1982, 1984a) were the first to report this phenomenon. In addition, Morgan and Morgan (1984b) also demonstrated that isoprenaline, papaverine, and forskolin, which are known to increase cellular cAMP levels, produced either an increase or no change in $[Ca^{2+}]_i$ in the face of arterial smooth muscle relaxation. On the other hand, nitrovasodilators such as nitroglycerine (Kobayashi et al., 1985) and sodium nitroprusside (Morgan and Morgan, 1984b; Karaki et al., 1988), which are believed to increase cellular cGMP levels (Schultz et al., 1977), and 8-bromo-cGMP (Kai et al., 1987), a slowly hydrolyzable cGMP analog, are reported to decrease $[Ca^{2+}]_i$ as well as relax the tissue. Interestingly, sodium nitroprusside was apparently more effective in relaxing vascular smooth muscle than in decreasing $[Ca^{2+}]_i$ similar to the effect of increasing cAMP (Morgan and Morgan, 1984b; Karaki et al., 1988). In any event, these studies using intact smooth muscle preparations clearly demonstrate that the force/$[Ca^{2+}]_i$ relationship can be modulated by either agonist stimulation or by a direct increase in second messengers.

Recent developments in the understanding of receptor signal transduction, especially the involvement of G proteins (Neer and Clapham, 1988), and the development of permeabilizing methods using α-toxin (Nishimura et al., 1988; Kitazawa et al., 1989) and β-escin (Kobayashi et al., 1989) have enabled us to demonstrate in permeabilized smooth muscle preparations that the force/Ca^{2+} relationship can be modulated. The results shown in Figures 2 and 3 as well as previous studies from our laboratory (Nishimura et al., 1989; Nishimura and van Breemen, 1989a,b) and Somlyo's laboratory (Kitazawa et al., 1989; Kobayashi et al., 1989; see also Kitazawa and Somlyo and Somlyo et al., this volume) clearly demonstrate that receptor stimulation significantly increases myofilament Ca^{2+} sensitivity of smooth muscle. Thus, it is evident that excitation-contraction coupling (or at least pharmacomechanical

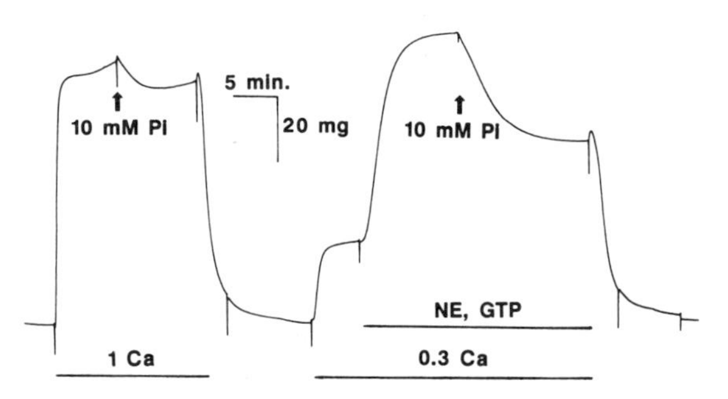

Figure 10. Effect of 10 mM P_i on the contractions induced by 1 μM Ca^{2+} alone or by 0.3 μM Ca^{2+} plus 10 μM NE plus 10 μM GTP. Ca^{2+} solutions were buffered by 10 mM EGTA. All contractions illustrated were performed using the same preparation. Traces are representative of three experiments.

coupling) of smooth muscle involves a change in the Ca^{2+} sensitivity of the contractile elements.

Involvement of G-Proteins and Protein Kinase C in the Increased Myofilament Ca^{2+} Sensitivity

Recent progress in the understanding of transmembrane signalling has shown that agonist binding to a receptor activates G proteins, which, in turn, stimulate the membrane-bound enzyme phospholipase C (PLC) (Gilman, 1987; Fain et al., 1988; Neer and Clapham, 1988). PLC catalyzes the hydrolysis of phosphatidyl-inositol 4,5-bisphosphate to yield two intracellular messengers, inositol-1,4,5-trisphosphate (IP_3) and diacylglycerol (DAG). IP_3 releases Ca^{2+} from the intracellular stores and DAG activates PKC. Protein phosphorylation catalyzed by PKC is postulated to modulate various Ca^{2+}-mediated processes such as exocytosis, cell proliferation and differentiation, membrane conductance and transport, and possibly smooth muscle contraction (Nishizuka, 1986).

The inhibitory effect of GDPβS on agonist induced contractions and the preferential requirement of GTP for agonist-induced contraction of the α-toxin permeabilized tissue shown in Figure 1, as well as in previous reports (Kitazawa et al., 1989; Nishimura et al., 1989, 1990), clearly demonstrates that the enhanced Ca^{2+} sensitivity of contraction following agonist activation is mediated through G proteins. The fact that the α-toxin permeabilized preparation can be activated by phorbol esters (Figure 6) with a similar increase in myofilament Ca^{2+} sensitivity as that elicited by agonist stimulation suggests that PKC may be intimately involved in this process(es). This possibility is further substantiated by the finding that H-7 and staurosporine apparently inhibit selectively the agonist induced enhanced force at any given [Ca^{2+}] (Fig. 4 and 5). Previous studies, using both permeabilized and intact tissues, have supported a role for PKC in the regulation of smooth muscle contraction and

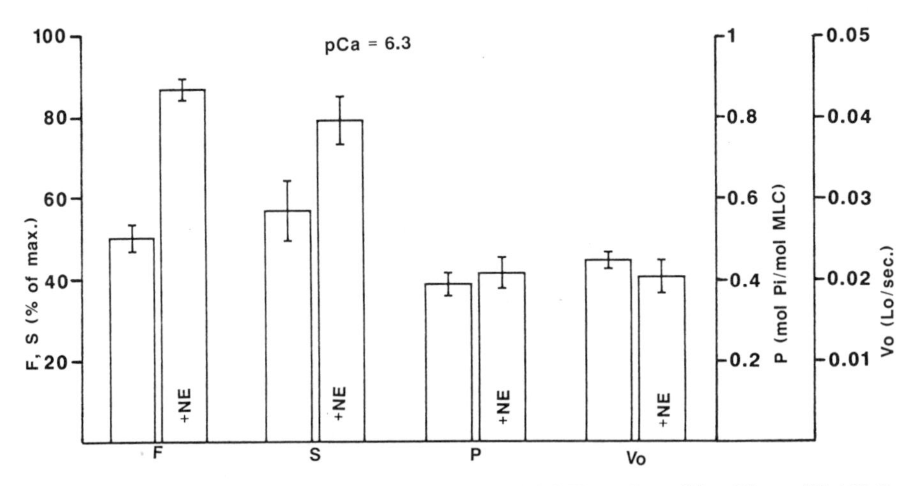

Figure 11. Effect of 10 μM norepinephrine plus 10 μM GTP on force (F), stiffness (S), MLC phosphorylation (P) and unloaded shortening velocity (V_o) at pCa = 6.3 buffered by 10 mM EGTA in permeabilized rabbit mesenteric artery. The measurement of these parameters were performed 20 min after activation by Ca^{2+} alone (open column) or by NE plus GTP (+NE). Values shown are means ± S.E.

more to the point, have supported the notion that PKC induced contractions require less Ca^{2+} than those elicited by membrane depolarization (Danthuluri and Deth, 1984; Rasmussen et al., 1984; Forder et al., 1985; Chatterjee and Tajeda, 1986; Nishimura and van Breemen, 1989a).

Modulation of Ca^{2+} Sensitivity by ADP and P_i

Studies using isolated contractile proteins have shown that actomyosin hydrolyzes ATP by a complex series of reactions (Eisenberg and Greene, 1980; Sleep and Hutton, 1980; Sellers, 1985), a simplified version of which is illustrated in Figure 12. Because AM-ADP and AM'-ADP are believed to be the force generating states in the crossbridge cycle, one would predict that increasing the concentration of ADP should increase the population of force generating crossbridges and result in an increase in force generated at any given intermediate Ca^{2+} concentration. In smooth muscle, this prediction has been supported by work from Kerrick and Hoar (1987). They suggested that the effect of increasing the [ADP] on contraction in smooth muscle would be independent of MLC phosphorylation levels and would slow ATP consumption and therefore, crossbridge cycling. This would result in a state of high maintained force but low levels of shortening velocity. Interestingly, these are properties characteristic of the "latch" state. The results in Figures 7 - 9 are also consistent with this prediction. These figures show that addition of 100 μM ADPβS or by the removal of CP, which would indirectly increase cellular [ADP], dramatically increases the Ca^{2+} sensitivity of α-toxin permeabilized smooth muscle. It is intriguing to speculate that the agonist induced enhanced myofilament Ca^{2+} sensitivity, or even the latch state, may be explained by this mechanism (Nishimura and van Breemen, 1989b).

If the Ca^{2+} sensitivity of force development could be increased by accumulating crossbridges in either the AM-ADP or AM'-ADP states through step "1" in Figure 12, then inhibition of ADP release from the AM-ADP complex (inhibition of step "2") should increase myofilament Ca^{2+} sensitivity. This has been shown to be the case by Lash et al. (1986) in a cell free, *in vitro* system. If a major step in the mechanism responsible for the agonist- or phorbol ester-induced enhancement of myofilament Ca^{2+} sensitivity is inhibition of ADP release, then the effect (relaxation) of added P_i should be greater for these contractions than for that induced by Ca^{2+} alone. This prediction is based on the assumption that contractile relaxation upon the

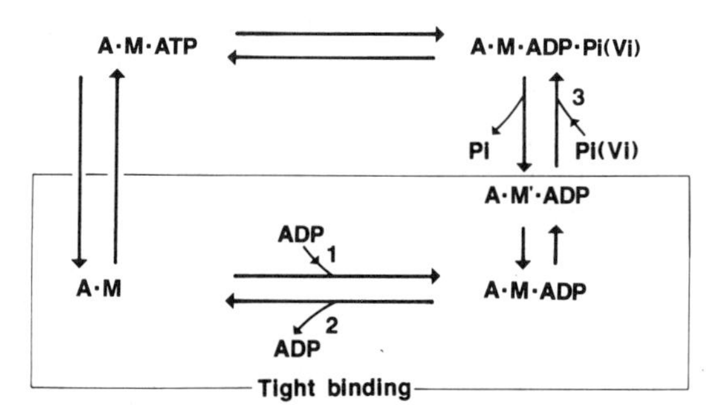

Figure 12. Simplified scheme of actomyosin ATPase activity. Abbreviations used are, A: actin, M: myosin, P_i: phosphate, V_i: vanadate.

addition of exogenous P_i is due to P_i binding to AM'-ADP (step "3" in Figure 12). The results shown in Figures 9 and 10 show a greater effect of P_i on the NE plus GTP induced contractions as compared to Ca^{2+} alone and support the hypothesis that inhibition of ADP release may be involved in the enhanced force response. Itoh et al. (1986) postulated that the inhibitory effect of P_i may be dependent on the rate of crossbridge cycling, such that the lower the cycling rate, the greater the inhibitory effect by P_i. The results obtained in Figures 9 and 10 would suggest that crossbridge cycling in response to 0.3 μM Ca^{2+} plus NE plus GTP was lower than that in response 1 μM Ca^{2+} alone, although the magnitude of steady state forces were similar. Estimates of crossbridge cycling rate at each of these conditions have shown that V_o increases only by increases in the [Ca^{2+}] and is unaffected by the addition of an agonist plus GTP (data not shown, but see Fig. 11 for a single [Ca^{2+}]). Therefore, it is possible that alterations in the release of ADP may play a role in the enhanced myofilament Ca^{2+} sensitivity following agonist activation.

MLC Phosphorylation and V_o

MLC phosphorylation is believed to be a primary step in the Ca^{2+}-dependent regulation of smooth muscle. Moreover, MLC phosphorylation levels have been shown by several laboratories to correlate with the level of force development in permeabilized tissues (Cassidy et al., 1981; Chatterjee and Murphy, 1983; Moreland and Murphy, 1986). It is therefore, of great interest to determine if the enhanced force at any given [Ca^{2+}] by NE plus GTP is accompanied by an increase in the level of MLC phosphorylation. Figure 11 clearly shows that, during steady state conditions (20 min of tissue activation), the increased force developed by stimulation with NE plus GTP, as compared to simply Ca^{2+} alone, was not accompanied by an increase in MLC phosphorylation. In addition, these data demonstrate the enhanced force is not associated with an increase in V_o. In this α-toxin permeabilized preparation, both MLC phosphorylation levels and V_o are increased in a Ca^{2+}-dependent fashion (data not shown). Moreover, stimulation with Ca^{2+} in the presence of GTPγS does significantly increase the level of MLC phosphorylation similar to the results obtained by Fujiwara et al. (1989), Kubota et al. (1990), and Kitazawa and Somlyo (this volume). However, it appears that although an increase in MLC phosphorylation may accompany the agonist and G-protein dependent enhancement of force, an increase in MLC phosphorylation, at least at steady state, is not required.

The data in Figure 11 clearly show that the agonist dependent force enhancement is associated with the recruitment of additional crossbridges. This is evident by the significant increase in stiffness which correlates with the level of force both in the presence and absence of NE plus GTP. These results, an increase in crossbridge activation without a concomitant increase in the level of MLC phosphorylation, suggest that MLC phosphorylation may not be the sole regulator of the smooth muscle contraction in the α-toxin permeabilized rabbit mesenteric artery. Based on several recent lines of evidence, we believe that the strongest candidates for a regulatory system which may function in parallel to MLC phosphorylation are the thin filament proteins of caldesmon (Sobue et al., 1981) and calponin (Takahashi et al., 1986; Winder et al., this volume). It will be of great interest to follow the results of future studies to determine if in fact these proteins are involved in the regulation of a smooth muscle contraction.

ACKNOWLEDGMENTS

We are particularly indebted to Dr. R. J. Hohman for supplying us with α-toxin. This work was supported by NIH grants HL 35657 (CvB) and HL 37956 (RSM) and a postdoctoral fellowship from the American Heart Association, Florida affiliate (JN). The authors also thank Ms. Gerry Trebilcock for her secretarial assistance.

REFERENCES

Adelstein, R. S., Conti, M. A., Hathaway, D. R., and Klee, C. B., 1978, Phosphorylation of smooth muscle myosin light chain kinase by the catalytic subunit of adenosine 3':5'-monophosphate-dependent protein kinase, *J. Biol. Chem.*, 253: 8347.

Aksoy, M. O., Mras, S., Kamm, K. E., and Murphy, R. A., 1983, Ca^{2+}, cAMP, and changes in myosin light chain phosphorylation during contraction of smooth muscle, *Am. J. Physiol.*, 245: C255.

Cassidy, P., Hoar, P. E., and Kerrick, W. G. L., 1979, Irreversible thiophosphorylation and activation of tension in functionally skinned rabbit ileum strips by [^{35}S]ATPγS, *J. Biol. Chem.*, 254: 11148.

Cassidy, P. S., Kerrick, W. G. L., Hoar, P. E., and Malencik, D. A., 1981, Exogenous calmodulin increases Ca^{2+} sensitivity of isometric tension activation and myosin phosphorylation in skinned smooth muscle, *Pflügers Arch.*, 392: 115.

Chatterjee, M. and Murphy, R. A., 1983, Calcium-dependent stress maintenance without myosin phosphorylation in skinned smooth muscle, *Science*, 221: 464.

Chatterjee, M. and Tejada, M., 1986, Phorbol ester-induced contraction in chemically skinned vascular smooth muscle, *Am. J. Physiol.*, 251: C356.

Danthuluri, N. R. and Deth, R. C., 1984, Phorbol ester-induced contraction of arterial smooth muscle and inhibition of α-adrenergic response, *Biochem. Biophys. Res. Commun.*, 125: 1103.

Dillon, P. F., Aksoy, M. O., Driska, S. P., and Murphy, R. A., 1981, Myosin phosphorylation and the cross-bridge cycle in arterial smooth muscle, *Science*, 211: 495.

Eisenberg, E. and Greene, L. E., 1980, The relation of muscle biochemistry to muscle physiology, *Ann. Rev. Physiol.*, 42: 293.

Fain, J. N., Wallage, M. A., and Wojcikiewicz, R. J. H., 1988, Evidence for involvement of guanine nucleotide-binding regulatory proteins in the activation of phospholipases by hormones, *FASEB J.*, 2: 2569.

Forder, J., Scriabine, A., and Rasmussen, H., 1985, Plasma membrane calcium flux, protein kinase C activation and smooth muscle contraction, *J. Pharmacol. Exp. Ther.*, 235: 267.

Fujiwara, T., Itoh, T., Kubota, Y., and Kuriyama, H., 1989, Effect of guanosine nucleotides on skinned smooth muscle tissue of the rabbit mesenteric artery, *J. Physiol.*, 408: 535.

Gaylinn, B. D. and Murphy, R. A., 1990, Quantitation of myosin 20 kD light chain phosphorylation using colloidal gold stained Western blots, *Biophys. J.*, 57: 165a.

Gilman, A. G., 1987, G proteins: Transducers of receptor-generated signals, *Ann. Rev. Biochem.*, 56: 615.

Hohman, R. J., 1988, Aggregation of IgE receptors induces degranulation in rat basophilic leukemia cells permeabilized with α-toxin from Staphylococcus aureus, *Proc. Nat'l. Acad. Sci. U.S.A.*, 85: 1624.
Itoh, T., Kanmura, Y., Kuriyama, H., and Sasaguri, T., 1985, Nitroglycerine- and isoprenaline-induced vasodilation: assessment from the actions of cyclic nucleotides, *Br. J. Pharmacol.*, 84: 393.
Itoh, T., Kanmura, Y., and Kuriyama, H., 1986, Inorganic phosphate regulates the contraction-relaxation cycle in skinned muscles of the rabbit mesenteric artery, *J. Physiol.*, 376: 231.
Itoh, T., Kubota, Y., and Kuriyama, H., 1988, Effects of phorbol ester on acetylcholine-induced Ca^{2+} mobilization and contraction in the porcine coronary artery, *J. Physiol.*, 397: 401.
Kai, H., Kanaide, H., Matsumoto, T., and Nakamura, M., 1987, 8-Bromoguanosine 3':5'-cyclic monophosphate decreases intracellular free calcium concentrations in cultured vascular smooth muscle cells from rat aorta, *FEBS Lett.*, 221: 284.
Kamm, K. E. and Stull, J. T., 1985, The function of myosin and myosin light chain kinase phosphorylation in smooth muscle, *Ann. Rev. Pharmacol. Toxicol.*, 25: 593.
Kamm, K. E. and Stull, J. T., 1989, Regulation of smooth muscle contractile elements by second messengers, *Ann. Rev. Physiol.*, 51: 299.
Karaki, H., Sato, K., Ozaki, H., and Murakami, K., 1988, Effects of sodium nitroprusside on cytosolic calcium level in vascular smooth muscle, *Eur. J. Pharmacol.*, 156: 259.
Kerrick, W. G. L. and Hoar, P. E., 1981, Inhibition of smooth muscle tension by cyclic AMP-dependent protein kinase, *Nature*, 292: 253.
Kerrick, W. G. L. and Hoar, P. E., 1987, Non-Ca^{2+}-activated contraction in smooth muscle, *in:* "Regulation and Contraction of Smooth Muscle", M. J. Siegman, A. P. Somlyo, and N. L. Stephens, eds., Alan R. Liss, New York, p. 437.
Kitazawa, T., Kobayashi, S., Horiuchi, K., Somlyo, A. V., and Somlyo, A. P., 1989, Receptor coupled, permeabilized smooth muscle: Role of the phosphatidylinositol cascade, G-proteins and modulation of the contractile response to Ca^{2+}, *J. Biol. Chem.*, 264: 5339.
Kobayashi, S., Kanaide, H., and Nakamura, M., 1985, Cytosolic free calcium transient in cultured smooth muscle cells: Microfluorometric measurements, *Science*, 229: 553.
Kobayashi, S., Kitazawa, T., Somlyo, A. V., and Somlyo, A. P., 1989, Cytosolic heparin inhibits muscarinic and α-adrenergic Ca^{2+} release in smooth muscle. Physiological role of inositol 1,4,5-trisphosphate in pharmacomechanical coupling, *J. Biol. Chem.*, 264: 17997
Kubota, Y., Kamm, K. E., and Stull, J. T., 1990, Mechanism of GTPγS-dependent regulation of smooth muscle contraction, *Biophys. J.*, 57: 163a.
Lash, J. A., Sellers, J. R., and Hathaway, D. R., 1986, The effects of caldesmon on smooth muscle heavy actomeromyosin ATPase activity and binding of heavy meromyosin to actin, *J. Biol. Chem.*, 261: 16155.
Moreland, R. S. and Murphy, R. A., 1986, Determinants of Ca^{2+}-dependent stress maintenance in skinned swine carotid media, *Am. J. Physiol.*, 251: C892.
Morgan, J. P. and Morgan, K. G., 1982, Vascular smooth muscle: The first recorded Ca^{2+} transients, *Pflügers Arch.*, 395: 75.

Morgan, J. P. and Morgan, K. G., 1984a, Stimulus-specific patterns of intracellular calcium levels in smooth muscle of the ferret portal vein, *J. Physiol.*, 351: 155.

Morgan, J. P. and Morgan, K. G., 1984b, Alteration of cytoplasmic ionized calcium level in smooth muscle by vasodilators in the ferret, *J. Physiol.*, 357: 539.

Neer, E. J. and Clapham, D. E., 1988, Roles of G protein subunits in transmembrane signalling, *Nature*, 333: 129.

Nishimura, J., Kolber, M., and van Breemen, C., 1988, Norepinephrine and GTP-γ-S increase myofilament Ca^{2+} sensitivity in α-toxin permeabilized arterial smooth muscle, *Biochem. Biophys. Res. Commun.*, 157: 677.

Nishimura, J. and van Breemen, C., 1989a, Direct regulation of smooth muscle contractile elements by second messengers, *Biochem. Biophys. Res. Commun.*, 163: 929.

Nishimura, J. and van Breemen, C., 1989b, Possible involvement of actomyosin ADP complex in regulation of Ca^{2+} sensitivity in α-toxin permeabilized smooth muscle, *Biochem. Biophys. Res. Commun.*, 165: 408.

Nishimura, J., Khalil, R. A., and van Breemen, C., 1989, Agonist-induced vascular tone, *Hypertension*, 13: 835.

Nishimura, J., Khalil, R. A., Drenth, J. P., and van Breemen, C., 1990, Evidence for increased myofilament Ca^{2+} sensitivity in norepinephrine-activated vascular smooth muscle, *Am. J. Physiol.*, 259: H2.

Nishizuka, Y., 1986, Studies and perspectives of protein kinase C, *Science*, 233: 305.

Rasmussen, H., Forder, J., Kojima, I., and Scriabine, A., 1984, TPA-induced contraction of isolated rabbit vascular smooth muscle, *Biochem. Biophys. Res. Commun.*, 122: 776.

Rüegg, J. C. and Paul, R. J., 1982, Vascular smooth muscle calmodulin and cyclic AMP-dependent protein kinase alter calcium sensitivity in porcine carotid skinned fibers, *Circ. Res.*, 50: 394.

Schultz, K. D., Schultz, K., and Schultz, G., 1977, Sodium nitroprusside and other smooth muscle-relaxants increase cyclic GMP levels in rat ductus deferens, *Nature*, 265: 750.

Sellers, J. R., 1985, Mechanism of the phosphorylation-dependent regulation of smooth muscle heavy meromyosin, *J. Biol. Chem.*, 260: 15815.

Sleep, J. A. and Hutton, R. L., 1980, Exchange between inorganic phosphate and adenosine 5'-triphosphate in the medium by actomyosin subfragment 1, *Biochemistry*, 19: 1276.

Sobue, K., Muramoto, Y., Fujita, M., and Kakiuchi, S., 1981, Purification of a calmodulin-binding protein from chicken gizzard that interacts with F-actin, *Proc. Nat'l. Acad. Sci. U.S.A.*, 78: 5652.

Somlyo, A. P., 1985, Excitation-contraction coupling and the ultrastructure of smooth muscle, *Circ. Res.*, 57: 497.

Takahashi, K., Hiwada, K., and Kokubo, T., 1986, Isolation and characterization of a 34000 dalton calmodulin- and F-actin-binding protein from chicken gizzard smooth muscle, *Biochem. Biophys. Res. Commun.*, 141: 20.

MYOSIN LIGHT CHAIN KINASE PHOSPHORYLATION: REGULATION OF THE Ca^{2+} SENSITIVITY OF CONTRACTILE ELEMENTS

James T. Stull, Malú G. Tansey, R. Ann Word, Yasutaka Kubota, and Kristine E. Kamm

The University of Texas Southwestern Medical Center
Department of Physiology
5323 Harry Hines Boulevard
Dallas, TX 75235

INTRODUCTION

Cyclic AMP-dependent protein kinase phosphorylates myosin light chain kinase purified from gizzard smooth muscle (Adelstein et al., 1978). In the absence of Ca^{2+}/calmodulin, 2 sites (sites A and B) are phosphorylated and the concentration of Ca^{2+}/calmodulin required for half-maximal activation (K_{CaM}) increases 10 fold (Conti and Adelstein, 1981). In the presence of Ca^{2+}/calmodulin, phosphate is incorporated into site B with no effect on myosin light chain kinase activity. As shown in Figure 1, site A is near the calmodulin binding domain whereas site B is more toward the c-terminus. Similar observations on kinase phosphorylation have been made with myosin light chain kinases purified from mammalian smooth muscles including trachea (Miller et al., 1983), stomach (Walsh et al., 1982), myometrium (Higashi et al., 1983), aorta (Vallet et al., 1981), and carotid artery (Bhalla et al., 1982). It has been proposed that phosphorylation of site A by cyclic AMP-dependent protein kinase could decrease the extent of myosin light chain kinase activation with a resultant decrease in myosin light chain phosphorylation and inhibition of contraction (Conti and Adelstein, 1981).

Purified myosin light chain kinase from smooth muscle is phosphorylated by other protein kinases (Table 1). Protein kinase C incorporates phosphate into two sites in gizzard myosin light chain kinase in the absence of Ca^{2+}/calmodulin (Ikebe et al., 1985; Nishikawa et al., 1985). Phosphorylation leads to a reduced affinity of the kinase for Ca^{2+}/calmodulin, a result similar to the effect produced by phosphorylation with cyclic AMP-dependent protein kinase. Calmodulin-dependent protein kinase II also phosphorylates myosin light chain kinase to a molar stoichiometry of 2.77 with an associated increase in K_{CaM} (Hashimoto and Soderling, 1990). Peptide mapping and sequence analysis showed that both calmodulin-dependent protein kinase II and cyclic AMP-dependent protein kinase phosphorylated the second of the two adjacent serines in phosphopeptide A which is located near the calmodulin-binding domain. Carbachol stimulation of muscarinic receptors could result in

Regulation of Smooth Muscle Contraction
Edited by R.S. Moreland, Plenum Press, New York, 1991

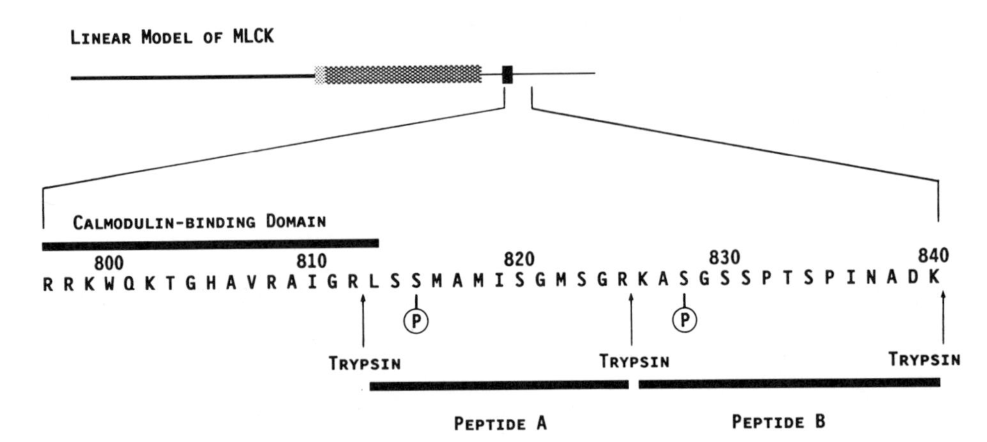

Figure 1. Sites of phosphorylation in smooth muscle myosin light chain kinase. In the linear model of myosin light chain kinase, the hatched regions represent a myosin light chain binding site and catalytic core. The solid bar represents the calmodulin binding domain. The structures of the 2 sites phosphorylated by cAMP-dependent protein kinase in phosphopeptides A and B are shown below.

Table 1. Phosphorylation of Smooth Muscle Myosin Light Chain Kinase

Kinase	Sites Phosphorylated		
	A	B	Other
cAMP-Dependent Protein Kinase	X	X	
cGMP-Dependent Protein Kinase		X	X
Protein Kinase C	X		X
Calmodulin-Dependent Protein Kinase II	X		X

activation of both protein kinase C and calmodulin-dependent protein kinase, whereas KCl depolarization would lead to activation of calmodulin-dependent protein kinase II. Finally, cGMP-dependent protein kinase phosphorylates smooth muscle myosin light chain kinase (Nishikawa et al., 1984). However, it does not phosphorylate site A and there is no change in the calmodulin activation properties (Table 1).

Protein kinases regulated by different second messengers phosphorylate smooth muscle myosin light chain kinase *in vitro*. From these biochemical observations, it is difficult to construct a simple cellular mechanism associating myosin light chain kinase phosphorylation to smooth muscle contraction. It has been shown that a 17-fold stimulation of cyclic AMP formation with forskolin resulted in myosin light chain kinase phosphorylation in ^{32}P-labelled tracheal smooth muscle (de Lanerolle et al., 1984). The sites of phosphorylation were not identified. However, treatment of tracheal smooth muscles with isoproterenol at a concentration sufficient for relaxation had no effect on the Ca^{2+}/calmodulin activation properties of myosin light chain kinase from those muscles (Miller et al., 1983). At a high concentration (5 μM) of isoproterenol, however, K_{CaM} increased slightly (less than 2-fold). Interestingly, K_{CaM}

increased more with carbachol or KCl treatments alone, both of which resulted in contraction (Miller et al., 1983). Although sites of phosphorylation in myosin light chain kinase were not identified, the conclusion was reached that β-adrenergic relaxation does not require an increase in K_{CaM} for myosin light chain kinase. The increase in K_{CaM} with agonist stimulation could be associated with activation of protein kinase C or calmodulin-dependent protein kinase II.

Myosin light chain kinase is dephosphorylated by an avian smooth muscle protein phosphatase (Pato and Adelstein, 1983). The enzyme consists of 3 distinct protein subunits with relative masses of 60, 55 and 38 kDa. The 38 kDa subunit is the catalytic subunit. The general biochemical properties of this enzyme are very similar to protein phosphatase type 2A (Cohen, 1989). When myosin light chain kinase was phosphorylated in sites A and B by cAMP-dependent protein kinase, both sites were dephosphorylated by the protein phosphatase in the absence of calmodulin. Interestingly, site A, but not site B, was dephosphorylated in the presence of calmodulin. This pattern of dephosphorylation in the presence or absence of calmodulin is opposite to the general pattern observed with protein kinases. It should also be noted that this protein phosphatase dephosphorylated isolated myosin light chain, but not intact myosin (Pato and Adelstein, 1983). Thus, it appears to be relatively specific for myosin light chain kinase.

We decided (1) to determine if myosin light chain kinase was phosphorylated in tracheal smooth muscle under conditions that activate different protein kinases, (2) to identify the number of sites phosphorylated in myosin light chain kinase by phosphopeptide mapping, and (3) to evaluate the functional significance of phosphorylation in terms of changes in calmodulin activation properties (Stull et al., 1990).

MYOSIN LIGHT CHAIN KINASE PHOSPHORYLATION

Treatment of bovine tracheal tissue with 3 μM isoproterenol resulted in no increase in myosin light chain phosphorylation or in force (Table 2). However, the extent of myosin light chain kinase phosphorylation increased from a control value of 0.86 mol of phosphate per mol of kinase to 1.28 and 2.56 by 1 min and 20 min, respectively. Carbachol treatment (0.1 μM) resulted in a transient increase in the extent of myosin light chain phosphorylation to a maximal value of 0.70 mol of phosphate per mol of light chain by 5 min while force increased to a sustained level. The extent of myosin light chain kinase phosphorylation also increased in the presence of carbachol to 1.94 mol of phosphate per mol of kinase by 25 min. The addition of isoproterenol to tissues that had been treated with carbachol for 5 min resulted in a prompt relaxation and decrease in myosin light chain phosphorylation. However, the extent of myosin light chain kinase phosphorylation was similar to values obtained with carbachol alone. Thus, the effects of isoproterenol and carbachol on myosin light chain kinase phosphorylation were not additive. Furthermore, there was no simple correlation between contractile force and myosin light chain kinase phosphorylation.

PDBu, an activator of protein kinase C, also stimulated myosin light chain kinase phosphorylation (Table 2). Treatment of tracheal tissues with 1 μM PDBu for 25 min resulted in an increase in myosin light chain kinase phosphorylation to 2.73 mol of phosphate per mol of kinase. Fractional force and myosin light chain phosphorylation increased slightly. These conditions lead to a predominant phosphorylation of protein kinase C phosphorylation sites in myosin light chain (Kamm et al., 1989) which has little apparent effect on smooth muscle contractile properties (Sutton and Haeberle, 1990). Treatment of tissues with a combination of carbachol and PDBu did not result in any greater increase in myosin light chain kinase phosphorylation compared to PDBu alone. Thus, the effects of PDBu and carbachol on kinase phosphorylation were not additive.

KCl stimulated myosin light chain phosphorylation and force development in tracheal smooth muscle (Table 2). KCl treatment also increased myosin light chain kinase phosphorylation to 1.60 mol of phosphate per mol of kinase.

These results show that treatment of tracheal smooth muscle with agents that activate different protein kinases stimulated myosin light chain kinase phosphorylation. However, there was no simple correlation between the extent of kinase phosphorylation and the contractile state. Since myosin light chain kinase can be phosphorylated at multiple sites by different protein kinases (Conti and Adelstein, 1981; Ikebe et al., 1985; Nishikawa et al., 1985; Hashimoto and Soderling, 1990), phosphopeptide mapping was used to identify specific sites of phosphorylation.

Table 2. Phosphorylation of Myosin Light Chain Kinase in ^{32}P-Labelled Tracheal Smooth Muscle

Treatment	Force	Phosphorylations: Light Chain	Phosphorylations: MLCK	Phosphorylations: Site A
	fractional	mol P/LC	mol P/MLCK	mol P/peptideA
Control	0.00	0.09 ± 0.02	0.09 ± 0.08	0.00
Carbachol[a]				
1 min	0.78	0.46 ± 0.01	1.28 ± 0.07	0.42
5 min	1.00	0.70 ± 0.06	1.66 ± 0.26	0.71
25 min	1.00	0.42 ± 0.01	1.94 ± 0.26	0.39
Iso[a]				
1 min	0.00	0.08 ± 0.02	1.28 ± 0.15	0.15
Carb (5 min)				
+ Iso (1 min)	0.10	0.16 ± 0.02	1.43 ± 0.13	0.33
PDBu[a]				
25 min	0.06	0.22 ± 0.03	2.73 ± 0.33	0.08
KCl[a]				
5 min	0.82	0.45 ± 0.03	1.60 ± 0.26	0.80

[a] Carbachol, 0.1 μM; isoproterenol, 3 μM; PDBu, 1 μM; KCl, 65 mM + 0.1 μM atropine

PHOSPHOPEPTIDE MAPPING

Myosin light chain kinase was immunoprecipitated from ^{32}P-labelled tracheal tissues and subjected to phosphopeptide mapping (Stull et al., 1990). Six phosphopeptides were consistently identified after treatment with various agents. Phosphopeptides A and B coincided to the phosphopeptides A and B identified in myosin light chain kinase diphosphorylated by cyclic AMP-dependent protein kinase; co-electrophoresis of digests of the purified myosin light chain kinase and kinase obtained from ^{32}P-labelled tissues resulted in coincident migrations of A and B phosphopeptides, respectively. Based on previous reports with the avian enzyme, phosphopeptide A was identified as the serine site phosphorylated near the calmodulin-binding domain (Conti and Adelstein, 1981; Lukas et al., 1986; Hashimoto and Soderling, 1990). Under control conditions, ^{32}P was found in peptides B, C, D, and F. Thus, there are multiple phosphorylation sites when the extent of myosin light chain kinase phosphorylation is 0.86 mol of phosphate per mol of kinase. Treatment of tracheal tissues with isoproterenol for 20 min results in 4 primary phosphopeptides from myosin light chain kinase: A, B, C and D. The relative amounts of ^{32}P in phosphopeptides A and B are less than C and D. PDBu treatment resulted in no significant ^{32}P incorporation into phosphopeptide A or B and the appearance of another phosphopeptide E.

Stimulation of muscarinic receptors for 25 min with carbachol resulted in significant ^{32}P incorporation into all six phosphopeptides. The combination of carbachol and isoproterenol gave a phosphopeptide pattern intermediate to the treatments with the individual agonists. Interestingly, phosphopeptide A was the most prominent ^{32}P-labelled peptide after treatment of tracheal smooth muscle with 65 mM KCl (Table 2).

The relative amounts of ^{32}P incorporated into ^{32}P-labelled peptides from myosin light chain kinase were analyzed (Table 2). Under control conditions there was no significant phosphorylation of peptide. With the addition of carbachol to tracheal smooth muscle, there was a sustained increase in myosin light chain kinase phosphorylation. However, the phosphorylation of peptide A appears to be transient with values reaching a maximum by 5 min and subsequently declining to lower levels by 25 min (Table 2). This pattern of peptide A phosphorylation is similar to the temporal pattern obtained with myosin light chain phosphorylation. Treatment of tracheal tissues with KCl also increased the extent of myosin light chain kinase phosphorylation with most of the radioactivity found in peptide A (Table 2). Assuming 1 phosphorylation site in peptide A, the calculated maximal extent of phosphorylation was 0.71 and 0.80 mol of phosphate per mol of peptide A for carbachol and KCl treatments, respectively.

There appears to be a positive relationship between the extent of Ca^{2+}-dependent myosin light chain phosphorylation and the extent of site A phosphorylation in myosin light chain kinase (Table 2). In tracheal smooth muscle depleted of Ca^{2+} by incubation in the presence of 5 mM EGTA (Ratz and Murphy, 1987), 0.1 μM carbachol did not result in the development of force nor myosin light chain kinase phosphorylation (data not shown). These results are consistent with the idea that a Ca^{2+}-dependent kinase may phosphorylate site A in myosin light chain kinase. If this is the case, then the positive correlation between site A phosphorylation and myosin light chain phosphorylation may be due to the Ca^{2+}-dependence of both reactions.

Myosin light chain kinase is phosphorylated to a high extent in site A in tracheal smooth muscle contracted with carbachol or KCl. Furthermore, there is a positive correlation between the extents of phosphorylation of myosin light chain and myosin light chain kinase. These data plus the inhibition of myosin light chain kinase phosphorylation at site A by EGTA treatment suggest that it is catalyzed by a Ca^{2+}-dependent protein kinase. Ca^{2+}/calmodulin-dependent protein kinase would be a likely candidate (Hashimoto and Soderling, 1990), because it also phosphorylates myosin light chain kinase at site A and this site is highly phosphorylated in contracting tissue. However, as noted by Hashimoto and Soderling (1990), when calmodulin is bound to myosin light chain kinase, phosphorylation of this regulatory site is blocked. The autophosphorylated, Ca^{2+}/calmodulin-independent form of the Ca^{2+}/calmodulin-dependent protein kinase II was used by these investigators for rapid phosphorylation without Ca^{2+}/calmodulin binding to myosin light chain kinase. Myosin light chain kinase could be phosphorylated by the Ca^{2+}/calmodulin-dependent protein kinase II during contraction if a large fraction of the myosin light chain kinase did not have calmodulin bound. Alternatively, a small population of nonbound kinase may be rapidly phosphorylated. During the time required for development of force, Ca^{2+}/calmodulin will associate and dissociate from myosin light chain kinase. While dissociated, the kinase may be phosphorylated, so that over some period of time substantial phosphorylation will be obtained.

The possibility must also be considered, however, that another Ca^{2+}/calmodulin-dependent protein kinase may phosphorylate myosin light chain kinase when Ca^{2+}/calmodulin is bound. It is known that a protein phosphatase dephosphorylates site A when Ca^{2+}/calmodulin is bound to myosin light chain kinase (Pato and Adelstein, 1983). Thus the phosphate moiety is not covered by calmodulin or otherwise totally inaccessible. Evidence has also been presented that site A is not necessary for the high affinity binding of Ca^{2+}/calmodulin to a 17 residue synthetic peptide representing the calmodulin binding domain of the avian myosin light chain kinase (Lukas et al., 1986). Thus the phosphorylatable serine in site A may not bind directly to Ca^{2+}/calmodulin and could be accessible for phosphorylation by an unidentified, Ca^{2+}-dependent protein kinase.

It is predicted that phosphorylation of site A in myosin light chain kinase increases K_{CaM} irrespective of the kinase that may phosphorylate this site (Kamm and Stull, 1985; 1989). Changes in the calmodulin activation properties were assessed by measurements of the kinase activity ratio in tissue homogenates. The ratio of enzyme activity at 1 μM Ca^{2+} to the activity at 100 μM Ca^{2+} in the presence of 400 nM calmodulin is quantitatively dependent upon the K_{CaM} value (Miller et al., 1983). When MLCK is phosphorylated the ratio of kinase activities at 1 and 100 μM Ca^{2+} decreases. There was a direct correlation between the myosin light chain kinase activity ratio and the extent of phosphorylation of peptide A in myosin light chain kinase (Stull et al., 1990). This relationship was observed with tissues treated with carbachol, KCl, isoproterenol, and PDBu. Thus, changes in K_{CaM} are probably determined by the extent of phosphorylation of site A.

The predicted effect of site A phosphorylation in myosin light chain kinase is to decrease the sensitivity of kinase activation to Ca^{2+}. Cytosolic Ca^{2+} concentrations were not measured in this study and therefore direct assessment of the functional consequences of myosin light chain phosphorylation are not possible. However, there are a number of recent physiological studies that have demonstrated a desensitization of contractile elements to cytosolic Ca^{2+} concentrations in contracting smooth muscles; i.e., there are time-dependent decreases in the ratio of force to cytosolic Ca^{2+} concentrations (Morgan and Morgan, 1984; Rembold and Murphy, 1988; Yagi et al., 1988; Himpens et al., 1989; Karaki, 1989). We propose that phosphorylation of myosin light chain kinase with a resultant increase in K_{CaM} could play an important role in this desensitization process. Direct measurements of cytosolic Ca^{2+} concentrations, myosin light chain kinase phosphorylation in site A, and myosin light chain phosphorylation will be needed to substantiate this hypothesis.

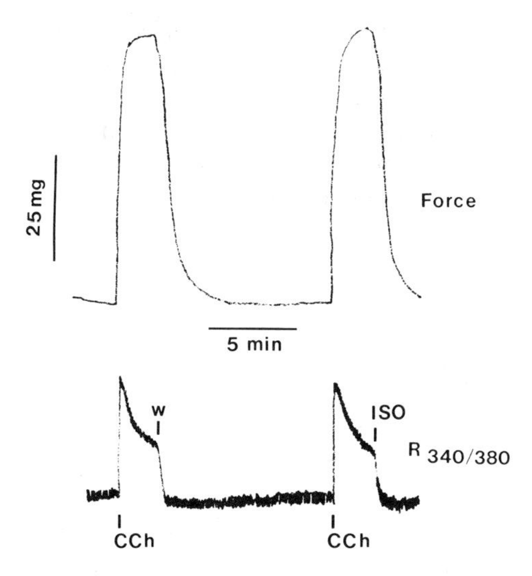

Figure 2. Tracings of fluorescence and force records from a Fura-2 loaded bovine tracheal smooth muscle strip. The upper trace shows contractile force while the lower trace shows the ratio of fluorescence intensity measured at excitation wavelengths of 340 and 380 nm. Contraction was elicited by 1 μM carbachol (CCh) which was followed by rinsing in PSS (W). After contraction in carbachol (second trace) 3 μM isoproterenol (ISO) was added.

Isoproterenol alone increased myosin light chain kinase phosphorylation, but only a small fraction of the radioactivity was incorporated into peptide A (Table 2). The maximal extent of ^{32}P incorporation was 0.20 mol of phosphate per mol of peptide A (20 min), a value significantly less than values obtained with carbachol or KCl treatments. The marked extent of peptide A phosphorylation induced by carbachol was inhibited by isoproterenol which may be related to a decrease in cytosolic Ca^{2+} concentration. As shown in Figure 2, the addition of isoproterenol to tracheal smooth muscle contracted in the presence of carbachol results in a prompt decrease in cytosolic Ca^{2+} concentration and in relaxation.

Why is cyclic AMP formation such a poor stimulus for site A phosphorylation? The sequence around the phosphorylation site (-R-X-X-S-X-) is a consensus sequence for Ca^{2+}/calmodulin-dependent protein kinase II, not for cyclic AMP-dependent protein kinase (-R-R-X-S-X-). Myosin light chain kinase is phosphorylated at a rate that is 1% or lower than the rates of phosphorylation of physiological substrates (Stull et al., 1986). Therefore, the rate of phosphorylation at site A by cyclic AMP-dependent protein kinase may not be sufficiently greater than the rate of dephosphorylation by a protein phosphatase; thus, only minor extents of phosphorylation can be achieved.

Activation of protein kinase C by PDBu resulted in a substantial increase in the extent of myosin light chain kinase phosphorylation, but interestingly no significant phosphorylation of peptide A (Table 2). Purified smooth muscle myosin light chain kinase is phosphorylated by protein kinase C (Ikebe et al., 1985; Nishikawa et al., 1985). Although there is disagreement about the sites of phosphorylation, both groups of investigators found that phosphorylation reduced the affinity of myosin light chain kinase for Ca^{2+}/calmodulin. The addition of PDBu, an activator of protein kinase C, to tracheal smooth muscle resulted in substantial phosphorylation of myosin light chain kinase. However, there was no phosphorylation of phosphopeptide A and there was no decrease in the myosin light chain kinase activity ratio. The possibility exists that protein kinase C may activate another kinase that phosphorylates myosin light chain kinase. However, the function of these phosphorylations is not clear since there is not a significant decrease in the myosin light chain kinase activity ratio.

SUMMARY

Purified myosin light chain kinase from smooth muscle is phosphorylated by cyclic AMP-dependent protein kinase, protein kinase C and the multifunctional calmodulin-dependent protein kinase II. Since phosphorylation in a specific site (site A) by any one of these kinases desensitizes myosin light chain kinase to activation by Ca^{2+}/calmodulin, kinase phosphorylation could play an important role in regulating smooth muscle contractility. This possibility was investigated in ^{32}P-labelled bovine tracheal smooth muscle. Treatment of tissues with carbachol, KCl, isoproterenol, or phorbol 12,13-dibutyrate increased the extent of kinase phosphorylation. Six primary phosphopeptides (A-F) of myosin light chain kinase were identified. Site A was phosphorylated to an appreciable extent only with carbachol or KCl, agents which contract tracheal smooth muscle. The extent of site A phosphorylation correlated to increases in the concentration of Ca^{2+}/calmodulin required for activation. These results show that cyclic AMP-dependent protein kinase and protein kinase C do not affect smooth muscle contractility by phosphorylating site A in myosin light chain kinase. It is proposed that phosphorylation of myosin light chain kinase in site A, perhaps by calmodulin-dependent protein kinase II, may play a role in reported desensitization of contractile elements in smooth muscle to activation by Ca^{2+}.

REFERENCES

Adelstein, R. S., Conti, M. A., Hathaway, D. R., and Klee, C. B., 1978, Phosphorylation of smooth muscle myosin light chain kinase by the catalytic subunit of adenosine 3':5'-monophosphate-dependent protein kinase, *J. Biol. Chem.*, 253: 8347.

Bhalla, R. C., Sharma, R. V., and Gupta, R. C., 1982, Isolation of two myosin light-chain kinases from bovine carotid artery and their regulation by phosphorylation mediated by cyclic AMP-dependent protein kinase, *Biochem. J.*, 203: 583.

Cohen, P., 1989, The structure and regulation of protein phosphatases, *Ann. Rev. Biochem.*, 58: 453.

Conti, M. A. and Adelstein, R. S., 1981, The relationship between calmodulin binding and phosphorylation of smooth muscle myosin kinase by the catalytic subunit of 3':5' cAMP-dependent protein kinase, *J. Biol. Chem.*, 256: 3178.

de Lanerolle, P., Nishikawa, M., Yost, D. A., and Adelstein, R. S., 1984, Increased phosphorylation of myosin light chain kinase after an increase in cyclic AMP in intact smooth muscle, *Science*, 223: 1415.

Hashimoto, Y. and Soderling, T. R., 1990, Phosphorylation of smooth muscle myosin light chain kinase by Ca^{2+}/calmodulin-dependent protein kinase-II. Comparative study of the phosphorylation sites, *Arch. Biochem. Biophys.*, 278: 41.
Higashi, K., Fukunaga, K., Matsui, K., Maeyama, M., and Miyamoto, E., 1983, Purification and characterization of myosin light-chain kinase from porcine myometrium and its phosphorylation and modulation by cyclic AMP-dependent protein kinase, *Biochim. Biophys. Acta*, 747: 232.
Himpens, B., Matthijs, G., and Somlyo, A. P., 1989, Desensitization to cytoplasmic Ca^{2+} and Ca^{2+} sensitivities of guinea-pig ileum and rabbit pulmonary artery smooth muscle, *J. Physiol.*, 413: 489.
Ikebe, M., Inagaki, M., Kanamaru, K., and Hidaka, H., 1985, Phosphorylation of smooth muscle myosin light chain kinase by Ca^{2+}-activated, phospholipid-dependent protein kinase, *J. Biol. Chem.*, 260: 4547.
Kamm, K. E. and Stull, J. T., 1985, The function of myosin and myosin light chain kinase phosphorylation in smooth muscle, *Ann. Rev. Pharmacol. Toxicol.*, 25: 593.
Kamm, K. E., Hsu, L.-C., Kubota, Y., and Stull, J. T., 1989, Phosphorylation of smooth muscle myosin heavy and light chains: Effects of phorbol dibutyrate and agonists, *J. Biol. Chem.*, 264: 21223.
Kamm, K. E. and Stull, J. T., 1989, Regulation of smooth muscle contractile elements by second messengers, *Ann. Rev. Physiol.*, 51: 299.
Karaki, H., 1989, Ca^{2+} localization and sensitivity in vascular smooth muscle, *Trends Pharmacol. Sci.*, 10: 320.
Lukas, T. J., Burgess, W. H., Prendergast, F. G., Lau, W., and Watterson, D. M., 1986, Calmodulin binding domains: Characterization of a phosphorylation and calmodulin binding site from myosin light chain kinase, *Biochemistry*, 25: 1458.
Miller, J. R., Silver, P. J., and Stull, J. T., 1983, The role of myosin light chain kinase phosphorylation in β-adrenergic relaxation of tracheal smooth muscle, *Mol. Pharmacol.*, 24: 235.
Morgan, J. P. and Morgan, K. G., 1984, Stimulus-specific patterns of intracellular calcium levels in smooth muscle of ferret portal vein, *J. Physiol.*, 351: 155.
Nishikawa, M., Shirakawa, S., and Adelstein, R. S., 1985, Phosphorylation of smooth muscle myosin light chain kinase by protein kinase C. Comparative study of the phosphorylated sites, *J. Biol. Chem.*, 260: 8978.
Pato, M. D. and Adelstein, R. S., 1983, Characterization of a Mg^{2+}-dependent phosphatase from turkey gizzard smooth muscle, *J. Biol. Chem.*, 258: 7055.
Ratz, P. H. and Murphy, R. A., 1987, Contributions of intracellular and extracellular Ca^{2+} pools to activation of myosin phosphorylation and stress in swine carotid media, *Circ. Res.*, 60: 410.
Rembold, C. M. and Murphy, R. A., 1988, Myoplasmic [Ca^{2+}] determines myosin phosphorylation in agonist-stimulated swine arterial smooth muscle, *Circ. Res.*, 63: 593.
Stull, J. T., Nunnally, M. H., and Michnoff, C. H., 1986, Calmodulin-dependent protein kinases, *in:* "The Enzymes", E. G. Krebs, P. D. Boyer, eds., Academic Press, Orlando, p. 113.
Stull, J. T., Hsu, L.-C., Tansey, M. G., and Kamm, K. E., 1990, Myosin light chain kinase phosphorylation in tracheal smooth muscle, *J. Biol. Chem.*, 265: 16683.

Sutton, T. A. and Haeberle, J. R., 1990, Phosphorylation by protein kinase C of the 20,000-dalton light chain of myosin in intact and chemically skinned vascular smooth muscle, *J. Biol. Chem.*, 265: 2749.

Vallet, B., Molla, A., and Demaille, J. G., 1981, Cyclic adenosine 3',5'-monophosphate-dependent regulation of purified bovine aortic calcium/calmodulin-dependent myosin light chain kinase, *Biochim. Biophys. Acta*, 674: 256.

Walsh, M. P., Hinkins, S., Flink, I. L., and Hartshorne, D. J., 1982, Bovine stomach myosin light chain kinase: Purification, characterization, and comparison with the turkey gizzard enzyme, *Biochemistry*, 21: 6890.

Yagi, S., Becker, P. L., and Fay, F. S., 1988, Relationship between force and Ca^{2+} concentration in smooth muscle as revealed by measurements on single cells, *Proc. Nat'l. Acad. Sci. U.S.A.*, 85: 4109.

MYOSIN HEAVY CHAIN ISOFORMS AND SMOOTH MUSCLE FUNCTION

Richard J. Paul, Timothy E. Hewett, and Anne F. Martin

Department of Physiology and Biophysics
Department of Pharmacology and Cell Biophysics
The University of Cincinnati
College of Medicine
Cincinnati, OH 45267

INTRODUCTION

Isoforms of striated muscle myosin have a long history. They have been extensively studied on biochemical and functional levels and more recently in terms of molecular biology. Under various environmental stresses the architecture of the contractile apparatus of striated muscle can be remodelled to meet the imposed demands.

A striking feature of smooth muscle is its functional diversity. While skeletal fast and slow muscle may differ in maximum shortening velocities and ATPase activities by two-fold (Ranatunga, 1982), smooth muscles can show a forty-fold range (Hellstrand and Paul, 1982). Moreover, smooth muscle is a very plastic tissue. Uterine muscle, for example, may triple its size over a short period of time in response to estrogen treatment (Csapo, 1949). The question arises as to whether any of these functional changes can be attributed to myosin isoforms, and if so, how is their expression controlled?

Isoforms of myosin heavy chains of smooth muscle do not have an extensive literature but have recently been a subject of considerable research interest, and concomitant controversy. Part of the controversy can be attributed to techniques. The pyrophosphate-polyacrylamide gel electrophoretic system of Hoh et al. (1977) has been widely used for separation of isoforms of myosin in many striated muscles. However, application of this technique to smooth muscle was associated with a number of artifacts, including inadequate separation, contamination with filamin (Rovner et al. 1986b), and differential mobilities attributable to the level of myosin phosphorylation (Trybus and Lowey, 1985; Persechini et al. 1986).

While the separation and identification of isoforms of native myosin still remains a goal (Takano-Ohmuro et al., 1983; Beckers-Bleukx and Marechal, 1985; Lema et al., 1986; Rovner et al., 1986b), there is considerable evidence for the presence of heavy chain isoforms. Two smooth muscle myosin heavy chain species have been identified by several laboratories on low density SDS gels (Rovner et al., 1986b; Kawamoto and Adelstein, 1987; Hewett and Martin, 1988). These have been termed SM1 (204 kDa) and SM2 (200 kDa) and have been distinguished from a non-muscle, platelet-type myosin isoform (196 kDa) through the use of isoform specific antibodies (Rovner et al. 1986a, Kawamoto and Adelstein 1987). Based on similar immunological evidence, a third potential smooth muscle myosin heavy chain (with an electrophoretic mobility slightly greater than SM2) has been reported in human smooth muscles (Sartore et al., 1989). Recently a full length cDNA for the myosin heavy chain of chicken gizzard (Yanagisawa et al., 1987) and a partial cDNA for the myosin heavy chain of rabbit uterine muscle have been cloned (Nagai et al., 1989). The evidence is consistent with two smooth muscle myosin heavy chain isoforms arising from alternative splicing at the C-terminus. This differs from skeletal muscle in which several distinct gene families for myosin are known to exist (Wydro et al. 1983).

FUNCTIONAL SIGNIFICANCE OF MYOSIN HEAVY CHAIN ISOFORM DISTRIBUTION

Given the Existence of Smooth Muscle Myosin Heavy Chain Isoforms, Do They Have Any Functional Significance?

Earlier reports suggested that SM1 and SM2 occurred in virtually equal proportions (Cavaille et al., 1986; Rovner et al., 1986b) and it was proposed that the heavy chains were always present as a heterodimer in the native myosin molecule. If this were the case, then the functional significance of myosin heavy chain isoforms would be obscure. However, a number of recent reports indicate that the proportions are unequal in a variety of tissues. The ratio of SM1/SM2 is reported to be about 2.5:1 in rat uterus (Hewett et al., 1988; Sparrow et al., 1988), 2.1:1 in rat tail artery (Packer, 1990), 1:1 in hog carotid artery (Rovner et al., 1986b), 0.7:1 in human trachea (Mohammad and Sparrow, 1989), and 0.6:1 in turkey gizzard (Kawamoto and Adelstein, 1987). These wide variations in tissue distribution of myosin heavy chain isoforms suggests the possibility of a functional role.

Given the Large Number of Factors Which May Underlie Functional Differences, How Can One Attribute Specific Differences to Myosin Isoform Distribution?

While the differences in isoform distribution are quite dramatic between, for example, turkey gizzard and rat uterus, any basis for associating this with functional differences would be highly speculative. The best approach available would appear to be a comparison of functional properties in the same tissue in which isoform distribution has been altered.

To this end we utilized a uterine model, capitalizing on the reported dramatic changes in uterine mass with estrogen treatment. The results are graphically summarized in Figure 1. Uteri from ovariectomized rats increased approximately 3-fold in mass following 3 days of β-estradiol treatment. Myosin heavy chain distribution was also significantly shifted from a SM1:SM2 ratio of 1.7 to 2.6 after estrogen treatment. We studied the mechanics in permeabilized fibers from these uteri, in order to avoid potential differences associated with activation processes (Meisheri et al., 1985). We found that active isometric force increased 3-fold in agreement with previous reported values and parallel increases in actomyosin content (Csapo, 1950). Importantly, in terms of isoform function, we found that the maximum velocity increased over 2-fold after estrogen treatment. Using standard ANOVA analysis, we found that the

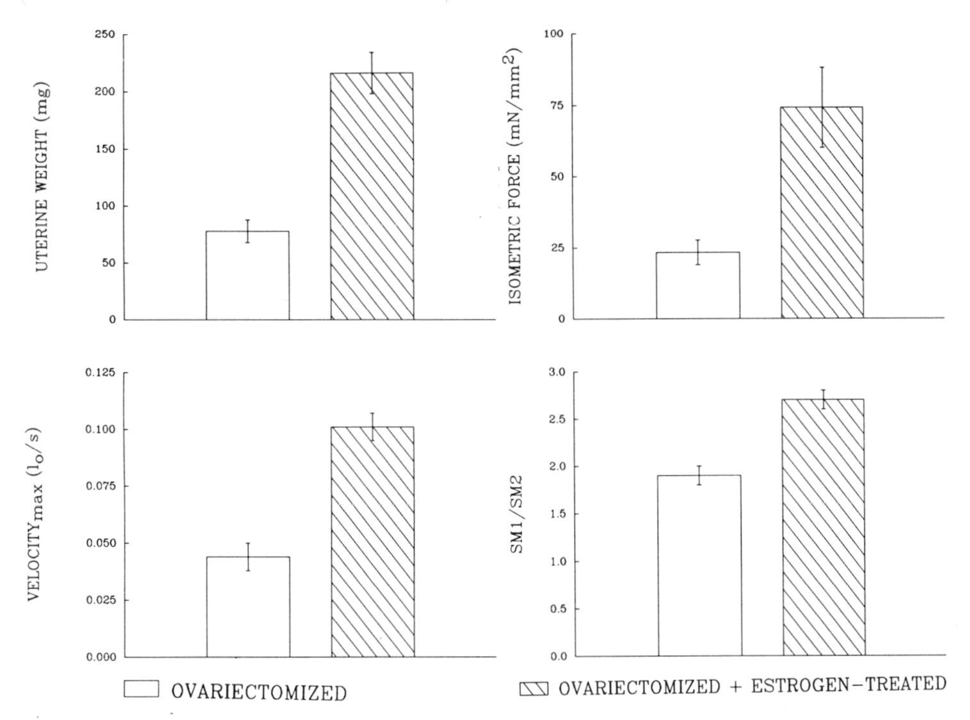

Figure 1. Changes in uterine mass, isometric force, unloaded shortening velocity, and myosin isoform distribution, following treatment of ovariectomized rats with β-estradiol. Female Sprague-Dawley rats (180 - 200g) were ovariectomized and two to four weeks post surgery were injected intramuscularly for 3 days with 2 μg/kg/day β-estradiol in sesame seed oil. Skinned fibers were prepared according to the method of Briggs (1963).

increase in velocity was significantly correlated with the %SM1 isoform, but not with the increase in force generating capacity. This would be anticipated if the increase in force reflected the increase in mass and actomyosin content, and velocity were related to isoform composition. The latter conclusion is tempting as striated muscle isoforms are associated with different ATPases and shortening velocities.

There is relatively little available data in which both myosin isoform distribution and mechanical parameters have both been measured. In one such study, Sparrow and colleagues (1988) compared changes in rat uterine muscle with pregnancy. They did not report significant changes in myosin heavy chain distribution. However, as shown in Figure 2, the relation between V_m and %SM1 was consistent with the changes we found using a different model system to alter myosin isoforms. On the other hand, Packer (1990) reported a small and non-significant decrease in the %SM1 isoform but it was associated with an increase (23%) in velocity in the caudal artery of SHR rats compared to that of WKY controls.

The changes observed in %SM1 content in these uterine muscle models would appear at first to be relatively small in comparison to the changes in velocity. The regression shown in Figure 2 indicates that an increase of 0.008 l_o/s is associated with each 1% increase in SM1. This is in fact consistent with results obtained in striated muscle, where, for example, Reiser et al. (1985) demonstrated a 0.022 l_o/s increase for each 1% increase in fast-type myosin heavy chain in single rabbit skeletal muscle fibers. Thus our data suggest that a

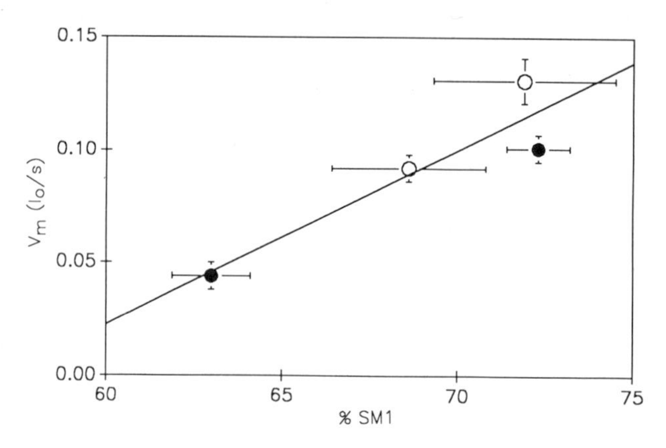

Figure 2. Relation between unloaded shortening velocity (V_m) and the percent SM1 myosin heavy chain isoform in skinned rat uterine muscle. Open circles: data from Sparrow et al. (1988), in which an increase in SM1 isoform was associated with pregnancy. Filled circles: data from this study, using uterine muscle from ovariectomized and ovariectomized plus estrogen treated rats. Correlation coefficient, r, = 0.93.

functional relation between myosin isoforms and dynamic mechanical parameters similar to that in skeletal muscle may also be present in smooth muscle. Further tests of this hypothesis rest on development of a model system for smooth muscle in which the myosin isoform distribution can be varied over a larger range.

Though interesting, correlations cannot be used to prove causality. If our hypothesis is correct, we would also anticipate isoform specific differences in actomyosin ATPase activities

Thus we have used this rat uterine model in corresponding studies of the enzymic properties of the myosin heavy chains. Our goal was to correlate changes in actomyosin ATPase with functional changes associated with changes in isoform distribution. However, even in the presence of an antiprotease cocktail, we noted that during the course of actomyosin preparation, there was an increase in the amount of material in the SM2 band compared to that observed in rapidly frozen tissue on low density SDS gels. To help assess the role of proteolysis in altering the electrophoretic mobility of myosin heavy chains and subsequent identification of isoforms, we have developed antibodies specific for SM1 and SM2. Based on the cDNA sequence (Yanagisawa et al., 1987; Nagai et al., 1989), differences in the protein composition of SM1 and SM2 have been identified at the C-terminal region of the molecule, with SM1 having an additional sequence of approximately 4 kDa. We thus made synthetic peptides corresponding to diverging sequences in this region and raised antibodies to these peptides. The resulting antibodies were monospecific, reacting only to their corresponding isoform on immunoblots of SDS gels. Using our isoform specific antibodies, we confirmed that a limited proteolysis of approximately 4 kDa at the C-terminal end of SM1 produced a cleaved SM1 (pSM1) which comigrated with SM2.

While such limited proteolysis is not unprecedented, it raises questions as to the unambiguous assignment of isoforms based only on electrophoretic migration patterns. However most surprisingly, the cleavage of this 4 kDa C-

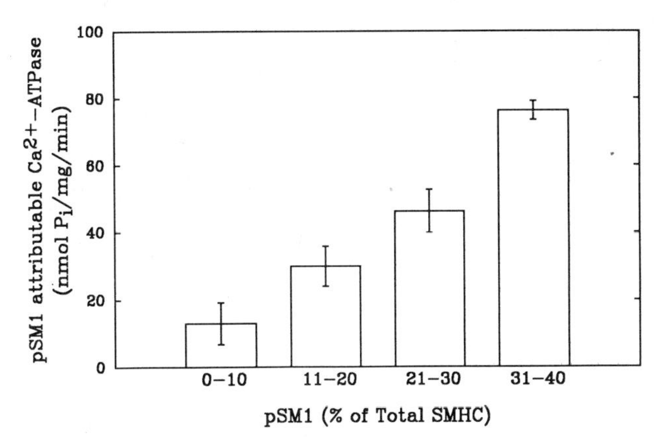

Figure 3. Relation between the actomyosin ATPase activity and cleaved SM1 myosin species (pSM1). The actomyosin (0.1 mg/ml) Ca^{2+}-ATPase activity was measured at 30° C in 500 mM KCl, 30 mM imidazole (pH 7.5), 5 mM ATP and 10 mM $CaCl_2$ as adapted from Martin et al. (1982). The percent of pSM1 in the actomyosin was calculated from the increase in the density of the material in the apparent SM2 (200 kDa) band on SDS gels in comparison to that observed in freeze-clamped uterine tissue, from which the actomyosin was prepared. SMHC refers to smooth muscle myosin heavy chain.

terminal tail-piece was associated with a dramatic increase in actomyosin ATPase activity (Hewett and Martin, 1988). As shown in Figure 3, the ATPase of actomyosin containing approximately 30 - 40% of the cleaved SM1, pSM1 species had an activity some two-fold greater than that corresponding to the native (uncleaved) myosin isoform distribution.

Given that the enzymic site for ATPase activity is at the N-terminal, globular region of the myosin molecule, a C-terminal cleavage which affected the ATPase activity was quite unexpected

However, a similar limited proteolysis could explain earlier data suggesting that limited proteolysis activates uterine ATPase activity (Needham and Williams, 1959). Our results suggest that a relatively small domain at the C-terminal end may modulate smooth muscle myosin ATPase activity. While the mechanism(s) for this activation is unknown, phosphorylation at the C-terminal end of non-muscle myosin is known to regulate filament formation and subsequent ATPase activity (Korn and Hammer, 1989). Whether this limited proteolysis represents a physiologically significant mechanism for activation in mammalian smooth muscle is unknown. However, our results would indicate that intramolecular interactions involving the C-terminal end of myosin are worth exploring as a regulatory mechanism.

SUMMARY

Using isoform specific antibodies we have verified the presence of two distinct muscle type myosin heavy chain isoforms in rat uterine muscle. We have shown that an endogenous protease can cleave a small 4 kDa region from the C-terminal of the SM1 isoform which generates a pSM1 species which comigrates with the SM2 isoform on low density SDS gels. While this cleavage can complicate isoform identification, more importantly, this cleavage was associated with a substantial increase in the actomyosin ATPase. Thus we have identified a domain at the C-terminal which may be involved in regulation of the ATPase activity. Interestingly, it is at this C-terminal, tail region of the smooth muscle myosin molecule where the only known isoform specific sequence differences are located. In skinned smooth muscle fibers of rat uterine muscle, we have also shown that differences in myosin heavy chain distribution, induced by β-estradiol treatment of ovariectomized rats, are correlated with changes in unloaded shortening velocity. Thus our work suggests that the functional significance of myosin heavy chain isoforms in smooth muscle may be similar to that observed in striated muscle.

ACKNOWLEDGMENTS

Supported in part by HL22619 and HL07571.

REFERENCES

Beckers-Bleukx, G. and Marechal, G., 1985, Detection and distribution of myosin isozymes in vertebrate smooth muscle, *Eur. J. Biochem.*, 152: 207.

Briggs, A. H., 1963, Characteristics of contraction in glycerinated uterine smooth muscle, *Am. J. Physiol.*, 204: 739.

Cavaille, F., Janmot, C., Ropert, S., and d'Albis, A., 1986, Isoforms of myosin and actin in human, monkey and rat myometrium. Comparison of pregnant and non-pregnant uterus proteins, *Eur. J. Biochem.*, 160: 507.

Csapo, A., 1949, Actomyosin content of the uterus, *Nature*, 162: 218.

Csapo, A., 1950, Actomyosin of the uterus, *Am. J. Physiol.*, 160: 46.

Hellstrand, P. J. and Paul, R. J., 1982, Vascular smooth muscle: Relations between energy metabolism and mechanics, *in:* "Vascular Smooth Muscle Metabolic, Ionic and Contractile Mechanisms", M. F. Crass III and C. D. Barnes, eds., Academic Press, New York, p. 1.

Hewett, T. E. and Martin, A. F., 1988, Changes in myosin heavy chains and ATPase activity in E_2 treated rat uterus, *Biophys. J.*, 53: 577a.

Hewett, T. E., Martin, A. F., and Paul, R. J., 1988, Alterations in mechanical parameters and myosin heavy chain species in E_2 treated rat uterus, *FASEB J.*, 2: A334.

Hoh, J. F. Y., McGrath, P. A., and Hale, P. T., 1977, Electrophoretic analysis of multiple forms of rat cardiac myosin: Effects of hypophysectomy and thyroxine replacement, *J. Mol. Cell. Cardiol.*, 10: 1053.

Kawamoto, S. and Adelstein, R. S., 1987, Characterization of myosin heavy chains in cultured aorta smooth muscle cells, *J. Biol. Chem.*, 262: 7282.

Korn, E. D. and Hammer III, J. A., 1989, Myosins of non-muscle cells, *Ann. Rev. Biophys. Biophys. Chem.*, 17: 23.

Lema, J. J., Pagani, E. D., Shemin, R., and Julian, F. J., 1986, Myosin isozymes in rabbit and human smooth muscles, *Circ. Res.*, 59: 115.

Martin, A. F., Pagani, E. D., and Solaro, R. J., 1982, Thyroxine induced redistribution of isoenzymes of rabbit ventricular myosin, *Circ. Res.*, 50: 117.

Meisheri, K. D., Rüegg, J. C., and Paul, R. J., 1985, Smooth muscle contractility: Studies on skinned fiber preparations, *in:* "Calcium and Smooth Muscle Contractility", E. E. Daniels and A. K. Grover, eds., Human Press, New York, p. 191.

Mohammad, M. A. and Sparrow, M. P., 1989, The distribution of heavy-chain isoforms of myosin in airways smooth muscle from adult and neonate humans, *Biochem. J.*, 260: 421.

Nagai, R., Kuro-o, M., Babij, P., and Periasamy, M., 1989, Identification of two types of smooth muscle myosin heavy chain isoforms by cDNA cloning and immunoblot analysis, *J. Biol. Chem.*, 264: 9734.

Needham, D. M. and Williams, J. M., 1959, Some properties of uterus actomyosin and myofilaments, *Biochem. J.*, 73: 171.

Packer C. S., 1990, Myosin heavy chain isoform pattern is not altered in arterial muscle from the spontaneously hypertensive rat, *in:* "Frontiers in Smooth Muscle Research", N. Sperelakis and J. D. Wood, eds., Wiley-Liss, New York, p. 575.

Persechini, A., Kamm, K. E., and Stull, J. T., 1986, Different phosphorylated forms of myosin in contracting tracheal smooth muscle, *J. Biol. Chem.*, 261: 6293.

Ranatunga, K. W., 1982, Temperature dependence of shortening velocity and rate of isometric tension development in rat skeletal muscle, *J. Physiol.*, 329: 465.

Reiser, P. J., Moss, R. L., Guilian, G. G., and Greaser, M. L., 1985, Shortening velocity and myosin heavy chains of developing rabbit muscle fibers, *J. Biol. Chem.*, 260: 9077.

Rovner, A. S., Murphy, R. A., and Owens, G. K., 1986a, Expression of smooth muscle and non-muscle myosin heavy chains in cultured vascular smooth muscle cell, *J. Biol. Chem.*, 261: 14740.

Rovner, A. S., Thompson, M. M., and Murphy, R. A., 1986b, Two different heavy chains are found in smooth muscle myosin, *Am. J. Physiol.*, 250: C861.

Sartore, S., De Marzo, N., Borrione, A. C., Zanellato, A. M. C., Saggin, L., Fabbri, L., and Schiaffino, S., 1989, Myosin heavy-chain isoforms in human smooth muscle, *Eur. J. Biochem.*, 179: 79.

Sparrow, M. P., Mohammad, M. A., Arner, A., Hellstrand, P., and Rüegg, J. C., 1988, Myosin composition and functional properties of smooth muscle from uterus of pregnant and non-pregnant rats, *Pflügers Arch.*, 412: 624.

Takano-Ohmuro, H., Obinata, T., Mikawa, T., and Masaki, T., 1983, Changes in myosin isozymes during development of chicken gizzard muscle, *J. Biochem.*, 93: 903.

Trybus, K. M. and Lowey, S., 1985, Mechanism of smooth muscle myosin phosphorylation, *J. Biol. Chem.*, 260: 15988.

Wydro, R. M., Nguyen, H. T., Gubits, R. M., and Nadal-Ginard, B., 1983, Characterization of sarcomeric myosin heavy chain genes, *J. Biol. Chem.*, 285: 670.

Yanagisawa, M., Hamada, Y., Katsuragawa, Y., Imamura, M., Mikawa, T., and Masaki, T., 1987, Complete primary structure of vertebrate smooth muscle myosin heavy chain deduced from its complementary cDNA sequence, *J. Mol. Biol.*, 198: 143.

MECHANICAL AND BIOCHEMICAL EVENTS DURING HYPOXIA-INDUCED RELAXATIONS OF RABBIT AORTA

Suzanne Moreland,[1,3] Ronald F. Coburn,[2] Carl B. Baron,[2] and Robert S. Moreland[2,3]

[1]Department of Pharmacology
Bristol-Myers Squibb Pharmaceutical Research Institute
Princeton, NJ 08543

[2]Department of Physiology
University of Pennsylvania
Philadelphia, PA 19104

[3]Bockus Research Institute
Graduate Hospital
Philadelphia, PA 19146

INTRODUCTION

Tissue hypoxia is involved both in chronic pathophysiological changes such as stroke, myocardial infarction, and renal ischemia and in acute physiological or environmental changes such as exposure to high altitudes. A current challenge in medicine and physiology is to better understand how cellular functions can be preserved during hypoxic events. One of the primary and initial aspects of cellular function affected by hypoxia is energy production and delivery. Energy delivery in cells is dependent on a variety of factors including the rates and sites of energy production, the transport of high energy compounds, and the local environment around energy-consuming reactions.

It has been known for some time that contraction of vascular smooth muscle is tightly coupled to energy delivery (for review see Paul, 1990). The primary factors involved in this energy delivery can be manipulated and studied in the vascular smooth muscle cell in an attempt to better understand the cellular response to tissue hypoxia.

Evidence that hypoxia-induced relaxations of vascular smooth muscle are due to a decrease in energy delivery to specific energy-dependent reactions involved in force maintenance has been summarized elsewhere (Scott and Coburn, 1989; Ishida and Paul, 1990; Marriott and Marshall, 1990a,b). As depicted in Figure 1, there are several energy-dependent reactions that might be involved in linking force with oxidative energy production. These include, but are not limited to, myosin light chain kinase, actin-activated myosin ATPase, ATP-dependent plasmalemmal ion channels and pumps, and reactions in the inositol phospholipid transduction mechanism. The object of this chapter is to

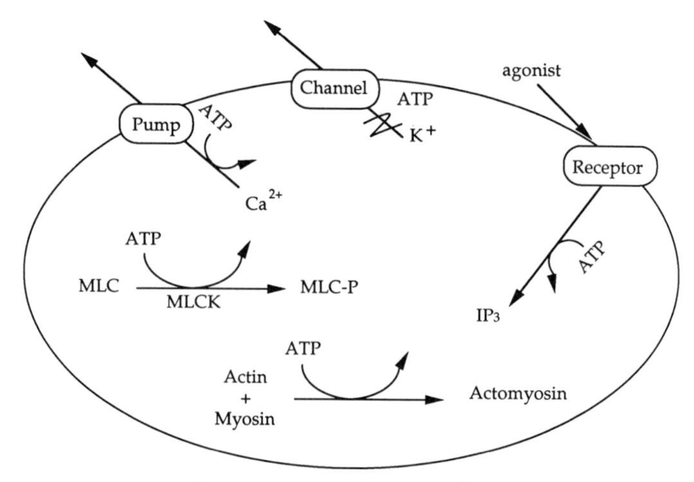

Figure 1. Model of a vascular smooth muscle cell showing possible sites of energy-dependent reactions affected by hypoxia.

review recent data obtained on the relationship between energy stores, force, and MLC phosphorylation (Coburn et al., 1991). In addition, new observations on the effect of glyburide on hypoxia-induced relaxations, O_2-induced redevelopment of force, and information on the stiffness and velocity during both relaxation and force redevelopment will be discussed. The central question asked is: Which of the ATP-dependent processes (ATP-dependent K^+ channels, myosin light chain kinase, and actin-activated myosin ATPase) could be involved in the coupling between energy production and force (oxidative metabolism-contraction coupling) in vascular smooth muscle.

MATERIALS AND METHODS

Male New Zealand white rabbits were sacrificed and the thoracic aortae were carefully removed, cleaned of fat and connective tissue, and mounted for isometric or isotonic recording in individual muscle chambers. The tissues were bathed in a bicarbonate buffered physiological salt solution (PSS) at 37° C and pH 7.4. The composition of the PSS was (in mM): 110 NaCl, 5.9 KCl, 26.1 $NaHCO_3$, 1.2 $MgCl_2$, 1.2 NaH_2PO_4, 2.0 $CaCl_2$, and 11.5 D-glucose.

All of the experiments in this study were conducted under conditions in which the tissue aeration was changed from a gas mixture containing O_2 to one in which N_2 was substituted for all O_2. Both gas mixtures contained equal CO_2 contents to maintain a constant pH. Following equilibration of the tissues and adjustment to L_{max}, the length for optimal active force development, contractions were elicited with either 30 μM norepinephrine or 50 mM KCl. For the biochemical determinations, rings of thoracic aortae were stimulated for 5 min under normoxic conditions, then hypoxia was induced. At various times during the stimulation, the chambers were rapidly dropped and the tissues freeze-clamped between large tongs precooled in liquid N_2. For the mechanical determinations, circumferential strips were stimulated for 15 min prior to induction of hypoxia. Force was recorded continuously in all experiments.

The freeze-clamped rings were analyzed for phosphocreatine (PCr) content by HPLC within 15 min of freezing (Scott et al., 1987). ATP determinations were performed on perchlorate homogenized samples stored at -60° C overnight. PCr and ATP concentrations are reported as mM cell water as described in Coburn et al. (1991). Free creatine was measured as described by Scott et al. (1987). Myosin light chain phosphorylation levels were determined by two-dimensional gel electrophoresis on tissue homogenates and quantitated by densitometric scans of the Coomassie blue stained gels as described previously (Aksoy et al., 1983; Moreland and Moreland, 1987). The data presented in this chapter on energy stores, force, and MLC phosphorylation may be found in greater detail in Coburn et al. (1991).

Isotonic shortening velocity was estimated during isotonic quick releases to afterloads of 0.15 times the force at the time of release using a computer controlled Cambridge Technology 300H servo-lever. The natural logarithm of the change in length between 0.5 and 1 sec of release was used to calculate velocity. Total dynamic stiffness was estimated by imposing a rapid 1% L_{max} stretch (>100 times maximal shortening velocity) at various times during tissue stimulation. At the end of each experiment, the tissues were exposed to PSS containing 2 mM EGTA for a minimum of 2 hours or until no response to either KCl or norepinephrine could be elicited. An identical stretch was then imposed on these calcium-depleted tissues and taken as passive stiffness. Passive stiffness was subtracted from all experimental stretches to yield active dynamic stiffness as described by Singer and Murphy (1987).

RESULTS AND DISCUSSION

A change in the aeration mixture from O_2 to N_2 was associated with a relatively rapid decrease in the PO_2 of the PSS bathing the tissues (Fig. 2). The PO_2 in the tissue baths, monitored by a polargraphic electrode, decreased to less than 20 mm Hg with a half-time of 35 sec. Within 50 sec, the PO_2 decreased

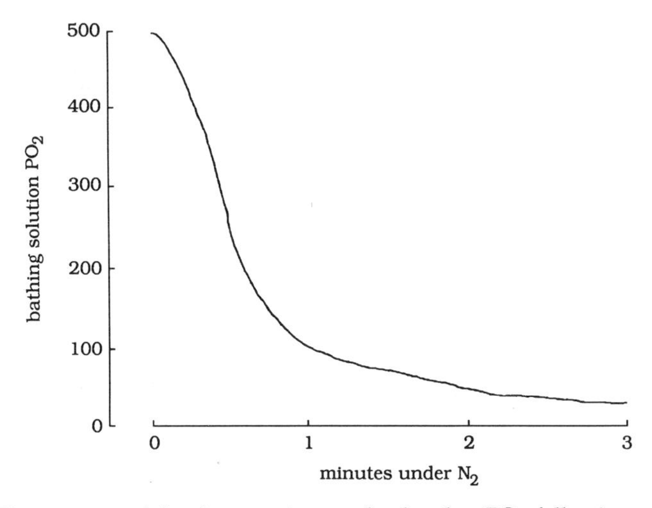

Figure 2. Time course of the decrease in muscle chamber PO_2 following a change in aeration from CO_2/O_2 to CO_2/N_2.

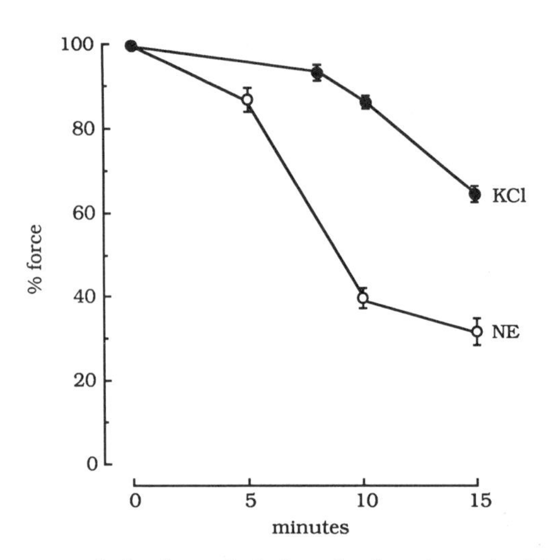

Figure 3. Time course of the hypoxia-induced relaxation of rabbit thoracic aorta. Tissues were stimulated with 30 μM norepinephrine (NE) or 50 mM KCl (KCl). Data are plotted as mean ± SEM; n = at least 5 tissues from different rabbits.

below the threshold for onset of relaxation. This rapid inhibition of oxidative energy production was brought about so that measurements could be made under conditions where total energy stores were slowly decreasing at a rate determined by energy consumption and glycolytic energy production. An alternative approach would have been to perform studies at different steady state levels of PO_2. Under those conditions, the outer layer of cells in the preparation may have remained normoxic while the anoxic core of the preparation would have increased in size as the PO_2 in the bathing solution decreased. Thus, the method we chose lead to less heterogeneity in the energy status of the cells.

Under hypoxic conditions, mean tissue PCr and ATP concentrations decreased at initial rates of 0.06 and 0.09 mM/min, respectively. Contractions elicited by norepinephrine relaxed rapidly and almost completely in response to hypoxia and therefore the decrease in PCr and ATP concentrations. In contrast, contractions generated in response to high extracellular concentrations of KCl relaxed more slowly and to a lesser extent (Fig. 3). For norepinephrine stimulated tissues, the threshold values for relaxation were 0.47 mM PCr and 1.10 mM ATP; 50% relaxation occurred at 0.32 and 0.58 mM PCr and ATP, respectively. For KCl induced contractions, the threshold levels were 0.24 mM PCr and 1.0 mM ATP; 50% relaxation occurred at less than 0.1 mM PCr and 0.1 mM ATP.

Our goal was to determine the energy-dependent reaction that links oxidative energy production to force. The data have allowed us to make some tentative conclusions about various energy-dependent reactions. We studied norepinephrine- and KCl-evoked force because different mechanisms are involved in force development and maintenance. The finding of markedly different threshold concentrations of PCr and ATP associated with relaxation of the two different types of contractions suggests different rate-limiting energy-dependent reactions are operating. Each of the energy-dependent reactions depicted in Figure 1 was examined for its potential role in the hypoxic relaxation.

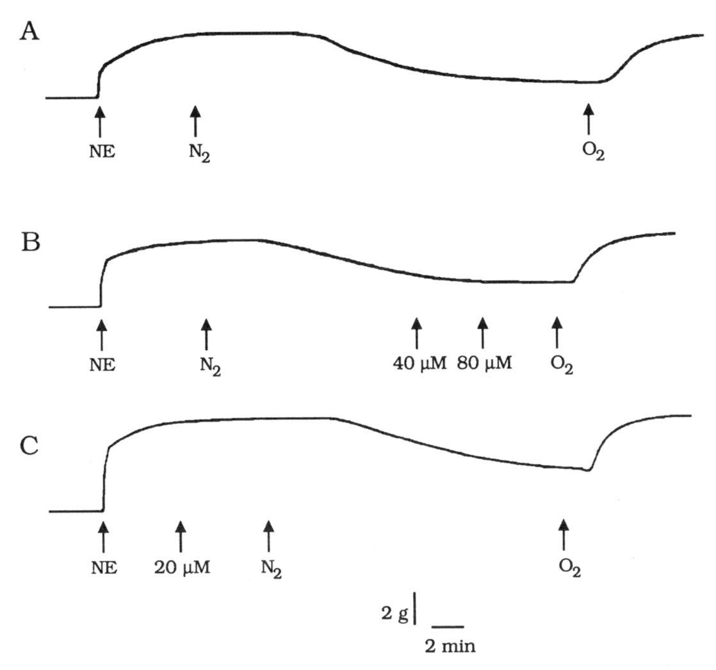

Figure 4. Effect of glyburide on hypoxia-induced relaxation. Tissues were stimulated with norepinephrine (NE), then, at the plateau of force, they were made hypoxic (N_2). Glyburide (G) was administered in the concentrations indicated either during (panel B) or before (panel C) hypoxia. Neither the hypoxic relaxation nor the redevelopment of force upon return to normoxia (O_2) was changed compared with control (panel A).

Energy-dependent Ion Channels and Pumps

It is unlikely that Ca^{2+}-ATPases in the plasma membrane and/or sarcoplasmic reticulum are involved in oxidative metabolism-contraction coupling. Inhibition of either of these calcium pumps would be expected to induce or augment contraction due to the resultant increase in intracellular Ca^{2+} concentration. These ATPase-linked pumps, also, have reportedly high affinities for ATP and are unlikely to be energy-limited under the conditions of these studies (Popescu and Ignat, 1983; Carafoli, 1984).

The ATP-dependent K^+ channels are, however, possible candidates for the energy-sensor. Activators of these channels have been shown to concentration-dependently relax contractions in response to norepinephrine, however KCl contractions are relatively resistant to their effects (Weir and Weston, 1986). Thus, activation of these channels in response to a decrease in intracellular ATP concentration would be expected to relax norepinephrine contractions and to have a much smaller effect on the KCl contractions. However, as shown in Figure 4, hypoxic relaxation of norepinephrine-induced force was not affected by glyburide, an inhibitor of these ATP-dependent K^+ channels in vascular smooth muscle (Standen et al., 1989). This finding argues against the operation of these channels during oxidative metabolism-contraction coupling. Further, O_2-evoked force redevelopment was not dependent on extracellular Ca^{2+} (data not shown). Thus, operation of ATP-dependent K^+ channels or energy-dependent Ca^{2+} influx such as occurs in some smooth muscles (Ashoori et al., 1984; Pierce, 1989) cannot be invoked to explain these findings.

Our data show that a glyburide-sensitive ATP-dependent channel is not involved in oxidative metabolism-contraction coupling. The finding that O_2-evoked contractions occur even at a very low extracellular Ca^{2+} concentration tends to exclude other ATP-dependent sarcolemmal channels as participating in oxidative energy-contraction coupling. Because it appears that ion pumps may not be involved in the hypoxia-induced relaxation and in oxidative metabolism-contraction coupling, the remainder of this discussion will focus more directly on the steps involved in contractile protein activation. These are myosin light chain kinase, actin-activated myosin ATPase, and the inositol phosphate cascade.

Myosin Light Chain Kinase

Phosphorylation of serine-19 in the 20,000 M_r myosin light chain is believed to play an important role in the regulation of vascular smooth muscle contraction (for review see Kamm and Stull, 1985), however, it does not act as a simple switch to control either force development or force maintenance (Moreland and Moreland, 1990). Hypoxia did not significantly affect the levels of myosin light chain phosphorylation during hypoxia-induced relaxation of the contractions in response to KCl. In contrast, hypoxia-induced relaxations of norepinephrine contractions were associated with a concomitant decrease in myosin light chain phosphorylation (data not shown).

The KCl data indicate that myosin light chain kinase is not inhibited even when tissue concentrations of PCr and ATP were each 0.1 mM, levels much lower than those associated with hypoxia-induced relaxation of force in response to norepinephrine. It appears that relaxation of KCl-evoked force is not due to energy-limitation of myosin light chain kinase. During norepinephrine-induced force, relaxations are associated with decreases in myosin light chain phosphorylation at mean tissue ATP and PCr levels considerably higher than values seen in the KCl-induced contractions. Therefore, myosin light chain kinase catalyzed phosphorylation of the myosin light chain is not likely to be the energy-limited step in hypoxia-evoked relaxations. (However, see Stull et al., this volume for discussion on alterations of myosin light chain kinase activity.) The decrease in myosin light chain phosphorylation seen during norepinephrine-stimulated force may be due to a decrease in intracellular Ca^{2+}.

Actin-activated Myosin ATPase

The primary utilizer of energy during the development and most likely maintenance of a smooth muscle contraction is the actin-activated myosin ATPase. Actin-activated myosin ATPase activity is the driving force for crossbridge cycling and is the biochemical correlate of isotonic shortening velocity (Bárány, 1967). The hypoxia-induced relaxation of both norepinephrine and KCl-induced contractions may be the result of direct inhibition of actin-activated myosin ATPase activity by the decrease in cellular ATP concentration. Hellstrand and Arner (1985) showed in permeabilized guinea pig taenia coli that crossbridge turnover was inhibited by 50% when the concentration of ATP in the bathing (intracellular) solution was 0.1 mM. In the experiments described herein, KCl-induced force was inhibited by 50% at an ATP concentration of 0.1 mM, suggesting that energy-limitations of the actin-

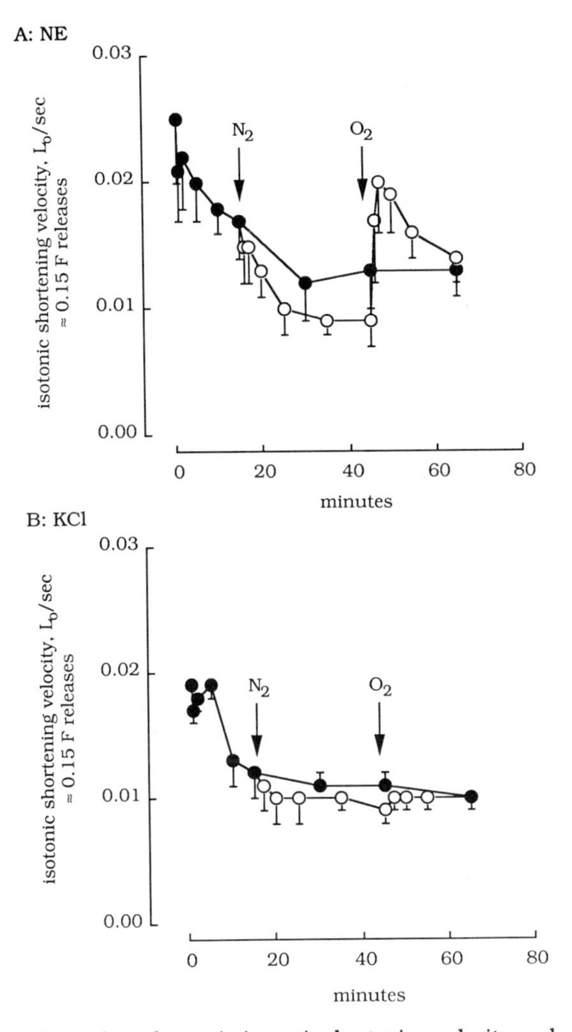

Figure 5. Time dependent change in isotonic shortening velocity under normoxic (filled symbols) and hypoxic (open symbols) conditions. A. Contraction in response to 30 µM NE. B. Contraction in response to 50 mM KCl. Data are plotted as mean ± SEM; n = at least 5 tissues from different rabbits.

activated myosin ATPase might be responsible for the hypoxic relaxations. This possibility is supported by the finding that estimates of isotonic shortening velocity determined during KCl-induced contractions tend to decrease during hypoxia (Fig. 5), but does not appear to account for the hypoxic relaxation of a norepinephrine-induced contraction. Isotonic shortening velocity and therefore crossbridge cycling rate decreases concomitantly with stress although as noted above, the concentration of ATP was 0.58 mM at half-maximal relaxation; significantly greater than the 0.1 mM necessary for energy-limitation at the actin-activated myosin ATPase. Depletion of cellular ATP concentrations, depending on the degree, can also produce a rigor state in smooth muscle (Somlyo et al., 1988). The decrease in stress following hypoxia may in fact be a high tension rigor state and not a true relaxation. If this were the case, then one would expect active stiffness, indicative of the number of

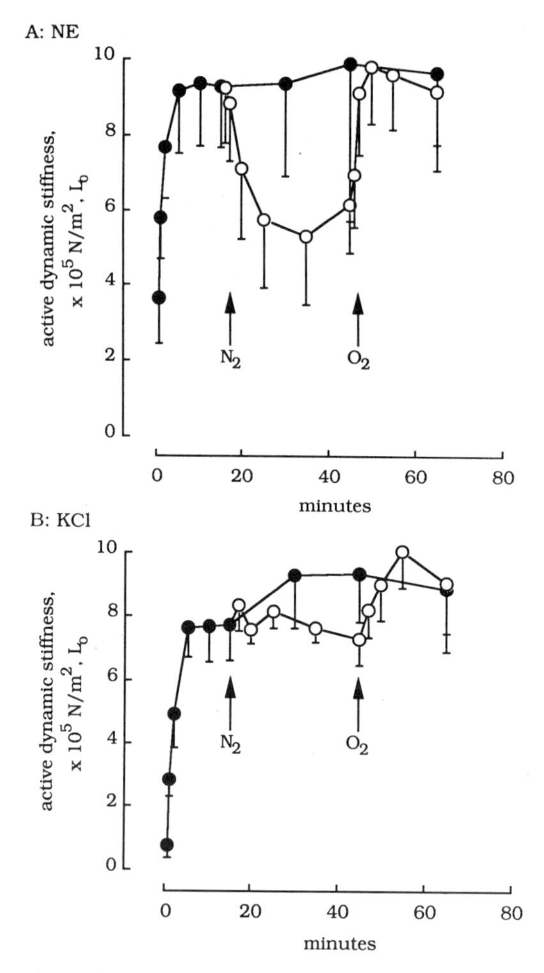

Figure 6. Time dependent change in active dynamic stiffness under normoxic (filled symbols) and hypoxic (open symbols) conditions. A. Contraction in response to 30 μM NE. B. Contraction in response to 50 mM KCl. Data are plotted as mean ± SEM; n = at least 4 tissues from different rabbits.

attached crossbridges, to remain high although stress decreases. Figure 6 shows that stiffness falls in relation to stress demonstrating that rigor bridges are not being formed during hypoxia induced relaxation of norepinephrine- or KCl-induced contractions.

Inositol Phosphate Metabolism

Pharmacomechanical coupling, or the activation of contraction by an agonist in the absence of a change in membrane potential, is generally accepted to be the mechanism by which norepinephrine activates smooth muscle (Somlyo and Somlyo, 1968). KCl contractions are dependent upon electromechanical and not pharmacomechanical coupling. The inositol phospholipid cascade has been implicated as an important player in pharmacomechanical coupling. Inositol 1,4,5-trisphosphate is thought to be the second messenger responsible for the release of Ca^{2+} from intracellular storage sites such as the sarcoplasmic reticulum (Berridge, 1984). The inositol phospholipid transduction system consumes relatively large amounts of ATP (Berridge, 1986) and so is a likely candidate for the energy-limiting step in norepinephrine contractions during hypoxia. That inositol phosphate metabolites are probably not important messengers in KCl contractions suggests an explanation for KCl contractions being relatively resistant to the effects of hypoxia. Coburn et al. (1988) have shown that the hypoxia-induced decrease in metabolic flux rate in norepinephrine-stimulated rabbit aorta was associated with a four- to five-fold increase in the phosphatidylinositol 4-phosphate pool size. This suggests that phosphatidylinositol 4-phosphate kinase is inhibited and rate-limiting under the conditions of these experiments. Thus, a decrease in energy delivery to phosphatidylinositol 4-phosphate kinase can be invoked to explain the large hypoxic relaxations observed in norepinephrine contracted tissues. The lack of importance of this enzyme during KCl contractions may account for the relative resistance of KCl contractions to hypoxia.

SUMMARY

Hypoxic relaxation of norepinephrine contractions of isolated rabbit aorta is rapid, whereas relaxation of KCl contractions is slower and blunted. The data given here suggest that with receptor-evoked contractions of rabbit aorta, the energy-limitation of ATP-dependent K^+ channels and other sarcolemmal channels, myosin light chain kinase, and actin-activated myosin ATPase are probably not involved in oxidative energy-contraction coupling. The data strongly support the hypothesis that the rate limiting, energy-dependent step is upstream to myosin light chain kinase, which is 50% inhibited at an ATP concentration of about 0.5 mM. This energy-dependent step may be in the inositol phospholipid transduction system, as we have previously postulated (Coburn et al., 1988). In contrast the energy-limited reaction during KCl contractions appears to be the actin-activated myosin ATPase which is 50% inhibited at a mean ATP concentration of about 0.1 mM.

ACKNOWLEDGMENTS

This study was supported in part by funds from HL 37956 (RSM) and HL 19737 (RFC).

REFERENCES

Aksoy, M. O., Mras, S., Kamm, K. E., and Murphy, R. A., 1983, Ca^{2+}, cAMP, and changes in myosin phosphorylation during contraction of smooth muscle, *Am. J. Physiol.*, 245: C255.

Ashoori, F., Takai, A., Tokuno, H., and Tomita, T., 1984, Effects of glucose removal and readmission on potassium contracture in the guinea-pig taenia coli, *J. Physiol.*, 356: 33.

Bárány, M., 1967, ATPase activity of myosin correlated with speed of muscle shortening, *J. Gen. Physiol.*, 50: 197.

Berridge, M. J., 1984, Inositol trisphosphate and diacylglycerol as second messengers, *Biochem. J.*, 220: 345.

Berridge, M. J., 1986, Agonist-dependent phosphoinositide metabolism. A bifurcating signal pathway, *in:* "New Insights into Cell and Membrane Transport Processes", G. Poste and S. T. Crooke, eds., Plenum Press, New York, p. 201.

Carafoli, E., 1984, Calmodulin-sensitive calcium-pumping ATPase of plasma membranes: Isolation, reconstitution, and regulation, *Fed. Proc.*, 43: 3005.

Coburn, R. F., Baron, C., and Papadopoulos, M. T., 1988, Phosphoinositide metabolism and metabolism-contraction coupling in rabbit aorta, *Am. J. Physiol.*, 255: H1476.

Coburn, R. F., Moreland, S., Moreland, R. S., and Baron, C. B., 1991, Rate limiting energy-dependent steps controlling oxidative metabolism-contraction coupling in vascular smooth muscle, *J. Physiol.*, in press.

Hellstrand, P. and Arner, A., 1985, Myosin light chain phosphorylation and the cross-bridge cycle at low substrate concentration in chemically skinned guinea pig taenia coli, *Pflügers Arch.*, 405: 323.

Ishida, Y. and Paul, R. J., 1990, Effects of hypoxia on high-energy phosphagen content, energy metabolism and isometric force in guinea-pig taenia caeci, *J. Physiol.*, 424: 41.

Kamm, K. E. and Stull, J. T., 1985, The function of myosin and myosin light chain kinase phosphorylation in smooth muscle, *Ann. Rev. Pharmacol. Toxicol.*, 25: 593.

Marriott, J. F. and Marshall, J. M., 1990a, Differential effects of hypoxia upon contractions evoked by potassium and noradrenaline in rabbit arteries in vitro, *J. Physiol.*, 422: 1.

Marriott, J. F. and Marshall, J. M., 1990b, Effects of hypoxia upon contractions evoked in isolated rabbit pulmonary artery by potassium and noradrenaline, *J. Physiol.*, 422: 15.

Moreland, S. and Moreland, R. S., 1987, Effects of dihydropyridines on stress, myosin phosphorylation, and V_0 in smooth muscle, *Am. J. Physiol.*, 252: H1049.

Moreland, S. and Moreland, R. S., 1990, Effects of calcium channel activators on contractile behavior in vascular smooth muscle, *in:* "Frontiers in Smooth Muscle Research", N. Sperelakis and J. D. Wood, eds., Wiley-Liss, New York, p. 525.

Paul, R. J., 1990, Smooth muscle energetics and theories of cross-bridge regulation, *Am. J. Physiol.*, 258: C369.

Pearce, W. J., 1989, Hypoxia inhibits calcium uptake in isolated cerebral arteries, *FASEB J.*, 3: A845.

Popescu, L. M. and Ignat, P., 1983, Calmodulin-dependent Ca^{2+}-pump ATPase of human smooth muscle sarcolemma, *Cell Calcium*, 4: 219.

Scott, D. P., Davidheiser, S., and Coburn, R. F., 1987, Effects of elevation of phosphocreatine on force and metabolism in rabbit aorta, *Am. J. Physiol.*, 253: H461.

Scott, D. P. and Coburn, R. F., 1989, Phosphocreatine and oxidative metabolism-contraction coupling in rabbit aorta, *Am. J. Physiol.*, 257: H597.

Singer, H. A. and Murphy, R. A., 1987, Maximal rates of activation in electrically stimulated swine carotid media, *Circ. Res.*, 60: 438.

Somlyo, A. V., Goldman, Y. E., Fujimori, T., Bond, M., Trentham, D. R., and Somlyo, A. P., 1988, Cross-bridge kinetics, cooperativity, and negatively strained cross-bridges in vertebrate smooth muscle. A laser-flash photolysis study, *J. Gen. Physiol.*, 91: 165.

Somlyo, A. V. and Somlyo, A. P., 1968, Electromechanical and pharmacomechanical coupling in vascular smooth muscle, *J. Pharmacol. Exp. Ther.*, 159: 129.

Standen, N. B., Quayle, J. M., Davies, N. W., Brayden, J. E., Huang, Y., and Nelson, M. T., 1989, Hyperpolarizing vasodilators activate ATP-sensitive K^+ channels in arterial smooth muscle, *Science*, 245: 177.

Weir, S. W. and Weston, A. H., 1986, The effects of BRL 34915 and nicorandil on electrical and mechanical activity and on ^{86}Rb efflux in rat blood vessels, *Br. J. Pharmacol.*, 88: 121.

CAN DIFFERENT FOUR-STATE CROSSBRIDGE MODELS EXPLAIN LATCH AND THE ENERGETICS OF VASCULAR SMOOTH MUSCLE?

Chi-Ming Hai,[1] Christopher M. Rembold,[2,3]
and Richard A. Murphy[2]

[1]Section of Physiology and Biophysics
Brown University
Providence, RI

[2]Department of Physiology
[3]Department of Internal Medicine (Cardiology)
University of Virginia Health Sciences Center
Charlottesville, VA 22908

INTRODUCTION

Ca^{2+} binding to troponin and the resulting thin filament conformational change acts as a switch in striated muscle that enables crossbridges to bind to actin and cycle (Rüegg, 1986). This allosteric "Ca^{2+}-switch" mechanism generates only two primary crossbridge states: free and attached (Huxley, 1957).

In smooth muscle, Ca^{2+}-calmodulin-dependent activation of myosin light chain kinase (MLCK) results in crossbridge phosphorylation. This phosphorylation of the 20 kDa myosin regulatory light chain appears to be essential for crossbridge attachment and cycling (Kamm and Stull, 1985; Hartshorne et al., 1989). However, phosphorylation is not a simple switch enabling crossbridge attachment because smooth muscles can maintain (Dillon et al., 1981) or slowly develop (Ratz and Murphy, 1987) high levels of stress (= force/cross-sectional area of muscle) at fairly low values of myoplasmic $[Ca^{2+}]$ and crossbridge phosphorylation. This state of high stress with reduced crossbridge cycling rates was termed "latch" (Dillon et al., 1981). The ability to reduce crossbridge cycling rates is potentially advantageous for muscle in the walls of hollow organs. Lower cycling rates decrease ATP usage when the smooth muscle is stabilizing organ dimensions against imposed loads such as blood pressure.

The simplest explanation for the mechanical properties of a smooth muscle during sustained contractions is to postulate that much of the force is generated by attached, dephosphorylated crossbridges or "latchbridges" (Dillon et al., 1981; Murphy, 1989). If the existence of latchbridges is accepted, the issue becomes one of explaining how they are formed and regulated. We originally thought that a second Ca^{2+}-dependent regulatory mechanism must be invoked to explain latch (Dillon et al., 1981), and numerous candidates have been

Regulation of Smooth Muscle Contraction
Edited by R.S. Moreland, Plenum Press, New York, 1991

proposed (see Kamm and Stull, 1989; Murphy, 1989; and other papers in this volume).

A plausible hypothesis for the formation of latchbridges is that MLCK and myosin light chain phosphatase (MLCP) can act on both free and attached crossbridges (Hai and Murphy, 1988a; Driska et al., 1989). If so, covalent regulation would yield four crossbridge states rather than two (see Fig. 1A). The two attached crossbridge states, AM and AM_p, are assumed to make equal individual contributions to force. Most attached crossbridges would be phosphorylated when myoplasmic [Ca^{2+}] and thus MLCK activity (K1 and K6) and phosphorylation values are high. At low levels of myoplasmic [Ca^{2+}] and phosphorylation, larger numbers of latchbridges (AM, Fig. 1) would be formed if MLCP is reasonably active (Driska et al., 1989). The proposed difference between AM_p and AM is that the detachment rate of the latchbridge (AM) is slower (K7 = 0.2 K4, Fig. 1A).

The predictive value of this four-state latchbridge hypothesis could be tested by mathematical modeling after resolving values for the seven rate constants from experimental data. Two approaches were used. The first approach (referred to as the "kinetic model") was to describe the contents of the four crossbridge species in four differential equations. This allowed fitting the time course of changes in phosphorylation and stress during contraction (Hai and Murphy, 1988a) and relaxation (Hai and Murphy, 1989a) induced by various stimuli, and predicting the steady-state dependence of stress on phosphorylation (Hai and Murphy, 1988a; Ratz et al., 1989). The second approach that allowed prediction of shortening velocity was based on that used by A. F. Huxley (1957) who showed that the two-state crossbridge model for skeletal muscle could predict its unique velocity-stress curve. We asked whether a similar approach modified for four crossbridge states could predict the phosphorylation-dependent family of velocity-stress relationships observed in the swine carotid media, as well as the steady-state dependencies of both stress and shortening velocity at zero load on phosphorylation (Hai and Murphy, 1988b). This more complex mathematical representation of the hypothesis diagrammed in Figure 1A will be referred to as the conformational model as it incorporates specific assumptions about the structure of the contractile unit (Hai and Murphy, 1988b). Applying both approaches to the scheme using the rate constants shown in Figure 1A gave good fits to experimental data (Panel A in Figs. 2, 3, 5, 6).

In an initial exploration of the energetics predictions of the four-state hypothesis, we estimated the steady-state ATP consumption used for phosphorylation and crossbridge cycling (Hai and Murphy, 1989b). The flux between crossbridge states was calculated using the four differential equations in the kinetic model. The unexpected result was a prediction that ATP consumption due to phosphorylation and dephosphorylation would be approximately six-fold higher than ATP consumption by crossbridge cycling. The implication was that the efficiency of the carotid media during shortening would be very low compared to skeletal muscle (Hai and Murphy, 1989b).

Paul (1990) pointed to the seemingly linear relationship between ATP consumption and stress as a hallmark of smooth muscle energetics and noted that this is not a prediction of the Hai and Murphy hypothesis. Paul proposed alternative sets of rate constants that favored cycling by phosphorylated crossbridges and minimized latchbridges as more consistent with the energetics properties of smooth muscle. These rate constants used with the kinetic model could approximate the time course of phosphorylation and stress in the electrically stimulated carotid media (Paul, 1990). Paul postulated that a high

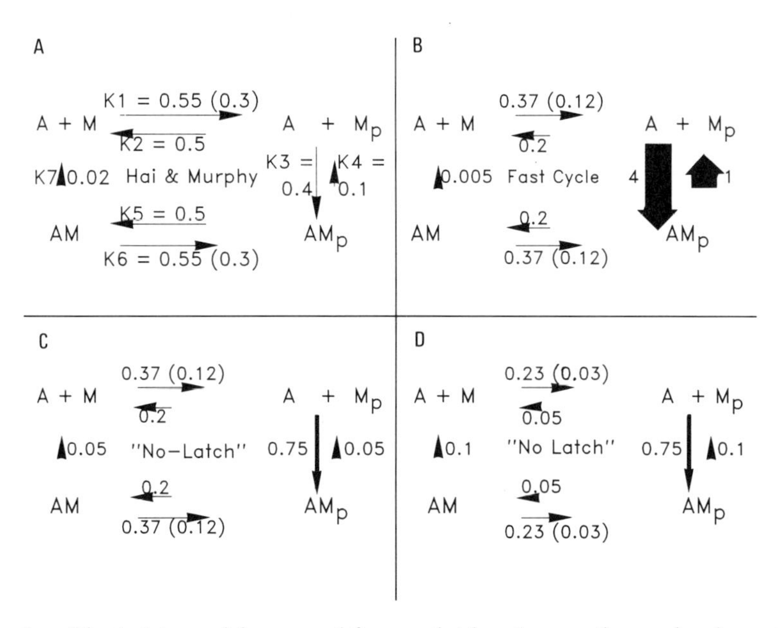

Figure 1. The 4-state model proposed for crossbridges in smooth muscle. A = actin in thin filament; M = myosin crossbridge; M_p = free, phosphorylated crossbridge; AM_p = attached, force generating, phosphorylated crossbridge; AM = attached, dephosphorylated, force-generating crossbridge or latchbridge. Values of the first order rate constants are in sec^{-1}. Their relative magnitudes are indicated by the size of the arrows. Panel A is the Hai and Murphy proposal. Panel B is Paul's fast cycling crossbridge model, while Panels C and D are Paul's two "no-latch" models where K7 = K4.

attachment to detachment rate for phosphorylated crossbridges may best explain the energetics of smooth muscle.

The aims of this study were to explore in more detail the predictions of Paul's alternative rate constants. How precisely do they describe the time course of phosphorylation and stress for the electrically stimulated carotid, and can they predict the responses to agonists? Can the alternative rate constants predict the steady-state dependencies of stress and shortening velocity at zero load on phosphorylation? Finally, we examined the quantitative predictions for ATP consumption in the carotid media using the conformational model.

MODEL COMPARISONS

The values of the rate constants in the Hai and Murphy hypothesis are shown in Figure 1A. They were resolved on the basis of optimizing the model's predictions to the data sets shown in Figures 2 and 3 (from Singer and Murphy, 1987). It is not known if crossbridge attachment alters the activities of MLCK or MLCP, nor could all seven rate constants be resolved from a data set. Therefore we assumed that K1 = K6 and K2 = K5. Similarly, the rate constants governing attachment and detachment of phosphorylated crossbridges could not be independently resolved. We set K3 = 4 x K4 to yield an assumed duty cycle where phosphorylated crossbridges would be attached 80% of the time (Hai and Murphy, 1988a). However, the value of the duty cycle has little effect on the model's predictions (Rembold and Murphy, 1990). The final value of K7

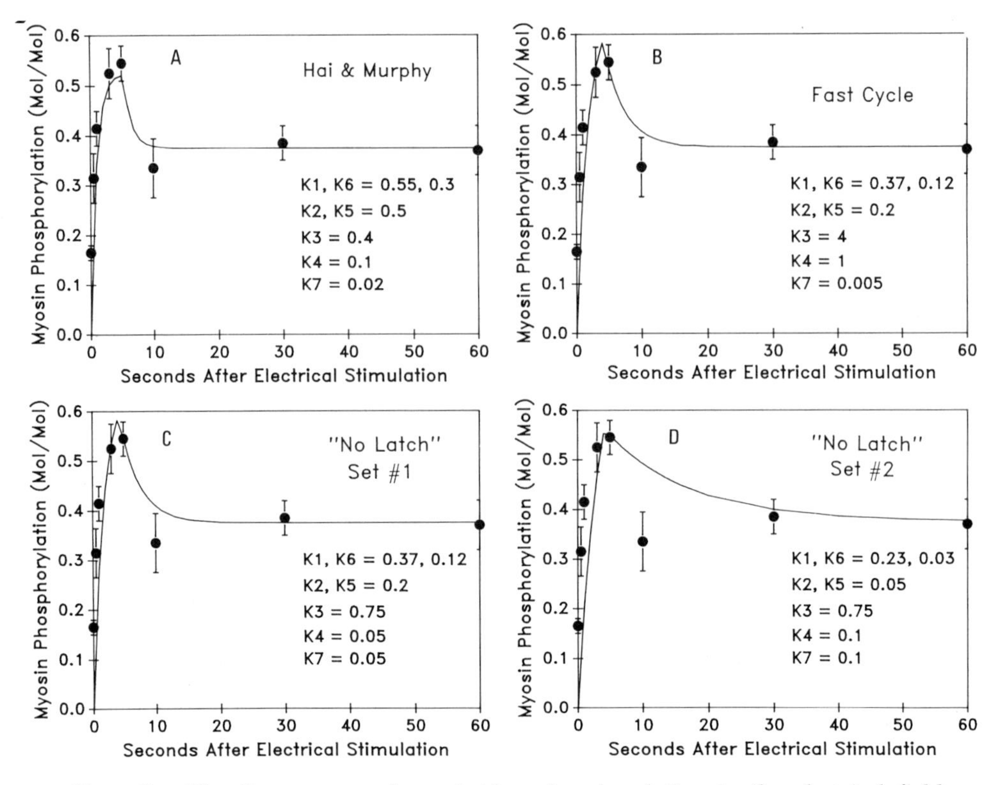

Figure 2. The time course of crossbridge phosphorylation in the electrical field stimulated swine carotid media (pretreated with phenoxybenzamine to block alpha receptors and with tetraethylammonium to reduce the potassium conductance, allowing the generation of action potentials). Solid circles ± 1 SEM are data from Singer and Murphy (1987). Lines are the model's predictions with the 4 sets of rate constants shown in the corresponding panels in Figure 1.

was selected after fitting other data sets (Hai and Murphy, 1989a). A key postulate of the hypothesis is that dephosphorylation of AM_p to form a latchbridge is associated with a slowing of its detachment rate (K7 = 0.2 x K4). Thus, phosphorylated crossbridges cycle faster than crossbridges that are dephosphorylated to form latchbridges. To reflect the initial transient in Ca^{2+}-induced phosphorylation the value of K1 (= K6) was increased from 0 to 0.55 sec^{-1} at time = 0, and reduced to a steady-state value of 0.3 sec^{-1} at 5 sec.

Paul's objective of fitting the data shown in Figures 2 and 3 with rate constants that partition most of the ATP consumption into crossbridge cycling rather than phosphorylation is illustrated in Figure 1B. The values of K3 and K4 governing turnover of phosphorylated crossbridges are increased by an order of magnitude with reductions in the rate constants for MLCK and MLCP activities (Paul, 1990).

Figures 1C and 1D were termed "no-latch" models (Paul, 1990) because the detachment rates of phosphorylated crossbridges and latchbridges are equal (K4 = K7). This designation is not strictly true since a pure "no-latch" model would have a linear dependence of stress on phosphorylation. However, the rate constants used in Figure 1D approximate such a state (see Fig. 5). The rate constants give a high duty cycle for crossbridges with attachment:detachment ratios of 15 and 7.5, respectively. If these alternative four-state models can predict the properties of the carotid media, then the concept of a latchbridge defined as a crossbridge with a reduced detachment rate may be unnecessary.

MODEL PREDICTIONS OF EXPERIMENTAL DATA

The Hai and Murphy rate constants were derived from the best fit to the data illustrated in Figures 2A and 3A. Thus, it is not surprising that the fit was optimal with these rate constants. However, the models assume instantaneous activation of MLCK to maximum levels. Therefore, the expected observation in a real muscle with delays reflecting the various steps involved in activation-contraction coupling is that increases in phosphorylation would follow model predictions rather than precede them (see Fig. 2B, C and D). The peak and steady-state phosphorylation predictions are similar for all four sets of rate constants.

All of Paul's rate constants predict more rapid increases in stress than was observed, but did yield accurate steady-state stress predictions (Fig. 3). Insofar as the considerable series elastic compliance in smooth muscle will slow measured rates of force development, the discrepancies between the initial rates of force development and the model's predictions might be explicable.

The ability of a model to predict one data set does not imply that it would be equally successful in predicting changes in phosphorylation or stress in response to other stimuli. We recently developed a rigorous test of the hypothesis that changes in phosphorylation as a result of changes in

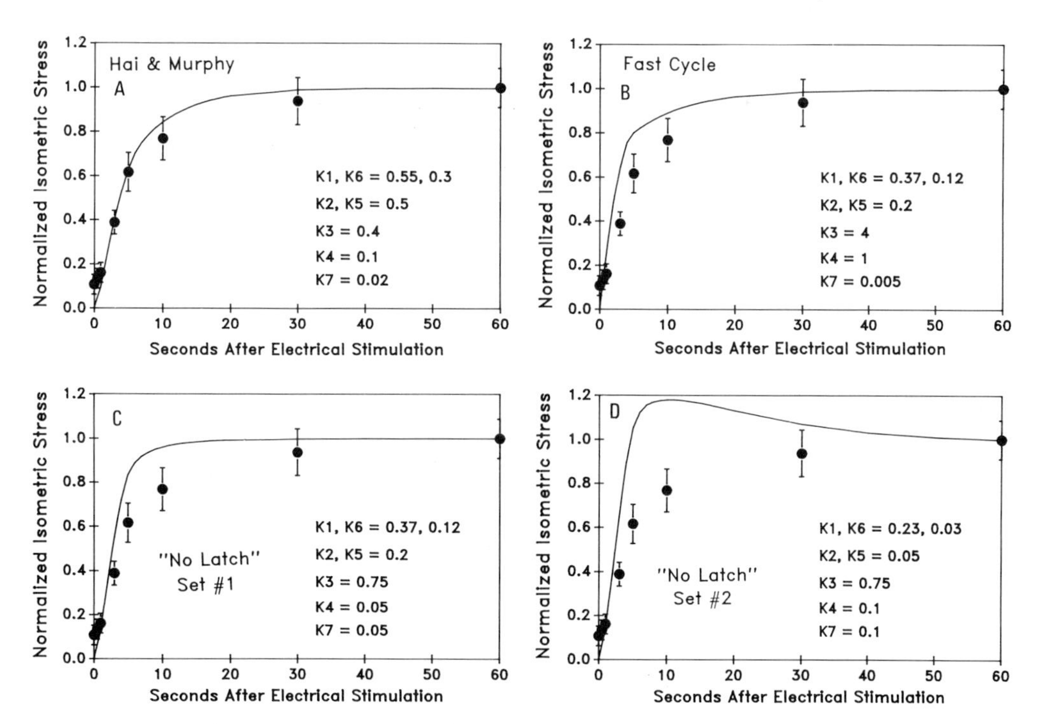

Figure 3. The time course of stress development in the same tissues described in Figure 2 (solid circles ± 1 SEM). The predictions of the four models in Figure 1 are given by the lines in the corresponding panels.

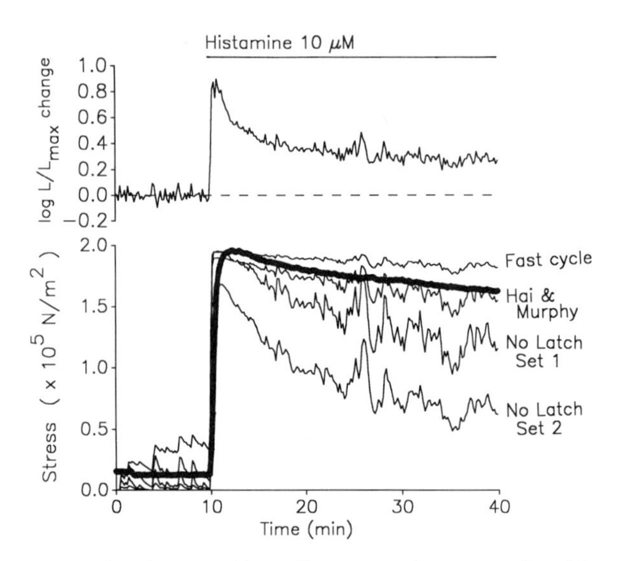

Figure 4. Response of swine carotid media preparation exposed to histamine. Upper panel is the measured, normalized light signal emitted from the aequorin-loaded tissues that is proportional to the myoplasmic [Ca^{2+}] (see Rembold and Murphy, 1988, for details of aequorin loading, signal normalization, calibration, etc.). Lower panel shows the corresponding measured isometric contraction (heavy line). The lighter, very irregular lines are the predicted stress time courses for each of the four sets of rate constants (Fig. 1) when the values of K1 (= K6) were set by the experimental Ca^{2+} signal in the upper panel. See text for details.

myoplasmic [Ca^{2+}] determine contraction and relaxation. We had previously determined the dependence of crossbridge phosphorylation on intracellular [Ca^{2+}] estimated using the photoprotein aequorin. With this relationship we could use the aequorin light signal to predict MLCK activity in agonist stimulated swine carotid media (Rembold and Murphy, 1990). The resulting time course for changes in K1 (= K6) were used with the kinetic model (Hai and Murphy, 1988a) to predict stress (Rembold and Murphy, 1990). Force development could be predicted using the light signal from aequorin-loaded tissues and the model with the rate constants shown in Figure 1A (Rembold and Murphy, 1990). The model does not make allowances for agonist diffusion times, tissue series elastic compliance, and activation processes. As a result stress develops more slowly than predicted by the model (Fig. 4). However, the Hai and Murphy rate constants predict stress values quite accurately when the Ca^{2+} signal is not changing rapidly (Fig. 4; Rembold and Murphy, 1990). Paul's "no-latch" rate constant sets underestimated the observed stress, while the "fast cycle" rate constants overestimated the stress induced by histamine. This was also true for the low level of tone in the unstimulated tissue.

An important test of the model is the ability to predict the steady-state dependence of stress on phosphorylation (Hai and Murphy, 1988a,b). The four differential equations of the kinetic model were used for the predictions illustrated in Figure 5, although the conformation model yields the same results. The data (Ratz et al., 1989) represent individual carotid media tissues activated by a variety of protocols. See McDaniel et al. (1990) for a more extensive data set. After corrections for artifacts that lead to elevated basal phosphorylation values, the Hai and Murphy model predicts a relationship that is centered in the experimental points (Figure 5A).

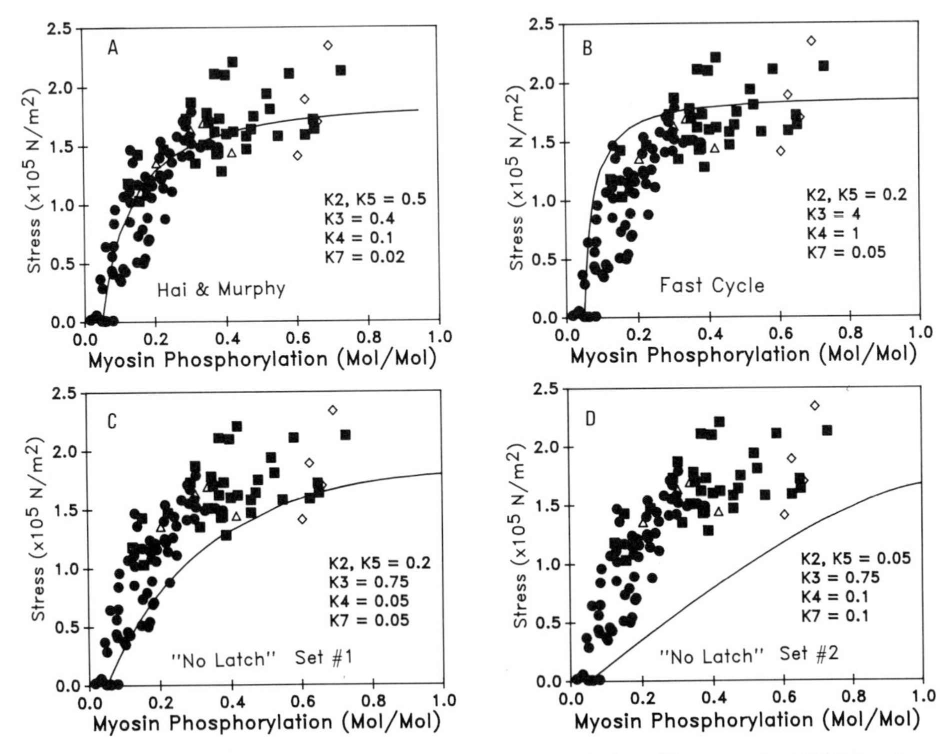

Figure 5. Model predictions (lines) of experimental data (Ratz et al., 1989) for the steady-state dependence of stress on phosphorylation in the swine carotid media. Symbols represent individual tissues activated by many different stimuli. Panels correspond to Figure 1.

None of the scenarios explored by Paul does a good job of predicting steady-state stress, at least at the comparatively low levels of phosphorylation that occur physiologically in sustained contractions (Fig. 5B, C, and D). Note that factors such as agonist diffusion, tissue series elastic compliance, or various activation reactions that slow force development should have no effect on the steady-state stress.

The simple kinetic model (Hai and Murphy, 1988a) used by Paul cannot predict shortening velocities. Therefore, we turned to the conformation model that can relate velocity-stress behavior to phosphorylation. As previously reported (Hai and Murphy, 1988b) this successfully fitted the experimental dependence of shortening velocity estimated for zero load on crossbridge phosphorylation (Fig. 6A, and see Rembold and Murphy, 1988, for a more extensive data set). None of Paul's sets of rate constants approximated the experimental findings (Fig. 6B, C, and D). As would be expected, adjusting the rate constants to favor a high crossbridge cycling rate leads to predictions of high shortening velocities that are not attained by the carotid media.

The calculation of a much higher rate of ATP consumption by the MLCK/MLCP system than by crossbridges during steady-state isometric contractions using the kinetic model (Hai and Murphy, 1989b) was unexpected, given the implied low efficiency of contraction in a working muscle. Since we shared Paul's (1990) reservations, we reanalyzed the energetics predictions not only for steady-state isotonic contractions, but also for isotonic contractions by coupling the four-state model with Huxley's (1957) concept of strain-dependent crossbridge cycling rates (Hai and Murphy, 1991). With this approach the estimated rate of ATP consumption due to crossbridge cycling was considerably higher (about five-fold) than the initial prediction using the simple kinetic flux analysis (Hai and Murphy, 1989b). The calculated steady-state ATP usage during isometric contractions would be partitioned almost equally between phosphorylation and crossbridge cycling according to this analysis (Hai and Murphy, 1991). We believe this is a more reasonable analysis since crossbridge cycling rates are considered to be strain-dependent (Eisenberg and Hill, 1985).

However, the model still predicts a highly nonlinear relationship between the rate of ATP consumption and steady-state stress (Fig. 7 and Paul, 1990). Clearly this is not consistent with the linear relationship suggested by experimental studies (Paul et al., 1976; Glück and Paul, 1977; Peterson, 1982; Arner and Hellstrand, 1983; Paul, 1989). Note, however, that the predicted relationship between ATP consumption and stress is approximately linear up to 25% or 30% phosphorylation (Fig. 7). The cited energetics studies were all performed under conditions when suprabasal phosphorylation was probably less than 30%. Thus, the experimental data might be expected to resemble a linear relationship. The model's prediction (Fig. 7) falls within the errors of a recent experimental study (Krisanda and Paul, 1988) using the swine carotid media in protocols where crossbridge phosphorylation levels are known and also provides an excellent fit to Peterson and Glück's data (see Paul, 1990) (Hai and Murphy, 1991).

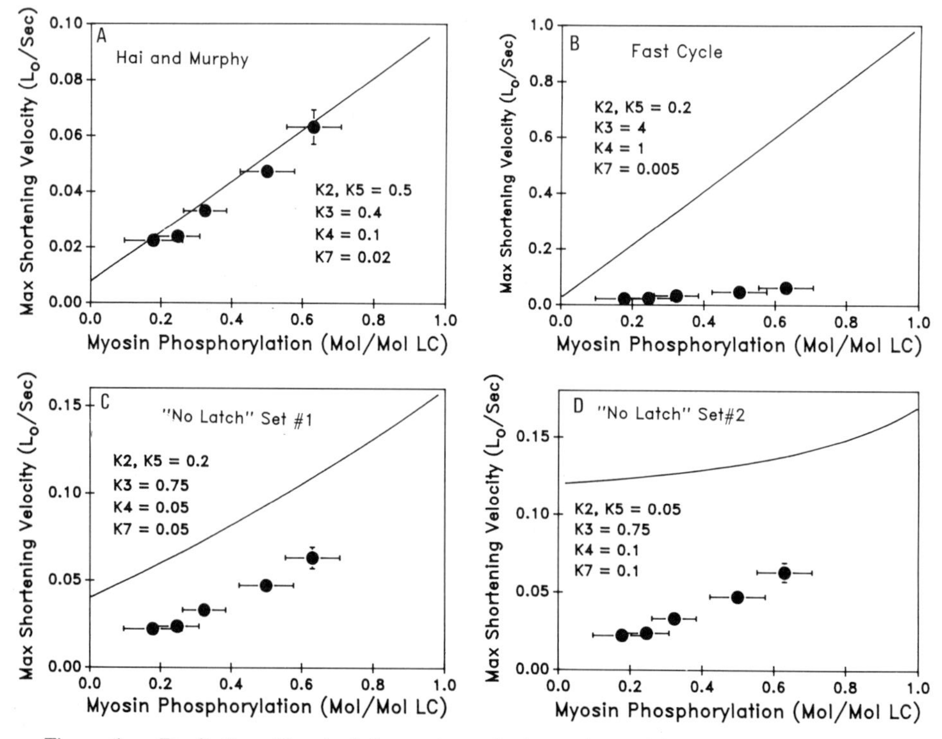

Figure 6. Predictions (lines) of shortening velocity estimated for zero load as a function of crossbridge phosphorylation (Hai and Murphy, 1988b). Experimental data are means ± 1 SEM from Aksoy et al. (1982) and Dillon and Murphy (1982). Note that changes in scale on the ordinates alter the position of the data set. Panels correspond to Figure 1.

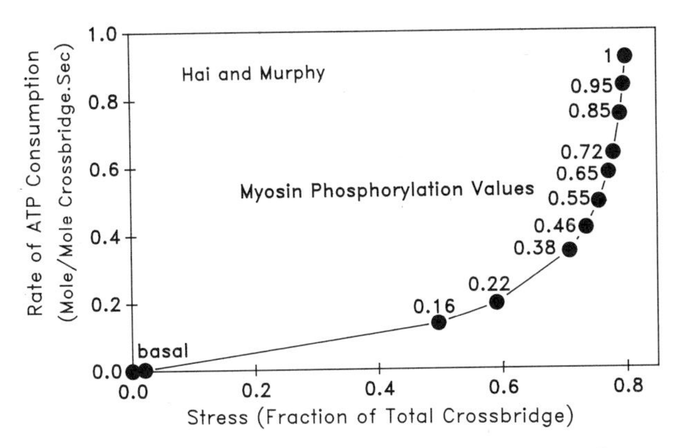

Figure 7. Model (Fig. 1A) predicted relationship between total suprabasal ATP consumption and stress for the carotid media. The solid circles represent corresponding phosphorylation values (in moles phosphate/mole 20 kDa regulatory light chain) and are not experimental points. See Hai and Murphy (1990) for a description of the modeling based on Huxley's (1957) approach.

DISCUSSION

Paul (1990) concludes that many sets of rate constants in a four-state model can explain experimental observations of myosin phosphorylation transients associated with monotonic stress development. Our more detailed analysis duplicates Paul's results in showing that a single time course of phosphorylation and stress can be approximated by alternative rate constants emphasizing crossbridge cycling and minimizing a slower cycle involving phosphorylation and dephosphorylation (Figs. 1, 2, and 3). However, such approximations are not equivalent to the fit achieved by the Hai and Murphy rate constants. When other data sets involving agonist stimulation are used, Paul's alternative scenarios are less satisfactory (Fig. 4). The most rigorous test of various models is their ability to predict the steady-state behavior of smooth muscle when the tissue should have a uniform level of phosphorylation. Only the original set of rate constants predicted steady-state stress or shortening velocity at zero load as a function of phosphorylation (Figs. 5 and 6).

Paul (1990) argues that it is unnecessary to postulate that most of the steady-state isometric ATP consumption is due to the activity of the MLCK/MLCP systems. We agree, although for somewhat different reasons. Calculating ATP consumption using the four differential equations to estimate crossbridge flux in the simple kinetic model appears to underestimate the turnover of phosphorylated crossbridges (Hai and Murphy, 1990). While the energy cost of covalent regulation in the tonic carotid media remains significant (about 50% of suprabasal ATP consumption during sustained isometric contractions) in the most recent analysis, overall ATP consumption may be minimized for a muscle where a high economy of force maintenance is more important than a high efficiency during shortening.

Perhaps the most important point underlying Paul's critique of the Hai and Murphy hypothesis is the prediction of a highly nonlinear relation between suprabasal ATP consumption and steady-state isometric stress (Fig. 7). Again, there is no disagreement about the model predictions with any of the tested rate constants, a result providing confidence in the mathematics and different computer programs. However, limitations in the available data preclude a firm decision. The model predicts that the relationship between suprabasal ATP consumption and stress is approximately linear at levels of steady-state phosphorylation typically elicited by agonists. The Hai and Murphy model did an excellent job of predicting ATP consumption in data sets available for the swine carotid media (Hai and Murphy, 1991).

Arner and Hellstrand (1983) reported a linear relationship between suprabasal ATP consumption and stress in intact rat portal vein preparations, although the slope was steeper early in a contraction (when phosphorylation levels would be higher). After skinning with the detergent Triton X-100, measured ATP consumption at various steady-state [Ca^{2+}] levels was dramatically altered to resemble the model's prediction shown in Fig. 7. Crossbridge phosphorylation was not measured in this early study, but phosphorylation can approach 100% in chemically skinned smooth muscle where high [Ca^{2+}] can be maintained and where MLCP activity may be low (Bialojan et al., 1985). It is likely that Arner and Hellstrand obtained much higher steady-state phosphorylation levels in the tissues after skinning, revealing a nonlinear relationship between ATP consumption and stress. An important caveat is that detergent skinned smooth muscle frequently does not exhibit latch and has a linear dependence of stress on phosphorylation (Tanner et al., 1988).

Data from intact tissues are needed. Paul (1990) emphasizes that quantitative modeling and computer simulation is a valuable technique for both formulating and testing hypotheses to explain experimental observations. Modeling is also useful for identifying important experiments that could test competing hypotheses. Our analysis indicates that one good example would be to measure suprabasal ATP consumption in the carotid media under conditions when steady-state phosphorylation values extend into the 30-60% range. This should reveal whether a linear or nonlinear relationship between ATP consumption and stress exists and provide an additional test of the four-state hypothesis.

ACKNOWLEDGMENTS

This study received support from NSF grant DCB8902438 (to C.-M.H.) and NIH grants 1RO1 HL38918 (to C.M.R.) and 5PO1 HL19242 (to R.A.M.). C. M. Rembold is a Lucille P. Markey Scholar with support from the Lucille P. Markey Charitable Trust. We are grateful for the editorial assistance of Kathy Dobbins.

REFERENCES

Aksoy, M. O., Murphy, R. A., and Kamm, K. E., 1982, Role of Ca^{2+} and myosin light chain phosphorylation in regulation of smooth muscle, *Am. J. Physiol.*, 242: C109.

Arner, A. and Hellstrand, P., 1983, Activation of contraction and ATPase activity in intact and chemically skinned smooth muscle of rat portal vein: dependence on Ca^{++} and muscle length, *Circ. Res.*, 53: 695.

Bialojan, C., Merkel, L., Rüegg, J. C., Gifford, D., and DiSalvo, J., 1985, Prolonged relaxation of detergent-skinned smooth muscle involves decreased endogenous phosphatase activity, *Proc. Soc. Exp. Biol. Med.*, 178: 648.

Dillon, P. F., Aksoy, M. O., Driska, S. P., and Murphy, R. A., 1981, Myosin phosphorylation and the cross-bridge cycle in arterial smooth muscle, *Science*, 211: 495.

Dillon, P. F. and Murphy, R. A., 1982, Tonic force maintenance with reduced shortening velocity in arterial smooth muscle, *Am. J. Physiol.*, 242: C102.

Driska, S. P., Stein, P. G., and Porter, R., 1989, Myosin dephosphorylation during rapid relaxation of hog carotid artery smooth muscle, *Am. J. Physiol.*, 256: C315.

Eisenberg, E. and Hill, T.L., 1985, Muscle contraction and free energy transduction in biological systems, *Science*, 227: 999.

Glück, E. and Paul, R. J., 1977, The aerobic metabolism of porcine carotid artery and its relationship to isometric force: Energy cost of isometric contraction, *Pflügers Arch.*, 370: 9.

Hai, C.-M. and Murphy, R. A., 1988a, Cross-bridge phosphorylation and regulation of latch state in smooth muscle, *Am. J. Physiol.*, 254: C99.

Hai, C.-M. and Murphy, R. A., 1988b, Regulation of shortening velocity by cross-bridge phosphorylation in smooth muscle, *Am. J. Physiol.*, 255: C86.

Hai, C.-M. and Murphy, R. A., 1989a, Cross-bridge dephosphorylation and relaxation in vascular smooth muscle, *Am. J. Physiol.*, 256: C282.

Hai, C.-M. and Murphy, R. A., 1989b, Crossbridge phosphorylation and the energetics of contraction in the swine carotid media, *in:* "Muscle Energetics", R. J. Paul, G. Elzinga, and K. Yamada, eds., Alan R. Liss, New York, p. 253.

Hai, C.-M. and Murphy, R. A., 1991, ATP consumption by smooth muscle as predicted by the four-state crossbridge model, *Biophys. J*, submitted.

Hartshorne, D. J., Ito, M., and Ikebe, M., 1989, Myosin and contractile activity in smooth muscle, *in:* "Calcium Protein Signalling", H. Hidaka, E. Carifoli, A. R. Means, T. Tanaka, eds., Plenum Press, New York, p. 269.

Huxley, A. F., 1957, Muscle structure and theories of contraction, *Prog. Biophys. Biophys. Chem.*, 7: 255.

Kamm, K. E. and Stull, J. T., 1985, The function of myosin and myosin light chain kinase phosphorylation in smooth muscle, *Ann. Rev. Pharmacol. Toxicol.*, 25: 593.

Kamm, K. E. and Stull, J. T., 1989, Regulation of smooth muscle contractile elements by second messengers, *Ann. Rev. Physiol.*, 51: 299.

Krisanda, J. M. and Paul, R. J., 1988, Dependence of force, velocity, and O_2 consumption on $[Ca^{2+}]_o$ in porcine carotid artery, *Am. J. Physiol.*, 255: C393.

McDaniel, N. L., Rembold, C. M., and Murphy, R. A., 1990, Covalent cross-bridge regulation in smooth muscle, *Ann. N.Y. Acad. Sci.*, 599: 66.

Murphy, R. A., 1989, Special topic: Contraction in smooth muscle cells, introduction, *Ann. Rev. Physiol.*, 51: 275.

Paul, R. J., Glück, E., and Rüegg, J. C., 1976, Cross bridge ATP utilization in arterial smooth muscle, *Pflügers Arch.*, 361: 297.

Paul, R. J., 1989, Smooth muscle energetics, *Ann. Rev. Physiol.*, 51: 331.

Paul, R. J., 1990, Smooth muscle energetics and theories of cross-bridge regulation, *Am. J. Physiol.*, 258: C369.

Peterson, J. W., 1982, Effect of histamine on the energy metabolism of K^+-depolarized hog carotid artery, *Circ. Res.*, 50: 848.

Ratz, P. H. and Murphy, R. A., 1987, Contributions of intracellular and extracellular Ca^{2+} pools to activation of myosin phosphorylation and stress in swine carotid media, *Circ. Res.*, 60: 410.

Ratz, P. H., Hai, C.-M., and Murphy, R. A., 1989, Dependence of stress on cross-bridge phosphorylation in vascular smooth muscle, *Am. J. Physiol.*, 256: C96.

Rembold, C. M. and Murphy, R. A., 1988, Myoplasmic $[Ca^{2+}]$ determines myosin phosphorylation in agonist-stimulated swine arterial smooth muscle, *Circ. Res.*, 63: 593.

Rembold, C. M. and Murphy, R. A., 1990, Latch-bridge model in smooth muscle: $[Ca^{2+}]_i$ can quantitatively predict stress, *Am. J. Physiol.*, 259: C251.

Rüegg, J. C., 1986, "Calcium in Muscle Activation", Springer-Verlag, Berlin.

Singer, H. A. and Murphy, R. A., 1987, Maximal rates of activation in electrically stimulated swine carotid media, *Circ. Res.*, 60: 438.

Tanner, J. A., Haeberle, J. R., and Meiss, R. A., 1988, Regulation of glycerinated smooth muscle contraction and relaxation by myosin phosphorylation, *Am. J. Physiol.*, 255: C34.

CALCIUM HOMEOSTASIS IN SINGLE INTACT SMOOTH MUSCLE CELLS

Edwin D. W. Moore, Peter L. Becker, Takeo Itoh,
and Fredric S. Fay

Program Molecular Medicine
University of Massachusetts Medical Center
373 Plantation Street
Worcester, MA 01605

INTRODUCTION

The force of contraction of smooth muscle cells, is a function of the Ca^{2+} ion activity of the myoplasm (Fay et al., 1974). There has therefore been a tremendous effort directed at understanding Ca^{2+} homeostatic mechanisms, with experiments ranging the spectrum from isolated genes to intact cells. The work of the scientists in this laboratory has concentrated on the latter, Ca^{2+} regulation in intact, single smooth muscle cells that have been enzymatically isolated from the stomach muscularis of the toad *Bufo marinus* (Fay et al., 1982). These experiments became possible only because of key technological developments of the last decade. In particular, it was the rational design of ion-sensitive fluorescent dyes that could be used ratiometrically to measure ion concentrations in living cells (Grynkiewicz et al, 1985) and the marriage of this technique with digital imaging microscopy (Fay et al, 1986), which enabled our experimental approach.

This paper will discuss three aspects of Ca^{2+} regulation in the intact smooth muscle cell. The first deals with the mechanism of action of the β-adrenergic relaxant isoproterenol (ISO) and the role played by the sodium gradient in this response. The second explores how the previous Ca^{2+} history of the myoplasm influences the time course of future Ca^{2+} transients. The third deals with the role played by calmodulin-dependent protein kinases both in linking Ca^{2+} to contractility as well as in feeding back on processes responsible for generation of the Ca^{2+} signal itself. We will first briefly discuss our methodology, but for greater detail, we refer the reader to a recent issue of Cell Calcium (Vol 11, 1990) that was devoted entirely to an extensive review of the use of ion-sensitive fluorescent dyes and their use with digital imaging microscopes.

METHODS

Figure 1 shows a schematic diagram of our dual wavelength micro-fluorimeter (DWM) (Yagi et al., 1988). The heart of the system is a Zeiss IM-35

Regulation of Smooth Muscle Contraction
Edited by R.S. Moreland, Plenum Press, New York, 1991

microscope equipped for epi-fluorescence, with quartz optics in the excitation path to optimize UV transmission. The filter wheel consists of two transmissive segments, centered at 340 and 380 nm, separated by two opaque segments the size of the illumination beam, so that the duty cycle is about 99%. The wheel rotation rate is variable, and controlled by the PDP 11/73. The image detector is a photomultiplier tube, the anode current is fed into a photon counting circuit and the values, after appropriate voltage discrimination, are stored on the computer. With this arrangement we can measure intracellular ion concentrations once every 7 msec., which is more than fast enough to follow changes in ion concentration within a smooth muscle cell during a contraction-relaxation cycle. Onto this system we have added the hardware necessary for voltage clamping and microinjection, and in addition, have developed a superfusion system so that the medium bathing the cell can be changed in < 1 sec. We use the fluorescent dyes fura-2, for measuring $[Ca^{2+}]_i$, and SBFI for measuring $[Na^+]_i$ (Minta and Tsien, 1989). Their spectral characteristics are virtually identical, so the same optical configuration can be used for both. The only significant difference between them is their intracellular calibration. In the toad stomach cells, the *in vivo* and the *in vitro* calibration of fura-2 are virtually identical, so that values of R_{max}, R_{min}, and β obtained from the free acid form of the dye *in vitro* can be used to convert intracellularly obtained ratios into calcium concentrations (Moore et al., 1990). This is not true of SBFI. We therefore produced an *in vivo* calibration curve for SBFI at the end of every experimental day (Williams and Fay, 1990).

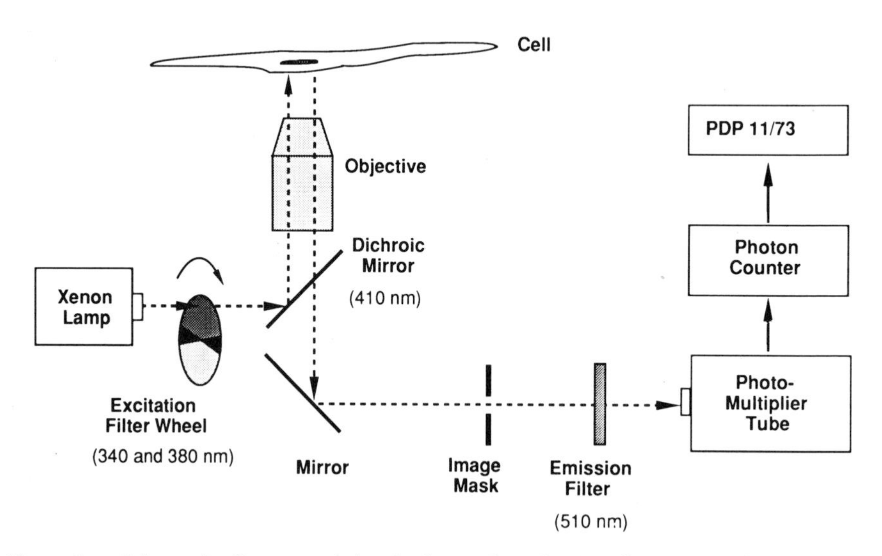

Figure 1. Schematic diagram of the dual wavelength microfluorimeter (DWM). The filter wheel contains two transmissive segments at 340 and 380 nm (bandpass 10 nm) separated by two opaque regions. The fluorescence emission was incident on the photocathode of a photomultiplier tube (Thorn EMI type 9813B), and PMT output was monitored with photon counting circuitry (E.G. & G. Ortec). Analysis of records was with the VAX 11-750. Objective is a Nikon UV-F 40X, NA=1.3.

ROLE OF THE Na^+ GRADIENT IN β-ADRENERGIC RELAXATION OF SMOOTH MUSCLE

The impetus for measuring $[Na^+]_i$ derives from the observations of Scheid et al. (1979). In that study, ISO produced an efflux of radiolabelled sodium from toad stomach cells. Since this response could be blocked with ouabain, it was suggested that β-agonists stimulate the Na^+/K^+-ATPase. The sequence of events following activation of a β-receptor are well known; there is activation of adenylate cyclase and the production of cAMP. cAMP dependent protein kinases then phosphorylate target proteins within that cell changing the cell's physiological state. The experiments of Scheid et al. (1979) indicate that the target protein in the toad stomach cell is the Na^+/K^+ ATPase, and suggested that there may be a reduction of $[Na^+]_i$. With the development of the Na^+-specific dye SBFI, we were able to directly test this hypothesis by pharmacologically targeting each of the steps following β-receptor activation. The results are presented in Table 1 and Figure 2.

Table 1. Mechanism of action of isoproterenol

Experiment	Change in $[Na^+]_i$ (in mM)	
100 μM isoproterenol	-7.0 ± 0.9	(7)
30 μM forskolin	-6.7 ± 3.1	(3)
5 mM 8-bromo-cyclic-AMP	-8.5	(2)
K^+-free saline	+10.6	(2)
K^+-free saline + 100 μM ISO	no change	
10 nM pindolol + 100 μM ISO	no change	

Summary of results from the investigation into the mechanism of action of ISO. Numbers in parentheses represent numbers of observations in that group; where appropriate standard errors are included.

In 27 cells the average $[Na^+]_i$ was 12.8 ± 4.1 mM. This is very close to the value reported with sodium-selective microelectrodes, and close to the value reported to be the K_m for sodium of the Na^+/K^+-ATPase of toad bladder (Geering and Rossier, 1979). Following application of 100 μM ISO, which is at the peak of its dose-response curve, $[Na^+]_i$ was reduced on average by 7 mM, and the response was complete in < 5 sec. The Na^+/K^+-ATPase can be inactivated by removing $[K^+]_o$ and this produced an increase in $[Na^+]_i$ since the passive inward sodium leak was no longer countered by the pump. This maneuver completely blocked the response to ISO, supporting the hypothesis that the reduction of $[Na^+]_i$ following ISO application was the result of stimulation of the Na^+/K^+ pump. This response could also be blocked with the non-selective β-antagonist pindolol, which by itself had no effect on $[Na^+]_i$. We were not therefore examining a non-specific action of the drug, occupation of the β-receptor was required. Forskolin is a relatively specific activator of adenylate cyclase, and 8-bromo-cAMP is a non-hydrolyzable analogue of cAMP. Both of these agents produced reductions in $[Na^+]_i$ comparable to that seen with ISO. These results suggest that occupation of the β-receptor stimulates the Na^+/K^+-ATPase through the activation of adenylate cyclase and cAMP. There are however a number of ways in which cAMP could stimulate activity of the enzyme. There could be a direct phosphorylation of either the α or β subunits

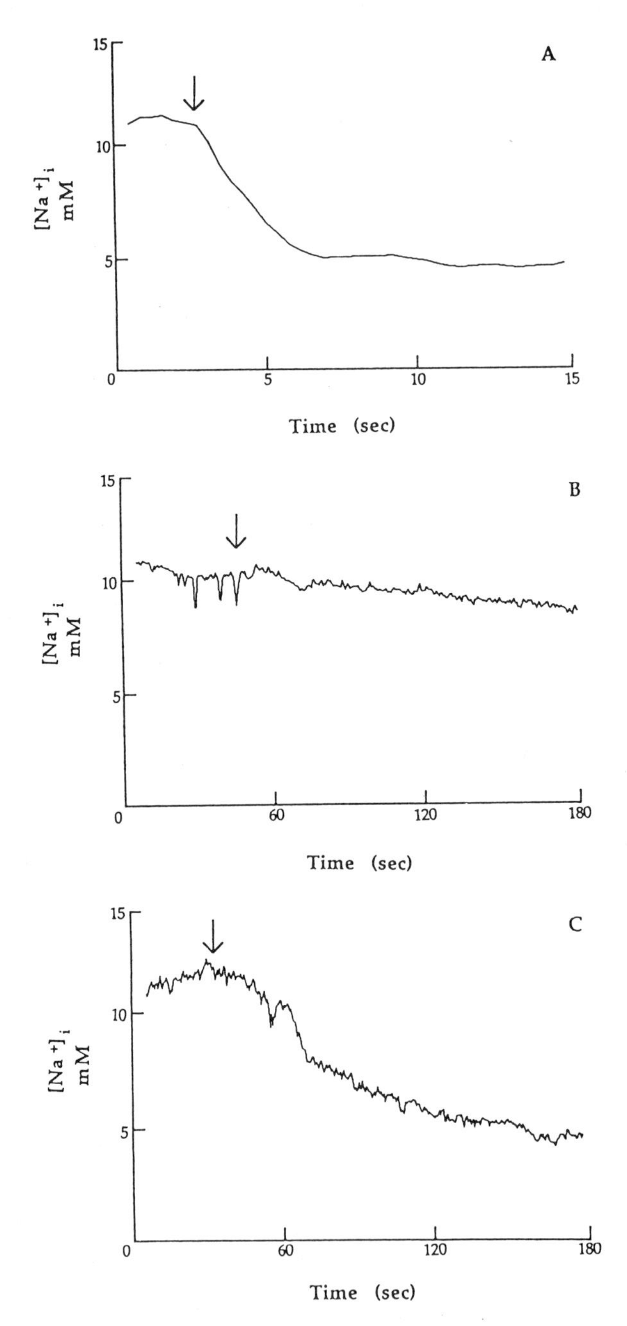

Figure 2. Records of $[Na^+]_i$ from the DWM. A) Cells were isolated as has been described, and maintained in an amphibian physiological salt solution (APS) of the following composition (in mM): 109 NaCl, 3 KCl, 0.56 Na_2HPO_4, 0.14 NaH_2PO_4, 20 $NaHCO_3$, 1.8 $CaCl_2$, 0.98 $MgSO_4$, 11.1 glucose, equilibrated with 95% O_2 - 5% CO_2, pH 7.4. At the point marked by the arrow the cell was superfused with APS containing 100 μM isoproterenol. $[Na^+]_i$ concentrations were recorded every 10 msec, the trace has been averaged with 50 msec intervals to reduce noise. B) After the response in A was recorded the shutter was closed and the cell was superfused with fresh APS for 3 minutes, followed by APS with 10 nM pindolol for 2 minutes. At the arrow 100 μM isoproterenol was re-introduced in the continued presence of pindolol. Note the different time scales in A and B. C) Five minutes after pindolol and isoproterenol were removed from the cell with fresh APS the cell was superfused, at the arrow, with 5 mM 8-Bromo-cAMP.

producing changes in K_m and/or V_{max}, there could simply be insertion of more enzymes into the plasma membrane from a readily accessible sub-plasmalemmal pool, or there could be stimulation indirectly through an intermediate protein similar to the stimulation by phospholamban of the sarcoplasmic reticular Ca^{2+}-ATPase in cardiac muscle. Our current experiments are addressing this issue.

Isoproterenol reduced $[Na^+]_i$ to a new steady state in an average of only 3 seconds. Cytoplasm does not have any appreciable sodium buffering capacity and thus the recorded changes in free ion also represent the change in total ion concentration (Harootunian et al., 1989). Given that the average surface area of a cell is about 5800 μm^2, and the rate of turnover of the Na^+/K^+-ATPase is 10000/min at rest and 74000/min after activation (Scheid et al., 1979) we have calculated that the density of Na^+/K^+-ATPase on the cell surface of a toad stomach smooth muscle cell must be about 350/μm^2 to account for the observed effect of isoproterenol on the Na^+ gradient. This estimate is within the range of pump densities measured on a variety of muscle cells (Hansen and Clausen, 1988).

How is the change in sodium concentration linked to contractility changes, if at all? We know that blockade of ISO with ouabain reduces the negative inotropic effect of ISO by roughly 50%, and that the decrease in $[Na^+]_i$ is complete before the effect of ISO on contractility is maximal. These data support the notion that the decrease in Na^+ is responsible, at least in part, for the decrease in contractility induced by ISO. But how is the change in the Na^+ gradient linked to a change in contractility? Several lines of indirect evidence suggest to us that $[Na^+]_i$ changes are linked to contractility via activation of the Na^+/Ca^{2+} exchanger. Given that $[Na^+]_i \approx 13$ mM, and $[Ca^{2+}]_i \approx 200$ nM, we calculate that the reversal potential of the Na^+/Ca^{2+} exchanger is -58 mV (Blaustein, 1974). The resting potential of these cells is -56 mV. Therefore at rest the exchanger is close to equilibrium. After application of ISO, activation of adenylate cyclase increases gK^+ which in conjunction with stimulation of the Na^+/K^+-ATPase hyperpolarizes the cell by about 11 mV (Yamaguchi et al., 1988). The reduction of $[Na^+]_i$ increases the reversal potential of the Na^+/Ca^{2+} exchanger to +19 mV, while the cell has hyperpolarized to $\approx$ -67 mV. With the reversal potential positive to the resting potential the exchanger would extrude Ca^{2+} at the expense of the increased transmembrane Na^+ gradient, and part of the negative inotropic effect of ISO would therefore be a reflection of the reduced Ca^{2+} availability. This hypothesis is supported by the observation that radiolabelled Ca^{2+} efflux is stimulated by ISO, and this stimulated efflux is blocked by ouabain (Scheid and Fay, 1984). A direct reduction of $[Ca^{2+}]_i$ by ISO has also been reported (Williams and Fay, 1986). Future experiments are directed toward establishing the relation between the effects of ISO on the Na^+ gradient and a decrease in Ca^{2+} availability following excitatory stimulation.

MYOPLASMIC Ca^{2+} HISTORY INFLUENCES Ca^{2+} METABOLISM

The second aspect of Ca^{2+} regulation in the intact cell that we will present is the effect which the previous Ca^{2+} history can have on the time course of future Ca^{2+} transients. We know from electrophysiological studies (Walsh and Singer, 1987) that the inward currents in smooth muscle cells isolated from the stomach of *Bufo marinus* are carried exclusively by Ca^{2+}.

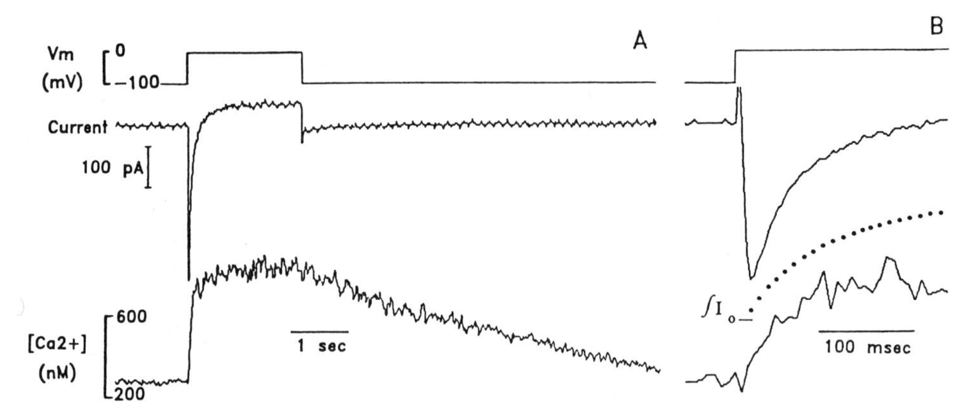

Figure 3. A) Recording of $[Ca^{2+}]_i$, membrane current, and command voltage during a 2 second voltage clamp pulse to 0 mV from a holding potential of -100 mV. The bath contained APS plus 20 mM Ca^{2+} and 10 mM TEA. The $[Ca^{2+}]_i$ record was smoothed using an averaging interval of 300 msec to reduce noise. B) Same records as Figure 3A, but displayed on a faster time scale. The Ca^{2+} trace has not been smoothed and in addition the integrated current is shown. The integral was calculated by measuring the area between the current trace and the current level at the end of the 2 second voltage command pulse. Reprinted from Becker et al., 1989 with permission.

Repolarization is largely provided by a Ca^{2+}-activated K^+ conductance. Therefore the membrane potential plays a direct role in determining the membrane Ca^{2+} permeability and $[Ca^{2+}]_i$. To isolate changes in membrane potential from the associated change in $[Ca^{2+}]_i$ we studied fura-2 loaded cells under voltage clamp conditions. Single cells were voltage clamped using micropipettes rather than patch pipettes to avoid dialyzing the internal contents of the cell against the patch electrode. The pipette contained 3 M CsCl to block the outward Ca^{2+}-activated K^+ currents, thereby facilitating assessment of the time course and magnitude of the inward Ca^{2+} currents.

The clamp protocol held the cells at -100 mV with 2 second depolarizations to 0 mV (Fig. 3A). As the cell depolarized there was an inward current of several hundred picoamps, and this current was accompanied by a large increase in $[Ca^{2+}]_i$. The increase in $[Ca^{2+}]_i$ is maintained for the 2 seconds that the cell was depolarized. After repolarization, $[Ca^{2+}]_i$ returned to baseline values at a rate that was about two orders of magnitude slower than its rate of increase. These events are depicted on a faster time scale in Figure 3B. As the inward current peaked, there was an increase in $[Ca^{2+}]_i$. The inset shows the integrated current as a function of time and the similarity between the integrated current and the rise in cytoplasmic Ca^{2+} indicates that the change in $[Ca^{2+}]_i$ is driven by the Ca^{2+} influx. In addition we have calculated from the integrated current that about 80 times as much Ca^{2+} enters the cell through the voltage activated Ca^{2+} channels than appears as free Ca^{2+} in the cytoplasm. This difference most probably reflects the Ca^{2+} buffering capacity of the myoplasm.

The voltage clamp protocol depicted in Figure 4A was used to study the rate of Ca^{2+} removal from the myoplasm. The cell was held at -100 mV, and depolarized to 0 mV for 300 msec. This was repeated 15 seconds later. Fifteen

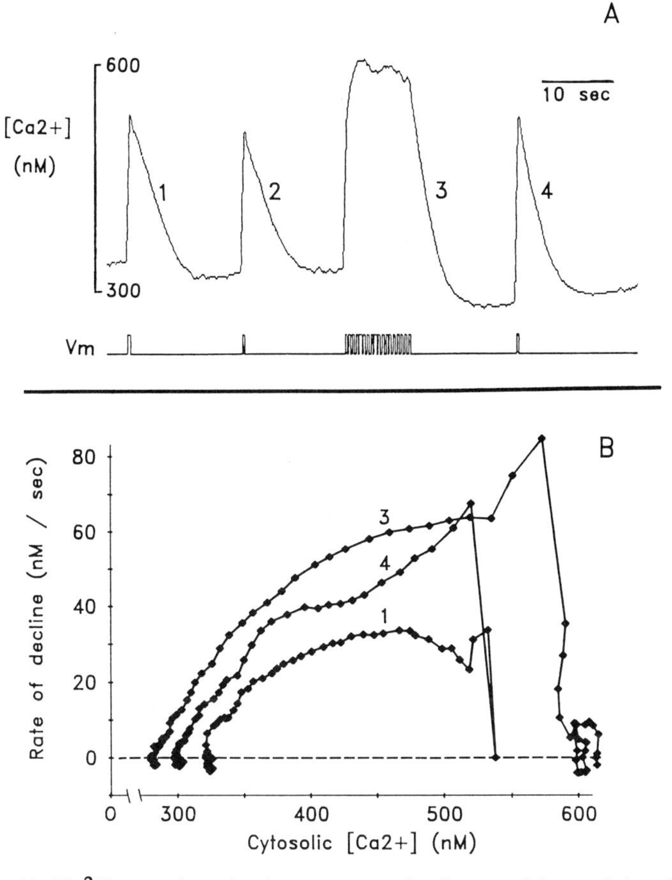

Figure 4. A) $[Ca^{2+}]_i$ records and voltage command pulses used for studying the effect of an elevation of $[Ca^{2+}]_i$ on the rate of decline of $[Ca^{2+}]_i$. Cell was clamped at -100 mV and depolarized to 0 mV for 300 msec in transients 1, 2, and 4. Nineteen such pulses separated by 200 msec were used to produce transient 3. Ca^{2+} transients were smoothed using an averaging interval of 300 msec to reduce noise. B) Rate of decline of $[Ca^{2+}]_i$ plotted as a function of $[Ca^{2+}]_i$ for transients 1, 3, and 4 from Figure 4A. Reprinted from Becker et al., 1989 with permission.

seconds after that the cell was depolarized by a train of 19 depolarizations, followed 15 seconds later by a final test depolarization. The first two Ca^{2+} transients were virtually identical. The third however peaked at a higher $[Ca^{2+}]_i$ usually around 800-1000 nM and there was an undershoot in $[Ca^{2+}]_i$ after the cell was repolarized. This was examined in greater detail in the diagrams presented in Figure 4B. This graph plots substrate concentration, Ca^{2+}, on the X-axis and the rate of Ca^{2+} removal on the Y-axis. Transients 1 and 2 were identical, so for clarity only the first is shown. The rate of Ca^{2+} removal plateaued at a relatively high value immediately following the stimulus, and its rate of removal decreased as the Ca^{2+} concentration fell. These data can be fit by a model with a passive constant leak and a pump exhibiting Michaelis-Menten properties with a K_m of several hundred nanomolar. This K_m range is close to the K_m values that have been reported for the Ca^{2+}-ATPase pumps, but much lower than the several μM K_m that has been reported for the smooth muscle Na^+/Ca^{2+} exchanger (Lucchesi et al., 1988). After the train of depolarizations however the Ca^{2+} peaked at a higher value and the rate of removal of Ca^{2+} was faster at every Ca^{2+} concentration; the Ca^{2+} returns to a level lower than it was before. These observations indicate that prolonged stimulation causes a persistent enhancement of Ca^{2+} removal mechanisms. In transient 4, the $[Ca^{2+}]_i$ achieved at the peak was no higher than in peaks 1 and 2, but the rate of removal of Ca^{2+} from the myoplasm is faster than in peak 1 at every Ca^{2+} concentration, and there was an undershoot of the basal $[Ca^{2+}]_i$, indicating that the Ca^{2+} removal mechanisms were still enhanced. By varying the time between the train of depolarizing pulses and the final test pulse, we have determined that this enhancement persists for approximately 30 seconds.

What might account for this persistent enhancement of Ca^{2+} removal processes? We know from the biochemical literature that both plasmalemmal and SR Ca^{2+}-ATPases can be stimulated by a Ca^{2+}-calmodulin dependent protein kinase (Wuytack et al., 1981; Lucchesi et al., 1988), and there are reports of phospholamban present in smooth muscle tissue (Raeymaekers and Jones, 1986). We believe that our voltage clamp data are best explained by Ca^{2+} stimulating, perhaps a slow turnover, Ca^{2+}/calmodulin-dependent protein kinase that phosphorylates a protein involved either directly or indirectly in Ca^{2+} pumping thereby enhancing Ca^{2+} removal (Becker et al., 1989).

CALMODULIN-DEPENDENT KINASE AND CONTRACTION

As indicated previously, contractility of smooth muscle cells is principally a function of the intracellular free calcium ion concentration (Fay et al., 1974). Current hypotheses suggest that cellular activation is accompanied by a rise in intracellular calcium which, through calcium/calmodulin, activates myosin light chain kinase (MLCK). MLCK phosphorylates the 20 kDa myosin light chains which in turn activates actomyosin ATPase to produce contraction (Adelstein and Eisenberg, 1980; Somlyo, 1985). This view however is not universally accepted. Some experimental observations have suggested that force production is possible in spite of low levels of phosphorylation, and calcium concentrations which are only slightly elevated above baseline values (Dillon et al., 1981; Siegman et al., 1984; Kamm and Stull, 1985; Walsh, 1987). While it is not questioned that Ca^{2+}/calmodulin will produce contraction through activation of MLCK, there have been suggestions that other, Ca^{2+}-independent, kinases can also phosphorylate myosin light chain and produce contraction, although these kinases have not been identified. It has also been suggested that there may be a thin filament associated regulatory mechanism, perhaps due to caldesmon (Walsh, 1987) or calponin (Winder and Walsh, 1989). With the recent availability of purified MLCK, purified Ca^{2+}-independent MLCK (IMLCK) (Ikebe et al., 1987), and the peptides corresponding to the autoinhibitory domain (SM1) and the Ca^{2+}/calmodulin binding domain (RS20) of MLCK (Lukas et al., 1986), we were able to probe the events linking Ca^{2+} to contraction in intact cells by following $[Ca^{2+}]_i$ and length changes in cells that were microinjected with these peptides (Itoh et al., 1989).

These experiments were done with fura-2 loaded cells on the DWM. Figure 5 shows the changes in $[Ca^{2+}]_i$ in response to microinjection of either Ca^{2+} or IMLCK, as well as the associated length changes. Following injection of Ca^{2+}, the intracellular Ca^{2+} concentration rises rapidly, and this rise in Ca^{2+} is

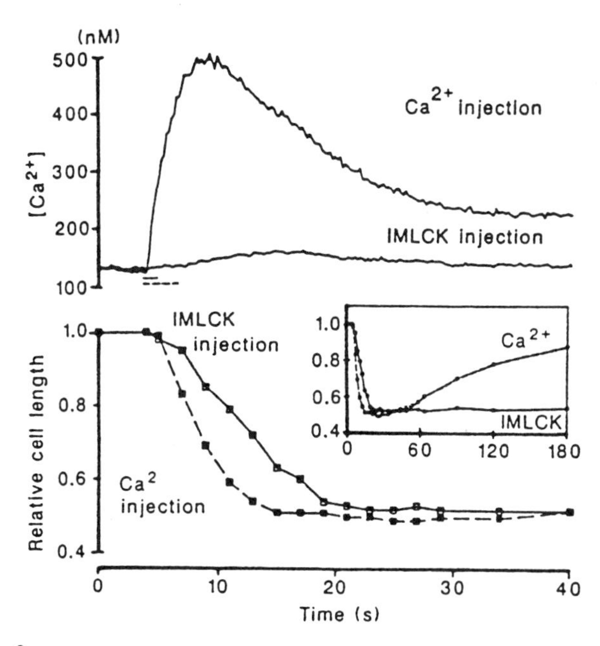

Figure 5. $[Ca^{2+}]_i$ and relative cell length following microinjection of either $CaSO_4$ (0.1 mM in 20 mM PIPES, pH 6.8) or IMLCK (1 mg/ml in 0.2 mM EGTA, 20 mM PIPES, pH 6.8) in 2 different cells. Cell length is expressed relative to that prior to injection. Lines below the Ca^{2+} traces indicate the duration of the microinjection; $CaSO_4$ solid, IMLCK dashed. Cells were impaled with borosilicate glass micropipettes (28 - 34 MΩ; 3 M KCl). Inset shows relative length changes over a longer time period. Cells were bathed in APS. Reprinted from Itoh et al., 1989 with permission.

associated with a shortening of the cell. When IMLCK is injected however there is virtually no change in $[Ca^{2+}]_i$ but the cell shortens; moreover the rate and magnitude of this shortening is indistinguishable from that produced by injection of Ca^{2+}. The estimated level of intracellular kinase activity attained by microinjection, 0.16 nmol phosphate/min/ml cell water, is comparable to the estimated peak endogenous level, 0.32 nmol phosphate/min/ml cell water. The inset of Figure 5 shows the length changes on a longer time scale. After injection of Ca^{2+} the cell eventually relaxes as $[Ca^{2+}]_i$ returns to baseline values. After injection of IMLCK however, the cell remains contracted, presumably because the intracellular kinase level remains high. In control experiments, either the injection buffer or boiled IMLCK was microinjected into the cell, and in both cases there was no change in either $[Ca^{2+}]_i$ or cell length. These results indicate that MLCK is itself sufficient to produce contraction of these smooth muscle cells.

In the next experiments we microinjected the peptides SM1, or RS20, then depolarized the cells with K^+-substituted APS to induce a contraction. Typical results are shown in Figure 6. The arrow in Figure 6 indicates the point of depolarization. In the control experiment $[Ca^{2+}]_i$ levels rose immediately following the application of K^+, the increase in Ca^{2+} was accompanied by a contraction, and the cell relaxed to pre-stimulus levels as $[Ca^{2+}]_i$ returned to baseline. In cells that were injected with the pseudosubstrate domain of IMLCK, the Ca^{2+} concentration following K^+ depolarization increased even further than in control cells. Not only the magnitude of the Ca^{2+} transient was increased, but the kinetics of the transient were grossly altered. Despite the fact that Ca^{2+} increased further, cell contraction was inhibited by about 60% following SM1 injection. A generally similar pattern was also seen following microinjection of the Ca^{2+}/calmodulin binding domain RS20; Ca^{2+} rose higher following depolarization yet contraction was significantly inhibited. The data from the control experiment and RS20 were obtained from the same cell. Therefore the difference in responses cannot be attributed to intercellular variability. In control experiments where there was a single, conservative, amino acid substitution on both peptides, no effect was observed on either the magnitude of the Ca^{2+} transient or on the associated contraction, and the responses were identical to uninjected cells. While contraction was not inhibited fully by injection of either SM1 or RS20 in these experiments, in experiments with skinned smooth muscle cells, where it is possible to have greater control over the intracellular concentration of peptide, it is possible to abolish the contraction in response to Ca^{2+}. *In vitro* assays indicate that SM1 does not inhibit either cAMP dependent protein kinase or protein kinase C at concentrations 100x greater than that needed to inhibit MLCK completely. SM1 will inhibit calmodulin dependent protein kinase II, but its potency as an inhibitor of calmodulin-dependent protein kinase II is slightly less than its potency to inhibit MLCK (Kargacin et al., 1990).

It is unlikely that the inhibitory effects of SM1 on contraction results from its ability to inhibit calmodulin-protein kinase II as a constitutively active fragment of this enzyme has no effect on contraction when applied to skinned

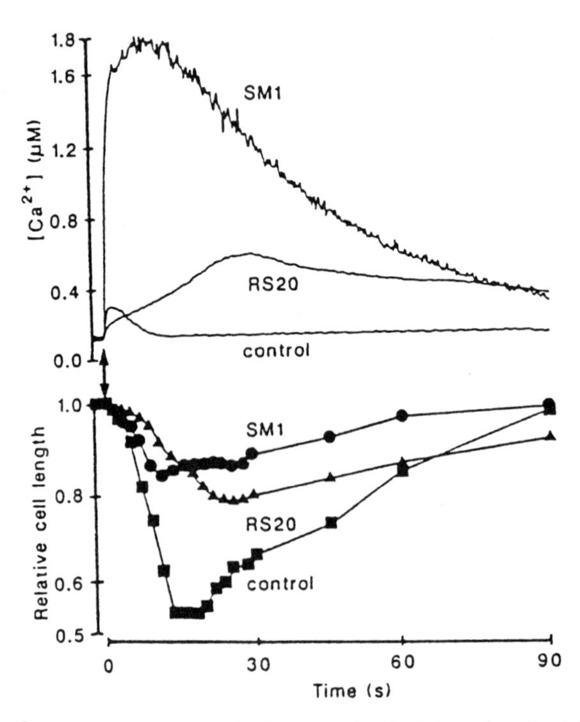

Figure 6. $[Ca^{2+}]_i$ and relative length changes of cells injected with either SM1 or RS20 and stimulated to contract with high K^+. Cells were injected with peptide 5 minutes prior to stimulation; K^+ substituted APS was applied at the vertical arrow. Estimated intracellular concentration of peptide was 35 µM. RS20 and control results were obtained from the same cell. Reprinted from Itoh et al., 1989 with permission.

cells either alone or in conjunction with IMLCK (Kargacin, Ikebe, and Fay, unpublished observations). The ability of SM1 and RS20 to inhibit the contraction induced by a rise in Ca^{2+} thus indicates that activation of MLCK is an obligatory step linking Ca^{2+} to contraction in smooth muscle. The observed effects of these peptides on the Ca^{2+} transient reveal an additional role for calmodulin and an as yet unspecified calcium/calmodulin-dependent protein kinase in feeding back in a negative manner on the processes responsible for generating the Ca^{2+} signal itself.

SUMMARY

We have demonstrated that ISO produces part of its negative inotropic action through activation of the plasmalemmal Na^+/K^+ pump, and reduction of $[Na^+]_i$. This action is mediated by the β-adrenergic receptor through activation of adenylate cyclase. The reduction of $[Na^+]_i$ is most probably translated to a change in the contractile state of the cell through activation of the Na^+/Ca^{2+} exchanger. While the exchanger is at equilibrium when the cell is at rest, after ISO it would extrude Ca^{2+} at the expense of the increased Na^+ gradient, resulting in a decreased Ca^{2+} availability and a reduction in the magnitude of subsequent contractions.

We have also seen that the previous calcium history of the myoplasm can influence the time course of future calcium transients. Prolonged large increases in $[Ca^{2+}]_i$ can accelerate the rate of its removal and depress basal $[Ca^{2+}]_i$ levels. This action is most probably mediated through a Ca^{2+}/calmodulin dependent protein kinase.

We have observed that MLCK is both necessary and sufficient to produce contraction of *Bufo marinus* stomach smooth muscle. There is also evidence that an as yet unidentified Ca^{2+}-calmodulin dependent protein kinase is acting to limit the magnitude and the duration of the Ca^{2+} transient by feeding back on processes involved in Ca^{2+} signal generation.

ACKNOWLEDGMENTS

This work was supported by a Muscular Dystrophy Post-Doctoral Fellowship to E.D.W.M., National Institutes of Health Fellowship to P.L.B., National Institutes of Health Grant (HL-14523) to F.S.F. We would like to extend our appreciation to Prof. Hirosi Kuriyama (Kyushu University, Fukuoka, Japan) for giving T.I. the opportunity for post-doctoral work.

REFERENCES

Adelstein, R. S. and Eisenberg, E., 1980, Regulation and kinetics of the actin-myosin-ATP interaction, *Ann. Rev. Biochem.*, 44: 921.

Becker, P. L., Singer, J. J., Walsh Jr., J. V., and Fay, F. S., 1989, Regulation of calcium concentration in voltage-clamped smooth muscle cells, *Science*, 244: 211.

Blaustein, M. P., 1974, The interrelationship between sodium and calcium fluxes across cell membranes, *Rev. Physiol. Biochem. Pharmacol.*, 70: 33.

Dillon, P. F., Aksoy, M. O., Driska, S. P., and Murphy, R. A., 1981, Myosin phosphorylation and the cross-bridge cycle in arterial smooth muscle, *Science,* 211: 495.

Fay, F. S., Fogarty, K., E., and Coggins, J. M., 1986, Analysis of molecular distribution in single cells using a digital imaging microscope, *in:* "Optical Methods in Cell Physiology", P. DeWeer and B. Salzberg, eds., John Wiley and Sons, New York, p. 51.

Fay, F. S., Hoffman, R., LeClaire, S., and Merriam, P., 1982, The preparation of individual smooth muscle cells from the stomach of *Bufo marinus, Methods Enzymol.,* 85: 284.

Fay, F. S., Shlevin, H. H., Granger, W. C., and Taylor, S. R., 1974, Aequorin luminescence during activation of single smooth muscle cells, *Nature,* 506: 280.

Geering, K. and Rossier, B. C., 1979, Purification and characterization of (Na^+ + K^+)-ATPase from toad kidney, *Biochim. Biophys. Acta,* 566: 157.

Grynkiewicz, G., Poenie, M, and Tsien, R., 1985, A new generation of Ca^{2+} indicators with greatly improved fluorescence properties, *J. Biol. Chem.,* 260: 3440.

Hansen, O. and Clausen, T., 1988, Quantitative determination of Na^+-K^+-ATPase and other sarcolemmal components in muscle cells, *Am. J. Physiol.,* 254: C1.

Harootunian, A. T., Kao, J. P. Y., Eckert, B. K., and Tsien, R. Y., 1989, Fluorescence ratio imaging of cytosolic free Na^+ in individual fibroblasts and lymphocytes, *J. Biol. Chem.,* 264: 19458.

Ikebe, M., Stepinska, M., Kemp, B. E., Means, A. R., and Hartshorne, D. J., 1987, Proteolysis of smooth muscle myosin light chain kinase, *J. Biol. Chem.,* 260: 13828.

Itoh, T., Ikebe, M., Kargacin, G. J., Hartshorne, D. J., Kemp, B. E., and Fay, F. S., 1989, Effects of modulators of myosin light-chain kinase activity in single smooth muscle cells, *Nature,* 338: 6211.

Kamm, K. E. and Stull, J. T., 1985, The function of myosin and myosin light chain kinase phosphorylation in smooth muscle, *Ann. Rev. Pharmacol. Toxicol.,* 25: 593.

Kargacin, G. J., Ikebe, M., and Fay, F. S., 1990, Peptide modulators of myosin light chain kinase affect smooth muscle cell contraction, *Am. J. Physiol.,* 259: C315.

Lucchesi, P., Cooney, R. A., Mangsen-Baker, T., Honeyman, W., and Scheid, C. R., 1988, Assessment of transport capacity of plasmalemmal Ca^{2+} pump in smooth muscle, *Am. J. Physiol.,* 255: C226.

Lukas, T. J., Burgess, W. H., Prendergast, F. G., Lau, W., and Watterson, D. M., 1986, Calmodulin binding domains: Characterization of a phosphorylation and calmodulin binding site from myosin light chain kinase, *Biochemistry,* 25: 1458.

Minta, A. and Tsien, R. Y., 1989, Fluorescent indicators for cytosolic sodium, *J. Biol. Chem.,* 264: 19449.

Moore, E. D. W., Becker, P. L., Fogarty, K. E., Williams, D. A., and Fay, F. S., 1990, Ca^{2+} imaging in single living cells: Theoretical and practical issues, *Cell Calcium,* 11: 157.

Raeymaekers, L. and Jones, L. R., 1986, Evidence for the presence of phospholamban in the endoplasmic reticulum of smooth muscle, *Biochim. Biophys. Acta,* 882: 258.

Scheid, C. R. and Fay, F. S., 1984, β-Adrenergic effects on transmembrane ^{45}Ca fluxes in isolated smooth muscle cells, *Am. J. Physiol.*, 246: C431.

Scheid, C. R., Honeyman, T. W., and Fay, F. S., 1979, Mechanism of β-adrenergic relaxation of smooth muscle, *Nature*, 277: 32.

Siegman, M. J., Butler, T. M., Mooers, S. U., and Michalek, A., 1984, Ca^{2+} can affect V_{max} without changes in myosin light chain phosphorylation in smooth muscle, *Pflügers Arch.*, 401: 385.

Somlyo, A. P., 1985, Excitation-contraction coupling and the ultrastructure of smooth muscle, *Circ. Res.*, 57: 497.

Walsh, Jr., J. V., and Singer, J. J., 1987, Identification and characterization of major ionic currents in isolated smooth muscle cells using the voltage-clamp technique, *Pflügers Arch.*, 408: 83.

Walsh, M. P., 1987, Caldesmon, A major actin- and calmodulin-binding protein of smooth muscle, *in:* "Regulation and Contraction of Smooth Muscle", M. J. Siegman, A. P. Somlyo, and N. L. Stephens, eds., Alan R. Liss, New York, p. 119.

Williams, D. A. and Fay, F. S., 1986, Calcium transients and resting levels in isolated cells as monitored with quin-2, *Am. J. Physiol.*, 250: C779.

Williams, D. A. and Fay, F. S., 1990, Intracellular calibration of the fluorescent calcium indicator fura-2, *Cell Calcium*, 11: 75.

Winder, S. J. and Walsh, M. P., 1989, Smooth muscle calponin, *Biochem. Soc. Trans.*, 17: 786.

Wuytack, F., De Schutter, G., and Casteels, R., 1981, Partial purification of the (Ca^{2+} and Mg^{2+})-dependent ATPase from pig smooth muscle and reconstitution of an ATP-dependent Ca^{2+}-transport system, *Biochem. J.*, 198: 265.

Yagi, S., Becker, P. L., and Fay, F. S., 1988, Relationship between force and Ca^{2+} concentration in smooth muscle as revealed by measurements on single cells, *Proc. Nat'l. Acad. Sci. U.S.A.*, 85: 4109.

Yamaguchi, H., Honeyman, T. W., and Fay, F. S., 1988, β-Adrenergic actions on membrane electrical properties of dissociated smooth muscle cells, *Am. J. Physiol.*, 254: C423.

PHARMACOMECHANICAL COUPLING: THE MEMBRANES TALK TO THE CROSSBRIDGES

Andrew P. Somlyo, Toshio Kitazawa, Sei Kobayashi,
Ming Cui Gong, and Avril V. Somlyo

Department of Physiology
University of Virginia School of Medicine
Charlottesville, VA 22908

INTRODUCTION

Excitation-contraction (E-C) coupling in smooth muscle (Table 1) is distinguished by the importance of pharmacomechanical coupling: a complex of signal transduction mechanisms that is not directly dependent on changes in membrane potential (Somlyo and Somlyo, 1968). In contrast, in striated muscle electromechanical coupling is the dominant E-C coupling mechanism. We shall emphasize pharmacomechanical coupling in this overview, not only because of its importance and our interest in it, but also because there has been major progress in our understanding of its molecular mechanisms, particularly the mechanisms of Ca^{2+} release and modulation of Ca^{2+}-sensitivity.

The sarcoplasmic reticulum (SR) is the most important physiological sink and source of activator Ca^{2+} in smooth muscle (A. V. Somlyo and Somlyo, 1971; Somlyo et al., 1971; Somlyo, 1980; Bond et al., 1984a; Kowarski et al., 1985; Somlyo and Somlyo, 1985). The SR can reduce cytoplasmic Ca^{2+} to cause relaxation, even when the calmodulin-regulated Ca-ATPase and the Na^{+}/Ca^{2+} exchanger are inactive (Bond et al., 1984a). A yet unsolved structural problem, related to the function of the SR in smooth muscle, is the need to reconcile the apparent continuity of the lumen of the extensively branching and inter-communicating network of peripheral and central SR and perinuclear space (Somlyo et al., 1971; Somlyo and Somlyo, 1975; Somlyo, 1980; Somlyo and Somlyo, 1991) with functional studies that imply the existence of pharmacologically separable intracellular Ca stores (Iino, 1986; 1989; 1990) that can be released by inositol 1,4,5-trisphosphate, but not by caffeine. Mitochondria, in smooth muscle as in other cells, accumulate Ca^{2+} only under abnormal conditions (Broderick and Somlyo, 1987; Somlyo et al., 1987).

In the face of voltage and ligand-gated Ca^{2+} influx (see below), energy dependent Ca^{2+} efflux also has to occur to maintain total cell calcium at steady state levels, and influx and efflux of Ca^{2+} across the plasma membrane can also contribute to, respectively, activation of contraction and relaxation. For the present, we can only reiterate our earlier conclusion that the "relative contributions of extracellular and intracellular Ca^{2+} to activation are not

Regulation of Smooth Muscle Contraction
Edited by R.S. Moreland, Plenum Press, New York, 1991

known, and may vary among different smooth muscles and mechanisms of activation". The experimental observation, that blockade of pharmacomechanical Ca^{2+} release (by heparin) inhibits the agonist-induced rise in $[Ca^{2+}]_i$ and contraction only in the absence, but not in the presence, of extracellular Ca^{2+} (Kobayashi et al., 1989), certainly suggests that (ileum) smooth muscle can utilize either source for activator Ca^{2+}.

Table 1. Excitation-Contraction Coupling in Smooth Muscle*

1. **ELECTROMECHANICAL COUPLING**

   A. **Ca^{2+} - influx though voltage-gated channels**

   B. Ca^{2+} - release from SR

2. **PHARMACOMECHANICAL COUPLING**

   A. **Ca^{2+} - release by messenger: $InsP_3$**

   B. **Ca^{2+} - sensitization/desensitization (regulation of kinase/phosphatase)**

   C. Ca^{2+} - influx through ligand-gated channels

*Large and bold print indicate quantitatively more significant and/or better established mechanisms.

ELECTROMECHANICAL COUPLING

Depolarization and action potentials cause an increase in cytoplasmic Ca^{2+}, while inhibitory electromechanical coupling involves the reduction of cytoplasmic Ca^{2+} through hyperpolarization of the plasma membrane. There is much evidence (Somlyo and Himpens, 1989) that depolarization can increase cytoplasmic Ca^{2+}. For example, intracellular Ca^{2+} transients can be clearly related to spontaneous, rhythmic surface membrane electrical activity (Fig. 1) (Himpens and Somlyo, 1988). The questions that still remain are: (1) the magnitude of the contribution of Ca^{2+} influx to electromechanical coupling and; (2) whether and by what mechanism depolarization and/or action potentials can release intracellular Ca^{2+}. The first question arises because of the relatively large buffering capacity of smooth muscle cytoplasm: during a maintained contraction induced by high K^+ and norepinephrine, total cytoplasmic Ca increases to approximately 200 µM (Bond et al., 1984b), far in excess of the Ca^{2+} current carried by an action potential (Johansson and Somlyo, 1980), and reflecting the concentration of cytoplasmic Ca^{2+} buffers. Concerning the second question, two lines of evidence suggest that depolarization may, in addition to promoting Ca^{2+} influx, also release intracellular Ca^{2+} and contribute to the increase in $[Ca^{2+}]_i$ during electromechanical coupling: high K^+ can release Ca^{2+} in cultured smooth muscle cells (Kobayashi et al., 1986), and spontaneous rhythmic contractions, associated with membrane electrical activity, occur in smooth muscles in Ca^{2+}-free solutions (Somlyo et al., 1971; Mangel et al., 1982). However, both of these observations are open to alternative interpretations. The replacement of Na^+ with K^+, used for depolarization,

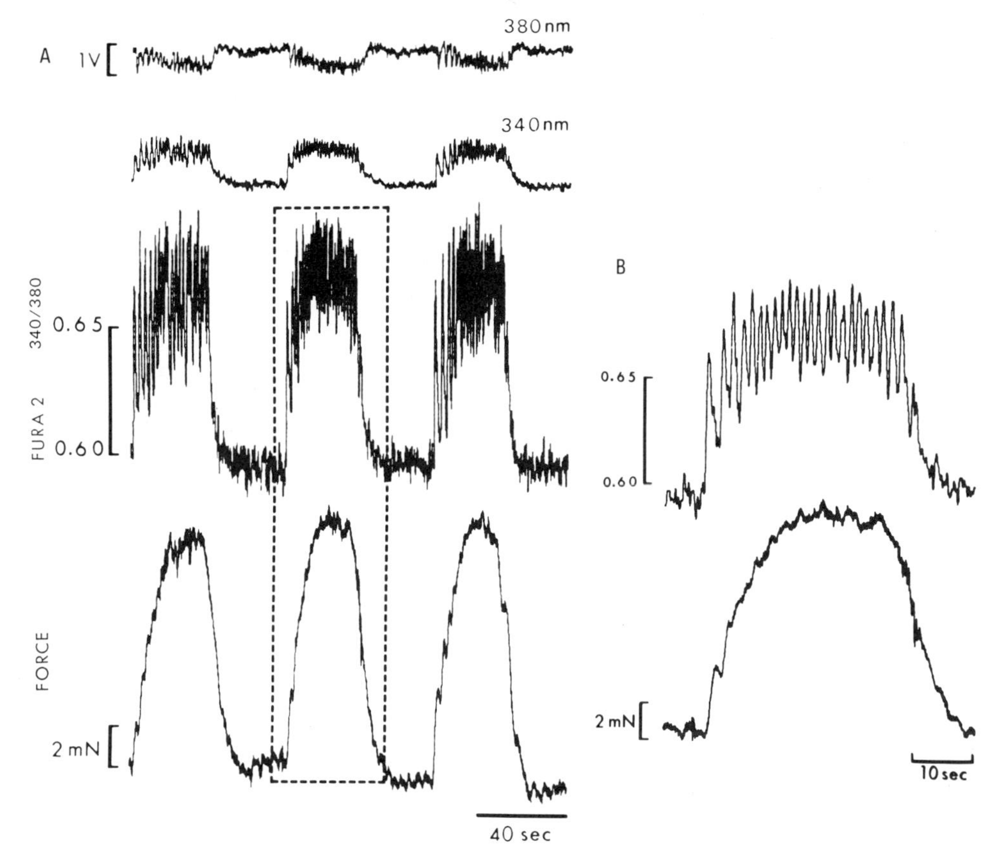

Figure 1. A: Spontaneous rhythmic contractions (lower trace) triggered by Ca^{2+} transients resembling action potentials superimposed on slow waves. The upper traces show fluorescence at 510 nm excited at, respectively 340 and 380 nm, above the ratio signal. The central wave shown in A is displayed on a faster time scale in B. Reprinted by permission from Himpens and Somlyo, 1988.

may activate phospholipase C (Smith et al., 1989) and, presumably, cause inositol 1,4,5-trisphosphate ($InsP_3$)-induced Ca^{2+} release. The rhythmic electrical activity (Mangel et al., 1982) accompanying the aforementioned rhythmic contractions in Ca^{2+}-free solutions, could be the result, rather than the cause, of intracellular Ca^{2+} release: oscillatory Ca^{2+} release from the SR can cause fluctuations in membrane potential, due to Ca^{2+}-induced increases in K^+ conductance (Desilets et al., 1989). The mechanism of "electromechanical" Ca^{2+} release, if this occurs, is even less well understood. In the presence of extracellular Ca^{2+}, the Ca^{2+} current carried by action potentials could trigger Ca^{2+}-induced Ca^{2+}-release (CICR), but the levels of Ca^{2+} required for CICR, at least in permeabilized smooth muscle, are relatively high (Iino, 1989). It is not known whether a fast rate of the increase in Ca^{2+} is also critical for CICR in smooth, as in cardiac, muscle (Fabiato, 1985). However, inhibition of CICR with procaine does not reduce Ca^{2+}-release ($t_{1/2} \approx$ 200ms at saturating [$InsP_3$]) by photo released $InsP_3$ (Somlyo et al., 1990a; 1990b) as would be expected in the presence of a "positive feedback" through CICR.

Considerable progress has been made in characterizing, by patch-clamp methods, the channels responsible for voltage-gated Ca^{2+} influx (Sperelakis and Ohya, 1989). Of the two main voltage-gated Ca^{2+}-channels, L- and T-type, the former are thought to play a greater role in E-C coupling, as T-type channels have a rather low conductance and are more readily inactivated. L-type channels have a higher conductance and are sensitive to inhibition by Ca^{2+}-channel blockers: verapamil, its cogeners and dihydropyridines (Bean et al., 1986; Benham and Tsien, 1987a). L-type channels are not fully inactivated by physiological levels of depolarization, thus permitting passage of a "window current" (Imaizumi et al., 1989) that could account for sustained Ca^{2+} influx during maintained depolarization. The effect of Ca^{2+} channel blockers in reducing cytoplasmic Ca^{2+} and relaxing smooth muscle (Fleckenstein 1983; Himpens and Somlyo, 1988) is consistent with a physiological role of these channels in E-C coupling. Voltage-gated Ca^{2+} channels can be modulated by agonists, but there remains a curious discrepancy between different experimental results: α-adrenergic agonists have been reported to both inhibit (Droogmans et al., 1987) or facilitate (Benham and Tsien, 1987a; 1988; Nelson et al., 1988; 1990) Ca^{2+} current through voltage-gated channels.

Electromechanical coupling probably makes a significant contribution to E-C coupling in phasic, spike generating smooth muscles, in which action potentials, spontaneous or evoked by agonists, trigger contractions (Bülbring, 1955; Somlyo and Somlyo, 1968; Bülbring and Szurszewski, 1974; Johansson and Somlyo, 1980; Kuriyama et al., 1982). It is less certain whether graded depolarization, in tonic smooth muscles, contributes to increases in $[Ca^{2+}]_i$ by opening voltage-gated channels: the few millivolts of depolarization by physiological concentrations of excitatory ligands may not reach the threshold required to open L-type channels, although it is possible that the threshold is more negative *in vivo* than found in patch-clamping studies. Quantitation of the contribution of voltage-gated Ca^{2+} channels to agonist-induced activation is further complicated by the fact that some Ca^{2+} channel blockers are α-adrenergic antagonists (Motulsky et al., 1983), and can also block intracellular Ca^{2+} release in permeabilized cells (Kobayashi et al., 1991) (see below).

Hyperpolarization can lead to relaxation through inhibition of action potentials and, to the extent that they are open, the closing of voltage-gated channels. The vasodilator effect of drugs like cromakalim and pinacidil, that open ATP-sensitive K^+ channels, has been ascribed to their hyperpolarizing action (Nelson et al., 1990; Videbaek et al., 1990). This may reflect inhibition of action potential firing in small resistance vessels, and does not necessarily indicate a pre-drug open state of L-type channels. The relaxant effect of the drugs on depolarized smooth muscle (Nelson et al., 1988) indicates that their action is not fully specific to the ATP-sensitive K^+ channels.

PHARMACOMECHANICAL COUPLING

Association With Electrical Events

Pharmacomechanical coupling was defined (Somlyo and Somlyo, 1968) as the collective of mechanisms that, in response to a ligand binding to its receptor, can cause contraction or relaxation without a necessary change in surface membrane potential. It was established through electrophysiological studies (Somlyo and Somlyo, 1968) that the voltage-independent action of drugs observed in muscles depolarized with high K^+ solutions (Evans et al., 1958), was not a "laboratory phenomenon", but a physiological mechanism of E-C coupling in smooth muscle. In the over twenty years since the introduction of the concept of pharmacomechanical coupling, very significant advances have been made in defining its molecular mechanisms and establishing it as a major component of physiological signal transduction in smooth muscle. Under physiological conditions (i.e., in a normally polarized smooth muscle), pharmacomechanical and electromechanical coupling can operate in parallel. For example, depolarization by excitatory agents can open and hyperpolarization by smooth muscle relaxants can close voltage-gated Ca^{2+} channels (see above), while their respective "pharmacomechanical" effects activate $InsP_3$-induced Ca^{2+}-release and modulate myosin light chain phosphorylation. Examples of "pure" pharmacomechanical coupling, changes in force in the absence of any change in membrane potential, have also been observed in some instances in normally polarized smooth muscles (Somlyo and Somlyo, 1968; Droogmans et al., 1977; Itoh et al., 1983; but cf. Trapani et al., 1981). This occasional "electromechanical dissociation" may be the result of experimental artifact, but can also be readily understood, by considering that, given the equilibrium potentials of these ions (Jones, 1980), an increase in P_{Na} or P_{Cl}, will cause depolarization and an increased P_K hyperpolarization: since agonists can change these ion permeabilities simultaneously, either directly or through second messengers, including Ca^{2+}, the effects of these conductance changes on the membrane potential can cancel each other (Somlyo and Somlyo, 1991). Furthermore, the absence of changes in membrane potential will have little or no effect on intracellular Ca^{2+} release or influx through

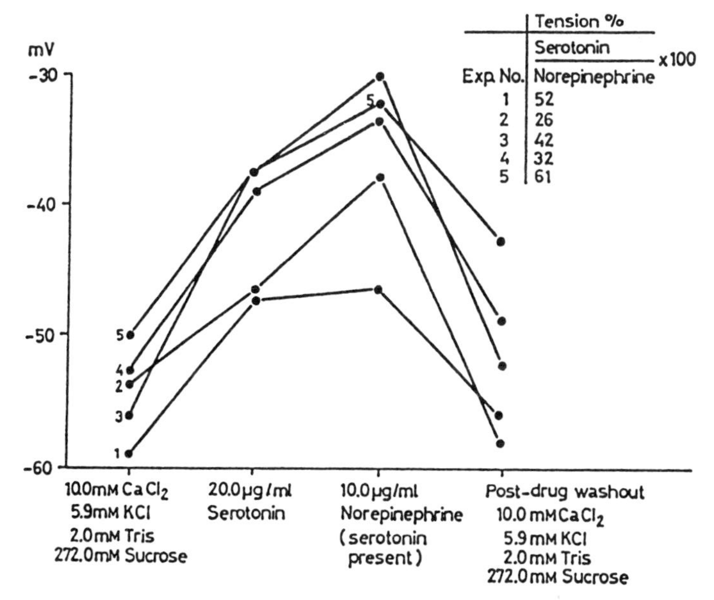

Figure 2. Unequal depolarization of rabbit main pulmonary artery in Na-free, 10.0 mM Ca solution. The points connected by one line represent membrane potentials of one vascular strip. Each point is the mean of 10 - 28 penetrations. The first and the last (post-drug washout) series of penetrations were obtained without drugs in Na-free solution. Note the greater maximal depolarization and contraction by norepinephrine than by serotonin. Reprinted by permission from A. P. Somlyo and Somlyo, 1971.

ligand-gated channels. Similarly, the hyperpolarizing effect of some (e.g. β-adrenergic) smooth muscle relaxants is dependent on external K^+ concentration (Somlyo et al., 1970; Somlyo et al., 1972), and the absence of hyperpolarization under some ionic conditions need not affect the "pharmacomechanical" relaxant effect of these agents (see below).

The interpretation of the effect of various agonists on the electrical activity of the plasma membrane is complicated by the fact that Ca^{2+} itself can open non-selective cation and Cl^- channels (Wasserman et al., 1986; Kaplan et al., 1987; Magliola and Jones, 1987; Aaronson and Jones, 1988). Because the Cl^- equilibrium potential (E_{Cl}) is generally more positive than the resting membrane potential in smooth muscle (Johansson and Somlyo, 1980), the opening of Cl^- channels by intracellularly released Ca^{2+} has a depolarizing effect. Therefore, it remains to be determined whether depolarization of smooth muscle in solutions containing Ca^{2+} as the sole external cation (Fig. 2) (A. P. Somlyo and Somlyo, 1971), was due to Ca^{2+} current through ligand-gated channels, Cl^- current through channels activated by Ca^{2+} released by $InsP_3$ (see below) or both. It is instructive to realize that unequal maximal depolarization by different agonists need not be the cause of the accompanying unequal maximal contractions, as thought earlier (Somlyo et al., 1969; A. P. Somlyo and Somlyo, 1971), but could be the result of unequal activation of the phosphatidylinositol cascade and intracellular Ca^{2+} release that, in turn, opened unequal numbers of Cl^- channels in the plasma membrane.

Mechanisms

The three potential pathways of pharmacomechanical coupling are: (1) $InsP_3$-induced Ca^{2+} release from the sarcoplasmic reticulum (SR); (2) Ca^{2+} influx through ligand-gated channels, and; (3) modulation of the Ca^{2+}-sensitivity of the contractile regulatory apparatus.

The increase in Ca^{2+} influx induced by agonists (Khalil and van Breemen, 1988; Karaki, 1989) supports the existence of ligand-gated channels, but does not exclude the possibility that this Ca^{2+} moves through voltage-gated channels (see below). The most direct evidence of ligand-gated Ca^{2+} influx was obtained recently through patch-clamp studies that revealed the existence of channels opened by extracellular ATP in vascular smooth muscle (Benham and Tsien, 1987b; Benham, 1989). These channels, like the muscarinic channels in ileum smooth muscle (Inoue and Isenberg, 1990) are non-selective, admitting both monovalent and divalent cations. Under physiological conditions (high Na^+, low Ca^{2+}) Ca^{2+} contributes only about 5% - 10% of the inward current that flows through ligand-gated channels (Benham, 1989), suggesting that, contrary to earlier expectations, these channels may have only a limited role in excitation-contraction coupling.

Ca^{2+}-Release: G-Protein-Coupled Phosphatidylinositol Cascade

The major physiological pathway of pharmacomechanical Ca^{2+}-release is the phosphatidylinositol cascade. Excitatory agents, acting on receptors coupled by G-proteins to phospholipase C, activate the latter (Baron et al., 1984; Abdel-Latif, 1986; Berridge, 1988) to hydrolyze phosphatidylinositol bisphosphate (PIP_2) to $InsP_3$ and diacylglycerol: the $InsP_3$ produced releases sufficient Ca^{2+}

from the SR to activate contraction (Somlyo et al., 1985; A. P. Somlyo et al., 1988; Somlyo et al., 1990b).

In no other system have criteria for $InsP_3$ as a physiological mediator been better documented than in smooth muscle. The effects of inhibition of G-protein activity with GDPβS (Kobayashi et al., 1988a; Kitazawa et al., 1989), the blockade of agonist-induced Ca^{2+}-release by the $InsP_3$ specific inhibitor, heparin (Kobayashi et al., 1988) and by the phospholipase C inhibitor, neomycin (Kobayashi et al., 1989) verify that interruption of any step of the phosphatidylinositol cascade prevents pharmacomechanical Ca^{2+} release. Several studies have also shown activation of phospholipase C by agonists (Abdel-Latif, 1986) and the rapid time course of $InsP_3$ production (Baron et al., 1984; Duncan et al., 1987; Miller-Hance et al., 1988; Chilvers et al., 1989). Finally, photolysis of caged phenylephrine in intact smooth muscle (A. P. Somlyo et al., 1988; Somlyo and Somlyo, 1990; Somlyo et al., 1990b) initiates contraction after a long lag (1 - 1.5s at room temperature $Q_{10} \approx 2.7$), comparable to that seen following activation by neuronal stimulation of arterial smooth muscle (Bevan and Verity, 1966) consistent with an enzymatic mechanism. In contrast, there is a very much shorter (about 30ms), concentration-dependent delay in Ca^{2+}-release upon photolysis of caged $InsP_3$ (Somlyo et al., 1990a; 1990b), followed by an additional delay (approx. 200 - 300ms) between Ca^{2+}-release and contraction (Walker et al., 1987; A. P. Somlyo et al., 1988; Somlyo and Somlyo, 1990). The latter delay is comparable to the one between depolarization-induced Ca^{2+} release and contraction (Himpens and Somlyo, 1988; Yagi et al., 1988), and likely reflects pre-phosphorylation reactions (A. V. Somlyo et al., 1988), such as the formation of active Ca^{2+}-calmodulin complexes and the activation of myosin light chain kinase by Ca-calmodulin, as well as light chain phosphorylation. The time sequence of these events is, therefore, consistent with a causal relationship between activation of receptors by agonists, production of $InsP_3$, $InsP_3$-induced Ca^{2+}-release and contraction (Walker et al., 1987; Somlyo and Himpens, 1989; Somlyo and Somlyo, 1990; Somlyo et al., 1990a)

The $InsP_3$ receptor has been isolated from smooth muscle (Chadwick et al., 1990): it is a protein of M_r 224,000 (SDS mol. wt.) and resembles the cloned cerebellar $InsP_3$ receptor (Furuichi et al., 1989; 1990). Receptors in SR vesicles incorporated into the lipid bilayers (Ehrlich and Watras, 1988; Chadwick et al., 1990) behave as channels, consistent with the kinetics of Ca^{2+} release following photolysis of caged $InsP_3$ (Somlyo and Somlyo, 1990; Somlyo et al., 1990a; 1990b). The $InsP_3$-receptor channel is modulated by adenine nucleotides: in their absence channel opening is very much reduced (Smith et al., 1985; Ehrlich and Watras, 1988). To a lesser extent, Ca^{2+} also modulates these channels (Iino, 1990). The $InsP_3$-induced Ca^{2+}-release shows no evidence of cooperative channel opening in smooth muscle (Somlyo and Somlyo, 1990; Somlyo et al., 1990a; 1990b).

In addition to the presence, in smooth muscles, of the enzymes (phosphatidylinositol kinases, phospholipase C) involved in the production of $InsP_3$ and its precursor, PIP_2, the major degradative pathway of $InsP_3$, $InsP_3$-ase [$K_D \sim 15 - 20\ \mu M$] is also present in smooth muscle (Walker et al., 1987; A. P. Somlyo et al., 1988). This is further evidence of the important physiological role of the phosphatidylinositol pathway in E-C coupling in smooth muscle, and contrasts with the low activity of $InsP_3$-ase (Walker et al., 1987; A. P. Somlyo et al., 1988) and low expression of the $InsP_3$-receptor (Furuichi et al., 1990) in striated muscles, in which $InsP_3$ does not open the "physiological"

Ca^{2+}-channel (Valdivia et al., 1990) nor plays a major physiological role in E-C coupling (Pape et al., 1988).

The G-Protein Coupled Modulation of the Ca^{2+} -Sensitivity of Force and Myosin Light Chain Phosphorylation

The Ca^{2+}-sensitivity of the contractile regulatory apparatus is not constant, but it can be modulated and the force output varied independently of free-Ca^{2+}. This component of pharmacomechanical coupling had been anticipated (Somlyo and Somlyo, 1968), but its experimental demonstration required recent advances in technology. Sensitization increases the force output for a given $[Ca^{2+}]_i$, desensitization decreases it. In intact smooth muscles, the sensitizing action of agonists is manifested by stimulus specific

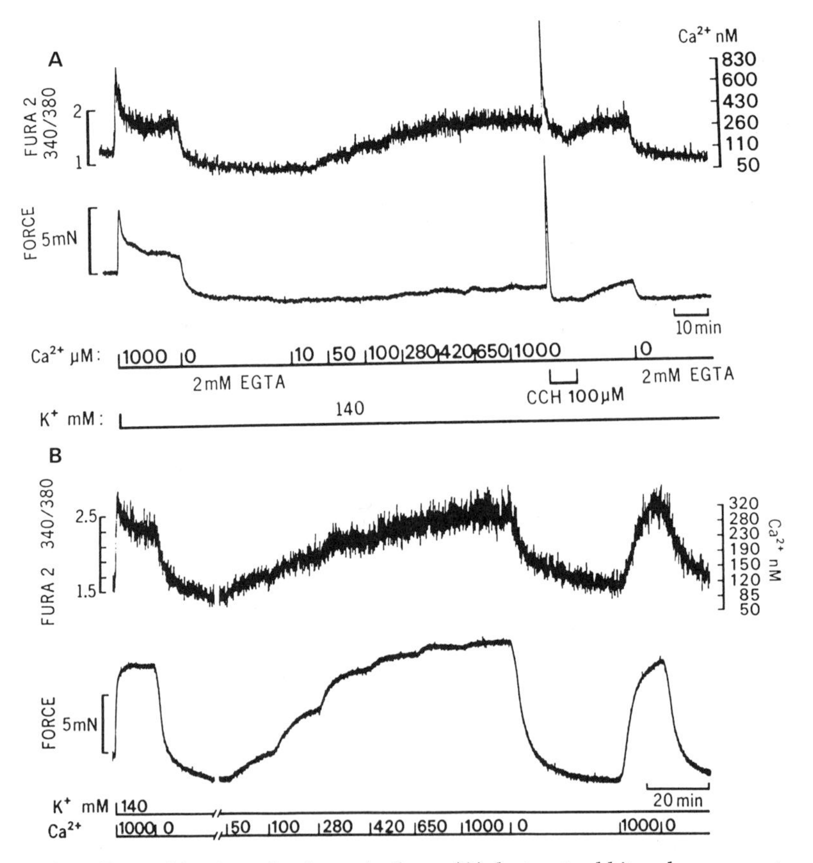

Figure 3. Desensitization of guinea pig ileum (A), but not rabbit pulmonary artery (B), to cytoplasmic Ca^{2+}. A) After the initial depolarization and incubation in Ca^{2+}-free, high K^+ solution, cumulative re-addition of extracellular Ca^{2+} to the ileum increases cytoplasmic Ca^{2+} to a level comparable to that observed during the maintained phase of the K^+ contraction, but force is markedly depressed. Addition of 100 µM carbachol evoked a rapid increase in $[Ca^{2+}]_i$ and contraction, indicating that the preceding low level of force was due to desensitization of the regulatory/contractile apparatus, rather than to damage. B) A similar experiment with rabbit pulmonary artery showed no evidence of desensitization. Reprinted by permission from Himpens et al., 1989.

differences in force/$[Ca^{2+}]_i$ ratios (Morgan and Morgan, 1984; Himpens and Casteels, 1987; Rembold and Murphy, 1988; Karaki, 1989; Himpens et al., 1990; Rembold, 1990): this ratio increases in proportion to the Ca^{2+}-sensitizing affect of a given agonist. The (partial) dissociation between force and $[Ca^{2+}]_i$ has been observed with both luminescent and fluorescent indicators, giving considerable reassurance that it is not an artifact of the Ca^{2+} indicators. Desensitization of intact smooth muscle to Ca^{2+} was first verified (Himpens et al., 1988; 1989) by the reduced force response to increases in $[Ca^{2+}]_i$ in intact, depolarized smooth muscle (Fig. 3).

Sensitization and desensitization to Ca^{2+} can also be demonstrated in permeabilized smooth muscles in which free Ca^{2+} is buffered with high concentrations of EGTA (see also Kitazawa and Somlyo, this volume). In smooth muscles that are permeabilized with α-toxin or with β-escin, receptors remain coupled to their effectors, and agonists (e.g. muscarinic, α-adrenergic, and eicosanoid) can induce contractions both through Ca^{2+}-release and through "sensitization" (Nishimura et al., 1988; Kitazawa et al., 1989; 1991; Kobayashi et al., 1989; Somlyo et al., 1989; Himpens et al., 1990; Kitazawa and Somlyo, 1990). To verify that increases in Ca^{2+} did not occur during sensitization-induced contractions some or all of the following precautions were taken in some of these experiments (Kitazawa et al., 1989; 1991; Somlyo et al., 1989; 1990b; Kitazawa and Somlyo, 1990): Ca^{2+} was buffered with 10 mM EGTA, all intracellularly releasable Ca^{2+} was removed with A23187, the diffuse intracellular distribution of EGTA, required for Ca^{2+} buffering at the myofibrils, was confirmed by electron probe microanalysis, and the absence of any change in cytoplasmic Ca^{2+} during agonist-induced contraction was verified with a fluorescent Ca^{2+} indicator. These studies, in our opinion, provide incontrovertible evidence of the modulation of Ca^{2+}-sensitivity of force development.

The Ca^{2+}-sensitizing effect of excitatory agonists is mediated by a G-protein: the increase in both force and 20 kDa myosin light chain (LC_{20}) phosphorylation are inhibited by GDPβS (Kitazawa and Somlyo, 1990, Kitazawa et al., 1991), and can also be mimicked by the non-hydrolyzable GTP analog, GTPγS (Fujiwara et al., 1989; Kitazawa et al., 1991).

Sensitization can be dissociated from agonist-induced Ca^{2+}-release, by Ca^{2+}-channel blockers (verapamil and nifedipine; Fig. 4), in permeabilized smooth muscles, in which the plasma membrane is "short-circuited" with staphylococcal α-toxin and, therefore, this dissociation can not be ascribed to inhibition of Ca^{2+}-influx through L-type Ca^{2+} channels. Dihydropyridines (e.g. nifedipine) have no α-blocking effect and GTPγS-induced Ca^{2+} release is also blocked by verapamil (Kobayashi et al., 1991), excluding the possibility that the effect is due to α_1-adrenergic block. Since Ca^{2+} channel blockers do not affect $InsP_3$ or caffeine-induced Ca^{2+} release, they probably act somewhere at the level of G-protein and phospholipase C. These and other experiments (Himpens et al., 1990) that show a lack of quantitative correlation between the Ca^{2+}-releasing and Ca^{2+}-sensitizing effects of different agonists, suggest that sensitization may not be mediated by the phosphatidylinositol cascade, and that Ca^{2+}-releasing and Ca^{2+}-sensitizing mechanisms are coupled to receptors by different G-proteins.

Desensitization to Ca^{2+} can also be demonstrated in permeabilized phasic smooth muscle: the fall in force, that, in phasic smooth muscles, follows the peak tension response to a step increase in Ca^{2+}, is clearly indicative of the time

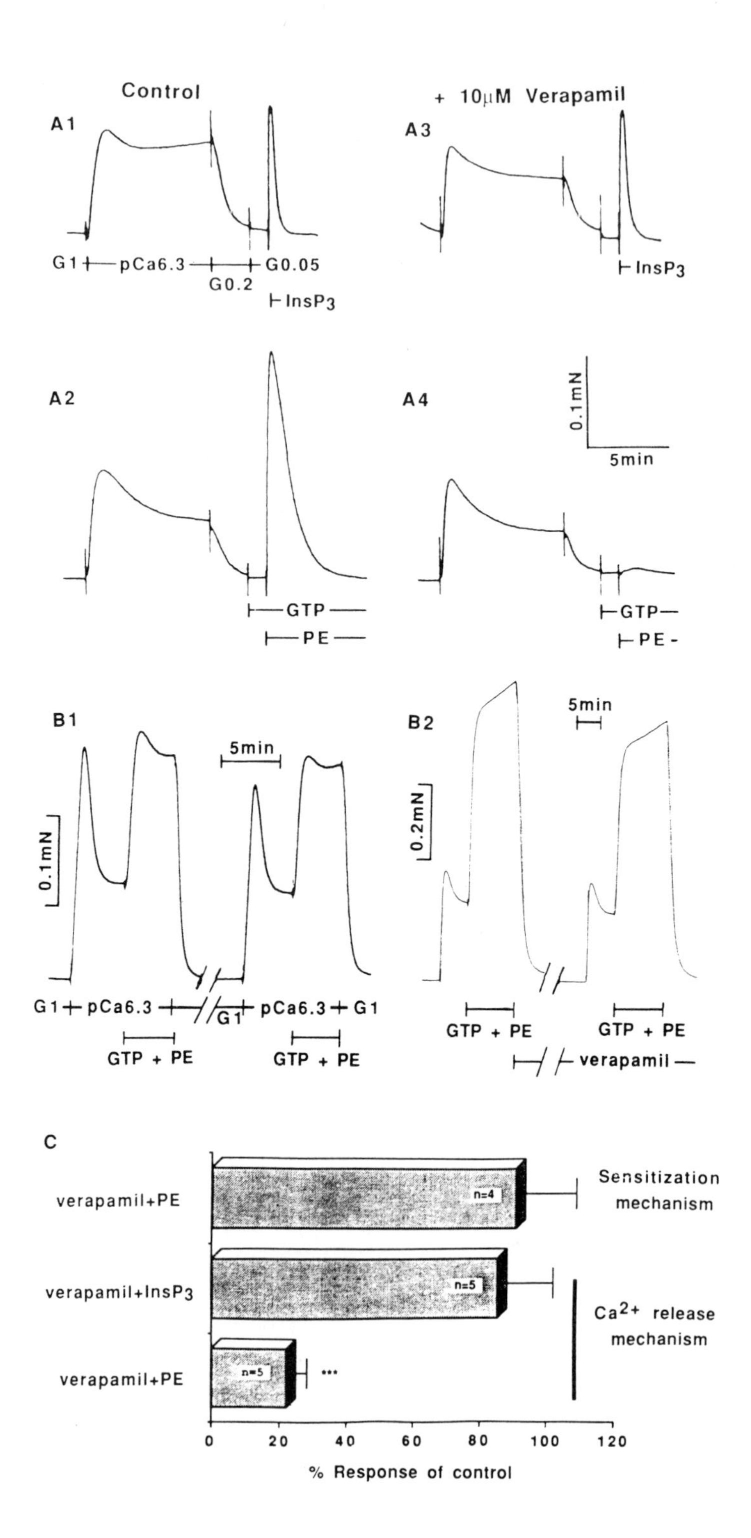
Control
+ 10μM Verapamil
A1
A3
G1 pCa6.3 G0.05
G0.2
InsP3
InsP3
A2
A4
0.1mN
5min
GTP
PE
GTP
PE
B1
B2
5min
5min
0.1mN
0.2mN
G1 pCa6.3 G1 pCa6.3 G1
GTP + PE
GTP + PE
GTP + PE
GTP + PE
verapamil
C
verapamil+PE
verapamil+InsP3
verapamil+PE
n=4
n=5
n=5
Sensitization mechanism
Ca2+ release mechanism
0
20
40
60
80
100
120
% Response of control

and/or Ca^{2+}-dependent reduction in the sensitivity of the contractile regulatory apparatus to Ca^{2+} (Somlyo et al., 1989; Kitazawa and Somlyo, 1990).

We have proposed a minimal mechanism of Ca^{2+}-sensitization/desensitization, based on modulation of the LC_{20} kinase/phosphatase ratio (Somlyo et al., 1989). The significant increase in LC_{20} phosphorylation, during Ca^{2+}-sensitization of the force response to agonists in permeabilized smooth muscles (Kitazawa and Somlyo, 1990; this volume; Kitazawa et al., 1991) and the decrease in LC_{20} phosphorylation during desensitization (Kitazawa and Somlyo, 1990) support this hypothesis.

The biochemical mechanism of desensitization, based on the above minimal scheme, could be the result of the time and/or Ca^{2+}-dependent inactivation of LC_{20} kinase or activation of a LC_{20} phosphatase. We know of no report of a LC_{20} phosphatase that is activated by Ca^{2+} and can dephosphorylate LC_{20}, but can not exclude the possibility of its occurrence. On the other hand, there is considerable evidence indicating that several kinases, kinase A, kinase G, kinase C, and Ca^{2+}-calmodulin kinase II can phosphorylate LC_{20} kinase and so reduce its affinity for Ca^{2+}-calmodulin (Adelstein et al., 1978; Conti and Adelstein, 1981; Ikebe et al., 1985; Nishikawa et al., 1985; Hashimoto and Soderling, 1990; Ikebe and Reardon, 1990). Such inhibition of LC_{20} kinase could be responsible for desensitization to Ca^{2+}. Relaxation of permeabilized smooth muscles by kinase A and by kinase G (Rüegg et al., 1981; Rüegg and Paul, 1982; Pfitzer et al., 1984; Sparrow et al., 1984) is *prima facie* evidence of such mechanisms of desensitization: the phasic decline in force at constant Ca^{2+} (Somlyo et al., 1989; Kitazawa and Somlyo, 1990), could be due to phosphorylation and inhibition of LC_{20} kinase by Ca^{2+}-calmodulin-sensitive kinase II (Kitazawa and Somlyo, 1990; Stull et al., 1990). Such a decrease in the kinase/phosphatase activity ratio would result in reduced LC_{20} phosphorylation and force. Support for, though not proof of, this mechanism is the conversion of the phasic decline in force to a tonic response when LC_{20} kinase activity is increased with calmodulin or LC_{20} phosphatases are inhibited with okadaic acid (Somlyo et al., 1989; Kitazawa and Somlyo, 1990). Finally, recent studies (Edelman et al., 1990) show that Ca^{2+}/calmodulin kinase II can

---

Figure 4. The effect of verapamil on the contractile responses mediated by Ca^{2+} release (A) and by "Ca^{2+}-sensitization" (B) in α-toxin permeabilized portal vein smooth muscle, to phenylephrine (PE) and $InsP_3$. A) The SR was loaded with Ca^{2+} at pCa 6.3 (buffered with 10 mM EGTA) for 7 min, followed by wash-out with 0.2 mM (G0.2) and 50 mM (G0.05) EGTA solutions, as shown in A1. The strips were incubated in normal relaxing solution with 1 mM EGTA (G1) for 3 - 10 min between each protocol. A1 and A2, control responses (before verapamil treatment) to 0.1mM $InsP_3$ (A1) and to 30 μM PE in the presence of added 0.05 mM GTP (A2). A3 and A4, responses in the presence of 10 μM verapamil to $InsP_3$ (A3) and to PE in the presence of added GTP (A4). In these experimental conditions, the free Ca^{2+} concentration is too low (≤ pCa 7.0) for the agonist to contract the muscle solely by increasing the Ca^{2+} sensitivity of the contractile apparatus (Kitazawa et al., 1989). B) The SR was depleted of calcium by 10 μM A23187, and free Ca^{2+} concentration was higher (pCa6.3) than in A. B1, control responses to 30 μM PE and 0.01 mM GTP, superimposed on highly buffered Ca^{2+} (pCa 6.3 with 10 mM EGTA)-induced contraction. B2, same protocol as B1, but in the presence of 100 μM verapamil. C) Summary of the experiments illustrated in A and B. Data are mean ± S.D.% response of control on ordinate: 100% = control response of the same strip (in the absence of verapamil). ***P <0.001. Note that verapamil inhibited by about 80% the PE-induced Ca^{2+} release, but had no effect on Ca^{2+}-sensitization or on $InsP_3$-induced Ca^{2+} release. Reprinted by permission from Kobayashi et al., 1991.

phosphorylate LC_{20}, albeit at a slower rate than LC_{20} kinase, and raise the possibility of this kinase also playing an auxiliary role in the regulation of tonic contraction.

TONIC AND PHASIC SMOOTH MUSCLES

The distinction between tonic and phasic smooth muscles (Somlyo and Somlyo, 1968; Somlyo et al., 1969) is also relevant when considering modulation of the Ca^{2+}-sensitivity of the contractile regulatory apparatus. We originally classified smooth muscles as tonic or phasic, to distinguish those smooth muscles that respond to excitatory agonists with graded depolarization and to high K^+ with maintained contractions from smooth muscles that generate action potentials and respond to high K^+-induced depolarization with contractions having a large, transient component. This classification is based on cellular properties, and should not be confused with the single unit versus multi-unit classification (Bozler, 1948), based on differences in intercellular conduction.

Differences in contractile regulation and crossbridge properties, at least on a quantitative level, also distinguish tonic from phasic smooth muscles. The contractile response of permeabilized tonic and phasic smooth muscles to a constant level of Ca^{2+} are also, respectively, tonic or phasic (see Kitazawa and Somlyo, this volume). The phasic response is the result of the aforementioned desensitization of phasic muscles to Ca^{2+}, and is mediated by rapid dephosphorylation (Himpens et al., 1988; Kitazawa and Somlyo, 1990). The inherent Ca^{2+}-sensitivity of tonic smooth muscles is also higher (Kitazawa et al., 1991; Kitazawa and Somlyo, this volume). Therefore, desensitization of the regulatory apparatus to Ca^{2+} can also contribute to the phasic property of some smooth muscles. Differences at the crossbridge level were revealed both in studies using caged ATP to circumvent diffusional delays (A. V. Somlyo et al., 1988; Horiuti et al., 1989) and through measurements of maximum shortening velocity (Paul et al., 1983). These studies indicate that the rate of crossbridge cycling is significantly lower in tonic than in phasic smooth muscle (Horiuti et al., 1989).

LATCH AND COOPERATIVITY

Any discussion of the control of contraction by the plasma membrane requires that we consider the possibility and consequences of crossbridge activity that is apparently autonomous of membrane control. We refer to contractile phenomena as "apparently autonomous", when they are controlled by mechanisms that, in the current state of our knowledge, cannot be ascribed to changes in Ca^{2+} or other mechanisms that modulate the LC_{20} kinase/phosphatase ratio. The possibility of Ca^{2+}-dependent mechanisms auxiliary to LC_{20} phosphorylation, in particular thin filament-mediated regulation, has been inferred largely from biochemical studies of actin-binding proteins. The physiological function of these proteins remains to be elucidated, and we will consider them outside the scope of this discussion. In any event, even "autonomous" mechanisms are initiated, even if not directly controlled (see below), by changes at the plasma membrane. The most widely recognized instance of apparent autonomy of crossbridges from membrane control is the quantitative dissociation between LC_{20} phosphorylation and force that occurs during the late, tonic phase of contraction in smooth muscle. In the following discussion of dissociation between these two events, we make no attempt to account for the possible existence of high levels of LC_{20} phosphorylation in the absence of force development (Gerthoffer, 1987; Tansey et al., 1990). Although potentially important, the difficulty of controlling the many variables affecting contraction in intact smooth muscle (e.g. ionic strength, pH, Ca^{2+}, ATP, etc.) complicate both the interpretation of such observations and the evaluation of their physiological significance.

The economy of maintained contraction in smooth muscles, at an energy cost much lower than that of force development and orders of magnitude below the energy cost of tetanus in striated muscle, is well known (Butler and Davies, 1980; Butler et al., 1983; Krisanda and Paul, 1984; Siegman et al., 1985). It was also recognized some time ago (Somlyo and Somlyo, 1967) that, when allowed to shorten during the tonic phase of contraction, vertebrate smooth muscle does not redevelop significant force, resembling the behavior of the catch muscles of invertebrates (Twarog, 1974). Following the discovery of the important role of LC_{20} phosphorylation in contractile regulation in smooth muscle (Gorecka et al., 1976; Small and Sobieszek, 1977), Murphy and his colleagues recognized that the high economy, low shortening velocity state of vertebrate smooth muscle is also characterized by a relatively low level of LC_{20} phosphorylation (Murphy, 1988; Hai and Murphy, 1989a; 1989b). They have referred to this "economic", low phosphorylation state as "latch", and ascribe it to the slow detachment of crossbridges that had been dephosphorylated while still in the attached state. A second mechanism, proposed by Paul (1990), ascribes the latch-state to a relatively high rate constant of attachment relative to detachment rates. A third mechanism, proposed by us (Himpens et al., 1988; A. V. Somlyo et al., 1988; Horiuti et al., 1989; Somlyo and Somlyo, 1990) assigns cooperativity as a major mechanism contributing to the low phosphorylation-high force state ("latch") in smooth muscle.

Cooperativity refers to the ability of an attached crossbridge to facilitate the attachment of another, detached crossbridge, without the necessary intervention of the normal regulatory mechanism. Cooperativity was originally characterized, in striated muscle, by A. Weber (Weber and Murray, 1973), who observed that rigor (ATP-free, attached) bridges could facilitate the attachment of other crossbridges, without the normally necessary intervention of the calcium switch on troponin/tropomyosin. Furthermore, micromolar ATP concentrations (lower than the concentration of crossbridges in the fiber) could induce contractions, in skinned skeletal muscle fibers in rigor, in the absence of Ca^{2+}; these contractions are attributed to reattachment of crossbridges, detached by ATP, through the aforementioned cooperative action of the remaining, attached rigor bridges (Goldman et al., 1984).

The first direct evidence of cooperativity in smooth muscle was the demonstration that photolysis of micromolar caged ATP (or CTP) in smooth muscle (Fig. 5) (A. V. Somlyo et al., 1988) caused contraction in the absence of Ca^{2+} and LC_{20} phosphorylation. Whether cooperativity in smooth muscle is propagated through thin (actin) or thick (myosin) filaments was not determined. We suggested that phosphorylated, attached crossbridges could, like rigor bridges, cooperatively facilitate the reattachment of unphosphorylated crossbridges, leading to a high force/phosphorylation ratio. Subsequent measurements of cytoplasmic Ca^{2+} and LC_{20} phosphorylation in intact smooth muscle revealed that force can be maintained at basal or near basal levels of phosphorylation, even in phasic (ileum) smooth muscles, possibly as the result

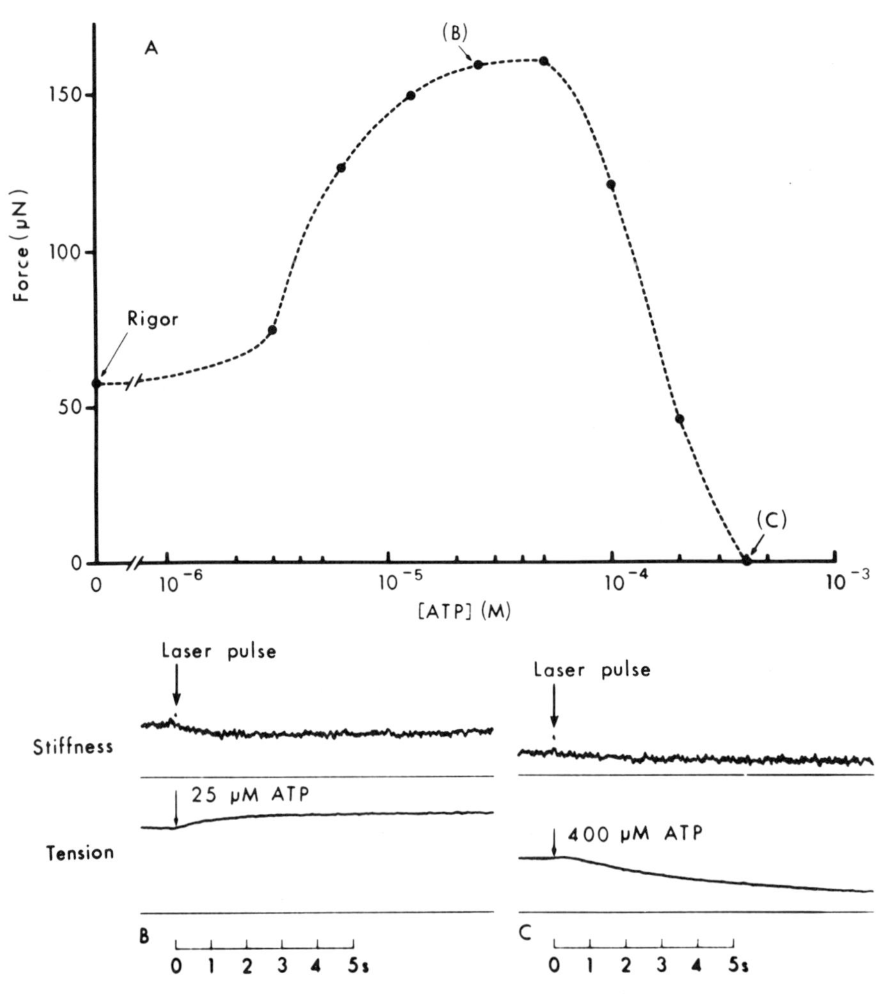

Figure 5. Tension vs [ATP] relationship for a single muscle strip, showing force development in the absence of Ca^{2+}. The muscle has been carried through the low-tension rigor series of solutions with an additional Mg^{2+}-free, 10 mM EDTA step in order to deplete ADP from the myofibrillar space. The Ca^{2+}-free rigor solution containing caged ATP did not include an ATP-regenerating system. Sequential increments of ATP in the myofilament space were obtained by varying the energy of the laser pulse. Stiffness and tension traces for points B and C are shown beneath in panels B and C. At concentrations of <50 μM ATP, the muscles contracted; at higher concentrations, the overall effect was a fall in tension. In-phase stiffness fell with all concentrations of ATP tested. The horizontal lines beneath the stiffness and tension transients in B and C represent the relaxed baseline levels. Reprinted by permission from A. V. Somlyo et al., 1988.

of cooperativity (Himpens et al., 1988). The major difference between our cooperative model and those of, respectively, Murphy (1988) and Paul (1990), is that cooperativity implies that unphosphorylated crossbridges can attach into a high force state while the models proposed by the latter authors do not include such reattachment. It is probable, however, that, as explicitly stated in Murphy's model, the detachment rate of unphosphorylated bridges is slower than that of regulated (i.e. phosphorylated) crossbridges. More recently, biochemical studies (Chacko and Eisenberg, 1990) have shown that rigor bridges can "cooperatively" enhance the ATPase activity of phosphorylated smooth muscle myosin in solution, but this study did not address the question of cooperative activation of unphosphorylated crossbridges. In motility assays, filaments that are copolymers of phosphorylated and unphosphorylated myosin can slide, while experiencing an additional load imposed by the unphosphorylated crossbridges (Sellers et al., 1985; Warshaw et al., 1990), suggesting to us that unphosphorylated crossbridges can cycle. We, although not necessarily the authors of this work, would consider this to be evidence of cooperativity.

The existence of sensitization, desensitization, and cooperativity, provide ample reason for the "loose coupling" between LC_{20} phosphorylation and force, particularly when correlations are attempted without taking into consideration the time dependence of these phenomena. Therefore, it would seem useful to consider (Fig. 6) a somewhat arbitrary division of the time course of contraction in smooth muscle into three components: (1) the early phase of activation and force development; (2) steady-state contraction having a variable duration, usually several minutes and depending on experimental conditions (e.g. temperature, level of activation) and the tonic and phasic properties of different smooth muscles; and (3) a late phase or "latch". During phase (1) there is a marked "excess" of LC_{20} phosphorylation relative to the amount of force developed. This early phosphorylation transient can be abolished by exposing smooth muscle to submaximal, rather than maximal, activation (Moreland et al., 1986), and the "discrepancy" can probably be accounted for by the lag between phosphorylation and force, secondary to series elastic contributions and possible internal, structural rearrangements between and within smooth muscle fibers. Accordingly, most authors relate phosphorylation to force during the second "steady-state". The difficulty in establishing

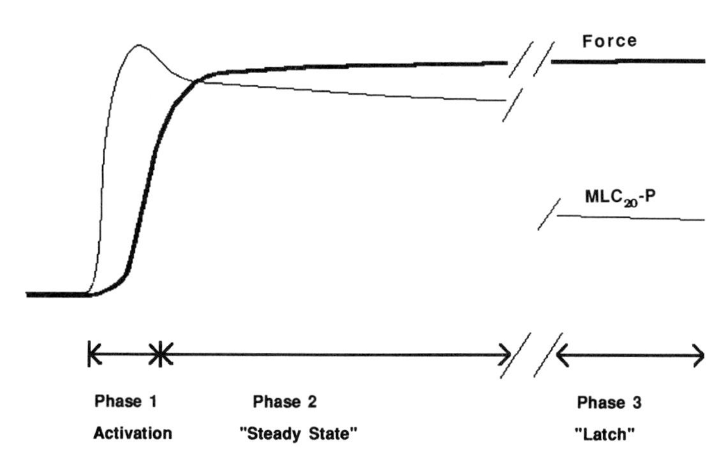

Figure 6. The three phases of contraction representative of the time dependence of the LC_{20} phosphorylation-tension relationship. During Phase I ("activation"), there is an "excess" of phosphorylation, presumably as a result of mechanical rearrangements. During Phase II ("steady-state"), a large number of cross-bridges are phosphorylated and rapidly cycling, and there is a strong correlation between phosphorylation and force, as seen in permeabilized smooth muscles. During Phase III ("latch"), a large number of crossbridges are cycling slowly due to cooperative reattachment, and the correlation between maintained force and phosphorylation is obscured. The time scale is arbitrary, and the duration of different phases is dependent on the, respective, tonic or phasic properties of different smooth muscles and on experimental conditions.

the extent and slope of the correlation originates from the definition of "steady-state" phase. The strong correlation between phosphorylation and force during this phase in permeabilized smooth muscles, is well-fitted with a quadratic function (Kitazawa et al., 1991). Clearly, if cooperativity plays a role in force maintenance, then its effect would be more appreciable and the correlation between phosphorylation and force less consistent during the latest stage of contraction. During this "latch" stage, progressive dephosphorylation of LC_{20} causes the initially graded levels of myosin phosphorylation in unequally stimulated muscles, to converge to low values (e.g. Fig. 3 in Ishikawa et al., 1988), at which time the relatively low precision of LC_{20} phosphorylation measurements obscures the correlation between phosphorylation and force evident during the earlier (e.g. Silver and Stull, 1984), "steady-state", of contraction. The force/phosphorylation ratio at this time would be a function of the time dependent dephosphorylation of myosin and of the possible redistribution of phosphorylated versus dephosphorylated crossbridges, into more or less "cooperative positions". A quantitative test of the validity of these suggestions will require determination of the minimum fraction of phosphorylated crossbridges sufficient for cooperative attachment of the number of non-phosphorylated crossbridges required for maintaining a given level of force. Therefore, whether cooperativity alone can account for the latter phases of "latch" during which, according to some reports, force is maintained at basal levels of myosin light chain phosphorylation, will also critically depend on the accuracy of phosphorylation measurements and the slope of the phosphorylation versus force relationship. Force maintenance through a cooperative mechanism would only be possible at a truly basal level of phosphorylation, if the thin filaments remained "regulated" in the presence of a "sufficient" number of attached crossbridges, regardless of whether the latter were phosphorylated. The kinetic constraints (e.g. attachment/detachment ratios) on this very speculative possibility remain to be explored. Regarding the existence of a "steady-state" (phase 2 of contraction), it is noteworthy that excellent correlation between LC_{20} phosphorylation and force has been observed in studies in permeabilized muscles (Tanner et al., 1988; Kitazawa et al., 1991; but cf. Kenney et al., 1990) in which the internal environment could be truly controlled and maintained in a steady-state.

REFERENCES

Aaronson, P. I. and Jones, A. W., 1988, Ca dependence of Na influx during treatment of rabbit aorta with NE and high K solutions, *Am. J. Physiol.*, 254: C75.

Abdel-Latif, A. A., 1986, Calcium-mobilizing receptors, polyphosphoinositides, and the generation of second messengers, *Pharmacol. Rev.*, 38: 227.

Adelstein, R. S., Conti, M. A., and Hathaway, D. R., 1978, Phosphorylation of smooth muscle myosin light chain kinase by the catalytic subunit of adenosine 3':5'-monophosphate-dependent protein kinase, *J. Biol. Chem.*, 253: 8347.

Baron, C. B., Cunningham, M., Strauss, J. F., and Coburn, R. F., 1984, Pharmacomechanical coupling in smooth muscle may involve phosphatidylinositol metabolism, *Proc. Nat'l. Acad. Sci. U.S.A.*, 81: 6899.

Bean, B. P., Sturek, M., Puga, A., and Hermsmeyer, K., 1986, Calcium channels in muscle cells isolated from rat mesenteric arteries: Modulation by dihydropyridine drugs, *Circ. Res.*, 59: 229.

Benham, C. D. and Tsien, R. W.. 1987a, Calcium-permeable channels in vascular smooth muscle: Voltage-regulated, receptor-operated, and leak channels, *in:* "Cell Calcium and the Control of Membrane Transport", L. J. Mandel and D. C. Eaton, eds., *Soc. Gen. Physiol.*, 40: 45.

Benham, C. D. and Tsien, R. W., 1987b, A novel receptor-operated Ca^{2+}-permeable channel activated by ATP in smooth muscle, *Nature*, 328: 275.

Benham, C. D. and Tsien, R. W., 1988, Noradrenaline modulation of calcium channels in single smooth cells from rabbit ear artery, *J. Physiol.*, 404: 767.

Benham, C. D., 1989, ATP-activated channels gate calcium entry in single smooth muscle cells dissociated from rabbit ear artery, *J. Physiol.*, 419: 689.

Berridge, M. J., 1988, Inositol lipids and calcium signaling, *Proc. R. Soc. Lond. B*, 234: 359

Bevan, J. A. and Verity, M. A., 1966, Postganglionic sympathetic delay in vascular smooth muscle, *J. Pharmacol. Exp. Ther.*, 152: 221.

Bond, M., Kitazawa, T., Somlyo, A. P., and Somlyo, A. V., 1984a, Release and recycling of calcium by the sarcoplasmic reticulum in guinea pig portal vein smooth muscle, *J. Physiol.*, 355: 677.

Bond, M., Shuman, H., Somlyo, A. P., and Somlyo, A. V., 1984b, Total cytoplasmic calcium in relaxed and maximally contracted rabbit portal vein smooth muscle, *J. Physiol.*, 357: 185.

Bozler, E., 1948, Conduction, automaticity and tonus of visceral smooth muscles, *Experientia*, 4: 213.

Broderick, R. and Somlyo, A. P., 1987, Calcium and magnesium transport *in situ* mitochondria: Electron probe analysis of vascular smooth muscle, *Circ. Res.*, 61: 523.

Bülbring, E., 1955, Correlation between membrane potential, spike discharge and tension in smooth muscle, *J. Physiol.*, 128: 200.

Bülbring, E. and Szurszewski, J. H., 1974, The stimulant action of noradrenaline (a-action) on guinea-pig myometrium compared with that of acetylcholine, *Proc. R. Soc. Lond. B*, 185: 225.

Butler, T. M. and Davies, R. E., 1980, High-energy phosphates in smooth muscle, *in:* "The Handbook of Physiology; The Cardiovascular System: Vascular Smooth Muscle", D. F. Bohr, A. P. Somlyo, and H. V. Sparks Jr., eds., American Physiological Society, Bethesda, p. 237.

Butler, T. M., Siegman, M. J., and Mooers, S. U., 1983, Chemical energy usage during shortening and work production in mammalian smooth muscle, *Am. J. Physiol.*, 244: C234.

Chacko, S. and Eisenberg, E., 1990, Cooperativity of actin-activated ATPase of gizzard heavy meromyosin in the presence of gizzard tropomyosin, *J. Biol. Chem.*, 265: 2105.

Chadwick, C. C., Saito, A. R., and Fleischer, S., 1990, Isolation and characterization of the inositol trisphosphate receptor from smooth muscle, *Proc. Nat'l. Acad. Sci. U.S.A.*, 87: 2132.

Chilvers, E. R., Challis, R. A. J., Barnes, P. J., and Nahorski, S. R., 1989, Mass changes of inositol (1,4,5)-trisphosphate in trachealis muscle following agonist stimulation, *Eur. J. Pharmacol.*, 164: 587.

Conti, M. A. and Adelstein, R. S., 1981, The relationship between calmodulin binding and phosphorylation of smooth muscle myosin kinase by the catalytic subunit of 3'-5' cAMP-dependent protein kinase, *J. Biol. Chem.*, 256: 3178.

Desilets, M., Driska, S. P., and Baumgarten, C. M., 1989, Current fluctuations and oscillations in smooth muscle cells from hog carotid artery: Role of the sarcoplasmic reticulum, *Circ. Res.*, 65: 708.

Droogmans, G., Raeymaekers, L., and Casteels, R., 1977, Electro- and pharmacomechanical coupling in the smooth muscle cells of the rabbit ear artery, *J. Gen. Physiol.*, 70: 129.
Droogmans, G., Declerck, I., and Casteels, R., 1987, Effect of adrenergic agonists on Ca^{2+}-channel currents in single vascular smooth muscle cells, *Pflügers Arch.*, 409: 7.
Duncan, R. A., Krabowski Jr., J. J., Davis, J. S., Polson, J. B., Coffey, R. G., Shimoda, T., and Szentivanyi, A., 1987, Polyphosphoinositide metabolism in canine tracheal smooth muscle (CTSM) in response to a cholinergic stimulus, *Biochem. Pharmacol.*, 36: 307.
Edelman, A. M., Lin, W.-H., Osterhout, D. J., Bennett, M. K., Kennedy, M. B., and Krebs, E. G., 1990, Phosphorylation of smooth muscle myosin by type II Ca^{2+}/calmodulin-dependent protein kinase, *Mol. Cell. Biol.*, 98: 87.
Ehrlich, B. E. and Watras, J., 1988, Inositol 1,4,5-trisphosphate activated by a channel from smooth muscle sarcoplasmic reticulum, *Nature*, 336: 583.
Evans, D. J. L., Schild, H. O., and Thesleff, S., 1958, Effects of drugs on depolarized plain muscle, *J. Physiol.*, 143: 474.
Fabiato, A., 1985, Time and calcium dependence of activation and inactivation of calcium-induced release of calcium from the sarcoplasmic reticulum of a skinned canine cardiac Purkinje cell, *J. Gen. Physiol.*, 85: 247.
Fleckenstein, A., 1983, "Calcium Antagonism in Heart and Smooth Muscle: Experimental Facts and Therapeutic Prospects", John Wiley and Sons, New York.
Fujiwara, T., Itoh, T., Kubota, Y., and Kuriyama, H., 1989, Effects of guanosine nucleotides on skinned smooth muscle tissue of the rabbit mesenteric artery, *J. Physiol.*, 408: 535.
Furuichi, T., Yoshikawa, S., Miyawaki, A., Wada, K., Maeda, N., and Mikoshiba, K., 1989, Primary structure and functional expression of the inositol 1,4,5-trisphosphate-binding protein P_{400}, *Nature*, 342: 32.
Furuichi, T., Shiota, C., and Mikoshiba, A., 1990, Distribution of inositol 1,4,5-trisphosphate receptor mRNA in mouse tissue, *FEBS Lett.*, 267: 85.
Gerthoffer, W. T., 1987, Dissociation of myosin phosphorylation and active tension during muscarinic stimulation of tracheal smooth muscle, *J. Pharmacol. Exp. Ther.*, 240: 8.
Goldman, Y. E., Hibberd, M. G., and Trentham, D. R., 1984, Relaxation of rabbit psoas muscle fibres from rigor by photochemical generation of adenosine 5'-trisphosphate, *J. Physiol.*, 354: 577.
Gorecka, A., Aksoy, M. O., and Hartshorne, D. J., 1976, The effect of phosphorylation of gizzard myosin on actin activation, *Biochem. Biophys. Res. Commun.*, 71: 325.
Hai, C.-M. and Murphy, R. A., 1989, Ca^{2+}, cross-bridge phosphorylation, and contraction, *Ann. Rev. Physiol.*, 51: 285.
Hai, C.-M., and Murphy, R. A., 1989, Crossbridge phosphorylation and regulation of the latch state in smooth muscle, *Am. J. Physiol.*, 254: C99.
Hashimoto, Y. and Soderling, T. R., 1990, Phosphorylation of smooth muscle myosin light chain kinase by Ca^{2+}/calmodulin-dependent protein kinase II: Comparative study of the phosphorylation sites, *Arch. Biochem. Biophys.*, 278: 41.
Himpens, B. and Casteels, R., 1987, Measurement by quin2 of changes of the intracellular calcium concentration in strips of the rabbit ear artery and of the guinea-pig ileum, *Pflügers Arch.*, 408: 32.

Himpens, B. and Somlyo, A. P., 1988, Free-calcium and force transients during depolarization and pharmacomechanical coupling in guinea-pig smooth muscle, *J. Physiol.*, 395: 507.

Himpens, B., Matthijs, G., Somlyo, A. V., Butler, T. M., and Somlyo, A. P., 1988, Cytoplasmic free calcium, myosin light chain phosphorylation, and force in phasic and tonic smooth muscle, *J. Gen. Physiol.*, 92: 713.

Himpens, B., Matthijs, G., and Somlyo, A. P., 1989, Desensitization to cytoplasmic Ca^{2+} and Ca^{2+} sensitivities of guinea-pig ileum and rabbit pulmonary artery smooth muscle, *J. Physiol.*, 413: 489.

Himpens, B., Kitazawa, T., and Somlyo, A. P., 1990, Agonist dependent modulation of Ca^{2+} sensitivity in rabbit pulmonary artery smooth muscle, *Pflügers Arch.*, 417: 21.

Horiuti, K., Somlyo, A. V., Goldman, Y. E., and Somlyo, A. P., 1989, Kinetics of contraction initiated by flash photolysis of caged adenosine trisphosphate in tonic and phasic smooth muscle, *J. Gen. Physiol.*, 94: 769.

Iino, M., 1986, Calcium dependent inositol trisphosphate-induced calcium release in the guinea-pig taenia caeci, *Biochem. Biophys. Res. Commun.*, 142: 47.

Iino, M., 1989, Calcium-induced calcium release mechanism in guinea pig *taenia caeci*, *J. Gen. Physiol.*, 94: 363.

Iino, M., 1990, Biphasic Ca^{2+} dependence of inositol 1,4,5-trisphosphate-induced Ca- release in smooth muscle cells of the guinea-pig *taenia caeci*, *J. Gen. Physiol.*, 95: 6.

Ikebe, M., Hartshorne, D. J., and Elzinga, M., 1987, Phosphorylation of the 20,000-dalton light chain of smooth muscle myosin by the calcium-activated, phospholipid-dependent protein kinase, *J. Biol. Chem.*, 262: 9569.

Ikebe, M. and Reardon, S., 1990, Phosphorylation of smooth myosin light chain kinase by smooth muscle Ca^{2+}/calmodulin-dependent multifunctional protein kinase, *J. Biol. Chem.*, 265: 8975.

Imaizumi, Y., Muraki, K., Takeda, M., and Watanabe, M., 1989, Measurement and stimulation of noninactivating Ca current in smooth muscle cells, *Am. J. Physiol.*, 256: C880.

Inoue, R. and Isenberg, G., 1990, Acetylcholine activates nonselective cation channels in guinea pig ileum through a G-protein, *Am. J. Physiol.*, 258: C1173.

Ishikawa, T., Chikiwa, T., Hagiwara, M., Mamiya, S., Saitoh, M., and Hidaka, H., 1988, ML-9 inhibits the vascular contraction via the inhibition of myosin light chain phosphorylation, *Mol. Pharmacol.*, 33: 598.

Itoh, T., Kuriyama, H., and Suzuki, H., 1983, Differences and similarities in the noradrenaline- and caffeine-induced mechanical responses in the rabbit mesenteric artery, *J. Physiol.*, 337: 609.

Johansson, B. and Somlyo, A., 1980, Electrophysiology and excitation-contraction coupling, *in:* "The Handbook of Physiology; The Cardiovascular System: Vascular Smooth Muscle", D. F. Bohr, A. P. Somlyo, and H. V. Sparks Jr., eds., American Physiological Society, Bethesda, p. 301.

Jones, A. W., 1980, Content and fluxes of electrolytes, *in:* "The Handbook of Physiology; The Cardiovascular System: Vascular Smooth Muscle", D. F. Bohr, A. P. Somlyo, and H. V. Sparks Jr., eds., American Physiological Society, Bethesda, p. 253.

Kaplan, J. H., Kennedy, B. G., and Somlyo, A. P., 1987, Calcium-stimulated sodium efflux from rabbit vascular smooth muscle, *J. Physiol.*, 388: 245.

Karaki, H., 1989, Ca^{2+} localization and sensitivity in vascular smooth muscle, *Trends Pharmacol. Sci.*, 10: 320.
Kenney, R. E., Hoar , P. E., and Kerrick, W. G. L., 1990, The relationship between ATPase activity, isometric force, and myosin light-chain phosphorylation and thiophosphorylation in skinned smooth muscle fiber bundles from chicken gizzard, *J. Biol. Chem.*, 265: 8642.
Khalil, R. A. and van Breemen, C., 1988, Sustained contraction of vascular smooth muscle: Calcium influx or C-kinase activation?, *J. Pharmacol. Exp. Ther.*, 244: 537.
Kitazawa, T., Kobayashi, S., Horiuti, T., Somlyo, A. V., and Somlyo, A. P., 1989, Receptor-coupled, permeabilized smooth muscle: Role of the phosphatidylinositol cascade, G-proteins, and modulation of the contractile response to Ca^{2+}, *J. Biol. Chem.*, 264: 5339.
Kitazawa, T. and Somlyo, A. P., 1990, Desensitization and muscarinic resensitization of force and myosin light chain phosphorylation to cytoplasmic Ca^{2+} in smooth muscle, *Biochem. Biophys. Res. Commun.*, 172: 1291.
Kitazawa, T., Gaylinn, B. D., Denney, G. H., and Somlyo, A. P., 1991, G-protein-mediated Ca^{2+} sensitization of smooth muscle contraction through myosin light chain phosphorylation, *J. Biol. Chem.*, 266: 1708.
Kobayashi, S., Kanaide, H., and Nakamura, M., 1986, Complete overlap of caffeine- and K^+ depolarization-sensitive intracellular calcium storage site in cultured rat arterial smooth muscle cells, *J. Biol. Chem.*, 261: 15709.
Kobayashi, S., Somlyo, A. P., and Somlyo, A. V., 1988, Guanine nucleotide- and inositol 1,4,5-trisphosphate-induced calcium release in rabbit main pulmonary artery, *J. Physiol.*, 403: 601.
Kobayashi, S., Somlyo, A. V., and Somlyo, A. P., 1988, Heparin inhibits the inositol 1,4,5-trisphosphate-dependent but not the independent calcium release induced by guanine nucleotide in vascular smooth muscle, *Biochem. Biophys. Res. Commun.*, 153: 625.
Kobayashi S., Kitazawa, T., Somlyo, A. V., and Somlyo, A. P., 1989, Cytosolic heparin inhibits muscarinic and α-adrenergic Ca^{2+} release in smooth muscle, *J. Biol. Chem.*, 264: 17997.
Kobayashi, S., Gong, M. C., Somlyo, A. V., and Somlyo, A. P., 1991, Ca^{2+}-channel blockers distinguish between G-protein-coupled, pharmacomechanical Ca^{2+} release and Ca^{2+} sensitization in smooth muscle, *Am. J. Physiol.*, 260: C364.
Kowarski, D., Shuman, H., Somlyo, A. P., and Somlyo, A. V., 1985, Calcium release by noradrenaline from central sarcoplasmic reticulum in rabbit main pulmonary artery smooth muscle, *J. Physiol.*, 366: 153.
Krisanda, J. M. and Paul, R. J., 1984, Energetics of isometric contraction in porcine carotid artery, *Am. J. Physiol.*, 246: C510.
Kuriyama, H., Ito, Y., Kitamura, K., and Itoh, T., 1982, Factors modifying contraction-relaxation cycle in vascular smooth muscles, *Am. J. Physiol.*, 243: H641.
Loirand, G., Pacaud, P., Mironneau, C., and Mironneau, J., 1990, GTP-binding proteins mediate noradrenaline effects on calcium and chloride currents in rat portal vein myocytes, *J. Physiol.*, 428: 517.
Magliola, L. and Jones, A. W., 1987, Depolarization-stimulated $^{42}K^+$ efflux in rat aorta is calcium- and cellular volume-dependent, *Circ. Res.*, 61: 1.

Mangel, A. W., Nelson, D. O., Rabovsky, J. L., Prosser, C. L., and Connor, J. A., 1982, Depolarization-induced contractile activity of smooth muscle in calcium-free solution, *Am. J. Physiol.*, 242: C36.

Miller-Hance, W. C., Miller, J. R., Wells, J. N., Stull, J. T., and Kamm, K. E., 1988, Biochemical events associated with activation of smooth muscle contraction, *J. Biol. Chem.*, 263: 13979.

Moreland, S., Moreland, R. S., and Singer, H. A., 1986, Apparent dissociation between myosin light chain phosphorylation and maximal velocity of shortening in KCl depolarized swine carotid artery: Effect of temperature and KCl concentration, *Pflügers Arch.*, 408: 139.

Morgan, J. P. and Morgan, K. G., 1984, Stimulus-specific patterns of intracellular calcium levels in smooth muscle of ferret portal vein, *J. Physiol.*, 351: 155.

Motulsky, H. J., Snavely, M. D., Hughes, R. J., and Insel, P. A., 1983, Interaction of verapamil and other calcium channel blockers with α_1- and α_2-adrenergic receptors, *Circ. Res.*, 52: 226.

Murphy, R. A., 1988, Special Topic: Contraction in smooth muscle cells, *Ann. Rev. Physiol.*, 51: 275.

Nelson, M. T., Standen, N. B., Brayden, J. E. and Worley III, J. F., 1988, Noradrenaline contracts arteries by activating voltage-dependent calcium channels, *Nature*, 336: 382.

Nelson, M. T., Patlak, J. B., Worley, J. F., and Standen, N. B., 1990, Calcium channels, potassium channels, and voltage dependence of arterial smooth muscle tone, *Am. J. Physiol.*, 259: C3.

Nishikawa, M., Shirokawa, S., and Adelstein, R. S., 1985, Phosphorylation of smooth muscle myosin light chain kinase by protein kinase C, *J. Biol. Chem.*, 260: 8978.

Nishimura, J., Kobler, M., and van Breemen, C., 1988, Norepinephrine and GTP-gamma-S increase myofilament Ca^{2+} sensitivity in alpha-toxin permeabilized arterial smooth muscle, *Biochem. Biophys. Res. Commun.*, 157: 677.

Pape, P. C., Konishi, M., Baylor, S. M., and Somlyo, A. P., 1988, Excitation-contraction coupling in skeletal muscle fibers injected with the $InsP_3$ blocker, heparin, *FEBS Lett.*, 235: 57.

Paul, R. J., Doerman, G., Zeugner, C., and Rüegg, J. C., 1983, The dependence of unloaded shortening velocity on Ca^{++}, calmodulin, and duration of contraction in "chemically skinned" smooth muscle, *Circ. Res.*, 53: 342.

Paul, R. J., 1990, Smooth muscle energetics and theories of crossbridge regulation, *Am. J. Physiol.*, 258: C369.

Pfitzer, G., Hofmann, F., DiSalvo, J., and Rüegg, J. C., 1984, cGMP and cAMP inhibit tension development in skinned coronary arteries, *Pflügers Arch.*, 401: 277.

Rembold, C. M. and Murphy, R. A., 1988, Myoplasmic $[Ca^{2+}]$ determines myosin phosphorylation in agonist-stimulated swine arterial smooth muscle, *Circ. Res.*, 63: 593.

Rembold, C., 1990, Modulation of the $[Ca^{2+}]$ sensitivity of myosin phosphorylation in intact swine arterial smooth muscle, *J. Physiol.*, 429: 77.

Rüegg, J. C., Sparrow, M. P., and Mrwa, U., 1981, Cyclic-AMP mediated relaxation of chemically skinned fibres of smooth muscle, *Pflügers Arch.*, 390: 198.

Rüegg, J. C. and Paul, R. J., 1982, Calmodulin and cyclic AMP-dependent protein kinase alter calcium sensitivity in porcine carotid skinned fibers, *Circ. Res.*, 50: 394.

Sellers, J. R., Spudich, J. A., and Sheetz, M. P., 1985, Light chain phosphorylation regulates the movement of smooth muscle myosin on actin filaments, *J. Cell. Biol.*, 101: 1897.

Siegman, M. J., Butler, T. M., and Mooers, S. U., 1985, Energetics and regulation of crossbridge states in mammalian smooth muscle, *Experientia*, 41: 1020.

Silver, P. J. and Stull, J. T., 1984, Phosphorylation of myosin light chain and phosphorylase in tracheal smooth muscle in response to KCl and carbachol, *Mol. Pharmacol.*, 25: 267.

Small, J. V. and Sobieszek, A., 1977, Ca-regulation of mammalian smooth muscle actomyosin via a kinase-phosphatase-dependent phosphorylation and dephosphorylation of the 20,000-M_r light chain of myosin, *Eur. J. Biochem.*, 76: 521.

Smith, J. B., Smith, L., and Higgins, B. L., 1985, Temperature and nucleotide dependence of calcium release by myo-inositol 1,4,5-trisphosphate in cultured vascular smooth muscle cells, *J. Biol. Chem.*, 260: 14413.

Smith, J. B., Dwyer, S. C., and Smith, L., 1989, Decreasing extracellular Na^+ concentration triggers inositol polyphosphate production and Ca^{2+} mobilization, *J. Biol. Chem.*, 264: 831.

Somlyo, A. P. and Somlyo, A. V., 1968, Electromechanical and pharmacomechanical coupling in vascular smooth muscle, *J. Pharmacol. Exp. Ther.*, 59: 129.

Somlyo, A. P. and Somlyo, A. V., 1971, Electrophysiological correlates of the inequality of maximal vascular smooth muscle contraction elicited by drugs, *in:* "Vascular Neuroeffector Systems", J. A. Bevan, R. F. Furchgott, and A. P. Somlyo, eds., Karger, Basel, p. 216.

Somlyo, A. P., Devine, C. E., Somlyo, A. P., and North, S. R., 1971, Sarcoplasmic reticulum and the temperature-dependent contraction of smooth muscle in calcium free solutions, *J. Cell Biol.*, 51: 722.

Somlyo, A. P., Somlyo, A. V., and Smiesko, V., 1972, Cyclic AMP and vascular smooth muscle, *in:* "Advances in Cyclic Nucleotide Research: Vol. I", R. Paoletti and G. A. Robinson, eds., Raven Press, New York, p. 175.

Somlyo, A. P. and Somlyo, A. V., 1975, Ultrastructure of smooth muscle, *in:* "Methods in Pharmacology, Vol. 3", E. E. Daniels and D. M. Paton, eds., Plenum Press, New York, p. 3.

Somlyo, A. P. and Somlyo, A. V., 1985, Excitation-contraction coupling and the ultrastructure of smooth muscle, *Circ. Res.*, 57: 497.

Somlyo, A. P., Somlyo, A. V., Bond, M., Broderick, R., Goldman, Y. E., Shuman, H., Walker, J. W., and Trentham, D. R., 1987, Calcium and magnesium movements in cells and the role of inositol trisphosphate in muscle, *in:* "Cell Calcium and Control of Membrane Support", L. J. Mandel and D. C. Eaton, eds., *Soc. Gen. Physiol.*, 42: 77.

Somlyo, A. P., Walker, J. W., Goldman, Y. E., Trentham, D. R., Kobayashi, S., Kitazawa, T., and Somlyo, A. V., 1988, Inositol trisphosphate, calcium and muscle contraction, *Phil. Trans. R. Soc. Lond. B*, 320: 399.

Somlyo, A. P. and Himpens, B., 1989, Cell calcium and its regulation in smooth muscle, *FASEB J.*, 3: 2266.

Somlyo, A. P., Kitazawa, T., Himpens, B., Matthijs, G., Horiuti, K., Kobayashi, S., Goldman, Y. E., and Somlyo, A. V., 1989, Modulation of Ca^{2+}-sensitivity and of the time course of contraction in smooth muscle: A major role of protein phosphatases?, *in:* "Adv. Prot. Phosphatases; Vol. 5", W. Merleude and J. DiSalvo, eds., Leuven University Press, Leuven, p. 181.

Somlyo, A. P. and Somlyo, A. V., 1990, Flash photolysis studies of excitation-contraction coupling, regulation, and contraction in smooth muscle, *Ann. Rev. Physiol.*, 52: 857.

Somlyo, A. P. and Somlyo, A. V., 1991, Smooth muscle structure and function, *in:* "The Heart and Cardiovascular System, Vol. 1", H. A. Fozzard, R. B. Jennings, E. Haber, A. M. Katz, and H. E. Morgan, eds., Raven Press, New York, in press.

Somlyo, A. V. and Somlyo, A. P., 1967, Active state and catch-like state in rabbit main pulmonary artery., *J. Gen. Physiol.*, 500: 168.

Somlyo, A. V., Vinall, P., and Somlyo, A. P., 1969, Excitation-contraction coupling and electrical events in two types of vascular smooth muscle, *Microvasc. Res.*, 1: 354.

Somlyo, A. V., Haeusler, G., and Somlyo, A. P., 1970, Cyclic adenosine-monophosphate: Potassium-dependent action on vascular smooth muscle membrane potential, *Science*, 169: 490.

Somlyo, A. V. and Somlyo, A. P., 1971, Strontium accumulation by sarcoplasmic reticulum and mitochondria in vascular smooth muscle, *Science*, 174: 955.

Somlyo, A. V., 1980, Ultrastructure of vascular smooth muscle, *in:* "The Handbook of Physiology; The Cardiovascular System: Vascular Smooth Muscle", D. F. Bohr, A. P. Somlyo, and H. V. Sparks Jr., eds., American Physiological Society, Bethesda, p. 33.

Somlyo, A. V., Bond, M., Somlyo, A. P., and Scarpa, A., 1985, Inositol trisphosphate-induced calcium release and contraction in vascular smooth muscle, *Proc. Nat'l. Acad. Sci. U.S.A.*, 82: 5231.

Somlyo, A. V., Goldman, Y. E., Fujimori, T., Bond, M., Trentham, D. R., and Somlyo, A. P., 1988, Cross-bridge kinetics, cooperativity and negatively strained cross-bridges in vertebrate muscle: A laser flash photolysis study, *J. Gen. Physiol.*, 91: 165.

Somlyo, A. V., Horiuti, K., Kitazawa, T., Trentham, D. R., and Somlyo, A. P., 1990a, Kinetics of $InsP_3$-induced Ca^{2+} release in smooth muscle isolated from guinea-pig portal vein, *J. Physiol.*, 429: 14P.

Somlyo, A. V., Kitazawa, T., Horiuti, K., Kobayashi, S. Trentham, D. R., and Somlyo, A. P., 1990b, Heparin-sensitive inositol trisphosphate signaling and the role of G-proteins in Ca^{2+} release and contractile regulation in smooth muscle, *in:* "Frontiers in Smooth Muscle Research", N. Sperelakis and J. D. Wood, eds. Wiley-Liss, New York, p. 167.

Sparrow, M. P., Pfitzer, G., Gagelmann, M., and Rüegg, J. C., 1984, Effects of calmodulin, Ca^{2+}, and cAMP protein kinase on skinned tracheal smooth muscle, *Am. J. Physiol.*, 246: C308.

Sperelakis, N. and Ohya, Y., 1989, Electrophysiology of vascular smooth muscle. Physiol. Pathophysiol. Heart, II. *Coronary Circ.*, 38: 773.

Stull, J. T., Hsu, L.-C., Tansey, M. G., and Kamm, K. E., 1990, Myosin light chain kinase phosphorylation in tracheal smooth muscle, *J. Biol. Chem.*, 265: 16683.

Tanner, J. A., Haeberle, J. R., and Meiss, R. A., 1988, Regulation of glycerinated smooth muscle contraction and relaxation by myosin phosphorylation, *Am. J. Physiol.*, 255: C34.

Tansey, M. G., Hori, M., Karaki, H., Kamm, K. E., and Stull, J. T., 1990, Okadaic acid uncouples myosin light chain phosphorylation and tension in smooth muscle, *FEBS Lett.*, 270: 219.

Trapani, A., Matsuki, N., Abel, P. W., and Hermsmeyer, K., 1981, Norepinephrine produces tension through electromechanical coupling in rabbit ear artery, *Eur. J. Pharmacol.*, 72: 87.

Twarog, B. M., 1974, Aspects of smooth muscle function in molluscan catch muscle, *Physiol. Rev.*, 56: 829.

Valdivia, C., Valdivia, H. H., Potter, B. V. L., and Coronado, R., 1990, Ca^{2+} release by inositol-trisphosphate in isolated triads of rabbit skeletal muscle, *Biophys. J.*, 57: 1233.

Videback, L. M., Aalkjaer, C., Hughes, A. D., and Mulvany, M. J., 1990, Effect of pinacidil on ion permeability in resting and contracted resistance vessels, *Am. J. Physiol.*, 259: H14.

Walker, J. W., Somlyo, A. V., Goldman, Y. E., Somlyo, A. P., and Trentham, D. R., 1987, Kinetics of smooth and skeletal muscle activation by laser pulse photolysis of caged inositol 1,4,5-trisphosphate, *Nature*, 327: 249.

Warshaw, D. M., Desrosiers, J. M., Work, S. S., and Trybus, K. M., 1990, Smooth muscle myosin cross-bridge interactions modulate actin filament sliding velocity in vitro, *J. Cell Biol.*, 8: 453.

Wasserman, A. J., McClellan, G., and Somlyo, A. P., 1986, Cellular and sub-cellular transport of sodium, potassium, magnesium and calcium in sodium loaded vascular smooth muscle: Electron probe analysis, *Circ. Res.*, 58: 790.

Weber, A. and Murray, J. M., 1973, Molecular control mechanisms in muscle contraction, *Physiol. Rev.*, 53: 612.

Yagi, S., Becker, P. L., and Fay, F. S., 1988, Relationship between force and Ca^{2+} concentration in smooth muscle as revealed by measurements on single cells, *Proc. Nat'l. Acad. Sci. U.S.A.*, 85: 4109.

CHARACTERISTICS OF THE VOLTAGE-DEPENDENT CALCIUM CHANNEL IN SMOOTH MUSCLE: PATCH-CLAMP STUDIES

Kenji Kitamura, Noriyoshi Teramoto, Masahiro Oike,
Zhiling Xiong, Shunichi Kajioka, Yoshihito Inoue,
Bernd Nilius, and Hirosi Kuriyama

Department of Pharmacology
Faculty of Medicine
Kyushu University
Fukuoka 812, Japan

INTRODUCTION

Visceral smooth muscle cells, including vascular smooth muscle cells, possess various types of Ca channels. The voltage-dependent Ca channel is commonly observed in many tissues and is thought to play an important role in the generation of action potentials. Neural transmitters, hormones, autacoids, peptides and other substances activate individual receptors and cause activation of the receptor-operated ion channels which are permeable to Na and Ca, and in some tissues, Cl ion. Thus, receptor activation may induce an influx of Ca via activation of the receptor-operated channel and voltage-dependent Ca channel, and also induce release of Ca from the sarcoplasmic reticulum (SR) via synthesis of inositol 1,4,5-trisphosphate (IP_3). In addition, the concentration gradient between extra- and intra-cellular Ca (2.5 mM and 100 nM, respectively) may promote the passive influx of Ca. However, analysis of this current has not yet been made in detail. In this chapter, we discuss mainly the features of the voltage-dependent Ca channel recorded from visceral smooth muscle cells using voltage- and patch-clamp procedures, and also compare their characteristics to those in cardiac muscle cells.

GENERAL FEATURES OF IONIC CURRENTS RECORDED FROM VISCERAL SMOOTH MUSCLE CELLS

Before describing the features of the voltage-dependent Ca channel, the features of K and Na currents, measured using voltage- and patch-clamp procedures, will be briefly introduced.

Regulation of Smooth Muscle Contraction
Edited by R.S. Moreland, Plenum Press, New York, 1991

General Features of the K Current

Using whole cell voltage- and patch-clamp procedures, the K channel in visceral smooth muscle cell membranes has been subdivided into more than ten different types from differences in biophysical parameters, ionic sensitivities, and drug actions. However, the conductance of the unitary K current varies with the experimental conditions (such as the ionic composition in the pipette and bath, the holding potential level, and bath temperature). For example, the K channel can be classified into Ca-dependent and independent K currents (Benham et al., 1985; Inoue et al., 1985, 1986; Benham and Bolton, 1986). Using the patch-clamp procedure, the former can be subtyped into large (KL, max-K, or BK), middle (KM), and small conductance (KS or SK) Ca-dependent K channels. The KL and KS are cytosolic Ca-dependent channels and the KM is the extracellular Ca-dependent one as estimated from experiments made using cell-free or cell-attached patch-clamp procedures (Inoue et al., 1986). The Ca-independent K channel is a small conductance K channel (10 - 20pS). The Ca-sensitive K channels have been classified using various drugs, as, for example, apamin-sensitive (KS) or -insensitive (KL) (Standen et al., 1989; Kajioka et al., 1990), charybdotoxin-sensitive (KL) or -insensitive (KS), tetraethylammonium (TEA)-sensitive (KL) or -less sensitive (KS), and 4-aminopyridine (4-AP)-sensitive (KS) or a -less sensitive (KL) (Kajioka et al., 1990). Intracellularly perfused ATP modifies the opening probability of the K channel and this ATP-sensitive K channel can be classified as Ca-sensitive (KS and KL) or -insensitive (KL). Whether the channel conductance is large or small, these ATP-sensitive channels are blocked by glibenclamide and are activated by nicorandil, cromakalim, or pinacidil (K channel openers, K channel modulator). The K channel can also be classified as voltage-dependent (delayed rectifier and 'A' current generating) or -independent (inward rectifying or anomalous rectifying). In some smooth muscles, agonists modify the receptor- or second messenger-mediated K channel (M or S channel), i.e. inhibit or activate the K channel opening (Mironneau and Savineau, 1980; Benham et al., 1987; Ohya et al., 1987; Okabe et al., 1987).

Using the whole-cell voltage-clamp procedure with prolonged depolarization, the macroscopic K current can be seen to consist of a transient Ca-sensitive K current which occurs just after the generation of the voltage-dependent Ca inward current, and subsequently a sustained Ca-independent K current and oscillatory Ca-dependent K currents (spontaneous transient outward current; STOCs) (Klöckner and Isenberg, 1985; Benham and Bolton, 1986; Ohya et al., 1987; Sakai et al., 1988; Bolton and Lim, 1989; Hume and Leblanc, 1989).

The transient outward current is sensitive to cytosolic Ca and has as a prerequisite the generation of the inward current. This Ca dependent K channel is not sensitive to apamin and is blocked by TEA. The nature of this channel suggests its involvement in the generation of the after-hyperpolarization of the action potential. When stored Ca in the SR is depleted, the generated inward current is not followed by the transient outward current (Benham et al., 1986; Ohya et al., 1987; Sakai et al., 1988; Bolton and Lim, 1989). Presumably, an influx of Ca activates the Ca-induced Ca release mechanism of the SR. Thus, Ca released from this storage site may contribute to the activation of the Ca-dependent K channel and thus generate the transient outward current.

The sustained outward current is composed of a Ca-less sensitive (or insensitive) K current and inactivation of this channel occurs with a very slow time course (Ohya et al., 1987). Distributions of Ca-less sensitive unitary K currents have been reported in smooth muscle cells (Benham and Bolton, 1983; Inoue et al., 1985).

The oscillatory outward currents occur with irregular amplitudes and frequencies, and these currents are blocked (after an initial transient enhancement) following depletion of stored Ca in the SR by application of caffeine, (Itoh et al., 1982; Ohya et al., 1987), ryanodine (Sutko and Kenyon, 1983; Hwang and van Breemen, 1987; Sakai et al., 1988), IP_3 (Suematsu et al., 1984; Ohya et al., 1988), or heparin (Kobayashi et al., 1988; Somlyo et al., 1988). Presumably Ca released from the SR by the Ca-induced Ca release mechanism (in the case of caffeine or ryanodine) or via IP_3-induced Ca release transiently enhances and then inhibits the generation of the oscillatory outward current by depletion of stored Ca (Ohya et al., 1987; Sakai et al., 1988). The general features of the transient and oscillatory outward currents correspond well with large conductance Ca-dependent K channels as measured using cell-attached or cell-free patch-clamp procedures (Sakai et al., 1988). More detailed studies of the correlation between the macroscopic current and unitary current are required.

General Features of the Na Current

A voltage-dependent Na current with fast inactivation can be recorded from the rabbit pulmonary artery and portal vein. In Ca-deficient, 2.5 mM Mn and 140 mM Na containing solution, the major part of the inward current ceases, but the transient fast Na inward current remains. This channel is activated at high membrane potential levels (-80 mV). However, at a holding potential of -60 mV, it is almost inactivated, as estimated from the steady state inactivation curve. The reversal potential level for the fast Na-current estimated from the current-voltage relationship is much lower than that of the Ca-current (+20mV vs +60mV; Okabe et al., 1988). This voltage-dependent fast Na channel recorded from the pulmonary artery and portal vein is very sensitive to tetrodotoxin (TTX; IC_{50} = 8.7 nM) and its TTX sensitivity is similar to that of the channel present in skeletal muscle and nerve cells (Hagiwara, 1983; Benoit et al., 1985), but greater than that of the channel present in cardiac muscle cells (IC_{50} of a few μM). However, Pidoplichko (1986) has reported that both TTX-sensitive and -less sensitive Na channels are present in cardiac muscle cells.

On the other hand, in cultured vascular smooth muscle cells and in the rat myometrium, while the presence of a TTX-sensitive Na channel has been reported, its sensitivity to TTX is much lower than that observed in the pulmonary artery (Amedee et al., 1986; Sturek and Hermsmeyer, 1987). Presumably, the nature of the Na channel varies by region and tissue. This Na current is enhanced, but the Ca current is inhibited by application of chloramine-T (0.3 mM; an agent that modifies Na current inactivation; Wong, 1984; Schmiedtmayer, 1985; Okabe et al., 1987). However, the physiological significance of this Na current is uncertain, because the resting membrane potential of most smooth muscle cells is about -70 to -50mV and, in this range, the Na channel is almost inactivated.

The Na current can also be recorded in unphysiological ionic environments. For example, in Ca-deficient solution (about 0.1 mM; EGTA not added, and the pipette solution containing high Cs), the Na current (a plateau formation in the action potential) occurs. The peak value of the current amplitude observed from the current-voltage relationship in Ca-deficient solution is shifted to the left (by about 10 mV) by comparison with that observed in 2.5 mM Ca containing solution (Krebs), and the reversal potential level shifts to a lower membrane potential level than that observed in Krebs solution. Generation of this current ceases on addition of dihydropyridine (DHP) derivative Ca antagonists (Ohya et al., 1986). Thus, the voltage-dependent Ca channel which is normally selectively permeable to Ca ions, may in abnormal ionic environments, allow other ions (Na and K) to permeate the membrane.

GENERAL FEATURES OF THE VOLTAGE-DEPENDENT CALCIUM CHANNEL IN SMOOTH MUSCLE

Macroscopic Ca Currents Recorded Using Whole-Cell Voltage-Clamp Procedure

To isolate the Ca current from the K current in freshly dispersed smooth muscle cells prepared using collagenase, high Cs is used in the pipette solution and 20 mM TEA added in the bath solution. When a depolarizing pulse is applied at a holding potential of -60 mV or -80 mV, a transient inward current is evoked. In the rabbit intestine, the inward current is elicited at levels of depolarization above -30 mV, and it reaches its maximum value at +10 mV depolarization. The amplitude of the inward current differs in smooth muscle cells prepared from different tissues. The polarity of the inward current is reversed at about +60 mV. The amplitude of the inward current depends on which tissue the cell is prepared from, and on the intra- and extracellular Ca (Ba) concentrations used. Thus, in the ileal smooth muscle cell, increased extracellular Ca enhances the amplitude, whereas increased intracellular Ca inhibits it, due to inactivation of the channel by Ca (the K_d value for Ca current inhibition is 100 nM). This Ca current is blocked in Mn and Ca-free solution but the amplitude is enhanced by replacement of Ca with Ba (Ohya et al., 1987: Aaronson et al., 1988). Macroscopic Ca currents have been recorded from various smooth muscle cells (Klöckner and Isenberg, 1985; Bean et al., 1986; Droogmans and Callewaert, 1986; Ganitkevich et al., 1986; Sturek and Hermsmeyer, 1986; Ohya et al., 1986, 1987; Okabe et al., 1987; Nakazawa et al., 1987; Toro and Stefani, 1987; Terada et al., 1987; Walsh and Singer, 1987; Yatani et al., 1987).

In some smooth muscle tissues, when the current-voltage relationship is observed at a holding potential of -80 mV, a small hump (second peak) can be observed at potentials more negative (-30 to -40 mV) than the peak amplitude (0 - 10mV). This small hump is not observed when the holding potential is kept at a more positive potential level (-60 mV). This hump current is resistant to Ca antagonists and is inactivated with a rapid time-course. On the other hand, the peak amplitude is markedly reduced by application of Ca antagonists and, moreover, inactivation of the current produced by depolarizing pulses occurs with a slow time-course. Therefore, in smooth muscle cells, the presence of two different Ca channels is recognized, as also observed in cardiac muscle cells (transient, T, and long lasting, L type). However, cell membranes in all smooth muscle tissues may not possess two different Ca channels; indeed, in some tissues, only the L type of Ca channel is present (rabbit small intestine, rabbit portal vein, etc). Furthermore, some tissues such as sphincter and dilator muscle cells of the iris do not possess the voltage-dependent Ca channel at all. In these cases, neurotransmitters initiate second messengers such as IP_3 (Suzuki et al., 1983; Yoshitomi and Ito, 1985), and thus mobilize the cytosolic Ca required for producing contraction. Furthermore, the distribution ratio between L and T type Ca channels may differ in different tissues. As described previously, inactivation of the Ca current occurs in L and T type channels with different time courses. Smooth muscle cells of the portal vein possess only the L-type, but the recovery from the inactivation requires a very long time course compared to that in cardiac muscle cells.

Unitary Ca Currents Recorded Using Patch-Clamp Procedures

In most smooth muscle cells, a unitary Ca current can be recorded using cell-free (fragmented) or cell-attached procedures. However, in the cell-free fragment, the unitary current rapidly ceases and this is called a 'run-down phenomenon'. Therefore, most experiments are carried out using the cell-attached patch-clamp procedure with Ba containing solution instead of Ca.

Individual investigators have measured the ionic conductance of the voltage-dependent Ca (Ba) channel, using different ionic conditions, holding potentials, and temperatures. Therefore, the values obtained are not strictly comparable.

From the results obtained by many investigators, voltage-dependent Ca channels of three different conductances can be identified in smooth muscle cell membranes, i.e. large (CaL; 25 - 30pS), middle (CaM; 12 - 15pS), and small (CaS; 7 - 9pS) conductance channels. This classification of the Ca channel differs from that made in neural cells, where the channels are termed L, T, or N (which is neither L nor T). In fact, the N type Ca channel has not yet been reported in smooth muscle cells. Only three different Ca channel distributions have been reported in the guinea-pig taenia coli, and other smooth muscle possesses either CaL and CaS, CaL and CaM, CaM and CaS, CaL alone, or CaM alone. The CaL is DHP-sensitive, whereas the CaS is insensitive. However, where CaL and CaM are present, the CaM is insensitive to DHP, but where CaM and CaS are present, the CaM is DHP-sensitive and CaS is insensitive. As yet, a distribution of CaS alone has not been reported. Upon application of nicardipine, opening of the large amplitude Ca channel current ceases but the small conductance Ca current remains unchanged. More details on the actions of Ca antagonists on channel activity will be provided later.

EFFECTS OF VARIOUS AGENTS AND MODULATING FACTORS ON THE VOLTAGE-DEPENDENT CALCIUM CHANNEL

Ionic channels, including the voltage-dependent Ca channel, in smooth muscle cell membranes are modified by various agents and modulating factors, such as endothelium-derived contracting and relaxing factors, neuro-transmitters, hormones, and peptides. The effects of these factors on the voltage-dependent Ca channel are described below.

The Effects of Endothelium-Derived Contracting and Relaxing Factors

Endothelial cells release various vasoconstricting and relaxing factors. The relaxing factors are: endothelium-derived relaxing factor (EDRF; Furchgott and Zawadzki, 1981; Vanhoutte et al., 1986) and hyperpolarizing factor (EDHF;

Komori and Suzuki, 1987a, 1987b), and prostaglandin I_2 (PGI_2; prostacycline; Moncada and Vane, 1977). The contracting factors are endothelin (Yanagisawa et al., 1988) and thromboxane A_2 (TXA_2; DeMay and Vanhoutte, 1985; Shirahase et al., 1988). Presumably, other substances may also be involved as regulating factors of the tone of vascular smooth muscle.

EDRF is thought to be NO derived from L-arginine in the endothelial cells and, in the cytosol of smooth muscle cells this factor activates the synthesis of cyclic GMP (Ignarro et al., 1987; Palmer et al., 1987; Ignarro, 1989). However, another possibility, namely the synthesis of S-nitrosocysteine, has been proposed (Myers et al., 1990). In general, cyclic GMP appears not to act directly on the voltage-dependent Ca channel, though according to one report dibutyryl cyclic GMP does inhibit the Ca channel in the guinea-pig mesenteric artery (Sperelakis and Ohya, 1990). The hyperpolarization of smooth muscle cell membranes induced by acetylcholine (ACh) has been noted by several investigators during observation of the effects of EDRF in intact tissues (endothelium present). However, such hyperpolarization is not observed in endothelium-denuded tissues. It is now clear that not all EDRFs contain the hyperpolarizing factor. In fact, EDHF is released from the endothelial cell independently of EDRF (Komori and Suzuki, 1987a). When ACh stimulates the muscarinic I (M_I) receptor it releases EDHF and when this agent stimulates the muscarinic II (M_{II}) receptor, it releases EDRF (Komori and Suzuki, 1987b; Chen and Suzuki, 1988; Nishiye et al., 1989). EDHF hyperpolarizes the membrane through activation of the Ca-independent K channel (glibenclamide-sensitive, ATP-sensitive K channel; Standen et al., 1988) and, as a consequence, the threshold required to activate the Ca channel is lowered. Thus, EDHF indirectly modifies the voltage-dependent Ca channel.

PGI_2 is reported to cause synthesis of cyclic AMP and hyperpolarize the membrane. Cyclic AMP initiates phosphorylation of the channel protein and activates the Ca-dependent K channel (Sadoshima et al., 1988; Kume et al., 1989). However, this agent has no direct effect on the voltage-dependent Ca channel. This action of cyclic AMP in smooth muscle differs from that in cardiac muscle cells.

Endothelin depolarizes the smooth muscle membrane by increasing mainly the Na ion permeability and also activates the voltage-dependent Ca channel (Nakao et al., 1989). Endothelin is released from endothelial cells by various stimulants and endothelin itself releases EDRF, PGI_2, and others from endothelial cells (Wright and Fozzard, 1988; Warner et al., 1989). Endothelin increases the open probabilities of the CaL and CaS unitary currents, as measured using cell-attached patch-clamp procedures. Endothelin augments the open probability of both CaL and CaS in the guinea pig portal vein. This action is postulated to be mediated through an unknown second messenger rather than by a direct action of endothelin on the channels.

TXA_2 directly modifies the smooth muscle membrane and produces contraction even though the membrane is only slightly depolarized (Makita, 1985). The ionic nature of the excitatory action of TXA_2 is not yet understood.

Effects of Neurotransmitters on the Voltage-Dependent Ca Channel

Neurotransmitters released from adrenergic and cholinergic nerves directly or indirectly regulate the voltage-dependent Ca channel as well as activating the receptor-operated ion channels. We do not intend to describe the receptor-operated ion channel (mainly cation) in detail, but only to describe relevant phenomena.

In intestinal smooth muscle cells, norepinephrine (NE) activates the α_1- and α_2-adrenoceptors. Alpha$_1$ adrenoceptors activate the Ca-dependent K channel which is inhibited by apamin, and is presumably not a large conductance Ca-dependent K channel. Activation of the α_2-adrenoceptor mainly increases Na channel activity and depolarizes the membrane. On the other hand, α_1-adrenoceptor activation depolarizes the vascular smooth muscle membrane also mainly via activation of the Na channel which may increase permeability to Na and also Ca (Bauer and Kuriyama, 1982; Bülbring and Tomita, 1987). However, detailed investigations using patch-clamp procedures have not yet been made.

NE is released from adrenergic nerve terminals with ATP as a co-transmitter (Burnstock, 1980). Actions of ATP on the purinergic II receptor have been investigated in detail on the rabbit ear artery and portal vein using voltage- and patch-clamp procedures (Benham and Tsien, 1987; Benham, 1990; Xiong et al., 1990). In the rabbit ear artery, ATP increases the open probability of the Ca current in Ca-rich media (in the absence of Na) and also increases the open probability of the Na channel in Na-rich media (absence of Ca). Benham and Tsien (1987) postulated from the above results that the permeability ratio of these two ions is 3:1. Furthermore, Benham (1989) measured the macroscopic Ca current and the Ca transient simultaneously from the same cell, and concluded that at least 10% of the total influx of inward current is carried by the Ca ion. In the rabbit portal vein, there are two ATP-induced inward currents, i.e. a rapidly developed transient (fast inactivation, within a few sec) inward current and a long lasting small amplitude inward current which occurs with or without the generation of the transient inward current. The former resembles the inward current recorded from the rabbit ear artery. In the case of the latter, ATP generates a Na current in Na-rich, Ca-free solution whereas, in Na-free solution containing 2.5 mM Ca, it generates mainly a Ca-dependent Cl current. Therefore, a minute amount of Ca influx may be enough to trigger the Cl efflux (Xiong et al., 1990). However, in physiological solutions, the ATP-induced inward current is smaller than the Na current or Cl current obtained in pathological ionic environments. This indicates that in physiological solutions, Na and Ca may compete at the mouth of the channel, and as a consequence, only a small amplitude inward current may occur.

ACh in the small intestine enhances the nonselective cation channel (mainly Na; Inoue et al., 1986), as also observed in the ear artery (Benham and Tsien, 1987; Benham, 1990). It is also reported that ACh causes depolarization of the membranes without increasing Cl channel activity. However, some neurotransmitters activate the receptor-operated anion (Cl) channel rather than the receptor-operated cation channel (Bolton and Large, 1986). The role of Cl current as well as the K, Ca, and Na currents should be considered in studies of the physiological function of the receptor-activated ionic current in some tissues.

NE and ATP act directly on the voltage-dependent Ca channel in vascular smooth muscle cells. However, the observed results are inconsistent. For example, NE shifts the threshold required to activate the voltage-dependent Ca channel and as a consequence, Ca channel activation is enhanced (Nelson et al., 1988). But it is also reported that NE inhibited activation of the Ca channel with a sequence reversed to the above results (Droogmans et al., 1987).

In smooth muscle cells of the portal vein, when low concentrations of ATP (below 100 μM) are applied, the Ca current evoked by depolarizing pulses is enhanced, but higher concentrations of ATP inhibit the current amplitude partly due to activation of the receptor-operated cation channel (Xiong et al., 1990). These ATP actions (either enhancement or inhibition) are potentiated by intracellular perfusion of GTPγS. Activation of the Ca channel by ATP may require activation of GTP-binding proteins (presumably the α-subunit). It is not yet completely clarified whether or not the same GTP-binding protein is required to activate both the voltage-dependent Ca channel and the receptor-operated cation channel.

The Effects of Ca Antagonists on the Voltage-Dependent Macroscopic Ca Current

Ca antagonists (Ca channel blockers) are classified into various types according to their chemical moieties. The most commonly used drugs are DHP derivatives, such as nifedipine, nicardipine, nitrendipine, and nisoldipine. At present, more than twenty different DHP derivatives have been introduced for the treatment of hypertension, angina pectoris, and other vascular diseases. In addition, milder Ca antagonists such as diltiazem (a benzothiazepine derivative), verapamil and gallopamil (phenylalkylamine derivatives), and flunarizine (a piperazine derivative) are also commonly used for treatment with the same purpose as the DHP derivatives (Fleckenstein, 1983; Godfraind et al., 1986).

When the effects of nifedipine (nicardipine), diltiazem, verapamil, and flunarizine are compared on the macroscopic Ca current evoked using the voltage-clamp procedure from the same tissue, some characteristic differences between the actions of individual derivatives on the channel activity can be seen (Terada et al., 1987a, 1987b, 1987c).

1) DHP is more effective than other Ca antagonists as estimated from the inhibition of the peak amplitude of the Ca current. The IC_{50} values for nicardipine, diltiazem, verapamil, and flunarizine are 24 nM, 1.4 μM, 1.3 μM, and 1.4 μM, respectively. These agents also act on the K channel (estimated from inhibition of the Ca-independent K current), the IC_{50} values for the above agents being 4.6 μM, 30 μM, 14 μM, and 5.8 μM, respectively. As a consequence, the ratios of the IC_{50} value for K current/ IC_{50} for Ca current are 190, 21, 11, and 4.1, respectively. This means that when relatively high concentrations of diltiazem, verapamil, and flunarizine are used, the inhibitory actions on the K channel may concomitantly involve depolarization of the membrane (Terada et al., 1987a, 1987b, 1987c).

2) Ca antagonists modify the inactivation curve, i.e. nicardipine and flunarizine shift the steady state inactivation curve to the left (the more polarized direction) as measured using the double-pulse method at a holding potential of -80 mV, but diltiazem and verapamil are less potent in shifting the voltage-dependent inactivation curve (Terada et al., 1987a, 1987b, 1987c).

3) When the "use-dependency" of Ca antagonists on the Ca current are observed, verapamil shows potent use-dependency, whereas flunarizine and diltiazem show moderate, and nicardipine shows virtually no effect. This means that nicardipine acts on the resting state and verapamil acts mainly on the open state of the channel (Terada et al., 1987a, 1987b, 1987c).

4) All Ca antagonists act on the L type Ca channel but not the T type Ca channel with the exception that flunarizine acts on both the L and T types of Ca channel in nerve cells (Terada et al., 1987a, 1987b, 1987c; Akaike et al., 1989).

5) Verapamil and gallopamil act from extracellular sites, so that intracellular perfusion of these agents does not modify the voltage dependent Ca channel (Ohya et al., 1987; Leblanc and Hume, 1989). These actions of phenylalkylamine derivatives differ from those observed in cardiac muscle (Nawrath et al., 1977; McDonald, 1984).

These effects of Ca antagonists on the voltage-dependent Ca channel using whole-cell voltage-clamp procedures seem to indicate that the mechanism of Ca antagonism differs for different agents.

Effects of Ca Antagonists and Ca Agonists on the Voltage-Dependent Unitary Ca Current

As described previously, Ca antagonists act on the CaL (L-subtype) and some CaM channels (Caffrey et al., 1986; Worley et al., 1986; Benham et al., 1987; Inoue et al., 1987; Kawashima and Ochi, 1987; Yatani et al., 1987; Yoshino et al., 1988; Inoue et al., 1990). It is clear that many DHP derivatives show both agonist and antagonistic actions. Some drugs show a potent agonistic, but little antagonistic action (Bay K 8644, YC 170, CGP 28-392, H160/51, 202-791) (Mannhold et al., 1982; Bechem and Schramm, 1987) whereas other drugs show a strong antagonistic, but little agonistic action (nitrendipine, nifedipine, benidipine, PN200-110, CV-4093 or FRC8653) (Hess et al., 1984; Brown et al., 1986; Okabe et al., 1987; Oike et al., 1990). These characteristics mostly depend on whether the drug is an (+)-R- or (-)-S-enantiomer. Phenyl-alkylamine derivatives are also reported to have such dual actions in cardiac and skeletal muscles (Pelzer et al., 1988).

Most DHP derivative antagonists reduce the open probability of the channel, whereas agonists increase the open probability and prolong the channel opening with no change in amplitude of the unitary current.

These actions of Ca antagonists and agonists on the CaL channel in smooth muscle have been analyzed using a hypothesis similar to that applied to cardiac muscle (Lee and Tsien, 1983; Hess et al., 1984; Sanguinetti and Kass, 1984; Brown et al., 1986; Triggle et al., 1986; Hering, 1987; Kass, 1987; Kass and Krafte, 1987; Pelzer et al., 1988). The modulated receptor hypothesis has been presented as an analogy of the action of local anesthetics (closed, open, and inactivated states), i.e. high affinity binding of these DHP derivatives to the Ca channel occurs in its inactive state. Further detailed experiments suggest that nifedipine acts not in the open state alone but in the closed, open, and inactivated states. Therefore, a nonmodulating receptor hypothesis (two closed, one open, and two inactivated states) and a subsequently modified nonmodulating receptor hypothesis have been introduced from the results obtained using DHP agonists and antagonists (Hess, 1984; Kokubun et al., 1986; Dolphin and Scott, 1989). However, detailed analysis of the action of Ca antagonists depending on the mode of the channel, using the unitary current recorded from smooth muscle cells, has not yet been made.

ROLES OF SECOND MESSENGERS (SIGNAL TRANSDUCTORS) ON THE VOLTAGE- DEPENDENT CALCIUM CHANNEL

It is now clear that many second messengers, namely cyclic GMP, cyclic AMP, IP_3, and diacylglycerol (DAG), are synthesized via activation of the agonist-receptor-GTP-binding protein-catalytic subunit (adenylate cyclase, guanylate cyclase, or phospholipase C)-complex. In addition, Ca is also categorized as a second messenger.

Cyclic AMP and cyclic GMP have no significant effects on the voltage-dependent Ca channel in many smooth muscle tissues, though there are reports from Sperelakis' group that permeable derivatives of cyclic GMP and cyclic AMP do inhibit the Ca channel (slow L-subtype; Sperelakis and Ohya, 1990). In cardiac muscle, intracellularly applied cyclic AMP accelerates channel opening of the voltage-dependent Ca channel without any change in unitary conductance, and it produces a positive inotropic action. However, this effect has not yet been confirmed in smooth muscle cells. Cyclic AMP increases Ca-dependent K channel activity (okadaic acid modulating phosphorylation site) in smooth muscles cells and induces hyperpolarization of the membrane (Kume et al., 1989).

IP_3 binds to an IP_3-binding protein and releases Ca from the SR (Berridge, 1984, 1986, 1988; Abdel-Latif, 1986; Shears, 1989). The primary structure of the IP_3-sensitive protein (receptor) has been elucidated (Takeshima et al., 1989; Supattapone et al., 1988; Furuichi et al., 1989) and the channel activity induced by IP_3 can be prevented by the application of heparin (Kobayashi et al., 1988; Somlyo et al., 1988). On the other hand, no direct effects of IP_3 on the voltage-dependent Ca channel are evident. Presumably released Ca augments the inactivation process of the voltage-dependent Ca channel and prevents the influx of Ca (Ohya et al., 1988). Except in the smooth muscle cell, IP_4 and IP_3 are thought to increase the pacemaker activity necessary for triggering the ionic channel activity (Irvine, 1989; Berridge and Irvine, 1989). However, in smooth muscle cells, IP_4 reduces the threshold for evoking the oscillatory outward current (generated by the Ca released from the SR), and this action is much weaker than that of IP_3 (Ohya et al., 1988). No evidence has been reported for activation of the voltage-dependent Ca channel induced by IP_4 in sarcolemma.

DAG, a co-product of IP_3 formation, is known to have multiple actions on physiological functions and is reported to act directly on the ionic channel (Nishizuka, 1984, 1986, 1988). However, activation of protein kinase C by DAG, phosphatidylserine, and Ca show no consistent effects on the Ca channel. Thus, activation of protein kinase C produces either activation or inhibition of the voltage-dependent Ca channel (Gleason and Flaim, 1986; Itoh et al., 1986; Litten et al., 1987; Vivaudou et al., 1988; Shearman et al., 1989). DAG is an unstable compound and therefore, various substances such as phorbol esters (Castagna et al., 1982; Ashendel et al., 1983; Kikkawa et al., 1983) are used instead. Phorbol esters also possess multiple actions. To clarify whether or not these multiple actions of phorbol esters are exactly the same as those of DAG, further investigations would be required.

When endothelin (Yanagisawa et al., 1988) is applied in the bath, the Ca channel opening is enhanced more than by its addition in the pipette. Therefore, Inoue et al. (1990) postulated that endothelin may not act directly on the channel but that an unknown second messenger synthesized by endothelin activation acts on the channel. This is unlikely to be IP_3, DAG, or cyclic nucleotides. Much the same suggestion has been made by Nelson et al. (1988) for the voltage-dependent Ca channel during bath application of NE. Presumably, other second messengers other than the established ones are involved.

In relation to the action of second messengers, the role of GTP-binding proteins have been elucidated on ionic channels (Brown and Birnbaumer,

1988; Rosenthal et al., 1988). In neurons, G_0 (brain GTP-binding protein) inhibits the voltage-dependent Ca channel (Heschler et al., 1987) whereas, in cardiac muscles G_S (cholera-toxin sensitive type) directly enhances the voltage-dependent Ca channel. In the guinea-pig portal vein, Ohya and Sperelakis (1988) reported that GTPγS activates the Ca channel (L-type). Xiong et al. (1990) also observed the enhancing effect of GTPγS on the ATP-operated cation channel in the rabbit portal vein, but this agent also accelerates the ATP-induced inhibition of the voltage-dependent Ca channel. Presumably, GTP-binding proteins have a role in regulating the voltage-dependent and receptor-operated channels. Further detailed experiments are required.

CONCLUSION

We have briefly described the biophysical features of the voltage-dependent Ca channel in relation to endothelium-derived factors and neurotransmitters. In addition, actions of Ca antagonists were briefly introduced.

The voltage-dependent Ca channel in smooth muscle cells possesses properties different from those in other excitable cells. The following are points of contrast between smooth and cardiac muscle cells.

a) In cardiac muscle cells, the L and T (in some cases, including S) subtypes of the voltage-dependent Ca channel are present whereas in smooth muscle, CaL (corresponds with L type), CaM, and CaS (corresponds with T type) are present.

b) In cardiac muscle cells, cyclic AMP augments activation of the voltage-dependent Ca channel (increases the open probability of the L type channel), but not in smooth muscle cells.

c) In cardiac muscle, gallopamil mainly acts from the internal side of the membrane (perfusion of this agent inside the cell more potently inhibits the channel than does bath application), whereas in smooth muscle cells, extracellular (bath) application is more effective than intracellular perfusion.

d) Activation and inactivation processes, and also their recovery processes require much longer in smooth muscle than in cardiac muscle. These phenomena may partly contribute to the slow rate of rise of the action potential (B. Nilius and K. Kitamura, personal communication).

e) In smooth muscle cells, Bay K 8644 increases the open probability of the CaL as also observed in cardiac muscle cells. However, in smooth muscle a shift of the channel state from mode 1 (short opening) to mode 2 (long opening) is also observed upon application of low frequency stimulation in the absence of the Ca agonist. Such a shift of mode by application of different frequencies of stimulations has not been reported in cardiac muscle cells.

f) In smooth muscle, intracellularly applied ATP activates the voltage-dependent Ca channel in a concentration dependent manner, and ATP prevents the 'run down' phenomenon. This action of intracellularly perfused ATP is much the same as that of cyclic AMP in cardiac muscles. This action of ATP is not mimicked by AMP-PNP. Such responses have been reported in cardiac muscle cells, too.

Presumably, even more contrasting features may remain to be discovered between the voltage-dependent Ca channels of these two tissues. A new line of investigation, using voltage- and patch-clamp procedures on the ion channel in dispersed smooth muscle cells, was started 10 years ago. However, there are still many unsolved problems before the exact nature of the voltage-dependent Ca channel can be determined. Investigations of the characteristics of purified channel proteins in relation to intact smooth muscle cells are awaited.

REFERENCES

Aaronson, P. I., Bolton, T. B., Long, R. J., and Mackenzie, I., 1988, Calcium currents in single isolated smooth muscle cells from the rabbit ear artery in normal-calcium and high-barium solution, *J. Physiol.*, 405: 57.

Abdel-Latif, A. A., 1986, Calcium-mobilizing receptors, polyphosphoinositides, and the generation of second messengers, *Pharmacol. Rev.*, 38: 227.

Akaike, N., Kostyuk, P. G., and Osipchuk, Y. V., 1989, Dihydropyridine-sensitive low-threshold calcium channels in isolated rat hypothalamic neurones, *J. Physiol.*, 412: 181.

Amedee, T., Renaud, J. F., Jmari, K., Lombert, A., Mironneau, J., and Lazdunski, M., 1986, The presence of Na^+ channels in myometrial smooth muscle cells revealed by specific neurotoxin, *Biochem. Biophys. Res. Commun.*, 137: 675.

Ashendel, C. L., Staller, J. M., and Boutwell, R. K., 1983, Identification of a calcium-and phospholipid-dependent phorbol ester binding activity in the solution fraction of mouse tissues, *Biochem. Biophys. Res. Commun.*, 111: 340.

Bauer, V. and Kuriyama, H., 1982, Evidence for non-cholinergic and non-adrenergic transmission in the guinea-pig, *J. Physiol.*, 330: 95.

Bechem, M. and Schramm, M., 1987, Calcium-agonists, *J. Mol. Cell. Cardiol.*, 19 (Suppl. 2): 63.

Benham, C. D., 1989, ATP-activated channels gate calcium entry in single smooth muscle cells dissociated from rabbit ear artery, *J. Physiol.*, 419: 689.

Benham, C. D. and Bolton, T. B., 1983, Patch-clamp studies of slow potential-sensitive potassium channels in longitudinal smooth muscle cells of rabbit jejunum, *J. Physiol.*, 340: 469.

Benham, C. D., Bolton, T. B., Lang, R. J., and Takewaki, T., 1985, The mechanism of action of Ba^{2+} and TEA on single Ca^{2+}-activated K^+ channels in arterial and intestinal smooth muscle cell membrane, *Pflügers Arch.*, 403: 120.

Benham, C. D., Hess, P., and Tsien, R. W., 1987, Two types of calcium channels in single smooth muscle cells from rabbit ear artery studied with whole-cell and single-channel recordings, *Circ. Res.*, 61 (Suppl 1): 10.

Benham, C. D. and Tsien, R. W., 1987, A novel receptor-operated Ca^{2+}-permeable channel activated by ATP in smooth muscle, *Nature*, 328: 275.

Benoit, E., Corbier, A., and Dubois, J.-M., 1985, Evidence for two transient sodium currents in the frog node of Ranvier, *J. Physiol.*, 361: 339.

Berridge, M. J., 1984, Inositol trisphosphate and diacylglycerol as second messengers, *Biochem. J.*, 220: 345.

Berridge, M. J., 1986, Intracellular signalling through inositol trisphosphate and diacylglycerol, *Biol. Chem. Hoppe-Seyler.*, 367: 447.

Berridge, M. J., 1988, Inositol trisphosphate-induced membrane potential oscillations in Xenopus oocytes, *J. Physiol.*, 403: 589.

Berridge, M. J. and Irvine, R. F., 1989, Inositol phosphates and cell signalling, *Nature*, 341: 197.

Bolton, T. B. and Large, W. A., 1986, Are junction potentials essential? Dual mechanism of smooth muscle cell activation by transmitter released from autonomic nerves, *Quart. J. Exp. Physiol.*, 71: 1.

Bolton, T. B. and Lim, S. P., 1989, Properties of calcium stores and transient outward currents in single smooth muscle cells of rabbit intestine, *J. Physiol.*, 409: 385.

Brown, A. M. and Birnbaumer, L., 1988, Direct G protein gating of ion channels, *Am. J. Physiol.*, 254: H401.

Brown, A. M., Kunze, D. L., and Yatani, A., 1986, Dual effects of dihydropyridines on whole cell and unitary calcium currents in single ventricular cells of guinea-pig, *J. Physiol.*, 379: 495.

Bülbring, E. and Tomita, T., 1987, Catecholamine action on smooth muscle, *Pharmacol. Rev.*, 39: 49.

Burnstock, G., 1980, Cholinergic and purinergic regulation of blood vessels, *in:* "Handbook of Physiology. The Cardiovascular System, section 2, vol. II", D. F. Bohr, A. P. Somlyo, H. V. Sparks, Jr., eds., Am. Physiol. Soc., Bethesda, p. 567.

Caffrey, J. M., Josephson, I. R., and Brown, A. M., 1986, Calcium channels of amphibian and mammalian aorta smooth muscle cells, *Biophys. J.*, 49: 1237.

Castagna, M., Takai, Y., Kaibuchi, K., Sano, K., Kikkawa, U., and Nishizuka, Y., 1982, Direct activation of calcium-activated, phospholipid-dependent protein kinase C by tumor-promoting phorbol esters, *J. Biol. Chem.*, 259: 7849.

Chen, G. and Suzuki H., 1988, Dissociation of the ACh-induced hyperpolarization and relaxation by methylene blue or haemoglobin in the rat main pulmonary artery, *Jpn. J. Pharmacol.*, 46: 184p.

DeMay, J. G. and Vanhoutte, P. M., 1983, Anoxia and endothelium-dependent reactivity of the canine femoral artery, *J. Physiol.*, 335: 65.

Dolphin, A. C. and Scott, R. H., 1989, Interaction between calcium channel ligands and guanine nucleotides in cultured rat sensory and sympathetic neurons, *J. Physiol.*, 413: 271.

Droogmans, G. and Callewaert, G., 1986, Ca^{2+}-channel current and its modification by the dihydropyridine agonist BAY K 8644 in the isolated smooth muscle cells, *Pflügers Arch.*, 406: 259.

Droogmans, G., Declerck, I., and Casteels, R., 1987, Effects of adrenergic agonists on Ca^{2+}-channel currents in single vascular smooth muscle cells, *Pflügers Arch.*, 409: 7.

Fleckenstein, A., 1983, History of calcium antagonists, *Circ. Res.*, 52: 3.

Furchgott, R. F. and Zawadzki, J. V., 1980, The obligatory role of endothelial cells in the relaxation of arterial smooth muscle by acetylcholine, *Nature*, 288: 373.

Furuichi, T., Yoshikawa, S., Miyawaki, A., Wada, K., Maeda, N., and Mikoshiba, K., 1989, Primary structure and functional expression of the inositol 1,4,5-trisphosphate-binding protein P_{400}, *Nature*, 342: 32.

Ganitkevich, V. YA., Shuba, M. F., and Sminov, S. V., 1986, Potential-dependent calcium inward current in a single isolated smooth muscle cell of the guinea-pig taenia coli, *J. Physiol.*, 380: 1.

Gleason, M. M. and Flaim, S. F., 1986, Phorbol ester contracts rabbit thoracic aorta by increasing intracellular calcium and by activating calcium influx, *Biochem. Biophys. Res. Comm.*, 138: 2362.

Godfraind, T., Miller, R., and Wibo, M., 1986, Calcium antagonism and calcium entry blockade, *Pharmacol. Rev.*, 38: 321.

Hagiwara, S., 1983, "Membrane potential-dependent ion channels in cell membrane. Phylogenic and developmental approaches", Raven Press, New York.

Hering, S., Beech, D. J., and Bolton, T. B., 1987, Voltage dependence of the actions of nifedipine and Bay K 8644 on barium currents recorded from single smooth muscle cells from rabbit ear artery, *Biomed. Biochem. Acta,* 467: S657.

Heschler, J., Tang, M., Jastorff, B., and Trautwein, W., 1987, On the mechanism of histamine induced enhancement of the cardiac Ca^{2+} current, *Pflügers Arch.,* 410: 23.

Hess, P., Lamsman, J. B., and Tsien, R. W., 1984, Different modes of Ca channel gating behaviour favoured by dihydropyridine Ca agonists and antagonists, *Nature,* 311: 538.

Hume, J. R. and Leblanc, N., 1989, Macroscopic K^+ currents in single smooth muscle cells of the rabbit portal vein, *J. Physiol.,* 413: 49.

Hwang, K. S. and van Breemen, C., 1987, Ryanodine modulation of ^{45}Ca efflux and tension in rabbit aortic smooth muscle, *Pflügers Arch.,* 408: 343.

Ignarro, L. J., 1989, Biological actions and properties of endothelium-derived nitric oxide formed and released from artery and vein, *Circ. Res.,* 65: 1.

Ignarro, L. J., Byrns, R. E., and Wood, K. S., 1987, Endothelium-dependent modulation of cGMP levels and intrinsic smooth muscle tone in isolated bovine intrapulmonary artery and vein, *Circ. Res.,* 60: 82.

Inoue, R., Kitamura, K., and Kuriyama, H., 1985, Two Ca-dependent K-channels classified by the application of tetraethylammonium distribute to smooth muscle membranes of the rabbit portal vein, *Pflügers Arch.,* 405: 173.

Inoue, R., Kitamura, K., and Kuriyama, H., 1987, Acetylcholine activates single sodium channels in smooth muscle cells, *Pflügers Arch.,* 410: 69.

Inoue, R., Okabe, K., Kitamura, K., and Kuriyama, H., 1986, A newly identified Ca^{2+} dependent K^+ channel in the smooth muscle cell membrane of single cells dispersed from the rabbit portal vein, *Pflügers Arch.,* 406: 138.

Inoue, Y., Oike, M., Nakao, K., Kitamura, K., and Kuriyama, H, 1990, Endothelin augments unitary Ca channel currents on the smooth muscle cell membrane of guinea-pig portal vein, *J. Physiol.,* 423: 171.

Inoue, Y., Xiong, Z., Kitamura, K., and Kuriyama, H., 1989, Modulation produced by nifedipine of the unitary Ba current of dispersed smooth muscle cells of the rabbit ileum, *Pflügers Arch.,* 414: 534.

Irvine, R. F., 1989, How do inositol 1,4,5-trisphosphate and inositol 1,3,4,5-tetrakisphosphate regulate intracellular Ca^{2+}?, *Biochem. Soc. Trans.,* 17: 6.

Itoh, T., Izumi, H., and Kuriyama, H., 1982, Mechanisms of relaxation induced by activation of β-adrenoceptors in smooth muscle cells of the guinea-pig mesenteric artery, *J. Physiol.,* 326: 475.

Itoh, T., Kanmura, Y., and Kuriyama, H., 1988, Inorganic phosphate regulates the contraction-relaxation cycle in skinned muscles of the rabbit mesenteric artery, *J. Physiol.,* 376: 231.

Kajioka, S., Oike, M., and Kitamura, K., 1990, Nicorandil opens a Ca-dependent and ATP-sensitive potassium channel in the smooth muscle cells of the rat portal vein, *J. Pharmacol. Exp. Ther.,* 254: 905.

Kass, R. S., 1987, Voltage-dependent modulation of cardiac calcium channel current by optical isomers of Bay K 8644: Implications for channel gating, *Circ. Res.,* 61: 1.

Kass, R. S. and Krafte, D. S., 1987, Negative surface charge density near heart calcium channels. Relevance to block by dihydropyridines, *J. Gen. Physiol.*, 89: 629.

Kawashima, Y. and Ochi, R., 1987, Two types of calcium channels in isolated vascular smooth muscles, *J. Physiol. Soc. Jpn*, 49: 369p.

Kikkawa, U., Takai, Y., Minakuchi, R., Inohara, S., and Nishizuka, Y., 1983, Protein kinase C as a possible receptor protein of tumor-promoting phorbol esters, *J. Biol. Chem.*, 258: 11442.

Klöckner, U. and Isenberg, G., 1985, Calcium current of cesium loaded isolated smooth muscle cells (unitary bladder of the guinea pig), *Pflügers Arch.*, 405: 340.

Kobayashi, S., Somlyo, A. P., and Somlyo, A. V., 1988, Heparin inhibits the inositol 1,4,5-trisphosphate-dependent, but not the independent, calcium release induced by guanine nucleotide in vascular smooth muscle, *Biochem. Biophys. Res. Comm.*, 153: 625.

Kokubun, S., Prod'hom, B., Becker, C., Porzzig, H., and Reuter, H., 1986, Studies on Ca channels in intact cardiac cells: Voltage-dependent effects and cooperative interactions of dihydropyridine enantiomers, *Mol. Pharmacol.*, 30: 571.

Komori, K. and Suzuki, H, 1987a, Electrical responses of smooth muscle cells during cholinergic vasodilation in the rabbit saphenous artery, *Circ. Res.*, 61: 586.

Komori, K. and Suzuki, H., 1987b, Heterogenous distribution of muscarinic receptors in the rabbit saphenous artery, *Br. J. Pharmacol.*, 92: 657.

Kume, H., Takai, A., Tokuno, M., and Tomita, H, 1989, Regulation of Ca^{2+}-dependent K^+-channel activity in tracheal myocytes by phosphorylation, *Nature*, 341: 152.

Leblanc, N. and Hume, J. R., 1989, D600 block of L-type Ca^{2+} channel in vascular smooth muscle cells: Comparison with permanently charged derivative, D890, *Am. J. Physiol.*, 257: C689.

Lee, K. S. and Tsien, R. W., 1983, Mechanism of calcium channel blockade by verapamil, D600, diltiazem, and nifedipine in single dialyzed heart cells, *Nature*, 302: 790.

Litten, R. Z., Suba, E. A., and Roth, B. L., 1987, Effects of a phorbol ester on rat aortic contraction and calcium influx in the presence or absence of Bay K 8644, *Eur. J. Pharmacol.*, 144: 185.

Makita, Y., 1984, Effects of adrenoceptor agonists and antagonists on smooth muscle cells and neuromuscular transmission in the guinea-pig renal artery and vein, *Br. J. Pharmacol.*, 80: 671.

Mannhold, R., Rodenkirchen, R., and Bayer, R., 1982, Qualitative and quantitative structure-activity relationships of specific Ca antagonists, *Prog. Pharmacol.*, 5: 25.

McDonald, T. F., Pelzer, D., and Trautwein, W., 1984, Cat ventricular muscle treated with D600: Characteristics of calcium channel block and unblock, *J. Physiol.*, 325: 217.

Mironneau, J. and Savineau, J.-P., 1980, Effects of calcium ions on outward membrane currents in rat uterine smooth muscle, *J. Physiol.*, 302: 411.

Moncada, S. and Vane, J. R., 1979, Pharmacology and endogenous roles of prostaglandin, endoperoxides, thromboxane A_2, and prostacyclin, *Pharmacol. Rev.*, 30: 293.

Myers, P. R., Minor, R. L., Guerra, R., Bates, J. N., and Harrison, D. G., 1990, Vasorelaxant properties of the endothelium-derived relaxing factor more closely resemble S-nitrosocysteine than nitric oxide, *Nature*, 345: 161.

Nakao, K., Inoue, Y., Oike, M., Kitamura, K., and Kuriyama, H., 1990, Mechanisms of endothelin-induced augmentation of the electrical and mechanical activity in rat portal vein, *Pflügers Arch.*, 415: 526.

Nakao, K., Okabe, K., Kitamura, K., Kuriyama, H., and Weston, A. H., 1988, Characteristics of cromakalim-induced relaxations in the smooth muscle cells of guinea-pig mesenteric artery, *Br. J. Pharmacol.*, 95: 785.

Nakazawa, K., Matsui, N., Shigenobu, K., and Kasuya, Y., 1987, Contractile response and electrophysiological properties in enzymatically dispersed smooth muscle cells of rat vas deferens, *Pflügers Arch.*, 408: 112.

Nawrath, H., Ten Eick, R. E., McDonald, T. F., and Trautwein, W., 1977, On the mechanism underlying the action of D600 on slow inward current and tension in mammalian myocardium, *Circ. Res.*, 40: 408.

Nelson, M. T., Standen, N. B., Brayden, J. E., and Worley III, J.F., 1988, Noradrenaline contracts arteries by activating voltage-dependent calcium channel, *Nature*, 336: 382.

Nishiye, E., Nakao, K., Itoh, T., and Kuriyama, H., 1989, Factors inducing endothelium-dependent relaxation in the guinea-pig basilar artery as estimated from the action of haemoglobin, *Br. J. Pharmacol.*, 96: 645.

Nishizuka, Y., 1984, The role of protein kinase C in cell surface signal transduction and tumor promotion, *Nature*, 308: 693.

Nishizuka, Y., 1986, Studies and prospectives of protein kinase C, *Science*, 233: 305.

Nishizuka, Y., 1988, The molecular heterogeneity of protein kinase C and its implications for cellular regulation, *Nature*, 344: 661.

Ohya, Y., Kitamura, K., and Kuriyama, H., 1987, Modulation of ionic currents in smooth muscle balls of the intestine by intercellularly perfused ATP and cyclic AMP, *Pflügers Arch.*, 408: 465.

Ohya, Y. and Sperelakis, N., 1988, Guanosine triphosphate dependent stimulation of L-type calcium channels of vascular smooth muscle, *Physiologist*, 31: A38.

Ohya, Y., Terada, K., Kitamura, K., and Kuriyama, H., 1986, Membrane currents recorded from a fragment of rabbit intestinal smooth muscle cells, *Am. J. Physiol.*, 251: C335.

Ohya, Y., Terada, K., Yamaguchi, K., Inoue, R., Okabe, K., Kitamura, K., Hirata, M., and Kuriyama, H., 1988, Effects of inositol phosphates on the membrane activity of smooth muscle cells of the rabbit portal vein, *Pflügers Arch.*, 412: 382.

Oike, M., Inoue, Y., Kitamura, K., and Kuriyama, H., 1990, Dual actions of FRC8653, a novel dihydropyridine derivative, on the Ba current recorded from the rabbit basilar artery, *Circ. Res.*, 67: 993.

Okabe, K., Kitamura, K., and Kuriyama, H., 1987, Features of 4-aminopyridine sensitive outward current observed in single smooth muscle cells from the rabbit pulmonary artery, *Pflügers Arch.*, 409: 561.

Okabe, K., Kitamura, K., and Kuriyama, H., 1988, The existence of a highly tetrodotoxin sensitive Na channel in freshly dispersed smooth muscle cells of the rabbit main pulmonary artery, *Pflügers Arch.*, 411: 423.

Palmer, R. M. J., Ferrige, A. G., and Moncada, S., 1987, Nitric oxide release accounts for the biological activity of endothelium-derived relaxing factor, *Nature*, 327: 524.

Pelzer, D., Cavalie A., Hofmann, F., Trautwein, W., and McDonald, T. F., 1988, Dual stimulating and inhibitory effects of the phenylalkylamine calcium antagonist D600 on cardiac and skeletal muscle calcium channels, *Pflügers Arch.*, 411: 39.

Pidoplichko, V. I., 1986, Two different tetrodotoxin-separable inward sodium currents in the membrane of isolated cardiomyocytes, *Gen. Physiol. Biophys.*, 6: 593.

Rosenthal, W., Heschler, J., Trautwein, W., and Schultz, G., 1988, Control of voltage dependent Ca^{2+} channels by G-protein-coupled receptor, *FASEB J.*, 2: 2784.

Sadoshima, J., Akaike, N., Tomoike, H., and Nakamura, M., 1988, Ca-activated K channel in cultured smooth muscle cells of rat aortic media, *Am. J. Physiol.*, 255: H410.

Sakai, T., Terada, K., Kitamura, K., and Kuriyama, H., 1988, Ryanodine inhibits the Ca-dependent KL current after depletion of Ca stored in smooth muscle cells of the rabbit ileal longitudinal muscle, *Br. J. Pharmacol.*, 95: 1089.

Sanguinetti, M. C. and Kass, R. S., 1984, Voltage-dependent block of calcium channel current in the calf cardiac purkinje fiber by dihydropyridine calcium channel antagonists, *Circ. Res.*, 55: 336.

Schmiedtmayer, J., 1985, Behaviour of chemically modified sodium channels in frog nerve supports a three-state model of inactivation, *Pflügers Arch.*, 404: 21.

Shearman, M. S., Sekiguchi, K., and Nishizuka, Y., 1989, Modulation of ion channel activity: A key function of the protein kinase C family, *Pharmacol. Rev.*, 41: 211.

Shears, S. B., 1989, Metabolism of the inositol phosphates produced upon receptor activation, *Biochem. J.*, 260: 313.

Shirahase, H., Usui, H., Kurahashi, K., Fujiwara, M., and Fukui, K., 1987, Possible role of endothelial thromboxane A_2 in the resting tone and contractile responses to acetylcholine and arachidonic acid in canine cerebral arteries, *Pharmacology*, 10: 517.

Somlyo, A. P., Walker, J. W., Goldman, Y. E., Trentham, D. R., Kobayashi, S., Kitazawa, T., and Somlyo, A. V., 1988, Inositol trisphosphate, calcium, and muscle contraction, *Phil. Trans. R. Soc. Lond. B*, 320: 399.

Sperelakis, N. and Ohya, Y., 1990, Cyclic nucleotide regulation of Ca^{2+} slow channels and neurotransmitter release in vascular muscle, *in:* "Frontiers in Smooth Muscle Research", N. Sperelakis and J. D. Wood, eds., Wiley-Liss, New York, p. 277.

Standen, N. B., Quayle, J. M., Davis, N. W., Brayden, J. E., Huang, Y., and Nelson, M. T., 1989, Hyperpolarizing vasodilators activate ATP-sensitive K^+-channels in arterial smooth muscle, *Science*, 245: 177.

Sturek, M. and Hermsmeyer, K., 1986, Calcium and sodium channels in spontaneously contracting vascular muscle cells, *Science*, 233: 475.

Suematsu, E., Hirata, M., Hashimoto, T., and Kuriyama, H., 1984, Inositol 1,4,5-trisphosphate releases Ca^{2+} from intracellular store sites in skinned single cells of porcine coronary artery, *Biochem. Biophys. Res. Comm.*, 120: 481.

Supattapone, S., Worley, P. F., Baraban, J. M., and Snyder, S. M., 1988, Solubilization, purification, and characterization of an inositol trisphosphate receptor, *J. Biol. Chem.*, 263: 1530.

Sutko, J. L. and Kenyon, J. L., 1983, Ryanodine modification of cardiac muscle responses to potassium-free solutions, *J. Gen. Physiol.*, 82: 385.

Suzuki, R., Osa, T., and Kobayashi, S., 1983, Cholinergic inhibitory response in the bovine iris dilator muscle, *Invest. Ophthalmol. Vis. Sci.*, 24: 760.

Takeshima, H., Nishimura, S., Matsumoto, T., Ishida, H., Kangawa, K., Minamino, N., Matsuo, H., Ueda, M., Hanaoka, M., Hirose, T., and Numa, S., 1989, Primary structure and expression from complementary DNA of skeletal muscle ryanodine receptor, *Nature,* 339: 439.

Toro, L. and Stefani, E., 1987, Ca^{2+} and K^+ currents in cultured vascular smooth muscle cells from rat aorta, *Pflügers Arch.,* 408: 417.

Terada, K., Kitamura, K., and Kuriyama, H., 1987a, Blocking actions of Ca^{2+} antagonists on the Ca^{2+} channels in the smooth muscle cell membrane of rabbit small intestine, *Pflügers Arch.,* 408: 552.

Terada, K., Nakao, K., Okabe, K., Kitamura, K., and Kuriyama, H., 1987b, Action of the 1,4-dihydropyridine derivative, KW-3049, on the smooth muscle membrane of the rabbit mesenteric artery, *Br. J. Pharmacol.,* 92: 615.

Terada, K., Ohya, Y., Kitamura, K., and Kuriyama, H., 1987c, Actions of flunarizine, a Ca^{++} antagonist, on ionic currents in fragmented smooth muscle cells of the rabbit small intestine, *J. Pharmacol. Exp. Ther.,* 240: 978.

Triggle, D. J., Skattebol, A., Rampe, D., Joslyn, A., and Gengo, P., 1986, Chemical pharmacology of Ca^{2+} channel ligands, *in:* "New Insight Into Cell and Membrane Transport Processes", G. Post and S. T. Crooke, eds., Plenum Press, New York, p. 125.

Vanhoutte, P. M., Rubanyi, G. M., Miller, J. M., and Houston, D. S., 1986, Modulation of vascular smooth muscle contraction by the endothelium, *Ann. Rev. Physiol.,* 48: 307.

Vivaudou, M. B., Clapp, L. H., Walsh Jr., J. V., and Singer, J. J., 1988, Regulation of one type of Ca^{2+} current in smooth muscle cells by diacylglycerol and acetylcholine, *FASEB J.,* 2: 2497.

Walsh Jr., J. V. and Singer, J. J., 1987, Identification and characterization of major ionic currents in isolated smooth muscle cells using the voltage-clamp technique, *Pflügers Arch.,* 408: 83.

Warner, T. D., De Nucci, G. R., and Vane, J. R., 1989, Rat endothelin is a vasodilator in the isolated perfused mesentery of the rat, *Eur. J. Pharmacol.,* 159: 325.

Wong, G. K., 1984, Irreversible modification of sodium channel inactivation in toad myelinated nerve fibres by the oxidant chloramine-T, *J. Physiol.,* 346: 127.

Worley III, J. F., Deitmer, J. W., and Nelson, M. T., 1986, Single nisoldipine-sensitive calcium channels in smooth muscle cells isolated from rabbit mesenteric artery, *Proc. Nat'l Acad. Sci. U.S.A.,* 83: 5746.

Wright, C. E. and Fozzard, J. R., 1988, Regional vasodilation is a prominent feature of the haemodynamic response to endothelin in anesthetized spontaneous hypertensive rats, *Eur. J. Pharmacol.,* 155: 201.

Xiong, Z. L., Kitamura, K., and Kuriyama, H., 1991, ATP activates a non-selective cation channel and modulates the voltage-dependent Ca channel in rabbit portal vein, *J. Physiol.,* submitted.

Yanagisawa, M., Kurihara, H., Kimura, S., Tomobe, Y., Kobayashi, M., Mitsui, Y., Yazaki, Y., Goto, K., and Masaki, T., 1988, A novel potent vasoconstrictor peptide produced by vascular endothelial cells, Nature, 332: 411.

Yatani, A., Seidel, C. L., Allen, J., and Brown, A. M., 1987, Whole-cell and single-channel calcium currents of isolated smooth muscle cells from saphenous vein, *Circ. Res.,* 60: 523.

Yoshino, M., Someya, T., Nishino, A., and Yabu, H., 1988, Whole-cell and unitary Ca channel currents in mammalian intestinal smooth muscle cells: Evidence for existence of two types of Ca channels, *Pflügers Arch.*, 411: 229.

Yoshitomi, T. and Ito, Y., 1986, Double reciprocal innervations in dog iris sphincter and dilator muscles, *Invest. Ophthalmol. Vis. Sci.*, 27: 83.

CHANGES IN CYTOPLASMIC CALCIUM INDUCED BY PURINERGIC P_{2X} RECEPTOR ACTIVATION IN VASCULAR SMOOTH MUSCLE CELLS AND SENSORY NEURONS

Christopher D. Benham, Muriel M. Bouvier,
and Martyn L. Evans

SmithKline Beecham Pharmaceuticals
The Pinnacles, Harlow
Essex CM19 5AD, England

INTRODUCTION

It is now widely accepted that ATP is an excitatory sympathetic co-transmitter (reviewed by Burnstock and Kennedy, 1986) which, on release generates fast excitatory junction potentials in arteries (Stjarne, 1986) and some other smooth muscles (Sneddon and Westfall, 1982). These contractile actions of ATP on smooth muscle appear to be due to activation of the P_{2x} subtype of purinoceptor (Burnstock and Kennedy, 1985) which gates a cation permeable channel. In sensory neurons (Krishtal et al., 1983), a very similar ATP receptor/channel also causes depolarization by the activation of inward currents. In vascular smooth muscle, ATP is generally thought to stimulate contraction by activating a mainly sodium permeable conductance that depolarizes the cell (Suzuki, 1985) and opens voltage-gated Ca^{2+} channels (Burnstock, 1988) allowing calcium entry. The advent of patch-clamp techniques has allowed detailed study of the ATP-activated channels in vascular smooth muscle revealing that the conductance is cation selective (Benham et al., 1987), and that the channels are closely coupled to the ATP receptor (Benham and Tsien, 1987). Results from current reversal potential measurements suggested that the channels are permeable to Ca^{2+} with a selectivity of three to one over Na^{+}. Bearing in mind the much higher concentration of Na^{+} present in extracellular saline, the calculated Ca^{2+} influx is less than 10% of the total ATP activated current. However, permeation calculations are subject to error, especially for Ca^{2+} permeable channels where the strict conditions of the Goldman equation are not fulfilled (Tsien et al., 1987).

Activation of calcium dependent K^{+} channels during responses to ATP suggested that this Ca^{2+} permeability might be physiologically significant (Benham et al., 1987). This can now be tested by making direct quantitative measurements of $[Ca^{2+}]_i$ using fluorescent dyes. We have used this technique to measure $[Ca^{2+}]_i$ in smooth muscle and neuronal cells during ATP

Regulation of Smooth Muscle Contraction
Edited by R.S. Moreland, Plenum Press, New York, 1991

application to see if Ca^{2+} influx through ATP gated channels was sufficient to elevate $[Ca^{2+}]_i$ and obtain quantitative data on the relationship between cation current and rise in $[Ca^{2+}]_i$. The results suggest that in the smooth muscle cells, about 10% of the current through ATP gated channels is carried by Ca^{2+} ions at -60 mV, sufficient to significantly elevate $[Ca^{2+}]_i$ in these single cells.

METHODS

Cell Preparation

Smooth muscle cells were enzymatically dissociated from the central ear arteries of adult New Zealand white rabbits as previously described (Benham and Bolton, 1986), and stored on glass coverslips at 4° C for use the same day. Sensory neurons were freshly dissociated from 2 - 4 day old rats (Forda and Kelly, 1985) and used within 24 hours before they had grown processes.

Membrane Current and Fluorescence Measurements

Cells were normally bathed in an extracellular solution containing (in mM): 130 NaCl, 5 KCl, 10 glucose, 1.5 $CaCl_2$, 1.2 $MgCl_2$, and 10 HEPES buffered to pH 7.3 with NaOH. For the $[Ca^{2+}]_i$ measurements, the pipette solution was made up in MilliQ water and in most experiments contained (in mM): 125 CsCl, 2 $MgCl_2$, 10 HEPES, and 0.1 indo-1 (K-salt, Molecular Probes) buffered to pH 7.2 with NaOH. No other Ca buffers were added to the pipette solution. In some experiments 125 mM KCl was substituted for CsCl.

Patch clamp experiments were performed using standard techniques (Hamill et al., 1981). Combined voltage clamp and fluorescence measurements were performed as described previously (Benham, 1989a). The apparatus has also been described elsewhere (Jacob and Benham, 1990).

Briefly, voltage-clamp and membrane current recording were made with standard patch clamp techniques using a List EPC-7 patch-clamp amplifier. $[Ca^{2+}]_i$ was estimated from indo-1 fluorescence by the ratio method using single wavelength excitation and dual emission (Grynkiewicz et al., 1985; Cobbold and Rink, 1987). Single cells loaded with indo-1 from the patch pipette were centered under a window created by a hole (diameter 20 microns) placed in the emission beam (Fig. 1). On breakthrough to whole cell mode, loading could be monitored from the emitted light which reached equilibrium at 3 - 5 min after breakthrough with counts of 10 to 50 times background (0.1 mM indo-1) and recording was started at this point. After background subtraction, $[Ca^{2+}]_i$ was estimated from the 405/480 ratio using a calibration for indo-1 determined within cells (Benham, 1989a).

RESULTS and DISCUSSION

Use of Ca^{2+}-Dependent K^+ Channels and Fluorescent Probes for Measuring $[Ca^{2+}]_i$

Indo-1 applied to the cell as free ions in the patch pipette (internal solution) diffuses into the cytoplasm (Fig. 1). Photometric measurements of the fluorescence yield a signal that provides an integrated measure of the free

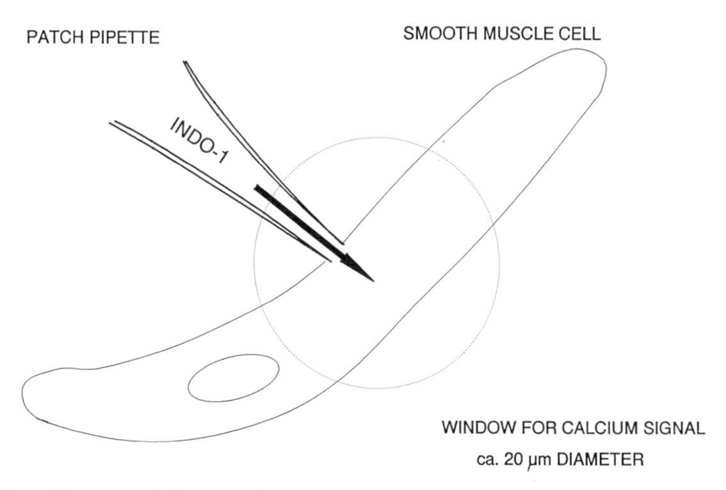

Figure 1. Diagram of the recording setup at the cellular level. Light is collected from a window of 20 μm in diameter. Smooth muscle cells were aligned so that a region of the cell that excluded the nucleus was viewed. Freshly dissociated sensory neurons were selected with cell bodies of about 20 μm diameter so as to fill this window.

Ca^{2+} concentration in the area of the cell under view. We routinely select a region of the smooth muscle cell that excludes the nucleus. The $[Ca^{2+}]_i$ signal will thus tend to be smoothed relative to a sharp signal in one part of the cell such as that caused by Ca^{2+} influx through membrane channels and resulting in a transiently high $[Ca^{2+}]_i$ in the immediate subsarcolemmal space.

Thus, photometric measurements might be expected to give quantitatively different results in some cases from those given by the widely used method of using Ca^{2+} activated K^+ channels as a measure of $[Ca^{2+}]_i$. Figure 2 shows $[Ca^{2+}]_i$ transients generated in the same cell by three different stimuli. As the cell was filled with a K^+ based internal solution, it was also possible to monitor K^+ channel activity. At the start of recording, (2 min after entering whole cell recording mode), spontaneous transient outward currents, (STOCS; Benham and Bolton, 1986) were observed. The largest of these had an amplitude of 175 pA. These transients are thought to reflect intermittent release of Ca^{2+} from internal stores, possibly triggered by the store being overloaded. Consistent with this idea, some of the larger transients were coincident with a $[Ca^{2+}]_i$ transient measured photometrically. Notice also that the fluorescence signal indicates that $[Ca^{2+}]_i$ falls steadily for the first 2 min of the trace with these transients superimposed on this declining base. As this basal level falls the large STOCS die out, so that just before caffeine was applied, the current trace shows only very small transient outward currents. $[Ca^{2+}]_i$ changes coincident with STOCS were seen in only two of ten cells where STOCS were recorded although the falling basal $[Ca^{2+}]_i$ was always seen. As we were only recording $[Ca^{2+}]_i$ from a region of the cell, we would expect that in some cells K^+ current activity represents changes in $[Ca^{2+}]_i$ that are not within the detection area for the fluorescence signal. Also, if the STOCS were very frequent, then individual $[Ca^{2+}]_i$ transients would tend to merge.

Caffeine application discharges internal Ca^{2+} stores resulting in a rise in $[Ca^{2+}]_i$. However, although the outward current signal was smaller than the

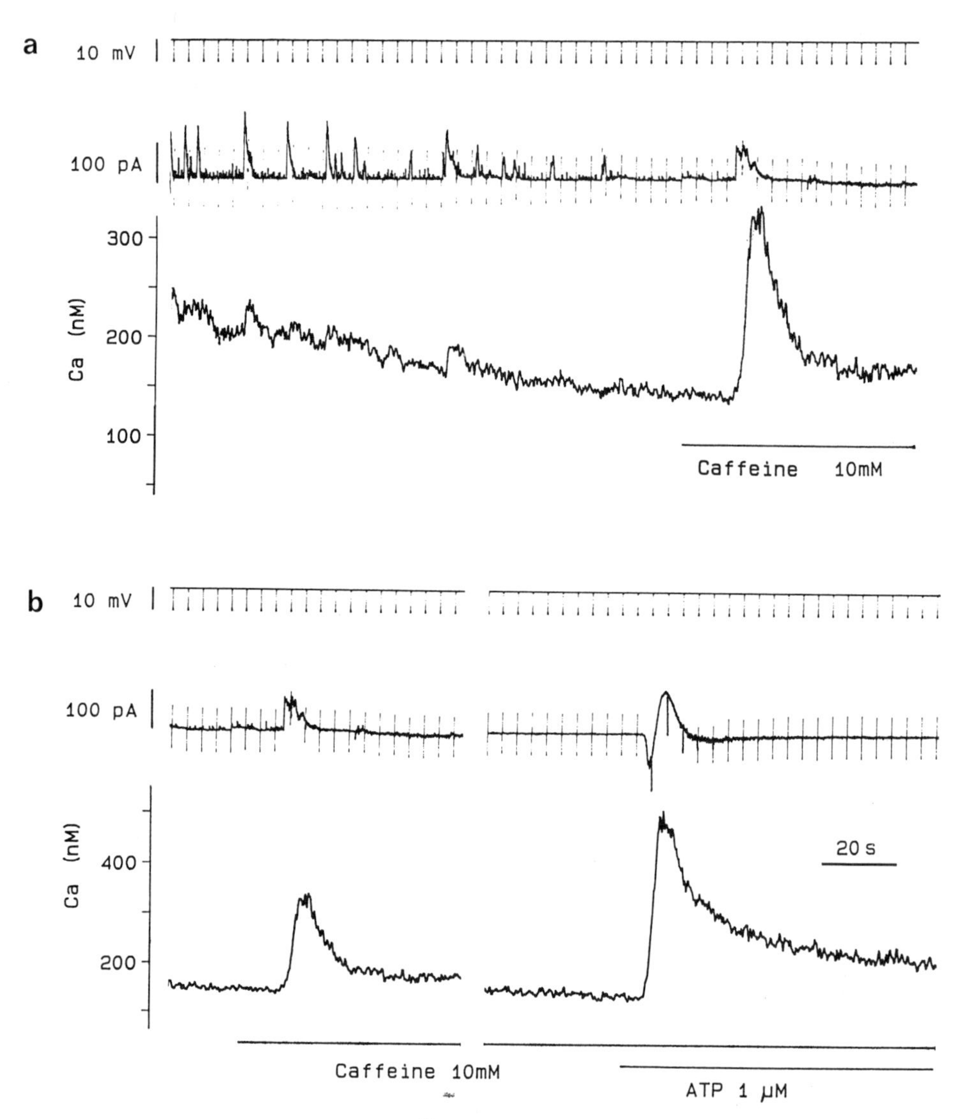

Figure 2. Outward currents and $[Ca^{2+}]_i$ signals generated by Ca store release and ATP gated current in an ear artery cell held under voltage clamp at -60 mV. In a and b, top trace is membrane potential, middle trace is membrane current, and lower trace is $[Ca^{2+}]_i$. a, Spontaneous transient outward currents followed by caffeine application. b, Caffeine response displayed on the same scale as the subsequent response to ATP in the same cell. The interruption in the record was 30s. (Fig. 2b is reprinted with permission from Benham, 1989b).

STOCS in amplitude, the rise in $[Ca^{2+}]_i$ was about 175 nM. This was much greater than the 20 - 40 nM signals associated with the STOCS. The much larger $[Ca^{2+}]_i$ rise in the caffeine response is not unexpected, as this treatment should release all the caffeine sensitive stores while the STOCS may represent localized release from individual storage compartments. However, it is surprising that caffeine did not generate a larger outward current in this cell. Clearly, the kinetics of release or the sites of release are such that the plasmalemmal K^+ channels are not exposed to the same transiently high Ca^{2+} signal as during STOC discharge.

Figure 2b shows the caffeine response redisplayed on the same scales as a subsequent response to ATP application in the same cell in the continued presence of caffeine. ATP activates an inward current followed by a net outward current of 110 pA, slightly greater than the 90 pA signal induced by caffeine. This was associated with a rise in $[Ca^{2+}]_i$ of nearly 400 nM. In this response the overlapping nature of the inward and outward currents make it impossible to measure their true peak values. However the inward current desensitized rapidly in these cells (Fig. 4), so the peak outward current was probably reduced rather little from the absolute value. The current response was biphasic in cells which showed a large rise in $[Ca^{2+}]_i$. Net outward currents were seen when the rise in $[Ca^{2+}]_i$ exceeded about 300 nM, close to the level of $[Ca^{2+}]_i$ expected to open significant numbers of Ca^{2+} activated K^+ channels at this membrane potential of -60 mV (Benham et al., 1986).

The relationship between the amplitude of the net outward current and the peak $[Ca^{2+}]_i$ signal in response to ATP application confirms this conclusion (Fig 3.) Data from 12 different cells showed that at -60 mV, the threshold for outward current was greater than 200 nM. $[Ca^{2+}]_i$ values of greater than 350 nM were accompanied by outward currents. Studies of the Ca^{2+} dependence of Ca^{2+} activated K^+ channels in vascular smooth muscle cells showed very low P_o (opening probability) at -60 mV when $[Ca^{2+}]_i$ was 200 nM. Increasing $[Ca^{2+}]_i$ to

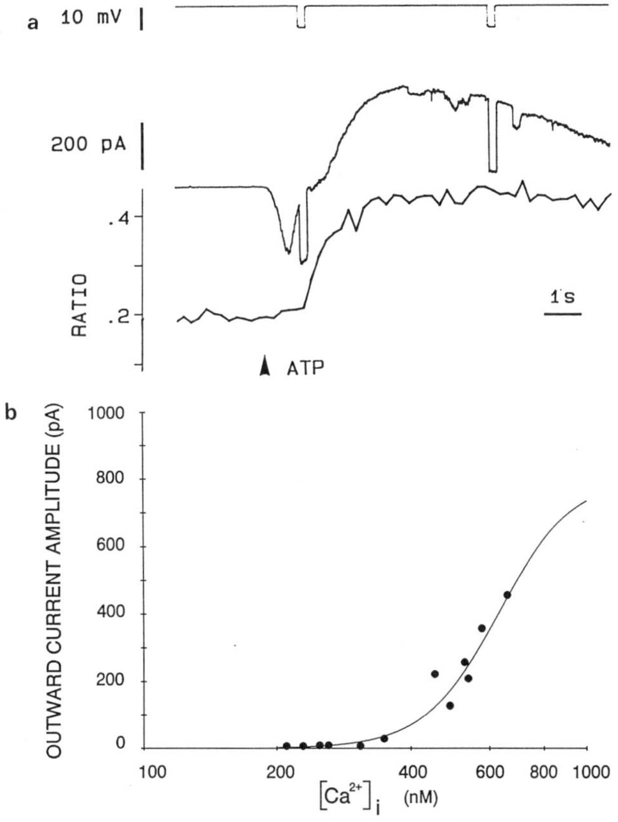

Figure 3. Relationship between outward currents and $[Ca^{2+}]_i$ signals generated by ATP. a, 10 µM ATP was applied from a puffer pipette. Traces as in Figure 2 except that the lower trace shows the indo-1 ratio rather than $[Ca^{2+}]_i$. b, The relationship between the peak outward current and peak $[Ca^{2+}]_i$ evoked by ATP at -60 mV in 12 ear artery cells.

1000 nM induced a near maximal increase in P_o. Thus, the two measuring systems are in good quantitative agreement for the $[Ca^{2+}]_i$ signals generated by ATP. This may in part be due to the fact that the Ca^{2+} influx is relatively slow allowing the whole cell Ca^{2+} signal to remain close to the subsarcolemmal Ca^{2+} concentration.

These data show that Ca^{2+} activated K^+ currents clearly can be used as a semi-quantitative indicator of $[Ca^{2+}]_i$. Appropriate choice of the membrane potential at which current is measured could be used to shift the sensitivity of the system into the range of $[Ca^{2+}]_i$ of interest. However, discrepancies may arise if the $[Ca^{2+}]_i$ signal is spatially localized and has a rapid onset. In this case, both systems may offer useful information. Using fluorescent Ca^{2+} indicators avoids the pitfalls of overlapping currents and particularly when studying Ca^{2+} influx processes, allows some quantitative analysis to be performed. In a later section results obtained with Cs^+ loaded cells to avoid this pitfall are discussed. Finally, it should be remembered that modulation of K^+ channel activity by other internal messengers could complicate analysis of outward current activity.

$[Ca^{2+}]_i$ Responses to ATP are Dependent on External Ca^{2+}

Figure 4a shows simultaneous recordings of membrane current and $[Ca^{2+}]_i$ from a cell bathed in saline containing 1.5 mM Ca^{2+}. Bath application of 10^{-6} M ATP to this cell held under voltage-clamp, activated an inward current which reached a peak of about 500 pA and then declined in the continued presence of ATP. The delay in onset of response to ATP was due to the dead space in the solution exchange system. Application of ATP by puffer pipette activates currents with minimal delay (Benham and Tsien, 1987). Hyperpolarizing voltage jumps to -80 mV were used to monitor the cell conductance and the access resistance. During the ATP response the voltage jumps evoked larger inward current steps, showing that there was an increase in membrane conductance. A rise in $[Ca^{2+}]_i$ from 140 nM to 300 nM was associated with this inward current which was also transient, but declined more slowly than the current. The maximum rate of rise of $[Ca^{2+}]_i$ coincided with the peak of the inward current as expected if Ca^{2+} influx through these channels was the cause of the rise.

Removal of extracellular Ca^{2+} did not inhibit the ATP activated inward current. Inward currents of similar amplitude were seen presumably due to monovalent cation movement through the ATP-activated channels (Benham and Tsien, 1987) (Fig. 4). However, biphasic responses with secondary outward currents were abolished and there was very little change in $[Ca^{2+}]_i$. The very small rise in $[Ca^{2+}]_i$ seen in two of these five cells (one of which is shown in Fig. 4b) could be due to residual Ca^{2+} in the bathing medium as no Ca^{2+} chelator was added. So, the rise in $[Ca^{2+}]_i$ seemed to be dependent on extracellular Ca^{2+} and was thus unlikely to be due to intracellular Ca^{2+} store release stimulated by generation of a second messenger such as inositol trisphosphate (IP_3). This is supported by the observation that ATP responses are unaffected by caffeine pretreatment to deplete Ca^{2+} stores or by norepinephrine pretreatment to activate phospholipase C and the IP_3 cascade. Effective voltage-clamp of these cells under these conditions (Benham and Tsien, 1988) means that it is unlikely that any depolarization occurs during the ATP response which might open voltage gated Ca^{2+} channels leading to Ca^{2+}

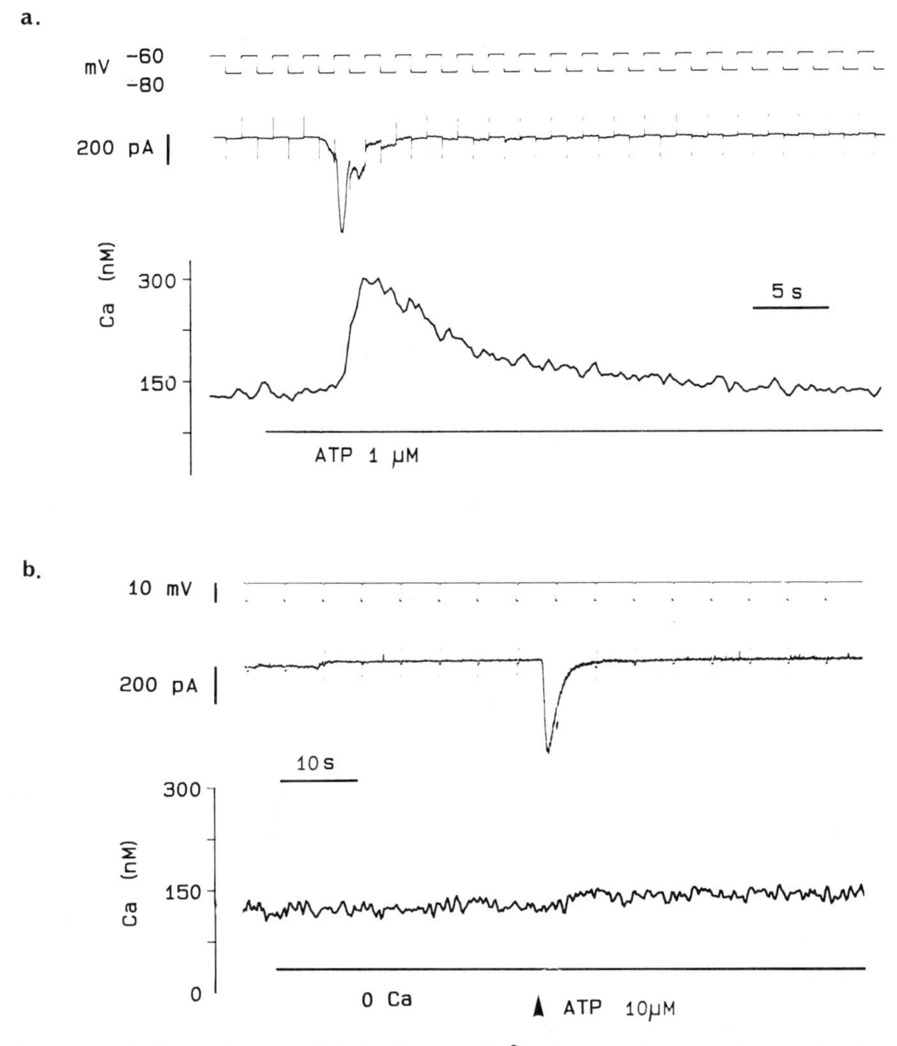

Figure 4. Bath application of ATP elevates $[Ca^{2+}]_i$ in vascular smooth muscle cells under voltage-clamp. a: Responses of a single rabbit ear artery cell to bath application of 10^{-6} M ATP (horizontal bar) in presence of 1.5 mM Ca^{2+}. Brief hyperpolarizing voltage steps were made throughout the recording to monitor cell input resistance and pipette series resistance (from Benham, 1989b with permission). b: Response of a cell bathed in nominally Ca^{2+}-free saline to application of ATP by pressure ejection from a micropipette. Note the similar magnitude and time course of the ATP activated current to that in Figure 3 but complete absence of a $[Ca^{2+}]i$ signal. Traces as for Figure 3a. (from Benham 1989b with permission).

influx through this route. The rise in $[Ca^{2+}]_i$ is also not blocked by dihydropyridine Ca^{2+} antagonists.

The ATP Evoked Rise in $Ca^{2+}{}_i$ is Voltage-Dependent

If the Ca^{2+} influx that was responsible for the rise in $[Ca^{2+}]_i$ was through the ATP gated channels and not through release of some unidentified store,

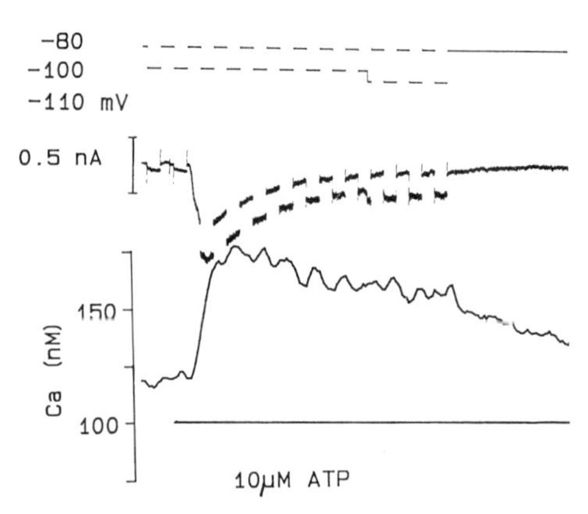

Figure 5. Voltage dependence of the ATP response in a sensory neuron. During the ATP response, $[Ca^{2+}]_i$ oscillations increase during the hyperpolarizing voltage jumps and decrease during the depolarization as the driving force is reduced.

then the $[Ca^{2+}]_i$ signal should be voltage-dependent such that at strongly depolarized potentials where the driving force for Ca^{2+} is much reduced there should be negligible Ca^{2+} influx. This was tested by applying ATP to cells held at depolarized potentials using CsCl filled cells so that no Ca^{2+} activated K^+ current was evoked. At a holding potential of +50 mV, calculations from constant field theory suggest that very little inward Ca^{2+} flux should occur through the ATP gated channels (Mayer and Westbrook, 1987). The channels should now carry outward current as the reversal potential was close to zero mV. As expected, ATP did evoke outward currents but with insignificant effects on $[Ca^{2+}]_i$ even though the magnitude of the outward currents suggested that the degree of ATP receptor activation was roughly comparable (Benham, 1989b).

An alternative demonstration of the increase in Ca^{2+} influx with increasing driving force through these cation channels could be achieved by varying membrane potential during the ATP current response. The ATP response declines too rapidly in vascular cells, but in sensory neurons, a similar conductance is activated by ATP that desensitizes less rapidly so current responses of tens of seconds may be recorded. Figure 5 shows membrane current and $[Ca^{2+}]_i$ responses to ATP in a rat sensory neuron held for 4 seconds alternately at -80 and -100 mV. The membrane potential changes had no effect on $[Ca^{2+}]_i$ in the unstimulated cell, but during the ATP response an oscillation was superimposed on the $[Ca^{2+}]_i$ response. This oscillation was of 3 - 5 nM amplitude and synchronized so that $[Ca^{2+}]_i$ rose during the periods at the more negative membrane potential. The peak change in $[Ca^{2+}]_i$ in this cell was about 50 nM. $[Ca^{2+}]_i$ responses in the sensory neurons were generally smaller than those seen in the vascular smooth muscle cells even though the magnitude of the currents was very similar.

A quantitative comparison between the magnitude of the $[Ca^{2+}]_i$ signal and the charge transferred in the current response in the two cell types confirms this observation (Fig 6). As expected, there is a positive correlation between the two signals in each cell, but the rise in $[Ca^{2+}]_i$ in the ear artery cells was about 10 fold greater than that in the neuronal soma. Measurements of the reversal potentials of the ATP gated current in the two cell types indicated that the permeation properties were very similar (Bouvier et al., 1990). This would

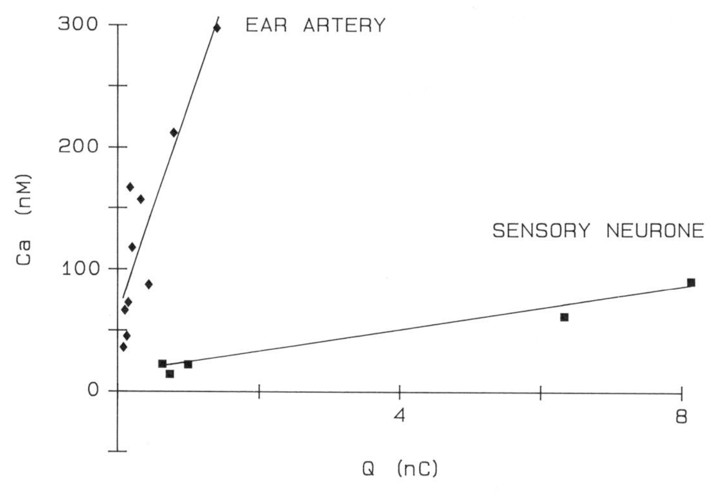

Figure 6. The relationship between the change in $[Ca^{2+}]_i$ and the charge transferred during the ATP response for ear artery smooth muscle and sensory neurons. Straight lines are drawn by eye. Data from ten smooth muscle cells and five neurons.

suggest that the same amount of Ca^{2+} is entering per unit charge. This suggests that the different relationship may be explained by the dissimilar buffering properties of the two types of cell. The results might also imply that the smooth muscle cells are less able to buffer this type of Ca^{2+} influx than the neuronal soma. We can speculate that the vascular cells are probably not physiologically exposed to large transient $[Ca^{2+}]_i$ loads such as might occur during a burst of action potentials in a neuron. Thus in contrast to the sensory neurons they may not be able to rapidly buffer the influx activated by ATP.

CONCLUSION

These data show that measurement of $[Ca^{2+}]_i$ with fluorescent indicators can be used to provide quantitative information about Ca^{2+} influx processes. The cation current activated by ATP in arterial smooth muscle and in sensory neurons admits sufficient Ca^{2+} to measurably raise cytoplasmic Ca^{2+} levels. The functional significance of this Ca^{2+} influx remains to be determined although in these cell types it would seem likely that the role of this pathway is secondary to influx of Ca^{2+} through voltage gated Ca^{2+} channels since the concomitant depolarization will reduce Ca^{2+} influx through the ATP gated channels. However, ATP gated channels do provide a pathway for Ca^{2+} entry that may be spatially distinct from that of other Ca^{2+} channels allowing Ca^{2+} entry targeted to particular parts of the cell.

Discrepancies between outward currents and $[Ca^{2+}]_i$ measured by fluorescence measurements suggest that $[Ca^{2+}]_i$ rises can be localized during STOC discharge in the ear artery cells and this is likely to be more important in neuronal processes. The growing use of Ca^{2+} imaging using confocal microscopy (Hernandez-Cruz et al., 1990) will provide a more powerful technique for studying these processes.

Finally, the technique provides a powerful means to investigate Ca^{2+} efflux processes in single cells that is a useful alternative to biochemical studies in broken cell preparations.

ACKNOWLEDGMENTS

Some of this work has recently been published in The Journal of Physiology (Benham, 1989b).

REFERENCES

Benham, C. D., 1989a, Voltage-gated and agonist mediated rises in intracellular Ca in rat clonal pituitary cells (GH3) held under voltage clamp, *J. Physiol.*, 415: 143.

Benham, C. D., 1989b, ATP-activated channels gate calcium entry in single smooth muscle cells dissociated from rabbit ear artery, *J. Physiol.*, 419: 689.

Benham, C. D. and Bolton, T. B., 1986, Spontaneous transient outward currents in single visceral and vascular smooth muscle cells of rabbit, *J. Physiol.*, 381: 385.

Benham, C. D., Bolton T. B., Byrne N. G., and Large, W. A., 1987, Action of extracellular adenosine triphosphate in single smooth muscle cells dispersed from the rabbit ear artery, *J. Physiol.*, 387: 473.

Benham, C. D., Bolton, T. B., Lang, R. J., and Takewaki, T., 1986, Calcium activated K-channels in single dispersed smooth muscle cells of rabbit jejunum and guinea-pig mesenteric artery, *J. Physiol.*, 371: 45.

Benham, C. D. and Tsien, R. W., 1987, Receptor-operated, Ca-permeable channels activated by ATP in arterial smooth muscle, *Nature*, 328: 275.

Benham, C. D. and Tsien, R. W., 1988, Noradrenaline modulation of calcium channels in single smooth muscle cells from rabbit ear artery, *J. Physiol.*, 404: 767.

Bouvier, M. M., Evans, M. L., Fowler, K., and Benham, C. D., 1990, Calcium influx induced by ATP receptor activation on neurones cultured from rat dorsal root ganglia, *J. Physiol.*, 424: 20P.

Burnstock, G., 1988, Sympathetic purinergic transmission in small blood vessels, *Trends Pharmacol. Sci.*, 9: 116.

Burnstock, G. and Kennedy, C., 1985, Is there a basis for distinguishing two types of P_2-purinoceptor?, *Gen. Pharmacol.*, 16: 433.

Burnstock, G. and Kennedy, C., 1986, A dual function for adenosine-5'-triphosphate in the regulation of vascular tone, *Circ. Res.*, 58: 319.

Cobbold, P. and Rink, T. J., 1987, Fluorescence and bioluminescence measurement of cytoplasmic free calcium, *Biochem. J.*, 248: 313.

Forda, S. R. and Kelly, J. S., 1985, The possible modulation of the development of rat dorsal root ganglion neurons by the presence of 5-HT containing neurones of the brainstem in dissociated cell culture, *Dev. Brain Res.*, 22: 55.

Grynkiewicz, G., Poenie, M., and Tsien, R. Y., 1985, A new generation of Ca indicators with greatly improved fluorescence properties, *J. Biol. Chem.*, 260: 3440.

Hamill, O. P., Marty, A., Neher, E., Sakmann, B., and Sigworth, F. J., 1981, Improved patch-clamp techniques for high resolution current recording from cells and cell free membrane patches, *Pflügers Arch.*, 391: 85.

Hernandez-Cruz, A., Sala, F., and Adams, P. R., 1990, Subcellular calcium transients visualised by confocal microscope in a voltage-clamped vertebrate neuron, *Science,* 247: 858.
Jacob, R. and Benham, C. D., 1990, Measuring cytoplasmic calcium in single living cells using fluorescent probes, *in:* "New Techniques of Optical Microscopy and Microspectrophotometry", R. J. Cherry, ed., Macmillan Press.
Krishtal, O. A., Marchenko, S. M., and Pidoplichko, V. I., 1983, Receptor for ATP in the membrane of mammalian sensory neurones, *Neurosci. Lett.,* 35: 41.
Mayer, M. L. and Westbrook, G. L., 1987, Permeation and block of N-methyl-D-aspartic acid receptor channels by divalent cations in mouse cultured central neurones, *J. Physiol.,* 394: 501.
Sneddon, P., Westfall, D. P., and Fedan, J. S., 1982, Co-transmitters in the motor nerves of the guinea-pig vas deferens: Electrophysiological evidence, *Science,* 218: 693.
Stjarne, L., 1986, New paradigm: Sympathetic transmission by multiple messengers and lateral interaction between monoquantal release sites? *Trends Neurosci.,* 9: 547.
Suzuki, H., 1985, Electrical responses of smooth muscle cells of the rabbit ear artery to adenosine triphosphate, *J. Physiol.,* 359: 401.
Tsien, R. W., Hess, P., McCleskey, E. W., and Rosenberg, R. L., 1987, Calcium channels : Mechanisms of selectivity, permeation and block, *Ann. Rev. Biophys. Biophys. Chem.,* 16: 265.

PURIFICATION AND RECONSTITUTION OF THE RYANODINE- AND CAFFEINE-SENSITIVE Ca^{2+} RELEASE CHANNEL COMPLEX FROM MUSCLE SARCOPLASMIC RETICULUM

Gerhard Meissner, F. Anthony Lai, Kristin Anderson, Le Xu, Qi-Yi Liu, Annegret Herrmann-Frank, Eric Rousseau, Rodney V. Jones, and Hee-Bong Lee

Department of Biochemistry and Biophysics
University of North Carolina
Chapel Hill, NC 27599

INTRODUCTION

Studies in our laboratory are directed towards obtaining a better understanding of the process of excitation-contraction coupling in muscle, primarily using purified membrane and protein preparations. In excitable tissues, release of Ca^{2+} ions can be triggered by a change in surface membrane potential (Ebashi, 1976), or it can occur via a chain of voltage-independent steps that involve agonist-induced formation of inositol 1,4,5-trisphosphate (IP_3) and its subsequent binding and activation of an intracellular membrane receptor/channel complex, the IP_3 receptor (Van Breemen and Saida, 1989). The voltage-dependent mechanism, commonly referred to as excitation - contraction (E-C) coupling, has been most extensively studied in skeletal and cardiac muscle. In striated muscle, rapid release of Ca^{2+} ions from the intracellular membrane compartment, sarcoplasmic reticulum (SR), is triggered by a surface membrane action potential that is thought to be communicated to the SR at specialized areas where the SR comes in close contact with tubular infoldings of the surface membrane (T-tubule). Spanning the gap between the two membrane systems are protein bridges (Peachey and Armstrong, 1983) which have been variously termed "feet" (Franzini-Armstrong, 1970), "bridges" (Somlyo, 1979), "pillars" (Eisenberg and Eisenberg, 1982), or "spanning" proteins (Caswell and Brandt, 1989), and are now believed to be identical with the ryanodine receptor or SR Ca^{2+} release channel complex (Fleischer and Inui, 1989; Lai and Meissner, 1989).

How the release of Ca^{2+} ions from SR is regulated in muscle has remained an enigma. One postulated mechanism, which appears to predominate in heart muscle, is the Ca^{2+} triggering mechanism with Ca^{2+} acting as a second messenger (Fabiato, 1983). According to this hypothesis, depolarization of the T-tubule allows a few Ca^{2+} ions to enter the muscle cell, most likely via a dihydropyridine sensitive Ca^{2+} channel located in the surface membrane

Regulation of Smooth Muscle Contraction
Edited by R.S. Moreland, Plenum Press, New York, 1991

(Näbauer et al., 1989; Tanabe, 1990). These Ca^{2+} ions then trigger the massive release of Ca^{2+} ions from SR by activating the intracellular SR Ca^{2+} release channel. A second mechanism, the mechanical coupling mechanism, suggests that the SR Ca^{2+} release channel is in direct contact with a voltage-sensing molecule in the T-tubule membrane (Schneider, 1981). Biophysical and pharmacological evidence (Rios and Pizarro, 1988) as well as cDNA expression studies (Tanabe et al., 1988) have indicated that the dihydropyridine (DHP) receptor is the voltage sensor in E-C coupling in skeletal muscle. Accordingly, protein conformational changes in the voltage sensor (DHP receptor; o in Fig. 1) occurring during T-tubule depolarization may be directly sensed by the feet and in turn cause an opening of a Ca^{2+} channel intrinsic to the SR feet (Lai and Meissner, 1989).

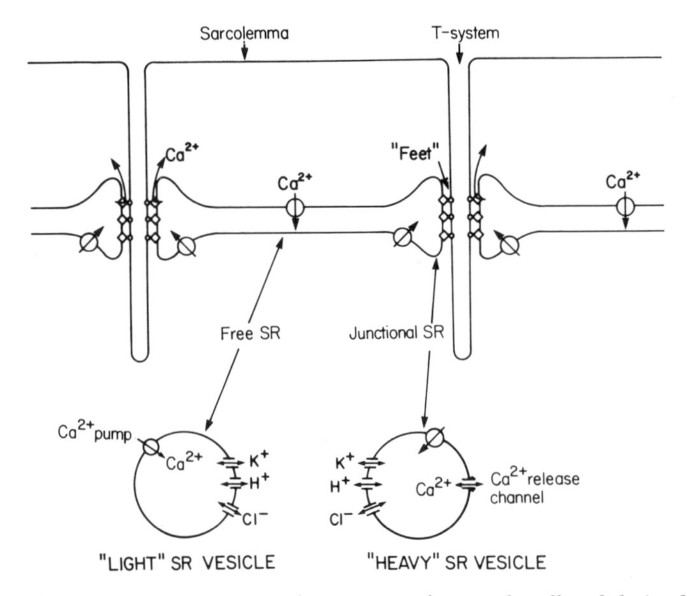

Figure 1. Schematic representation of a segment of a muscle cell and derived SR vesicle fractions. Indicated are dihydropyridine receptor (o) in T-system membrane, and Ca^{2+} pump, Ca^{2+} release channel (◊) and monovalent ion channels in SR membrane.

Fragmentation of the SR during homogenization of muscle and subsequent fractionation by differential and density gradient centrifugation has yielded distinct types of vesicles called "light" and "heavy" SR with respect to their sedimentation properties (Meissner, 1975; Smith et al., 1986a). Light vesicles correspond to the free or nonjunctional SR, whilst heavy SR are predominantly derived from the terminal cisternae, junctional region of SR (Fig. 1). In our studies, we use heavy, junctionally-derived SR vesicle fractions to purify and characterize a Ca^{2+}-gated, ryanodine- and methylxanthine-sensitive 30S Ca^{2+} release channel complex in skeletal (Lai et al., 1988) and cardiac (Anderson et al., 1989) muscle. More recently, partially purified 30S channel complexes have been also obtained from smooth muscle (Herrmann-Frank et al., 1990).

RESULTS AND DISCUSSION

Purification of 30S Ryanodine Receptor Complex

The SR Ca^{2+} release channel complex is isolated from "heavy" SR membrane fractions prepared from rabbit skeletal or canine heart muscle homogenates. Membranes are isolated in the presence of protease inhibitors, and "heavy" vesicles containing the ryanodine-sensitive Ca^{2+} release channel protein are enriched by differential and sucrose gradient centrifugation (Meissner, 1984; Meissner and Henderson, 1987). The membrane-bound Ca^{2+} release channel is solubilized in the presence of the channel-specific probe [^{3}H]-ryanodine, using the zwitterionic detergent Chaps and high ionic strength, then purified as a 30S protein complex by centrifugation through a linear sucrose gradient (Fig. 2). Density centrifugation results in efficient separation of the 30S ryanodine receptor complex from other solubilized SR proteins due to its faster sedimentation rate. The sucrose gradient centrifugation procedure is relatively simple and straightforward and has been applied in our laboratory to isolate the 30S ryanodine receptor from rabbit skeletal muscle (Lai et al., 1988), canine cardiac muscle (Anderson et al., 1989), and bovine brain (Lai et al., 1990), as well as to identify a 30S channel complex from canine and porcine aorta (Herrmann-Frank et al., 1990). SDS polyacrylamide gel electrophoresis has shown that the skeletal muscle ryanodine receptor peak fractions contain a single polypeptide with an apparent M_r of ~350,000-450,000 (Fig. 3; Imagawa et al., 1987; Inui et al., 1987; Lai et al., 1988; Anderson et al., 1989) and a calculated M_r ~560,000 as determined by cloning and sequencing the complementary

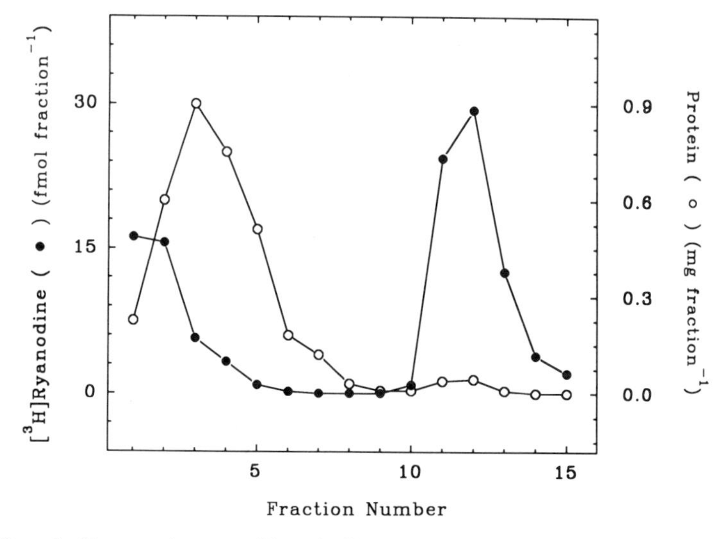

Figure 2. Sedimentation profile of Chaps-solubilized heavy SR proteins and ryanodine receptor. Heavy SR membranes were solubilized in Chaps, centrifuged through linear sucrose gradients and fractionated. Fractions were analyzed for protein and radioactivity content. The majority of the solubilized proteins sediment between fractions 1-6, whereas the [^{3}H]-ryanodine receptor peak is present in fractions 11-13 comigrating with a small protein peak. The radioactivity remaining at the top of the gradient (fractions 1-4) represents unbound [^{3}H]-ryanodine. (in modified form from Lai et al., 1988).

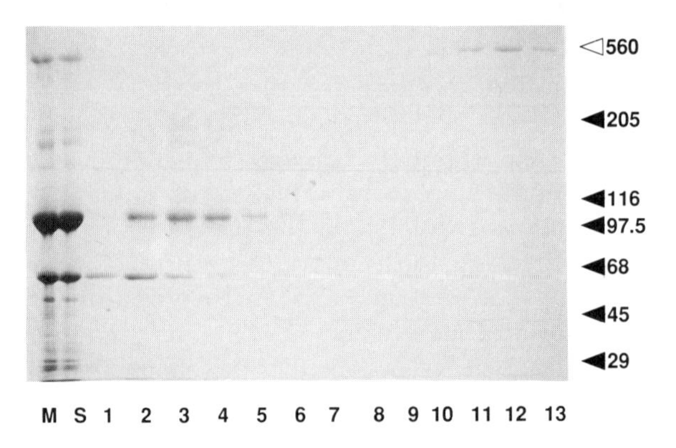

Figure 3. SDS polyacrylamide gel analysis. Heavy SR membranes (lane 1, 30 µg protein) were solubilized in Chaps (lane 2, 28 µg) and centrifuged through linear sucrose gradients as described in Figure 2. An aliquot (30 µl) of each gradient fraction was then analyzed by SDS-gel electrophoresis on a 3-12% linear polyacrylamide gradient gel and stained with Coomassie blue. The top gradient fractions contain the major heavy SR proteins, Ca-ATPase and calsequestrin. Fractions 11-13 specifically contain a single polypeptide which comigrates with the ryanodine receptor peak (Fig. 2). Molecular weight standards (x 10^{-3}) are shown to the right (closed arrowheads). Open arrowhead represents the predicted molecular weight of the ryanodine receptor polypeptide from cDNA sequence analysis. (in modified form from Lai et al., 1988).

DNA of the skeletal and cardiac muscle ryanodine receptors (Takeshima et al., 1988; Zorzato et al., 1989; Otsu et al., 1990). The skeletal and cardiac M_r 560,000 polypeptides are arranged in the form of tetrameric, four leaf clover-like structures, as revealed by electron microscopy of negatively stained samples (Lai et al., 1988; Wagenknecht et al., 1989). Cross-linking studies (Lai et al., 1989) and scanning transmission electron microscopy, which indicated a mass of about 2 million for the purified ryanodine receptor (Saito et al., 1989), have also suggested a tetrameric assembly of the M_r ~560,000 polypeptide. These structures are morphologically identical to the previously described protein bridges ("feet") that span the transverse tubular-SR junctional gap (Lai et al., 1988).

Purified Ryanodine Receptor Complex Contains an Intrinsic Ca^{2+} Channel Activity

The Chaps-solubilized 30S ryanodine receptor complex, purified in the absence of [^{3}H]-ryanodine, upon reconstitution into Müeller-Rudin type lipid bilayers, displays a channel activity typical of the SR Ca^{2+} release channel (Lai et al., 1988; Smith et al., 1988; Anderson et al., 1989). The skeletal and cardiac muscle receptors readily insert into the bilayer in a symmetrical monovalent cation medium and conduct monovalent cations more efficiently than Ca^{2+} ions. For these reasons, the pharmacology and kinetics of activation and inactivation of single purified channels have been primarily determined in our laboratory in a K^+ or Na^+ medium (Fig. 4). If desired, a Ca^{2+}-conducting channel can be obtained by perfusing the cis chamber with an impermeable cation such as $Tris^+$ and the trans chamber with a $CaCl_2$ or Ca-Hepes buffer, with Ca^{2+} serving as the current carrier. The effects of endogenous regulatory

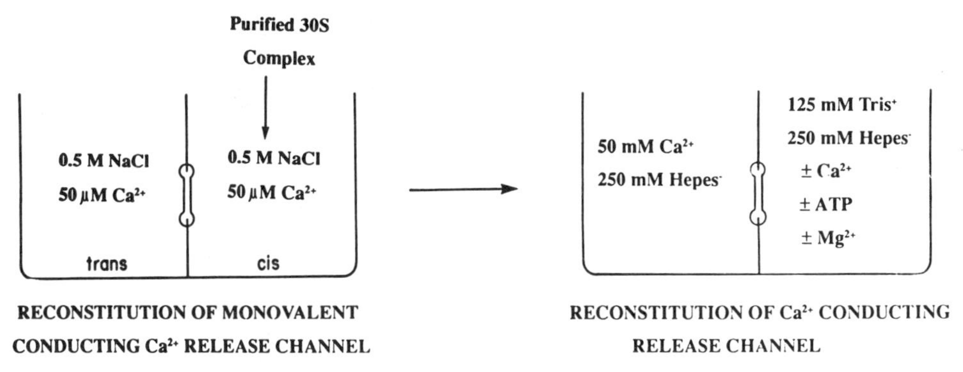

Figure 4: Reconstitution of monovalent and Ca^{2+} ion conducting Ca^{2+} release channel. The trans and cis sides of the bilayer are equivalent to the lumenal and myoplasmic sides of the SR membrane, respectively.

ligands such as Ca^{2+}, ATP, and Mg^{2+}, or drugs, are determined by varying their concentrations in the cis chamber of the bilayer set-up.

In Figure 5, a single skeletal channel, purified in the absence of [^{3}H]-ryanodine, was reconstituted in symmetric 0.5 M NaCl containing 6 µM free Ca^{2+}. Single channel conductance was 600 pS with 500 mM Na^+ as the current carrier. As observed for the native channel, channel activity was decreased by lowering the free Ca^{2+} concentration in the cis chamber (SR myoplasmic side) by the addition of EGTA, and again increased by the addition of 2 mM ATP cis. Perfusion of the trans chamber (SR lumenal side) with a 50 mM Ca^{2+} solution, and cis chamber with the slowly permeating cation Tris, resulted in a Ca^{2+} conductance of 90 pS and E_{rev} ~30 mV, which under the same buffer conditions are typical for the native channel (Smith et al., 1986b).

Single channel reconstitution studies with purified channel proteins can be ambiguous since a miniscule contaminant, undetectable by more conventional techniques such as SDS-gel analysis, may give rise to the observed channel activity. It is therefore important to correlate microscopic single channel studies (Fig. 5) with some macroscopic measurements. One such parameter has been determination of [^{3}H]-ryanodine binding to the native and purified receptor. The SR Ca^{2+} release channel interacts in a highly specific and characteristic manner with the neutral plant alkaloid, ryanodine. In vesicle-ion flux studies this drug displays a dual effect in that nanomolar concentrations appear to lock the channel into an open configuration, whereas at concentrations above 10 µM, ryanodine appears to completely close the channel (Fig. 6).

Studies with membrane-bound and detergent-solubilized preparations have suggested that ryanodine exerts its activating and inhibitory action by binding in a Ca^{2+}-dependent manner to high and low affinity sites of the SR Ca^{2+} release channel, respectively (Fig. 7; Lai et al., 1989). Taking into account a calculated molecular weight of 560,000, a specific activity of ~500 pmol ryanodine per mg protein to the purified skeletal and cardiac receptors corresponded to approximately one high-affinity site per 30S tetramer (Lai et al., 1988; Anderson et al., 1989). Recently, more extensive binding studies with the membrane-bound skeletal receptor have indicated the presence of both high and low affinity sites with a ratio of 1:3 (Lai et al., 1989).

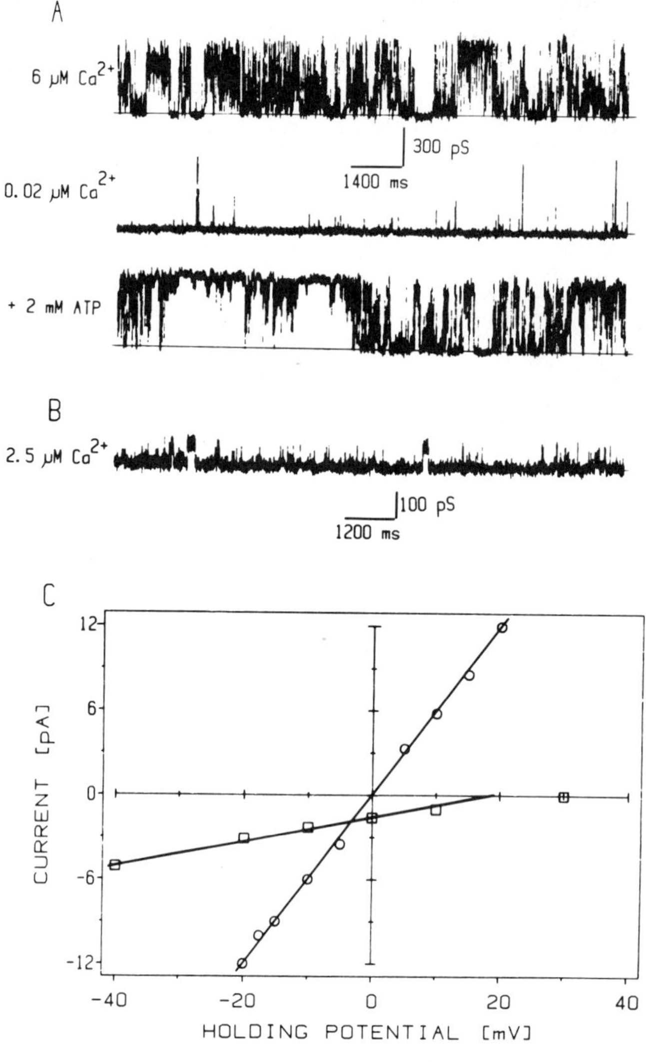

Figure 5. Reconstitution of the 30S ryanodine receptor complex into planar lipid bilayers. A: Single channel currents, shown as upward deflections, were recorded in symmetric 0.5 M NaCl, 10 mM NaPIPES, pH 7 with 6 μM free Ca^{2+} cis (top trace), with 0.02 μM free Ca^{2+} cis (second trace), or with 0.02 μM free Ca^{2+} plus 2 mM ATP in the cis chamber (third trace). Holding potential (HP): -15 mV. B: Single-channel current recorded after perfusion with 50 mM $Ca(OH)_2$/250 mM HEPES, pH 7.4, 10% glycerol trans, and 125 mM Tris/250 mM HEPES, pH 7.4, 10% glycerol, plus 2.5 μM free Ca^{2+} cis. HP: 0 mV. C: Current-voltage relationship for recordings A (top trace) and B. Values of unit conductance: γ, 595 pico Siemens (pS) with 0.5 M Na^+ (○); and γ, 91 pS with 50 mM Ca^{2+} (□) as the conducting ion. Recordings were filtered at 300 Hz and sampled at 2 kHz. (with permission from Lai et al., 1988).

The effects of ryanodine on a single K^+-conducting, purified skeletal Ca^{2+} release channel are shown in Figure 8. Several minutes after the addition of μM ryanodine, the channel entered into a subconductance state with an open probability close to unity. In Figure 8, a relatively high ryanodine concentration of 30 μM was used to reduce the time required to observe the otherwise very slow interaction of ryanodine with the channel (Meissner, 1986b). Upon further addition of millimolar ryanodine cis, the channel's subconductance state abruptly disappeared and the channel entered into a fully closed state. The similar effects of ryanodine on the purified, bilayer-reconstituted SR Ca^{2+} release channel and in vesicle $^{45}Ca^{2+}$-flux studies, together with [^{3}H]-ryanodine binding studies, has suggested that the Ca^{2+} release channel is identical with the 30S complex comprised of four polypeptides of M_r ~560,000 rather than a spurious, minor component co-purifying with the ryanodine receptor (Lai and Meissner, 1989; but see also Zaidi et al., 1989). More recently, functional expression of full-length rabbit skeletal ryanodine receptor cDNA in CHO cells has provided additional evidence that the 30S ryanodine receptor complex indeed contains an intrinsic Ca^{2+} channel activity (Penner et al., 1989).

SR Ca^{2+} Release Channel is a Ligand-Gated Channel

In our laboratory, the mechanism of skeletal and cardiac muscle Ca^{2+} release has been primarily studied using isolated SR membrane and protein preparations and by applying two complementary techniques. In the first one, microscopic Ca^{2+} currents are recorded through single Ca^{2+} release channels which are obtained by incorporation of SR vesicles (Rousseau et al., 1986; Smith et al., 1986b) or purified channel preparations (Fig. 5) into planar lipid bilayers. In the other one, macroscopic $^{45}Ca^{2+}$ efflux is measured from actively or passively loaded vesicles using a rapid mixing, quench, and filtration protocol (Meissner et al., 1986). Although single channel measurements provide more direct information, an advantage of the vesicle flux technique is that it more readily yields representative data by averaging the kinetic behavior of a large number of channels.

Table 1. Ca^{2+} release properties of skeletal and cardiac Ca^{2+} release vesicles

Additions to release medium	$^{45}Ca^{2+}$ Efflux: Skeletal	$^{45}Ca^{2+}$ Efflux: Cardiac
	$t_{1/2}$ (s)	
2×10^{-9} M Ca^{2+}	8	25
2×10^{-9} M Ca^{2+}, 5 mM ATP	0.06	12
2×10^{-9} M Ca^{2+}, 5 mM ATP, 5 mM Mg^{2+}	20	20
10^{-5} M Ca^{2+}	0.6	0.02
2×10^{-6} M Ca^{2+}, 5 mM AMP-PCP	0.01	0.01
10^{-5} M Ca^{2+}, 5 mM AMP-PCP, 5 mM Mg^{2+}	0.09	0.03
10^{-5} M Ca^{2+}, 1 mM Mg^{2+}	15	0.25
10^{-3} M Ca^{2+}	10	0.1
10^{-5} M Ca^{2+}, 2 μM calmodulin	1.2	0.1

Skeletal and cardiac heavy SR vesicles were passively loaded with 1 mM $^{45}Ca^{2+}$ and diluted into release media containing the indicated concentrations of free Ca^{2+}, Mg^{2+}, and adenine nucleotide. Release rates were determined with the use of a chemical quench apparatus and by filtration (Meissner et al., 1986; Meissner, 1986a). Ca^{2+} permeable vesicles released half their $^{45}Ca^{2+}$ stores within the indicated times.

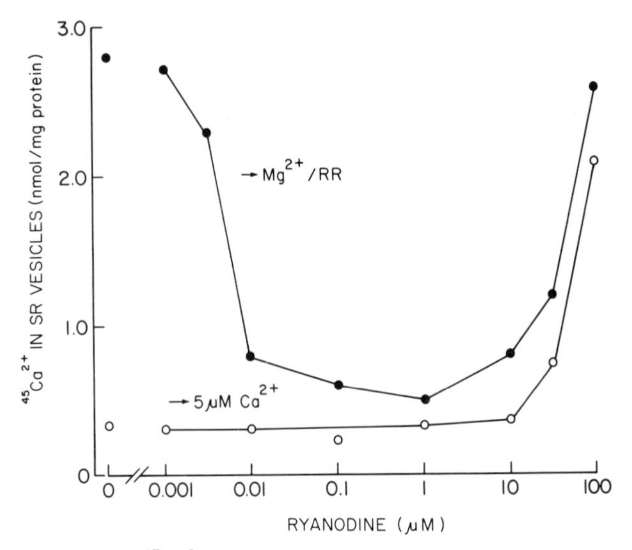

Figure 6. Dependence of $^{45}Ca^{2+}$ efflux on ryanodine concentration. Skeletal muscle SR Ca^{2+} release vesicles were incubated for 45 min at 37° C with 0.1 mM free $^{45}Ca^{2+}$ and the indicated concentrations of ryanodine. Amounts of $^{45}Ca^{2+}$ retained by the vesicles in Ca^{2+} release inhibiting (10 mM Mg^{2+}, 10 μM ruthenium red (RR, ●)) and activating (5 μM free Ca^{2+} (○)) media were determined by filtration (with permission from Meissner, 1986b).

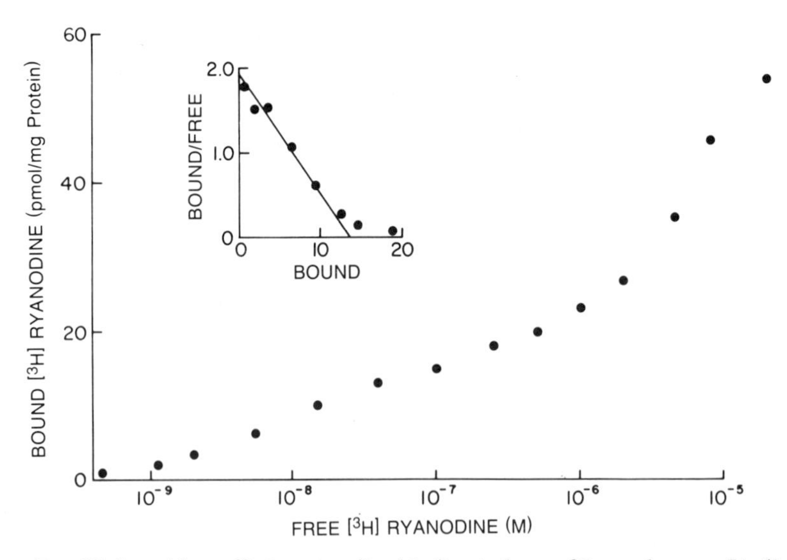

Figure 7. High and low affinity ryanodine binding to heavy SR membranes. Binding of [^{3}H]-ryanodine to heavy SR membrane was determined as described (Lai et al., 1989). Scatchard analysis (inset) reveals a curvilinear slope indicating the presence of both high affinity (K_D = 7 nM), and low-affinity sites (adapted from Lai et al., 1989).

Table 1 shows that skeletal and cardiac SR vesicle-$^{45}Ca^{2+}$ release is activated in a similar, although not identical, manner by micromolar Ca^{2+} and millimolar adenine nucleotide. In vesicle flux (Table 1) and single channel measurements (Smith et al., 1986b) the presence of both ligands was required to optimally activate the channel. The presence of millimolar Mg^{2+} and Ca^{2+} resulted in reduced $^{45}Ca^{2+}$ release rates. Another endogenous inhibitor is calmodulin, which reduced $^{45}Ca^{2+}$ efflux rates (Table 1) and channel open times (Smith et al., 1989) 2-6 fold. Ca^{2+}, ATP, and Mg^{2+} regulate SR Ca^{2+} release in a cooperative or noncooperative manner, depending on the experimental conditions (Meissner et al., 1986). The presence of cooperatively coupled domains within the tetrameric 30S channel complex has been also indicated in [^{3}H]-ryanodine binding and single channel studies. [^{3}H]-ryanodine binding revealed an allosteric interaction between high and low affinity sites (Lai et al., 1989). Single channel recordings of the purified 30S complex have shown the presence of multiple conductance states within the 30S tetramer. Up to four approximately equivalent conductances have been reported with either Na^{+} or Ca^{2+} as the conducting ion (Liu et al., 1989). In a recent, rare single channel recording, at least six distinct conductances can be discerned (Fig. 9). Again, cooperative interactions within the channel complex must be postulated to occur, in order to account for the appearance of the multiple conductance states.

Activation of SR Ca^{2+} Release Channel by Methylxanthines

Among a large number of drugs that have been identified to modulate SR Ca^{2+} release (Palade et al., 1989), caffeine is one of the best known because of its extensive use in the assessment of SR function in controlling myoplasmic Ca^{2+} (Endo, 1977; Stephenson, 1981; Fabiato, 1983). Figures 10 and 11 illustrate that the effectiveness of caffeine in stimulating SR Ca^{2+} release is dependent on the free Ca^{2+} concentration and the presence of MgATP in the release medium. In the absence of caffeine, a maximal release rate with a first-order rate constant of about 1 s^{-1} was obtained in release media containing 1-20 μM free Ca^{2+} (Fig. 10). Ca^{2+} release was slow at both nanomolar and millimolar Ca^{2+} concentrations, which has suggested that the Ca^{2+} release channel contains high-affinity activating and low-affinity inhibitory Ca^{2+} binding sites. Addition of 20 mM caffeine resulted in about a 10-fold increase of the Ca^{2+} release rate at nanomolar Ca^{2+} concentrations. In the presence of micromolar and millimolar free Ca^{2+}, caffeine was less effective, increasing the release rate by a factor of only about 1.5. Figure 11 shows that at nanomolar free Ca^{2+}, the effectiveness of caffeine in stimulating ^{45}Ca release was several fold increased by the addition of 5 mM MgATP to the release medium. A similar potentiating effect has been observed when the nucleotide complex was added to release media containing micromolar concentrations of free Ca^{2+} (Meissner et al., 1986).

Single channel measurements have revealed that caffeine exerts its Ca^{2+} releasing effects in skeletal and cardiac muscle by increasing the open time of the SR Ca^{2+} release channel without changing the unit conductance (Rousseau et al., 1988; Rousseau and Meissner, 1989). In Figure 12, a single skeletal Chaps-solubilized channel, purified in the form of a 30S complex in the absence of [^{3}H]-ryanodine, was recorded in a symmetrical 500 mM NaCl, 6 μM free Ca^{2+} medium. In the upper trace, a Na^{+}-conducting channel activity (570 pS with

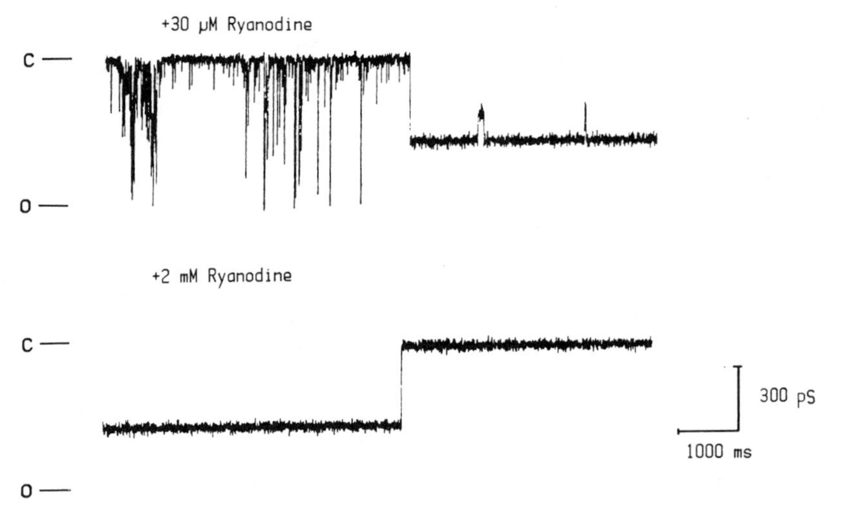

Figure 8. Effect of ryanodine on a single reconstituted, purified Ca^{2+} release channel. Shown are single-channel recordings of K^+ current of purified 30S channel complex incorporated into a planar lipid bilayer in symmetric 250 mM KCl buffer with 50 μM free Ca^{2+}. Unitary conductance = 700 pS, Holding potential = 20 mV. Upper trace shows appearance of subconducting state with open probability (P_o) ~1, following several minutes after cis addition of 30 μM ryanodine. An additional, infrequently observed substate can also be noticed. Lower trace illustrates the sudden transition from the subconductance state to a fully closed state within one minute after cis addition of 2 mM ryanodine. Bars on left represent the open (o) and closed (c) channel (from Lai et al., 1989, by permission).

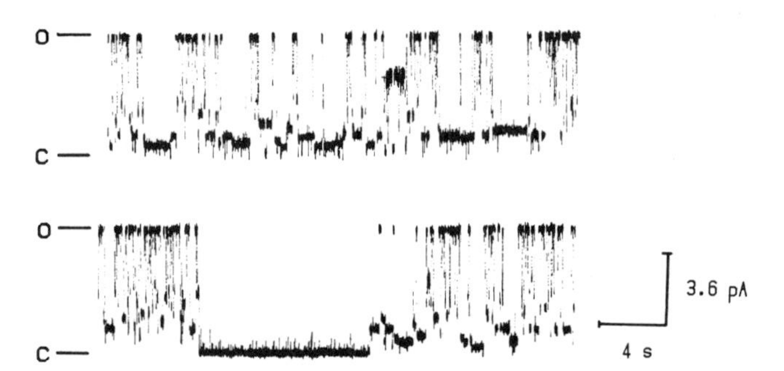

Figure 9. Multiple conductance levels of K^+-conducting skeletal muscle Ca^{2+} release channel. Single channel activity of purified 30S channel complex was recorded in symmetric 250 mM KCl, with 50 μM free Ca^{2+} and 5 mM MgATP cis. Maximal conductance was 470 pS. Bars on left represent the maximally open (o) and closed (c) channel.

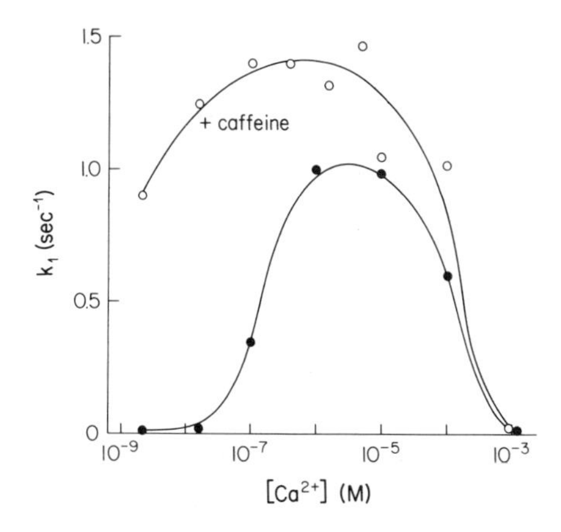

Figure 10. Dependence of $^{45}Ca^{2+}$ efflux on Ca^{2+} concentration in the presence and absence of caffeine. Skeletal muscle SR vesicles were passively loaded with 1 mM $^{45}Ca^{2+}$ and diluted into media containing the indicated concentrations of free Ca^{2+} (●,○). $^{45}Ca^{2+}$ efflux in the presence of 20 mM caffeine (○) was determined by initially lowering free Ca^{2+} before adding caffeine in a second mixing step (Meissner et al., 1986). The first-order rate constants of $^{45}Ca^{2+}$ efflux from vesicles containing the Ca^{2+} release channel are shown (from Rousseau et al., 1988, by permission).

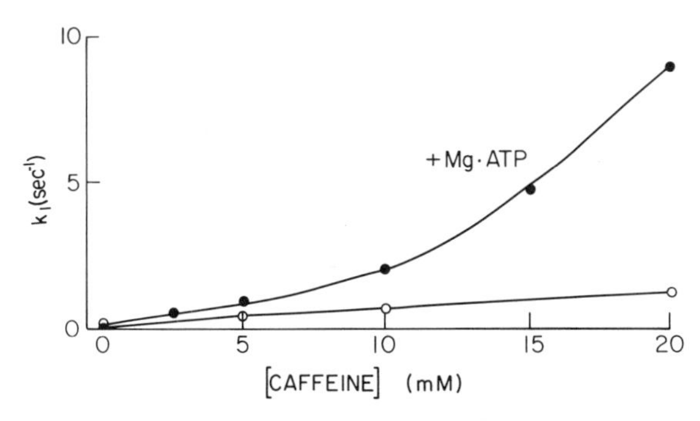

Figure 11. Caffeine concentration dependence of $^{45}Ca^{2+}$ efflux in the presence and absence of Mg^{2+} and ATP. Skeletal muscle SR vesicles loaded with 1 mM $^{45}Ca^{2+}$ were initially diluted into media containing 4 x 10^{-9} free Ca^{2+} using a rapid mixing apparatus (Meissner et al., 1986). After a second mixing step, first-order rate constants of $^{45}Ca^{2+}$ efflux were determined in media containing 2 x 10^{-9} M free Ca^{2+} and the indicated concentrations of caffeine with no addition (○) or with 5 mM Mg^{2+} plus 5 mM ATP (●). (from Rousseau et al., 1988, by permission).

500 mM Na^{+} as the permeant cation) was observed, which was decreased upon lowering the free Ca^{2+} concentration from 6 μM to 40 nM (trace B) and again increased by the addition of 10 mM caffeine to the nM Ca^{2+} medium (trace C of Fig. 12).

Single channel measurements of Figures 8 and 12 show that the two Ca^{2+} releasing drugs ryanodine and caffeine affect the SR Ca^{2+} release channel in quite different ways. Micromolar concentrations of ryanodine induce the formation of a channel state characterized by long open time intervals and a reduced conductance. Furthermore, the ryanodine-modified channel becomes insensitive to Ca^{2+}, Mg^{2+}, and ATP, which all greatly affect the gating behavior of the unmodified channel. In contrast, caffeine activates the channel by increasing channel open time without significantly affecting single channel conductance (Fig. 12), or eliminating sensitivity to Mg^{2+} and ATP (Rousseau et al., 1988).

Evidence for a Ryanodine- and Caffeine-Sensitive Channel in Smooth Muscle

More recently, we have also obtained evidence for the presence of a 30S cation conducting channel complex in canine and porcine aorta (Herrmann-Frank et al., 1990). Microsomal proteins were solubilized in Chaps and centrifuged through linear sucrose gradients. As has been observed for the skeletal and cardiac muscle ryanodine receptors, a single [^{3}H]-ryanodine receptor peak with an apparent sedimentation coefficient of ~30S was obtained (Table 2). Incorporation of the vascular 30S protein fractions into planar bilayers induced the formation of a monovalent ion and Ca^{2+} ion conducting channel which was activated by Ca^{2+} and caffeine and inhibited by ruthenium red. However, in our preliminary studies we were not able to observe the formation of a subconducting state upon the addition of μM ryanodine, although mM ryanodine closed the channel. A second noticeable difference has been that the maximal conductance observed in 250 mM K^+ buffer was half that of the main conductance of the skeletal muscle channel. The appearance of multiple subconductances present within the channel tetramer is a frequently observed phenomenon upon reconstitution of the detergent solubilized, purified channel (Fig. 9; Liu et al., 1989). It may, therefore, be that further studies will also reveal a conductance level of 700-800 pS for the vascular channel in 250 mM K^+ medium, as observed for the skeletal muscle 30S channel complex. Several of the presently known properties of the skeletal, cardiac, and smooth ryanodine receptor-Ca^{2+} release channel complexes, as determined in our laboratory, are summarized in Table 2.

Table 2. Properties of K^+- and Ca^{2+}-conducting channel complexes from skeletal, cardiac and vascular muscle

	Skeletal	Cardiac	Vascular
Apparent sedimentation coefficient	30S	30S	30S
Main conductance			
In 50 mM Ca^{2+}	100 pS	75 pS	~100 pS
In 250 mM K^+	750 pS	N.D.	360 pS
Regulation			
Activation by μM Ca	Yes	Yes	Yes
Activation by mM caffeine	Yes	Yes	Yes
Inhibition by μM ruthenium red	Yes	Yes	Yes
Modification by ryanodine to			
subconducting state	Yes	Yes	Not Observed
closed state	Yes	N.D.	Yes

N.D. = not determined

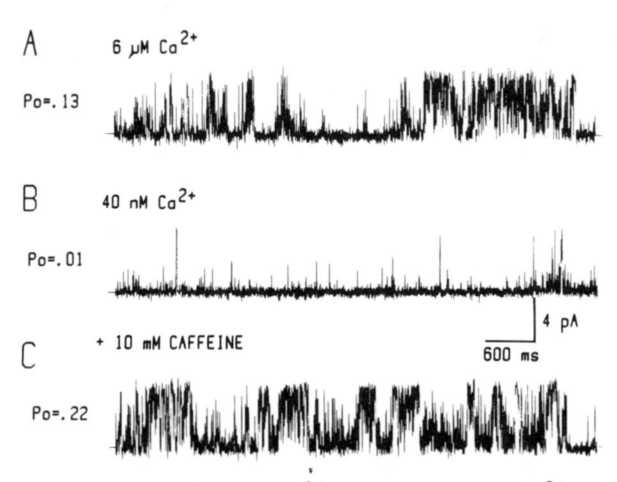

Figure 12. Activation of skeletal muscle Ca^{2+} release channel by Ca^{2+} and caffeine. The 30S purified channel complex was reconstituted in its soluble form into a planar lipid bilayer (Fig. 4). Single channel activity was recorded in symmetric 500 mM NaCl, 6 μM free Ca^{2+} (A), after cis free Ca^{2+} was decreased to 40 nM by the addition of EGTA (B), and after the addition of 10 mM caffeine to the 40 nM free Ca^{2+} cis medium (C). Holding potential = -10 mV. The unitary conductance of the Ca^{2+}- and caffeine-activated channel was 570 pS (from Rousseau et al., 1988, by permission).

ACKNOWLEDGMENTS

This work was supported by USPHS and MDA grants.

REFERENCES

Anderson, K., Lai, F. A., Liu, Q. Y., Rousseau, E., Erickson, H. P., and Meissner, G., Structural and functional characterization of the purified cardiac ryanodine receptor-Ca^{2+} release channel complex, 1989, *J. Biol. Chem.*, 264: 1329.

Caswell, A. H. and Brandt, N. R., 1989, Triadic proteins of skeletal muscle, *J. Bioenerg. Biomembr.*, 21: 149.

Ebashi, S., 1976, Excitation-contraction coupling, *Ann. Rev. Physiol.*, 38: 293.

Eisenberg, B. R. and Eisenberg, R. S., 1982, The T-SR junction in contracting single muscle fibers, *J. Gen. Physiol.*, 79: 1.

Endo, M., 1977, Calcium release from the sarcoplasmic reticulum, *Physiol. Rev.*, 57: 71.

Fabiato, A., 1983, Calcium-induced release of calcium from the cardiac sarcoplasmic reticulum, *Am. J. Physiol.*, 245: C1.

Fleischer, S. and Inui, M., 1989, Biochemistry and biophysics of excitation-contraction coupling, *Ann. Rev. Biophys. Chem.*, 18: 333.

Franzini-Armstrong, C., 1980, Structure of sarcoplasmic reticulum, *Fed. Proc.*, 39: 2403.

Herrmann-Frank, A., Darling, E., and Meissner, G., 1990, Single channel measurements of the Ca^{2+}-gated ryanodine-sensitive Ca^{2+} release channel of vascular smooth muscle, *Biophys. J.*, 57: 156a.
Imagawa, T., Smith, J. S., Coronado, R., and Campbell, K. P., 1987, Purified ryanodine receptor from skeletal muscle sarcoplasmic reticulum is the Ca^{2+}-permeable pore of the calcium release channel, *J. Biol. Chem.*, 262: 16636.
Inui, M., Saito, A., and Fleischer, S., 1987, Isolation of the ryanodine receptor from cardiac sarcoplasmic reticulum and identity with the feet structures, *J. Biol. Chem.*, 262: 15637.
Lai, F. A., Erickson, H. P., Rousseau, E., Liu, Q. Y., and Meissner, G., 1988, Purification and reconstitution of the calcium release channel from skeletal muscle, *Nature*, 331: 315.
Lai, F. A. and Meissner, G., 1989, The muscle ryanodine receptor and its intrinsic Ca^{2+} channel activity, *J. Bioenerg. Biomembr.*, 21: 227.
Lai, F. A., Misra, M., Xu, L., Smith, H. A., and Meissner, G., 1989, The ryanodine receptor-Ca^{2+} release channel complex of skeletal muscle sarcoplasmic reticulum: evidence for a cooperatively coupled, negatively charged homotetramer, *J. Biol. Chem.*, 264: 16776.
Lai, F. A., Xu, L., and Meissner, G., 1990, Identification of a ryanodine receptor in rat and bovine brain, *Biophys. J.*, 57: 529a.
Liu, Q. Y., Lai, F. A., Rousseau, E., Jones, R. V., and Meissner, G., 1989, Multiple conductance states of the purified calcium release channel complex from skeletal sarcoplasmic reticulum, *Biophys. J.*, 55: 415.
Meissner, G., 1975, Isolation and characterization of two types of sarcoplasmic reticulum vesicles, *Biochim. Biophys. Acta*, 389: 51.
Meissner, G., 1984, Adenine nucleotide stimulation of Ca^{2+}-induced Ca^{2+} release in sarcoplasmic reticulum, *J. Biol. Chem.*, 259: 1365.
Meissner, G., 1986a, Evidence of a role for calmodulin in the regulation of calcium release from skeletal muscle sarcoplasmic reticulum, *Biochemistry*, 25: 244.
Meissner, G., 1986b, Ryanodine activation and inhibition of the Ca^{2+} release channel of sarcoplasmic reticulum, *J. Biol. Chem.*, 261: 6300.
Meissner, G., Darling, E., and Eveleth, J., 1986, Kinetics of rapid Ca^{2+} release by sarcoplasmic reticulum. Effects of Ca^{2+}, Mg^{2+}, and adenine nucleotides, *Biochemistry*, 25: 236.
Meissner, G. and Henderson, J. S., 1987, Rapid calcium release from cardiac sarcoplasmic reticulum vesicles is dependent on Ca^{2+} and is modulated by Mg^{2+}, adenine nucleotide, and calmodulin, *J. Biol. Chem.*, 262: 3065.
Näbauer, M., Callewaert, G., Cleemann, L., and Morad, M., 1989, Regulation of calcium release is gated by calcium current, not gating charge, in cardiac myocytes, *Science*, 244: 800.
Otsu, K., Willard, H. F., Khanna, V. K., Zorzato, F., Green, N. M., and MacLennan, D. H., 1990, Molecular cloning of cDNA encoding the Ca^{2+} release channel (ryanodine receptor) of rabbit cardiac muscle sarcoplasmic reticulum, *J. Biol. Chem.*, 265: 13472.
Palade, P., Dettbarn, C., Brunder, D., Stein, P., and Hals, G., 1989, Pharmacology of calcium release from sarcoplasmic reticulum, *J. Bioenerg. Biomembr.*, 21: 295.

Peachey, L. D. and Franzini-Armstrong, C., 1983, Structure and function of membrane systems of skeletal muscle cells, *in:* "Skeletal Muscle", Handbook of Physiology, Sect. 10, L. D. Peachey, R. H. Adrian, S. R. Geiger eds., American Physiological Society, Bethesda, p. 23.

Penner, R., Neher, E., Takeshima, H., Nishimura, S., and Numa, S., 1989, Functional expression of the calcium release channel from skeletal muscle ryanodine receptor cDNA, *FEBS Lett.*, 259: 217.

Rios, E. and Pizarro, G., 1988, Voltage sensors and calcium channels of excitation-contraction coupling, *News Physiol. Sci.*, 3: 223.

Rousseau, E., Smith, J. S., Henderson, J. S., and Meissner, G., 1986, Single channel and $^{45}Ca^{2+}$ flux measurements of the cardiac sarcoplasmic reticulum calcium channel, *Biophys. J.*, 50: 1009.

Rousseau, E., LaDine, J., Liu, Q. Y., and Meissner, G., 1988, Activation of the Ca^{2+} release channel of skeletal muscle sarcoplasmic reticulum by caffeine and related compounds, *Arch. Biochem. Biophys.*, 267: 75.

Rousseau, E. and Meissner, G., 1989, Single cardiac sarcoplasmic reticulum Ca^{2+}-release channel: activation by caffeine, *Am. J. Physiol.*, 256: H328.

Saito, A., Inui, M., Wall, J. S., and Fleischer, S., 1989, Mass measurement of the feet structures/calcium release channel of sarcoplasmic reticulum by scanning transmission electron microscopy (STEM), *Biophys. J.*, 55: 206a.

Schneider, M. F., 1981 Membrane charge movement and depolarization-contraction coupling, *Ann. Rev. Physiol.*, 43: 507.

Smith, J. S., Coronado, R., and Meissner, G., 1986a, Single-channel calcium and barium currents of large and small conductance from sarcoplasmic reticulum, *Biophys. J.*, 50: 921.

Smith, J. S., Coronado, R., and Meissner, G., 1986b, Single measurements of the calcium release channel from skeletal muscle sarcoplasmic reticulum. activation by Ca^{2+} and ATP and modulation by Mg^{2+}, *J. Gen. Physiol.*, 88: 573.

Smith, J. S., Imagawa, T., Ma, J., Fill, M., Campbell, K. P., and Coronado, R., 1988, Purified ryanodine receptor from rabbit skeletal muscle is the calcium-release channel of sarcoplasmic reticulum, *J. Gen. Physiol.*, 92: 1.

Smith, J. S., Rousseau, E., and Meissner, G., 1989, Calmodulin modulation of single sarcoplasmic reticulum Ca^{2+}-release channels from cardiac and skeletal muscle, *Circ. Res.*, 64: 352.

Somlyo, A. V., 1979, Bridging structures spanning the junctional gap at the triad of skeletal muscle, *J. Cell Biol.*, 80: 743

Stephenson, E. W., 1981, Activation of fast skeletal muscle: Contributions of studies on skinned fibers, *Am. J. Physiol.*, 240: C1.

Takeshima, H., Nishimura, S., Matsumoto, T., Ishida, H., Kangawa, K., Minamino, N., Matsuo, H., Ueda, M., Hanaoka, M., Hirose, T., and Numa, S., 1989, Primary structure and expression from complementary DNA of skeletal muscle ryanodine receptor, *Nature*, 339: 439.

Tanabe, T., Beam, K. G., Powell, J. A., and Numa, S., 1988, Restoration of excitation-contraction coupling and slow calcium current in dysgenic muscle by dihydropyridine receptor complementary DNA, *Nature*, 336: 134.

Tanabe, T., Mikami, A., Numa, S., and Beam, K. G., 1990, Cardiac-type excitation-contraction coupling in dysgenic skeletal muscle injected with cardiac dihydropyridine receptor cDNA, *Nature*, 344: 451.

Van Breemen, C., and Saida, K., 1989, Cellular mechanisms regulating [Ca^{2+}] in smooth muscle, *Ann. Rev. Physiol.*, 51: 315.
Wagenknecht, T., Grassucci, R., Frank, J., Saito, A., Inui, M., and Fleischer, S., 1989, Three-dimensional architecture of the calcium channel/foot structure of the sarcoplasmic reticulum, *Nature*, 338: 167.
Zaidi, N. F., Lagenaur, C. F., Hilker, R. J., Xiong, H., Abramson, J. J., and Salama, G., 1989, Disulfide linkage of biotin identifies a 106-kDa Ca^{2+} release channel in sarcoplasmic reticulum, *J. Biol. Chem.*, 264: 21737.
Zorzato, F., Fujii, J., Otsu, K., Phillips, M., Green, N. M., Lai, F. A., Meissner, G., and MacLennan, D. H., 1990, Molecular cloning of cDNA encoding human and rabbit forms of the Ca^{2+} release channel (ryanodine receptor) of skeletal muscle sarcoplasmic reticulum, *J. Biol. Chem.*, 265: 2244.

CYTOSOLIC CALCIUM ION REGULATION IN CULTURED ENDOTHELIAL CELLS

Rachel E. Laskey, David J. Adams, Sherry Purkerson,
and Cornelis van Breemen

Department of Molecular and Cellular Pharmacology
University of Miami School of Medicine
Miami, FL 33101

INTRODUCTION

Endothelial cells profoundly affect the cardiovascular system by interacting with the blood at the luminal surface and with the underlying smooth muscle of the media. Endothelial secretions carry out multiple and sometimes opposing functions. For example, thrombotic and antithrombotic, proliferative and antiproliferative, as well as vasodilatory and vasoconstrictor substances have been identified with the endothelium.

Impairment of endothelial function, particularly production of endothelium-derived relaxing factor (EDRF), is associated with various pathological conditions such as coronary vasospasm (Ganz and Alexander, 1985), increased peripheral resistance and hypertension (Lockette et al., 1986), and diabetes (Oyama et al., 1986). EDRF has been identified as nitric oxide (NO). Its physiological importance in maintaining the proper balance of vascular resistance has recently been demonstrated with *in vivo* studies in animals and in man in which injections of N-monomethyl-L-arginine (L-NMMA), an inhibitor of NO production, led to a powerful and prolonged vasoconstriction (Aisaka et al., 1989; Rees et al., 1989; Vallance et al., 1989). This is quite impressive in that it reveals the presence of a dynamic vasodilator tone within the cardiovascular system.

The synthesis and/or release of EDRF and other endothelial-derived secretions such as prostacyclin (PGI_2) are closely associated with the intracellular free calcium ion concentration, $[Ca^{2+}]_i$. As in many other cell types, $[Ca^{2+}]_i$ plays an important second messenger role. The level of free calcium within the endothelial cell regulates a range of cellular responses although its regulatory role in the production of the vasoconstrictor peptide endothelin (ET-1) is currently unknown. The regulation of intracellular calcium is not clearly delineated in endothelial cells, yet its importance as a second messenger is unquestioned.

That increases in $[Ca^{2+}]_i$ are necessary for production of EDRF is based on several observations. First, a number of investigators have shown that the absence of extracellular Ca^{2+} rapidly and completely inhibits relaxations of

Regulation of Smooth Muscle Contraction
Edited by R.S. Moreland, Plenum Press, New York, 1991

blood vessels in response to endothelial-activating agents such as acetylcholine, bradykinin, and thrombin (Furchgott and Zawadski, 1980; Singer and Peach 1982; Long and Stone 1985). In addition, the calcium ionophore A23187 which facilitates Ca^{2+} entry into the cell evokes endothelium-dependent relaxations (Furchgott and Zawadski, 1980). More recently, Palmer et al. (1988) identified a cytoplasmic divalent cation- and NADPH-dependent enzyme system in endothelial cells that releases NO by converting L-arginine to L-citrulline. Meyer et al. (1989) have subsequently shown this synthetic pathway to be Ca^{2+} dependent.

Calcium as a second messenger, is involved in coupling stimulus-secretion mechanisms in response to various endothelial activating agents. The versatility of the calcium signalling system may depend on its inherent complexity. For example, EDRF secretion occurs with basal and stimulated levels of $[Ca^{2+}]_i$. In addition, two types of Ca^{2+} signals have been recognized: 1) the $[Ca^{2+}]_i$ can be set at any value within the range of 50 to 1000 nM (Adams et al., 1989a) and 2) the $[Ca^{2+}]_i$ can oscillate at a range of frequencies and amplitudes (Jacob et al., 1988; Sage et al., 1989). This complexity may allow for maximal informational encoding of the Ca^{2+} signal. The most commonly described Ca^{2+} signal in endothelial cells in response to agonist stimulation consists of two components. Initially, a maximal Ca^{2+} spike (500 - 1000 nM from a basal level of 75 - 150 nM) occurs and then falls within two minutes to an elevated plateau (200 - 300 nM). The initial spike, refractory to the removal of extracellular Ca^{2+}, is associated with release of Ca^{2+} from the intracellular stores of the endoplasmic reticulum, while the plateau is highly dependent upon extracellular Ca^{2+} entry. Figure 1 illustrates the $[Ca^{2+}]_i$ response to agonist stimulation of cultured endothelial cells loaded with fura-2, showing the dependence of the plateau phase on extracellular Ca^{2+} availability. This maintained phase of Ca^{2+} entry is essential to the continued release of EDRF from stimulated endothelial cells. Changes in the extracellular Ca^{2+} concentration around the physiological range have been recently shown to modulate the synthesis of NO by the vascular endothelium and consequently, vascular tone (Lopez-Jaramillo et al., 1990).

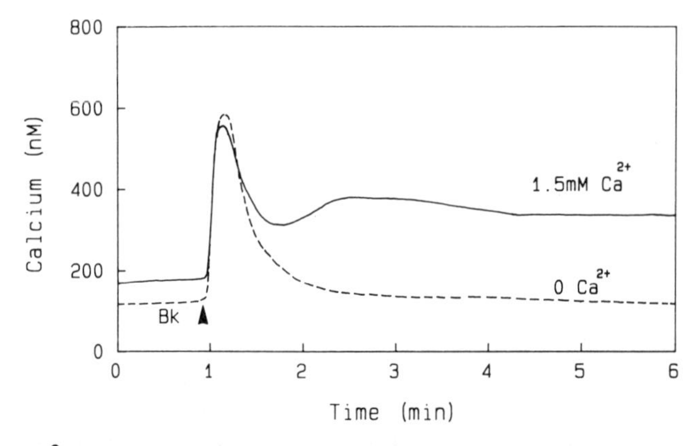

Figure 1. $[Ca^{2+}]_i$ response of cultured endothelial cells stimulated with bradykinin (Bk, 200 nM) in normal physiological saline solution (PSS) containing 1.5 mM Ca^{2+} (solid line) or in Ca^{2+}-free PSS (dashed line). (From Laskey et al., 1990, by permission).

At least four separate routes for the supply of activating Ca^{2+} to the cytosol have been proposed: 1) a Ca^{2+} 'leak' pathway, 2) receptor-operated non-selective cation channels (ROCs), 3) stretch-activated cation channels, and 4) IP_3-activated calcium channels in the endoplasmic reticulum. These mechanisms involving Ca^{2+} transport will be discussed below.

CALCIUM ENTRY

EDRF release is mainly dependent upon the availability of extracellular Ca^{2+} although a transient component is related to Ca^{2+} mobilization upon receptor activation. A basal rate of EDRF release occurs in the absence of agonist-stimulated Ca^{2+} entry and is reduced by the removal of extracellular Ca^{2+}. An inherent Ca^{2+} 'leak' influx driven by the electrochemical gradient for Ca^{2+} can account for this basal release of EDRF. This Ca^{2+} leak appears to be unregulated, but nevertheless vital to modulation of systemic blood pressure. Removal of extracellular Ca^{2+} leads to a sharp decrement of EDRF production *in vitro* (Luckhoff et al., 1988; Laskey et al., 1990). Furthermore, interference with basal EDRF release *in vivo* by using L-NMMA causes a concomitant rise in blood pressure (Aisaka et al., 1989). Vascular smooth muscle cells also exhibit Ca^{2+} 'leak' which contributes to myogenic tone (van Breemen et al., 1986). At "rest", this myogenic tone is balanced by the Ca^{2+}-dependent basal release of EDRF from the endothelium. In addition to providing Ca^{2+} for basal EDRF release, the Ca^{2+} leak pathway contributes to maintaining the plateau phase of the agonist-stimulated increase in cytosolic Ca^{2+} concentration and in refilling intracellular Ca^{2+} pools. The nature of the Ca^{2+} 'leak' pathway is not known. It may represent spontaneous openings of plasmalemmal ion channels and transporters, or the permeability may be a consequence of perturbation of the phospholipid bilayer by the insertion of proteins. Some surface "structural" Ca^{2+} may be internalized as a consequence of membrane turnover. The fact that changes in extracellular pH modulate Ca^{2+} leak may suggest that protonation of surface phospholipids and integral membrane proteins may play a role in Ca^{2+} entry via this route. Johns et al. (1987) observed a basal $^{45}Ca^{2+}$ influx into unstimulated endothelial cells which was reduced by membrane depolarization. Depolarization diminishes the driving force for Ca^{2+} entry through the leak pathway by reducing the electrochemical gradient. The rate of $^{45}Ca^{2+}$ leak, corresponding to 16 pmol.10^6 cells.s^{-1}, was reduced by approximately 15% in an isotonic KCl external solution. Recent studies in our laboratory indicate that a large receptor-independent $^{45}Ca^{2+}$ influx in porcine aortic endothelial cells is inhibited by external La^{3+} and reduced in isotonic KCl. Figure 2 illustrates the magnitude of the $^{45}Ca^{2+}$ influx in unstimulated and ATP-stimulated cultured endothelial cells and the inhibition of the ATP-induced $^{45}Ca^{2+}$ entry by the antagonists, SK&F 96365 and La^{3+}.

Ca^{2+} entry appears to be critical in regulating endothelial cell secretory activity. Agonist-stimulation of endothelial cells results in a secondary maintained phase of cytosolic Ca^{2+} elevation that is related to Ca^{2+} entry across the plasma membrane. Electrophysiological and $^{45}Ca^{2+}$ flux studies indicate that this Ca^{2+} entry is mediated by ROCs rather than voltage-dependent calcium channels (Johns et al., 1987; Bregestovski et al., 1988; Lodge et al., 1988; Schilling et al., 1988).

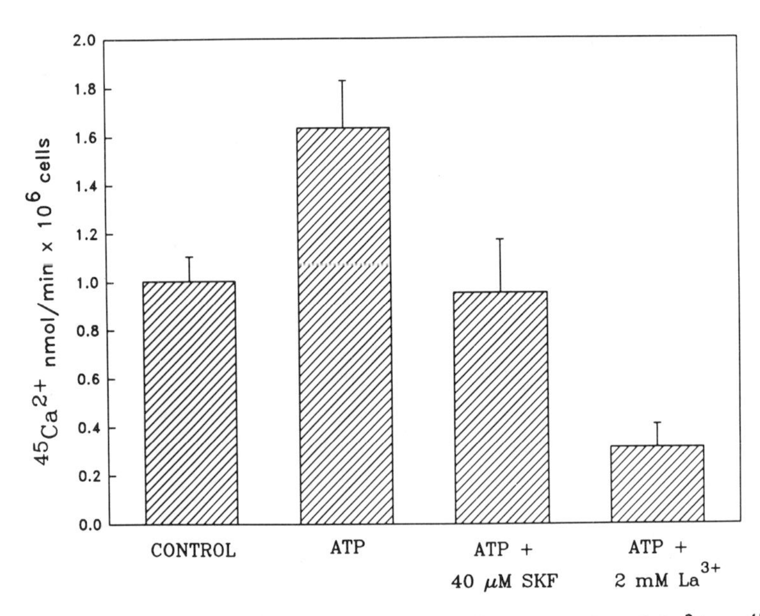

Figure 2. The effects of the receptor-antagonist, SK&F 96365 and La^{3+} on $^{45}Ca^{2+}$ uptake in resting and ATP-stimulated cultured endothelial cells from pig aorta. $^{45}Ca^{2+}$ uptake was measured by exposing confluent endothelial cell monolayers to $^{45}Ca^{2+}$-labelled PSS with or without 40 μM SK&F 96365 or 2 mM $LaCl_3$ for 5 min.

Agonist-induced release of EDRF *in vitro* is reduced in high K^+ solution and is not inhibited by calcium channel antagonists suggesting that voltage-gated calcium channels are not involved in excitation-secretion coupling for EDRF release (Spedding et al., 1986; Jayakody et al., 1987). Subsequently, calcium channel antagonists were shown to have no effect on either basal $^{45}Ca^{2+}$ influx (Whorton et al., 1984) or fura-2 fluorescence in cultured porcine aortic endothelial cells at concentrations sufficient to block voltage-gated calcium channels. Similarly, stimulated $^{45}Ca^{2+}$ influx due to bradykinin or ATP was not blocked by calcium channel antagonists. Membrane depolarization produced a decrease in $^{45}Ca^{2+}$ uptake into cultured endothelial cells (Johns et al., 1987; Schilling, 1989; Luckhoff and Busse, 1990b). In studies using the fluorescent Ca^{2+} indicator, fura-2, depolarization by raising the extracellular $[K^+]$ reduced the resting $[Ca^{2+}]_i$ and also decreased the plateau phase of $[Ca^{2+}]_i$ associated with agonist stimulation (Laskey et al., 1990). The absence of voltage-dependent calcium channels was also directly demonstrated by the observed decline in resting $[Ca^{2+}]_i$ upon depolarization in voltage-clamped cultured endothelial cells loaded with Indo-1 (Cannell and Sage, 1989).

The presence of receptor-operated channels is supported by the finding that a number of endothelium-dependent agonists increase $^{45}Ca^{2+}$ influx (Whorton et al., 1984; Johns et al., 1987). Thrombin, bradykinin, and histamine evoke a depolarizing inward current associated with an increase in membrane conductance in voltage-clamped cultured endothelial cells (Johns et al., 1987; Bregestovski et al., 1988; Lodge et al., 1988; Takeda and Klepper, 1990). The agonist-activated inward currents were non-selective to monovalent and divalent cations, reversed at 0 mV, and the current amplitude was sensitive to the extracellular Na^+ and Ca^{2+} concentrations (Johns et al., 1987; Bregestovski et

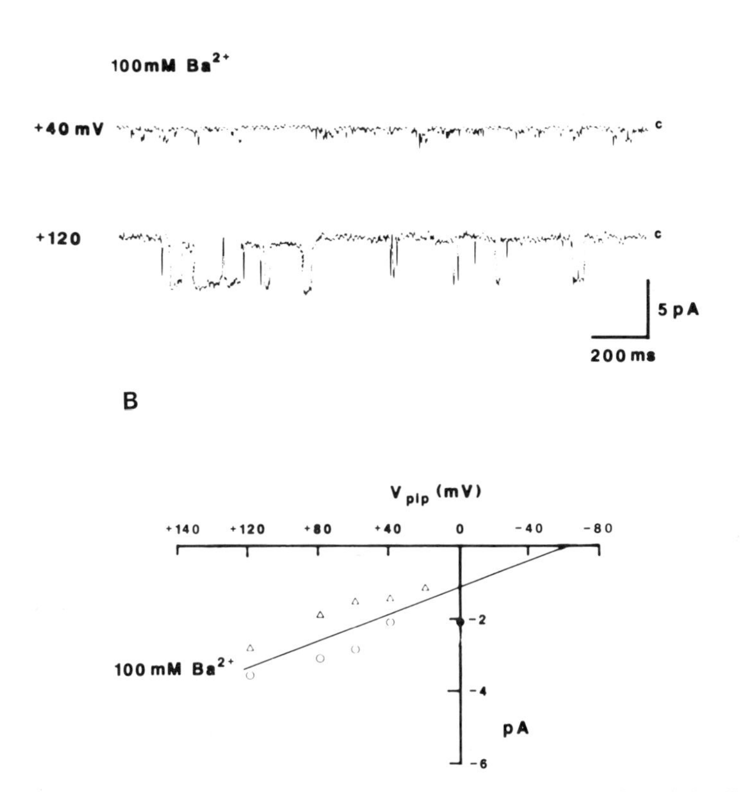

Figure 3. Single channel currents evoked by thrombin in cultured endothelial cells from bovine pulmonary artery. (A) Unitary currents recorded from a cell-attached membrane patch with a pipette solution containing 0.1 U/ml thrombin in isotonic (100 mM) $BaCl_2$ (22° C). (B) Unitary current amplitude plotted as a function of potential applied to the patch pipette. The resting potential of the cell was -60 mV. (From Adams et al., 1989a, by permission).

al., 1988). In cell-attached membrane patches from bovine pulmonary artery endothelial cells, thrombin- and bradykinin-activated channels with a single channel conductance of 15 - 20 pS in isotonic Ba^{2+} were observed (Lodge et al., 1988; Adams et al., 1989a). In excised membrane patches from human umbilical vein endothelial cells, histamine activates a non-selective cation channel with a single channel conductance of 8 pS in 110 mM Ca^{2+} (pipette) and 140 mM K^+ (bath) (Nilius, 1990). This non-selective cation channel was sufficient to induce a sustained agonist-mediated Ca^{2+} influx. Although a Ca^{2+} permeable receptor-operated non-selective cation channel has been described in vascular endothelial cells, the link between the receptor-ligand binding and ionic channel activation remains to be elucidated.

Another pathway for Ca^{2+} entry into endothelial cells which may be of physiological importance in the regulation of blood vessel tone, is the mechanosensitive or stretch-activated channel. Non-selective cation channels activated by negative pressure applied to the pipette have been reported in cell-attached patches in cultured endothelial cells from porcine aorta (Lansman et al., 1987; Adams et al., 1989a). A single channel conductance of 19 pS was obtained in the presence of isotonic Ca^{2+} and together with an estimated P_{Ca}/P_{Na} of 6, suggests that in physiological solutions Ca^{2+} may contribute to an inward current and hence cell depolarization in response to stretch.

Although ionic channels that permit Ca^{2+} entry are not voltage-gated, membrane potential (E_m) plays a role in regulating Ca^{2+} entry by affecting the electrochemical gradient (E_m - E_{Ca}) for Ca^{2+} movement. It is important then to consider other ion channels that modulate membrane potential since they indirectly influence EDRF release by affecting Ca^{2+} influx. Potassium channels constitute the predominant ionic permeability of cultured endothelial cells. The resting E_m of cultured endothelial cells is depolarized when the extracellular K^+ concentration ($[K^+]_o$) is raised (52 mV change in E_m per 10-fold change in $[K^+]_o$) (see Adams et al., 1989a). Inwardly rectifying K^+ channels have been reported in cultured endothelial cells (Colden-Stanfield et al., 1987; Johns et al., 1987; Takeda et al, 1987; Olesen et al., 1988a; Cannell and Sage, 1989; Silver and Decoursey, 1990) and may contribute to the resting membrane potential. Olesen et al. (1988b) have shown that shear force due to endothelial superfusion also activates an inward rectifying K^+ conductance. The resulting hyperpolarization increases the cytosolic $[Ca^{2+}]$ which in turn would stimulate EDRF secretion and vasodilation. There are at least three other types of K channels found in endothelial cells: Ca^{2+}-activated K channels (Suave et al., 1988; Colden-Stanfield et al., 1990), transient outward K channels (Takeda et al., 1987; Silver and Decoursey, 1990), and acetylcholine-activated K channels (Olesen et al., 1988a). Acetylcholine-activated K channels lead to hyperpolarization and a subsequent increase in the electrochemical gradient for Ca^{2+} entry through leak channels. Ca^{2+}-activated K currents can be indirectly activated by endothelium-dependent vasodilators such as ATP and bradykinin (Suave et al., 1988; Colden-Stanfield et al., 1990; Rusko et al., 1991) to hyperpolarize the cell and thus maintain the electrical driving force for Ca^{2+} entry.

The above evidence leads to the concept that Ca^{2+} entry into endothelial cells is mediated by the Ca^{2+} leak, ROCs, and in some instances, stretch-activated channels. None of these mechanisms are activated by membrane depolarization which reduces Ca^{2+} entry due to a decrease in the electrical driving force for Ca^{2+}. In order to maintain homeostasis, this Ca^{2+} entry is balanced by active Ca^{2+} extrusion from the cells. To date, minimal attention has been focused on this aspect of endothelial Ca^{2+} regulation. Present evidence suggests that Ca^{2+} extrusion is principally mediated by an ATP-sensitive Ca^{2+} pump (Hagiwara et al., 1983). Although a Na^+-Ca^{2+} exchange mechanism exists in cultured endothelial cells (Adams et al., 1989b), it appears to affect neither resting Ca^{2+} levels nor bradykinin-stimulated increase in $[Ca^{2+}]_i$ (Laskey et al., 1990). However, the Na-Ca exchanger could accelerate the decline in $[Ca^{2+}]_i$ upon removal of agonist (Laskey and van Breemen, 1989).

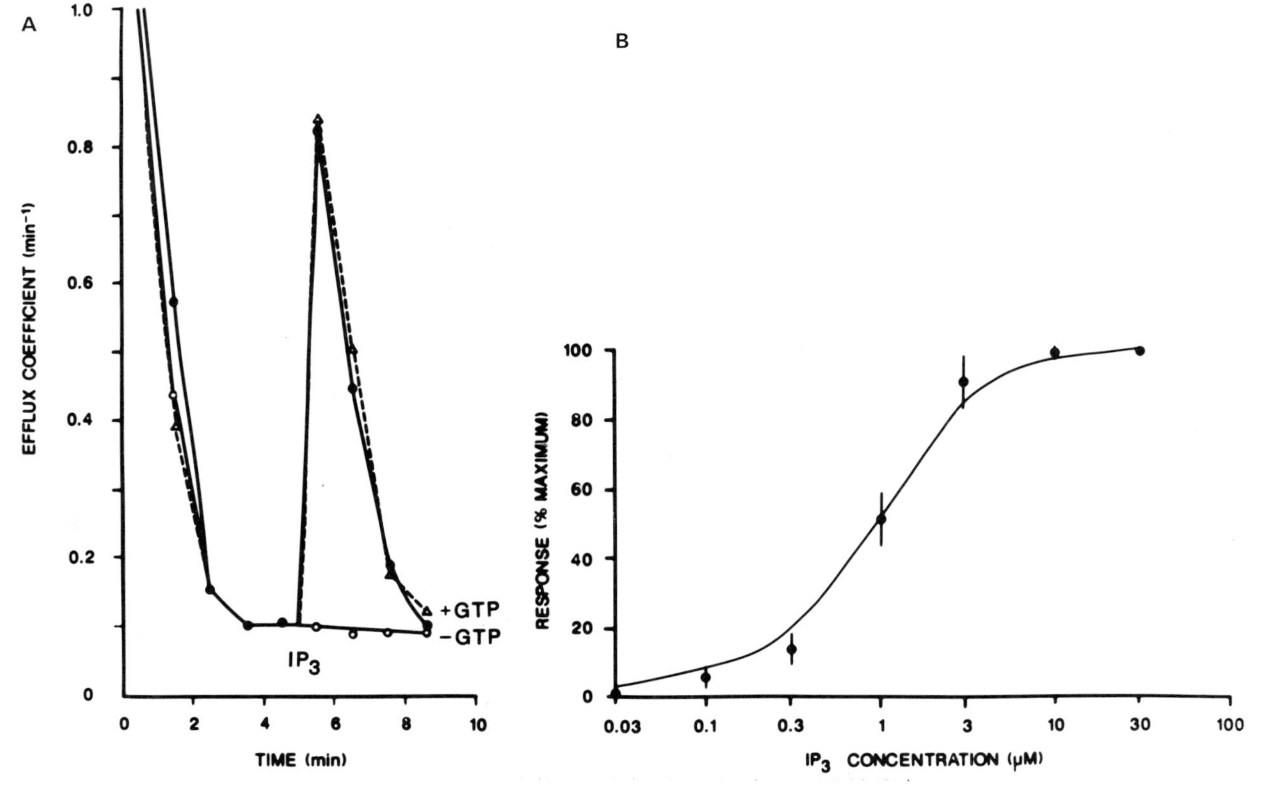

Figure 4. (A) IP_3-induced $^{45}Ca^{2+}$ efflux from saponin-permeabilized endothelial cells. The cells were incubated for 20 min at a pCa of 6 and then $^{45}Ca^{2+}$ efflux was measured in Ca^{2+}-free solution. The effects of 100 μM GTP (○), 10 μM IP_3 (●), and 100 μM GTP + 10 μM IP_3 (Δ) were tested 5 - 6 min after the start of the efflux period. (B) Dose-response curve for IP_3-induced release of $^{45}Ca^{2+}$ from saponin-permeabilized endothelial cells. The different concentrations of IP_3 were introduced 6 - 7 min after the start of the efflux period. The values (mean ± SEM, n = 4) represent the % of the IP_3-sensitive $^{45}Ca^{2+}$ released by each concentration of IP_3 with respect to the maximum released by 30 μM IP_3. The curve fitted to these data is for an adsorption isotherm with a K_d of 1 μM. (From Freay et al., 1989, by permission).

Ca^{2+} RELEASE FROM INTRACELLULAR STORES

A second mechanism by which receptor activation can lead to an increase in cytosolic Ca^{2+} is via an initial transient release from intracellular stores. As in other cells, the intracellular Ca^{2+} stores of endothelial cells are comprised of endoplasmic reticulum and mitochondria. $^{45}Ca^{2+}$ flux measurements in saponin-permeabilized endothelial cells, in which the endoplasmic reticulum (ER) remains intact, showed that the ER contains more than sufficient Ca^{2+} for cell activation and takes up this Ca^{2+} with a high affinity in an ATP-dependent manner (Freay et al., 1989). The mitochondrial pool has a much greater capacity but much lower pump affinity for Ca^{2+}. The cytosolic second messenger, inositol 1,4,5-trisphosphate (IP_3), was effective in releasing the ER Ca^{2+} with a high affinity (K_d = 1 μM) but had no effect on the mitochondrial Ca^{2+} uptake. Several additional lines of evidence indicate that IP_3 is the relevant intracellular messenger for agonist-induced Ca^{2+} release. The EC_{50} for $^{45}Ca^{2+}$ release by bradykinin in intact cells was similar to the EC_{50} for IP_3 synthesis, and direct application of IP_3 to permeabilized cells elicited $^{45}Ca^{2+}$ release (Fig. 4). Lambert et al. (1986) have shown that bradykinin activates phospholipase C (PLC) and that the dose-response of this function correlates with the dose-response for Ca^{2+} release measured by tracer efflux. Calcium mobilization from the ER thus appears to involve a chemical cascade stemming from receptor - guanine nucleotide binding protein (G protein) activation of phospholipase C-mediated phosphatidylinositol 4,5-bisphosphate (PIP_2) hydrolysis, yielding IP_3 and diacylglycerol (DAG). The IP_3 sensitive Ca^{2+} pools of the ER required refilling before a subsequent stimulated efflux could occur. This refilling is highly dependent upon the availability of extracellular Ca^{2+} although the exact route of Ca^{2+} entry into the ER is not clear. A recent study by Luckhoff and Busse (1990a) confirms the idea that refilling of intracellular stores must occur indirectly via the cytosol rather than by a more direct route from the extracellular space.

$[Ca^{2+}]_i$ OSCILLATIONS

Oscillations in $[Ca^{2+}]_i$ of endothelial cells were initially reported in single cultured human umbilical vein endothelial cells (Jacob et al., 1988). These oscillations appearing as a ramp and spike of fura-2 fluorescence, were insensitive to elevated extracellular $[K^+]$ and were slowly abolished by removal of extracellular Ca^{2+}. Subsequently, synchronous $[Ca^{2+}]_i$ oscillations were described in confluent endothelial cell monolayers from bovine pulmonary artery (Sage et al., 1989) and human umbilical vein (Neylon and Irvine 1990). Several recent reviews have speculated on the possible role and mechanism(s) underlying single cell Ca^{2+} oscillations (Berridge 1988; Jacob, 1990).

In confluent bovine endothelial monolayers loaded with fura-2, bradykinin elicited synchronous sinusoidal oscillations superimposed on an elevated level of $[Ca^{2+}]_i$. Since these $[Ca^{2+}]_i$ oscillations were highly dependent upon the level of confluency, it was proposed that the development of gap junctions was necessary to permit cell to cell communication and thus synchronize the response to agonists. Both dye and electrical coupling have been demonstrated in endothelial cells *in situ* (Beny and Gribi, 1989). Laskey et al. (1990), also using fura-2 loaded endothelial monolayers, found that under conditions which slightly depolarize the cells (i.e., 0 mM K^+, or ouabain) bradykinin stimulated synchronous oscillations in a much more reproducible manner allowing a more comprehensive study of this phenomenon. This method of producing $[Ca^{2+}]_i$ oscillations by combining agonist stimulation with depolarization was first reported for fibroblasts by Harootunian et al. (1988). Figure 5 depicts $[Ca^{2+}]_i$ oscillations induced in bovine atrial endothelial cells upon stimulation with bradykinin in the absence of external K^+. These oscillations were dependent on Ca^{2+} influx as the removal of extracellular Ca^{2+} abruptly disrupted the oscillations while readdition of the Ca^{2+} restored the oscillatory signal. The oscillations could occasionally be induced in normal K^+ at low extracellular $[Ca^{2+}]$ (<300 nM). This observation led to the idea that a critical range of Ca^{2+} entry is necessary for the production of $[Ca^{2+}]_i$ oscillations. Furthermore, it was found that the $[Ca^{2+}]_i$ oscillations were sensitive to temperature. $[Ca^{2+}]_i$ oscillations in confluent endothelial monolayers were absent at room temperature (22° C), however, as the temperature of the experimental chamber was slowly raised, oscillations began to occur around 35° C

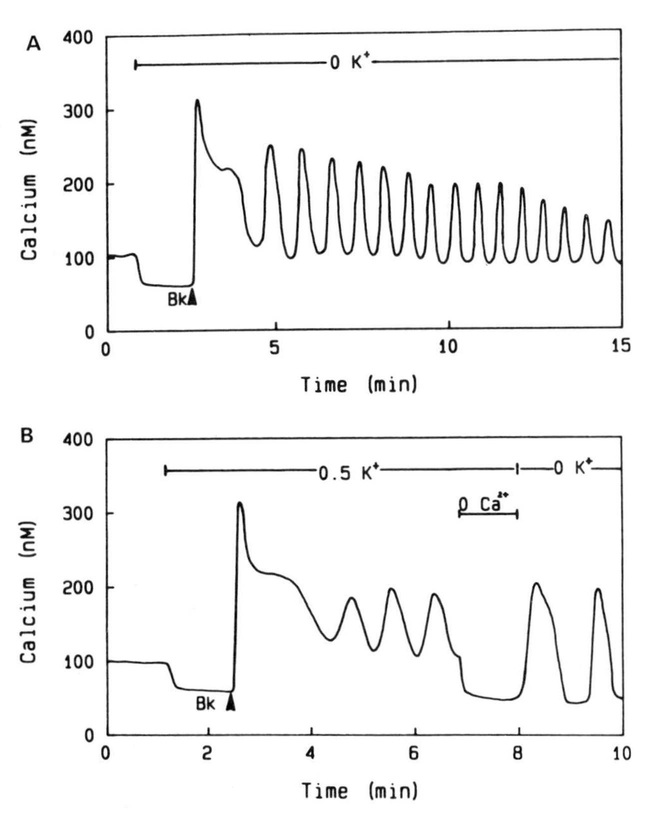

Figure 5. (A.) Cytosolic calcium oscillations in response to bradykinin (Bk, 200 nM) in K^+-free PSS. (B.) Oscillations were dependent upon the extracellular $[K^+]$ and upon the availability of extracellular Ca^{2+}. (From Laskey et al., 1990, by permission).

In order to explore the possibility that gap junctions are involved in the synchronization of the $[Ca^{2+}]_i$ signals, n-octanol, which disrupts electrical coupling, was applied to the endothelial monolayers. In all cases this procedure abolished the oscillations in $[Ca^{2+}]_i$ without apparent morphological damage to the endothelium, suggesting the spread of either electrical or chemical signals through gap junctions during the oscillation of $[Ca^{2+}]_i$.

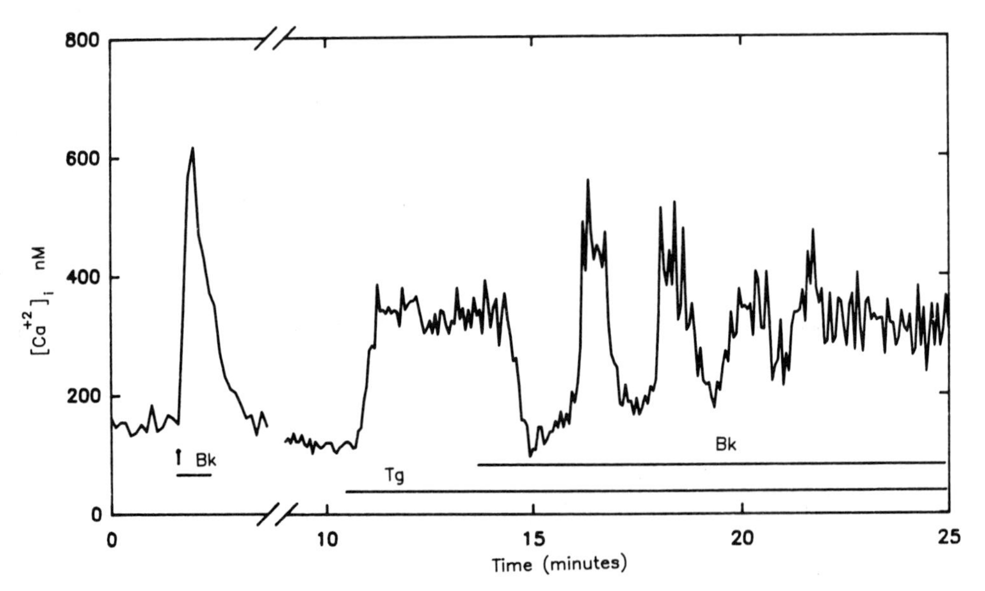

Figure 6. A brief exposure to bradykinin (Bk; 200 nM) elicited a reproducible transient $[Ca^{2+}]_i$ spike. Pretreatment with thapsigargin (1 μM) raised the basal $[Ca^{2+}]_i$ and prevented a subsequent Bk-induced transient. Oscillations in $[Ca^{2+}]_i$ began with a downward deflection upon Bk stimulation.

In an effort to determine the possible mechanism(s) underlying the observed oscillations, several agents which affect Ca^{2+} entry were studied on cultured endothelial cells. SK&F 96365 (40 μM), a putative antagonist for receptor-operated cation channels, consistently abolished the $[Ca^{2+}]_i$ oscillations. La^{3+}, a non-specific inhibitor of Ca^{2+} influx, also completely abolished the oscillations. These observations suggest that the oscillations are dependent upon Ca^{2+} entry via receptor-operated channels with a possible contribution by the Ca^{2+} leak pathway. To explore the contribution of the endoplasmic reticulum to $[Ca^{2+}]_i$ oscillations, endothelial cells were incubated in the presence of thapsigargin, a sesquiterpine lactone. Thastrup et al. (1989) reported that thapsigargin selectively inhibited ER Ca^{2+} uptake without acting on a plasma membrane component. Addition of thapsigargin to bradykinin-stimulated endothelial cells in the absence of extracellular K^+ caused an increase in the $[Ca^{2+}]_i$ level without inhibiting the fluctuations in $[Ca^{2+}]_i$.

Inhibition of refilling of the internal Ca^{2+} store by thapsigargin was demonstrated by the following protocol. It was observed that bradykinin could repeatedly stimulate the release of Ca^{2+} from intracellular stores when administered for a period of 30 sec followed by at least 5 min recovery in physiological saline. Pretreatment of the monolayer with 1 μM thapsigargin elevated $[Ca^{2+}]_i$ and prevented the transient $[Ca^{2+}]_i$ response to bradykinin administration (Fig. 6). Oscillations in $[Ca^{2+}]_i$ still occurred in spite of the removal of ER Ca^{2+} transport. Thus it appears that Ca^{2+} movement through the ER is not directly involved in the observed oscillations, and that the more likely site of regulation for these oscillations is at the level of the plasma membrane. It was previously reported by Sage and coworkers (1989) that phorbol esters block oscillations, suggesting a role for protein kinase C in their production. This effect may have been due to an inhibitory role of PKC on Ca^{2+} entry.

The membrane potential in confluent endothelial cell monolayers was monitored using the anionic potential sensitive dye, bis-(1,3-dibutylbarbituric acid)trimethine oxonol ($DiBAC_4(3)$) with an excitation wavelength of 490 nm and emission at 520 nm. Simultaneous recording of membrane potential and the fura-2 fluorescence (excitation wavelength of 340 nm), have revealed an interesting correlation. The E_m oscillated over a very narrow range with the same frequency as the fluctuations in fura-2 signal. Correlating the periods of these two oscillations indicated that depolarization corresponded to increases in $[Ca^{2+}]_i$. If variations in E_m were responsible for the fluctuating cation entry, depolarization would have reduced the driving force for Ca^{2+} resulting in a decrease in Ca^{2+} entry as previously described. Thus the observed increase in $[Ca^{2+}]_i$ coincident with depolarization supports the idea of fluctuations in the open probability of ROCs as a source of the oscillations in $[Ca^{2+}]_i$. The postulated mechanism by which Ca^{2+} entry is caused to fluctuate may be periodic inhibition of ROCs by a DAG-PKC pathway or changes in $[Ca^{2+}]_i$ near the plasma membrane which may have a subsequent effect on membrane potential.

At present, little is known regarding the functional aspects of this phenomenon. Oscillations in Ca^{2+} signalling within the endothelial cells may represent an important mechanism for regulating vasomotion in an environment subject to pulsatile blood flow. This mechanism may be useful in preventing "steal" of blood flow within various vascular beds. A role for the endothelium in the generation of oscillatory tension development in hamster aorta has been proposed by Jackson (1988). Segal and Duling (1987) demonstrated that locally applied autocoids elicited a vasodilatory response that propagated along vessels in the absence of blood flow. A mechanism involving intercellular communication may underlie this phenomenon.

SUMMARY

Mechanisms involved in Ca^{2+} homeostasis and stimulus-secretion coupling in cultured endothelial cells in response to humoral and physical stimuli include passive leak, activation of ion channels, and chemical second messengers. Calcium entry is controlled by receptor activation, passive leak, and mechanical stretch. The rate at which Ca^{2+} enters the cell through these pathways is dependent on the transmembrane potential which governs the electrochemical gradient for Ca^{2+} and which is set by participation of various K channels.

REFERENCES

Adams, D. J., Barakeh, J., Laskey, R., and van Breemen, C., 1989a, Ion channels and regulation of intracellular calcium in vascular endothelial cells, *FASEB J.*, 3: 2389.

Adams, D. J., van Breemen, C., Cannell, M. B., and Sage, S. O., 1989b, Na^{+}-Ca^{2+} exchange in fura-2 loaded cultured bovine pulmonary artery endothelial monolayers, *J. Physiol.*, 418: 183P.

Aisaka, K., Gross, S. S., Griffith, O. W., and Levi, R., 1989, N^{G}-methylarginine, an inhibitor of endothelium-derived nitric oxide synthesis, is a potent pressor agent in the guinea pig: Does nitric oxide regulate blood pressure in vivo?, *Biochem. Biophys. Res. Commun.*, 160: 881.

Beny, J. L. and Gribi, F., 1989, Dye and electrical coupling of endothelial cells in situ, *Tissue and Cell*, 21: 797.

Berridge, M., 1990, Calcium oscillations, *J. Biol. Chem.*, 265: 9583.

Bregestovski, P., Bakhramov, A., Danilov, S., Moldobaeva, A., and Takeda, K., 1988, Histamine-induced inward currents in cultured endothelial cells from human umbilical vein, *Br. J. Pharmacol.*, 95: 429.

Cannell, M. B. and Sage, S. O., 1989, Bradykinin-evoked changes in cytosolic calcium and membrane currents in cultured bovine pulmonary artery endothelial cells, *J. Physiol.*, 419: 555.

Colden-Stanfield, M., Schilling, W. P., Ritchie, A. K., Eskin, S. G., Navarro, L. T., and Kunze, D., 1987, Bradykinin-induced increases in cytosolic calcium and ionic currents in cultured bovine aortic endothelial cells, *Circ. Res.*, 61: 632.

Colden-Stanfield, M., Schilling, W. P., Passani, L. D., and Kunze, D. L., 1990, Bradykinin-induced potassium current in cultured bovine aortic endothelial cells, *J. Membr. Biol.*, 116: 227.

Freay, A., Johns, A., Adams, D. J., Ryan, U. S., and van Breemen, C., 1989, Bradykinin and inositol 1,4,5-trisphosphate-stimulated calcium release from intracellular stores in cultured bovine endothelial cells, *Pflügers Arch.*, 414: 377.

Furchgott, R. F. and Zawadski, J. V., 1980, The obligatory role of endothelial cells in the relaxation of arterial smooth muscle by acetylcholine, *Nature*, 288: 373.

Ganz, P. and Alexander, R. W., 1985, New insights into the cellular mechanisms of vasospasm, *Am. J. Cardiol.*, 56: 11E.

Hagiwara, H., Ohtsu, Y., Shimonaka, M., and Inada, Y., 1983, Ca^{2+}- or Mg^{2+}-dependent ATPase in plasma membrane of cultured endothelial cells from bovine carotid artery, *Biochim. Biophys. Acta*, 734: 133.

Harootunian, A. T., Kao, J. P. Y., and Tsien, R. Y., 1988, Agonist-induced calcium oscillations in depolarized fibroblasts and their manipulation by photoreleased $Ins(1,4,5)P_3$, Ca^{2+} and Ca^{2+} buffer, *Cold Spring Harbor Symp. Quant. Biol.*, 8: 935.

Jackson, W. F., 1988, Oscillations in active tension in hamster aortas: Role of the endothelium, *Blood Vessels*, 25: 144.

Jacob, R., 1990, Calcium oscillations in electrically non-excitable cells, *Biochim. Biophys. Acta*, 1052: 427.

Jacob, R., Merritt, J. E., Hallam, T. J., and Rink, T. J., 1988, Repetitive spikes in cytoplasmic calcium evoked by histamine in human endothelial cells, *Nature,* 335: 40.

Jayakody, R. L., Kappagoda, C. T., Senaratne, M. P. J., and Sreeharan, N., 1987, Absence of effect of calcium antagonists on endothelium-dependent relaxation in rabbit aorta, *Br. J. Pharmacol.,* 91: 155.

Johns, A., Lategan, T., Lodge, N., Ryan, U., van Breemen, C., and Adams, D. J., 1987, Calcium entry through receptor-operated channels in bovine pulmonary artery endothelial cells, *Tissue and Cell,* 19: 733.

Lambert, T. L., Kent, R. S., and Whorton, A. R., 1986, Bradykinin stimulation of inositol polyphosphate production in porcine aortic endothelial cells, *J. Biol. Chem.,* 261: 15288.

Lansman, J. B., Hallam, T. J., and Rink, T. J., 1987, Single stretch-activated ion channels in vascular endothelial cells as mechanotransducers?, *Nature,* 325: 811.

Laskey, R. E. and van Breemen, C., 1989, Lack of evidence for active Na^+/Ca^{2+} exchange in cultured endothelial cells, *FASEB J.,* 3: A879.

Laskey, R. E., Adams, D. J., Johns, A., Rubanyi, G. M., and van Breemen, C., 1990, Membrane potential and Na^+-K^+ pump activity modulate resting and bradykinin-stimulated changes in cytosolic free calcium in cultured endothelial cells from bovine atria, *J. Biol. Chem.,* 265: 2613.

Lockette, W. E., Otshuba, Y., Carretero, O. A., 1986, Endothelium-dependent relaxation in hypertension, *Hypertension,* 8(Suppl. II): II-61.

Lodge, N. J., Adams, D. J., Johns, A., Ryan, U. S., and van Breemen, C., 1988, Calcium activation of endothelial cells, *in:* "Resistance Arteries", W. Halpern, ed., Perinatology Press, Ithaca, p. 152.

Long, C. J. and Stone, T. W., 1985, The release of endothelium-derived relaxant factor is calcium dependent, *Blood Vessels,* 22: 205.

Lopez-Jaramillo, P., Gonzalez, M. C., Palmer, R. M. J., and Moncada, S., 1990, The crucial role of physiological Ca^{2+} concentrations in the production of endothelial nitric oxide and the control of vascular tone, *Br. J. Pharmacol.,* 101: 489.

Luckhoff, A. and Busse, R., 1990a, Refilling of endothelial calcium stores without bypassing the cytosol, *FEBS Lett.,* 275: 108.

Luckhoff, A. and Busse, R., 1990b, Activators of potassium channels enhance calcium influx into endothelial cells as a consequence of potassium currents, *Naunyn-Schmiedeberg's Arch. Pharmacol.,* 342: 94.

Luckhoff, A., Pohl, U., Mulsch, A., and Busse, R., 1988, Differential role of extra- and intracellular calcium on EDRF and prostacyclin from cultured endothelial cells, *Br. J. Pharmacol.,* 95: 189.

Meyer, B., Schmidt, K., Humbert, R., and Bohme, E., 1989, Biosynthesis of endothelium-derived relaxing factor: A cytosolic enzyme in porcine aortic endothelial cells Ca^{2+}-dependently converts L-arginine into an activator of soluble guanylyl cyclase, *Biochem. Biophys. Res. Commun.,* 164: 678.

Neylon, C. B. and Irvine, R. F., 1990, Synchronized repetitive spikes in cytoplasmic calcium in confluent monolayers of human umbilical vein endothelial cells, *FEBS Lett.,* 275: 173.

Nilius, B., 1990, Permeation properties of a non-selective cation channel in human vascular endothelial cells, *Pflügers Arch.,* 416: 609.

Olesen, S. P., Davies, P. F., and Clapham, D. E., 1988a, Muscarinic-activated K^+ current in bovine aortic endothelial cells, *Circ. Res.*, 62: 1059.

Olesen, S. P., Clapham, D. E., and Davies, P. F., 1988b, Haemodynamic shear stress activates a K^+ current in vascular endothelial cells, *Nature*, 331: 168.

Oyama, Y., Kawasaki, H., Hattori, Y., and Kauno, M., 1986, Attenuation of endothelium-dependent relaxation in aorta from diabetic rat, *Eur. J. Pharmacol.*, 131: 75.

Palmer, R. M., Ashton, D. S., and Moncada, S., 1988, Vascular endothelial cells synthesize nitric oxide from L-arginine, *Nature*, 333: 664.

Rees, D. D., Palmer, M. J., and Moncada, S., 1989, Role of endothelium-derived nitric oxide in the regulation of blood pressure, *Proc. Nat'l. Acad. Sci. U.S.A.*, 86: 3375.

Rusko, J., Tanzi, F., van Breemen, C., and Adams, D. J., 1991, Gating and block of Ca^{2+}-activated potassium channels in native endothelial cells from rabbit aorta, *Biophys. J.*, 59: 81a.

Sage, S. O., Adams, D. J., and van Breemen, C., 1989, Synchronized oscillations in cytoplasmic free calcium concentration in confluent bradykinin-stimulated bovine pulmonary artery endothelial cell monolayers, *J. Biol. Chem.*, 264: 6.

Schilling, W. P., Ritchie, A. K., Navarro, L. T., and Eskin, S. G., 1988, Bradykinin-stimulated calcium influx in cultured bovine aortic endothelial cells, *Am. J. Physiol.*, 255: C219.

Segal, S. S. and Duling, B. R., 1987, Propagation of vasodilation in resistance vessels of the hamster: Development and review of a working hypothesis, *Circ. Res.*, 61: 1120.

Singer, H. A. and Peach, M. J., 1982, Calcium- and endothelial-mediated vascular smooth muscle relaxation in rabbit aorta, *Hypertension*, 4(Suppl. II): II-19.

Silver, M. R. and Decoursey, T. E., 1990, Intrinsic gating of inward rectifier in bovine pulmonary artery endothelial cells in the presence or absence of internal Mg^{2+}, *J. Gen. Physiol.*, 96: 109.

Suave, R., Parent, L., Simoneau, C., and Roy, G., 1988, External ATP triggers a biphasic activation process of a calcium-independent K^+ channel in cultured bovine aortic endothelial cells, *Pflügers Arch.*, 412: 469.

Spedding, M., Schini, V., Schoeffter, P., and Miller, R. C., 1986, Calcium channel activation does not increase release of endothelial-derived relaxant factors (EDRF) in rat aorta although tonic release of EDRF may modulate calcium channel activity in smooth muscle, *J. Cardiovasc. Pharmacol.*, 8: 1130.

Takeda, K. and Klepper, M., 1990, Voltage-dependent and agonist-activated ionic currents in vascular endothelial cells: A review, *Blood Vessels*, 27: 169.

Takeda, K., Schini, V., and Stoeckel, H., 1987, Voltage-activated potassium, but not calcium currents in cultured bovine aortic endothelial cells, *Pflügers Arch.*, 410: 385.

Thastrup, O., Dawson, A., Scharff, O., Foder, B., Cullen, P., Drobak, B., Bjerrum, P., Christensen, S., and Hanley, N., 1989, Thapsigargin, a novel molecular probe for studying intracellular calcium release and storage, *Agents and Actions*, 27: 17.

Vallance, P., Collier, J., and Moncada, S., 1989, Effects of endothelium-derived nitric oxide on peripheral arteriolar tone in man, *Lancet*, 2: 997.

Van Breemen, C., Cauvin, C., Johns, A., Leijten, P., and Yamamoto, H., 1986, Ca^{2+} regulation of vascular smooth muscle, *Fed. Proc.*, 45: 2746.
Whorton, A. R., Willis, C. E., Kent, R. S., and Young, S. L., 1984, The role of calcium in the regulation of prostacyclin synthesis by porcine aortic endothelial cells, *Lipids*, 19: 17.

ALTERED EXCITATION-CONTRACTION COUPLING IN HYPERTENSION: ROLE OF PLASMA MEMBRANE PHOSPHOLIPIDS AND ION CHANNELS

Robert H. Cox[1] and Thomas N. Tulenko[2]

[1]Bockus Research Institute
The Graduate Hospital
Philadelphia, PA 19146

[2]Department of Physiology
Medical College of Pennsylvania
Philadelphia, PA 19129

INTRODUCTION

Established hypertension is characterized by an elevation of peripheral resistance (Frohlich, 1973). A variety of *in vivo* studies in human as well as animal models of hypertension have demonstrated augmented responsiveness of arterial smooth muscle to contractile agents (Triggle, 1989). However, identification of the cellular mechanisms responsible for this increased responsiveness has proven elusive. Most older *in vitro* studies of maximum contractile responses as well as of the sensitivity of isolated vascular smooth muscle to agonists failed to reveal augmented responses in hypertensive arteries (Cox, 1989; Mulvany, 1989; Triggle, 1989). Folkow proposed the hypothesis that increased arterial wall thickness which impinged on the lumen was responsible for an increased geometric component of peripheral resistance as well as the augmented *in vivo* smooth muscle responsiveness to agonists in hypertension (Folkow, 1973). The latter was thought to be the result of an amplifying effect of the increased wall thickness being translated into augmented resistance responses to smooth muscle activation (Folkow, 1973). This hypothesis was generally accepted as a reconciliation of the results reported up to the mid 1970s.

In the late '70s, the development of new experimental techniques led to the demonstration of augmented contractile responses in genetic as well as experimental models of systemic hypertension in both large and small arteries. Figure 1 shows a comparison of maximum values of active stress for carotid and tail arteries from 20-week old Wistar-Kyoto (WKY) and spontaneously hypertensive rats (SHR) plotted as a function of normalized muscle length (Cox, 1981). These results showed that augmented contractile responses are present in distributing arteries of hypertensive compared to normal subjects. Other studies documented similar differences in small resistance arteries between WKY and SHR (Mulvany and Halpern, 1977) as well as similar differences in DOCA-salt and renal stenosis hypertension (Cox, 1982a; 1982b).

Regulation of Smooth Muscle Contraction
Edited by R.S. Moreland, Plenum Press, New York, 1991

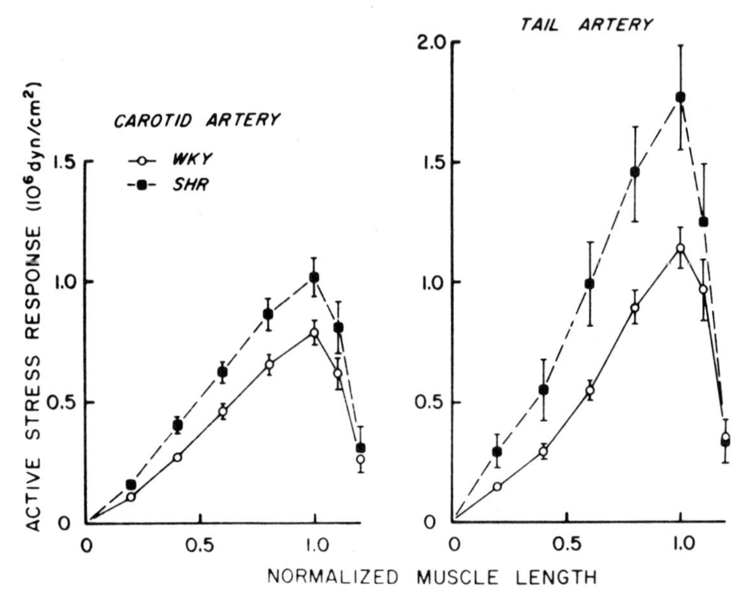

Figure 1. Comparison of active wall stress to 140 mM K^+ for intact cylindrical segments of carotid and tail arteries of WKY and SHR. Wall diameter was normalized by dividing by the value associated with maximum active stress. Symbols represent mean ± 1 SEM of data. Reprinted from Cox, 1981 by permission of the American Heart Association.

The augmented active stress was found to be the result of medial hypertrophy and an increase in the smooth muscle cell volume per unit length of hypertensive arteries. When normalized on the basis of medial cell volume, maximum values of active cell stress were generally similar between normotensive and hypertensive animals. Additional studies showed that an increase in the actin:myosin ratio also occurred in arteries from hypertensive animals (Chacko and Cox, 1981). These results suggest that the latter changes represent an increase in the cellular actin content with no change in the amount of myosin per cell.

Other studies demonstrated more sensitive contractile responses to agonists in small resistance vessels of hypertensive animals, sites at which peripheral resistance is determined. An example of such differences is shown in Figure 2. Under conditions in which innervation to small mesenteric arteries was intact, there were no significant differences in norepinephrine (NE) sensitivity between arteries from controls and hypertensive animals. When NE uptake into sympathetic nerve terminals in the tissues was blocked, significantly higher sensitivity to NE was found in hypertensive arteries (Mulvany and Nyborg, 1980; Whall et al., 1980). These results in the late '70s and early '80s demonstrated that differences exist in maximal contractile capacity as well as sensitivity to agonist activation in arteries of genetic and experimental hypertensive subjects that could contribute to the augmented *in vivo* sensitivity of vascular resistance to vasoactive agents.

In general, contraction in vascular smooth muscle is the result of an increase in the free [Ca^{2+}] in the cytoplasm which activates the contractile apparatus (Somlyo, 1985). The augmented contractile sensitivity to agonists demonstrated in hypertensive arteries could be the result of increased Ca^{2+}

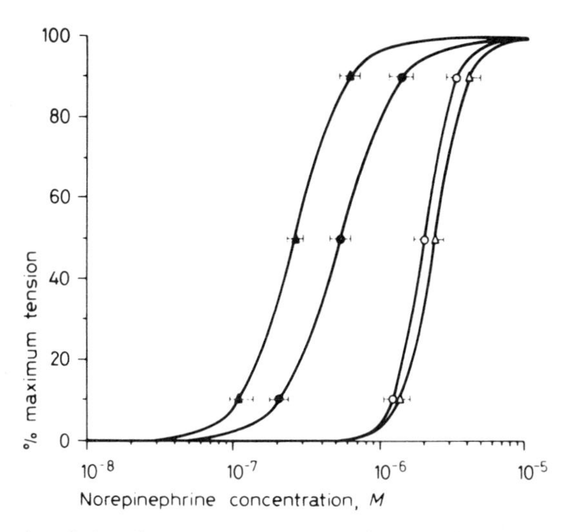

Figure 2. Norepinephrine dose-response curves obtained from mesenteric resistance arteries of WKY (circles) and SHR (triangles), in the innervated (open symbols) and denervated (filled symbols) conditions. Symbols represent mean ± 1 SEM. Arteries were denervated by treatment with 6-hydroxydopamine. Reprinted from Whall et al., 1980 by permission.

sensitivity of the contractile apparatus or increased free cytoplasmic Ca^{2+} levels. Prevailing opinion held that no changes in Ca^{2+} sensitivity of the contractile proteins exist in hypertension. For example, McMahon and Paul (1985) compared the Ca^{2+} sensitivity of the thoracic aorta of control and aldosterone-hypertensive rats after chemical disruption of the surface membrane with the detergent Triton X-100. The response to Ca^{2+} under these conditions represents an intrinsic response of the contractile proteins independent of Ca^{2+} handling by the surface membrane and intracellular structures. Their results shown in Figure 3 demonstrated no difference in Ca^{2+} sensitivity between control and hypertensive subjects. Results of this sort have been cited to support the hypothesis that changes in the contractile proteins do not occur in hypertension. These results as well as other direct evidence suggested that a defect in Ca^{2+} regulation by vascular smooth muscle may occur in hypertension. In support of this conclusion, increased cytoplasmic levels of Ca^{2+} have been reported under resting and agonist-activated conditions in hypertensive smooth muscle cells (Erne and Hermsmeyer, 1989; Sugiyama et al., 1990; Morgan and Papageorgiou, this volume).

MEMBRANE CHANGES

Over the last 20 years, a growing body of evidence has documented a wide variety of membrane related changes in hypertension including receptors (Graham et al., 1982), ion channels (Jones, 1973; Bing et al., 1986; Rusch and Hermsmeyer, 1988) and transport molecules (Matlib et al., 1985; Ives, 1989). One of the earliest demonstrations of such alterations was that of an increase in the permeability of hypertensive smooth muscle to ^{42}K or its equivalent ^{86}Rb (Jones, 1973). An example of this difference is shown in Figure 4, where efflux rate constants are represented as a function of time under basal conditions as well as in response to challenge by increasing concentrations of NE. For each condition, larger efflux rate constants were found in SHR aorta compared to WKY. Equivalent measurements in small mesenteric arteries as well as tail arteries from the SHR demonstrated qualitatively similar differences. This increase in ^{42}K or ^{86}Rb efflux was found to be partially calcium dependent since blockers of calcium influx decreased the difference between normals and hypertensives (Jones and Smith, 1986).

Additional studies have demonstrated that not only do changes occur in potassium channels in arterial smooth muscle membranes in hypertension, but in all ion channels (Jones, 1980). Since all ion channels contribute to the determination of the membrane potential, it could be expected that differences in this quantity might exist in hypertensive smooth muscle cells. Stekiel et al. (1986) have demonstrated that such differences do exist. Figure 5 summarizes values of membrane potential measured in small resistance arteries of 4 - 5 week-old and 12 - 15 week-old WKY and SHR *in situ*. In young animals, no significant differences were found between WKY and SHR, and local

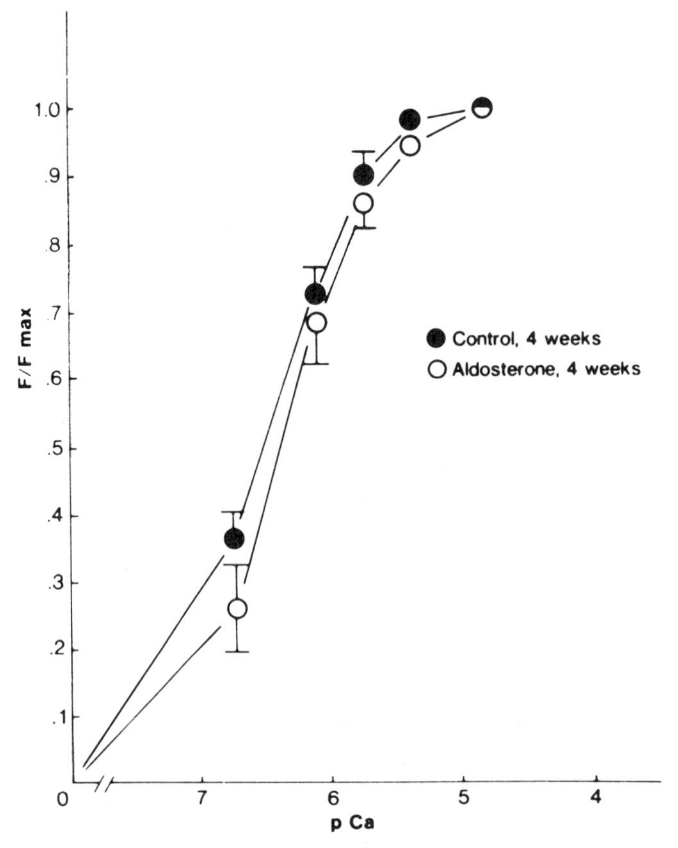

Figure 3. Calcium-dose response relationship in chemically skinned aortic rings from control (●, n=9) and aldosterone hypertensive rats (○, n=8). Isometric tension responses have been normalized to the force obtained at pCa = 4.9 in each vessel [C (n=9), 4.08 ± 0.36; AHR (n=8), 3.99 ± 0.35 mN]. Symbols represent means ± 1 SEM. Reprinted from McMahon and Paul, 1985 by permission of the American Heart Association.

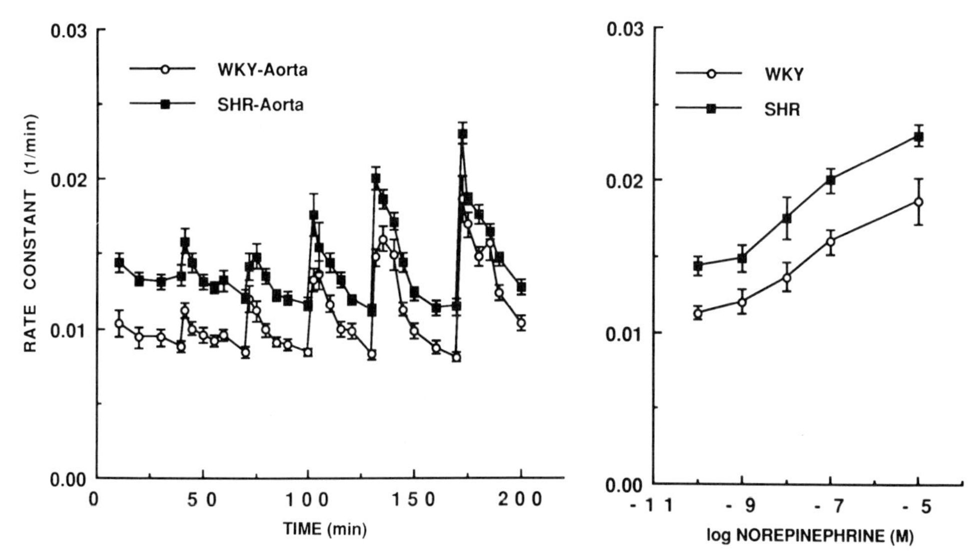

Figure 4. Effects of norepinephrine (NE) on rate of ^{86}Rb efflux from segments of thoracic aorta of 20 week old WKY and SHR. After 40 minutes of basal efflux, segments were exposed to increasing doses of norepinephrine from 0.1 nM to 10 μM in separate exposures followed by rinsing in normal buffer. Symbols are means ± 1 SEM. Panel on the right shows dose-responses relations of peak efflux response to NE for WKY and SHR groups.

sympathectomy produced no significant changes in values of membrane potential. On the other hand, in the older animals values of resting membrane potential were significantly depolarized in SHR compared to WKY. Sympathectomy in older animals produced a hyperpolarization of membrane potential in both WKY and SHR to values that were not significantly different between the two groups. These results show that in the older SHR altered ion conductances exist such that membranes are relatively depolarized and that this difference is, in part, related to the sympathetic innervation to vascular smooth muscle. Since potassium conductance is increased, these results suggest that even larger increases in Na^+ and/or Cl^- conductance must occur to produce the observed membrane depolarization (Stekiel et al., 1986).

CONTRACTILE PROTEINS

As indicated above, older studies had suggested that no differences exist in the properties of the contractile proteins in normal and hypertensive subjects. However, over the last few years, evidence has begun to accumulate which suggests that such alterations do exist in hypertrophied smooth muscle. Figure 6 shows a comparison of force-velocity relations of caudal artery smooth muscle from Wistar-Kyoto and SHR by Packer and Stephens (1985). Maximum values of velocity of shortening were significantly elevated in the SHR compared to WKY. On the basis of studies performed in striated muscle (Bárány, 1967; Carey et al., 1979), these results could be interpreted to suggest an augmented ATPase activity of myosin in the caudal artery cells of the SHR. While differences in composition of actin and intermediate filaments have

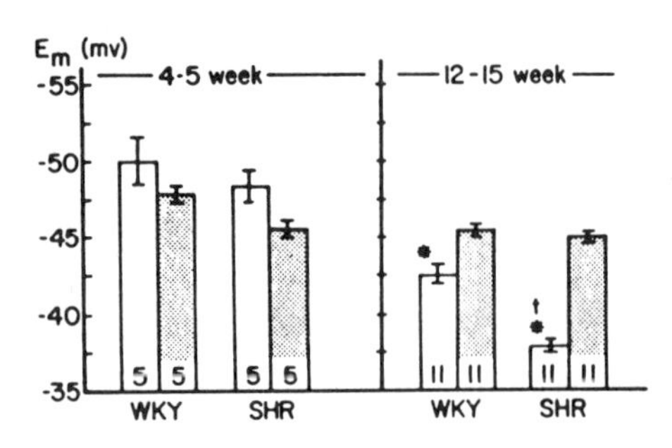

Figure 5. In situ membrane potentials (E_m) of small mesenteric arteries in 4-5 wk-old and 12-15 wk old SHR and WKY. Open bars: innervated arteries; shaded bars: arteries sympathetically denervated locally by pretreatment with 6-hydroxydopamine. Bars represent mean ± SEM of E_m in mV while numbers at base represent group size. * indicates values of E_m in 12-15 wk-old WKY and SHR that were significantly less than values in 4-5 wk-old group. † indicates values of E_m in SHR that differed from WKY. Reprinted from Stekiel et al., 1986 by permission.

been demonstrated in hypertension-associated hypertrophy (Owens, 1991), direct evidence in support such a suggestion has been slow in coming.

A study by Upadhya et al. (1986) using femoral arteries from renal hypertensive dogs has provided this evidence. No differences in alpha, beta and gamma isoforms of actin were found between control and hypertensive arteries. Differences were observed in tropomyosin chains, however, with an increase in the relative amount of the beta compared to alpha chain content in hypertensive arteries. Electrophoresis of intact myosin performed under non-denaturing conditions demonstrated a second band of myosin in hypertensive arterial smooth muscle which accounted for approximately 20% of the total myosin in the tissue. Following limited proteolysis with trypsin, peptide maps of myosin from control and hypertensive arteries demonstrated qualitative differences in various peptide bands. This suggests compositional differences in myosin in these two groups of animals. Functional studies of the ATPase activity of myosin were also performed using these tissues. While no significant differences were found in calcium- or magnesium-activated ATPase activity, potassium-EDTA activated ATPase activity was significantly higher in myosin from hypertensive femoral arteries compared to controls. The results of these studies suggest that alterations in the ATPase activity of myosin may be causally associated with renal hypertension in the dog, and they support the studies of augmented velocity of shortening in hypertensive arterial smooth muscle.

Several additional reports have recently been made regarding changes in the functional characteristics of myosin in different models of smooth muscle hypertrophy. Estrogen treatment in the rat was found to be associated with an increase in the ratio of SM1 to SM2 myosins in uterine smooth muscle determined by SDS-PAGE electrophoresis (Paul et al., this volume). In addition, augmented force-velocity relations were found in hypertrophied uterine smooth muscle in this study, suggesting increased ATPase activity. In studies on femoral arteries from renal hypertensive dogs mentioned above, there were no significant changes in the ratio of SM1 to SM2. In hypertrophied rabbit urinary bladder secondary to urethral obstruction, Kim et al. (1990) found a decrease in the ratio of SM1 to SM2 myosins with no change in ATPase

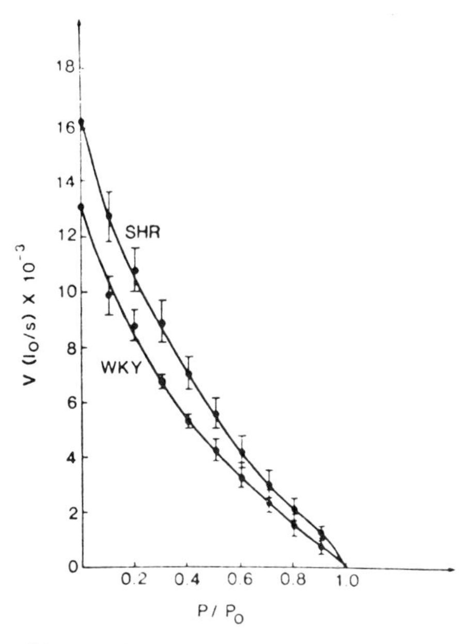

Figure 6. Summary of force-velocity relations from caudal arterial muscle of WKY (n = 6) and SHR (n = 5). Symbols are means while vertical lines represent ± 1 SEM. Reprinted from Packer and Stephens, 1985 with permission.

activity. The results of these three studies suggest no correlation between changes in myosin heavy chain composition and myosin ATPase activity in hypertrophied smooth muscle. It appears, therefore, that the heavy chain does not directly regulate ATPase activity of hypertrophied smooth muscle as occurs in striated muscle (Pagani and Julian, 1984).

EXCITATION-CONTRACTION COUPLING HYPOTHESIS

Signal transduction mechanisms involving the plasma membrane of cells involve protein movement (interactions) within the plane of the membrane (Benga and Holmes, 1984). In the process, proteins associate and dissociate to produce the interactions required for signal transduction to occur. The mobility of proteins within the bulk lipid phase is under constraints determined in part by the membrane's physical state, that is, by its fluidity or microviscosity (Quinn, 1980). Two primary determinants of membrane physical state are: a) the order characteristics of phospholipid fatty acyl chains, including the degree of unsaturation and chain length; and b) the free cholesterol content of the membrane (Quinn, 1980; van Blitterswijk et al., 1987).

Work in several cell systems including vascular smooth muscle has shown that enrichment of the cell plasma membrane with free cholesterol produces a number of functional cellular changes including: a) increased membrane microviscosity (decreased fluidity) (Medow and Segal, 1987; Gleason and Tulenko, 1989); b) increased Ca^{2+} influx (Locher et al., 1984; Bialecki and

Tulenko, 1989; Gleason and Tulenko, 1989); c) alterations in ATP-sensitive and Ca^{2+}-activated K^+ channels (Rock and Tulenko, 1991); and d) decreases in Ca^{2+}-ATPase (Madden et al., 1981) and Na^+/K^+-ATPase activity (Broderick et al., 1989). Acute cholesterol enrichment of rabbit carotid arteries produced an increase in sensitivity to norepinephrine (NE) activation which was completely reversed by organic Ca^{2+} channel blockers (Bialecki and Tulenko, 1989). This increase in NE sensitivity was associated with an increase in basal and NE activated ^{45}Ca influx (Bialecki and Tulenko, 1989) which was also inhibited by organic Ca^{2+} channel blockers. Similar results have also been obtained using rabbit aortic smooth muscle cells in culture (Gleason and Tulenko, 1989). Acute cholesterol enrichment also increased basal and NE-activated ^{86}Rb efflux in cultured cells (Tulenko et al., 1990) which was also inhibited by diltiazem. Since NE-activated ^{86}Rb efflux is thought to reflect the activity of Ca^{2+}-sensitive K^+ channels, these results suggest that altered Ca^{2+} and K^+ channel function as well as increased cellular $[Ca^{2+}]$ occurs with acute cholesterol enrichment. Taken together, these results suggest that acute cholesterol enrichment of the plasma membrane of smooth muscle cells causes an organic Ca^{2+} channel blocker-dependent increase in sensitivity to NE that may be the result of an alteration in ion channel properties (Tulenko et al., 1990). That enrichment with cholesterol directly causes these alterations is confirmed by the observation that these changes are reversible with removal of excess cholesterol (Gleason and Tulenko, 1989). These results support the concept that the plasma membrane bilayer structure is sensitive to changes in its lipid composition (Schnitzky and Barenholz, 1978) and these changes have the capacity to alter the functional properties of membrane-associated proteins including ion channels, and as a consequence cell and tissue function.

Based upon this intriguing line of thought we offer the following hypothesis to explain these findings: changes in the composition of vascular smooth muscle membranes produce alterations in the molecular structure and lipid dynamics of the membranes which in turn alters the expression of the intrinsic characteristics of membrane bound proteins including ion channels and transport molecules. These changes result in alterations in excitation-contraction coupling, leading to altered sensitivity to contractile agonists in arterial smooth muscle cells.

Table 1. Summary of Phospholipid Analysis of Thoracic Aorta From 20 Week WKY and SHR (n = 4)

PHOSPHOLIPID	WKY		SHR	
(% Total)	Mean	SE	Mean	SE
Diphosphatidylglycerol (DPG)	5.9	1.5	1.3	1.3
Lysophosphatidylcholine (LPC)	14.0	1.3	18.9	3.7
Phosphatidylethanolamine (PE)	6.8	2.5	5.2	3.0
Phosphatidylcholine (PC)	18.5	9.1	14.0	5.6
Phosphatidic Acid (PA)	4.2	3.0	4.8	2.8
Sphingomyelin (SM)	15.6	1.4	25.3	5.7
Phosphatidylserine (PS)	13.7	4.6	6.8	3.0
Phosphatidylinositol (PI)	11.5	3.4	17.8	2.6

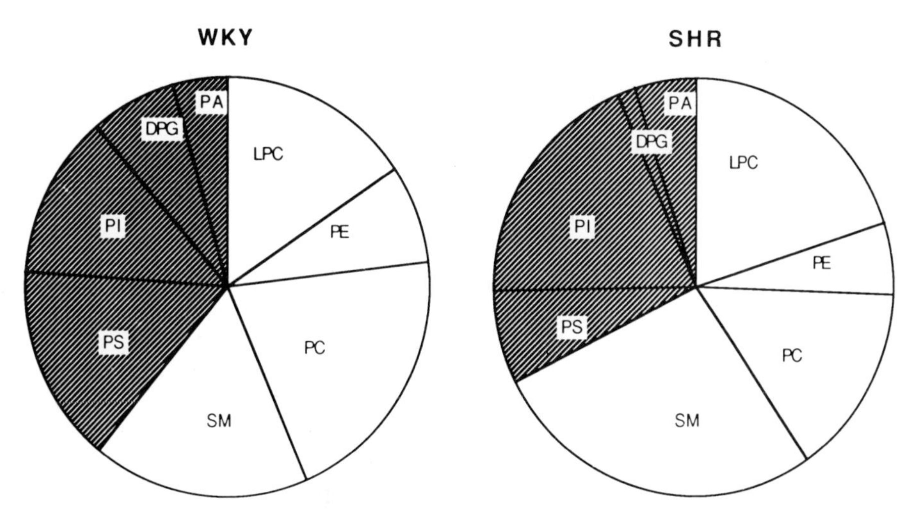

Figure 7. Summary of phospholipid profile of thoracic aorta from WKY and SHR. Anionic phospholipids are represented by shaded areas and neutral phospholipids by clear areas. Phospholipids are as follows: LPC, lysophosphatidylcholine; PE, phosphatidylethanolamine; PC, phosphatidylcholine; SM, sphingomyelin; PS, phosphatidylserine; PI, phosphatidylinositol; DPG, diphosphatidylglycerol; and PA, phosphatidic acid.

MEMBRANE PHOSPHOLIPIDS

Are there any experimental data to support this hypothesis? Studies performed by us have demonstrated a remodelling of the phospholipid profile of cell membranes from the thoracic aorta of the spontaneously hypertensive rat (SHR) compared to its Wistar-Kyoto control (WKY) as shown in Table 1. One of the most striking findings of this study is a decrease in total anionic phospholipid content expressed as a percentage of total phospholipid as shown in Figure 7. Anionic phospholipids on the outer leaflet of the plasma membrane are the sites to which divalent cations bind which is thought to stabilize the permeability of the plasma membrane and modulate the gating characteristics of ion channels. Stabilization of the phospholipid matrix by calcium binding has been suggested (Hauser and Phillips, 1979) and therefore may play an important role in altered smooth muscle reactivity in hypertension (Webb and Bohr, 1978). Thus, a decrease in anionic binding sites would suggest a decrease in the stabilizing effect of divalent ions on smooth muscle membrane permeability supporting the hypothesis offered by Webb and Bohr (1978) to explain altered excitation-contraction coupling in hypertension. In addition, some other important changes in specific phospholipid species can be noted. We observed an increase in sphingomyelin and a decrease in phosphatidylcholine content in SHR compared to WKY. Since sphingomyelin is a rigidizing phospholipid while phosphatidylcholine is a fluidizing phospholipid, changes in the ratio of sphingomyelin to phosphatidylcholine content would be expected to alter membrane fluidity (Borochov et al., 1977). This ratio is larger in the SHR as a result of both an increase in the rigidizing phospholipid (sphingomyelin) and a decrease in the fluidizing phospholipid (phosphatidylcholine) suggesting a decreased membrane fluidity. In addition, phosphatidylinositol content is significantly increased (55%) in the SHR aorta compared to the WKY. Phosphatidylinositol plays an important role in membrane signal transduction mechanisms involved in the mobilization of calcium in cellular stimulus response coupling (Nishizuka, 1984).

These results suggest a decrease in calcium binding sites as well as a rigidizing effect of the phospholipid differences on the smooth muscle cell membrane in the SHR. How could changes in anionic phospholipids influence the properties of ion channels and excitation-contraction in arterial smooth muscle? An example of such an effect is demonstrated by an experiment in cellular electrophysiology. Figure 8 shows the effects of varying extracellular calcium on calcium currents measured in single, freshly dissociated, voltage-clamped rabbit portal vein smooth muscle cells. To the left are (from top to bottom) the voltage clamp protocol and current tracings for 2, 5, and 10 mM external [Ca^{2+}]. These currents were obtained with depolarizations from a holding potential of -60 mV to +40 mV in steps of 10 mV for 75 msec. The peak currents at each voltage for the three extracellular calcium concentrations are plotted in the graph to the right as current-voltage (I-V) relations. Increasing extracellular calcium results in an increase in the maximum current and a shift to the right in the voltage-dependent activation of the current by approximately 10 mV per doubling of extracellular calcium.

This effect of extracellular Ca^{2+} on Ca^{2+} currents is the result of several actions of the ion. First, increasing the external concentration of Ca^{2+} increases the driving force leading to an increase in the maximum current. The increase in external [Ca^{2+}] increases binding to the cell surface shifting the voltage-

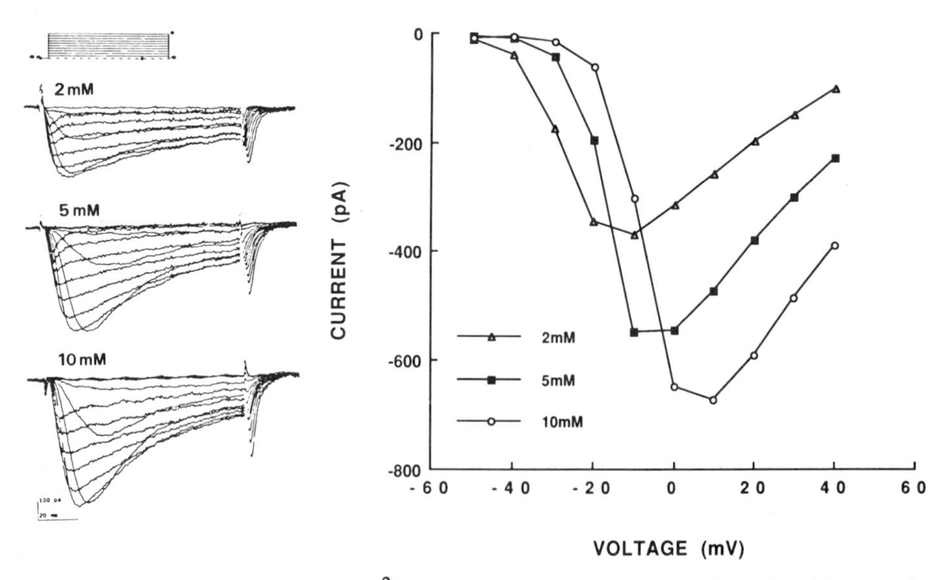

Figure 8. Effects of external [Ca^{2+}] on current-voltage relations of single rabbit portal vein myocyte under voltage clamp conditions. On the top left is voltage step protocol from a holding potential of -60 mV in 10 mV, 75msec steps to +40 mV. Below are current responses at 2, 5 and 10 mM $[Ca^{2+}]_o$. Panel on the right shows current-voltage relations for peak currents recorded at each voltage. Increasing $[Ca^{2+}]_o$ shifts the I-V relation, the activation voltage, and the reversal potential to the right as well as increases the maximum current. Cell capacitance was 56.3 pF.

dependency of activation and inactivation processes to the right. The net result of these changes is that the voltage range over which significant window currents occur shifts to the right as shown in Figure 9. This figure shows steady-state inactivation and activation of the Ca^{2+} current describing the availability relation (h_∞) and the channel conductance, respectively, at 2, 5, and 10 mM $[Ca^{2+}]_o$. This effect of extracellular Ca^{2+} is mediated by virtue of a surface charge effect of the extracellular divalent cations in conjunction with anionic charges (binding sites) on the smooth muscle cell plasma membrane (Kass and Krafte, 1987). This surface charge effect modulates the voltage-dependent gating characteristics of the Ca^{2+} channel producing the shift in the voltage range for window currents.

These results demonstrate that modulation of the gating characteristics of ion channels, in this example by varying extracellular calcium, can have large effects on the voltage-dependence of ionic currents. Changes in surface charge as a result of changes in the relative content of anionic versus neutral phospholipids would also be expected to have significant effects on the voltage-dependency of ion currents within this same voltage range, the lower range of which coincide with values recorded from arterial smooth muscle *in vivo*. These results also support the idea that changes in ionic current can occur as a result of altered gating characteristics of the ion channel independent of changes in the characteristics of the channel *per se*, supporting the hypothesis that modulation of ion channel function can occur through alterations in membrane composition.

POTASSIUM CHANNELS

Potassium channels also play an important role in excitation-contraction coupling in vascular smooth muscle. Since the permeability characteristics of the plasma membrane of muscle cells is dominated by that of potassium, resting membrane potential is strongly effected by changes in potassium

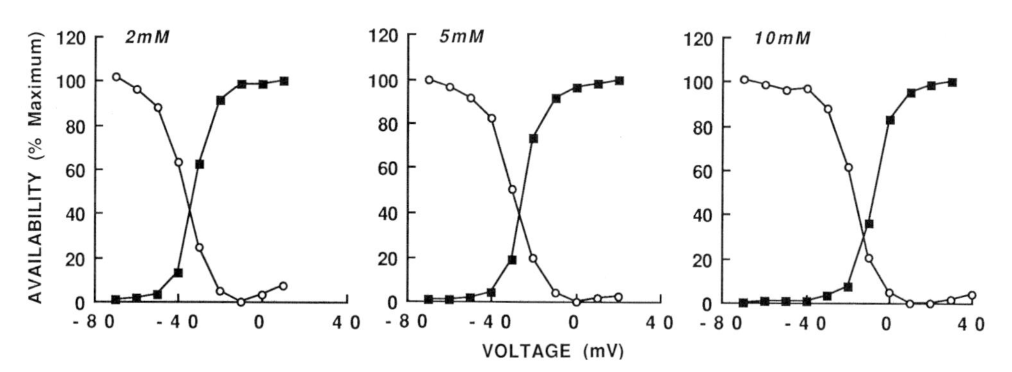

Figure 9. Comparison of calcium channel availability relations as a function of external $[Ca^{2+}]$. Open symbols show steady-state inactivation relation as a function of voltage while closed symbols show activation relation of the channels. The panels are from top to bottom for 2, 5 and 10 mM $[Ca^{2+}]_o$. The area under the intersection of the two curves represents the voltage range for "window currents". With increasing $[Ca^{2+}]_o$, the voltage range over which window currents are active shifts to higher voltages.

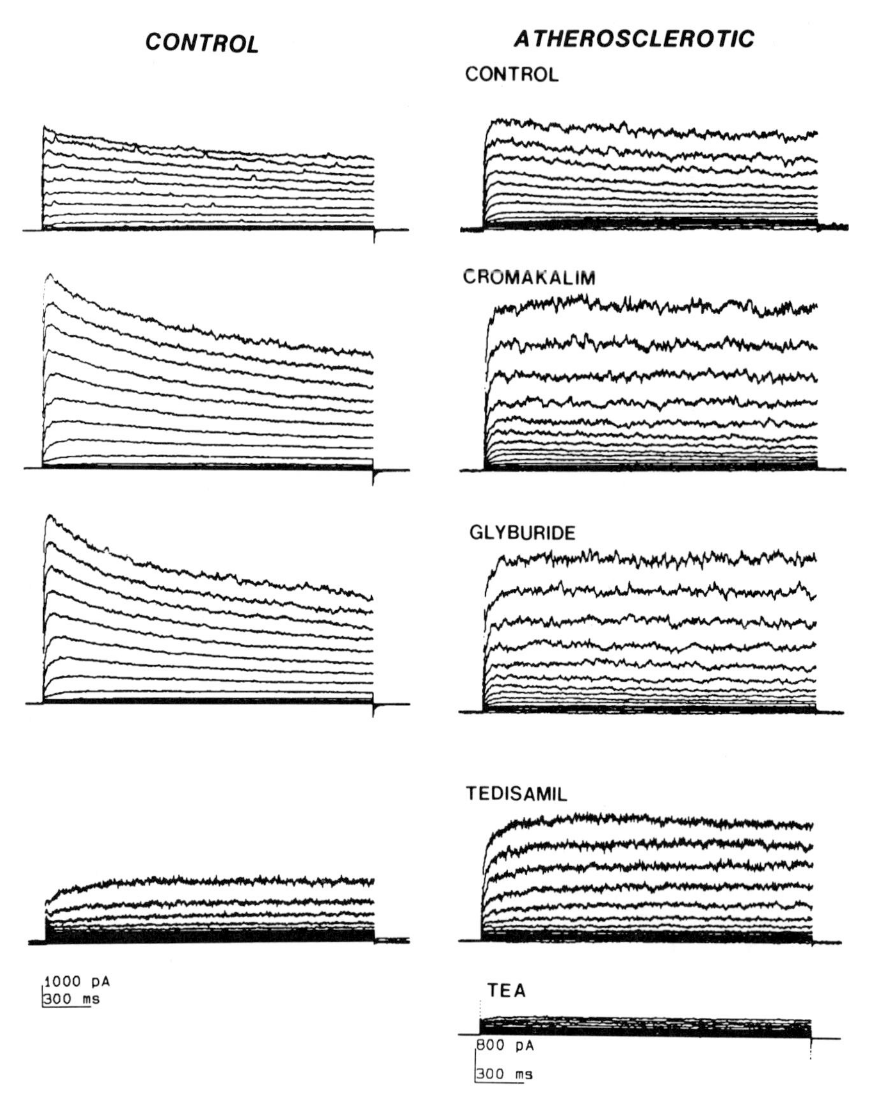

Figure 10. Comparison of potassium currents recorded from single portal vein myocytes from control and dietary atherosclerotic rabbits. Currents were recorded with an external solution consisting of 135 mM NaCl, 5 mM $CaCl_2$, 1 mM $MgCl_2$, 10 mM HEPES and 10 mM glucose, and pipet solution of 140 mM KCl, 10 mM NaCl, 5 mM MgATP, 14 mM EGTA, and 10 mM HEPES. Following control measurements responses to the cumulative addition of the following were made: 2 µM cromakalim, 2 µM glyburide, 10 µM tedisamil, and 20 mM TEA.

channel permeability. Thus, potassium channel permeability plays an important role in determining membrane potential and the degree of activation of calcium channels in vascular smooth muscle cells.

In addition to hypertension as mentioned above, other physiological and pathological conditions are associated with alterations in potassium channel properties in arterial smooth muscle. Thus, growth and development, aging, diabetes, atherosclerosis, and obesity, conditions associated with changes in the functional properties of potassium channels and their regulation, would be expected to be associated with altered excitation-contraction coupling. Alterations in agonist sensitivity for vascular smooth muscle contraction have been reported in these conditions (Hruza and Zweifach, 1967; Cohen and Berkowitz, 1976; Fleisch, 1980; White and Carrier, 1988; Cox and Kikta, 1989; Henry, 1990).

Dietary hypercholesterolemia (experimental atherosclerosis) is another condition associated with alterations in K^+ channel properties (Rock and Tulenko, 1991). In carotid arteries from atherosclerotic rabbits an increase in ^{86}Rb basal efflux, and a decrease in the augmentation of ^{86}Rb efflux by the putative potassium channel activators cromakalim and pinacidil have been reported (Rock and Tulenko, 1991). Changes in the characteristics of potassium channels measured under voltage clamp conditions in single cells from the portal vein and renal artery have also been observed in dietary atherosclerosis as shown in Figure 10. Under control conditions in cells from normal animals, a slow voltage-dependent inactivation of an outward current was observed during long voltage steps. Qualitatively similar alterations were present in cells from atherosclerotic animals, but augmented rectification was also observed. In response to the K^+ channel activator cromakalim, potassium currents were increased in both control and atherosclerotic cells but with the manifestation of different inactivation characteristics. In cells from control animals, cromakalim augmented the transient nature of these potassium currents increasing peak current as well as its inactivation. In atherosclerotic cells, on the other hand, the peak current was also increased by cromakalim but the transient nature of currents was reduced, producing a nearly non-inactivating current. Glyburide (1 μM) had small inhibitory effects in both control and atherosclerotic cells. Tedisamil, a drug known to block transient outward and delayed rectifier channels in ventricular myocytes (Dukes and Morad, 1990), exerted large effects in control portal vein cells, but only modest inhibitory effects in atherosclerotic cells. Residual currents in atherosclerotic cells were completely inhibited by 10 mM TEA.

SESSION ON "LESSONS FROM ADAPTATION"

In this chapter, we have reviewed some of the changes in arterial smooth muscle associated with the development of hypertension-associated hypertrophy, and have presented a hypothesis integrating current data and concepts that we believe provides an explanation for the altered excitation-contraction coupling that occurs in hypertension. These changes represent one example of a more general capability of blood vessels, ie. their ability to adapt on altered functional demand (Johansson, 1984). Adaptation in smooth muscle can be defined as a change in structure, function, or form that produces better adjustment to its environment. Other examples of adaptation are changes in blood vessels associated with growth and development, or in uterine smooth muscle associated with pregnancy. Changes in functional properties that accompany adaptation are usually the result of changes in cellular composition secondary to changes in cellular synthesis or degradation processes.

It would seem intuitive that adaptation of smooth muscle-containing structures represents a response of the tissue to optimize some measure(s) of performance. At the present time we do not understand which measures of performance are optimized by such adaptive changes. Furthermore, it is not clear what signals provide the information that modifies cellular processes altering composition. However, detailed studies of these processes should provide an opportunity to gain insights into the mechanisms involved. An understanding of these mechanisms is a prerequisite to the development of a rationale strategy to prevent or reverse the adverse consequences of adaptive changes in smooth muscle that may have pathological consequences.

The general sequence of events in excitation-contraction coupling in smooth muscle cells is now reasonably well established. Figure 11 shows a schematic representation of these excitation-contraction coupling events. Neural, humoral, and chemical signals impinge on the arterial smooth muscle cell where they are integrated at the membrane level. As a result of this integration, excitation-contraction coupling events are initiated that result in changes in cytoplasmic free calcium. These changes lead to an interaction between actin and myosin, and ultimately in the mechanical responses associated with contraction.

Details of changes in the sequence of events associated with excitation-contraction coupling will be discussed in more detail in the chapters to follow. Membrane mechanisms in hypertension will be discussed by Bohr et al. Morgan and Papageorgiou will describe changes in the regulation of cytoplasmic free calcium in hypertrophic smooth muscle cells. Alterations in actin and myosin isoforms associated with hypertrophy will be described by Seidel et al. and the consequences of asthma-induced adaptation on contractile apparatus mechanical function will be discussed by Stephens et al.

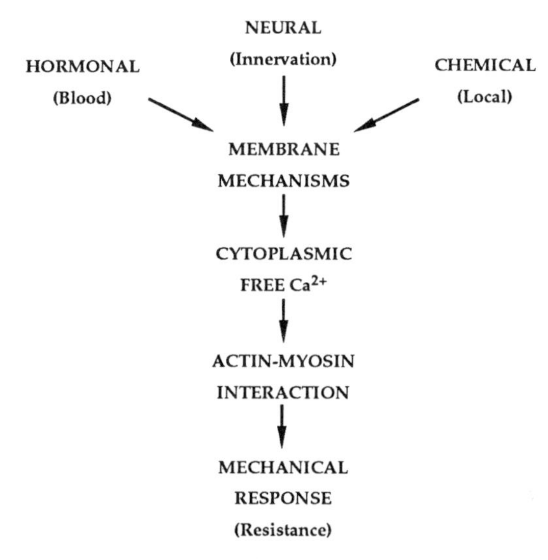

Figure 11. Schematic representation of excitation-contraction coupling in vascular smooth muscle. Various neural, hormonal and chemical signals are integrated at the level of the plasma membrane. These signals ultimately determine the degree of activation of the contractile proteins, actin and myosin, through an effect on cytoplasmic calcium levels thereby determining the mechanical response of the blood vessel.

ACKNOWLEDGMENTS

The authors wish to express their sincere appreciation to Ms. Elaine Veit, Sue VanBuren and Elizabeth Cannon for their tireless participation in the work performed in their laboratories. This work was supported in part by NIH Grants HL27846, HL30496 and HL39388, and by grants from the Southeastern Pennsylvania Affiliate and Delaware Chapter of the American Heart Association.

REFERENCES

Bárány M., 1967, ATPase activity of myosin correlated with speed of muscle shortening, *J. Gen. Physiol.*, 50: 197.

Benga, G. and Holmes, R. P., 1984, Interactions between components in biological membranes and their implications for membrane function, *Prog. Biophys. Mol. Biol.*, 43: 195.

Bialecki R. and Tulenko T. N., 1989, Excess membrane cholesterol alters calcium channel activity in smooth muscle of rabbit carotid artery, *Am. J. Physiol.*, 257: C306.

Bing, R. F., Heagerty, P. M., Thurston, H., and Swales, J. D., 1986, Ion transport in hypertension: Are changes in the cell membrane responsible?, *Clin. Sci.*, 71: 225.

Borochov, H., Zahler, P., Wilbrandt, W., and Shinitzky, M., 1977, Effect of phosphatidylcholine to sphingomyelin mole ratio on the dynamic properties of sheep erythrocytes membrane, *Biochim. Biophys. Acta*, 470: 382.

Broderick, R., Bialecki, R., and Tulenko, T. N., 1989, Cholesterol-induced changes in arterial sensitivity to adrenergic stimulation, *Am. J. Physiol.*, 257: H170.

Carey, R. A., Bove, A. A., Coulson, R. L., and Spann, J. F., 1979, Correlation between cardiac muscle myosin ATPase activity and velocity of shortening, *Biochem. Med.*, 21: 235.

Chacko, S. and Cox, R. H., 1981, Contractile protein content of arteries from normotensive (WKY) and spontaneously hypertensive rats (SHR), *Fed. Proc.*, 40: 575.

Cohen, M. L. and Berkowitz, B. A., 1976, Vascular contraction: Effect of age and extracellular calcium, *Blood Vessels*, 13: 139.

Cox, R. H., 1981, Basis for the altered arterial wall mechanics in the spontaneously hypertensive rat, *Hypertension*, 3: 485.

Cox, R. H., 1982a, Time course of arterial wall changes with DOCA plus salt hypertension in the rat, *Hypertension*, 4: 27.

Cox, R. H., 1982b, Changes in arterial wall properties during development and maintenance of renal hypertension, *Am. J. Physiol.*, 242: H477.

Cox, R. H., 1989, Mechanical properties of arteries in hypertension, *in*: "Blood Vessel Changes in Hypertension: Structure and Function Vol. 1", R. M. K. W. Lee, ed., CRC Press, Boca Raton, p. 65.

Cox, R. H., Katzka, D., and Morad, M., 1990, Characteristics of calcium current in single isolated rabbit portal vein myocytes, *Biophys. J.*, 57: 526a.

Cox, R. H., Katzka, D., and Morad, M., 1990, Characteristics of inactivation of calcium currents in freshly isolated rabbit portal vein myocytes, *FASEB J.*, 4: A441.

Cox, R. H. and Kikta, D. C., 1989, Comparison of norepinephrine (NE) activated ^{86}Rb efflux in arteries from genetically obese Zucker rats, *FASEB J.*, 3: A1009.

Dukes, I. and Morad, M., 1990, Tedisamil inactivates transient outward K^+ current in rat ventricular myocytes, *Am. J. Physiol.*, 257: H1746.

Erne, P. and Hermsmeyer, K., 1989, Intracellular vascular muscle Ca^{2+} modulation in genetic hypertension, *Hypertension*, 14: 145.

Fleisch, J. H., 1980, Age-related changes in the sensitivity of blood vessels to drugs, *Pharmacol. Ther.*, 8: 477.

Folkow, B., Hallback, M., Lundgren, Y., Sivertsson, R., and Weiss, L., 1973, Importance of adaptive changes in vascular design for establishment of primary hypertension, studied in man and in spontaneously hypertensive rats, *Circ. Res.*, 32(Suppl I): 2.

Frohlich, E. D., 1973, Clinical significance of hemodynamic findings in hypertension, *Chest*, 64: 94.

Gleason, M. M. and Tulenko, T., 1989, Excess cholesterol alters calcium fluxes and membrane fluidity in cultured arterial smooth muscle cells, *Circulation*, 80: II-63.

Graham, R. M., Pettinger, W. A., Sagalowsky, A., Brabson, J., and Gandler, T., 1982, Renal alpha-adrenergic receptor abnormality in the spontaneously hypertensive rat, *Hypertension*, 4: 881.

Hauser, H. and Phillips, M. C., 1979, Interactions of the polar head groups of phospholipid bilayer membranes, *in:* "Progress in Surface Membrane Science", D. A. Cadenheaad and J. F. Danielli, eds., Academic Press, New York, p. 297.

Henry, P. D., 1984, Hyperlipidemic arterial dysfunction, *Circulation*, 81: 697.

Hermsmeyer, K., 1984, Altered arterial muscle ion transport mechanism in the spontaneously hypertensive rat, *J. Cardiovasc. Pharmacol.*, 6: Sl0.

Hruza, Z. and Zweifach, B. W., 1967, Effect of age on vascular reactivity to catecholamines in rats, *J. Gerontology*, 22: 469.

Ives, H. E., 1989, Ion transport defects in hypertension. Where is the link?, *Hypertension*, 14: 590.

Johansson, B., 1984, Different types of smooth muscle hypertrophy, *Hypertension*, 6 (Suppl III): III-64.

Jones, A. W., 1973, Altered ion transport in vascular smooth muscle from spontaneously hypertensive rats. Influences of aldosterone, norepinephrine and angiotensin, *Circ. Res.*, 33: 563.

Jones, A. W., 1980, Content and fluxes of electrolytes, *in:* "The Handbook of Physiology; The Cardiovascular System: Vascular Smooth Muscle", D. F. Bohr, A. P. Somlyo, and H. V. Sparks Jr., eds., American Physiological Society, Bethesda, p. 253.

Jones, A. W. and Smith, J. M., 1986, Altered Ca-dependent fluxes of ^{42}K in rat aorta during aldosterone-salt hypertension, *in:* "Recent Advances in Arterial Disease: Atherosclerosis, Hypertension and Vasospasm", T. N. Tulenko and R. H. Cox, eds., Alan R. Liss, New York, p. 265.

Kass, R. S. and Krafte, D. S., 1987, Negative surface charge density near heart calcium channels. Relevance to block by dihydropyridines, *J. Gen. Physiol.*, 89: 629.

Kim, Y. S., Samuel, M., Levin, R. M., and Chacko, S., 1990, Characteristics of the contractile and cytoskeletal proteins of the hypertrophied urinary bladder smooth muscle, *J. Urology*, 143: 35A.

Locher, O. H., Neyes, L., Stimple, M., Kuffer, B., and Vetter, W., 1984, The cholesterol content of the human erythrocyte influences calcium influx through the channel, *Biochem. Biophys. Res. Comm.*, 124: 822.

Madden, T. D., King, M. D., and Quinn, P. J., 1981, The modulation of Ca^{++}-ATPase activity of sarcoplasmic reticulum by membrane cholesterol - the effect of enzyme coupling, *Biochim. Biophys. Acta*, 641: 265.

Matlib, M. A., Schwartz, A., and Yamori, Y., 1985, A Na^{+}-Ca^{2+} exchange process in isolated sarcolemmal membranes of mesenteric arteries of WKY and SHR, *Am. J. Physiol.*, 249: C166.

McMahon, E. G. and Paul, R. J., 1985, Calcium sensitivity of isometric force in intact and chemically skinned aortas during the development of aldosterone-salt hypertension in the rat, *Circ. Res.*, 56: 427.

Medow, M. S. and Segal, S, 1987, Age-related changes in fluidity of rat renal brush border membrane vesicles, *Biochem. Biophys. Res. Comm.*, 142: 849.

Mulvany, M. J. and Nyborg, N., 1980, An increased calcium sensitivity of mesenteric resistance vessels in young and adult spontaneously hypertensive rats, *Br. J. Pharmacol.*, 71: 585.

Mulvany, M. J., 1989, Contractile properties of resistance vessels related to cellular function, *in:* "Blood Vessel Changes in Hypertension: Structure and Function, Vol. 1", R. M. K. W. Lee, ed., CRC Press, Boca Raton, p. 1.

Mulvany, M. J. and Halpern, W., 1977, Contractile properties of small resistance vessels in spontaneously hypertensive and normotensive rats, *Circ. Res.*, 41: 19.

Nishizuka, Y., 1984, Turnover of inositol phospholipids and signal transduction, *Science*, 225: 1365.

Owens, G. K., 1991, Role of contractile agonists in growth regulation of vascular smooth muscle cells, *in:* "Cellular and Molecular Mechanisms in Hypertension", R. H. Cox, ed., Plenum Press, New York, in press.

Packer, C. S. and Stephens, N. L., 1985, Force-velocity relationships in hypertensive arterial smooth muscle, *Can. J. Physiol. Pharmacol.*, 63: 669.

Pagani, E. D. and Julian, F. J., 1984, Rabbit papillary muscle myosin isoenzymes and the velocity of muscle shortening, *Circ. Res.*, 54: 586.

Quinn, P. J., 1980, The fluidity of cell membranes and its regulation, *Prog. Biophys. Mol. Biol.*, 38: 1.

Rock, D. E. and Tulenko, T. N., 1991, Excess membrane cholesterol and atherosclerosis impair ATP-dependent K^{+} channel activation in arterial smooth muscle cells, *FASEB J.*, 5: A532.

Rusch, N. J. and Hermsmeyer, K., 1988, Calcium currents are altered in the vascular muscle cell membrane of spontaneous hypertensive rats, *Circ. Res.*, 63: 997.

Schnitzky, M. and Barenholz, Y., 1978, Fluidity parameters of lipid regions determined by fluorescence polarization, *Biochim. Biophys. Acta.*, 515: 367.

Somlyo, A. P., 1985, Excitation-contraction coupling and the ultrastructure of smooth muscle, *Circ. Res.*, 57: 497.

Stekiel, W. J., Contney, S. J., and Lombard, J. H., 1986, Small vessel membrane potential, sympathetic input, and electrogenic pump rate in SHR, *Am. J. Physiol.*, 250: C547.

Sugiyama, T., Yoshizumi, M., Takaku, F., and Yazaki, Y., 1990, Abnormal calcium handling in vascular smooth muscle cells of spontaneously hypertensive rats, *J. Hypertension*, 8: 369.

Triggle, C. R., 1989, Reactivity and sensitivity changes of blood vessels in hypertension, *in:* "Blood Vessel Changes in Hypertension: Structure and Function, Vol. 1", R. M. K. W. Lee, ed., CRC Press, Boca Raton, p. 25.

Tulenko, T. N., Bialecki, R., Gleason, M., and D'Angelo, G., 1990, Ion channels, membrane lipids and cholesterol: A role for membrane lipid domains in arterial function, *in:* " Potassium Channels: Basic Function and Therapeutic Aspects", T. Colatsky, ed., Alan R. Liss, New York, p. 187.

Tulenko, T. N., Rabinowitz, J. L., Cox, R. H., and Santamore, W. P., 1988, Altered Na^+/K^+-ATPase, cell Na^+ and lipid profiles in canine arterial wall with chronic cigarette smoking, *Eur. J. Biochem.*, 20: 285.

Tulenko, T. N., Lapatofsky, D., and Cox, R. H., 1988, Alterations in membrane phospholipid bilayer composition with age in the Fisher 344 rat, *Physiologist*, 31: A138.

Upadhya, A., Fariel, M. R., Bagshaw, R. J., Cox, R. H., and Chacko, S., 1986, Alteration of the contractile proteins of the arterial muscle in hypertension, *Fed. Proc.*, 45: 1074.

van Blitterswijk, W. B., van der Meer, B., and Hilkmann, H., 1987, Quantitative contributions of cholesterol and the individual classes of phospholipids and their degree of fatty acyl (un)saturation to membrane fluidity measured by fluorescence polarization, *Biochemistry*, 26: 1746.

Webb, R. C. and Bohr, D. F., 1978, Mechanism of membrane stabilization by calcium in vascular smooth muscle, *Am. J. Physiol.*, 235: C227.

Whall Jr., C. W., Myers, M. M., and Halpern, W., 1980, Norepinephrine sensitivity, tension development and neuronal uptake in resistance arteries from spontaneously hypertensive and normotensive rats, *Blood Vessels*, 17: 1.

White, R. E. and Carrier, G. O., 1988, Enhanced alpha-adrenergic neuroeffector system in diabetes: importance of calcium, *Am. J. Physiol.*, 255: H1036.

MANY MEMBRANE ABNORMALITIES IN HYPERTENSION RESULT FROM ONE PRIMARY DEFECT

David F. Bohr, Philip B. Furspan, and Anna F. Dominiczak

Department of Physiology
University of Michigan
School of Medicine,
Ann Arbor, MI 48109

INTRODUCTION

Ever since experimental tools have been available to study cellular mechanisms responsible for hypertension, special attention has been given to the possible role of an abnormality of the cell membrane in this process. This focus evolved not only because all regulatory processes of the cell must be mediated through a cell membrane function, but also because specific clues suggested that the cell membrane might be at fault. Both experimental and clinical hypertension were shown to be influenced by salt intake, implying that some subjects had a defect in regulating salt metabolism, perhaps a problem of cell membrane permeability to sodium or chloride. This idea was supported by the observation that mineralocorticoid excess, which was known to alter membrane permeability to sodium, caused hypertension.

Based on these clues many physiological studies have been undertaken to determine whether functional differences could be observed that would reflect membrane abnormalities in vascular muscle from hypertensive animals. Increases in responsiveness to many constrictor agonists have been reported (Holloway and Bohr, 1973; Moreland et al., 1984; Webb and Bohr, 1984; Turla and Webb, 1987; Bruner et al., 1989; Joshua and Bohr, submitted). These include norepinephrine, angiotensin, serotonin, endothelin, phorbol ester (an activator of protein kinase C) and Bay K 8644 (a calcium channel opener). From the extensiveness and diversity of these agonists it was inferred that the vascular smooth muscle membrane is generally more excitable in hypertension. Another type of vascular smooth muscle activity that appears to be a characteristic of hypertension is its oscillatory activity. Webb and his associates (Lamb et al., 1985) have found that this abnormality reflects differences in membrane permeability to potassium and to calcium and is driven by changes in the membrane potential. These investigators have recently observed this oscillatory activity in vascular smooth muscle from patients with essential hypertension (personal communication).

Even when vascular smooth muscle from the hypertensive animal does not evidence this oscillatory activity, it may be unique in that it maintains a

Regulation of Smooth Muscle Contraction
Edited by R.S. Moreland, Plenum Press, New York, 1991

myogenic tone when mounted in a physiological salt solution in which normal vascular smooth muscle remains completely relaxed (Noon et al., 1978; Suzuki et al., 1979; Soltis and Bohr, 1987). This maintained tone of vascular smooth muscle from hypertensive animals is the result of an influx of extracellular calcium through a membrane that is abnormally permeable to this ion.

This observation introduces what is probably the most relevant abnormality of specific electrolyte transport systems in vascular smooth muscle in hypertension. Several different types of observations have documented this abnormality of calcium channels. Van Breemen et al. (1986) observed an increase in ^{45}Ca efflux from vascular smooth muscle of spontaneously hypertensive rats (SHR) through three types of channels: leak, voltage sensitive, and receptor operated. Aoki and Asano (1986) and Storm et al. (1990) reported that the voltage sensitive calcium channels were more sensitive than normal to the calcium agonist Bay K 8644. Rush and Hermsmeyer (1988) using patch clamp techniques observed a greater activity of the tonic calcium channel in vascular smooth muscle from hypertensive rats.

INDIVIDUAL TRANSPORT SYSTEMS

The literature related to abnormalities of ion transport systems and permeability associated with hypertension is extensive: among the transport processes affected are the Na^+-H^+ antiporter, Na^+-Ca^{2+} exchange, Ca^{2+}, Mg^{2+} ATPase, Na^+-K^+ ATPase, and Na^+, K^+, Ca^{2+} and Cl^- permeabilities (see Fig. 1).

The hypothesis that an abnormality in Na^+-H^+ exchange might play a role in the pathogenesis of hypertension was first proposed (Aaronson, 1983) after Canessa et al. (1980) found that erythrocytes from patients with essential hypertension have increased Na^+-Li^+ exchange activity when compared to normotensive subjects. Subsequent studies have reported similar differences in the activity of the Na^+-H^+ exchanger in smooth muscle cells from stroke-prone SHR (SHRSP) compared to those from Wistar-Kyoto (WKY) rats (Berk et al., 1989) and in platelets from hypertensive patients compared to normal control subjects (Livne et al., 1987).

Thompson et al. (1990) investigated the activity of the Na^+-Ca^{2+} exchanger in rings from tail arteries of SHRSP and WKY rats. After the rings were contracted with 1 μM norepinephrine, they placed them in calcium-free

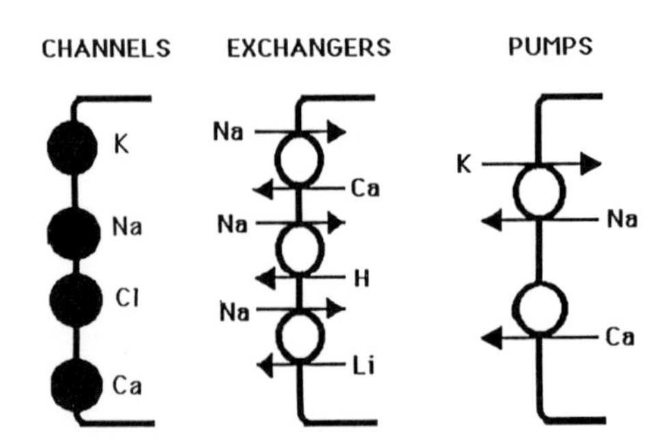

Figure 1. Electrolyte transport systems found to be abnormal in hypertension. Here they are categorized into channels, exchangers and active pumps.

physiological salt solution containing vanadate (to inhibit the calcium extrusion pump), ryanodine (to inhibit calcium uptake by the sarcoplasmic reticulum), and no sodium (to inhibit Na^+-Ca^{2+} exchange). To test the activity of the Na^+-Ca^{2+} exchanger, sodium was added back to the bath solution and the time to 50% relaxation was measured. The arteries from the SHRSP took more than twice as long to achieve 50% relaxation as did those from the WKY rats. The authors concluded that the activity of the Na^+-Ca^{2+} exchanger is depressed in tail arteries of SHRSP.

Cell membrane Ca^{2+}, Mg^{2+} ATPase activity has been reported to be altered in both essential and experimental hypertension. Takaya et al. (1990) studied a variety of parameters of the Ca^{2+}, Mg^{2+} ATPase activity in platelets isolated from essential hypertensives and normotensive controls. The maximal initial reaction velocity (V_{max}) of the ATPase was significantly lower in the platelets from the hypertensives compared to those from the normotensives (14.99 ± 1.71 nmol inorganic phosphate (P_i)/mg protein/min and 27.54 ± 4.37 nmol P_i/mg protein/min, respectively).

A greater relaxation in response to potassium in tail arteries from SHR, 1K-1C renal hypertensive rabbits, and DOCA-hypertensive pigs has been interpreted to reflect greater activity of vascular smooth muscle cell membrane Na^+-K^+ ATPase in the hypertensive animals (Webb and Bohr, 1979; Hagen et al., 1982; Webb, 1982). This interpretation is supported by a study by Allen and Seidel (1977), who reported that Na^+-K^+ ATPase activity in microsomal fractions from aortas of SHR was increased (28 moles P_i/mg protein/hr) as compared with that in aortic microsomal fractions of normotensive rats (18 moles P_i/mg protein/hr).

Increased membrane permeability to sodium has been reported for a variety of cell types in both essential and experimental hypertension (for review, see Furspan and Bohr, 1988). In a study from our laboratory (Furspan and Bohr, 1985), we measured net Na^+ influx at 4° C in lymphocytes from rats with either genetic (SHRSP), mineralocorticoid (DOCA), or renal (two kidney-one clip) hypertension. Lymphocytes from the SHRSP and DOCA rats exhibited elevated net Na^+ influx (43 and 35%, respectively) when compared to appropriate controls.

Jones et al. (1973; 1975) have reported greater membrane permeability to potassium in large and small blood vessels from SHR and from rats with DOCA hypertension than in those from normotensive controls. We (Furspan and Bohr, 1985) found that net potassium efflux at 4° C is greater in lymphocytes from SHRSP than in those from WKY.

Evidence for increased cell membrane permeability to calcium associated with genetic hypertension has been reported for lymphocytes. Bruschi et al. (1986) used the intracellular fluorescent calcium indicator quin2 to measure the rate of accumulation of calcium in vanadate-treated (calcium extrusion pump inhibitor) and calcium-depleted (incubation in calcium-free solution) lymphocytes after calcium was added back to the solution. Intracellular calcium concentration increased at a faster rate in the lymphocytes from SHR than in those from WKY. This observation suggests a greater membrane permeability to calcium associated with this type of hypertension.

The evidence for changes in membrane permeability to chloride associated with hypertension is sparse, mainly consisting of Jones's demonstration of increased turnover of this ion in vascular smooth muscle from SHR and DOCA hypertensive rats (Jones, 1973; Jones and Hart, 1975).

A case has been made for the primacy of many of these abnormalities in the pathogenesis of hypertension. Although it is possible that such a multiplicity of primary membrane defects are present in hypertension, there is strong evidence to suggest that these alterations in membrane transport are secondary to underlying changes in the physicochemical properties of the cell membrane. More specifically, Bing et al. (1986) have proposed that the disturbances in ion transport and in other membrane functions which contribute to the increased vascular reactivity associated with hypertension are related to alterations in cell membrane lipids.

Recognizing the diversity of the integral proteins that have been shown to function abnormally in hypertension it seems unlikely that there should be this many defects in the expression of the individual proteins involved. Recently, the possibility has been recognized that the primary defect may reside in the matrix in which all of the transport proteins operate. This is the lipid bilayer of the cell membrane.

The fluid mosaic model of the cell membrane, proposed by Singer and Nicolson in 1972, comprises protein macromolecules embedded in a lipid bilayer. The role of the cell membrane is to maintain a permeability barrier and to transmit information. Phospholipids constitute the predominant structural element of membranes. The hydrophobic character of fatty acids of the phospholipids make membranes impermeable to hydrophilic molecules. The integral proteins span the membrane completely, thus a localized event at the outer surface of the membrane can be coupled with an event at the inner side. This property permits integral proteins to play an important role as membrane receptors or channels, regulating agonist-cell interactions and the transport of ions across the membrane (Carruthers and Melchior, 1986; Yeagle 1989). The regulation of these functions of the integral proteins by membrane lipids has become a subject of many current studies. Evidence has been presented that Na^+-K^+ ATPase activity is inhibited by high levels of membrane cholesterol in several cell types (Yeagle, 1983; 1989; Yeagle et al., 1988). This inhibition of the electrogenic sodium pump has been demonstrated as a result of excess cholesterol in the diet or *in vitro* enrichment of cell membranes using cholesterol-rich liposomes. Broderick et al. (1989) have demonstrated a five-fold increase in norepinephrine sensitivity in the rabbit femoral arteries perfused with the cholesterol-enriched liposomes compared with control arteries perfused with the liposome medium without added cholesterol. Bialecki and Tulenko (1989) presented evidence that the increased contractile sensitivity to norepinephrine in cholesterol-enriched smooth muscle is dependent on increased Ca^{2+} influx, and Locher et al. (1984) showed a direct relationship between the cholesterol-to-phospholipid molar ratio in red blood cell and increased $^{45}Ca^{2+}$ influx. Other direct effects attributed to cholesterol include: reduced membrane fluidity (Criado et al., 1982; Yeagle, 1989), altered activity of certain membrane receptors (Criado et al., 1982), and reduced calcium ATPase activity of sarcoplasmic reticulum membranes (Madden et al., 1979). Specific phospholipids of the bilayer have also been shown to regulate membrane protein function (Yeagle, 1989). Phosphatidylethanolamine stimulates the activity of Ca^{2+} ATPase possibly through interactions between the bilayer surface and the extramembranous portions of the calcium pump protein (Yeagle and Sen, 1986).

The negatively-charged phospholipids such as phosphatidic acid and cardiolipin have been investigated as calcium binding sites in membranes

(Zolese and Curatola, 1989). Phase separation, bilayer-non-bilayer transition, changes of membrane permeability and fusion have all been shown to result from calcium interactions with the membrane (Weiss, 1986; Zolese and Curatola, 1989). Moreover, calcium binding to vascular smooth muscle membranes is related to availability of negatively-charged phospholipids (Weiss, 1986).

Indirect manipulation of membrane lipids by changing dietary intake of fatty acids has provided a considerable body of evidence demonstrating the effect of such changes on membrane transport systems, especially the various ATPase "pumps" (for review, see Wahle, 1983). In a recent study, Swanson et al. (1989) described the effects of changes in dietary lipid intake on Ca^{2+}, Mg^{2+} ATPase of cardiac tissue in mice. A diet high in n-3 polyunsaturated fatty acids led to an increase in the n-3/n-6 fatty acid ratio of cell membranes and a corresponding decrease in sarcoplasmic reticulum ATPase activity and calcium transport.

Although the hypotensive effects of diets high in polyunsaturated fatty acids have been observed in both experimental (MacDonald et al., 1980; Hoffman et al., 1985) and human hypertension (Rao et al., 1981) there have been relatively few attempts to relate this effect on blood pressure to altered ion transport effects of this diet. In a study of normotensive men, Heagerty et al. (1986b) found that a diet high in linoleic acid led to a significant rise in the mean total and ouabain-sensitive sodium efflux of leukocytes and a small decrease in blood pressure. Quite different results were reported by Murray et al. (1986) in a study utilizing the spontaneously hypertensive rat. They found that a diet high in polyunsaturated fatty acids led to a significantly higher blood pressure than the diet low in these fatty acids and that this greater blood pressure was associated with a lower ouabain-sensitive efflux rate constant in thymocytes. Kawahara et al. (1990) recently reported that the ability of a diet high in linoleic acid to attenuate the development of deoxycorticosterone acetate-salt hypertension was associated with altered electrolyte content and fluxes in the aorta, lymphocytes, and erythrocytes.

Strong evidence that the altered membrane environment associated with hypertension is responsible for an alteration in an ion transport process is provided in a study by Adeoya et al. (1989). Although they found that Ca^{2+}-dependent ATP hydrolysis activity of erythrocyte ghost membranes was reduced in SHR compared to WKY rats, there was no significant difference in this parameter between SHR and WKY after detergent solubilization of these membranes. The authors interpret this result to indicate that the reduced activity of Ca^{2+}, Mg^{2+} ATPase in intact SHR membrane is due to alterations in the lipid bilayer.

During the past decade the observation has been well documented that the cell membrane of various tissues from hypertensive animals bind less calcium than do tissues from normotensive controls (Kwan et al., 1979; Postnov et al., 1980; Devynck et al., 1981; 1982). The consequences of altered binding of calcium to cell membranes are well recognized in relation to the phenomenon of membrane stabilization (Rothstein, 1968). Webb and Bohr (1978) observed that elevations of extracellular calcium concentrations above physiological levels caused relaxation of vascular smooth muscle. They attributed this effect of calcium to its ability to "stabilize" the cell membrane. Holloway and Bohr (1973) reported that this relaxing effect of high calcium concentrations was less effective in vascular smooth muscle from renal, DOCA, and genetic hypertensive rats than in their normotensive controls. Hansen and Bohr (1975) suggested that this difference was due to the impaired calcium binding of the membrane in hypertension. Alterations in membrane bound calcium also affects ion permeability. Jones and Hart (1975) and Furspan and Bohr (1986) studied the effect of varying extracellular calcium concentrations on potassium efflux from vascular smooth muscles and lymphocytes, respectively. They found that as extracellular calcium concentration was increased potassium efflux decreased, i.e., membrane stabilization by calcium decreased membrane permeability to potassium. In addition, it was noted that this effect was less pronounced (potassium efflux is greater) in cells from hypertensive rats than those from controls. The plasma membrane of these tissues from the hypertensive rats is less well stabilized by calcium than is that from the normotensive controls.

Some speculation about the molecular basis of this abnormality has focussed on the possibility that plasma membranes from hypertensive animals possessed less of a calcium-binding protein than those from normotensive controls. This hypothesis appeared to be confirmed by Kowarski et al. (1986) who reported that various tissues from SHR contained less of an "integral membrane binding protein" (IMCAL) that binds calcium with high affinity than do those from WKY rats. The affinity of this protein for calcium, however, was too high: the binding sites would be saturated at physiological concentrations of extracellular calcium. Indeed, a relatively minor role for proteins in the binding of calcium to cell membranes was suggested in a study of calcium binding to isolated rat liver membranes (Shlatz and Marinetti, 1972). It was concluded that of proteins, phospholipids, and neuraminic acid, "phospholipids are the most important components which bind Ca^{2+}", especially at low affinity sites. Consequently, we now believe that the alteration in membrane calcium-binding ability associated with experimental and essential hypertension is related to an altered characteristic of the lipid bilayer.

Recent studies have documented that one or more of the known second messenger systems (cyclic nucleotides, inositol trisphosphate, Ca^{2+}, diacylglycerol, protein kinase C, prostaglandins) may be altered in hypertension (for review see Hamet and Tremblay, 1989). Below we have presented a few examples of such abnormalities.

The signal initiated by the binding of an agonist to its specific membrane receptor is transmitted by G-proteins to membrane-associated phospholipases (Neer and Clapham, 1988). The activated phospholipase C hydrolyses phosphatidylinositol 4,5 bisphosphate (PIP_2) forming two products that have second messenger functions in the cell: inositol trisphosphate (IP_3) which enters the cytosol and elicits the release of Ca^{2+} from sarcoplasmic reticulum and diacylglycerol which remains within the cell membrane and increases the affinity of protein kinase C for calcium (Berridge and Irvine, 1984; Rasmussen et al., 1987; van Breemen and Saida, 1989). Increased turnover rates of phosphoinositides have been demonstrated in erythrocyte membranes (Koutouzov et al., 1983), platelets (Koutouzov et al., 1987), and aortae (Heagerty et al., 1986a; Turla and Webb, 1990) from genetically hypertensive rats. Turla and Webb (1990) described increased accumulation of inositol phosphates following treatment with serotonin or norepinephrine in aortae from stroke-prone SHR. In addition, the enhanced activity of the protein kinase C branch of the phosphoinositide system has been reported in genetic hypertension (Turla and Webb, 1987). It is of interest that diacylglycerol is capable not only of activating protein kinase C, but also of disrupting the lipid bilayer of the membranes in which it is produced (Siegel et al., 1989; Yeagle 1989). This is dictated by the ability of diacylglycerol to reduce dramatically the temperature of the lamellar-hexagonal (II) phase transition of the lipid bilayer. The formation of the hexagonal (II) phase results in transient destabilization of the bilayer structure, which could be related to changes in transmembrane ionic fluxes, cell pH, differentiation, and growth. Indeed, diacylglycerol formation is increased in erythrocytes of SHR (Kato and Takenawa, 1987). More recently, Okumura et al. (1990) have presented evidence for increased 1,2-diacylglycerol production in thoracic aortae of 4 week old SHR as compared to their normotensive controls. The classical pathway involving PIP_2 as the sole source of diacylglycerol has been challenged recently and evidence has been presented for the agonist-induced generation of diacylglycerol from phosphatidylcholine (Besterman et al., 1986; Cabot et al., 1988).

The physical state or "fluidity" of the phospholipid bilayer is a useful index of the influence that this matrix may have on the integral proteins. This fluidity, measured as lateral movement of fluorescent dyes in the membrane, is reduced in erythrocytes and platelets from hypertensive patients (Orlov and Postnov, 1982; Naftilan et al., 1986) and in erythrocytes from the SHR (Orlov et al., 1982). In conjunction with reduced fluidity, platelet (Naftilan et al., 1986) and erythrocyte (Ollerenshaw et al., 1987) membranes contained less unsaturated fatty acids. Nara et al. (1986) studied lipid composition in cultured smooth muscle cells from SHRSP. They found decreased cholesterol:phospholipid ratio and decreased levels of linoleic, arachidonic, and eicosapentaenoic acid in phospholipids extracted from cells from SHRSP as compared to those from WKY. Moreover, Marche and his coworkers (Remmal et al., 1988; Marche et al., 1990) reported markedly increased phosphatidylcholine turnover in quiescent platelets from SHR, SHRSP, and Dahl salt-sensitive rats as compared to their normotensive control strains. The enhanced turnover of this most abundant membrane phospholipid, also implicated in signal transduction and diacylglycerol formation, may contribute to a primary membrane abnormality in genetic hypertension.

SUMMARY

Evidence has been presented that:

1.) Changes in lipid bilayer alter the function of integral membrane proteins.

2.) There is less calcium bound to the plasma membrane in hypertension.

3.) Structural and functional abnormalities of the lipid bilayer have been reported in genetic hypertension.

We hypothesize that multiple abnormalities of membrane transport systems in hypertension are secondary to an inherent abnormality of the lipid bilayer in which these transport proteins reside.

ACKNOWLEDGMENTS

This work was done during the tenure of a British-American Research Fellowship of the American Heart Association and the British Heart Foundation (AFD). It was also supported by the National Heart, Lung, and Blood Institute grant HL-18575.

REFERENCES

Adeoya, A. S., Norman, R. I., and Bing, R. F., 1989, Erythrocyte membrane calcium adenosine 5'-triphosphatase activity in the spontaneously hypertensive rat, *Clin. Sci.*, 77: 395.

Allen, J. C. and Seidel, C., 1977, EGTA stimulated and ouabain inhibited ATPase of vascular smooth muscle, *in:* "Excitation-Contraction Coupling in Smooth Muscle", R. Casteels, T. Goodfraind, and J. C. Rüegg, eds., Elsevier/North Holland, Amsterdam, p. 211.

Aoki, K. and Asano, M., 1986, Effect of BAY K 8644 and nifedipine on femoral arteries of spontaneously hypertensive rats, *Br. J. Pharmacol.*, 88: 221.

Aaronson, P. S., 1983, Red cell sodium-lithium countertransport and essential hypertension, *N. Eng. J. Med.*, 307: 317.

Berk, B. C., Vallega, G., Muslin, A. J., Gordon, H. M., Canessa, M., and Alexander, R. W., 1989, Spontaneously hypertensive rat vascular smooth muscle cells in culture exhibit increased growth and Na^+-H^+ exchange, *J. Clin. Invest.*, 83: 822.

Berridge, M. J. and Irvine, R. F., 1984, Inositol triphosphate, a novel second messenger in cellular signal transduction, *Nature*, 312: 315.

Besterman, J. M., Duronio, V., and Cuatrecasas, P., 1986, Rapid formation of diacylglycerol from phosphatidylcholine: A pathway for generation of a second messenger, *Proc. Nat'l. Acad. Sci. U.S.A.*, 83: 6785.

Bialecki, R. A. and Tulenko, T. N., 1989, Excess membrane cholesterol alters calcium channels in arterial smooth muscle, *Am. J. Physiol.*, 257: C306.

Bing, R. F., Heagerty, A. M., Thurston, H., and Swales, J. D., 1986, Ion transport in hypertension: Are changes in the cell membrane responsible?, *Clin. Sci.*, 71: 225.

Broderick, R., Bialecki, R., and Tulenko, T. N., 1989, Cholesterol-induced changes in rabbit arterial smooth muscle sensitivity to adrenergic stimulation, *Am. J. Physiol.*, 257: H170.

Bruner, C. A., Webb, R. C., and Bohr, D. F., 1989, Vascular reactivity and membrane stabilizing effect of calcium in spontaneously hypertensive rats, *in:* "Calcium in Essential Hypertension", K. Aoki and E. D. Frolich, eds., Academic Press, Tokyo, p. 275.

Bruschi, G., Bruschi, M. E., Cavatorta, A., and Borghetti, A., 1986, The mechanism of Ca increase in blood cells of spontaneously hypertensive rats. *J. Cardiovasc. Pharmacol.*, 8 (Suppl 8): S139.

Cabot, M. C., Welsh, C. J., Cao, H., and Chabbott, H., 1988, The phosphatidylcholine pathway of diacylglycerol formation stimulated by phorbol diester occurs via phospholipase D activation, *FEBS Lett.*, 233: 153.

Canessa, M., Adragna, N., Solomon, H. S., Connolly, T., and Tosteson, D. C., 1980, Increased sodium-lithium countertransport in red cells of patients with essential hypertension, *N. Eng. J. Med.*, 302: 772.

Carruthers, A. and Melchior, D. L., 1986, How bilayer lipids affect membrane protein activity, *Trends Biochem. Sci.*, 11: 331.

Criado, M., Eibl, H., and Barrantes, F. J., 1982, Effects of lipid on acetylcholine receptor: Essential need for cholesterol for maintenance of agonist-induced state transition in lipid vesicles, *Biochemistry*, 21: 362.

Devynck, M. A., Pernollet, M. G., Nunez, A. M., and Meyer, P., 1981, Analysis of calcium handling in erythrocyte membranes of genetically hypertensive rat, *Hypertension*, 3: 397.

Devynck, M. A., Pernollet, M. G., Nunez, A, M., Aragon, I., Montenay-Garestier, T., Helene, C., and Meyer, P., 1982, Diffuse structural alteration in cell membranes of spontaneously hypertensive rats, *Proc. Nat'l. Acad. Sci. U.S.A.*, 79: 5057.

Furspan, P. B. and Bohr, D. F., 1985, Lymphocyte abnormalities in three types of hypertension in the rat, *Hypertension*, 7: 860.

Furspan, P. B. and Bohr, D. F., 1986, Calcium related abnormalities in lymphocytes from genetically hypertensive rats, *Hypertension*, 8 (Suppl II): II-123.

Furspan, P. B. and Bohr, D. F., 1988, Cell membrane permeability in hypertension, *Clin. Physiol. Biochem.*, 6: 122.

Hagen, E. C., Johnson, J. C., and Webb, R. C., 1982, Ouabain binding and potassium relaxation in aortae from renal hypertensive rabbits, *Am. J. Physiol.*, 12: H896.

Hamet, P. and Tremblay, J., 1989, Abnormalities of second messenger systems in hypertension, *in:* "Blood cells and arteries in hypertension and atherosclerosis", P. Meyer and P. Marche, eds., Raven Press, New York, p. 171.

Hansen, T. R. and Bohr, D. F., 1975, Hypertension, transmural pressure, and vascular smooth muscle response in rats, *Circ. Res.*, 36: 590.

Heagerty, A. M., Ollerenshaw, J. D., and Swales, J. D., 1986a, Abnormal vascular phosphoinositide hydrolysis in the spontaneously hypertensive rat, *Br. J. Pharmacol.*, 89: 803.

Heagerty, A. M., Ollerenshaw, J. D., Robertson, D. I., Bing, R. F., and Swales, J. D., 1986b, Influence of dietary linoleic acid on leucocyte sodium transport and blood pressure, *Br. Med. J.*, 293: 295.

Hoffman, P., Taube, C., and Heinroth-Hoffman, I., 1985, Antihypertensive action of dietary polyunsaturated fatty acids in spontaneously hypertensive rats, *Arch. Int. Pharmacodyn. Ther.*, 276: 222.

Holloway, E. T. and Bohr, D.F., 1973, Reactivity of vascular smooth muscle in hypertensive rats, *Circ. Res.*, 33: 678.

Jones, A. W., 1973, Altered ion transport in vascular smooth muscle from spontaneously hypertensive rats. Influences of aldosterone, norepinephrine, and angiotensin, *Circ. Res.*, 33: 563.

Jones, A. W. and Hart, R. G., 1975, Altered ion transport in aortic smooth muscle during deoxycorticosterone acetate hypertension in the rat, *Circ. Res.*, 37: 333.

Joshua, I. G. and Bohr, D. F., Increased vascular reactivity to endothelin in genetically hypertensive rats, *Hypertension*, submitted.

Kato, H. and Takenawa, T., 1987, Phospholipase C activation and diacylglycerol kinase inactivation lead to an increase in diacylglycerol content in spontaneously hypertensive rat, *Biochem. Biophys. Res. Commun.*, 146: 1419.

Kawahara, J., Sano, H., Yoshihisa, K., Hattori, K., Miki, T., Suzuki, H., and Fukuzaki, H., 1990, Dietary linoleic acid prevents the development of deoxycorticosterone acetate-salt hypertension, *Hypertension*, 15 (Suppl I): I-81.

Koutouzov, S., Marche, P., Girad, A., and Meyer, P., 1983, Altered turnover of polyphosphoinositides in the erythrocyte membrane of the spontaneously hypertensive rat, *Hypertension*, 5: 409.

Koutouzov, S., Remmal, A., Marche, P., and Meyer, P., 1987, Hypersensitivity of phospholipase C in platelets of spontaneously hypertensive rats, *Hypertension*, 10: 497.

Kowarski, S., Cowen, L. A., and Schachter, D., 1986, Decreased content of integral membrane calcium-binding protein (IMCAL) in tissues of the spontaneously hypertensive rat, *Proc. Nat'l. Acad. Sci. U.S.A.*, 83: 1097.

Kwan, C. Y., Belbeck, L., and Daniel, E. E., 1979, Abnormal biochemistry of vascular smooth muscle plasma membrane as an important factor in the initiation and maintenance of hypertension in rats, *Blood Vessels*, 16: 259.

Lamb, F. S., Myers, J. H., Hamlin, M. N., and Webb, R. C., 1985, Oscillatory contractions in tail arteries from genetically hypertensive rats, *Hypertension*, 7 (Suppl. I): I-25.

Livne, A., Veitch, R., Grinstein, S., Balfe, J. W., Marquez-Julio, A., and Rothstein, A., 1987, Increased platelet Na^{+}-H^{+} exchange rates in essential hypertension: Application of a novel test, *Lancet*, 1: 553.

Locher, R., Neyses, M., Stimpel, M., Kuffer, B., and Vetter, W., 1984, The cholesterol content of the human erythrocyte influences calcium influx through the channel, *Biochem. Biophys. Res. Commun.*, 124: 822.

MacDonald, M. C., Kline, R. L., and Mogenson, G. J., 1980, Dietary linoleic acid and salt-induced hypertension, *Can. J. Physiol. Pharmacol.*, 59: 872.

Madden, T. D., Chapman, D., and Quinn, P. J., 1979, Cholesterol modulates activity of calcium-dependent ATPase of the sarcoplasmic reticulum, *Nature*, 279: 538.

Marche, P., Limon, I., Blanc, J., and Girard, A., 1990, Platelet phosphatidylcholine turnover in experimental hypertension, *Hypertension*, 16: 190.

Moreland, R. S., Lamb, F. S., Webb, R. C., and Bohr, D. F., 1984, Functional evidence for increased sodium permeability in aortas from DOCA hypertensive rats, *Hypertension*, 6 (Suppl I): I-88.

Murray, G. E., Nair, R., and Patrick, J., 1986, The effect of dietary polyunsaturated fat on cation transport and hypertension in the rat, *Br. J. Nutrition*, 56: 587.

Naftilan, A. J., Dzau, V. J., and Loscalzo, J., 1986, Preliminary observations on abnormalities of membrane structure and function in essential hypertension, *Hypertension*, 8 (Suppl II): II-174.

Nara, Y., Sato, T., Mochizuki, S., Mano, M., Horie, R., and Yamori, Y., 1986, Metabolic dysfunction in smooth muscle cells of spontaneously hypertensive rats, *J. Hypertension*, 4 (Suppl III): S105.

Neer, E. J. and Clapham, D. E., 1988, Role of G protein subunits in transmembrane signalling, *Nature*, 333: 129.

Noon, J. P., Rice, P. I., and Baldassani, R. J., 1978, Calcium leakage as a cause of high resting tension in vascular smooth muscle from spontaneously hypertensive rat, *Proc. Nat'l. Acad. Sci. U.S.A.*, 75: 1605.

Okumura, K., Kondo, J., Shirai, Y., Muramatsu, M., Yamada, Y., Hashimoto, H., and Ito, T.. 1990, 1,2-diacylglycerol content in thoracic aorta of spontaneously hypertensive rats, *Hypertension*, 16: 43.

Ollerenshaw, J. D., Heagerty, A. M., Bing, R. F., and Swales, J. D., 1987, Abnormalities of erythrocyte membrane fatty acid composition in human essential hypertension, *J. Human Hypertension*, 1: 9.

Orlov, S. N. and Postnov, Y. V., 1982, Ca^{2+} binding and membrane fluidity in essential and renal hypertension, *Clin. Sci.*, 63: 281.
Orlov, S. N., Gulak, P. V., Litvinov, I. S., and Postnov, Y. V., 1982, Evidence of altered structure of the erythrocyte membrane in spontaneously hypertensive rats, *Clin. Sci.*, 63: 43.
Postnov, Y. V., Orlov, S. N., and Pokudin, N. J., 1980, Decrease of calcium binding by red blood cell membrane in spontaneously hypertensive rats and in essential hypertension, *Pflügers Arch.*, 385: 191.
Rao, R. H., Rao, V. B., and Stikantia, S. G., 1981, Effect of polyunsaturated-rich vegetable oils on blood pressure in essential hypertension, *Clin. Exp. Hypertension*, 3: 27.
Rasmussen, H., Tukuwa, Y., and Park, S., 1987, Protein kinase C in the regulation of smooth muscle contraction, *FASEB J.*, 1: 177.
Remmal, A., Koutouzov, S., and Marche, P., 1988, Enhanced turnover of phosphatidylcholine in platelets of hypertensive rats. Possible involvement of a phosphatidylcholine-specific phospholipase C, *Biochim. Biophys. Acta*, 690: 236.
Rothstein, A., 1968, Membrane phenomena, *Ann. Rev. Physiol.*, 30: 15.
Rush, N. J. and Hermsmeyer, K., 1988, Calcium currents are altered in the vascular muscle cell membrane of spontaneously hypertensive rats, *Circ. Res.*, 63: 997.
Shlatz, L. and Marinetti, G. V., 1972, Calcium binding to the rat liver plasma membrane, *Biochim. Biophys. Acta*, 290: 70.
Siegel, D. P., Banschbach, J., Alford, D., Ellens, H., Lis, L. J., Quinn, P. J., Yeagle, P. L., and Bentz, J., 1989, Physiological levels of diacylglycerols in phospholipid membranes induce membrane fusion and stabilize inverted phases, *Biochemistry*, 28: 3703.
Singer, S. J. and Nicolson, G. L., 1972, The fluid mosaic model of the structure of cell membranes, *Science*, 175: 720.
Soltis, E. E. and Bohr, D. F., 1987, Vascular reactivity in the spontaneously hypertensive stroke-prone rat: Effect of antihypertensive treatment, *Hypertension*, 9: 492.
Storm, D. S., Turla, M. B., Todd, K. M., and Webb, R. C., 1990, Calcium and contractile responses to phorbol esters and the calcium channel agonist, Bay K 8644, in arteries from hypertensive rats, *Am. J. Hypertension*, 3: 245S.
Suzuki, A., Yanagawa, T., and Tajiri, T., 1979, Effects of some smooth muscle relaxants on the tonus and on the actions of contractile agents in isolated aorta of SHRSP, *Jpn. Heart J.*, 20 (Suppl 1): 219.
Swanson, J. E., Lokesh, B. R., and Kinsella, J. E., 1989, Ca^{2+}-Mg^{2+} ATPase of mouse cardiac sarcoplasmic reticulum is affected by membrane n-6 and n-3 polyunsaturated fatty acid content, *J. Nutrition*, 119: 364.
Takaya, J., Lasker, N., Bamforth, R., Gutkin, M., Byrd, L. H., and Aviv, A., 1990, Kinetics of Ca^{2+}-ATPase activation in platelet membranes of essential hypertensives and normotensives, *Am. J. Physiol.*, 258: C988.
Thompson, L. E., Rinaldi, G. J., and Bohr, D. F., 1990, Decreased activity of the sodium-calcium exchanger in tail artery of stroke-prone spontaneously hypertensive rats, *Blood Vessels*, 27: 197.
Turla, M. B. and Webb, R. C., 1987, Enhanced vascular reactivity to protein kinase C activators in genetically hypertensive rats, *Hypertension*, 9 (Suppl III): III-150.
Turla, M. B. and Webb, R. C., 1990, Augmented phosphoinositide metabolism in aortas from genetically hypertensive rats, *Am. J. Physiol.*, 258: H173.

Van Breemen, C., Leijten, P., Yamamoto, H., Aaronson, P., and Cauvin, C., 1986, Calcium activation of vascular smooth muscle, *Hypertension,* 8 (Suppl II): II-89.

Van Breemen, C. and Saida, K., 1989, Cellular mechanisms regulating $[Ca^{2+}]_i$ in smooth muscle, *Ann. Rev. Physiol.,* 51: 315.

Wahle, K. W. J., 1983, Fatty acid modification and membrane lipids, *Proc. Nutrition Soc.,* 42: 273.

Webb, R. C., 1982, Potassium relaxation of vascular smooth muscle from DOCA hypertensive pigs, *Hypertension,* 4: 609.

Webb, R. C. and Bohr, D. F., 1984, The membrane of the vascular smooth muscle cell in experimental hypertension and its response to serotonin. *in:* "Smooth Muscle Contraction", N. L. Stephens, ed., M. Dekker, Inc., New York, p. 485.

Webb, R. C. and Bohr, D. F., 1978, Mechanism of membrane stabilization by calcium in vascular smooth muscle, *Am. J. Physiol.,* 235: C227.

Webb, R. C. and Bohr, D. F., 1979, Potassium relaxation of vascular smooth muscle from spontaneously hypertensive rats, *Blood Vessels,* 16: 71.

Weiss, G. B., 1986, Phospholipids, calcium binding and arterial smooth muscle membranes, *in:* "Recent Advances in Arterial Diseases: Atherosclerosis, Hypertension and Vasospasm", T. N. Tulenko and R. H. Cox, eds., Alan R. Liss, New York, p. 123.

Yeagle, P. L., 1983, Cholesterol modulation of $(Na^+ + K^+)$ ATPase ATP hydrolyzing activity in human erythrocyte, *Biochim. Biophys. Acta,* 727: 39.

Yeagle, P. L., 1989, Lipid regulation of cell membrane structure and function, *FASEB J.,* 3: 1833.

Yeagle, P. L. and Sen, A., 1986, Hydration and the lamellar to hexagonal II phase transition of phosphatidylethanolamine, *Biochemistry,* 25: 7518.

Yeagle, P. L., Young, J., and Rice, D., 1988, Effects of cholesterol on (Na^+, K^+) ATPase ATP hydrolyzing activity in bovine kidney, *Biochemistry,* 27: 6449.

Zolese, G. and Curatola, G., 1989, Ca^{2+} interaction with phospholipid bilayers studied by multifrequency phase fluorometry, *Biosci. Rep.,* 9: 497.

$[Ca^{2+}]_i$ DISTRIBUTION AND SIGNALLING IN VASCULAR HYPERTROPHY

Kathleen G. Morgan and Panos Papageorgiou

Departments of Medicine and Physiology
Harvard Medical School
Beth Israel Hospital
Boston, MA 02214

INTRODUCTION

We were initially attracted to the field of vascular hypertrophy by two themes in the hypertrophy literature. First, an extensive literature suggests that, in many different experimental models of hypertension, the consequent hypertrophy is associated with an increased contractile sensitivity to pharmacologic agents (e.g., Mulvany et al., 1981; Field and Soltis, 1985). Many authors have suggested that calcium handling might be altered in these hypertrophied vessels (for review, see Morgan and Suematsu, 1990). Alterations in calcium handling have been measured in platelets from hypertensive patients (Ernie et al., 1984) and in cultured cells from hypertensive animals (Nabika et al., 1985; Sugiyama et al., 1986), but a direct measurement of elevated intracellular ionized calcium $[Ca^{2+}]_i$ levels in contractile hypertrophic vessels had not yet been made. Second, there is a separate literature suggesting, but again not directly showing, that growth of cultured cells requires changes in $[Ca^{2+}]_i$ handling (Metcalfe et al., 1986). Being influenced by these two lines of investigation, we became interested in investigating $[Ca^{2+}]_i$ levels and the changes in $[Ca^{2+}]_i$ signalling in contractile hypertrophic vascular smooth muscle cells.

METHODS

A variety of methods were used in these studies and each will be discussed briefly below.

Calcium Indicators

Two Ca indicators were used in this study. The rationale for the use of two indicators is based on the fact that all Ca indicators have the potential to produce artifactual signals and also that the calibration of steady state levels in terms of absolute $[Ca^{2+}]_i$ can be misleading. If two markedly different types of indicators can be shown to give the same $[Ca^{2+}]_i$ profile, then there can be an increased level of confidence in the results.

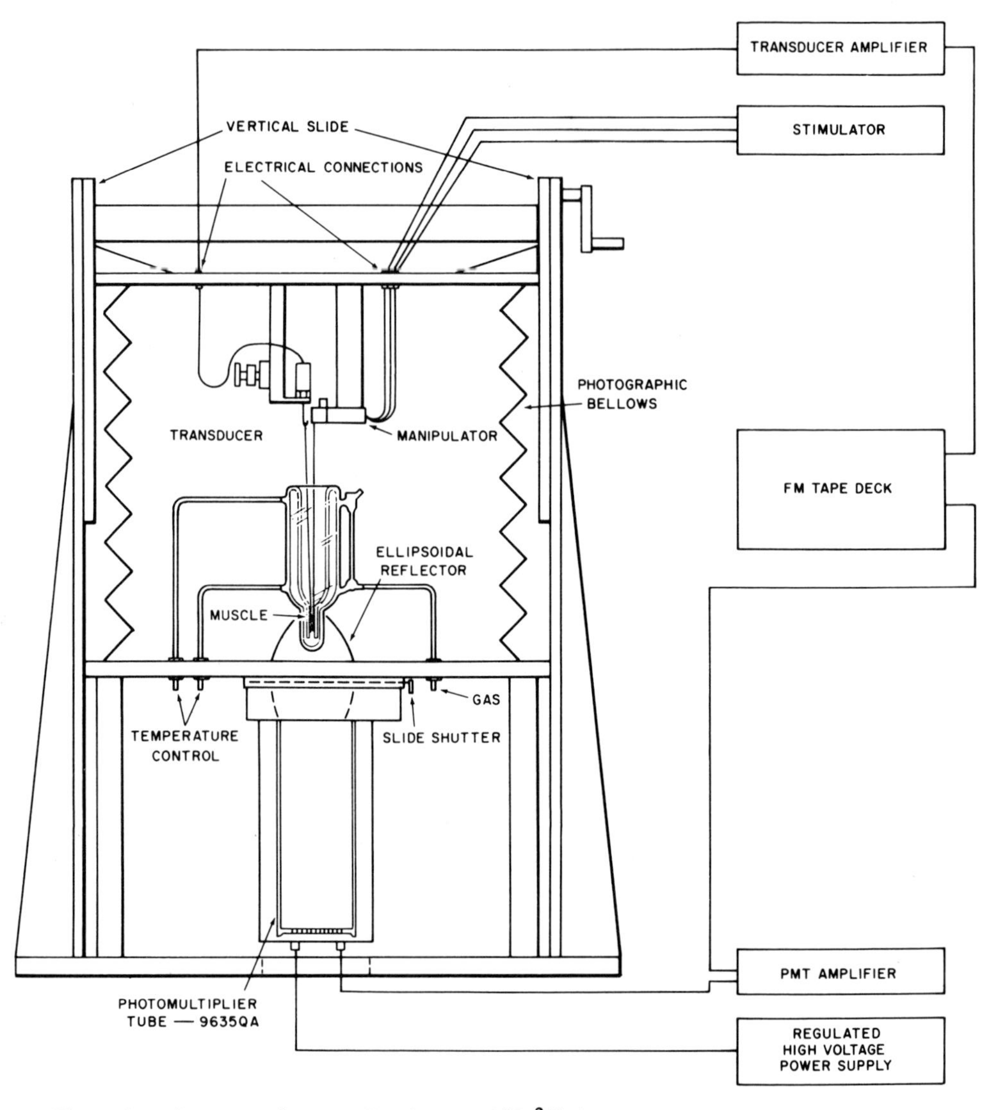

Figure 1. Apparatus for recording force and $[Ca^{2+}]_i$ from aequorin-loaded muscle strips. Modified with permission from Papageorgiou, 1990.

Aequorin. We have used aequorin primarily in intact strips of vascular smooth muscle. Aequorin is loaded into the preparation conveniently with a chemical loading procedure we have previously described (Morgan and Morgan, 1982). Since aequorin is a luminescent rather than fluorescent calcium indicator, a lamp is not necessary in the apparatus, but since it is a low light level calcium indicator, the light collection apparatus must be extremely efficient. The apparatus we use (Fig. 1) is modified from one initially developed in the laboratory of Dr. John Blinks at the Mayo Clinic. All mechanical connections are made in a light-tight manner by a skilled machinist. The entire apparatus is surrounded by a photographic bellows, and a pair of ellipsoidal mirrors are used to enhance light collection. A photomultiplier tube specially selected for low dark current (EMI 9635QA) is used. Since light is collected from all surfaces of the muscle and we restrict our preparations to thin, translucent preparations, it has been determined that there are no movement artifacts associated with aequorin signals during muscle contraction (Housmans et al., 1983).

The ongoing luminescence obtained from the muscle can be normalized by a lysis signal (L_{max}) obtained on application of Triton X-100 at the end of the experiment. L_{max} is proportional to the total aequorin content, thus the ratio of L/L_{max} at any point in time gives an index of $[Ca^{2+}]_i$. The absolute value is determined by comparison to an *in vitro* calibration curve (Blinks et al., 1982). Aequorin luminescence is proportional to the 2.5^{th} power of the $[Ca^{2+}]_i$ over essentially all of the physiologic range of $[Ca^{2+}]_i$ levels in vascular cells. Thus, aequorin is an extremely sensitive calcium indicator, but the nonlinear responsiveness should be kept in mind in interpreting transient events, since focal areas of $[Ca^{2+}]_i$ within a single cell will dominate the average cellular signal in a disproportionate manner. For this reason, we generally restrict calibration to steady state events.

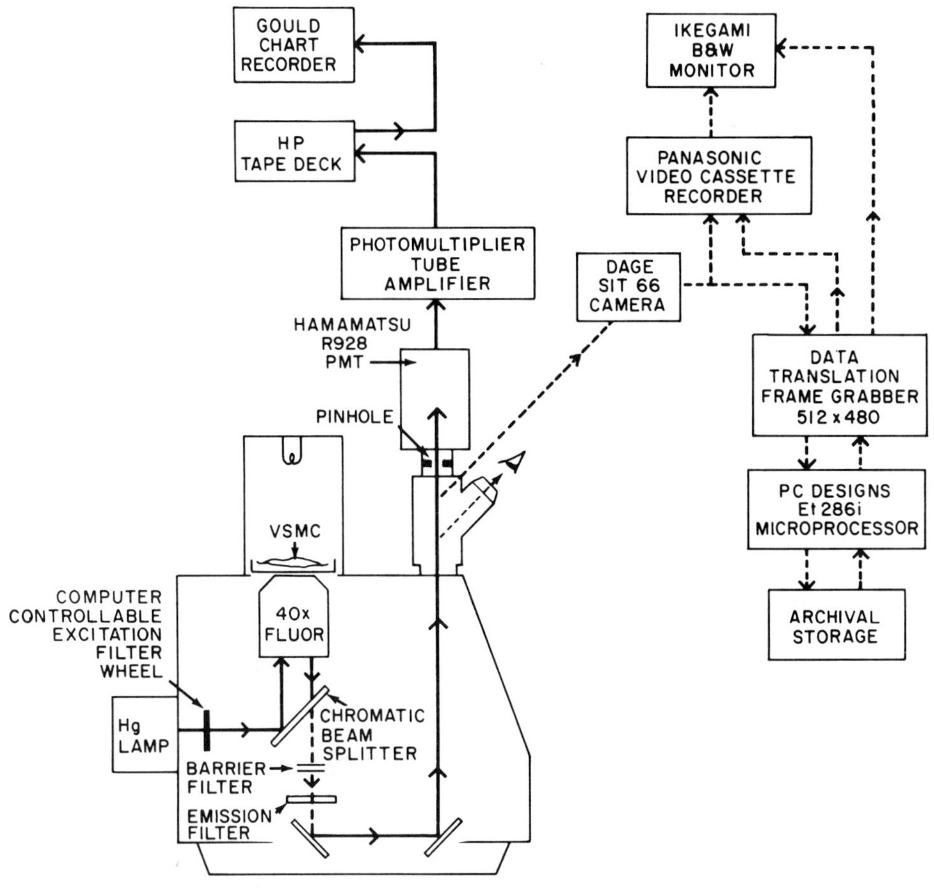

Figure 2. Apparatus for recording cell length and $[Ca^{2+}]_i$ from fura-2-loaded cells. Reproduced with permission from Papageorgiou, 1990.

Fura-2. We have used the fluorescent Ca indicator fura-2 exclusively in single cells, to avoid the additional technical challenges that can be encountered in the use of fluorescent indicators in intact muscle strips (Himpens and Somlyo, 1988). With the use of digital image analysis, many of the potential problems involved in the use of fura-2 as a Ca indicator can be avoided. The apparatus we use is shown in Figure 2. A mercury lamp is used for excitation, and it has been found empirically that the best wavelengths for ratio pairs are 350 and 390 nm in this apparatus. Interference filters with a 10-12 nm band width are used in combination with a 420 nm chromatic beam splitter and a long pass barrier filter of 485 nm. Images were obtained either with a 40X (N.A. 0.85) or a 100X (N.A. 1.3) objective. The entire image of the fluorescent cells can be recorded on the SIT camera and stored on VCR tape for later analysis in a computer equipped with Data Translation imaging boards. In some experiments, continuous photomultiplier tube recordings of isolated 2 μm diameter circular areas within single cells were recorded on magnetic tape and observed in strip chart records. With the use of imaging, however, the problem of organelle loading of the Ca indicator can be evaluated and "hot spots" can be specifically avoided. We have found that, in fresh tissues, it is generally possible, with short loading times, to obtain preparations with essentially homogeneous loading of the cytoplasm. We have also found that this is much more difficult in cultured cells. Other problems, such as partial de-esterification of fura-2-AM (Scanlon et al., 1987) and the generation of Ca-insensitive bleached intermediates (Becker and Fay, 1987) are more difficult to handle, and we have generally used an *in vivo* calibration of each individual cell with the use of 4-bromo-A23187 to make the cell membrane hyperpermeable, followed by the exposure to solutions of known $[Ca^{2+}]$ concentration (Peeters et al., 1987).

Cell Dissociation Methods

We have described a method for isolating healthy, contractile, pharmacologically normal cells from the ferret portal vein, using a mixture of collagenase, elastase, and trypsin inhibitor (DeFeo and Morgan 1985a). It was found to be essential to avoid centrifugation or aspiration of the tissue through wide-bore pipettes, because these procedures, while they increase cell yield, also cause irreversible damage to mammalian vascular cells. In the rat studies, a similar method was used for the aorta, but it was found to be far more difficult to isolate cells from the thick, elastic aorta than from the relatively thin portal vein. The details of the method are given elsewhere (Papageorgiou and Morgan, 1991b). The main modification was to eliminate the mincing of the tissue into small pieces, and to take an intact cylinder of the thoracic aorta, remove loose connective tissue and endothelium, invert the aorta, and then tie it at the ends. The inverted aortae were incubated at 34° C under 95% O_2 and 5% CO_2 and gently agitated in a Ca^{2+}-Mg^{2+}-free Hanks solution, containing collagenase, elastase, and trypsin inhibitor. The modifications resulted in an improved quality of the cells with a relatively large proportion of relaxed-looking cells at rest and responsive cells after addition of agonists, but still gave a very low yield, with only approximately two healthy cells per optical field at 400X. The mean length of the aortic cells was 90 ± 15 μm, whereas that in the

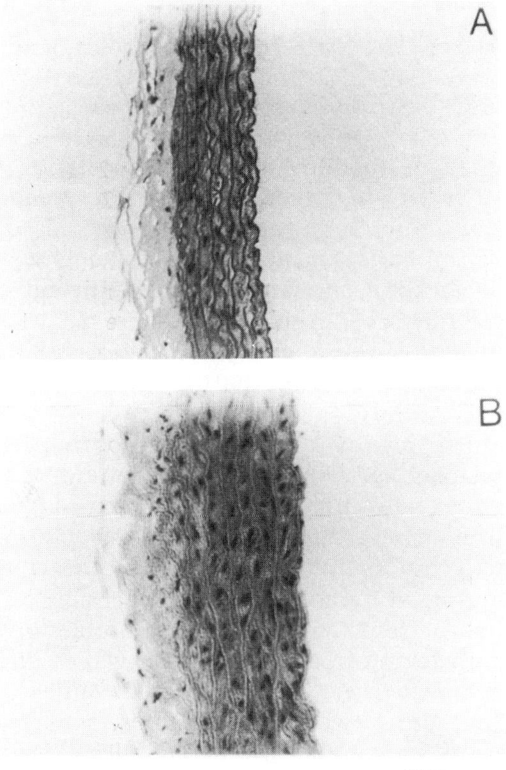

Figure 3. Haematoxylin-eosin stained sections of age-matched control (A) and hypertrophied (B) rat aorta. Reproduced with permission from Papageorgiou, 1990.

portal vein cells was 130 ± 3.3 μm (DeFeo and Morgan, 1985a). We cannot rule out that the difference is due to some damage in the aortic cells from the isolation procedure, and this has to be kept in mind in interpretation of results in all isolated cell experiments. However, we know from intact tissue studies that the ferret and rat aortae possess considerable tone at rest compared to the ferret portal vein (DeFeo and Morgan, 1985b; Papageorgiou and Morgan, 1991b), and this would be expected to result in some shortening of the cells at rest. Similarly, it is possible that the cells simply are shorter in the vessels *in situ*.

Aortic Coarctation Model

The abdominal aortae of male Sprague-Dawley rats were partially ligated between the renal arteries in order to induce medial hypertrophy. Sixty day old animals were used and followed after ligation for 3 - 4 weeks, after which their blood pressure was measured from the left carotid artery, confirmed to be elevated, and their thoracic aortae were removed. Age-matched animals served as controls. Medial hypertrophy of the thoracic aortae from the experimental animals compared to the controls was confirmed on haematoxylin and eosin stained sections (Papageorgiou and Morgan, 1991a) (Fig. 3).

EFFECTS OF HYPERTROPHY ON BASAL $[Ca^{2+}]_i$

Cytoplasmic $[Ca^{2+}]_i$

Digital images were obtained at the two excitation wavelengths from single freshly isolated cells from the rat thoracic aorta loaded with fura-2. After appropriate background subtraction, the 350 image was divided by the 390 image on a pixel-by-pixel basis, and then measurements were taken at a (non-nuclear) cytoplasmic location showing a low coefficient of variation. The cytoplasmic resting $[Ca^{2+}]_i$ level in the normal aortic cells had a mean value of 201 nM, whereas the cells from the hypertrophic aortae had a resting $[Ca^{2+}]_i$ of 324 nM (Papageorgiou and Morgan, 1991b). The resting values from the normal aortic cells were very similar to what we have observed in other vascular cells previously; however, the levels in the hypertrophic aortic cells are far higher than we have seen in any other smooth muscle cell type.

Because of uncertainties involved in calibration of fura-2 signals, as well as the possibility of cell damage during enzymatic isolation, we confirmed these findings using intact strips of rat aorta loaded with aequorin. The resting $[Ca^{2+}]_i$ value as determined by aequorin in normal aortic strips was 215 nM, whereas the strips taken from the hypertrophic aortae loaded with aequorin gave a resting calcium level of 340 nM. These values were not statistically significantly different from the respective fura-2 values (Papageorgiou and Morgan, 1991b). Thus, using two vastly different methods, the same finding was obtained, i.e., cells from hypertrophic aortae had significantly elevated basal $[Ca^{2+}]_i$ levels.

Nuclear-Cytoplasmic Calcium Gradients?

Although imaging was not possible in the intact aequorin-loaded strips, by the use of digital image analysis we were able to compare fura-2 signals from the isolated cells at different points in space within the cell. In the normal aortic cells we found no significant difference between cytoplasmic and nuclear $[Ca^{2+}]_i$ levels. The nuclear area was loosely defined as the central part of the cell, having the widest diameter. Previous electron micrographs have confirmed the nucleus to be identified by this definition. The $[Ca^{2+}]_i$ level in the "nuclear" area in the normal aortic cells was not significantly different from the non-nuclear area. This is in contrast to our past studies on ferret portal vein cells (Papageorgiou and Morgan, 1990). In those studies, normal portal vein cells were shown by digital image analysis to have an apparent nuclear-cytoplasmic $[Ca^{2+}]_i$ gradient at rest. We also demonstrated that the apparent gradient could be abolished by the application of caffeine, suggesting that the appearance of a gradient was an artifact due to the presence of fura-2 in the perinuclear sarcoplasmic reticulum.

In contrast to the normal rat aortic cells, the hypertrophic rat aortic cells displayed an apparent nuclear-cytoplasmic gradient. Although the cytoplasmic values were elevated, the nuclear values were elevated significantly further, to a value of 410 nM. And, in contrast to what was seen in the ferret portal vein cells, this gradient persisted after application of caffeine to deplete the perinuclear sarcoplasmic reticulum (Papageorgiou and Morgan, 1991b). Thus, this suggests that the nuclear envelope, in spite of the existence of nuclear pores (which can pass molecules of 200Å size at appropriate times in the cell cycle; Feldherr and Akin, 1990), is capable of generating a gradient in $[Ca^{2+}]_i$.

This has been previously suggested by Fay and coworkers (Williams et al., 1985; Williams et al., 1987) and is consistent with reports of ion-specific membrane channels in the nuclear envelope (Mazzanti et al., 1990). Although we remain skeptical that the nuclear envelope, which anatomically appears quite leaky, could generate significant concentration gradients, the implications of this finding, if true, may be significant with respect to mechanisms of Ca^{2+}-induced regulation of gene transcription.

AGONIST RESPONSES IN HYPERTROPHIED CELLS

On stimulation with an elevated potassium solution, the change in $[Ca^{2+}]_i$ in response to the depolarizing stimulus was not significantly different between the normal and hypertrophic tissues. However, the absolute level of $[Ca^{2+}]_i$ was significantly higher in the depolarized hypertrophic cells than in the depolarized control cells because of the significantly elevated basal $[Ca^{2+}]_i$ (Papageorgiou and Morgan, 1991b).

Very different findings were observed in response to α-agonists in these cells. The control rat aortic strips loaded with aequorin and the isolated cells loaded with fura-2, both showed pronounced contractions to phenylephrine, but displayed little or no change in $[Ca^{2+}]_i$, either transient or steady state. The hypertrophic strips and cells, in contrast, showed a pronounced spike-like initial elevation in $[Ca^{2+}]_i$ in response to phenylephrine. This finding suggests that the pathway of signalling was altered by the hypertrophy process (Papageorgiou and Morgan, 1991b). The increase in $[Ca^{2+}]_i$ signalling in the hypertrophied cells may explain both the decreased ED_{50} (i.e., increased sensitivity to α-agonists) and the increased speed of contraction, which we and others have observed in hypertrophic tissues in response to α-agonists.

HETEROGENEITY OF HYPERTROPHIED CELLS

Vascular Cells

It has previously been reported that hypertrophied rat aortae contain a significantly increased population of polyploid cells (Owens et al., 1981; Owens and Schwartz, 1983; Goldberg et al., 1984). Using ethidium dimer fluorescence to quantitatively measure an index of nuclear volume, we have confirmed that the nuclear volumes of cells from hypertrophic aortae were not normally distributed, and that whereas approximately 12% of the normal cells appear to be tetraploid, in the hypertrophic aortae the tetraploid population represents approximately 48%, with an additional 28% being octaploid (Papageorgiou and Morgan, 1991b). These percentages are in good agreement with past reports of hypertrophic aortae from other laboratories.

Using this method, we were able to co-load cells with ethidium dimer and fura-2 to study $[Ca^{2+}]_i$ levels and signalling in cells having an identified nuclear volume. It was found that the diploid normal and hypertrophic cells had indistinguishable $[Ca^{2+}]_i$ levels. In both the normal and hypertrophic aortic cells, the tetraploid cells showed a significantly increased $[Ca^{2+}]_i$ level compared to the diploid cells. The octaploid hypertrophied cells had an elevated $[Ca^{2+}]_i$ indistinguishable from that of the tetraploid cells (Papageorgiou and Morgan, 1991b). Thus, it appears that on the transition to

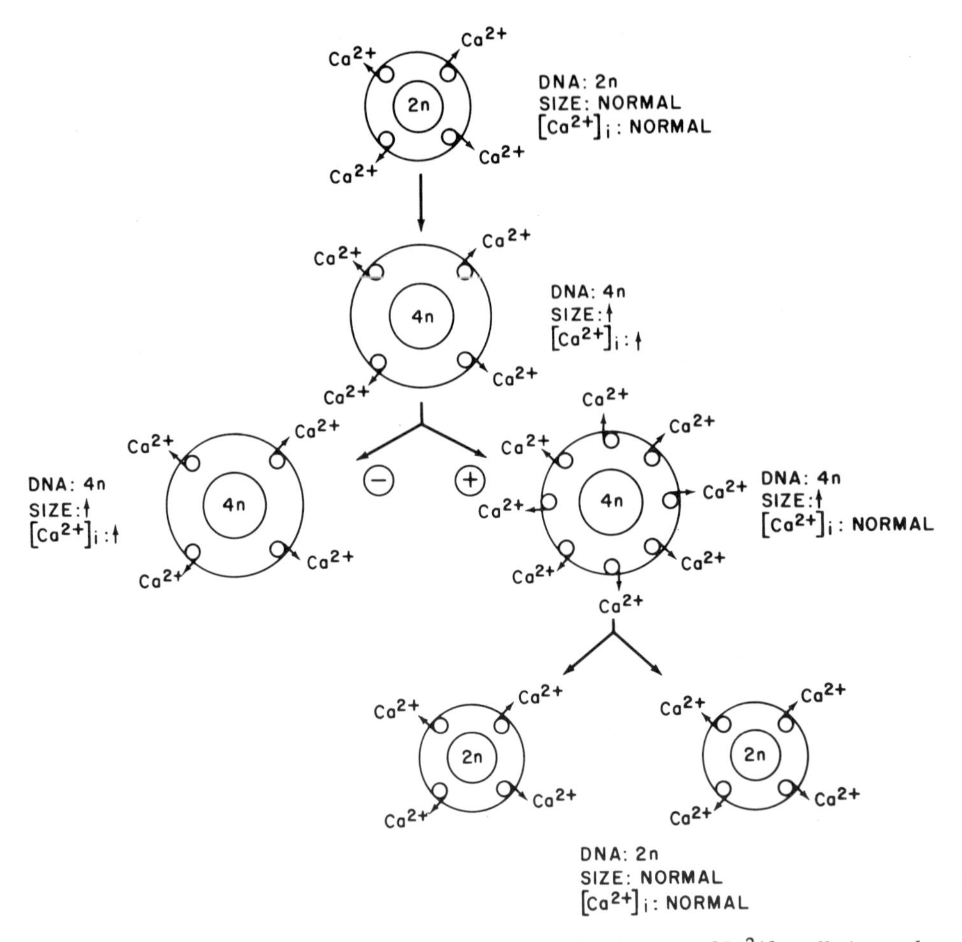

Figure 4. Possible scheme to explain the relationship between $[Ca^{2+}]_i$, cell size and ploidy. Reproduced with permission from Papageorgiou, 1990.

tetraploidy there is an increase in $[Ca^{2+}]_i$, both in the normal and the hypertrophic vessels, but because of the larger population of polyploid cells in the hypertrophic vessels, the mean $[Ca^{2+}]_i$ level is significantly elevated in the hypertrophic vessels compared to the controls. It is worth noting that the hypertrophic diploid cells did have a significantly increased cell volume compared to the control cells, even though the resting $[Ca^{2+}]_i$ was not significantly elevated. This suggests (but certainly does not prove) that the increase in cell volume preceded the increase in $[Ca^{2+}]_i$ levels.

The mechanism for the increase in basal $[Ca^{2+}]_i$ is unknown at the present time. An increase in cell size normally occurs when cells become committed to proceeding through the G_1-to-S transition and begin DNA synthesis. We have wondered whether this increase in cellular size could cause an increase in basal $[Ca^{2+}]_i$ because of the decreased surface membrane:volume ratio, as illustrated in Figure 4. A perturbation of the optimal surface membrane-to-volume ratio could decrease the efficacy of membrane-bound Ca^{2+}-homeostatic mechanisms, thus leading potentially to an increased $[Ca^{2+}]_i$. It is known that an increased $[Ca^{2+}]_i$ prevents the formation of the mitotic apparatus and inhibits the progression through mitosis (Hafner and Petzelt, 1987; McIntosh and Koonce, 1989). Thus, the cell remains hypertrophic and tetraploid with an increased $[Ca^{2+}]_i$. Cells undergoing normal mitosis must be capable of generating compensatory mechanisms to bring the cell $[Ca^{2+}]_i$ back to normal, allowing mitosis to proceed and to lead to the generation of an increased number of cells rather than cells of increased size.

Alternatively, the increased basal $[Ca^{2+}]_i$ levels could be the result of an alteration in membrane Ca pumps or possibly an increase in the size of the sarcoplasmic reticulum volume in these growing cells. Consistent with the latter point is the finding that the hypertrophied cells had an increased transient $[Ca^{2+}]_i$ spike in response to some agonists, possibly due to IP_3-mediated release of Ca from intracellular sites. Arguing against this concept is the fact that angiotensin-induced responses showed a decreased $[Ca^{2+}]_i$ spike amplitude in the hypertrophied cells.

Liver Cells

The polyploid rat aortic muscle cells described above were derived from a cell population developed as a response to a pathophysiological perturbation, i.e., increased blood pressure. In order to examine whether the relationship between $[Ca^{2+}]_i$ and nuclear volume as well as cellular volume would hold in the absence of any apparent physical stimuli, normal rat liver cells were also studied. Liver cells *in situ* are known to display various degrees of ploidy (Brodsky and Uryvaeva, 1977).

The same method of co-loading with fura-2 and ethidium dimer was used to correlate $[Ca^{2+}]_i$ and nuclear volume in individual cells. The resting $[Ca^{2+}]_i$ was found to be 245 ± 41 nM (n=10). A histogram analysis of nuclear

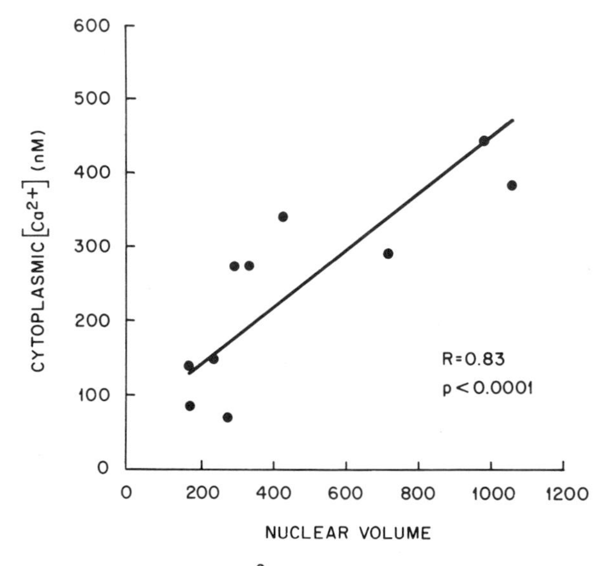

Figure 5. Relationship between $[Ca^{2+}]_i$ and nuclear volume in normal liver cells. Reproduced with permission from Papageorgiou, 1990.

volume values revealed that the sample was not normally distributed. This suggested the presence of liver cells with varying degrees of ploidy. A scatter plot of $[Ca^{2+}]_i$ as a function of nuclear volumes (Fig. 5) showed a highly significant linear correlation (R=0.83). Similar significant correlations were present in plots of $[Ca^{2+}]_i$ vs cellular volumes and cellular volumes vs nuclear volumes.

CONCLUSIONS

In conclusion, there are a number of alterations in Ca regulatory mechanisms in hypertrophied cells. In these studies, aortic coarctation induced hypertension was used to trigger the generation of hypertrophy. These results do not elucidate the mechanism of hypertension, but the secondary changes, linked to the hypertrophy would be expected to contribute to the disease process of ongoing hypertension, when it exists.

The increased basal $[Ca^{2+}]_i$ levels as well as the increased signalling in response to α-agonists would tend to make the cells more irritable, more prone to contraction, and perhaps explains the decreased ED_{50} to alpha agonists; but, interestingly, we have also observed that the maximal force per cross sectional area in response to various agonists is either unchanged or significantly decreased, depending on the agonist used (Papageorgiou and Morgan, 1991a). This may suggest that there is a less efficient force-generating mechanism, perhaps as a result of the increased intermediate filament lattice, which has been described by others in hypertrophic cells (Berner et al., 1981), or perhaps due to decreased force generation by polyploid cells. It is known that smooth muscle cells are generally non-contractile in culture. Whether polyploid cells generate normal or decreased contractile force remains to be demonstrated.

REFERENCES

Becker, P. L. and Fay, F. S., 1987, Photobleaching of fura-2 and its effect on determination of calcium concentrations, *Am. J. Physiol.*, 253: C613.

Berner, P. F., Somlyo, A. V., and Somlyo, A. P., 1981, Hypertrophy-induced increase of intermediate filaments in vascular smooth muscle, *J. Cell Biol.*, 88: 96.

Blinks, J. R., Wier, W. G., Hess, P., and Prendergast, F. G., 1982, Measurement of Ca^{2+} concentrations in living cells, *Prog. Biophys. Mol. Biol.*, 40: 1.

Brodsky, W. Y. A. and Uryvaeva, I. V., 1977, Cell polyploidy: Its relation to tissue growth and functioning, *Int. Rev. Cytol.*, 50: 275.

DeFeo, T. T. and Morgan, K. G., 1985a, Responses of enzymatically isolated mammalian vascular smooth muscle cells to pharmacological and electrical stimuli, *Pflügers Arch.*, 404: 100.

DeFeo, T. T. and Morgan, K. G., 1985b, Calcium-force relationships as detected with aequorin in two different vascular smooth muscles of the ferret, *J. Physiol.*, 369: 269.

Ernie, P., Bolli, P., Burgesser, E., and Buhler, F. R., 1984, Correlation of platelet calcium with blood pressure, *N. Eng. J. Med.*, 310: 1084.

Feldherr, C. M. and Akin, D., 1990, The permeability of the nuclear envelope in dividing and nondividing cell cultures, *J. Cell Biol.*, 111: 1.

Field, F. P. and Soltis, E. E., 1985, Vascular reactivity in the spontaneously hypertensive rat: Effect of high pressure stress and extracellular calcium, *Hypertension*, 7: 228.

Goldberg, I. D., Shapiro, H. M., Zoller, L. C., Myrick, K., Levenson, S. E., and Christenson, L., 1984, Isolation and culture of a tetraploid subpopulation of smooth muscle cells form the normal rat aorta, *Science*, 226: 559.

Hafner, M. and Petzelt, C., 1987, Inhibition of mitosis by an antibody to the mitotic calcium transport system, *Nature*, 330: 264.

Himpens, B. and Somlyo, A. P., 1988, Free-calcium and force transients during depolarization and pharmacomechanical coupling in guinea-pig smooth muscle, *J. Physiol.*, 395: 507.

Housmans, P. R., Lee, N. K. M., and Blinks, J. R., 1983, Active shortening retards the decline of the intracellular calcium transient in mammalian heart muscle, *Science*, 221: 159.

Mazzanti, M., DeFelice, L. J., Cohen, J., and Malter, H., 1990, Ion channels in the nuclear envelope, *Nature*, 343: 764.

McIntosh, J. R. and Koonce, M. P., 1989, Mitosis, *Science*, 246: 622.

Metcalfe, J. C., Moore, J. P., Smith, G. A., and Hesketh, T. R., 1986, Calcium and cell proliferation, *Br. Med. Bull.*, 42: 405.

Morgan, J. P. and Morgan, K. G., 1982, Vascular smooth muscle: The first recorded Ca^{2+} transients, *Pflügers Arch.*, 395: 75.

Morgan, K. G. and Suematsu, E., 1990, The role of calcium in vascular smooth muscle tone, *Am. J. Hypertension*, 3: 291S.

Mulvany, M. J., Korsgaard, N., and Nyborg, N., 1981, Evidence that the increased calcium sensitivity of resistance vessels in spontaneously hypertensive rats is an intrinsic defect of their vascular smooth muscle, *Clin. Exp. Hypertension*, 3: 749.

Nabika, T., Velletri, P. A., Beaven, M. A., Endo, J., and Lovenberg, W., 1985, Vasopressin-induced calcium increases in smooth muscle cells from spontaneously hypertensive rats, *Life Sci.*, 37: 579.

Owens, G. K., Rabinovitch, P. S., and Schwartz, S. M., 1981, Smooth muscle cell hypertrophy versus hyperplasia in hypertension, *Proc. Nat'l. Acad. Sci. U.S.A.*, 78: 7759.

Owens, G. K. and Schwartz, S. M., 1983, Vascular smooth muscle cell hypertrophy and hyperploidy in the Goldblatt hypertensive rat, *Circ. Res.*, 53: 491.

Papageorgiou, P., 1990, "Alterations in excitation-contraction coupling in hypertrophic vascular smooth muscle", Harvard Univ. Press, Cambridge.

Papageorgiou, P. and Morgan, K. G., 1990, The nuclear-cytoplasmic [Ca^{2+}] gradient in single mammalian vascular smooth muscle cells, *Proc. Soc. Exp. Biol. Med.*, 193: 331.

Papageorgiou, P. and Morgan, K. G., 1991a, Increased Ca^{2+} signalling following alpha adrenoceptor activation in vascular hypertrophy, *Circ. Res.*, 68: 1080.

Papageorgiou, P. and Morgan, K. G., 1991b, Intracellular free Ca^{2+} is elevated in hypertrophic aortic muscle from hypertensive rats, *Am. J. Physiol.*, 260: H507.

Peeters, G. A., Hlady, V., Bridge, J. H. B., and Barry, W. H., 1987, Simultaneous measurement of calcium transients and motion in cultured heart cells, *Am. J. Physiol.*, 253: H1400.

Scanlon, M., Williams, D. A., and Fay, F. S., 1987, A Ca^{2+}-insensitive form of fura-2 associated with polymorphonuclear leukocytes, *J. Biol. Chem.*, 262: 6308.

Sugiyama, T., Yoshizumi, M., Takaku, F., Urabe, H., Tsukakoshi, M., Kasuya, T., and Yazaki, Y., 1986, The elevation of the cytoplasmic calcium ions in vascular smooth muscle cells in SHR -- Measurement of the free calcium ions in single living cells by laser microfluorospectrometry, *Biochem. Biophys. Res. Commun.*, 141: 340.

Williams, D. A., Becker, P. L., and Fay, F. S., 1987, Regional changes in calcium underlying contraction of single smooth muscle cells, *Science*, 235: 1644.

Williams, D. A., Fogarty, K. E., Tsien, R. Y., and Fay, F. S., 1985, Calcium gradients in single smooth muscle cells revealed by the digital imaging microscope using fura-2, *Nature*, 318: 558.

CONTROL AND FUNCTION OF ALTERATIONS IN CONTRACTILE PROTEIN ISOFORM EXPRESSION IN VASCULAR SMOOTH MUSCLE

Charles L. Seidel, David Rickman, Heidi Steuckrath,
Julius C. Allen, and Andrew M. Kahn[1]

Baylor College of Medicine
Department of Medicine
Houston, TX, 77030

[1]University of Texas Health Science Center, Houston
School of Medicine
Houston, TX 77225

INTRODUCTION

The use of autologous saphenous veins for reconstruction of occluded, atherosclerotic peripheral and coronary arteries is a surgical procedure widely used to relieve the symptoms of compromised arterial flow. Independent of graft location, failure within the first week results primarily from thrombus formation while late failure (months to years) can be due to intimal and medial hypertrophy and/or the re-occurrence of atherosclerosis (Unni et al., 1974; Lawrie et al., 1976; Gunthaner et al., 1979; Tracey et al., 1979).

The wall hypertrophy associated with late graft failure can be either localized or diffuse (Dilley et al., 1988) and most likely results from the combined effects of chemical and physical factors acting on the wall. Dobrin et al. (1989) has shown in canine vein grafts that intimal hyperplasia was associated with low shear stress which may permit increased adherence of monocytes and platelets to the vessel wall with the subsequent release of chemical mitogens and chemoattractants. In contrast, medial hypertrophy was associated with increased wall strain. Strain could initiate hypertrophy by compression of the vasa vasorum causing ischemia, a known initiator of medial hypertrophy in grafts (Brody et al., 1972) or by acting directly on smooth muscle cells to stimulate protein synthesis through unknown mechanisms (Seidel and Schildmeyer, 1987).

Because a decrease in force generation was observed in association with venous hypertrophy produced by portal vein coarctation (Johansson, 1976), we determined the contractile ability of hypertrophied canine saphenous vein grafts. These studies (Seidel et al., 1984) indicated a reduction in active force development that was agonist independent and was accompanied by a net increase in contractile protein content. These observations suggested that the reduction in active stress was not the result of an agonist specific alteration in

Regulation of Smooth Muscle Contraction
Edited by R.S. Moreland, Plenum Press, New York, 1991

excitation-contraction coupling but, because of the increase in contractile protein content, may be due to quantitative and qualitative changes in contractile protein expression.

Berner et al (1981) observed that in coarcted portal veins, smooth muscle cell hypertrophy was associated with an increase in the number of thin and intermediate filaments without a parallel increase in thick filaments. Such an alteration in filament content may contribute to the reduced force generation observed in this model (Johansson, 1976). These changes are in contrast to those in coarcted arterial tissue where contractile filament content increased in parallel with smooth muscle cell hypertrophy (Olivetti et al., 1980) and no change in contractile ability was observed (Bevan et al., 1975). This suggests that venous hypertrophy can result in reduced force development possibly through altered contractile protein production.

Primary cultures of canine saphenous vein smooth muscle cells (Seidel et al., 1988; 1989) maintained in a serum free medium express the same myosin isoforms present in intact vein. When the medium is changed to one containing serum, the expression of a myosin heavy chain isoform is induced that is immunologically and electrophoretically similar to the myosin of platelets and endothelial cells. With increased duration of serum exposure, this non-muscle myosin heavy chain (nmMHC) becomes the sole form of myosin present in the cells. Since vein grafts undoubtedly have altered endothelial cell function, they are most likely exposed to serum mitogens and therefore, the smooth muscle cells may express the nmMHC isoform. If the expression of this heavy chain is observed in vein grafts, then cultured saphenous vein smooth muscle cells could be used to determine if the presence of non-muscle myosin is associated with altered cell function.

The purpose of this work was to quantitatively and qualitatively characterize the myosin isoforms in canine saphenous vein grafts and to determine if changes in myosin isoform expression observed in grafts, when induced in cultured saphenous vein smooth muscle cells, were associated with alterations in contraction or migration.

METHODS

Characterization of Myosin Isoforms in Vein Grafts

Saphenous veins were used to replace equal lengths of femoral arteries in dog as previously described (Seidel et al., 1984). One month after surgery, patent grafts were removed, flushed of blood, cleaned of adhering connective tissue, and homogenized in a solution containing 25mM sodium-phosphate (pH 7), 1% beta-mercaptoethanol, and 1% sodium dodecylsulfate (Seidel et al., 1984). The insoluble material was pelleted by centrifugation and the myosin isoforms in the supernatant separated on 5% polyacrylamide-1% SDS gels (Seidel et al., 1988). The types of myosin isoforms present in graft extracts were characterized by comparing their mobility to the myosin present in similar extracts from unoperated saphenous veins as well as to purified human platelet myosin heavy chain. Quantitative differences were assessed by densitometric scanning of Coomassie blue stained gels (Seidel and Murphy, 1979) with the proportion of each heavy chain expressed relative to the total heavy chain present.

Contraction and Migration of Single Cells

Primary cultures of canine saphenous vein smooth muscle cells were prepared by enzyme dispersion as described previously (Seidel et al., 1988) and placed in either the serum-free medium described by Libby and O'Brien (1984) to maintain the *in vivo* myosin heavy chain isoform pattern or in medium containing 10% fetal bovine serum (FBS) to stimulate non-muscle myosin expression (Seidel et al., 1989). Cells to be used for contractile studies were seeded into 30 mm petri dishes coated with type I collagen prepared from rat tail tendon, while cells to be used for migration studies were seeded into plastic tissue culture flasks. Cells remained in culture for at least four days before study in order to induce the desired myosin isoform change.

To study the contractile ability of cells, 30 mm culture dishes containing cells were placed on the heated (37° C) stage of a Nikon inverted, phase contrast microscope. The culture medium was replaced with a solution containing 136mM NaCl, 4mM KCl, 2mM $CaCl_2$, 1mM $MgSO_4$, 5mM glucose, and 10mM HEPES (pH 7.4 at 37° C). After 45 minutes, a field of at least ten cells was photographed at 200X and then serotonin was added or the solution was changed to one in which the KCl concentration was increased to 100 mM by equal molar substitution for NaCl. Ten minutes later, a second photograph was taken of the same field. The length of the longest axis of the same ten cells was determined before and after exposure to the contractile agent, the percent change in length calculated, and averaged for all ten cells. This average was taken as the response of that particular culture preparation.

To determine the ability of the cells to migrate in response to a chemoattractant gradient, cells were removed from the culture dish with a dilute trypsin-EDTA solution, spun out of solution, and resuspended in Dulbecco's Modified Eagles Medium (DMEM). Aliquots of the cell suspension were placed in the upper wells of a 12 well Neuroprobe^R Chemotaxis chamber at a density of 10^5 cells per well. The upper wells were separated from the lower wells by a gelatin coated polycarbonate membrane with 8 µm pores. In the lower wells was placed either DMEM containing 10% FBS to act as a chemoattractant or DMEM alone as a control for random migration. After five hours in a tissue culture incubator, the chamber was disassembled, the cells on the upper membrane surface were removed by drawing the membrane over the edge of a razor blade and those on the lower surface stained. The number of stained cells was determined in three random, non-overlapping areas of the membrane that had separated the upper and lower wells. This number represented the number of cells that had penetrated the membrane toward the lower well. To correct for random migration, the cell counts obtained when FBS was present in the lower well were expressed relative to the number of cells that had crossed the membrane when DMEM alone was in the lower well. This ratio is defined as the migration index (MI).

RESULTS

Saphenous Vein Grafts

The total myosin heavy chain content of one month saphenous vein grafts was not different from that of control vessels (Graft: 1.3 ± 0.9 mg/gww, n = 4; Control: 1.8 ± 0.5 mg/gww, n = 8). Grafts contained the two muscle specific myosin heavy chain isoforms present in unoperated veins (SMHC-1, SMHC-2)

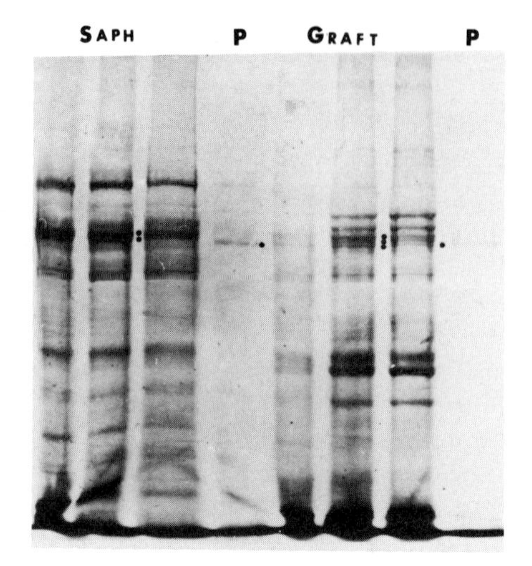

Figure 1. A 5% polyacrylamide gel illustrating the myosin heavy chain isoforms (indicated with dots) in extracts from a control canine saphenous vein (Saph) and a one month vein graft (Graft). Lanes labeled P contain purified platelet myosin. Note that the graft contains a third heavy chain that co-migrates with platelet myosin. Within each group, protein loading increases from left to right.

as well as a third heavy chain with electrophoretic mobility identical to that of platelet myosin heavy chain, nmMHC (Fig. 1). Of the total heavy chain present (Fig. 2), 26% was the nmMHC isoform. Relative to the total muscle specific myosin heavy chains, the proportions of SMHC-1 and SMHC-2 were unchanged by grafting (Fig. 2). If nmMHC did not function in force generation, its presence would contribute to the reduction in active force per total myosin observed in one month grafts (Seidel et al., 1984).

Single Cells

Cultured saphenous vein cells upon exposure to serum mitogens undergo a shift in myosin heavy chain expression similar to that seen in one month grafts. Muscle specific isoforms decrease without changing their relative proportions while the nmMHC isoform increases (Seidel et al., 1989). This suggests that cultured cells can serve as models of smooth muscle cells in grafts and therefore can be used to study the relationship between myosin isoform expression and cell function.

Cultured saphenous vein smooth muscle cells maintained in the absence of serum express only the muscle specific myosin isoforms characteristic of the intact vessel (Seidel et al., 1989). Such cells shortened upon exposure to serotonin (10^{-5} M) or to 100 mM KCl (SFM in Fig. 3).

Cells maintained in serum containing medium however, undergo a shift in myosin isoform expression. By the fifth day in culture over 50% of the myosin is the nmMHC isoform and at confluence (8 - 10 days) this isoform accounts for all the myosin heavy chain (Seidel et al., 1989). During this culture interval, such cells did not shorten when exposed to serotonin or to KCl at concentrations that contracted cells in serum-free medium (FBS in Fig. 3). Even increasing the serotonin concentration ten fold did not initiate shortening (0.0 ±

2.6%, n = 4). These data suggest that contractile function is compromised in association with the shift from muscle to non-muscle myosin isoforms.

The contractile response of cells maintained in the presence and absence of serum are in contra distinction to the migratory response of cells maintained under similar conditions. When cells were maintained in serum for four days to induce a shift to the non-muscle myosin isoform, approximately ten times as many cells migrated in response to the chemoattractant effect of serum as migrated in response to DMEM (Fig. 4). However, if the cells had been maintained in serum-free medium to retain the expression of muscle specific myosin isoforms, migration was not stimulated by serum (Fig. 4).

Because maintaining cells in serum can induce changes important for the migratory response that may not include alterations in myosin isoform expression, experiments were performed to further test the importance of the expression of specific myosin isoforms. Previous studies indicated that those cells that did not attach to the culture dish two days after seeding in serum containing medium, also did not express nmMHC but retained only the muscle specific isoforms (Seidel et al., 1989). If culturing in the presence of serum initiates cell changes necessary for migration which do not include nmMHC

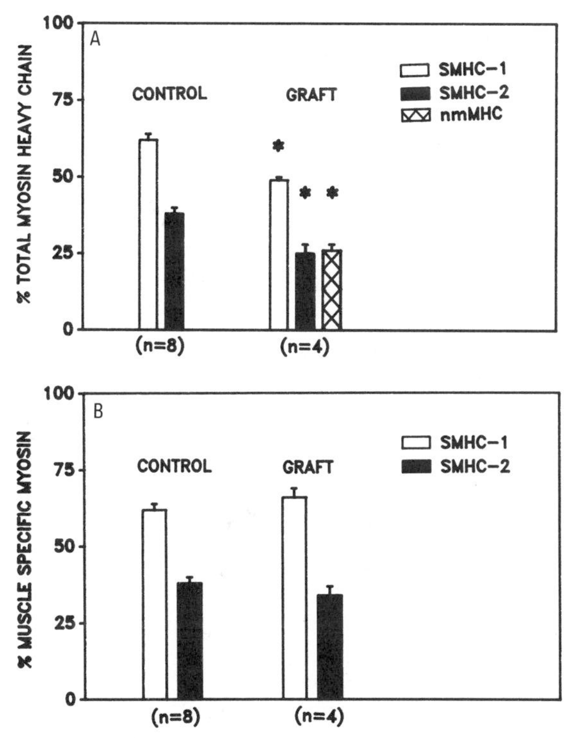

Figure 2. Relative proportion of three myosin heavy chain isoforms in canine saphenous veins and one month vein grafts. a: Myosin heavy chain isoforms; Control and one month grafts: The three isoforms are expressed relative to the total amount of myosin heavy chain present. The proportion of muscle specific isoforms (SMHC-1 and SMHC-2) are significantly lower in the graft whereas the proportion of the non-muscle isoform (nmMHC) is greater in the graft. b: Muscle specific myosin isoforms; Control and one month grafts. The individual muscle specific isoforms are expressed relative to the total amount of muscle myosin. The proportions of SMHC-1 and SMHC-2 are not changed in the graft. All results are mean ± SEM.

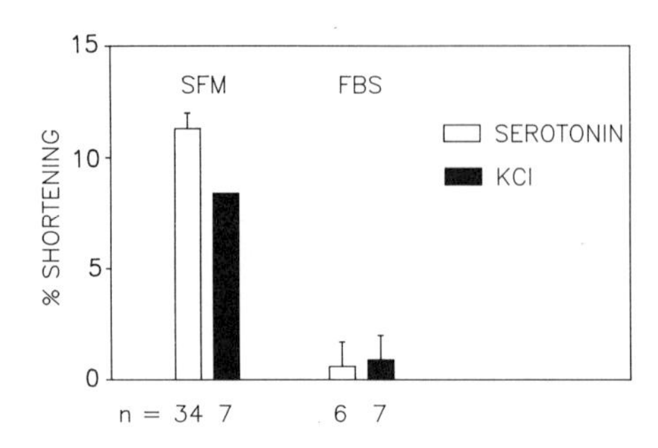

Figure 3. Contractile response to serotonin and KCl. The contractile response of cultured saphenous vein smooth muscle cells maintained for at least five days in a serum free medium (SFM) or in a medium containing serum (FBS). Only cells maintained in SFM shortened in response to serotonin (10^{-5} M) or KCl (100 mM). All results are mean ± SEM.

expression, unattached cells should be able to migrate. Migratory ability was determined for attached (expressing nmMHC) and unattached cells from the same cultures after two days in medium containing serum. Only attached cells migrated (Attached: MI = 8.6; Unattached: MI = 1.2; n = 2) in response to serum. This suggests that the induction of nmMHC is important for migration.

DISCUSSION

Saphenous veins used to replace segments of femoral arteries hypertrophy and have an impaired maximum force generating ability. Impaired contractile function occurs even though net contractile protein content increases with graft duration in parallel with continued wall hypertrophy (Seidel et al., 1984). Present data (Figs. 1 and 2) indicate that grafts at one month contain a non-muscle isoform of myosin heavy chain which comprises 26% of the total myosin heavy chain present. Primary cultures of canine saphenous vein smooth muscles cells can be induced to undergo a shift in myosin isoform expression upon serum exposure similar to that observed in grafts (Seidel et al., 1984; 1988). When the contractile ability of such cells is compared to that of cells in which only the muscle specific isoforms of myosin are present, reduced shortening ability is observed (Fig. 3). These results are consistent with the hypothesis that arterial grafting of the saphenous vein induces a shift in myosin expression from the muscle to the non-muscle isoform which results in a reduction in the force generating ability of the smooth muscle cells and the graft in general.

The stimulus for the shift in myosin isoform expression in vein grafts is not known. Since such a shift can be induced in cultured saphenous vein smooth muscle cells by serum, serum may play a similar role in the graft if the barrier function of the endothelium is altered. Endothelial damage has been associated with early (weeks) graft failure, however, the endothelium has been shown to be intact in long term grafts that have failed because of intimal hyperplasia and medial hypertrophy (Spray and Roberts, 1977). Whether the presence of non-muscle myosin is a property of grafts only early in the adaptive response prior to the reestablishment of the endothelium or is maintained in grafts of long duration is not known.

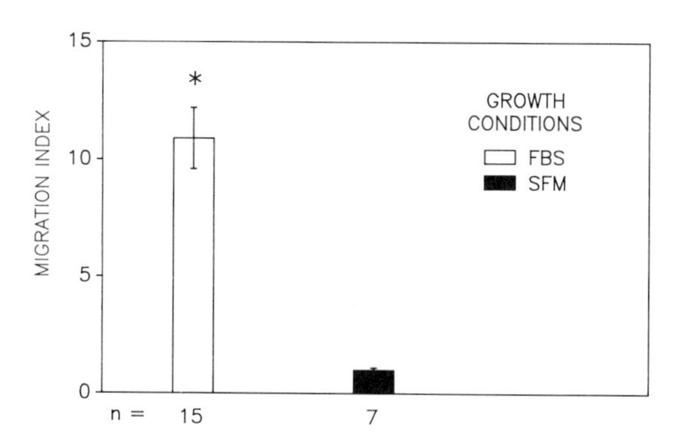

Figure 4. Migratory response to 10% serum. The migratory response of cells maintained for four days in either a serum-free medium (SFM) or in medium containing serum (FBS). 10% serum was used as the chemoattractant. Migration index is the ratio of the number of cells migrating in response to serum relative to the number migrating in the absence of serum. Only cells maintained in FBS migrated. All values are mean ± SEM.

From a quantitative perspective, the shift from muscle to non-muscle myosin isoforms can not completely account for the magnitude of force reduction observed in one month grafts. At the end of one month, maximum force generation is reduced approximately 70% (Seidel et al., 1984), however, only 26% of the total myosin heavy chain has shifted to the non-muscle isoform. Since the reduction in maximum contractile response is agonist independent (Seidel et al., 1984), general changes in the force generating system of the graft must have occurred. It is possible that the organization of the smooth muscle cells within the graft has been altered so that active force generation is not effectively transmitted in a circumferential direction. The observed increase in graft collagen content (Seidel et al., 1984) may contribute to altered force transmission.

From the study of cultured cells, the shift to the non-muscle myosin isoform and the loss of contractile ability is associated with the appearance of chemoattractant induced migration (Fig. 4). Migration depends upon the expression of non-muscle myosin since cells that contain only the muscle specific isoforms, whether they were maintained in the absence (serum-free medium) or in the presence (unattached cells) of serum, do not respond to a chemoattractant (see Results).

The explanation for why the muscle and non-muscle myosin isoforms are associated with contraction and migration, respectively, is not known. Myosin light chain phosphorylation regulates actin activation of smooth muscle and non-muscle myosin isoforms (Sellers and Umemoto, 1989) and therefore a fundamental change in the activation mechanism may not be involved. The mechanism of cell migration has been most extensively studied in non-muscle cells where it has been shown to involve the coordinated interaction of actin and myosin in ways that differ from their interaction in contraction (DeBiasio et al., 1988; McKenna et al., 1989). By following the movement of fluorescently labeled myosin in the lamellipodia of migrating fibroblasts it was determined that myosin was absent from the initial developing membrane protrusion and only appeared after the protrusion was established. Myosin however, was concentrated at the base of the protrusion in association with actin. Actin was present in both developing and established protrusions. From this work it has been suggested that protrusions result from the combined effects of: (1) actin polymerization in the leading edge of the protrusion driving the membrane forward and (2) turgor pressure developed by the interaction of actin and myosin at the base of the protrusion (DeBiasio et al., 1988). This process requires extensive polymerization-depolymerization and movement of myosin within the cell. The absence of organized thick myosin containing filaments is a characteristic of cultured smooth muscle cells that have lost the ability to contract (Chamley-Campbell et al., 1981; Sasaki et al., 1989). We suggest that the migratory ability of smooth muscle cells results from the expression of a myosin isoform, non-muscle myosin, that is better suited for the dynamic structural reorganizations required for migration.

The present study suggests that nmMHC expression is a necessary component for cell migration, however, other changes may also be required. One such change may include actin. Alterations in actin isoforms have been observed to occur in arterial tissue in association with intimal hyperplasia (Kocher et al., 1984) as well as in culture following serum exposure (Fatigati and Murphy, 1984; Barja et al., 1986; Owens et al., 1986). As with myosin, actin expression can be shifted from the muscle specific to the non-muscle isoforms, therefore, this shift may also contribute to the change in smooth muscle function from contractile to motile. Changes in venous smooth muscle actin isoforms during wall hypertrophy or in culture have not been described. A comparison of the relative amounts of the three actin isoforms in different smooth muscle tissues (Fatigati and Murphy, 1984) indicated that in venous tissue the proportion of the muscle specific alpha actin isoform was less than that in arterial tissue, whereas the gamma isoforms (smooth muscle plus non-muscle) were greater. The beta non-muscle was comparable in both. Because of these differences between arterial and venous tissue, it is possible that venous tissue may not undergo actin isoform shifts during hyperplasia and therefore a change in actin isoforms may not be required to enable venous smooth muscle cells to migrate.

Smooth muscle cell migration from the media of the vessel contributes significantly to intimal thickening observed after endothelial denudation of arterial tissue (Clowes and Schwartz, 1985) and may play an important part in the thickening that occurs in vein grafts. An understanding of the mechanism of smooth muscle migration would reveal points for intervention to reduce intimal thickening. If changes in contractile protein isoform expression is important for migration, preventing such changes may be therapeutically valuable. Alternatively, blocking the effect of chemoattractants may also be practical. In this regard, it has been shown that intimal thickening due to endothelial denudation was prevented in thrombocytopenic rats even though the same level of medial smooth muscle cell proliferation was achieved (Fingerle et al., 1989). It was concluded that intimal thickening was prevented because the absence of platelet derived chemoattractants prevented smooth muscle cell migration into the subintimal space.

It is also possible that nmMHC is involved in cellular functions in addition to migration. In non-muscle cells, myosin has been shown to be involved in cytokinesis because cell division is prevented if actin-myosin interaction is blocked (Pollard et al., 1990). It may play a similar role in vascular smooth muscle, however work from our laboratory suggests that proliferation can occur in the absence of nmMHC expression (Seidel et al., 1989; 1990). Another role for nmMHC may be in vesicular movement within the cell. Smooth muscle cells expressing nmMHC are in the "synthetic" state (Chamley-Campbell et al., 1981) characterized by the presence of large amounts of rough endoplasmic reticulum, free ribosomes, and Golgi complexes. In this state the cells synthesize and secrete the extracellular matrix proteins, collagen and elastin. It is possible that nmMHC may be involved in the transport of vesicles containing these proteins to the sarcolemma for secretion.

In conclusion, vein grafts undergo a reduction in contractile function that may be due in part to a shift from the muscle specific to the non-muscle isoforms of myosin. The presence of the non-muscle myosin isoform in cultured venous smooth muscle cells is associated with a decreased ability to contract and an increased ability to migrate. The increased migratory ability of cells that have undergone a shift in myosin isoform expression may contribute to the intimal thickening common to various vascular pathologies.

ACKNOWLEDGMENTS

This research was supported by NIH grant HL34280. We thank Corneille Smith for manuscript preparation.

REFERENCES

Barja, F., Coughlin, C., Belin, D., and Gabbiani, G., 1986, Actin isoform synthesis and mRNA levels in quiescent and proliferating rat aortic smooth muscle cells *in vivo* and *in vitro*, *Lab. Invest.*, 55: 226.

Berner, P. F., Somlyo, A. V., and Somlyo, A. P., 1981, Hypertrophy-induced increase of intermediate filaments in vascular smooth muscle, *J. Cell Biol.*, 88: 96.

Bevan, J. A., Bevan, R. D., Chang, P. C., Pegram, B. L., Purdy, R. E., and Su, C., 1975, Analysis of changes in reactivity of rabbit arteries and veins two weeks after induction of hypertension by coarctation of the abdominal aorta, *Circ. Res.*, 37: 183.

Brody, W. R., Kosek, J. C., and Angell, W. W., 1972, Changes in vein grafts following aorto-coronary bypass induced by pressure and ischemia, *J. Thorac. Cardiovasc. Surg.*, 64: 847.

Chamley-Campbell, J. H., Campbell, G. R., and Ross, R., 1981, Phenotype-dependent response of cultured aortic smooth muscle to serum mitogens, *J. Cell Biol.*, 89: 379.

Clowes, A. W. and Schwartz, S. M., 1985, Significance of quiescent smooth muscle migration in the injured rat carotid artery, *Circ. Res.*, 56: 139.

DeBiasio, R. L., Wang, L.-L., Fisher, G. W., and Taylor, D. L., The dynamic distribution of fluorescent analogues of actin and myosin in protrusions at the leading edge of migrating Swiss 3T3 fibroblasts, *J. Cell Biol.*, 107: 2631.

Dilley, R. J., McGeachie, J. K., and Prendergast, F. J., 1988, A review of the histologic changes in vein-to-artery grafts, with particular reference to intimal hyperplasia, *Arch. Surg.*, 123: 691.

Dobrin, P. B., Littooy, F. N., and Endean, E. D., 1989, Mechanical factors predisposing to intimal hyperplasia and medial thickening in autogenous vein grafts, *Surgery*, 105: 393.

Fatigati, V. and Murphy, R. A., 1984, Actin and tropomyosin variants in smooth muscles: dependence of tissue type, *J. Biol. Chem.*, 259: 14383.

Fingerle, J., Johnson, R., Clowes, A. W., Majesky, M. W., and Reidy, M. A., 1989, Role of platelets in smooth muscle cell proliferation and migration after vascular injury in rat carotid artery, *Proc. Nat'l. Acad. Sci. U.S.A.*, 86: 8412.

Guthaner, D. F., Robert, E. W., Alderman, E. L., and Wexler, L., 1979, Long-term serial angiographic studies after coronary artery bypass surgery, *Circulation*, 60: 250.

Johansson, B., 1976, Structural and functional changes in rat portal veins after experimental portal hypertension, *Acta Physiol. Scand.*, 98: 381.

Kocher, O., Skalli, O., Bloom, W. S., and Gabbiani, G., Cytoskeleton of rat aortic smooth muscle cells: Normal conditions and experimental intimal thickening, *Lab. Invest.*, 50: 645.

Lawrie, G. M., Lie, J. T., Morris, G. C., and Beazley, H. L., 1976, Vein graft patency and intimal proliferation after aortocoronary bypass: Early and long term angiopathologic correlations, *Am. J. Cardiol.*, 38: 856.

Libby, P. and O'Brien, K. V., 1984, The role of protein breakdown in growth, quiescence, and starvation of vascular smooth muscle cells, *J. Cell. Physiol.*, 118: 317.

McKenna, N. M., Wang, Y.-L., and Konkel, M. E., 1989, Formation and movement of myosin-containing structures in living fibroblasts, *J. Cell Biol.*, 109: 1163.

Olivetti, G., Anversa, P., Melissari, M., and Loud, A. V., 1980, Morphometry of medial hypertrophy in the rat thoracic aorta, *Lab. Invest.*, 42: 559.

Owens, G. K., Loeb, A., Gordon, D., and Thompson, M. M., 1986, Expression of smooth muscle-specific alpha-isoactin in cultured vascular smooth muscle cells: Relationship between growth and cytodifferentiation, *J. Cell Biol.*, 102: 343.

Pollard, T. D., Satterwhite, L., Cisek, L., Corden, J., Sato, M., and Maupin, P., 1990, Actin and myosin biochemistry in relation to cytokinesis, *Ann. N. Y. Acad. Sci.*, 582: 120.

Sasaki, Y., Uchida, T., and Sasaki, Y., 1989, A variant derived from rabbit aortic smooth muscle: Phenotype modulation and restoration of smooth muscle characteristics in cells in culture, *J. Biochem.*, 106: 1009.

Seidel, C. L. and Murphy, R. A., 1979, Changes in rat aortic actomyosin content with maturation, *Blood Vessels* 16: 98.

Seidel, C. L., Lewis, R. M., Bowers, R., Bukoski, R. D., Kim, H-S., Allen, J. C., and Hartley, C., 1984, Adaptation of canine saphenous veins to grafting: correlation of contractility and contractile protein content, *Circ. Res.*, 55: 102.

Seidel, C. L., White, V., Wallace, C., Amann, J., Dennison, D., Schildmeyer, L. A., Vu, B., Allen, J. C., Navarro, L., and Eskin, S., 1988, Effect of seeding density and time in culture on vascular smooth muscle cell proteins, *Am. J. Physiol.*, 254: C235.

Seidel, C. L., Wallace, C. L., Dennison, D. K., and Allen, J. C., 1989, Vascular myosin expression during cytokinesis, attachment, and hypertrophy, *Am. J. Physiol.*, 256: C793.

Seidel, C. L. and Schildmeyer, L. A., 1987, Vascular smooth muscle adaptation to increased load, *Ann. Rev. Physiol.*, 49: 489.

Seidel, C. L., Allen, J. C., Wallace, C. L., Jemelka, S. K., Navran, S. S., and Dennison, D. K., 1990, Relationship between cell Na, Na pump number, myosin expression and proliferation in cultured canine vascular smooth muscle cells, *in:* "Molecular Biology of the Cardiovascular System", R. Roberts, M. D. Schneider, eds., Wiley-Liss, New York, p. 269.

Sellers, J. R. and Umemoto, S., 1989, Effect of multiple phosphorylations on movement of smooth muscle and cytoplasmic myosin, *in:* "Calcium Protein Signalling", H. Hidaka, E. Carafoli, A. R. Means, T. Tanaka, eds., Plenum Press, New York, p. 299.

Spray, T. L. and Roberts, W. C., 1977, Changes in saphenous veins used as aortocoronary bypass grafts, *Am. Heart J.*, 94: 500.

Tracy, R. E., Strong, J. P., Toca, V. T., and Lopez, C. R., 1979, Atheronecrosis and its fibroproliferative base and cap in the thoracic aorta, *Lab. Invest.*, 41: 546.

Unni, K. K., Kottke, B. A., Titus, J. L., Frye, R. L., Wallace, R. B., and Brown, A. L., 1974, Pathologic changes in aortocoronary saphenous vein grafts, *Am. J. Cardiol.*, 34: 526.

BIOPHYSICAL AND BIOCHEMICAL PROPERTIES OF "ASTHMATIC" AIRWAY SMOOTH MUSCLE

Newman L. Stephens, He Jiang and Chun Y. Seow

Department of Physiology
Faculty of Medicine
University of Manitoba
770 Bannatyne Avenue
Winnipeg, Manitoba, R3E 0W3, Canada

INTRODUCTION

It is fair to say that the lessons learned from basic asthma research have not afforded any major insights into fundamental mechanisms of smooth muscle mechanical function. However they have helped to focus on the fact that very specialized, basic mechanical processes are affected in "asthma". It is the use of analytic procedures developed in basic research that has helped sharpen our insight into asthma. Hopefully results obtained with these newer methods will lead to elucidation of the pathogenesis of asthma.

Perhaps one positive lesson that has been learned is that mechanical studies lead to the formulation of well-focussed biochemical research hypotheses. Without the former the latter would be merely fishing expeditions.

Prior to proceeding *in media res*, it must be pointed out that though the mandate for this conference is smooth muscle contraction, the liberty has been taken to deal with some aspects of the process of relaxation since it could be important in the pathogenesis of asthma, as it is, indeed, in hypertension.

Finally to place this presentation in its proper context it must be pointed out that our whole focus has been on what is clinically termed the early asthmatic response. This occurs an hour or two after exposure of a sensitized individual to specific antigen. The bronchospasm involves mainly the central bronchi (3rd to 5th or 6th generation) and mural inflammation and edema are not marked features. Diagnosis is made by recognizing reduction in specific airway conductance. The other phase of the disease is the late asthmatic response which occurs 6-8 hours after exposure to antigen. Typically the child wakes up in the early hours of the morning with respiratory distress. This phase is characterized by obstruction to ventilation in the peripheral airways (1 mm in diameter) and results not in alterations of airway conductance but of dynamic compliance. Infiltration with inflammatory cells, and edema of the wall of the airway are the hallmarks of the disease. The brunt of chronic

Regulation of Smooth Muscle Contraction
Edited by R.S. Moreland, Plenum Press, New York, 1991

disease is on similar airways but chronic inflammatory cells, such as lymphocytes, basophils, eosinophils, and macrophages make their appearance. A host of chemotactic factors, liberated from mast cells, make their appearance and several transmitters such as the leukotrienes, substance P, platelet activating factor, and NPY to name a few, are also involved. These do not seem to be involved to a major extent in the acute asthmatic response. The importance of chronic disease and the late asthmatic response is shown by the fact that more than 90% of the research in the field is devoted to it. In short, our research is directed at a clinically less important phase of the disease, but, because it occurs first, it may be more relevant to pathogenesis.

MODELS OF ASTHMA

Deficiencies in Animal Models

No adequate animal model of asthma exists at this time and for this reason the word asthma, strictly cannot be used in animal work. Either quotation marks - "asthma" - should be used, or, what is better, is to refer to it as allergic bronchospasm. Human asthma is characterized by non-specific factors in addition to allergy; the latter is not a *sine qua non.* Examples of the former are emotional stress, exercise, exposure to cold, respiratory infections, certain industrial occupations, and familial incidence. Furthermore, spontaneous attacks which are typical of human asthma have not been documented in animals. A final point to be made in this regard is that the allergic state in an animal model must be IgE based and not IgG, since human asthma is based on IgE.

Airway Smooth Muscle Hyperreactivity

It is perhaps best, at this juncture, to provide some definitions.

The term hyperreactivity is essentially a clinical one and does not exactly conform to a pharmacological definition; partly this is due to the fact that complete dose-response (methacholine or histamine) curves are not obtainable in severe asthmatics. Because of this, what the clinicians refer to as hyperreactivity is a combination of a leftward and upward shift (increased slope) of the curve. For *in vitro* work the practice described by Armour et al. (1988) is employed. It requires a dose-response curve that displays saturation; leftward shifts are termed increases in sensitivity, and changes in the slope of the linear part of the curve are recognized as alterations in reactivity.

Models of Acquired "Asthma"

Canine. The most effective are those in which immunization with either ovalbumin or ragweed pollen commences from the day of birth. Such a method has been described by our group (Kepron et al., 1977; Antonissen et al., 1979). Highly sensitized (IgE based) animals are obtained with hyperresponsive airways, hyperresponsiveness encompassing both increases in sensitivity and reactivity.

The adult ascaris-sensitized dog is a poor model, as antibody titres and airway responses are very variable and poor.

Guinea-pig. This model has been employed by Souhrada and Souhrada (1981). Immunization is carried out with intraperitoneal injections of ovalbumin and pertussis vaccine. The airway smooth muscle appears much more reactive than that of other models; however the immunity is based on IgG antibodies which makes it less desirable.

Horse. The condition of 'heaves' in horses bears striking similarity to human asthma. Our colleague, Dr. E.A. Kroeger, has conducted several studies on airway smooth muscle from horses with heaves and shown that it is hyperreactive. One unique feature and advantage is that the smooth muscle can be stimulated directly by electrical means. This is not so for muscle from practically all other species, where in the absence of cholinergic stimulation i.e. by acetylcholine release from the electrically stimulated nerves, no stimulation is possible.

Models of Inbred Hyperreactivity

The Basenji Greyhound. This model was developed by Carol Hirshman (Hirshman and Dorones, 1986) and has provided useful data relating to the influence of inbreeding on the airway hyperreactivity response.

The Polygenic Rat Model of Inbred Hyperreactivity. This was developed by Holme and Piechuta (1981) by sensitizing Sprague-Dawley rats and inbreeding the hyperreactors. By the 3rd generation 90% of the rats were hyperreactive. By arrangement with Merck, Frosst Inc. Canada, we have been able to obtain such rats and have shown that similar changes occur in airway smooth muscle from such rats as occur in the canine acquired model.

Rat Acetylcholine Hyperreactivity Model. This was developed by Levitt and Mitzner (1988) and is essentially a monogenic model in which the sole alteration is in acetylcholine receptor activity.

Model of Combined Disease

Since human asthma likely incorporates inbred and acquired factors a model incorporating these would be very useful. As it is possible to sensitize with ovalbumin the inbred hyperreactive rat model of Holme and Piechuta (1981), the desired model is at hand. It will be interesting to determine whether the effects of the mentioned factors are additive or interactive.

All our studies have been carried out on the canine model (Kepron et al., 1977; Antonissen et al., 1979) in which immunization of the pups begins on the day of birth.

ISOMETRIC FORCE (P_0) DEVELOPMENT

Popularity of Use; Shortcomings

For *in vitro* study of smooth muscle function, as for example in asthma, isometric measurement of force is time honored. The basic reason is that technically it is much easier to record isometric force data since all that is required is an isometric transducer. Isotonic measurements require more complex equipment. The shortcomings of isometric studies are as follows:

a) Isometric measurements yield data only permitting evaluation of tissue (or wall) stiffness, and not directly relating to shortening or constriction, which are the mechanical parameters of relevance.
b) Our studies with sensitized airway smooth muscle (Stephens et al., 1988) have shown that alterations in maximum isometric force development (P_o) are relatively late indicators of disease and are preceded by those in maximum isotonic shortening capacity (Δl_{max}) and velocity (V_o).
c) In support of P_o studies it must be conceded that crossbridge cycling is involved in this mode of contraction as it is in isotonic. However, the energetics of the two are quantitatively dissimilar in that isotonic shortening utilizes more energy which may, in turn, impose limits on physiological function. Hence isometric contraction may provide an overestimate of functional capacity.

Isometric studies are, at best those of steady state and are essentially static. The rate of tension development has been used as an indicator of shortening velocity. However it does not provide a measure of V_o which is the best indicator of isotonic shortening velocity and also provides an index of actomyosin ATPase activity. Furthermore it provides very little information about shortening ability since the maximum shortening that can occur is limited by the magnitude and properties of the muscle's series elastic component (SEC) and in any case is entirely an internal shortening. It cannot, for example, tell us much about the operation of the internal resistance to shortening (Stephens and Seow, 1987) which is an important component of shortening *in vivo.*

Normalization of P_o

In recent years considerable interest has developed in the effects of maturation of airway smooth muscle (Armour et al., 1984; Shioya et al., 1987; Mapp et al., 1989; Halayko et al., 1990). This has entailed comparison of contractile properties of airway smooth muscle at different loci and at different ages. Since the absolute force developed isometrically is dependent on muscle thickness which varies with age, to enable comparison of contractility, a suitable normalizing function is required. Some of these are as follows:
a) Measured P_o as a fraction of the maximum force developed in response to KCl. This is a very appropriate device and has only one drawback; in a poor muscle preparation the force developed may be only 10% of that expected. This would of course be normalized with respect to 80 mM KCl. While the ratio is quite valid, it should be remembered that one is working essentially at a low point on the response curve and results obtained may not apply at high points.
b) Maximum isometric force (P_o) expressed with respect to gram wet weight of tissue. This is an erroneous device as the force developed by a muscle only depends upon the actomyosin crossbridges in its cross-sectional area, and not on the bridges present in the remainder of the tissue. The latter are in series with each other, and therefore only pull on each other and do not transduct force to the outside world.
c) Maximum isometric force normalized with respect to cross-sectional area. This is probably the best normalization in current use. It converts raw force into tissue stress units. However it is only valid when applied to homogeneous structures, such as steel wires or rubber bands. Such homogeneity does not exist in biological tissues and its use there is only a first approximation. In skeletal muscle where the cross-section consists almost entirely of muscle cells, one can meaningfully talk about muscle stress. However in smooth muscle, where connective tissue can constitute more than

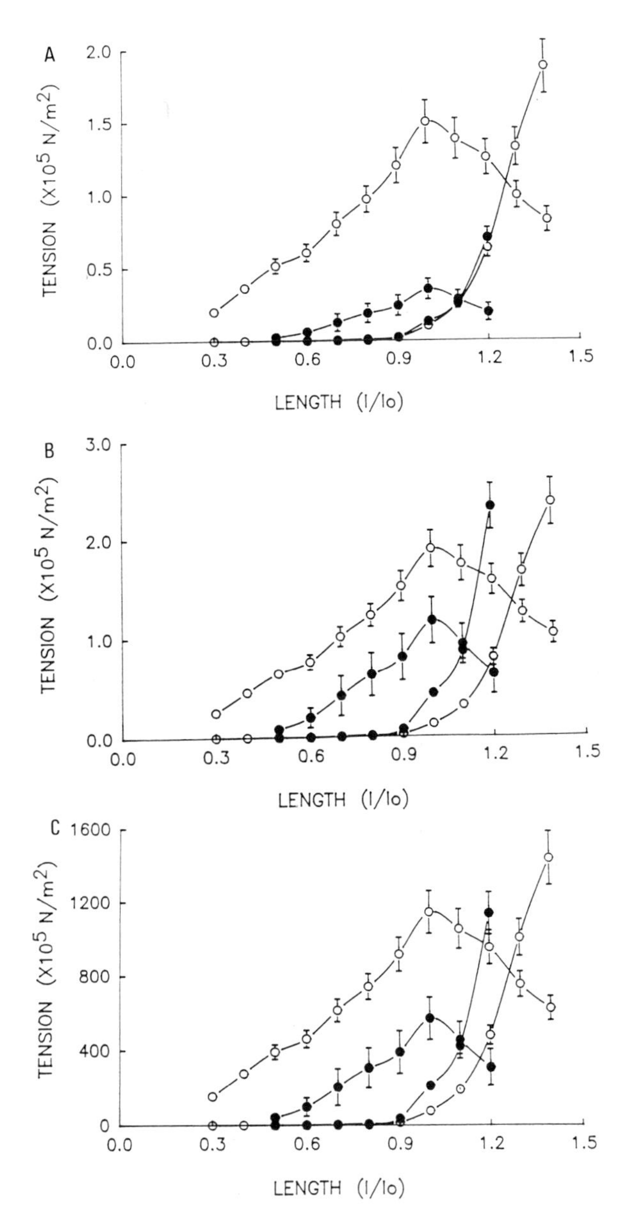

Figure 1. Length-tension relationships of tracheal (open circles) and bronchial (filled circles) smooth muscle computed with different methods of normalization. A: The active and resting tensions normalized with respect to total cross sectional area of the strip were plotted against relative lengths of the tracheal (TSM) and bronchial (BSM) muscle strips. B: The tensions were normalized with respect to the actual muscle proportions in the muscle strips. C: The tensions were normalized with respect to the myosin heavy chain content of the tissue cross-section. While the initial difference in unadjusted tension between the TSM and BSM was 5.24 fold, after normalization with respect to tissue cross-sectional area (tissue stress) this difference was reduced to 4.24 fold; after normalization with respect to muscle cross-sectional area it was further reduced to 1.60. Normalization with respect to myosin cross-section as a fraction of the total tissue cross-section yielded a difference of 2.16 fold.

50% of the cross-section, muscle stress carries less validity. In such a case force must by normalized with respect to the cross-sectional area of muscle cell present in the cross-section of the total tissue which yields a measure of muscle cell stress. This is not the end of the story however as the structures actually producing force are the actomyosin crossbridges present in a half sarcomere. Therefore accurate normalization should be with respect to the amount of contractile protein present in the tissue cross-section. An indirect measure of this may be obtained by measuring the quantity of these proteins in unit weight of tissue. Under ideal circumstances this should be the same as the proportion in the tissue cross-section. At any rate this provides a measure of contractile protein (actomyosin) stress.

Figure 1A-C shows stress-length curves for bronchial and tracheal smooth muscle. Whole tissue stress (tension/cm^2), muscle cell stress, and actomyosin stress are depicted. While the difference in maximum stress for whole tissue of the tracheal and bronchial smooth muscle shows a 5.24 fold difference; it drops to 4.24 fold after appropriate normalization, which demonstrates the crucial importance of normalization in assessing mechanical function in smooth muscle. We also noted that the content of actomyosin in cells from normal tracheal and bronchial smooth muscle was the same. This permits one to normalize with respect to muscle cross-sectional area alone. However the important caveat is that the actomyosin content must be the same for the muscles being compared. This cannot be taken for granted as we have recently obtained preliminary data to show that actomyosin content in sensitized airway smooth muscle is greater than that in control.

We have estimated actomyosin from band patterns in polyacrylamide (4%-20% gradient) gels. As stated before this measures the total actomyosin in the muscle. If the tissue is completely homogeneous then from the total tissue actomyosin the quantity of actomyosin in the cross-section can be computed. A more direct way, and one which we propose to use in the near future, is to carry out quantitative immunohistochemistry of muscle cross-sections treated with antibodies to myosin and actin.

One other feature worth mentioning is that the magnitude of force development depends upon the cycling rate of crossbridges. In airway smooth muscle where the bulk of developed isometric force results from the activity of latchbridges, the greater force seen (*vis-a-vis* that developed by normally cycling crossbridges) is due to the lesser velocity of the former. Presumably this results in a longer period over which actin and myosin remain attached and thus develop greater force.

Use of *in vivo* Airway Conductance Studies

In clinical work specific airway conductance in peak expiratory flow rate is measured in quantitating bronchoconstriction. This is fortunate because these measurements provide information relating to narrowing of airways which is the important variable to consider in diagnosing and managing asthma. The only shortcoming is that in cases of moderately severe or severe asthma, it is not possible to elicit a complete dose-response which precludes analysis in terms of sensitivity and reactivity. At any rate it is interesting to observe that while the majority of *in vitro* work is based on isometric force measurement which does not afford much insight into mechanisms of allergic bronchospasm, clinicians employ a parameter that does.

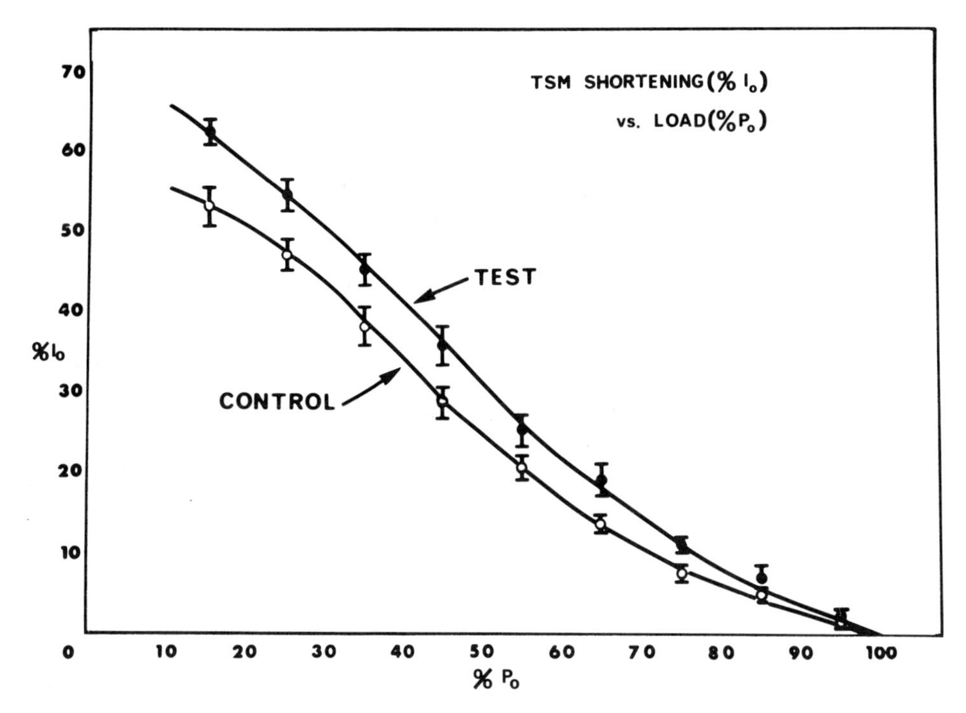

Figure 2. Isometric length-tension curves of canine tracheal smooth muscle, showing derived maximum shortening capacity (Δl_{max}) for a muscle shortening isotonically from optimal length (l_{max}, synonymous with l_o) carrying a load equal to resting tension at l_{max}.

SHORTENING PARAMETERS

From all that we have stated so far it is clear that with respect to studying bronchoconstriction *in vitro,* the analogous measurement to make is that of isotonic shortening. As elastic loading of the muscle progressively increases with shortening, due to the recoil of the cartilage and connective tissues, perhaps it would be better to study auxotonic shortening. However this has not been attempted yet and the work to be reported below is entirely restricted to isotonic shortening.

Maximum isotonic shortening (Δl_{max}) is the maximum shortening the airway smooth muscle is capable of, starting from its optimal length (l_o) and carrying a load equivalent to that it normally carries *in vivo.* The expectation is that it is increased in asthma. Δl_{max} can be measured in two ways.

Derived Isometric Measurement of Δl_{max}

From conventional length-tension curves of the type shown in Figure 2, the value of Δl_{max} can be deduced. The horizontal line indicates the capability of the muscle to shorten isotonically. This of course is an over-estimate since it has been derived from static measurements.

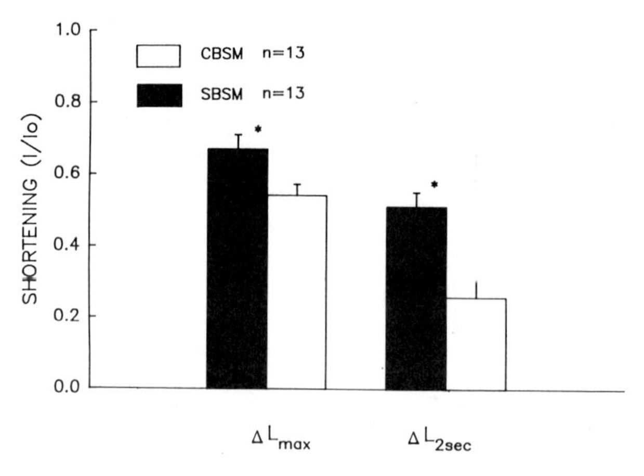

Figure 3. Maximum shortening capacities of sensitized and control bronchial smooth muscle; SBSM showed greater shortening capacity (Δl_{max}) than that of CBSM. The bulk of this shortening was effected at 2 seconds.

Isotonic Measurements

In vivo isotonic shortening would be less because of the extra energy utilized for shortening purposes since friction and inertia would require expenditure of energy to overcome these forces and therefore leaving less energy for shortening itself. Values of Δl_{max} obtained either from isometric or isotonic measurements are greater in sensitized airway smooth muscle as compared to control. Data for isotonic shortening is shown in Figure 3. This we feel is the most important piece of mechanical datum we have been able to obtain with respect to elucidation of the pathogenesis of asthma. This also was the first report in the literature that the intrinsic mechanical properties of the smooth muscle itself had become altered in asthma. Prior to this, the general notion was that the muscle itself was quite normal but was driven harder by increased amounts of liberated neurotransmitters.

Mechanisms of Increased Δl_{max}. Macklem (personal communication) has pointed out that the fact that airway smooth muscle can shorten by almost 80% of l_o, suggests that most people would be very prone to asthma unless there was some physiological process that cut short maximum shortening. He thought that reflex activation of the nonadrenergic noncholinergic inhibitory system (Coburn and Tomita, 1973; Richardson, 1979) whose putative transmitter is vasoactive intestinal polypeptide (VIP) was responsible. Impaired function of this system could result in asthma. Though undoubtedly present in human airways, its inhibitory effect is very weak and not sufficient to prevent the abnormally large degree of shortening. A second theory that he put forward was that normally the bronchi cannot narrow to extreme degrees because of the restraining effect of the lung parenchyma, the so-called interdependence effect. In asthma it is postulated that the compliance of the lung parenchyma increases and the restraint is removed from the airway, i.e. there is uncoupling of the parenchyma from the airway. A third notion, put forward by us (Stephens, 1987), is that as smooth muscle shortens it undergoes reduced activation and hence bronchoconstriction becomes limited. Siegman and colleagues have reported similar findings (Siegman et al., 1985). Her data showed that the innermost cells in a strip of smooth muscle remain unactivated. The mechanism for this is not known for smooth muscle. Rudel and Taylor (1974; 1989) have shown reduced activation in single fibers of skeletal muscle. Their data suggest that in shortening the thickness of the muscle increases and the T-tubules become distorted. Because of this, propagation of the action potential is defective, excitation-contraction coupling is impaired, and reduced activation develops. Failure of this mechanism would result in asthma. However, as smooth muscle does not possess T-tubules such an explanation is not tenable. Figure 3 shows another interesting and hitherto unreported feature, *viz.*, that while the total shortening at the end of the contraction (at 10 seconds from stimulus onset) is indeed increased, 75% of this is effected within the first 2 seconds which indicates that it is only the normally cycling crossbridges that are involved. Statistical analysis revealed that the shortening developed between 2 and 10 seconds was the same for the sensitized and control bronchial smooth muscle, i.e. that the latchbridges are unaffected.

The second shortening parameter that needs to be dealt with is that of maximum velocity. This is discussed below.

ALTERED COMPLIANCE OF THE INTERNAL RESISTANCE TO SHORTENING (IRS)

It is fairly evident that intracellular and intercellular structural constraints could limit the extent of shortening developed normally. Increases in the compliance of the structures responsible could result in the increased bronchoconstriction of asthma. That such a structure likely exists can be demonstrated by a simple experiment. If an electrically stimulated muscle is allowed to shorten maximally in isotonic mode, and if at the peak of this shortening the stimulus is withdrawn, the muscle re-elongates exactly to its starting length. Conversely, if it were forcibly stretched and the stimulus turned off at peak stretch the muscle would relax back to original length. Cardiac muscle physiologists have demonstrated the existence of an internal resistor and suggested its recoil is responsible in part, for filling of the ventricle in diastole.

Finally it must be pointed out that the internal resistor is probably part of the muscle's parallel elastic component (PEC), which is generally delineated by the muscle's passive length-tension curve. The internal resistor limits

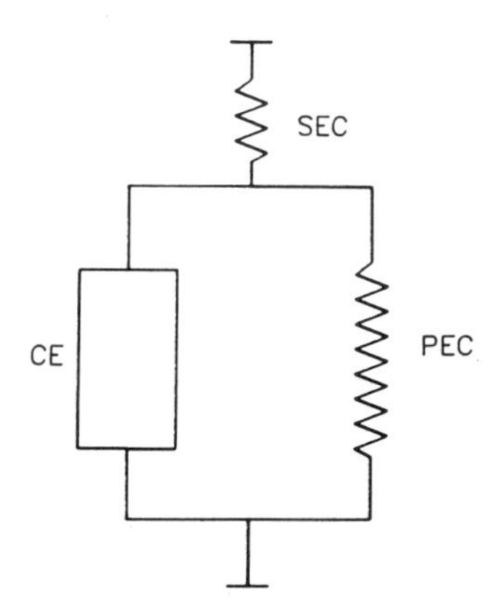

Figure 4. Voigt model for muscle. SEC = series elastic component; CE = contractile element; PEC = parallel elastic component.

shortening only when it is stressed in compression mode. The conventional PEC is usually studied in extension mode.

The Methodology for Measuring the Internal Resistance to Shortening

The rationale behind this is best demonstrated with the help of the Voigt model shown in Figure 4. The model choice is arbitrary. In tracheal smooth muscle (TSM) PEC is stressed very little at l_o since resting tension is very low. With only a little shortening, tension falls to zero and the PEC spring adopts its neutral position. Further shortening results in compression of the PEC which sets a limit on the shortening ability of the muscle. It would also internally load the contractile element (CE) whose velocity would therefore also drop. Employing these considerations, the mechanical properties of the internal resistance to shortening (IRS) can be measured; the only assumption to be made is that PEC compression loads the CE in the same way that the SEC loads it when an external load is applied. The methods requires the combination of the force-balance equation and A. V. Hill's equation for the force-velocity relation for muscle (1938). In the former $F(t) = KX + \alpha V + mA$ where F(t) is force as a function of time, KX, αV, and mA represent elastic, viscous, and inertial forces, respectfully. In isotonic mode, SEC length is constant and KX is therefore equivalent to P and F(t) can be replaced by $P_o(t)$. In smooth muscle, where velocity is very low, mA is usually ignored. The equation then becomes $P_o(t) = P + \alpha V$. Taking time into consideration, the Hill equation may be rearranged to $P_o(t) = P + [(P+a)/b]V$, where a and b are asymptotes of the

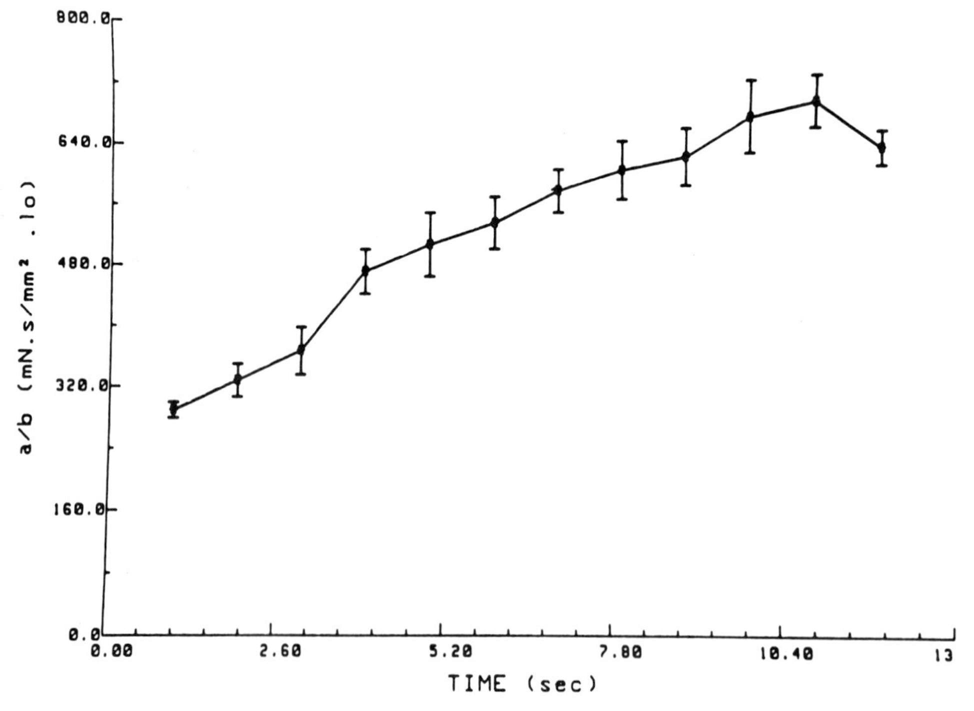

Figure 5. Internal resistance, a/b, as a function of time; a and b are constants derived from hyperbolic force-velocity curves elicited by quick-release experiments at 1 second intervals during an isometric contraction. The constant a has units of force while b has units of velocity. (From Stephens and Seow, 1987, by permission)

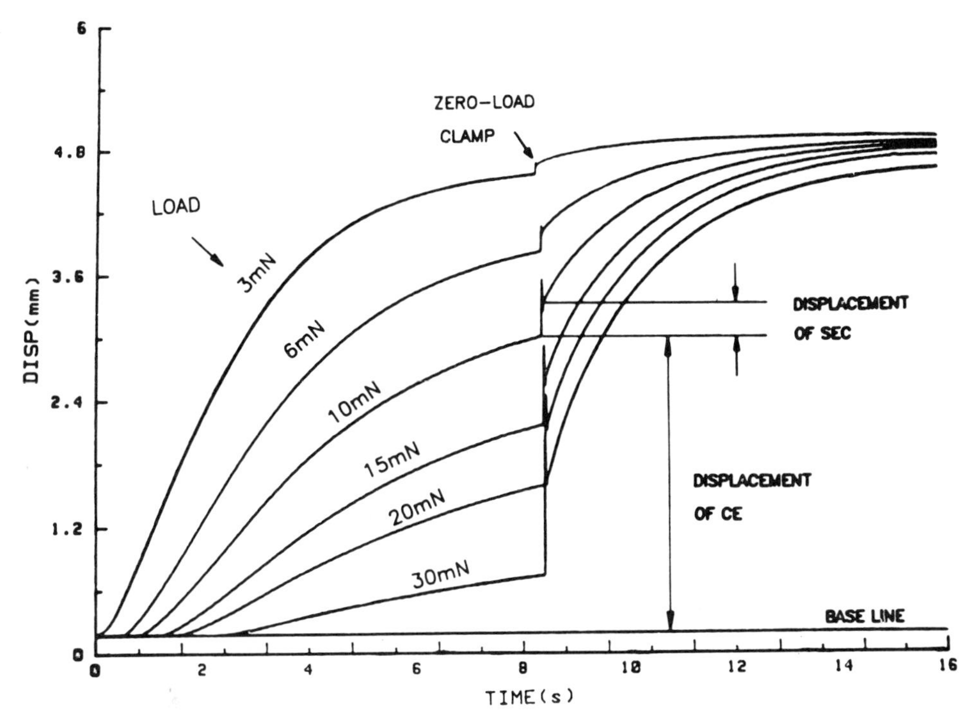

Figure 6. Displacement curves showing zero-load clamps applied to isotonic contractions under different loads. (From Stephens and Seow, 1987, by permission)

hyperbolic force-velocity curve with units of force and velocity, respectively. The two equations can then be written thus:

$P + \alpha V = P + [(P+a)/b]V$: From this, α, which we are terming as an index of internal resistance to shortening, may be solved for: $\alpha = (P+a)/b$; when $P = 0$; $\alpha = a/b$.

Experimentally values of a/b can be obtained from force-velocity curves elicited at consecutive second intervals. Figure 5 shows a plot of a/b versus time; it is clear that a/b increases with time, from which we conclude that internal resistance to shortening increases with time and reduces velocity.

We have recently described a method to directly measure the tension-compression properties of the PEC based on the theoretical considerations mentioned above (Stephens and Seow, 1987). Zero load clamps were applied at the same instant in time in the course of several isotonic contractions all starting at different lengths determined by different preloads. Records from a typical experiment are shown in Figure 6. A plot of zero load velocities, computed from the records, versus contractile element length is shown in Figure 7. In Figure 8 this curve is superimposed on a force-velocity curve elicited from the same muscle. The abscissa represents load for the force-velocity curve and contractile element length for the length-velocity phase plane. At iso-velocity points the corresponding length and load parameters must represent tension-compression points for the PEC. A curve is shown in Figure 9.

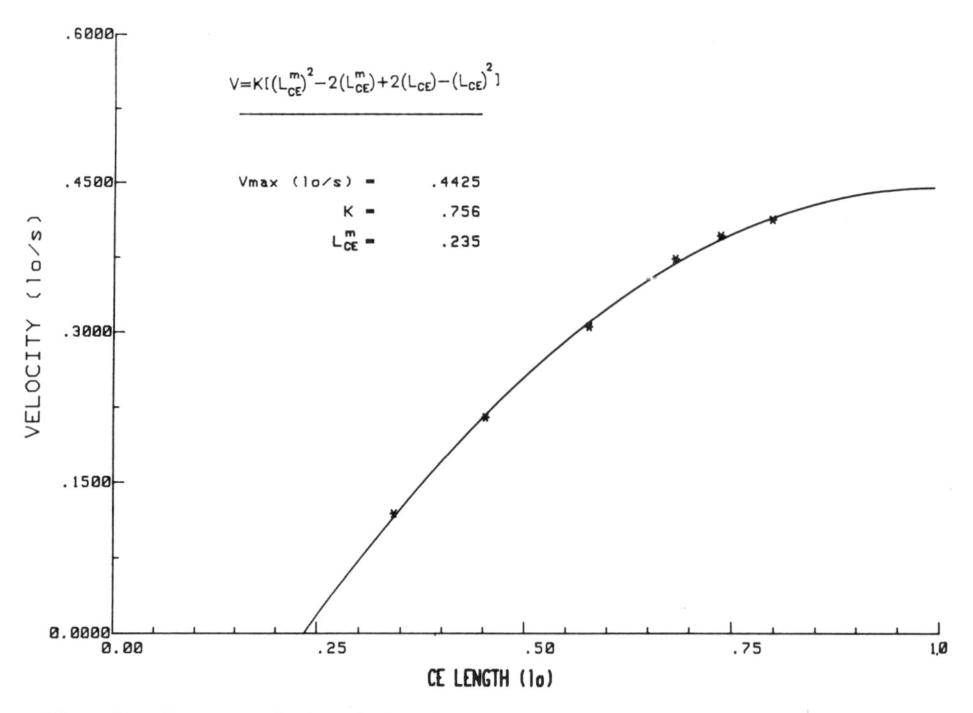

Figure 7. V_o versus CE length data fitted by a parabolic function: $V_o = V_{max}(1-[(1-l_{CE})/(1-l_{CE}M)]^2$, where V_{max} is the maximum velocity at zero-load which occurs at l_o; l_{CE} is the contractile element length; and $l_{CE}M$ is the maximally contracted contractile element length at zero load (From Stephens and Seow, 1987, by permission).

Comparison of Compliance of IRS of Airway Smooth Muscle from Control and Sensitized Dogs

A similar study was conducted in sensitized TSM. Curves for control and sensitized muscles are shown in Figure 10. It is clear that the sensitized muscles possess a more compliant PEC (or IRS) which could account for the increased shortening ability of this muscle.

Speculations on the Causes of Increased Compliance of IRS

There are two possible causes for the increased compliance.

Intercellular and Interfascicular Collagen and Elastin. This can be tested for by examining single airway smooth muscle cells. These would necessarily be free of connective tissue, and, if alterations in compliance still persisted they would have to be stemming from the IRS.

Intracellular Structures such as are made up of Cytoskeletal Proteins. Among these are desmin, vimentin, α–actinin, and others that have been studied by Rasmussen et al. (1987). Changes in these cytoskeletal elements may result in increased compliance of the PEC. Whether these changes are determined by altered states of phosphorylation remains to be seen.

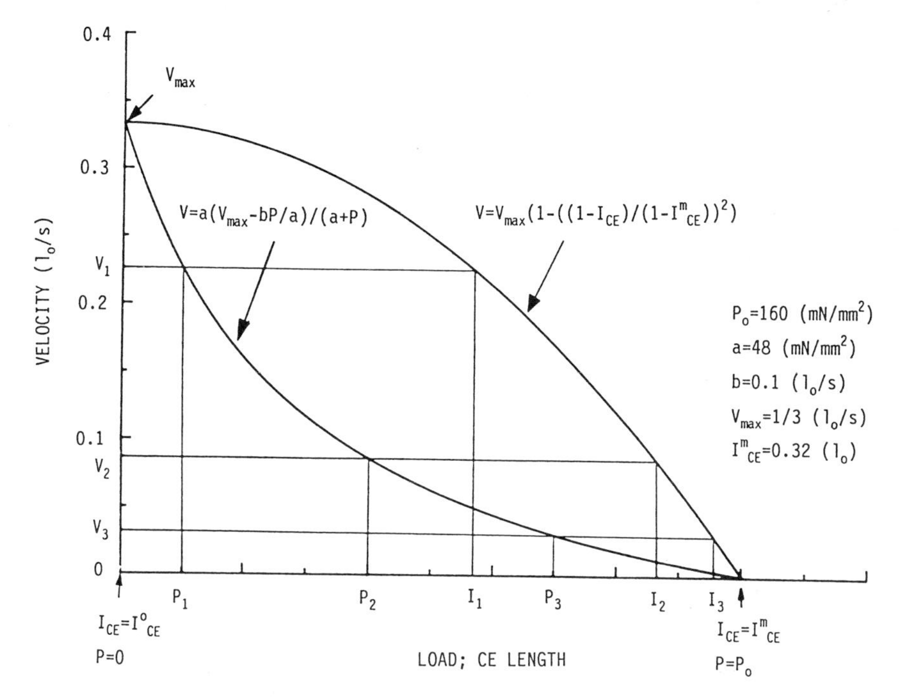

Figure 8. Illustration of the method for obtaining the PEC's tension-compression curve (From Stephens and Seow, 1987, by permission).

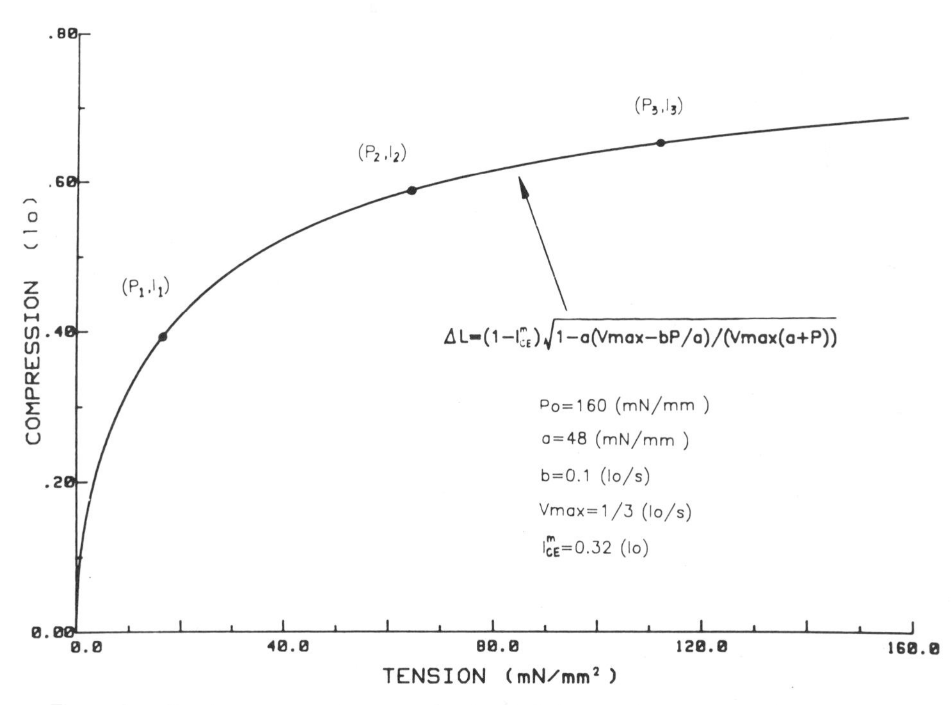

Figure 9. Tension-compression curve for the PEC obtained from Figure 8. (From Stephens and Seow, 1987, by permission).

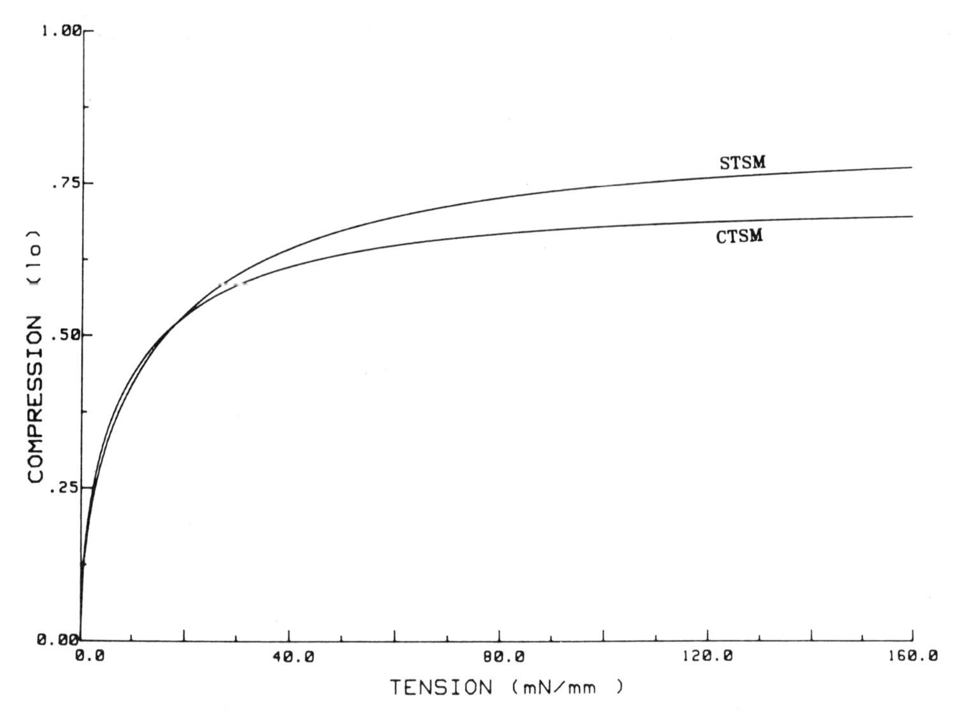

Figure 10. Tension-compression curves of sensitized and control canine trachealis.

The role of the cytoskeleton in the pathogenesis of asthma and hypertension could turn out to be important. Somlyo et al. (1984) have reported that in portal veins from animals made hypertensive by partial coarctation of the portal-anterior mesenteric vein, the very first change that is seen is a major increase in cytoskeletal elements.

DOES V_0 PLAY ANY ROLE IN THE PATHOGENESIS OF ASTHMA?

Though changes in resistance to airflow in airways and to blood in arteries and veins must depend upon shortening of the mural smooth muscle, sufficient attention has not been devoted to shortening. In studies of striated muscle both P_0 and V_0 have been extensively studied. In smooth muscle however, mainly P_0 has been investigated.

The role of V_0 is of predominant importance in the case of cardiac muscle. Stroke output must depend upon adequate muscle shortening. However, because the heart muscle can only contract in twitch mode, time becomes a limiting factor. Hence if the velocity of shortening is low, adequate ejection cannot be affected.

The Two Second Limitation

Initially we had thought that as airway smooth muscle can be easily tetanized and has a contraction time of about 10 seconds, velocity would not limit maximum shortening capacity. However, our measurements have shown that 80% of the muscle's shortening is achieved within 2 seconds, which is when normally cycling crossbridges are active. Hence reduced velocity of shortening could seriously reduce shortening. Therefore in airway smooth muscle, velocity is an important regulator of shortening capacity. Our studies have shown that both V_o and Δl_{max} (due to crossbridge activity; mainly latchbridges) after the first 2 seconds of stimulation are not different between sensitized and control TSM.

Biochemical Mechanisms of Increased V_o of STSM

V_o, unloaded shortening velocity, is an interesting parameter as it is an index of actin-activated myosin Mg^{2+}-ATPase activity. Bárány (1967) has shown a linear relationship between these two variables for a variety of muscles.

Table 1. ATPase Activity of Myofibrils

nmol P_i, mg myosin^{-1}, min^{-1}

STSM	499.29 ± 11.49	(n=7)
CTSM	345.00 ± 6.81	(n=7)

STSM = sensitized tracheal smooth muscle; CTSM = control tracheal smooth muscle.

Myofibrillar ATPase. We have measured this employing a technique reported by Sobieszek and Bremel (1975). Table 1 shows that the values for myofibrillar ATPase in STSM are significantly greater than those for CTSM.

Myosin Heavy Chain (MHC) Isozymes. In an attempt to determine the mechanism underlying the increased ATPase activity, we undertook a study of MHC isozymes of the TSM. Hoh et al. (1978) have shown that for cardiac muscle, three MHC isozymes are present which they termed V_1, V_2, and V_3. Isozyme V_1 of greatest molecular weight and ATPase activity, was seen when V_o was elevated as for example after treatment with thyroxine. Isozyme V_3 possessed the lowest molecular weight, lowest ATPase activity, and was associated with the lowest V_o.

We carried out SDS-polyacrylamide gel (4%) analysis and demonstrated the presence of two isozymes, of molecular weights 204,000 and 200,000, respectively. Using specific polyclonal antibodies, kindly given to us by Dr. R. Adelstein, and Western blot analysis we confirmed the two bands were indeed those of myosin heavy chains. Occasionally a third band is also seen (molecular weight 196,000 daltons). Western blots using platelet myosin have shown it was a non-muscle myosin isozyme. We usually only see this third band when we use a heavy crude extract load for the gels. Under these conditions the separation between the other two muscle myosin bands is lost, which, for most of our experiments, forces us to use light loads. At any rate we have found no difference in band pattern with sensitization. Figure 11 shows isozyme patterns for STSM and CTSM. Western blots and densitometrograms are also shown.

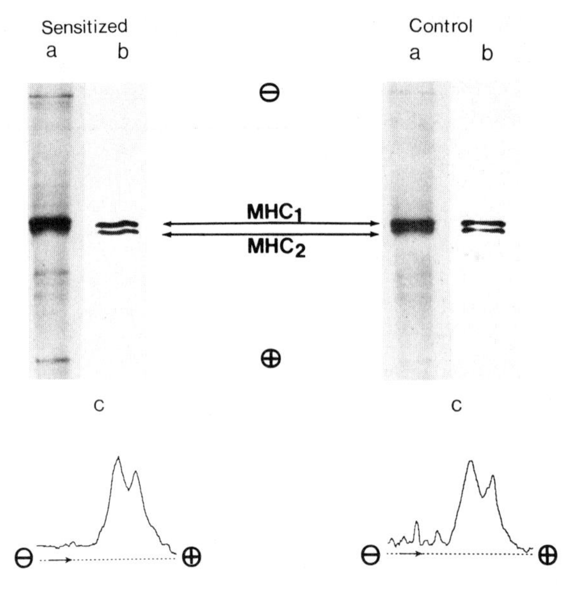

Figure 11. Separation of myosin heavy chain isoforms from STSM and CTSM using 4% SDS-PAGE. a: Silver stained gels; total crude protein loaded was 3 µg. b: Western blots from the 4% SDS-PAGE gels, total crude protein loaded was 0.4 µg. Primary antibody was specific for smooth muscle myosin heavy chains (rabbit, anti-chicken gizzard immunoglobulin G (IgG). c: Laser densitometrograms of silver stained gels shown in a.

Myosin light Chain (MLC) Phosphorylation. Because myosin isozyme distribution patterns are unchanged in the STSM as compared to the CTSM, the remaining possibility to account for the increased myofibrillar ATPase activity was increased phosphorylation of MLC. Two-dimensional gel analysis revealed such was the case as shown in Figure 12.

MLC Kinase. The logical next step was to examine the MLC kinase as it is this enzyme that may be responsible for the increased MLC phosphorylation in the sensitized TSM. Decreased activity of the MLC phosphatase could also yield the same result, but its study has been left to the near future. Using antibodies supplied us by Dr. Mary Pato, we have demonstrated that MLCK content in STSM is greater than that in CTSM. This is shown in Figure 13. Studies of the activity of the enzyme represent another promissory note.

Future Studies

These will primarily involve gene expression.

mRNA-MLCK. We have initiated studies of gene expression of MLC kinase using cDNA probes (chicken gizzard kinase) kindly provided by Dr. Vince Guerriero. With extracts from chicken gizzard and esophagus we have shown that the probes are very effective in identifying messenger RNA for MLCK. However, we have not as yet succeeded in identifying mRNA in canine TSM. Either activity across species is a problem - this seemed unlikely as we felt that MLCK sequences would be highly conserved - or we need to explore a range of stringency conditions. These explorations are underway.

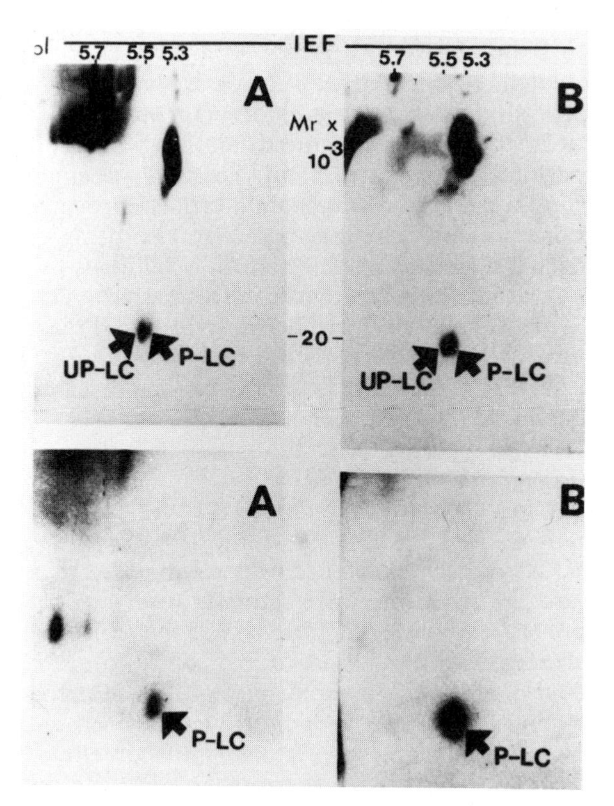

Figure 12. Autoradiograms showing phosphorylated 20 kDa myosin light chain (MLC) separated by 2-dimensional electrophoresis from crude extracts of control (A) and sensitized (B) TSM. Whole cell homogenates were incubated in a reaction mixture (in mM: 0.5 $CaCl_2$; 2 $MgCl_2$; 1 cysteine, 60 KCl, 40 imidazole, pH 6.9), containing ouabain (1 mM), sodium azide (10mM) and 1 mM [γ-^{32}P] ATP for 12 seconds. Top: Coomassie blue stained gels; Bottom: corresponding autoradiogram profiles. Phosphorylated (P-LC) and unphosphorylated (UP-LC) 20 kDa myosin light chains are indicated by arrows.

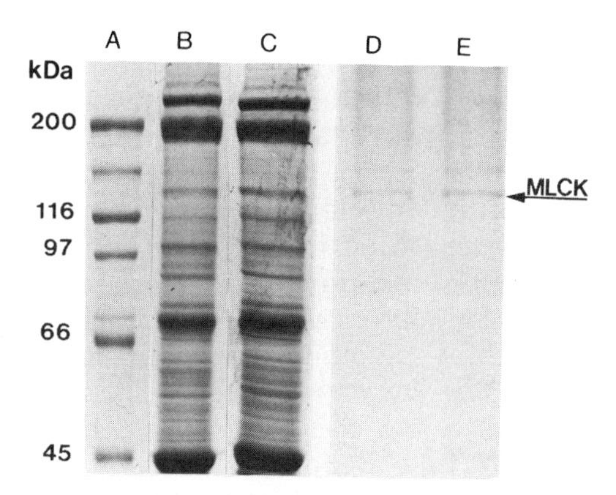

Figure 13. Identification of myosin light chain kinase (MLCK) in gel profiles of STSM and CTSM crude extracts separated by 7.5% SDS-PAGE. A: Molecular weight markers; B and C: Coomassie blue stained gel lanes, showing control and sensitized tracheal muscle band patterns respectively; crude protein load 10 µg; D and E: Western blots of CTSM and STSM samples, respectively; crude protein load was 8 µg for each; primary antibody used was specific for smooth muscle myosin light chain kinase (rabbit, anti-chicken gizzard immunoglobulin G). Antibody was a gift from Dr. Mary Pato of the University of Saskatchewan.

Restriction Fragment Length Polymorphism (RFLP). On a different tack we have also initiated studies of gene linkage analysis. These are running *pari passu* with similar studies that we are carrying out with respect to essential hypertension. In brief, we have identified families with asthma occurring in at least two generations. A positive family history, documented evidence of clinical attacks, and a positive methacholine bronchoprovocation test are being used as phenotypic markers. Blood samples are also being collected. From the white cells, nuclear DNA will be subjected to RFLP analysis. Patterns obtained from asthmatics and non-asthmatics will be compared. The co-segregation of polymorphism with the phenotypic markers will be followed. The hope is to eventually determine - admittedly a very long shot - what particular gene is responsible and what gene product is involved. The disease is likely polygenic in origin which compounds the problem further.

DEVELOPMENT OF A BRONCHIAL SMOOTH MUSCLE PREPARATION

For the last 20 years we have used the canine TSM as a model for central bronchi which are the flow limiting segments for ventilation and which are the airways responsible for the bronchospasm of the early asthmatic response.

As the objective of our research has been to study smooth muscle contractility in such a way as to make meaningful deductions about subcellular mechanisms of contraction, the bronchus was not suitable because of the presence of multiple cartilaginous plaques that contributed additional and undesirable friction and inertia. Since we believed at the time that the airway smooth muscle was directly attached to the plaques, there seemed to be no way of obtaining a strip of BSM free of injury. One of us (H. J.) noted in micrographs of the bronchus published 40 years ago (see von Hayek, 1990), that the muscle was not directly inserted into the superjacent plaques but via a layer of very loose connective tissue. He was therefore encouraged to try to dissect out intact muscle.

Length-tension curves for such a prepared strip of BSM have been shown above in Figure 1. Analysis showed that mean P_0 when converted to stress units was significantly increased in the cartilage-free preparation compared to the stress measured in a cartilage-intact strip. This was surprising as isometric force or stress is essentially a static measurement. The increment is probably more apparent than real and arises from the fact that the weight of the tissue is less after cartilage is removed. As it is from this weight and l_0 that the cross-sectional area of the muscle is computed, it is this decrease in weight which probably results in the greater stress.

There was no significant difference in the stress developed by control (CBSM) or sensitized (SBSM)- bronchial smooth muscle which confirms what we have reported for tracheal smooth muscle (Antonissen et al., 1979).

Force-velocity studies employing abrupt load-clamp techniques have elicited curves that conform to a rectangular hyperbola (Jiang and Stephens, 1990) and are fitted by the Hill equation (Hill, 1938). The maximum velocity of shortening of the sensitized BSM is higher than that of the TSM and also higher than that of the control BSM. By employing zero load clamps (which provide measures of maximum velocity of shortening (V_0)) at consecutive 1 second intervals during the course of an isometric contraction we have shown (Jiang and Stephens, 1990) that normally cycling and latchbridges are both present in BSM. Studies of sensitized BSM have just been initiated.

From the same data, we have also been able to compute the maximum shortening capacity (Δl_{max}) of the muscle. Δl_{max} for a muscle shortening from l_o and bearing a preload equal to the resting tension at l_o, was greater in the strip from which cartilaginous plaques had been removed.

This confirms that the removal of the cartilage plaques had not damaged the smooth muscle. It also led us to give up the TSM as a model for studying BSM now that the latter is directly available. While this is true for biophysical studies, for biochemical the TSM enjoys the convenience that more starting material can be obtained from it.

STUDIES OF INBRED HYPERREACTIVITY OF AIRWAY SMOOTH MUSCLE

Though not conclusive, there is fairly strong evidence that a congenital component is inherent in the pathogenesis of human asthma. For several members in a family to be attacked is quite common. With the exception of a few studies (Holme and Piechuta, 1981; Hirshman and Dorones, 1986; Levitt and Mitzner, 1988) the vast majority of others have dealt exclusively with models of acquired disease.

We have been able to obtain, from Merck Frosst Inc. Canada, Sprague-Dawley rats with inbred hyperreactive airways. Fisher rat airways are the best controls for the hyperreactive strain. The Sprague-Dawley rats are immunized with ovalbumin and pertussis vaccine, and then challenged with specific antigen. About 45% of the animals hyperreact with dyspnea. Inbreeding of rats selected from this group results in 90% of the animals developing hyperreactive airways in the F2 generation. The rapid development of a high prevalence of hyperreactivity suggests that changes in the expression of a few genes is all that is required.

Length-tension, force-velocity, and zero load clamp studies have been carried out on tracheal smooth muscle from control and test rats. As the muscle strips were about 1.5 to 2.0 mm in length, special techniques were developed to obtain valid measurements (Xu et al., 1990). Essentially these were modifications of methods employed by investigators of single skeletal muscle fibres (Chiu et al., 1987). A preliminary report of our results has been made (Xu et al., 1990). The mechanical data are qualitatively very similar to those of the canine TSM and BSM. Quantitative differences are that V_o is greater which is perhaps not unexpected. However, maximum developed stress is about 50% less than that of canine muscle. We have no explanation for this. All we can say is, *pro tem*, that this is probably due to muscle damage as V_o and Δl_{max} are decidedly normal.

STUDIES OF HUMAN AIRWAY SMOOTH MUSCLE

This is, of course, the only tissue in which definitive studies for the elucidation of allergic or nonspecific hyperreactivity should be carried out. However there are serious practical problems. Firstly, adequate numbers of tissue samples are hard to come by since, fortunately, the mortality rate is low. Secondly, whether obtained post-mortem or post-surgical resection, the tissue inevitably deteriorates before *in vitro* studies can be carried out. Yet studies of human tissue are a *sine qua non* since qualitative differences exist between it and animal models. For example, while a nonadrenergic, noncholinergic inhibitory nervous system is present in humans (Richardson, 1979), it is absent in dogs.

The only solution seems to be to obtain punch biopsy material obtained during bronchoscopy. Miniaturizing the method we have described above for rat TSM, may suffice to obtain reliable data for human bronchial smooth muscle.

RELAXATION OF AIRWAY SMOOTH MUSCLE

Relaxation of smooth muscle is a process that has been relatively neglected, and such studies as exist, have not probed basic mechanisms to the same depth as in skeletal muscle (Jewell and Wilkie, 1960). Shibata and Cheng (1977) have reported that vascular smooth muscle from hypertensive rats shows an increased relaxation time which could contribute to the magnitude of the hypertension.

We have conducted a pilot study of relaxation in airway smooth muscle with the objective of developing a valid index that would provide information about the status of the muscles' contractility during relaxation.

As isometric relaxation provides information mainly about the waning of muscle stiffness we did not think it was a good index of relaxation (vasodilation or bronchodilatation *in vivo*) which is really re-elongation of a shortened muscle.

We have measured isotonic relaxation following a preloaded isotonic contraction and computed the half-time. This could be used to compare isotonic relaxation in control and sensitized airway smooth muscle. However, this parameter is load dependent. To eliminate load we have computed half-times for relaxation following a series of contractions with different loads. The half-times were plotted against their loads and a $t_{1/2}$ obtained for zero load by fitting a curve to the plot and estimating the value from the resultant equation.

One other problem is that relaxation at different loads is a function of contractile element length which must vary as a result of the varying loads. To eliminate this, the muscle was allowed to shorten isotonically carrying a preload equal to the resting tension at l_o. At a given moment a series of abrupt loads were applied in consecutive contractions. Simultaneously with the load clamp, the electrical muscle stimulus was also turned off. Each isotonic relaxation showed a rapid transient followed by a slow transient, the former represented the abrupt elastic re-elongation of the SEC, the latter representing re-elongation of the contractile element. The latter always commenced from the same initial contractile element length. Zero load half-time computed from such curves was therefore independent of initial CE length and of load. We feel that this is the appropriate parameter to use in comparing relaxation of control and sensitized airway smooth muscles. Once the pilot study is completed the studies involving the airways from sensitized animals will commence.

With respect to mechanisms, we have identified three phases during isotonic relaxation. These follow naturally from the reverse sigmoid curve shape of the record. The first phase which is curvilinear is probably due to reduction in the rate of ATP hydrolysis. The second phase is linear and while we do not know the precise mechanism our speculation is that ATP hydrolysis is waning even more and the load on the muscle is able to re-elongate the muscle more rapidly. Whether the recoil of the compressed internal resistor is contributing to re-elongation is conjectural. Since we have found that the compliance of this resistor is increased with sensitization we expect analysis to show a slower linear phase of relaxation in sensitized muscle. The final phase is again curvilinear with the concavity being upward. The curvilinearity is only seen at low loads. Preliminary experiments on relaxation show that active contraction redevelops during this phase of contraction; this would result in slowing of re-elongation. Time resolved studies of ATP hydrolysis using fast freezing techniques will help determine whether the muscle has become active again.

MISCELLANY

A host of mechanisms remain that have not been dealt with in this chapter but which should be mentioned. An itemized list of these is follows.

Properties of the Membrane

This includes both electrical and biochemical. Alterations in sensitivity of the type reported in sensitized airway smooth muscle could stem from changes in resting membranes potential and threshold. Souhrada and Souhrada (1981) have shown that sensitization of guinea-pig airway smooth muscle results in hyperpolarization. However, challenge with specific antigen leads to depolarization above the level of the normal resting membrane potential. This could account for the increased sensitivity of sensitized ASM.

In sensitized canine TSM we do not observe any hyperpolarization and find instead a 5-7 mV depolarization from the control value of 50 mV ± 4.2 (SE).

Considerable interest has developed recently in the electrophysiological mechanism of smooth muscle relaxation. In this connection the role of Ca^{2+}-dependent and ATP-dependent K^+ channels is being studied to determine whether the changes seen in hypersensitive ASM result from alterations in K^+ conductance. No clear cut decision has as yet been arrived.

Excitation-Contraction Coupling (ECC)

With respect to normal mechanisms, it has been shown that ECC is of two types, electromechanical and pharmacomechanical (Somlyo and Somlyo, 1968). While several studies have elucidated mechanisms of ECC in normal smooth muscle, relatively few have been undertaken to determine what changes occur in sensitized airway smooth muscle. In pharmacomechanical coupling, ligand binds to membrane receptor and through G protein interaction in which attached GDP dissociates to be replaced by GTP leading to dissociation of the α, β, and γ subunits of the G protein. The dissociated subunit activates phospholipase C which results in the production of inositol triphosphate (IP_3) which in turn releases Ca^{2+} from intracellular stores. The reaction also produces diacylglycerol (DAG) which activates protein kinase C (PKC) a powerful second messenger.

Little information is currently available concerning coupling of receptors to effector mechanisms in ASM. However, the role of inositol phospholipid is undeniably important especially with respect to pharmacomechanical coupling.

With respect to electromechanical coupling, membrane depolarization induced by electrical stimulation or ligand binding activates a variety of Ca^{2+} channels, resulting in an increase in myoplasmic calcium with subsequent contraction; receptor-mediated events that result in membrane depolarization and voltage-operated Ca^{2+} channels are also involved.

Altered Sensitivity of the Contractile and Regulatory Proteins

Recent evidence (Somlyo et al., 1989) indicates that contraction may be regulated by alterations in sensitivity of the contractile proteins. It is likely that allergic bronchospasm could be due to increased sensitivity but, as yet, no data dealing with this issue have been reported.

The Role of the Epithelial Cell

Just as in blood vessels where endothelial cells secrete endothelium derived relaxing factor (EDRF; Furchgott and Zavodzki, 1980) and a powerful agonist called endothelin (ET; Yanagisawa et al., 1988), in bronchi, epithelial cells are said to play a similar role (Aizawa et al., 1988). The secreted EDRF and ET relax and contract, respectively, airway smooth muscle. Whether disturbances occur in their activities in asthma has not yet been reported.

Cell Adhesion Molecules (ICAM-1)

What the role of ICAM-1 is in the pathogenesis of the early asthmatic response is the subject of recent research. A recent report (Wagner et al., 1990) indicates that ICAM-1 may be an important adhesion molecule which is secreted from the surface of the epithelial cell. Evidently, receptors for this molecule exist on eosinophils and interaction between the two is responsible for the directed migration of the eosinophil into the site of the immune response.

Other adhesion molecules and their receptors probably exist for lymphocytes and macrophages which facilitate migration of these cells to the immune response site.

The Role of Inflammation

Considerable infiltration of the bronchial walls and parenchyma with chronic inflammatory cells is well established. This leads to thickening of the wall which contributes to increased reactivity via the operation of the so-called geometric factor. These changes occur in the late asthmatic response and in clinically chronic asthma. At least in airway smooth muscle from a dog which has only been sensitized and never challenged at a time when both V_0 and Δl_{max} have increased, we have not detected any signs of infiltration or of smooth muscle hypertrophy or hyperplasia after examining many histological sections. We believe therefore that inflammation is a secondary phenomenon.

The Role of the Blood Vessels in Asthma

The idea that changes in blood vessels in the wall of the airways lead to edema and inflammation of the wall with increased resistance to airflow is currently under investigation (Long et al., 1988; Wanner et al., 1988). We have shown that the sensitization process, not unexpectedly, is not confined to the airways. We have reported that the pulmonary blood vessels exhibit a strong Schultz-Dale response (Kong and Stephens, 1981) as does the saphenous vein (unpublished) that the hepatic vein, splenic artery, terminal ileum, and upper ureter are sensitized. Fortunately the antigen does not normally get to these sites so disease is not manifested, apart from the rare anaphylactic shock.

Therefore it is possible that intramural vessels are sensitized and their response to antigen contributes to rapid airway narrowing and asthma. No one has yet suggested that asthma is primarily a vascular disease, but the notion is not beyond the realm of possibility.

CONCLUSION

It is clear that a wide variety of structural and functional adaptations develop in antigen sensitized and challenged airway smooth muscle. Our studies indicate that in the early stage of disease, where only sensitization has occurred, changes in mechanical properties occur in the absence of any structural change. Nor do we see any inflammation, which, is in any case, the hallmark of chronic disease. What the relationship of the acute asthmatic response is to the late chronic state is not known.

Current studies indicate that changes in myosin light chain kinase content are linked to, and could be responsible for, the increased shortening velocity of the sensitized muscle. This increase is only seen in normally cycling crossbridges. Latchbridges appear unaffected in all the models we have examined to date. Perhaps one answer to the question of how is smooth muscle regulated, is an answer obtained from some of the adaptations seen in asthma. Identification of the type of crossbridge that is affected in asthma is crucially important as it points to the direction in which the next line of research should be directed. Derangement of normally cycling bridges suggests that the activity of the phosphorylating mechanism should be studied. Had changes in latchbridges been detected, studies of myosin light chain phosphatase activity would have been mandated.

With respect to determining the cause of the increased shortening capacity of sensitized muscle, studies of the biochemistry (which may ultimately come to include molecular biology) and biophysics of the cytoplasm of airway smooth muscle are evidently very important.

Finally, we believe that gene linkage analysis which includes delineation of inheritance patterns of phenotypic markers and cosegregation of patterns of restriction fragment length polymorphism will be of major importance in the study of asthma.

ACKNOWLEDGMENTS

Supported by an operating grant from the Council for Tobacco Research Inc. USA. H. J. and C. Y. S. were both supported by separate Medical Research Council of Canada Fellowships.

REFERENCES

Aizawa H., Miyazaki, N., Shigematsu, N., and Tomooka, M., 1988, A possible role of airway epithelium in modulating hyperresponsiveness, *Br. J. Pharmacol.,* 93: 139.

Antonissen, L. A., Mitchell, R. W., Kroeger, E. A., Kepron, W., Tse, K. S., and Stephens, N. L., 1979, Mechanical alterations of airway smooth muscle in a canine asthmatic model, *J. Appl. Physiol.*, 46: 681.

Armour, C. L., Black, J. L., Berend, N., and Woolcock, A. J., 1984, The relationship between bronchial hyperresponsiveness to methacholine and airway smooth muscle structure and reactivity, *Resp. Physiol.*, 58: 223.

Armour, C. L., Black, J. L., and Johnson, P. R. A., 1988, A role for inflammatory mediators in airway hyperresponsiveness, *in:* "Mechanisms in Asthma", C. L. Armour and J. L. Black, eds., Alan R. Liss, New York, p. 99.

Bárány, M., 1967, ATPase activity of myosin correlated with speed of muscle shortening, *J. Gen. Physiol.*, 50: 197.

Chiu, Y. C., Ballon, E. W., and Ford, L. E., 1987, Force, velocity, and power changes during normal and potentiated contractions of cat papillary muscle, *Circ. Res.*, 60: 446.

Coburn, R. F. and Tomita, T, 1973, Evidence of nonadrenergic noncholinergic inhibitory nerves in the guinea pig trachealis muscle, *Am. J. Physiol.*, 224: 1072.

Furchgott, R. F. and Zavodzki, J. V., 1980, Acetylcholine relaxes arterial smooth muscle by releasing a relaxing substance from endothelial cells, *Fed. Proc.*, 39: 581.

Guerriero, V., Russo, M. A., Olson, N. J., Putkey, J. A., and Means, A. R., 1986, Domain organization of chicken gizzard myosin light chain kinase deduced from a cloned cDNA, *Biochemistry*, 25: 8372.

Halayko, A., Jiang, H., Rao, K., and Stephens, N. L., 1990, Muscle cellularity and contractile proteins in canine airway smooth muscle, *FASEB J.*, 4: A444.

Hill, A. V., 1938, The heat of shortening and the dynamic constants of muscle, *Proc. R. Soc. Lond. B*, 126: 136

Hirshman, C. A. and Dorones, H., 1986, Airways responses to methacholine and histamine in Basenji greyhounds and other purebred dogs, *Resp. Physiol.*, 63: 339.

Hoh, Y. H., McGrath, P. A., and Hale, P. T., 1978, Electrophoretic analysis of multiple forms of rat cardiac myosin: effect of hypophysectomy and thyroxine replacement, *J. Mol. Cell Cardiol.*, 10: 1053.

Holme, G. and Piechuta, H., 1981, The derivation of an inbred line of rats which develop asthma-like symptoms following challenge with aerosolized antigen, *Immunology*, 42: 19.

Jewell, B. R. and Wilkie., D. R., 1960, The mechanical properties of relaxing muscle, *J. Physiol.*, 152: 30.

Jiang, H. and Stephens, N. L., 1990, Contractile properties of bronchial smooth muscle with and without cartilage, *J. Appl. Physiol.*, 69: 120.

Kepron, W., James, J. M., Kirk, B., Sehon, A. H., and Tse, K. S., 1977, A canine model for reaginic hypersensitivity and allergic bronchoconstriction, *J. Allergy Clin. Immunol.*, 59: 64.

Kong, S. and Stephens, N. L., 1981, Pharmacological studies of sensitized canine pulmonary blood vessels, *J. Pharmacol. Exp. Ther.*, 219: 551.

Levitt, R. C. and Mitzner, W., 1988, Expression of airway hyperreactivity to acetylcholine as a simple autosomal recessive trait, *FASEB J.*, 2: 2605.

Long, W. M., Yerger, L. D., Martinez, H., Codias, E., Sprung, C. L., Abraham, W. M., and Wanner, A., 1988, Modification of bronchial blood flow during allergic airway responses, *J. Appl. Physiol.*, 65: 272.

Mapp, C. E., Chitano, P., DeMarzo, N., DiBlasi, P., Saetta, M., DiStefano, A., Bosco, V. M., Allegra, L., and Fabbri, L. M., 1989, Response to acetyl-

Mapp, C. E., Chitano, P., DeMarzo, N., DiBlasi, P., Saetta, M., DiStefano, A., Bosco, V. M., Allegra, L., and Fabbri, L. M., 1989, Response to acetylcholine and myosin content of isolated canine airways. *J. Appl. Physiol.*, 67: 1331.

Rasmussen, H., Takuwa, Y., and Park, S., 1987, Protein kinase C in the regulation of smooth muscle contraction, *FASEB J.*, 1: 177.

Richardson, J. B., 1979, Nerve supply to the lungs, *Am. Rev. Resp. Dis.*, 119: 785.

Rudel, R. and Taylor, S. R., 1989, Striated muscle fibers; Facilitation of contraction at short lengths by caffeine, *Science*, 172: 389.

Shibata, S. and Cheng, J. R., 1977, Relaxation of vascular smooth muscle in spontaneously hypertensive rats, *Blood Vessels*, 14: 247.

Shioya, T., Pollack, E. R., Munoz, N. M., and Leff, A. R., 1987, Distribution of airway contractile responses in major resistance airways of the dog, *Am. J. Pathol.*, 129: 102.

Siegman, M. J., Davidheiser, S., Butler, T. M., and Mooers, S. U., 1985, What is the length-tension relation in smooth muscle, *Fed. Proc.*, 44: 456A.

Sobieszek, A. and Bremel, R., 1975, Preparation and properties of vertebrate smooth muscle myofibrils and actomyosin, *Eur. J. Biochem.*, 55: 49.

Somlyo, A. P., Kitazawa, T., Himpens, B., Matthijs, G., Horiuti, K., Kobayashi, S., Goldman, Y. S., and Somlyo, A. V., 1989, Modulation of Ca^{2+}-sensitivity and of the time course of contraction in smooth muscle: A major role of protein phosphatases, *in:* "Advances in Protein Phosphatases, Vol. 5", W. Merleude and J. DiSalvo, eds., Leuven University Press, Leuven, p. 181.

Somlyo, A. V. and Somlyo, A. P., 1968, Electromechanical and pharmacomechanical coupling in vascular smooth muscle, *J. Pharmacol. Exp. Ther.*, 159: 129.

Somlyo, A. V., Bond, M., Beaner, P. F., Ashton, F. T., Holtzen, H., and Somlyo, A. P., 1984, The contractile apparatus of smooth muscle: An update, *in:* "Smooth Muscle Contraction", N. L. Stephens, ed., M. Dekker Inc., New York.

Souhrada, M. and Souhrada, J. F., 1981, Reassessment of electrophysiological and contractile characteristics of sensitized airway smooth muscle, *Resp. Physiol.*, 46: 17.

Stephens, N. L., 1987, State of art: Airway smooth muscle, *Am. Rev. Resp. Dis.*, 135: 960.

Stephens, N. L. and Seow, C. Y., 1987, Smooth Muscle Contraction: Mechanisms of Crossbridge Slowing, *in:* "Regulation and Contraction of Smooth Muscle", M. J. Siegman, A. P. Somlyo and N. L. Stephens, eds., Alan R. Liss, New York, p. 357.

Stephens, N. L., Kong, S. K., and Seow, C. Y., 1988, Increased shortening of Airway Smooth Muscle, *in:* "Mechanisms in Asthma", C. L. Armour and J. L. Black, eds., Alan R. Liss, New York, p. 231.

Taylor, S. R., 1974, Decreased activation in skeletal muscle fibers at short lengths, *in:* "The Physiologic Basis of Starling's Law of the Heart", R. Porter, D. W. Fitzsimmons, eds., Elsevier/Excerpta Medica North Holland, London.

Von Hayek, H., 1990, "The Human Lung", Hafner Press, New York.

Wagner, C. D., Gundel, R. H., Reilly, P., Haynes, N., Letts, L. G., and Rothlein, R., 1990, Intercellular adhesion molecule-1 (ICAM-1) in the pathogenesis of asthma, *Science*, 247: 456.

Wang, Z., Seow, C. Y., Kepron, W., and Stephens, N. L., 1990, Mechanical alterations in sensitized canine saphenous vein, *J. Appl. Physiol.*, 69:171.

Wanner, A., Barker, J. A., Long, W. M., Mariassy, A. T., and Chediak, A. D., 1988, Measurement of airway mucosal perfusion and water volume with an inert soluble gas, *J. Appl. Physiol.*, 65: 264.

Xu, J., Stephens, N. L., Ford-Hutchinson, A., Jones, T., and Piechuta, H., 1990, Mechanical properties of inbred hyperreactive rat trachealis, *FASEB J.*, 4: A269.

Yanagisawa, M., Kurchaya, H., and Kimura, S., Tomobe, Y., Kobayashi, M., Mitsui, Y., Yazaki, Y., Goto, K., and Masaki, T., 1988, A novel potent vasoconstrictor peptide produced by vascular endothelial cells, *Nature*, 322: 411.

PURIFICATION OF PROTEIN KINASE C AND IDENTIFICATION OF ISOZYMES IN VASCULAR SMOOTH MUSCLE

Charles M. Schworer and Harold A. Singer

Weis Center for Research
Geisinger Clinic
Danville, PA 17822

INTRODUCTION

Although protein kinase C (PKC) is known to be present in vascular smooth muscle (VSM), its physiologic significance is not understood. Among its actions, PKC has been proposed to regulate tonic contractile responses (Rasmussen et al., 1987) and mediate, at least in part, intracellular responses to growth stimuli (Woodgett et al., 1987). Functional roles for PKC have been inferred largely from studies using pharmacological activators of the kinase (phorbol esters) and putative selective inhibitors (e.g. H-7, staurosporine). These studies are indirect and interpretation of them must be tempered by the possibility that phorbol ester-induced activation of PKC is not equivalent to physiological agonist-induced activation and that currently available kinase inhibitors are not completely selective (for review, see Woodgett et al., 1987). Another approach used has been to predict and observe the effects of phorbol ester-stimulated down regulation of PKC activity (Rozengurt et al., 1983; Coughlin et al., 1985).

Attempts to define the role of protein kinase C in tissues are further complicated by the fact that PKC is known to be a complex family of isozymes with as many as seven potential subtypes comprising the total intracellular pool of the enzyme (Kikkawa et al., 1989). Some of the subtypes have been shown to have a distinct subcellular distribution (Leach et al., 1989; Rogue et al., 1990), which may vary depending on the activation state of the cell (Leach et al., 1989). Also, some of the different isozymes may be down-regulated to varying degrees following prolonged phorbol ester exposure (Isakov et al., 1990; Godson et al., 1990). These observations imply that the individual PKC subtypes may have unique physiological roles. In order to understand the function of PKC in vascular smooth muscle, it is essential to know what PKC subtypes are present. We have purified PKC from swine carotid arteries and have investigated its subtype composition.

EXPERIMENTAL PROCEDURES

Purification of Swine Carotid Artery PKC

The protocol is based on procedures used previously for the purification of brain PKC (Walton et al., 1987; Wooten et al., 1987). Medial strips of swine carotid arteries were quick-frozen and powdered in liquid nitrogen. 25 g of the frozen, powdered tissue were homogenized in 5 volumes of buffer H (20 mM Tris (pH 7.4), 2 mM EDTA, 2 mM EGTA, 0.25 M sucrose, 0.5 mM phenylmethylsulfonyl fluoride (PMSF), 10 μg/ml leupeptin, 20 μg/ml trypsin inhibitor, and 1 mM dithiothreitol (DTT)) by means of a Polytron (two 30 sec bursts). Three additional 25 g amounts of tissue were processed in an identical manner. The homogenates were combined and centrifuged at 10,000 x g for 20 min. The supernatant was saved; the pellets were resuspended in a total of 150 ml of buffer H, homogenized as above, and re-centrifuged. These supernatants were combined with the original supernatant and centrifuged at 100,000 x g for 40 min. The 100,000 x g supernatant was loaded onto a DEAE-Sephacel column (5 x 10.2 cm), and the column was washed with buffer A (20 mM Tris (pH 7.4), 2 mM EDTA, 2 mM EGTA, 0.1 mM PMSF, 5 μg/ml leupeptin, 1 mM DTT). PKC activity was eluted with a 0 to 0.3 M NaCl gradient (12.5 ml/fraction). The peak of the phospholipid-dependent histone III-S phosphorylation activity was pooled and subjected to 75% ammonium sulfate precipitation. The pellet was resuspended in buffer A, then dialyzed against more buffer A. This pool of PKC activity was adjusted to 1 M NaCl and loaded onto a phenyl-Sepharose CL-4B column (1 x 8 cm) equilibrated in buffer A containing 1 M NaCl. PKC was eluted with a 1 M to 0 M NaCl gradient (2 ml/fraction) followed by additional washing with buffer A. Fractions containing PKC activity were pooled and loaded onto a column of protamine-agarose (0.7 x 3.9 cm) equilibrated in buffer A containing 0.1 M NaCl. After the column was washed with buffer A plus 0.5 M NaCl, PKC activity was eluted with buffer A plus 1.5 M NaCl (1 ml/fraction). This final pool of PKC activity was dialyzed in buffer A plus 20% glycerol and stored at -70° C. PKC activity was stable for at least 4 months.

Standard Assay for PKC Activity

The reaction mixture contained 50 mM HEPES (pH 7.4), 10 mM MgAcetate, 1.1 mM $CaCl_2$, approximately 0.8 mM EDTA plus EGTA (carry-over from buffer H or buffer A), 0.12 mg/ml histone (type III-S), 200 μM [γ-^{32}P]-ATP (400-500 cpm/pmol), and either plus or minus 0.1 mg/ml phosphatidylserine and 0.01 mg/ml dioleoylglycerol. The 25 μl reaction was incubated at 30° C for 3 min. 15 μl of the reaction were spotted onto 1.5 x 2.5 cm pieces of Whatman P81 phosphocellulose chromatography paper (Roskoski, 1983). The papers were washed in 75 mM phosphoric assay to remove free [γ-^{32}P]-ATP. The amount of radioactivity in the histone III-S that remained bound to the papers was determined by liquid scintillation counting.

Electrophoresis and Immunoblotting

Gel electrophoresis was carried out in the presence of SDS according to Laemmli's procedure for discontinuous polyacrylamide gels (Laemmli, 1970). Gels of 9% acrylamide were run using the Hoefer Mighty Small II apparatus. Protein bands were detected by either silver (Wray et al., 1981) or Coomassie blue staining. For immunoblotting, methods described by Harlow and Lane

(1988) were followed. Proteins in the gels were electrophoretically transferred onto Immobilon (Millipore) membranes by wet transfer at 70 V for 2 hr in 25 mM Tris, 192 mM glycine. The transferred proteins were incubated with anti-PKC antibody overnight at 4° C using either mouse monoclonal antibody (raised against bovine brain PKC) that recognizes both PKC subtypes II and III but not I (Amersham) or mouse monoclonal antibodies (developed with rabbit brain PKC) that recognize specific PKC isozymes, i.e., subtypes I, II, or III (Seikagaku Kogyo, Tokyo). Immunodetection was accomplished by either of two methods. One method involved a subsequent incubation with goat anti-mouse IgG conjugated with horseradish peroxidase (Bio-Rad). The color reaction used 4-chloro-1-naphthol (Bio-Rad HRP color development reagent) as substrate. The alternative and more sensitive method utilized secondary incubation with [^{125}I]-goat anti-mouse IgG (Dupont, New England Nuclear) and subsequent autoradiography of the transfer membrane.

Hydroxylapatite Chromatography

Swine carotid artery PKC activity (20-40 units) that had been purified through the DEAE-Sephacel and ammonium sulfate precipitation steps was loaded onto a hydroxylapatite column (DNA-grade Bio-Gel HTP, 0.7 x 7.8 cm) equilibrated in buffer B that contained 20 mM KPO_4 (pH 7.5), 1 mM EDTA, 1 mM EGTA, 10% (v:v) glycerol, 0.1 mM PMSF, 10 µg/ml leupeptin, and 1 mM DTT. The column was run at a flow rate of 9.2 ml/hr, and 0.9 ml fractions were collected. PKC activity was eluted with a 70 ml gradient from 20 mM KPO_4 to 150 mM KPO_4.

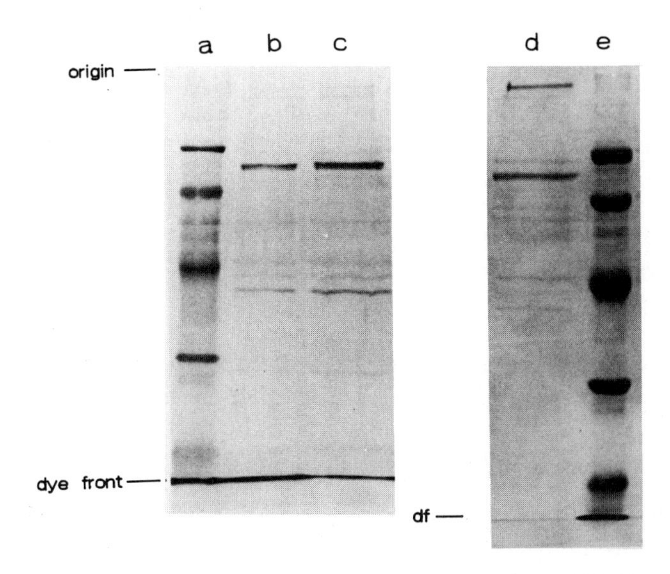

Figure 1. Gel electrophoresis of purified swine carotid artery PKC. Protein bands were detected using either silver stain (lanes a-c) or Coomassie blue (lanes d, e). The protein M_r markers (lanes a and e) are: phosphorylase b (94,000), bovine serum albumin (67,000), ovalbumin (43,000), carbonic anhydrase (30,000), and soybean trypsin inhibitor (20,100). The loadings of purified PKC are: 0.6 µg (lane b), 1.2 µg (lane c), and 3 µg (lane d).

Other Materials and Methods

[γ-^{32}P]-ATP was purchased from Amersham. DEAE-Sephacel and phenyl-Sepharose CL-4B were obtained from Pharmacia. Protamine-agarose, leupeptin, soybean trypsin inhibitor, PMSF, L-α-phosphatidyl-L-serine, 1,2-dioleoyl-sn-glycerol (C18:1,[cis]-9), histone (type III-S), and bovine brain myelin basic protein were purchased from Sigma. [Ser^{25}]-protein kinase C (19 - 31) was obtained from Peninsula Laboratories. Syntide-2 was provided by T. Soderling (Vanderbilt University, Nashville). Protein concentrations were determined by the Bradford assay (Bradford, 1976) using bovine serum albumin as the standard.

RESULTS

The summary of a recent purification of PKC from swine carotid artery smooth muscle is given in Table I. These numbers are representative of three separate preparations of PKC. Based on the PKC activity in the 100,000 x g supernatant, a 7500-fold purification was attained with a yield of about 20 percent. Analysis of the most highly purified PKC (through the protamine-agarose step) by gel electrophoresis (Fig. 1) indicated that the major constituent of the most highly purified PKC activity pool was a band of 82 kDa that represented approximately 70-75% of the total protein in the sample.

To verify that this band was actually PKC, we utilized immunoblotting to examine the cross-reactivity of the 82-kDa protein to antibody that recognizes brain PKC. For these experiments we used the mouse monoclonal anti-PKC antibody that recognizes both PKC subtypes II and III. Swine carotid artery PKC was tested at three stages of purification: after the DEAE-Sephacel column, after the phenyl-Sepharose column, and after the protamine-agarose column. These samples were compared to a partially purified rabbit brain PKC. For each sample only a single band cross-reacted with the antibody (data not shown). The bands co-migrated with an M_r of 82,000 (data not shown).

We tested the ability of the purified swine carotid artery PKC to phosphorylate some additional protein and synthetic peptide substrates (Fig. 2). The values in the figure are the levels of phosphate incorporation attained after 1 min of incubation. As might be expected, the phosphorylation of histone was completely dependent on the presence of Ca^{2+} and phospholipid (phosphatidylserine plus dioleoylglycerol). The same was true for myelin basic protein except for the assay in the presence of added phospholipid but no Ca^{2+} where phosphate incorporation above basal levels was detected. On the other hand, when either of the synthetic peptides was used as substrate, relatively high levels of phosphate incorporation were observed in the absence of Ca^{2+} and/or phospholipid. At 1 min, Ca^{2+} plus phospholipid still had a stimulatory effect, but beyond that time of incubation the differences between the four conditions were lesser in magnitude, and in some cases the values for all four conditions were nearly equal (data not shown).

In some tissues PKC exists as a number of isozymes (Kikkawa et al., 1989). We were interested in examining swine carotid artery PKC to determine what subtypes are expressed. For these experiments we used the mouse monoclonal anti-PKC antibodies that supposedly recognize the specific PKC subtypes I, II, or III. The results of our initial experiments with these antibodies indicated that subtype III was the major isozyme present in fractions of purified

Table 1. Purification of PKC from swine carotid artery.

PROCEDURE	VOLUME (ml)	UNITS/ML	TOTAL UNITS (nmol/min)	TOTAL PROTEIN (mg)	UNITS/MG	YIELD (%)	PURIFICATION (fold)
HIGH-SPEED SUPERNATANT	570	0.30	169	4708	0.036	100	1
DEAE-SEPHACEL	114	1.06	121	225	0.54	71.5	15
75% AMMONIUM SULFATE	7.8	11.9	93	102	0.91	54.8	25
PHENYL-SEPHAROSE	16.4	3.63	59	5.6	10.6	35.0	295
PROTAMINE-AGAROSE	1.85	18.0	33	0.12	277	19.6	7500

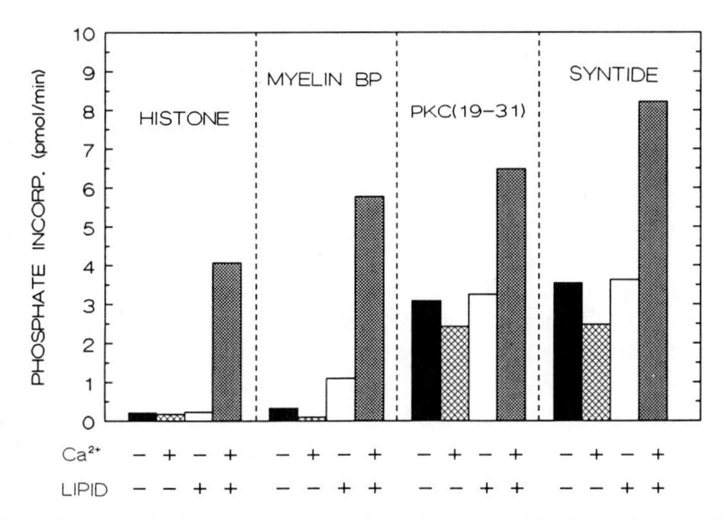

Figure 2. Phosphorylation of proteins and synthetic peptides by swine carotid artery PKC. The protocol used to measure the phosphorylation of these substrates was the same as the standard PKC assay (see "Experimental Procedures") except that $CaCl_2$, when present, was at a concentration of 0.24 mM and the duration of incubation was only 1 min. The "-Ca^{2+}" assays contained 1 mM EGTA but no added $CaCl_2$. Substrates were present at the following concentrations: 0.12 mg/ml histone, 0.30 mg/ml myelin basic protein (MYELIN BP), 0.01 mM [Ser^{25}]-protein kinase C peptide 19 - 31 (PKC(19 - 31)), and 0.024 mM syntide-2 (SYNTIDE).

PKC from VSM but that there was also a significant amount of subtype II (data not shown). We wanted to verify this observation in some alternative, independent way and decided to utilize hydroxylapatite chromatography. Hydroxylapatite chromatography is a widely accepted method to separate brain PKC subtypes (Huang et al., 1986). When we loaded a sample of partially purified swine carotid artery PKC (through the DEAE-Sephacel and

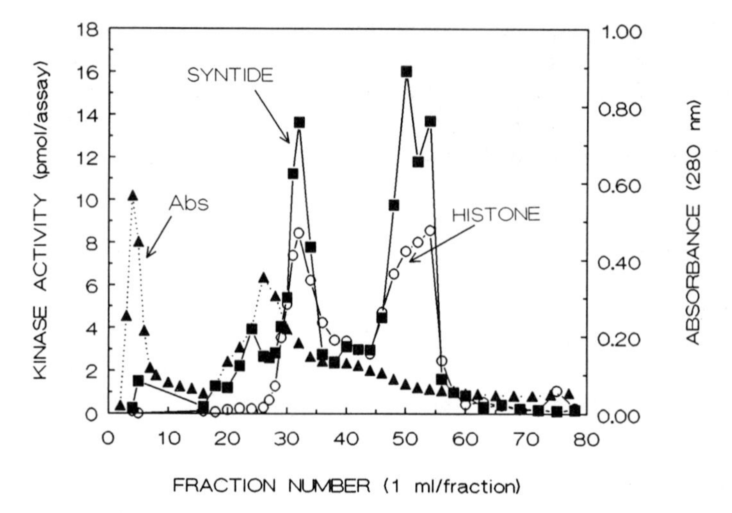

Figure 3. Hydroxylapatite chromatography of swine carotid artery PKC. 36 units of swine carotid artery PKC were loaded onto the column and eluted with a gradient of 20 - 150 mM KPO_4. Fractions were assayed for histone phosphorylation (+Ca^{2+}, +phospholipid) and syntide-2 phosphorylation (-Ca^{2+}, -phospholipid).

ammonium sulfate steps) onto hydroxylapatite and eluted the kinase activity with a gradient of KPO_4, we observed that the PKC activity was resolved into two distinct peaks of phospholipid-dependent histone kinase activity that coincided with two peaks of Ca^{2+}- and phospholipid-independent syntide-2 kinase activity (Fig. 3). We estimated that the first peak of PKC activity eluted at about 60 mM KPO_4 and the second peak over a range of about 92 - 95 mM KPO_4. No phosphorylation of histone by any fractions was observed in the absence of Ca^{2+} and phospholipid. When the fractions of kinase activity were analyzed by immunoblotting using the subtype II- and subtype III-specific antibodies, we observed that the fractions comprising the first peak of PKC activity cross-reacted with the subtype II antibody but not the subtype III antibody, while the fractions comprising the second peak cross-reacted with subtype III antibody but not subtype II antibody (Fig. 4).

DISCUSSION

Our results indicate that there are multiple isozymes of PKC in swine carotid artery smooth muscle. These findings are in contrast to a recent report describing PKC purified from bovine aorta (Watanabe et al., 1989). Those investigators detected only PKC subtype III in either the soluble or particulate fraction using a subtype-specific antibody, and they observed only a single, broad peak of PKC activity when either of the fractions were subjected to hydroxylapatite chromatography. No explanation is readily apparent that can account for these disparate findings. It is possible that there is a real difference in PKC composition between bovine aorta and swine carotid artery. On the other hand, the discrepancy could have resulted from differences in the methods used to process the tissue. Other reports of smooth muscle PKC purified from bovine aorta (Dell et al., 1988) or rabbit iris (Howe and Abdel-

Latif, 1988) did not attempt to identify PKC isozymes. Additional comparative studies may be required to clarify this issue.

Assuming specificity of the antibodies used and homology between swine and rabbit PKC isozymes, it would appear that both types II and III PKC are expressed in swine carotid artery VSM. However, because of the number and similarity of PKC isozymes, we need to verify our findings with additional subtype-specific antibodies. One of the more recently discovered PKC subtypes (δ, ε, or ζ) (Ono et al., 1988) or an as yet unidentified subtype may be cross-reacting or go undetected with these antibodies. We have considered the possibility that a minor component of Ca^{2+}-independent PKC (δ, ε, or ζ) is also expressed in swine carotid artery. Konno et al. (1989) have shown that PKC-ε elutes from hydroxylapatite very closely to the rabbit brain subtype II isozyme. Therefore, if subtype II is present, PKC-ε may not be resolved by hydroxylapatite chromatography. Also, myelin basic protein, but not histone, was a good substrate for PKC-ε (Schaap and Parker, 1990). Phosphorylation by PKC-δ, ε, or ζ is unique in that it is phospholipid-dependent but Ca^{2+}-independent. We observed a small amount of phospholipid-dependent, Ca^{2+}-independent phosphorylation of myelin basic protein, a finding consistent with expression of one of these isozymes. Immunoblotting with specific antibodies, as well as additional biochemical characterization, will be required to determine whether any of these forms is present.

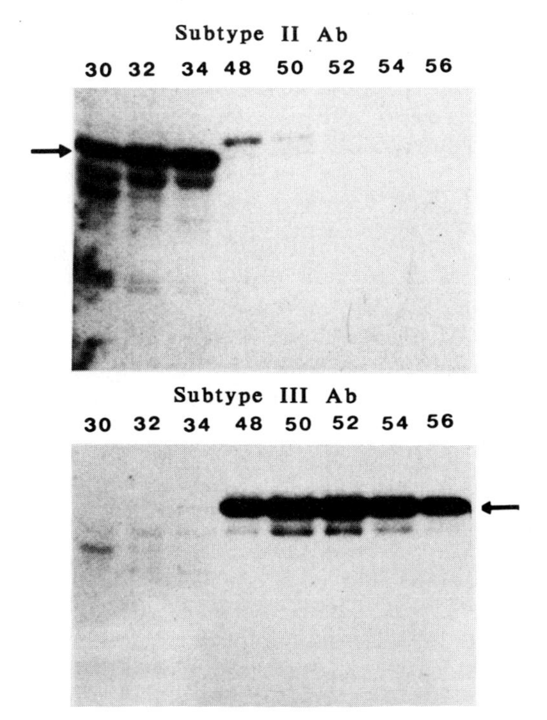

Figure 4. Immunoblotting of hydroxylapatite fractions. Aliquots (0.30 ml) of fractions from the hydroxylapatite chromatography (Fig. 3) were adjusted to 10% trichloroacetic acid to precipitate proteins. The precipitate was resuspended in electrophoresis sample buffer, and identical samples were loaded onto parallel acrylamide gels. After electrophoresis the samples were analyzed by immunoblotting utilizing antibodies specific for PKC subtypes II or III. Cross-reactivity was detected by secondary incubation with [^{125}I]-goat anti-mouse IgG and autoradiography. The numbers above the lanes refer to fractions from the column (Fig. 3). Arrows mark the position of 82-kDa PKC.

The fact that significant phosphorylation of the synthetic peptides was observed in the absence of any activators draws into question the validity of using either of these peptides for the quantitative determination of PKC activity in crude fractions of cells or tissues. It would be difficult to assess phospholipid-dependent activity (i.e., activity measured in the presence of phospholipid minus activity in the absence of phospholipid) since any observed phospholipid-independent activity could be the result of phosphorylation by either PKC or some other phospholipid-independent kinase. The ability of PKC to phosphorylate these peptides in the absence of activators is also of interest from a mechanistic standpoint. A pseudosubstrate domain of PKC has been proposed to interact with the active site in such a way that substrates can not gain access to that site. Activation would allow the active site to disengage and interact with substrates. Our observations suggest that this interaction may exclude large proteins but not small peptides such as PKC(19 - 31) or syntide-2 that can interact with the active site and become phosphorylated in the absence of activators. If so, it would be informative to define the size limit or conformational constraints that exclude proteins so that it could be determined if any potential *in vivo* substrates are small enough to be constitutively phosphorylated by PKC.

If multiple subtypes of PKC actually exist in VSM cells, then the possibility of distinct subcellular compartmentalization for each subtype must be considered. Rogue et al. (1990) purified PKC from rat liver nuclei and determined that only isozyme II was present. Studies are underway in our own laboratory using immunofluorescent techniques to determine if the multiple PKC subtypes have a distinct subcellular localization in cultured VSM cells.

REFERENCES

Bradford, M. M., 1976, A rapid and sensitive method for the quantitation of microgram quantities of protein utilizing the principle of protein-dye binding, *Anal. Biochem.*, 72: 248.

Coughlin, S. R., Lee, W. M. F., Williams, P. W., Giels, G. M., and Williams, L. T., 1985, c-myc gene expression is stimulated by agents that activate protein kinase C and does not account for the mitogenic effect of PDGF, *Cell*, 43: 243.

Dell, K. R., Walsh, M. P., and Severson, D. L., 1988, Characterization of bovine aortic protein kinase C with histone and platelet protein P47 as substrates, *Biochem. J.*, 254: 455.

Godson, C., Weiss, B. A., and Insel, P. A., 1990, Differential activation of protein kinase C is associated with arachidonate release in Madin-Darby canine kidney cells, *J. Biol. Chem.*, 265: 8369.

Harlow, E. and Lane, D., 1988, Immunoblotting, *in:* "Antibodies: A laboratory manual", Cold Spring Harbor Laboratory, Cold Spring Harbor, New York, p. 471.

Howe, P. H. and Abdel-Latif, A. A., 1988, Purification and characterization of protein kinase C from rabbit iris smooth muscle, *Biochem. J.*, 255: 423.

Huang, K.-P., Nakabayashi, H., and Huang, F. L., 1986, Isozymic forms of rat brain Ca^{2+}-activated and phospholipid-dependent protein kinase, *Proc. Nat'l. Acad. Sci. U.S.A.*, 83: 8535.

Isakov, N., McMahon, P., and Altman, A., 1990, Selective post-transcriptional down-regulation of protein kinase C isoenzymes in leukemic T cells chronically treated with phorbol ester, *J. Biol. Chem.*, 265: 2091.

Kikkawa, U., Kishimoto, A., and Nishizuka, Y., 1989, The protein kinase C family: Heterogeneity and its implications, *Ann. Rev. Biochem.*, 58: 31.

Konno, Y., Ohno, S., Akita, Y., Kawasaki, H., and Suzuki, K., 1989, Enzymatic properties of a novel phorbol ester receptor/protein kinase, nPKC, *J. Biochem.*, 106: 673.

Leach, K. L., Powers, E. A., Ruff, V. A., Jaken, S., and Kaufmann, S., 1989, Type 3 protein kinase C localization to the nuclear envelope of phorbol ester-treated NIH 3T3 cells, *J. Cell Biol.*, 109: 685.

Laemmli, U., 1970, Cleavage of structural proteins during the assembly of the head of bacteriophage T4, *Nature*, 227: 680.

Ono, Y., Fujii, T., Ogita, K., Kikkawa, U., Igarashi, K., and Nishizuka, Y., 1988, The structure, expression, and properties of additional members of the protein kinase C family, *J. Biol. Chem.*, 263: 6927.

Rasmussen, H., Takuwa, Y., and Park, S., 1987, Protein kinase C in the regulation of smooth muscle contraction, *FASEB J.*, 1: 177.

Rogue, P., Labourdette, G., Masmoudi, A., Yoshida, Y., Huang, F. L., Huang, K.-P., Zwiller, J., Vincendon, G., and Malviya, A. N., 1990, Rat liver nuclei protein kinase C is the isozyme type II, *J. Biol. Chem.*, 265: 4161.

Roskoski, Jr., R., 1983, Assays of protein kinase, *Methods Enzymol.*, 99: 3.

Rozengurt, E., Rodriguez-Pena, M., and Smith, K., 1983, Phorbol ester, phospholipase C, and growth factors rapidly stimulate the phosphorylation of a M_r 80,000 protein in intact quiescent 3T3 cells, *Proc. Nat'l. Acad. Sci. U.S.A.*, 80: 7244.

Schaap, D. and Parker, P. J., 1990, Expression, purification, and characterization of protein kinase C-ε, *J. Biol. Chem.*, 265: 7301.

Walton, G. M., Bertics, P. J., Hudson, L. G., Vedvick, T. S., and Gill, G. N., 1987, A three-step purification procedure for protein kinase C: characterization of the purified enzyme, *Anal. Biochem.*, 161: 425.

Watanabe, M., Hachiya, T., Hagiwara, M., and Hidaka, H., 1989, Identification of type III protein kinase C in bovine aortic tissue, *Arch. Biochem. Biophys.*, 273: 165.

Woodgett, J. R., Hunter, T., and Gould, K. L., 1987, Protein kinase C and its role in cell growth, *in:* "Cell Membranes: Methods and Reviews, Vol. 3", E. Elson, W. Frazier, and L. Glaser, eds., Plenum Press, New York, p. 215.

Wooten, M. W., Vandenplas, M., and Nel, A. E., 1987, Rapid purification of protein kinase C from rat brain. A novel method employing protamine-agarose affinity column chromatography, *Eur. J. Biochem.*, 164: 461.

Wray, W., Boulikas, T., Wray, V. P., and Hancock, R., 1981, Silver staining of protein in polyacrylamide gels, *Anal. Biochem.*, 118: 197.

POLYLYSINE: AN ACTIVATOR OF SMOOTH MUSCLE CONTRACTILITY

Pawel T. Szymanski and Richard J. Paul

Department of Physiology and Biophysics
University of Cincinnati
College of Medicine
231 Bethesda Avenue
Cincinnati, OH 45267

INTRODUCTION

Polycationic amines are known to affect a wide variety of biochemical reactions and have thus been used as probes of mechanism in cell-free systems and isolated cell fractions (Ahmed et al., 1986; Gröschel-Stewart et al., 1989). Polycations are of particular interest to the study of the mechanisms of regulation of contractility in smooth muscle for they are reported to be modulators of phosphoprotein phosphatases (DiSalvo et al., 1985). Phosphatases play a major role in the control of the level of myosin light chain phosphorylation, which in smooth muscle is the primary transduction mechanism for regulation of contractility (Hartshorne, 1987). A specific phosphatase inhibitor would be of interest not only in terms of the mechanism of crossbridge regulation, but also in terms of identification of the nature of the physiologically relevant phosphatase in smooth muscle. We have thus studied the effects of polylysine on the contractility and ATPase activity of "skinned" (permeabilized) smooth muscle fibers and ATPase activity of smooth muscle actomyosin. For comparison we have also studied the effects of polylysine on actomyosin prepared from skeletal muscle.

We report here that polylysine can elicit a contraction in skinned taenia coli smooth muscle and activates the ATPase of actomyosin prepared from bovine aorta. In contrast, polylysine does not stimulate skeletal muscle actomyosin ATPase and indeed inhibits its Ca^{2+}-activated, Mg^{2+}-ATPase activity. Our preliminary experiments indicate that the effects of polylysine are not mediated through alteration of myosin light chain phosphorylation. Thus polylysine may be a useful probe not only for the differences in smooth and skeletal actomyosin but also for the mechanism whereby myosin phosphorylation activates smooth muscle.

Regulation of Smooth Muscle Contraction
Edited by R.S. Moreland, Plenum Press, New York, 1991

METHODS

Skinned Taenia Coli Fiber Preparation

Guinea pig taenia coli fiber bundles were permeabilized using a combination of Triton X-100 extraction and freeze glycerinization (Rüegg and Paul, 1982).

Preparation of Native Actomyosin

Ca^{2+}-sensitive, phosphatase-free actomyosin (Erdodi et al., 1988) was prepared from bovine aorta (Litten et al., 1977). Briefly, actomyosin was extracted in 80 mM KCl, 18 mM MOPS (pH 7.0), 4 mM EGTA, 4 mM ATP, 1 mM NaN_3, 1 mM DTT, and dialyzed against 25 mM KCl, 4.4 mM MOPS (pH 7.0), 4 mM $MgCl_2$, 0.5 mM DTT and 0.2 mM EGTA. Membrane fragments were removed by extensive washes in 1% Triton X-100.

Mechanical Studies with Skinned Fibers

Isometric force in skinned fibers (5 mm long, 100-200 μm thick) was measured at 25° C using an AME-801 (SensoNor a.s., Horten, Norway) force transducer in total volume of 0.3 ml. "Relaxing solution" contained 20 mM MOPS (pH 6.7), 4 mM EGTA, 1 mM ATP, 10 mM $MgCl_2$, 24 mM KCl, 1 mM NaN_3, 0.5 mM DTT, and 0.1 μM calmodulin; and "contracting solution" also contained 3.6 mM $CaCl_2$, for a calculated Ca^{2+} concentration of 14.9 μM (Godt and Maughan, 1988). All solutions contained an ATP regenerating system consisting of 10 mM phosphocreatine and 10 U/ml creatine phosphokinase. Polylysine here refers to 10 - 13 kDa poly-*l*-lysine hydrobromide obtained from Sigma (P-6516). A control contraction/relaxation cycle was performed before any test measurements, and the reported values for isometric force were normalized to the maximum isometric force in this control cycle.

ATPase Activity Measurements

ATPase activity in smooth (100 - 200 μg/ml) and skeletal (10 - 20 μg/ml) muscle actomyosin was measured at 25° C in a final volume of 0.6 ml of either relaxing or contracting solutions. All solutions contained an ATP regenerating system, consisting of 10 mM phosphoenolpyruvate and 20 μg/ml pyruvate kinase. ATPase activity was determined by measuring the rate of ADP production, linked stoichiometrically to the pyruvate produced by the regenerating system. The pyruvate in turn was coupled to the production of NAD by including 50 μg/ml lactate dehydrogenase and 50 μM NADH (Lowry and Passonneau, 1972). The rate of ATP hydrolysis was measured spectrofluorimetrically as a decrease of NADH fluorescence.

Protein Determination

Protein concentration was determined using the procedure of Bradford (1976) with plasma gamma globulin as a standard.

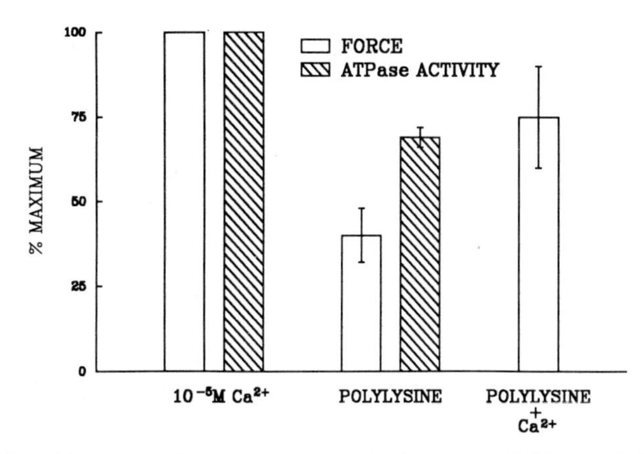

Figure 1. The effects of polylysine on isometric force and ATPase activity in skinned fibers of guinea pig taenia coli. The absolute values for fiber ATPase were 2.17 ± 0.16 and 4.38 ± 0.41 μM/min/mg dry weight for basal and Ca^{2+}-stimulated conditions respectively. Isometric force development ranged from 2 to 7 mN.

Statistical Analysis

Standard ANOVA or the Student's t-test, as appropriate, were used for statistical analysis with $p < 0.05$ taken as an indication of a statistically significant difference.

RESULTS AND DISCUSSION

The addition of 10 - 20 μM polylysine to relaxing solution ($[Ca^{2+}] < 10^{-9}$ M) elicited a contraction in skinned fibers, 40% ± 8% (n = 6) of maximally stimulated contraction ($[Ca^{2+}]$ = 14.9 μM). As shown in Figure 1, polylysine also stimulated the ATPase activity in fibers in parallel to the increases in isometric force. Polylysine also activated the actomyosin ATPase of smooth muscle in a dose-dependent manner. In contrast, the basal ATPase of skeletal muscle actomyosin was not affected by polylysine (Fig. 2). Moreover, addition of polylysine to skeletal muscle actomyosin, previously activated by Ca^{2+}, resulted in an inhibition of the actin-activated, Mg^{2+}-ATPase.

The activation of smooth muscle actin-myosin interaction by polylysine is consistent with the hypothesis that the effects of polylysine are mediated by phosphatase inhibition. In these skinned guinea pig taenia coli preparations, the addition of ATPγS in the absence of Ca^{2+} is sufficient to elicit a slowly developing contracture, similar to that induced by polylysine (Strauss and Paul, unpublished observations). ATPγS is known to be a substrate for myosin light chain kinase, but is a poor substrate for phosphatases (Sherry et al., 1978). Thus the contracture elicited by ATPγS indicates that there is an intrinsic level of kinase activity in these fibers at low Ca^{2+} levels (< 10^{-9} M). Thus the contracture and increased ATPase activity induced by polylysine could be attributable to the unmasking of this intrinsic kinase activity by inhibition of the phosphatase.

To test this hypothesis, we measured the level of phosphorylation of the 20 kDa myosin light chain in skinned fibers using an IEF-immunoblot assay

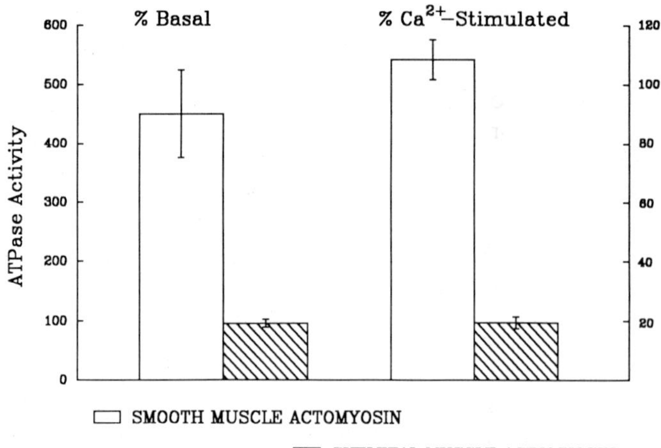

Figure 2. The effects of polylysine (10 μM) on skeletal and smooth muscle actomyosin ATPase. Control ATPase activities (no polylysine) are 100%; the left ordinate scale refers to the basal activities and the right ordinate to the Ca^{2+}-stimulated activity. The absolute values for basal activities were 2.68 ± 0.2 nmole/min/mg (n = 6) and 27.3 ± 4.2 nmole/min/mg (n = 5) for smooth and skeletal muscle, respectively. The Ca^{2+}- stimulated ATPase activities were 9.90 ± 0.78 nmole/min/mg (smooth muscle; n = 6) and 277.2 ± 10.3 nmole/min/mg (skeletal muscle; n = 5).

(Obara and de Lanerolle, 1989). Surprisingly, no significant changes from the basal level of myosin light chain phosphorylation could be detected in the presence of polylysine (10 μM). It is possible that the effects of polylysine are mediated by causing a net increase in phosphorylation of other proteins. As a further control, we measured the incorporation of ^{32}P associated with polylysine stimulation in smooth muscle actomyosin using autoradiography of SDS gels. Again, there were no detectable increases in phosphorylation of any protein induced by polylysine.

If the effects of polylysine are not mediated by phosphorylation, then what is the basis for its activation of smooth muscle actin-myosin interaction? And why does this differ from skeletal muscle actomyosin? The effects of polylysine on skinned fibers were not readily reversible, suggesting a mechanism involving strong binding. These effects could be completely abolished by heparin, a highly negatively charged heteropolysaccharide. Thus the observed effects are specific to polylysine and not ascribable to the counterion, bromide, or potential impurities in the polylysine. The effects are also not simply a charge or ionic strength effect as 100 mM lysine did not stimulate the bovine aortic actomyosin ATPase activity.

The mechanism(s) by which polylysine activates smooth muscle actin-myosin interaction is unknown. However, activation of both contraction in skinned fibers and actomyosin ATPase in the absence of phosphorylation is not unprecedented (Ikebe et al., 1984; Wagner et al., 1987; Moreland and Moreland, 1989). A contraction in the absence of Ca^{2+} can be induced by high Mg^{2+} concentrations (Ikebe et al., 1984) and, in terms of the time course and maximal forces attained, is similar to that elicited by polylysine.

Two very general classes of mechanisms can be invoked to explain these observations. One class might involve polylysine binding in a region of the molecule which alters the position of the myosin light chains. This could result in a deinhibition by a mechanism similar to that proposed for the effects of phosphorylation of myosin light chains (Kohama et al., 1987).

An alternative hypothesis would involve filament formation. The basis for this class of mechanism rests on the observation that in some non-muscle systems, phosphorylation of myosin heavy chains at the c-terminal region of the molecule leads to filament formation and activation of ATPase activity (Korn and Hammer, 1989). Furthermore, based on available sequence data (Nagai et al., 1989), the c-terminal region of smooth muscle myosin is highly negatively charged. Thus binding of polylysine to this region could conceivably lead to filament formation and subsequent increase in ATPase activity (Trybus, 1989). In preliminary experiments, we have observed that polylysine increases the sedimentation of myosin, indicating some form of aggregation may be involved. As suggested by Hartshorne and colleagues (1987), this might initially involve a conversion from the 10S to 6S configuration of myosin, which is a prerequisite for filament formation. Clearly more experimentation is required to further define the mechanism(s) underlying the polylysine effects.

In summary, polylysine activates both contraction in skinned smooth muscle fibers and actomyosin ATPase in the absence of Ca^{2+} and myosin light chain phosphorylation. Significantly, polylysine inhibited skeletal muscle actomyosin ATPase activity, the opposite to that in observed in smooth muscle. Thus polylysine is a useful and specific probe of activation in smooth muscle.

ACKNOWLEDGMENTS

This work was supported in part by HL22619 and HL23240.

REFERENCES

Ahmed, K., Goueli, S. A., and Williams-Ashman, H. G., 1986, Mechanisms of significance of polyamine stimulation of various protein kinase reaction, *Adv. Enzyme Reg.*, 25: 401.

Bradford, M., 1976, A rapid and sensitive method for quantitation of microgram quantities of protein utilizing the principle of protein-dye binding, *Anal. Biochem.*, 72: 248.

DiSalvo, J., Gifford, D., and Kokkinakis, A., 1985, Heat-stable regulatory factors are associated with polycation-modulable phosphatases, *Adv. Enzyme Reg.*, 23: 103.

Erdodi, F., Rokolya, A., Bárány, M., and Bárány, K., 1988, Phosphorylation of the 20,000 Da myosin light chain isoforms of arterial smooth muscle by myosin light chain kinase and protein kinase C, *Arch. Biochem. Biophys.*, 266: 583.

Godt, R. E. and Maughan, D. W., 1988, On the composition of the cytosol of relaxed skeletal muscle of the frog, *Am. J. Physiol.*, 254: C591.

Gröschel-Stewart, U., Lehman-Klose, S., Magel, E., and Baumgart, P., 1989, Histones bind to contractile proteins and inhibit actomyosin ATPase, *Biochem. Internat'l.*, 18: 519.

Hartshorne, D. J., 1987, Biochemistry of the contractile process in smooth muscle, *in:* "Physiology of the Gastrointestinal Tract, Second edition", L. R. Johnson, ed., Raven Press, New York, p. 423.

Ikebe, M., Barsotti, R. J., Hinkins, S., and Hartshorne, D. J., 1984, Effects of magnesium chloride on smooth muscle actomyosin adenosine-5'-triphosphatase activity, myosin conformation and tension development in glycerinated smooth muscle fibers, *Biochemistry,* 23: 5062.

Kohama, K., Kohama, T., and Kendrick-Jones, J., 1987, Effect of N-ethylmaleimide on Ca-inhibition of Physarum myosin, *J. Biochem.,* 102: 17.

Korn, E. D. and Hammer III, J. A., 1989, Myosin of non-muscle cells, *Ann. Rev. Biophys. Biophys. Chem.,* 17: 23.

Litten, R. Z., Solaro, J. R., and Ford, G. D., 1977, Properties of the calcium sensitive components of bovine arterial actomyosin, *Arch. Biochem. Biophys.,* 182: 24.

Lowry, O. H. and Passonneau, J. V., 1972, Measurement of enzyme activities with pyridine nucleotides, *in:* "A Flexible Systems of Enzymatic Assays", O. H. Lowry and J. V. Passonneau, eds., Academic Press, New York, p. 93.

Moreland, R. S. and Moreland, S., 1989, Regulation by calcium and myosin phosphorylation of the development and maintenance of stress in vascular smooth muscle, *in:* "Muscle Energetics", R. J. Paul, G. Elzinga, and K. Yamada, eds., Alan R. Liss, New York, p. 317.

Nagai, R., Kuro-o, M., Babij, P., and Periasamy, M., 1989, Identification of two types of smooth muscle myosin heavy chain isoforms by cDNA cloning and immunoblot analysis, *J. Biol. Chem.,* 264: 9734.

Obara, K. and de Lanerolle, P., 1989, Isoproterenol attenuates myosin phosphorylation and contraction of tracheal muscle, *J. Appl. Physiol.,* 66: 2017.

Rüegg, J. C. and Paul, R. J., 1982, Vascular smooth muscle: Calmodulin and cyclic AMP-dependent protein kinase alter calcium sensitivity in porcine carotid skinned fibers, *Circ. Res.,* 50: 394.

Sherry, J. M. F., Gorecka, A., Aksoy, M. O., Dabrowska, R., and Hartshorne, D. J., 1978, Roles of calcium and phosphorylation in the regulation of gizzard myosin, *Biochemistry,* 17: 4411.

Trybus, K., 1989, Filamentus smooth muscle is regulated by phosphorylation, *J. Cell Biol.,* 109: 2887.

Wagner, J., Pfitzer, G., and Rüegg, J. C., 1987, Calcium/calmodulin activation of gizzard skinned fibers at low levels of myosin phosphorylation, *in:* "Regulation and Contraction of Smooth Muscle", M. J. Siegman, A. P. Somlyo, and N. L. Stephens, eds., Alan R. Liss, New York, p. 427.

INCREASED ATPase ACTIVITY AND MYOSIN LIGHT CHAIN KINASE (MLCK) CONTENT IN AIRWAY SMOOTH MUSCLE FROM SENSITIZED DOGS

Kang Rao, He Jiang, Andrew J. Halayko, Nan Pan,
Wayne Kepron, and Newman L. Stephens

Department of Physiology,
Faculty of Medicine,
University of Manitoba,
Winnipeg, Manitoba R3E 0W3 Canada

INTRODUCTION

The bronchial hyperreactivity associated with acute asthma appears to be a manifestation of alterations in the regulatory pathways governing airway smooth muscle function (Stephens et al., 1988; Jiang et al., 1990; Kong et al., 1990). We have developed a canine model of asthma (Kepron et al., 1977), in which newborn dogs are sensitized to ragweed pollen using an intraperitoneal immunization regime.

Shortening capacity and zero-load shortening velocity of sensitized tracheal and bronchial smooth muscle (TSM and BSM) were increased compared to littermate controls (Antonissen et al., 1979; Antonissen et al., 1980; Stephens et al., 1988; Jiang et al., 1990). In sensitized TSM and pulmonary artery this increase in shortening velocity is a correlate of both increased levels of 20 kDa regulatory myosin light chain phosphorylation and increased myofibrillar ATPase activity (Kong et al., 1986; Kong et al., 1990). These results implicate myosin light chain kinase (MLCK) and/or the pathways regulating its activity as being essential to the mechanism underlying enhanced dynamic mechanical properties in ragweed pollen sensitized airway smooth muscle.

We have studied the effects of sensitization on the content of MLCK as well as myosin and actin in TSM and BSM. In addition, we have quantified the activity of these proteins in myofibrillar preparations using Mg^{2+}-ATPase assays. We also used peptide mapping techniques to identify potential changes in the primary sequence of MLCK.

METHODS

ATPase activity was determined as reported by us (Stephens et al., 1991). After obtaining dog lungs, tracheal and bronchial smooth muscle samples for ATPase determination were dissected promptly and stored at -70° C. Myofibrils were prepared as previously described (Sobieszek and Bremel, 1975). Myofibril homogenization buffer was composed of (in mM): imidazole, 20 (pH 6.9); KCl, 60; cysteine, 1; $MgCl_2$, 1. Triton X-100 (5%) was added into the initial homogenization solutions to eliminate contaminating ATPases of membranous origin (Solaro et al., 1971). For determination of ATPase activity, each assay tube contained 40 - 60 mg of myofibrils. Mg^{2+}-ATP was added to the tube to initiate the reaction, which was allowed to proceed for 30 sec prior to stopping with the addition of trichloroacetic acid to 10% v/v. The concentration of P_i liberated was determined as described by Lanzetta et al. (1979) with some alterations (Itaya and Ui, 1966; Anner and Moosmayer, 1975). The rates of ATP hydrolysis were normalized to myofibrillar myosin heavy chain content.

Myosin heavy chain (MHC), actin, and MLCK content of whole tissue and myofibrillar preparations were determined using quantitative sodium dodecyl sulfate polyacrylamide gel electrophoresis (SDS-PAGE). Homogenates of tracheal and bronchial smooth muscle tissues and myofibrils were prepared in denaturing sample buffer (Laemmli, 1970). Individual proteins were separated using 4 - 20% linear gradient SDS-PAGE (for MHC and actin) or 7.5% SDS-PAGE (MLCK) and their identities confirmed using specific polyclonal antibodies and Western blotting techniques. Gels were stained with Coomassie blue R-250, dried, and protein content estimated using two dimensional laser densitometric analysis. Bovine serum albumin (BSA) was used as a standard on each gel and contents of MHC, actin, and MLCK extrapolated after correcting for individual differences in dye binding capacities.

Peptide mapping techniques were applied to analyze the primary structures of myosin light chain kinase as described by Cleveland et al. (1977). Bands representing MLCK were cut from 7.5% SDS gels and applied to a 4 - 20% SDS gel. The protein was subsequently digested in the stacking gel of the second gel with either chymotrypsin or Staphylococcus aureus V8 protease.

RESULTS

The Mg^{2+}-ATPase activity of sensitized bronchial smooth muscle (SBSM) myofibrils was significantly greater than that of the control whether normalized to myofibrillar weight (Figure 1a) or myofibrillar myosin heavy chain (MHC) content (Figure 1b).

The pattern of protein bands seen on SDS gels was considerably different between whole tissue homogenates and myofibrillar homogenates. This is indicative of preferential loss of protein during the preparation of myofibrillar samples from whole tissues. The identities of bands specific for MHC, MLCK (Fig. 2), and actin were confirmed immunochemically in Western blots. The estimated M_r for MLCK was 138 kDa. Comparison of peptides generated by Cleveland digestion of purified proteins and specific gel bands indicated the bands were not contaminated with proteins other than MLCK, MHC, or actin.

The content of MLCK was increased in whole tissue homogenates from sensitized tracheal smooth muscle (STSM) and in SBSM compared with their control counterparts (Fig. 3). The ratio of MHC:MLCK is considerably lower in BSM than in TSM in whole tissues indicating a relatively higher proportion of MLCK in BSM (Table 1). Actin and myosin contents were higher in SBSM tissue than in control BSM, however, this difference was not detected in tracheal tissue homogenates (Fig. 3).

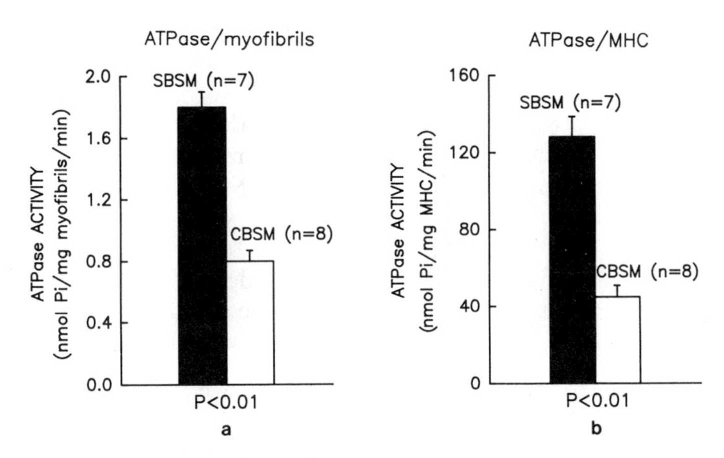

Figure 1. ATPase activity of myofibrils from control (C) and sensitized (S) bronchial and tracheal myofibril preparations.

Table 1. Molar Ratios of Contractile Proteins in Whole Tissue Homogenates

	STSM	CTSM	SBSM	CBSM
MHC:MLCK	1.42	2.30	0.64	0.64
actin:MHC	14.2	15.1	18.0	18.9

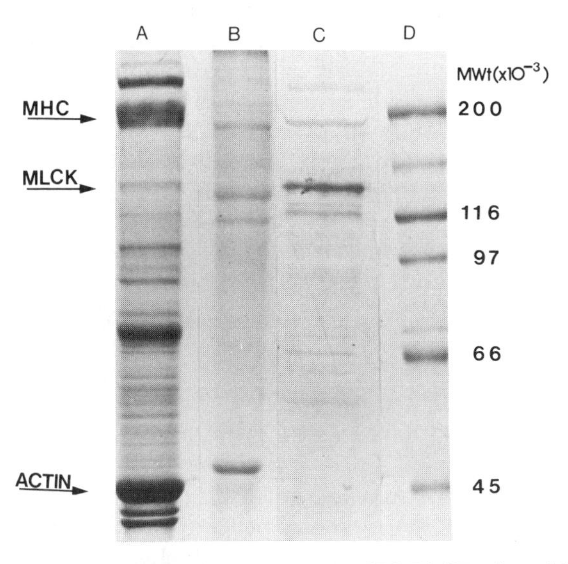

Figure 2. Typical stained patterns seen on 4 - 20% SDS-PAGE gels and Western blot. A. Crude homogenate of sensitized tissue; B. Protein extract from sensitized myofibrillar total sample; C. Western blot using antibodies specific for smooth muscle MLCK and MHC (Antibodies were gifts from Dr. M. Pato, Univ. Saskatchewan and Dr. M. Walsh, Univ. Calgary, respectively); D. Molecular weight markers.

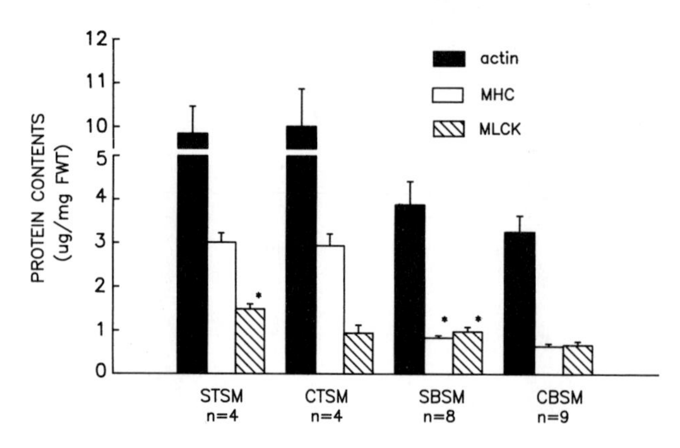

Figure 3. Content of MLCK, MHC, and actin in sensitized and control smooth muscle from the trachea and bronchi. STSM and CTSM: sensitized and control tracheal smooth muscle; SBSM and CBSM: sensitized and control bronchial smooth muscle.

Table 2. Molar Ratio of Contractile Proteins in Myofibril Preparations

	STSM	CTSM	SBSM	CBSM
MHC:MLCK	0.26	0.25	0.15	0.14
actin:MHC	18.7	16.4	16.0	19.1

The content of MLCK was higher in the myofibrils prepared from sensitized airway smooth muscle compared to myofibrils prepared from control tissue. This is similar to the results for whole tissue homogenates. In addition, actin and MHC contents were higher in myofibril preparations from sensitized tracheal and bronchial smooth muscle compared to their control counterparts (Fig. 4). The ratio of MHC:MLCK was greatly reduced in all myofibril preparations compared to the MHC:MLCK seen for corresponding whole tissue extracts (Table 2). This ratio, MHC:MLCK, is lower in bronchial myofibrils than in tracheal smooth muscle myofibrils (Table 2), a pattern also seen for whole tissue extracts (Table 1).

Peptide maps generated for MLCK from sensitized and control bronchial smooth muscle were not different regardless of which protease was employed in the digestion.

DISCUSSION

In sensitized canine airway smooth muscle, the content of some major contractile proteins, myosin, actin, and MLCK, appear to be increased but the isometric force development is known to be similar to that of controls (Jiang et al., 1990; Jiang et al., this volume). The functional relevance of the increased myosin and actin contents in sensitized airway smooth muscle is intriguing. One could speculate that the increased myosin and actin are in a non-polymerized form, which do not contribute to the force production. A second

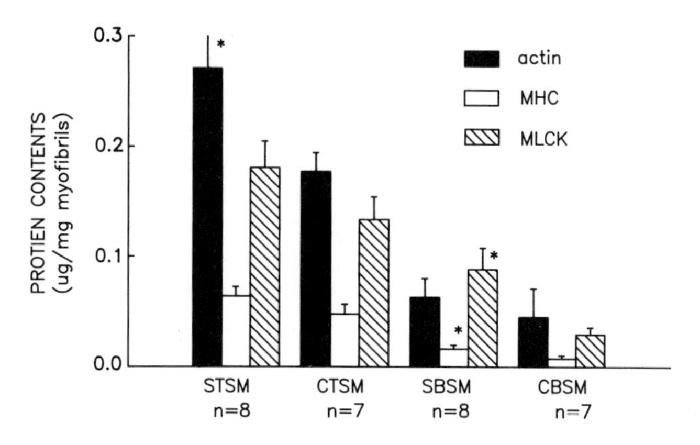

Figure 4. Content of MLCK, MHC, and actin in myofibrillar preparations of sensitized and control smooth muscle from the trachea and bronchi.

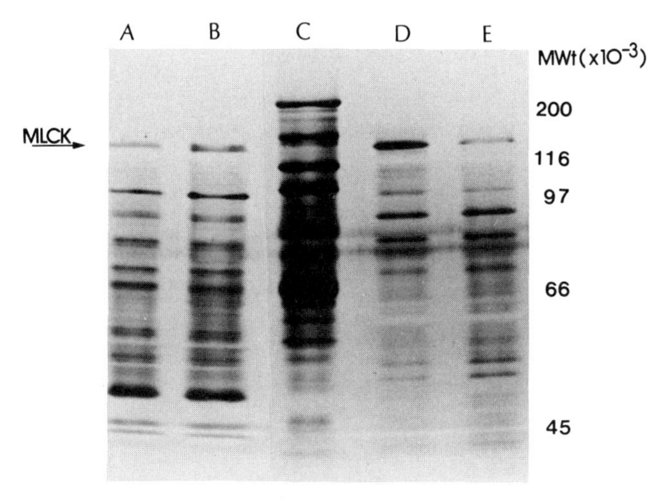

Figure 5. Cleveland peptide mapping of MLCK from sensitized and control bronchial smooth muscle using chymotrypsin and Staphylococcus aureus V8 protease. Peptides were separated by 4 - 20% SDS-PAGE. A. MLCK from SBSM cut with V8 protease; B. MLCK from CBSM cut with V8 protease; C. High molecular weight markers; D. MLCK from SBSM cut with Chymotrypsin; E. MLCK from CBSM cut with chymotrypsin.

thought might be that the myofilaments have been made longer because of more myosin and actin. This might explain the increased shortening capacity and unchanged force production that has been noted in sensitized smooth muscle strips (Jiang et al., 1990; Jiang et al., this volume). Had the increased myosin and actin formed more myofilaments in the sensitized group however, we would have expected to have measured more force when normalized with respect to muscle tissue.

The increased isotonic shortening velocity and shortening capacity have been reported in trachealis (Antonissen et al., 1979; Mitchell and Stephens, 1983), pulmonary artery (Kong and Stephens, 1983), saphenous vein (Wang et al., 1990), and bronchial smooth muscle (Jiang et al., this volume) from sensitized dogs. These changes are mainly observed in the mechanical properties of early normally cycling crossbridges (Jiang et al., this volume). It has been shown that in skeletal muscle, zero load shortening velocity represents crossbridge cycling, therefore, correlates with actomyosin ATPase activity (Bendall, 1964; Bárány, 1967). We confirm that ATPase activity is significantly higher in sensitized airway smooth muscle myofibril preparation than in controls as predicted by the increased zero load shortening velocity in the sensitized group. The elevated shortening velocity and capacity found in our sensitized airway smooth muscles (Stephens et al., 1988; Jiang et al., this volume) could play a role in the acute asthmatic attack.

It is generally accepted that in smooth muscle the actomyosin Mg^{2+}-ATPase activity is regulated by the phosphorylation of myosin light chain (Sobieszek, 1977; Persechini and Hartshorne, 1981; Aksoy et al., 1983). We have shown that the level of myosin light chain phosphorylation is increased in sensitized canine trachealis as compared to controls (Kong et al., 1990). This accounts for the increased myofibrillar ATPase activity of the sensitized tracheal and bronchial smooth muscle. Phosphorylation of 20 kDa myosin light chain (20 kDa-MLC) is catalyzed by Ca^{2+} -and calmodulin-dependent MLCK (for detailed discussion, see Kamm and Stull, 1985). In the present study, we have revealed that in sensitized airway smooth muscle, MLCK content is increased (Fig. 4). This suggests that the elevated levels of 20 kDa-MLC phosphorylation seen in airway smooth muscle from sensitized animals (Kong et al., 1990) may be the result of an increase in the enzyme responsible for 20 kDa-MLC phosphorylation, namely MLCK. The primary structure of MLCK seems to be unchanged in sensitized tissues, as suggested in our peptide mapping study. However, the question of whether the specific activity of MLCK is somehow altered during the sensitization process needs to be addressed. Of equal importance will be to determine whether calmodulin or MLC specific phosphatase content or function is affected by sensitization.

The ATPase activity of skeletal muscle myofibril preparations was first measured by Bendall (1961; 1964). Smooth muscle myofibrillar preparations represent a melange of regulatory and contractile proteins such as MHC, MLC, actin, tropomyosin, calmodulin, caldesmon, and MLCK (Sobieszek and Bremel, 1975; Murakami and Uchida, 1985). Sobieszek and Bremel (1975) have suggested that myofibril preparations from vertebrate smooth muscles are well suited to biochemical studies of the smooth muscle regulation. We have found, however, that the relative contents of actin and myosin and MLCK in myofibrillar preparations are not necessarily the same as those in tissues from which they are made. Therefore, we suggest that actomyosin specific ATPase activities in total tissue homogenates and myofibrillar preparations be compared to avoid artifactual misrepresentations.

CONCLUSIONS

1. The content of the major contractile proteins, myosin and actin was increased in sensitized bronchial smooth muscle compared to their littermate controls.
2. Myosin light chain kinase content is increased in sensitized airway smooth muscle compared to control tissue. However, the primary structure of the enzyme appears to be unaltered.
3. Changes in the content of major contractile proteins in smooth muscle after sensitization suggest that expression of some major, muscle-specific genes may be affected in the process.

4. Increased myosin light chain kinase content in sensitized smooth muscle may be part of the fundamental mechanism underlying increases seen in the myofibrillar ATPase activity of sensitized airway smooth muscle.
5. The relative proportions of the major catalytic and regulatory enzymes responsible for ATP hydrolysis during crossbridge cycling in myofibril preparations are not the same as the proportions seen in the whole airway smooth muscle tissue from which they are made. Hence, enzyme activity in whole tissue homogenates should be studied instead of or in conjunction with myofibril preparations to avoid artifactual results.

ACKNOWLEDGMENTS

This work was supported by an operating grant from the Medical Research Council of Canada. H. Jiang is the recipient of a fellowship from Medical Research Council of Canada.

REFERENCES

Aksoy, M. O., Mras, S., Kamm, K. E., and Murphy., R. A., 1983, Ca^{2+}, cAMP, and changes in myosin phosphorylation during contraction of smooth muscle, *Am. J. Physiol.*, 245: C255.

Anner, B. and Moosmayer, 1975, Rapid determination of inorganic phosphate in biological systems by a highly sensitive photometric method, *Anal. Biochem.*, 65: 305.

Antonissen, L. A., Mitchell, R. W., Kroeger, E. A., Kepron, W., Tse, K. S., and Stephens, N. L., 1979, Mechanical alterations of airway smooth muscle in a canine asthmatic model, *J. Appl. Physiol.*, 46: 681.

Antonissen, L. A., Mitchell, R. W., Kroeger, E. A., Kepron, W., Tse, K. S., Stephens, N. L., and Bergen, J., 1980, Histamine pharmacology in airway smooth muscle from a canine model of asthma, *J. Pharmacol. Exp. Ther.*, 213: 150.

Bárány, M., 1967, ATPase activity of myosin correlated with speed of muscle shortening, *J. Gen. Physiol.*, 50: 197.

Bendall, J. R., 1961, A study of the kinetics of the fibrillar adenosine triphosphatase of rabbit skeletal muscle, *Biochem. J.*, 81: 520.

Bendall, J. R., 1964, The myofibrillar ATPase activity of various animals in relation to ionic strength and temperature, *in:* "Biochemistry of Muscle Contraction", J. Gergely, ed., Little Brown, Boston, p. 448.

Cleveland, D. W., Fischer, S. G., Kirschner, N. M., and Laemmli, U. K., 1977, Peptide mapping by limited proteolysis in sodium dodecyl sulfate and analysis by gel electrophoresis, *J. Biol. Chem.*, 252: 1102.

Itaya, K. and Ui, M., 1966, A new micromethod for the colorimetric determination of inorganic phosphate, *Clin. Chem. Acta*, 14: 361.

Jiang, H., Rao, K., Halayko, A. J., Kepron, W., and Stephens, N. L., 1990, Mechanical alterations in sensitized (S) canine bronchial smooth muscle (BSM), *FASEB J.*, 4: A444.

Kamm, K. E. and Stull, J. T., 1985, The function of myosin and myosin light chain kinase phosphorylation in smooth muscle, *Ann. Rev. Pharmacol. Toxicol.*, 25: 593.

Kepron, W., James, J. M., Kirk, B., Sehon, A. H., and Tse, K. S., 1977, A canine model for reaginic hypersensitivity and allergic bronchoconstriction, *J. Allergy Clin. Immunol.*, 59: 64.

Kong, S. K., Shiu, R. P. C., and Stephens, N. L., 1986, Studies of myofibrillar ATPase in ragweed-sensitized canine pulmonary smooth muscle, *J. Appl. Physiol.*, 60: 92.

Kong, S. K., Halayko, A. J., and Stephens, N. L., 1990, Increased myosin phosphorylation in sensitized canine tracheal smooth muscle, *Am. J. Physiol.*, 259: L53.

Laemmli, U. K., 1970, Cleavage of structural proteins during the assembly of the head of bacteriophage T4, *Nature*, 27: 680.

Lanzetta, P. A., Alvarez, L. J., Reinach, P. S., and Candia, O. A., 1979, An improved assay for nanomole amounts of inorganic phosphate, *Anal. Biochem.*, 100: 95.

Mitchell, R.W. and Stephens, N. L., 1983, Maximum shortening velocity of smooth muscle: Zero load-clamp vs afterloaded method, *J. Appl. Physiol.*, 55: 1630.

Murakami, U. and Uchida, K., 1985, Contents of myofibrillar proteins in cardiac, skeletal and smooth muscle, *J. Biochem.*, 98: 187.

Persechini, A. and Hartshorne, D. J., 1981, Phosphorylation of smooth muscle myosin: evidence for cooperativity between the myosin head, *Science*, 213: 1383.

Sobieszek, A. and Bremel, R. D., 1975, Preparation and properties of vertebrate smooth muscle myofibrils and actomyosin, *Eur. J. Biochem.*, 55: 49.

Sobieszek, A., 1977, Ca-linked phosphorylation of a light chain of vertebrate smooth muscle myosin, *Eur. J. Biochem.*, 73: 477.

Solaro, R. J., Pang, D. C., and Briggs, F. N., 1971, The purification of cardiac myofibrils with Triton X-100, *Biochim. Biophys. Acta*, 245: 259.

Stephens, N. L., Kong, S. K., and Seow, C. Y., 1988, Mechanisms of increased shortening of sensitized airway smooth muscle, *in:* "Mechanisms of Asthma: Pharmacology, Physiology, and Management", C. L. Armour, J. L. Black, eds., Alan R. Liss, New York, p. 231.

Stephens, N. L., Halayko, A. J., and B. Swynghedauw, B., 1991, Myosin heavy chain isoform distribution in normal and hypertrophied rat aorta smooth muscle, *Can. J. Physiol. Pharmacol.*, 69: 8.

ISOFORMS (CONFORMATIONS?) OF TURKEY GIZZARD MYOSIN LIGHT CHAIN KINASE: SEPARATION BY ANION EXCHANGE HIGH PERFORMANCE LIQUID CHROMATOGRAPHY

Louise M. Garone

Medical Biotechnology Center and
Department of Biological Chemistry
University of Maryland at Baltimore School of Medicine
660 W Redwood Street
Baltimore, Maryland 21201

INTRODUCTION

Myosin light chain kinase illustrates well how an enzyme simultaneously affects and is affected by interlocking pieces in the puzzle of smooth muscle regulation. Coincident with binding calcium-calmodulin (Ca^{2+}-CaM), myosin light chain kinase (MLCK) is apparently released from pseudosubstrate inhibition and allowed to phosphorylate its substrate, the regulatory myosin light chain (MLC) (Pearson et al., 1988). These events at least facilitate (Sobieszek, 1977; Persechini et al., 1981), and probably trigger, (Itoh et al., 1989) the initiation of contraction in smooth muscle.

Factors in the biological regulation of MLCK activity, other than CaM, are still only poorly understood. One of the most perplexing aspects of this regulation is the role of posttranslational modification of MLCK, specifically phosphorylation, in modulating interactions between MLCK, MLC, and CaM. Adelstein et al. (1978) showed that turkey gizzard MLCK (tgMLCK) could be phosphorylated by cAMP-dependent protein kinase and that this phosphorylation decreased the enzyme's catalytic activity. Later, Conti and Adelstein (1981) demonstrated that cAMP-dependent protein kinase (A-kinase) could phosphorylate two sites on tgMLCK, one of which was only accessible in the absence of CaM. This led to the hypothesis that A-kinase phosphorylation of MLCK functioned as a receptor-mediated inhibition of smooth muscle contraction. Although this hypothesis is no longer widely supported, the most recent data on MLCK phosphorylation by Ca^{2+}/CaM-dependent protein kinase II are still consistent with a modulatory role for the structural effect of phosphorylation of MLCK on this enzyme's protein-protein interactions (Hashimoto and Soderling, 1990; Ikebe and Reardon, 1990).

There would seem to be two major problems to be overcome in understanding the relationship between posttranslational modification of MLCK, MLCK conformation and catalytic activity, and MLCK participation in protein-protein interactions. The first is the plethora of kinases capable of phosphorylating MLCK, and the second is the possibility that phosphorylation at more than just two sites is functionally significant (Shoemaker et al., 1990). In order to develop an accurate structural model of MLCK it will be necessary to determine the basal phosphorylation state of this protein and the changes which occur in that state under different physiological conditions.

In this paper I report on the finding of a "cluster" of HPLC peaks in electrophoretically-pure tgMLCK. In addition, I propose a possible explanation for why the cluster was not resolved previously and speculate as to the chemical nature of the resolved proteins.

METHODS AND RESULTS

Fresh turkey (hen or tom) gizzards were obtained from Round Hill Foods (New Oxford PA) and stored at -80° C. Reagents were purchased from Sigma except when otherwise indicated. [γ-^{32}P]-ATP was purchased from New England Nuclear. Bovine testis calmodulin (CaM) was the generous gift of Dr. Robert F. Steiner of the Department of Chemistry and Biochemistry, University of Maryland Baltimore County. The HPLC system used contained a Glenco SV-3 injector, a Waters M45 pump, a Waters M6000A pump, a Waters M 660 gradient maker, a Waters M-480 variable-wavelength absorbance detector and a Linear dual channel recorder.

MLCK Assay

MLCK activity was determined as described by Corbin and Reimann (1974) using Whatman 3MM paper. The assay conditions were: ~10 μM myosin light chain, 20 mM Tris (pH 7.3), 10 mM $MgCl_2$, 2 mM $CaCl_2$ (or 5 mM EGTA), 0.1 μM CaM, 0.1 mM ATP (specific activity 0.3 Ci/mmol), 0.2 mg/ml bovine serum albumin (BSA), and 1 nM MLCK. Both CaM and MLCK were diluted in 1 mg/ml BSA and added last to initiate the reaction. Under certain circumstances (very pure myosin light chain substrate) it is possible to "turn off" the enzyme in the presence of Ca^{2+} by using only diluent in place of CaM. The "blank" for background determination was an assay in which diluent replaced enzyme.

MLCK Isolation

The protocol for isolating tgMLCK was designed to combine aspects of the tgMLCK purification strategies developed by Adelstein and Klee (1981) and Walsh et al. (1983). The approach of Walsh and co-workers (1983), Affi Gel Blue affinity chromatography followed by DEAE Sephacel anion exchange chromatography, provided a good yield and was more convenient than using gel filtration columns. Walsh and co-workers (1983), however, had to perform "reverse" ammonium sulfate fractionation to avoid degradation of the MLCK by endogenous proteolytic enzymes. Adelstein and Klee (1981) on the other hand, obtained undegraded MLCK using standard ammonium sulfate precipitation by including a "cocktail" of protease inhibitors. It therefore seemed appropriate to incorporate protease inhibitors into the chromatographic protocol. The inhibitors selected were phenylmethylsulfonyl fluoride (PMSF), streptomycin sulfate (SS), N-tosyl-L-phenylalanine

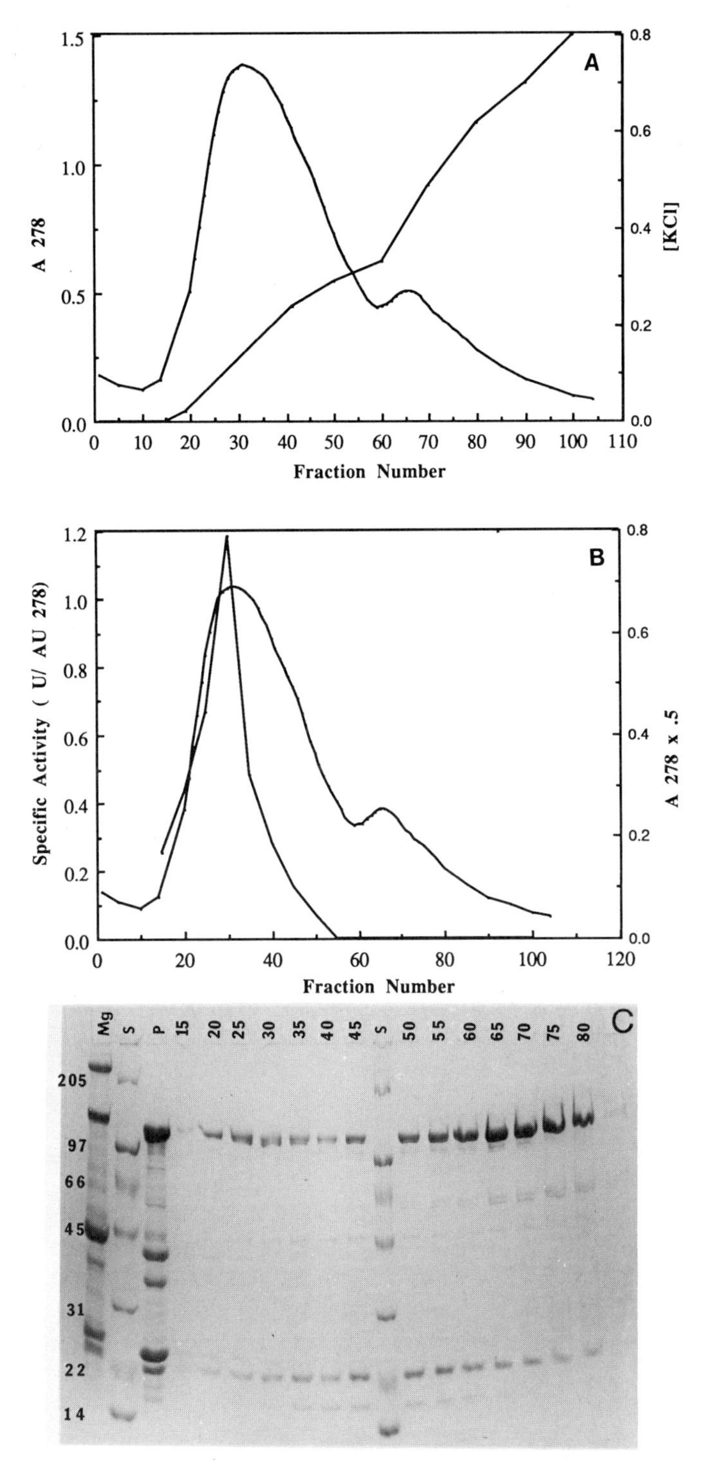

Figure 1. A: Absorbance and salt concentrations of Affi Gel Blue column fractions. B: MLCK activity of AGB fractions. The bulk of MLCK activity eluted at 0.02 M < [KCl] < 0.24 M. Fractions 20- 40, inclusive, were pooled for anion exchange chromatography. C: SDS PAGE profile of AGB fractions. This gel and all other gels stained in Coomassie Blue R (0.2% dissolved in 50% methanol, 10% acetic acid) for 1 h followed by 30 min "fast" destaining in 50% methanol, 10% acetic acid and 4 h "slow" destaining in 10% methanol, 10% acetic acid.

chloromethyl ketone (TPCK), Nα-benzoyl-L-arginine methyl ester (BAME), and soybean trypsin inhibitor (STI). This procedure has been used for about two years in our laboratory to obtain tgMLCK for spectroscopic and kinetic experiments (Garone, 1991; Garone et al., 1991) and routinely provides a good separation of MLCK from caldesmon.

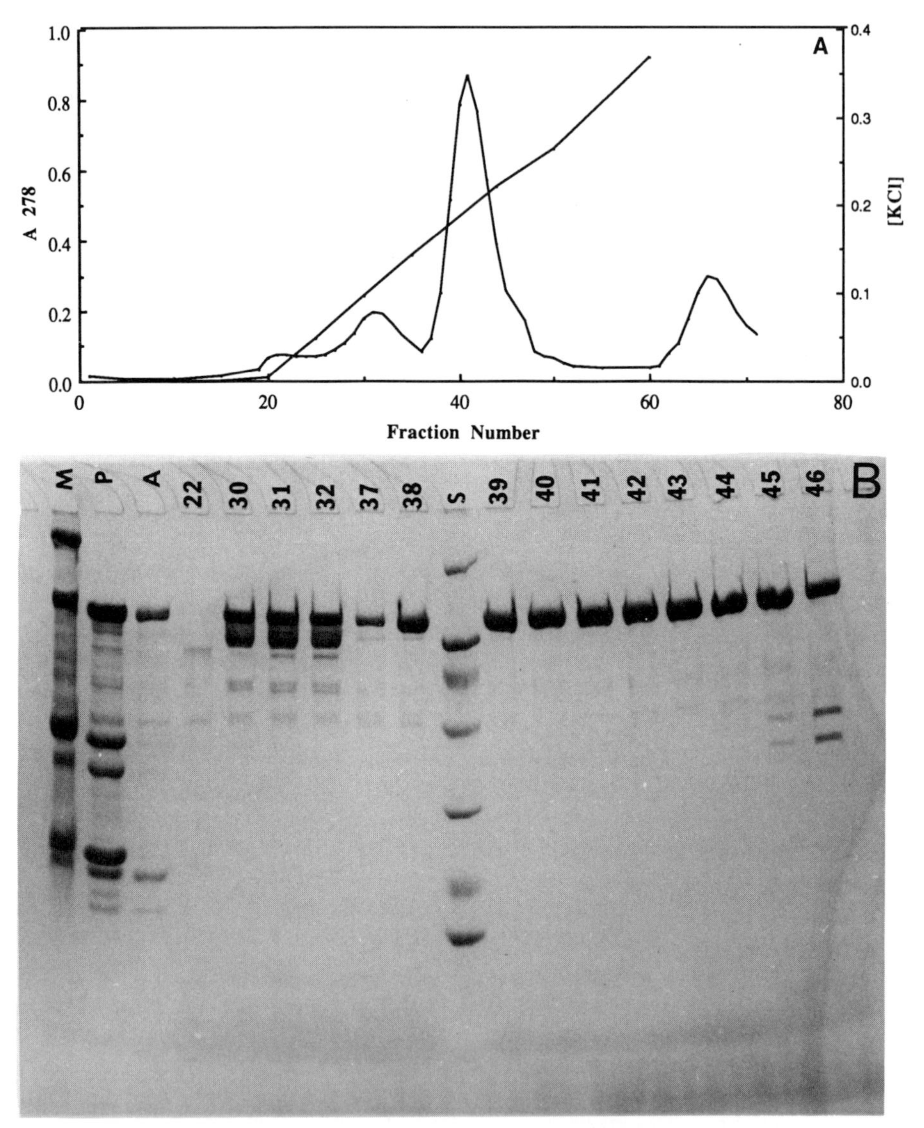

Figure 2. A: Absorbance and salt concentration of DEAE Sephacel fractions. MLCK eluted at 0.14 < [KCl] < 0.22 M. B: SDS PAGE profile of DEAE Sephacel fractions. Fractions 38 - 40 were pooled (9 ml) as DEAE A and fractions 41 - 43 were pooled as DEAE B. "Early" and "late" pools of tgMLCK activity, DEAE A and DEAE B, respectively, were maintained separate as a precautionary measure because a protein of $M_r \approx 20{,}000$ daltons contaminated the "late" tgMLCK fractions in some preparations. The contaminant was absent from this preparation, but when present can be removed by passage over a short G-100 column. Unfortunately, the contaminant's presence does not seem to be correlated with any other known factor in the protocol.

All manipulations of tissue, and conventional chromatography, were conducted at 4° C. Fourteen thawed gizzards were cleaned of fat and connective tissue and ground to yield ~250 g of smooth muscle mince (good results have been obtained with tissue frozen at -80° C as little as three days and as long as four months). The mince was homogenized in a commercial Waring blender at high speed in 4 volumes (w/vol ratio) of 20 mM Tris (pH 7.5), 40 mM KCl, 1 mM $MgCl_2$, 1 mM DTT, 5 mM EGTA, 0.5% Triton X-100, 75 µg/ml PMSF, 100 µg/ml SS, 10 µg/ml STI, 10 µg/ml TPCK, 10 µg/ml BAME. The homogenate was centrifuged at 13,000 x g (r_{max}) for 15 min and the supernatant discarded. The pellet was re-homogenized in 1.0 l of buffer as described above, but without detergent, and the centrifugation repeated. Two detergent-free washes were performed.

Remaining intact tissue was removed from the pellet and the washed myofibrils were suspended (Waring blender at high speed) in 500 ml of 40 mM Tris (pH 7.5), 80 mM KCl, 30 mM $MgCl_2$, 5 mM EGTA, 1 mM DTT, protease inhibitors. The myofibril suspension was then homogenized (Potter-Elvjham) with an additional 500 ml of buffer and centrifuged 30 min at 13,000 x g. The supernatant was filtered through glass wool and the pellet saved for later use in the preparation of myosin light chains.

After measuring the volume of "magnesium extract" (Mg Ext) and taking aliquots of the solution for activity determination and electrophoresis, the solution was brought to 40% saturation with solid ammonium sulfate (226 g/l). The solution was stirred at 4° C for 15 min, then centrifuged for 30 min at 13,000 x g. The supernatant was filtered through glass wool and brought to 60% saturation with solid ammonium sulfate (363 g/l) based on the original volume of the Mg Ext. The 40-60% precipitation was centrifuged 30 min at 13,000 x g and the supernatant discarded. The pellets were dissolved in a minimal volume of Affi Gel Blue (AGB) buffer and dialyzed against 2.0 l of: 20 mM K_2HPO_4, 2.5 mM EGTA, 2.5 mM EDTA, 1 mM DTT, 0.02% NaN_3, 75 µg/ml PMSF, 10 µg/ml BAME, and 10 µg/ml TPCK (note the absence of SS and STI), pH 8.0. The dialysate was loaded onto the AGB column at 0.5 ml/min. This and other flow rates were maintained with a Glenco pump, originally used for HPLC. The loaded column was washed with 150 ml of AGB buffer and eluted with a gradient of KCl from 0 to 0.8 M. Fractions were collected using an LKB Superrak Model 2211. Conductance (at 25° C) and A_{278} of the gradient fractions were monitored (Figure 1A) and selected fractions were tested for MLCK activity (Figure 1B). The same fractions were also evaluated by polyacrylamide gel (5-20% gradient, continuous Tris buffer system, pH 8.8) electrophoresis in the presence of 2 mM EGTA and 0.1% SDS (Figure 1C). Fractions containing the MLCK peak were pooled (AGBB) and dialyzed against 2.0 l of DEAE Sephacel buffer consisting of: 20 mM Tris (pH 7.5), 2.5 mM EGTA, 2.5 mM EDTA, 1 mM DTT, 75 µg/ml PMSF, 10 µg/ml BAME, and 10 µg/ml TPCK. A small (~20 ml bed volume) column of DEAE Sephacel was loaded with the MLCK peak eluted from the AGB column and washed with 150 ml of buffer. It was eluted with a 0 - 0.4 M KCl gradient. Fractions were collected and analyzed as described for AGB chromatography except that only the peak MLCK-containing fractions, identified by electrophoretic profile, were assayed (Figure 2). MLCK was stored at -80° C in the column elution buffer and dialyzed overnight against an appropriate buffer for the experimental protocol. Results from this isolation and purification are summarized, quantitatively, in the first five entries of Table 1.

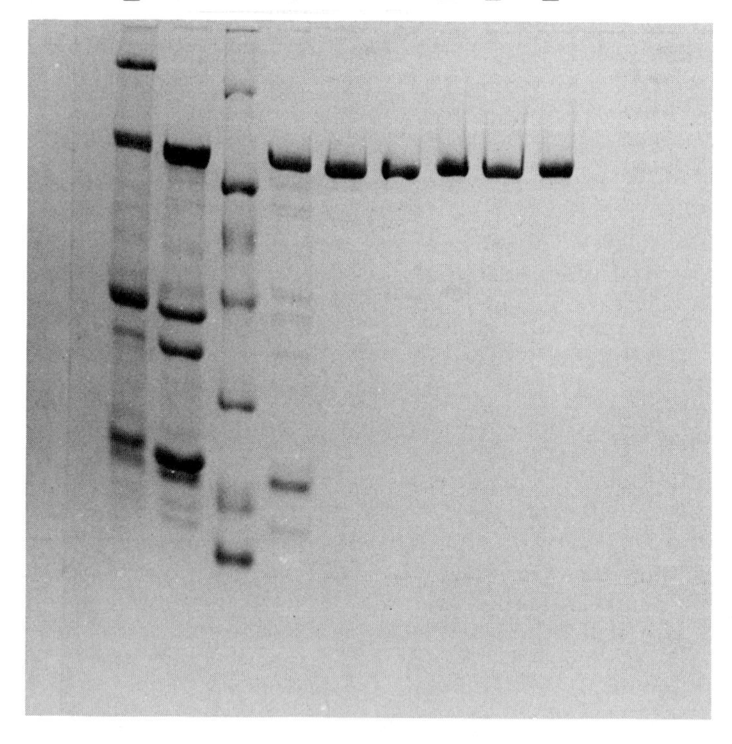

Figure 3. SDS PAGE illustration of tgMLCK isolation and purification stages. Quantity of protein loaded per lane is given in AU_{278} (volume in ml x A_{278}) 1: Mg Ext, 0.023; 2: 40-60% ammonium sulfate cut, 0.023; 3: Molecular weight standards as in Figure 1; 4: Affi Gel Blue pool, 0.006; 5: DEAE A, 0.0025; 6: DEAE B, 0.0036; 7: (H) B1, 0.0021; 8: (H) B2, 0.0023; and 9: (H) B3, 0.0023.

Anion Exchange HPLC of "Pure" tgMLCK

Figure 4 is a chromatogram of 600 µg of tgMLCK subjected to anion exchange HPLC (Bio-Gel TSK DEAE-5-PW, 7.5 x 75 mm). A 1.0 ml aliquot of fraction DEAE B was dialyzed overnight at 4° C against solvent A composed of: 50 mM MOPS, 0.1 M KCl, 0.5 mM DTT, 0.1 mM $CaCl_2$, pH 7.5. Solvent B was solvent A brought to 0.4 M in KCl after equilibration to room temperature. Following injection of MLCK, the column was washed with solvent A (1 ml/min) prior to initiating the salt gradient elution. This was performed at 25° C (similar results were obtained at 4° C). The protein in the first three peaks (labeled B1, B2 and B3) is shown on the gel in Figure 3. The M_r of these fractions are indistinguishable from that of the original fraction (DEAE B). MLCK activity of these fractions were determined, the results of which are shown in Table 1. The specific activity decreased with increasing retention time (anionic character). The activity ratio (defined as the rate of phosphorylation obtained by activation with .001 µM CaM divided by that with .1 µM CaM) was 0.756 for B1 and 0.570 for B3. This decrease in activity ratio and specific activity with increasing anionic character suggests the possibility that

Ser^{814} phosphorylation may be involved (sequence numbering of Olson et al., 1990). This site is susceptible to A-kinase catalyzed phosphorylation which is inhibited in the presence of CaM (Stull et al., this volume).

Table 1. Summary of tgMLCK Activity During Purification and After Anion Exchange HPLC.

Fraction	A_{278}	Volume (ml)	Activity (U/ml)	Spec. Activity (U/AU_{278})	Total Activity U
Mg Ext.*	3.06	920	0.136	0.044	124.8
P60*	11.50	62	1.950	0.170	120.9
AGBB*	1.19	53	0.337	0.283	17.9
DEAE A**	0.51	9	0.835	1.637	7.5
DEAE B**	0.71	9	1.108	1.539	10.0
(H) B1***	0.14	-	0.254	1.814	-
(H) B2***	0.20	-	0.176	0.880	-
(H) B3***	0.13	-	0.108	0.831	-

* = Abbreviation defined in text. ** = Abbreviation defined in legend to Figure 2. *** = Abbreviation defined in legend to Figure 4.

DISCUSSION

The multiple peaks of MLCK activity separated by HPLC was surprising when first noted. I had previously subjected tgMLCK to isoelectric focussing (Righetti and Drysdale ,1970; O'Farrell, 1975) which indicated that the isoelectric point was 6.8. This was reasonably consistent with the pI for tgMLCK reported earlier (6.5) by Walsh et al. (1984) and, within the limits of resolution, indicated that the protein focussed as a single band. One possibility for the multiple peaks could have been attributed to proteolysis. However, there was little quantitative difference between chromatographic runs conducted at 4° and 25° C. In addition, the lack of difference with respect to apparent molecular weight would argue against enzyme degradation or autolysis. Clearly, minor proteolysis cannot be completely ruled out, but it is likely that the presence of protease inhibitors during isolation and purification would effectively minimize contributions of this factor to the observed results. Microheterogeneity due to genetic variance inherent in the turkeys (14 animals were represented in a single preparation) also seems improbable. Commercially-raised turkeys are highly inbred. These animals only reproduce by artificial insemination making it likely that all of the animals used on a given day were the offspring of a single father and a group of sisters. High levels of consanguinity decrease the likelihood of finding a large number of true genetic isoforms in a population sample as small as 14 animals.

It would be attractive to suppose that the HPLC peaks correspond to MLCK phosphorylated at a successively larger number of sites by the action of protein kinases *in vivo*, but certain disquieting facts must be rationalized before this can be proven to be the case. Endogenous protein phosphatases in smooth

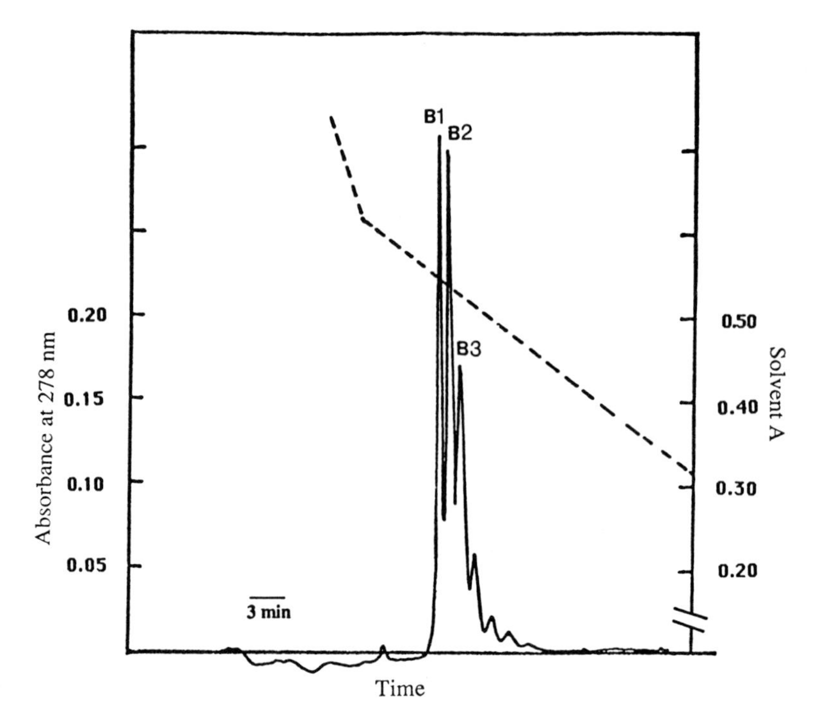

Figure 4. Anion exchange HPLC of 1 ml of DEAE B (≈ 600 μg tgMLCK). The design of this elution was 0% B, 3 min; 0-40% B, 9 min; 40-70% B, 30 min. Full scale (100 divisions) is equal to absorbance 0.5 at 278 nm. Peaks labelled B1, B2, and B3 correspond to ((H) for HPLC) B1, (H) B2, and (H) B3 in Table I and Figure 3.

muscle (SMP) dephosphorylate MLCK, presumably without regulatory control. Since there were no particular precautions built into the experimental protocol to inhibit phosphatase activity it seems fair to assume that basal levels of phosphatase activity were present. In particular, SMP-I is known to cleave phosphate from one or both of the A-kinase sites on tgMLCK, depending on the conditions of incubation (Pato and Adelstein, 1983). Moreover, isolated SMP-I is better able to dephosphorylate MLCK after the phosphatase has been denatured by freezing and thawing. If freezing and thawing induce catalytic subunit dissociation of the phosphatase equally well *in situ* as they do following isolation it would imply that storage at -80° C should have augmented SMP-I activity in the homogenates of these gizzards. Thus, if the resolved peaks did correspond to phosphorylation states of tgMLCK these probably would not involve the well-characterized A-kinase sites. The lower activity ratio of B3, as compared with that of B1, would also now need to be explained by some phenomenon similarly unrelated to phosphorylation at Ser^{814}.

The recent work by Dalla Libera and his colleagues (Dalla Libera et al., 1990) may have direct bearing on this problem. They purified MLCK from the gizzard of a single animal by a procedure in which conventional chromatography was altogether circumvented in favor of anion exchange HPLC. Unfortunately, because their chromatograms were generated by an extremely steep gradient elution (0 to 0.5 M NaCl over 20 min) it would be impossible to resolve MLCK pI isoforms, if they were present. Other recent work with bearing on the problem involves analysis of the MLCK cDNA from chick embryo fibroblasts. Not less than 36 distinct potential phosphorylation sites were identified within an amino acid sequence identical (except at one residue) to that predicted by the chicken gizzard cDNA of Olson et al. (1990). Sixteen additional potential phosphorylation sites were found in the 285 residues not contained in the chicken gizzard MLCK cDNA. Since MLCK appears to contain a large number of phosphorylation sites, if high resolution HPLC were to reveal multiple forms of MLCK from a preparation such as Dalla Libera's (Dalla Libera et al., 1990) further investigation would certainly be warranted.

ACKNOWLEDGMENTS

Dr. Robert F. Steiner was of great assistance in the early stages of this work and Dr. John H. Collins was of equivalent help in the later stages. I am grateful for both of their contributions.

REFERENCES

Adelstein, R. S. and Klee, C. B., 1981, Purification and characterization of smooth muscle myosin light chain kinase, *J. Biol. Chem.*, 256: 7501.

Adelstein, R. S., Conti, M. A., Hathaway, D. R., and Klee, C. B., 1978, Phosphorylation of smooth muscle myosin light chain kinase by the catalytic subunit of adenosine 3':5'-monophosphate-dependent protein kinase, *J. Biol. Chem.*, 253: 8347.

Conti, M. A. and Adelstein, R. S., 1981, The relationship between calmodulin binding and phosphorylation of smooth muscle myosin light chain kinase by the catalytic subunit of 3':5' cAMP-dependent protein kinase, *J. Biol. Chem.*, 256: 3178.

Corbin, J. D. and Reimann, E. M., 1974, Assay of cyclic AMP-dependent protein kinase, *Methods Enzymol.*, 38: 287.

Dalla Libera, L., Cavallini, L., Fasolo, M., Cavanni, P., Ratti, E., and Gaviraghi, G., 1990, Smooth muscle myosin light chain kinase: Rapid purification by anion exchange high performance liquid chromatography, *Biochem. Biophys. Res. Comm.*, 167: 1249.

Garone, L., 1991, A comparison of the steady state kinetics of MLCK-catalyzed phosphoryl transfer to the gizzard myosin RLC and to a synthetic polypeptide substrate, *Biochem. Biophys. Res. Commun.*, submitted.

Garone, L., Albaugh, S. A., and Steiner, R. F., 1991, The secondary structure of turkey gizzard myosin light chain kinase and the nature of its interaction with calmodulin, *Biopolymers*, 30: 1139.

Hashimoto, Y. and Soderling, T. R., 1990, Phosphorylation of smooth muscle myosin light chain kinase by Ca^{2+}/calmodulin-dependent protein kinase II: Comparative study of the phosphorylation sites, *Arch. Biochem. Biophys.*, 278: 41.

Ikebe, M. and Reardon, S., 1990, Phosphorylation of smooth muscle myosin light chain kinase by smooth muscle Ca^{2+}/calmodulin-dependent multifunctional protein kinase, *J. Biol. Chem.*, 265: 8975.

Ikebe, M., Maruta, S., and Reardon, S., 1989, Location of the autoinhibitory region of smooth muscle myosin light chain kinase, *J. Biol. Chem.*, 264: 6967.

Itoh, T., Ikebe, M., Kargacin, G. J., Hartshorne, D. J., Kemp, B. E., and Fay, F. S., 1989, Effects of modulators of myosin light chain kinase activity in smooth muscle cells, *Nature,* 338: 164.

O'Farrell, P. H., 1975, High resolution two dimensional electrophoresis of proteins, *J. Biol. Chem.,* 250: 4007.

Pato, M. D. and Adelstein, R. S., 1983, Purification and characterization of a multisubunit phosphatase from turkey gizzard smooth muscle, *J. Biol. Chem.,* 258: 7047.

Pearson, R. B., Wettenhall, R. E. H., Means, A. R., Hartshorne, D. J., and Kemp, B. E., 1988, Autoregulation of enzymes by pseudosubstrate prototypes, *Science,* 241: 970.

Persechini, A., Mrwa, U., and Hartshorne, D. J., 1981, Effect of phosphorylation on the actin-activated ATPase activity of myosin, *Biochem. Biophys. Res. Comm.,* 98: 800.

Righetti, P. and Drysdale, J. W., 1970, Isoelectric focusing in polyacrylamide gels, *Biochim. Biophys. Acta,* 236: 17.

Shoemaker, M. O., Lau, W., Shattuck, R., Kwiatkowski, A. P., Matrisian, P. E., Guerra-Santos, L., Wilson, E., Lukas, T., Van Eldik, L., and Watterson, D. W., 1990, Use of DNA sequence and mutant analyses and antisense oligodeoxynucleotides to examine the molecular basis of nonmuscle myosin light chain kinase autoinhibition, calmodulin recognition and activity, *J. Cell Biol.,* 111: 1107.

Sobieszek, A., 1977, Ca^{2+}-linked phosphorylation of a light chain of vertebrate smooth muscle myosin, *Eur. J. Biochem.,* 73: 477.

Walsh, M. P., Cavadore, J.-C. Vallet, B., and Demaille, J., 1980, Calmodulin-dependent myosin light chain kinases from cardiac and smooth muscle: A comparative study, *Can. J. Biochem.,* 58: 299-308, 1980.

Walsh, M. P., Hinkins, S., Dabrowska, R. and Hartshorne, D. J., 1983, Smooth muscle myosin light chain kinase, *Methods Enzymol.,* 99: 279.

CLONING PHOSPHOLAMBAN cDNA FROM RAT AORTIC SMOOTH MUSCLE

Kwang S. Hwang and Bernardo Nadal-Ginard

Laboratory of Molecular and Cellular Cardiology and
Howard Hughes Medical Institute, Children's Hospital
Harvard Medical School
300 Longwood Avenue
Boston, MA 02115

INTRODUCTION

Phospholamban (PLN, Kirchberger and Tada, 1976), a pentameric integral protein ($M_r \approx 6$ kDa) of sarcoplasmic reticulum (SR), is an endogenous inhibitor of the SR Ca^{2+}-ATPase. PLN plays a crucial role in excitation-contraction coupling in the heart. Phosphorylation of PLN catalyzed by cyclic AMP-dependent protein kinase activates the SR Ca^{2+}-ATPase in myocardial cells (Tada and Katz, 1982) resulting in the rapid removal of Ca^{2+} ions from the myoplasm to lower the intracellular Ca^{2+} concentration ($[Ca^{2+}]_i$) during each excitation-contraction cycle. The cDNA of PLN has been cloned from the canine heart (Fujii et al., 1987), and subsequently from both cardiac and slow-twitch skeletal muscle of the rabbit (Fujii et al., 1988).

In vascular smooth muscle, the cDNA sequence for PLN has not been reported, and evidence for the presence of PLN in smooth muscle is scarce. In porcine stomach and in rabbit and canine aorta, components ($M_r \approx 22$ kDa and 11 kDa) from the SR were identified as PLN (Raeymaekers and Jones, 1986). Those peptides showed the same characteristics as those of cardiac PLN: phosphorylation by catalytic subunit of cAMP-dependent protein kinase, similar electrophoretic mobility in SDS-PAGE, and cross-reactivity with antibodies raised against purified canine cardiac PLN. Furthermore, PLN was immunolocalized in canine visceral and vascular smooth muscle by labelling gently disrupted tissues with an affinity-purified PLN polyclonal antibody and indirect immunogold, using preembedding techniques (Ferguson et al., 1988). The SR of smooth muscle cells was specifically labelled with patches in a nonuniform fashion. Recently, cloning of cDNA for PLN from porcine stomach (Verboomen et al., 1989) has been reported. PLN from smooth muscle has been shown to be phosphorylated by cAMP-dependent protein kinase, Ca^{2+}- calmodulin-dependent protein kinase (Raeymaekers and Jones, 1986), and cGMP-dependent protein kinase (Raeymaekers et al., 1988).

PLN regulates cardiac SR Ca^{2+}-ATPase, whose cDNA in rabbit is identical to that of slow-twitch skeletal muscle, coding 997 amino acids

Regulation of Smooth Muscle Contraction
Edited by R.S. Moreland, Plenum Press, New York, 1991

(MacLennan et al., 1985). SR Ca^{2+}-ATPase in vascular smooth muscle and non-muscle cells is alternatively spliced with 49 amino acids, which are substituted for 4 amino acids of the cardiac form at the carboxyl terminus (Lytton and MacLennan, 1988). Since the smooth (and non-) muscle SR Ca^{2+}-ATPase shares the majority (993 amino acids) of the cardiac Ca^{2+}-ATPase, we can hypothesize that the SR Ca^{2+}-ATPase of smooth (and non-) muscle is modulated by PLN through an identical mechanism as in cardiac cells.

In order to elucidate the PLN modulation of SR Ca^{2+}-ATPase of vascular smooth muscle, we cloned the cDNA encoding for PLN from rat aortic smooth and cardiac muscles, and analyzed the RNAs from the rat cell types. In this communication, we report evidence for the presence of rat PLN in heart, aorta, and L_6E_9. Rat PLN cDNA in the aorta is identical to that of the heart, suggesting similar PLN modulation in these cells. Quantitative analysis of its distribution, however, shows that the vascular smooth muscle cells express less PLN than the myocardial cells. These data suggest that PLN regulates SR Ca^{2+}-ATPase in rat vascular smooth muscle, as in the heart, which is critical for the rapid mobilization of Ca^{2+} in and out of the SR.

MATERIALS AND METHODS

Recombinant DNA techniques were performed essentially according to Molecular Cloning (Maniatis, et al., 1982). The rat cardiac cDNA library in λgt10, constructed from mRNAs of Sprague-Dawley rats, was kindly donated by Dr. B. Scott. The rat aortic cDNA library in λgt11 from Sprague-Dawley rats was purchased from Clontech. A synthetic oligo DNA (PA59, 5' GACGTGCTTGTT-GAGGCATTTCAATGGTTGAAGCTCTTCTAATAGCAGAGCGAGTGAGG 3'), complimentary to the coding region of canine cDNA (see Fig. 1), was 5' end labelled with [γ-^{32}P]-ATP (Amersham) by T4 polynucleotide kinase (New England BioLabs) and used for screening. From 1 x 10^6 plaques of the rat cardiac library, five clones were detected and the longest clone, RC623 (Fig. 1), was subcloned in pGEM-3Z (Promega) and further characterized. RC623 was sequenced according to the manufacturer's suggested protocol, with primers in vector and internal sequence, and by subcloning. Sequencing of cDNA was carried out using Sequenase (USB) with [α-^{35}S]-dATP(Amersham). Sequences were analyzed using a Digital Vax computer and the University of Wisconsin Genetics Computer Group software.

RC623 was identified as a cDNA for PLN by confirming the deduced amino acid sequence, and was used for further screening of the aortic library and RNA analysis. RC623 (Fig. 1) was uniformly labelled with [α-^{32}P]-dCTP (Amersham) by Klenow (New England BioLabs). 1 x 10^6 plaques of the rat aortic library were screened with uniformly labelled RC623 (Fig. 1) yielding a clone RAN18. RAN18 was subcloned and analyzed in the same manner.

RNAs were prepared from various cultured cells and tissues of female Sprague-Dawley rats (190 - 210g, Charles River) using guanidine thio-cyanate/CsCl method and analyzed by S1 nuclease protection assay. S1 nuclease assays were performed by hybridization of 30 µg of each RNA with 5' end labelled BspMI fragment of RC623 (number 83 - 381) in hybridizing solutions (80% formamide with 400 mM NaCl, 0.05% SDS, 1 mM EDTA, 10 mM PIPES, pH 6.4) at 42° C overnight, followed by digestion with 300 units of S1 nuclease (Sigma) in 30 mM NaAc, 300 mM NaCl, 3 mM $ZnSO_4$, pH 4.5 at 25° C for one hr. These were separated on PAGE and exposed on XAR5 (Eastman-Kodak) with an intensifying screen.

RESULTS

Figure 1A shows the cloning strategy using the reported nucleotide sequence of PLN from the canine heart (Fujii, et al., 1987). A synthetic oligo DNA, PA59, complimentary to the open reading frame of canine PLN, was 5' end labelled with [γ-^{32}P]-ATP and used to screen the rat cardiac library in λgt10, yielding five clones. RC623 (700 bp, Fig. 1B), the longest one, was subcloned in pGEM-3Z and sequenced (Fig. 1C) with primers in the sequencing vector, internal primers, and by subcloning. We identified RC623 as a PLN cDNA (see below, Figs. 2 and 3) from the comparison to that of canine PLN. Uniformly labelled RC623 was used to screen the rat aortic library in λgt11, producing RAN18 (400 bp). RAN18 was sequenced (Fig. 1C) in the same manner.

The nucleotide sequence of RAN18 from the rat aorta was identical to that of RC623 from the heart. Homology of the nucleotide sequence between the canine PLN and the rat RC623 or RAN18 (Fig. 2) was 78.8%, including untranslated regions (400bp). There was 86.2% homology in the open reading frame of 52 amino acids. This resulted in a single conserved mutation at the second position (see below, Fig. 3). From this comparison, we concluded that RC623 and RAN18 were PLN cDNAs.

The nucleotide sequences of PLNs reported from various species and their deduced amino acid sequences were compared in Figure 3. They shared 86.2 - 88.7% of the nucleotide sequence with a single amino acid mutation.

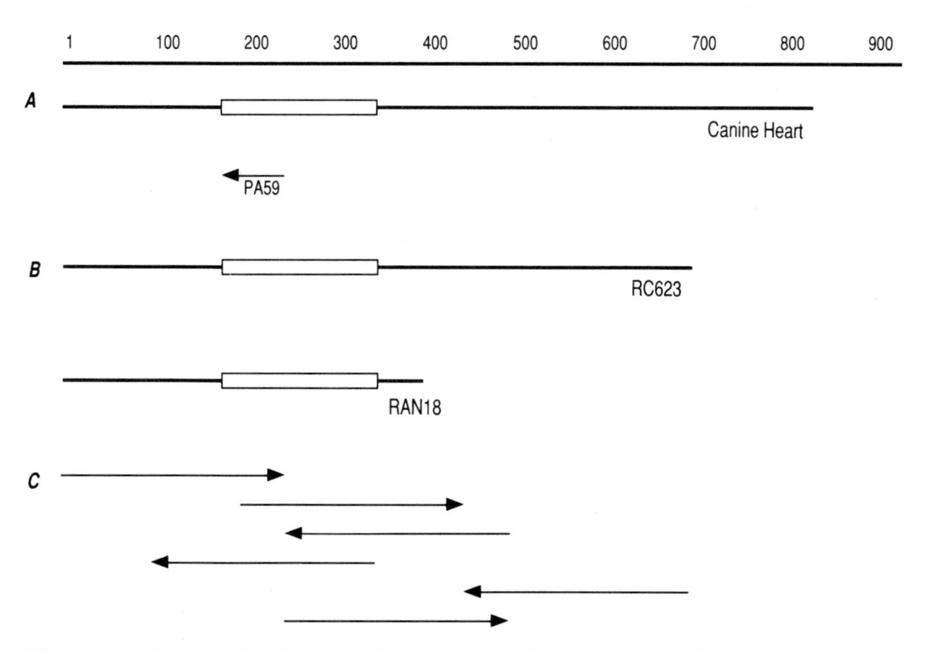

Figure 1. Strategy for cloning and sequencing of rat phospholamban. a: A synthetic oligo DNA (PA59) complimentary to the coding region of canine PLN cDNA. b. Rat PLN cDNAs from cardiac (RC623) and aortic smooth muscle cells (RAN18). c. Sequencing strategy for RC623 and RAN18.

```
                       .         .         .         .         .
RC623     32   AGACTTCACACAACTA..AACAGTCTGCATTGTGACGATCACAGAAGCCA   79
               |||||||| ||||||   || | | |  ||||| |  ||||||  |||||
CANINE    32   AGACTTCATACAACTCACAATACTTTATATTGTAATCATCACAAGAGCCA   81
                       .         .         .         .         .
RC623     80   AGGCCTCCTAAAAGGAGACA........GCTCGCGTTTGGCTGCCTGCTG  121
               ||||  |||||||| ||| |        |||| | |||||| ||
CANINE    82   AGGCTACCTAAAAGAAGAGAGTGGTTGAGCTCACATTTGGCCGC......  125
                       .         .         .         .         .
RC623    122   TCAACTTTTTATCTTTCTCTTGACTACTTAAAAAAGACTT..GTCTTCCT  169
                || ||||||| ||||||||| || | |||||    ||||  | ||||||
CANINE   126   .CAGCTTTTTACCTTTCTCTTCACCATTTAAA....ACTTGAGACTTCCT  170
                       .         .         .         .         .
RC623    170   ACTTTTGTCTTCCTGGCATCATGGAAAAAGTCCAATACCTTACTCGCTCG  219
                ||      |||||||  ||||||| |||||||||||||| ||| ||||
CANINE   171   GCT......TTCCTGGGGTCATGGATAAAGTCCAATACCTCACTCGCTCT  214
                       .         .         .         .         .
RC623    220   GCTATCAGGAGAGCCTCGACTATTGAAATGCCCCAGCAAGCGCGTCAGAA  269
               ||||| || ||||| || || ||||||||||| || ||||| ||||| ||
CANINE   215   GCTATTAGAAGAGCTTCAACCATTGAAATGCCTCAACAAGCACGTCAAAA  264
                       .         .         .         .         .
RC623    270   CCTCCAGAACCTCTTTATCAATTTCTGTCTCATCTTGATATGTCTGCTGC  319
                || |||||||| ||||| |||||||||||||| || ||||||||  ||
CANINE   265   TCTTCAGAACCTATTTATAAATTTCTGTCTCATTTTAATATGTCTCTTGT  314
                       .         .         .         .         .
RC623    320   TGATCTGCATCATTGTGATGCTTCTGTGACCAGCTGTCGCCA.CCGCAGA  368
               ||||||||||||||||||||||||| |||    |||  || | |  |||
CANINE   315   TGATCTGCATCATTGTGATGCTTCTCTGAAGTTCTGCTGCAATCTCCAGT  364
                       .         .         .         .
RC623    369   CCTGCACCATGCCAACGTCAGCTTACAACCTGGCCCTCCACCACGAGGG   417
                 |||| | || || | ||| ||||  | ||| |   ||| | |||||
CANINE   365   GATGCAACTTGTCACCATCAACTTAATATCTGCCATCCCATGAAGAGGG   413
```

Figure 2. Identification of rat phospholamban. The rat nucleotide sequence of RC623 (top) was aligned to that of canine phospholamban (bottom) for the best match. Dots were inserted for best alignment and vertical lines indicate the identical nucleotides. The overall homology (400bp including untranslated regions) was 78.9% with 86.2% in the open reading frame, from ATG (number 190) to TGA (number 348) of RC623.

Most of the mismatches in the nucleic acid sequence appeared at the third bases of each amino acid codons, which conserved the amino acid sequence except for the second position. Glutamic acid was used in rat and rabbit (GAA and GAG, respectively) while aspartic acid (GAT) was encoded in canine and porcine. Since both amino acids were negatively charged, PLN with either amino acid could interact with SR Ca^{2+}-ATPase in the same way, with a possible modification of its binding site(s). Mutation of two consecutive nucleotides for amino acid position 42 and 43, however, was tolerated by the variation in leucine codons at both the first and third bases.

Distribution of rat PLN was studied by S1 nuclease protection assay of rat RNAs prepared from various tissues (Fig. 4) and cultured cells (Fig 5). A protected band of 298bp was seen in RNAs from tissues of heart (8 hr.) and aorta (24 hr.). The protection was not detected in either spleen or liver, while there were trace levels in RNAs from the kidney and brain after long exposure (not shown). The relative amount of PLN RNAs from the aorta (weak after 24 hr.), however, was less than that obtained from the heart (strong after 8 hr.). In order to eliminate the possible contamination of vascular muscle in each tissue, RNAs from cultured cells (Fig. 5) of rat clones were tested. RNAs from L_6E_9 (myogenic) and A_7R_5 (aorta) showed protection, while RNAs from the kidney and fibroblast did not.

```
                        10                  30                  50
          1     MetGluLysValGlnTyrLeuThrArgSerAlaIleArgArgAlaSerThrIleGluMet    20
RAT       1     ATGGAAAAAGTCCAATACCTTACTCGCTCGGCTATCAGGAGAGCCTCGACTATTGAAATG    60
RABBIT    1          G     T        C        T     A  A  G     A  C             60
PORCINE   1          T              C        T     T  A     T  A  C             60
CANINE    1          T              C        T     T  A     T  A  C             60

                        70                  90                 110
          21    ProGlnGlnAlaArgGlnAsnLeuGlnAsnLeuPheIleAsnPheCysLeuIleLeuIle    40
RAT       61    CCCCAGCAAGCGCGTCAGAACCTCCAGAACCTCTTTATCAATTTCTGTCTCATCTTGATA    120
RABBIT    61      T  A     A     A              A                               120
PORCINE   61      T  A     A     A     T        A                       A       120
CANINE    61      T  A     A     A  T  T        A     A              T  A       120


                       130                 150
          41    CysLeuLeuLeuIleCysIleIleValMetLeuLeuEnd      53
RAT       121   TGTCTGCTGCTGATCTGCATCATTGTGATGCTTCTGTGA     159
RABBIT    121        C                 C  C        C        159    88.7%
PORCINE   121     C  CT       T        C           C        159    86.8%
CANINE    121        CT  T                         C        159    86.2%
```

Figure 3. Sequence comparison of phospholamban reported from various species. The rat phospholamban protein sequence (top) with its nucleotides (second) was aligned to the altered nucleotides of rabbit (third), porcine (fourth) and canine (bottom). Note that the second amino acid, glutamic acid in rat (GAA) and rabbit (GAG) PLN, respectively, is mutated to aspartic acid (GAT) in both porcine and canine PLN.

DISCUSSION

The data clearly demonstrate that rat aortic smooth muscle cells express PLN, as do cardiac cells. However, the magnitude of expression is less than that of cardiac cells. This is based, in part, on the following evidence. First, cDNA for PLN (RAN18) was obtained from the rat aortic library (Figs. 1 and 3). Importantly, the nucleotide sequence was identical to that (RC623) obtained from the rat cardiac library. Second, the S1 nuclease protection assay showed the same protected band of expected size (Figs. 4 and 5) in the rat RNAs from the aortic (both the tissue in Fig. 4 and its cultured cells in Fig. 5), cardiac, and myogenic cells. Third, the quantitative PLN level in S1 nuclease protected band of aortic smooth muscle cells was lower than that of cardiac muscle cells.

Even though the level of PLN expression was different among cell types (Figs. 4 and 5), identical PLN, both at the nucleotide and amino acid levels, was expressed in rat aorta (RAN18) and heart (RC623). This finding remains to be verified in other species. The S1 nuclease protection seen with cultured myogenic cells in this study (L_6E_9, Fig. 5) is consistent with the findings of Fujii et al., (1988) using both cardiac and slow twitch skeletal muscle cells in slow muscles. The trace level of nuclease protection seen with RNAs from the kidney tissue, but not from its cultured cells, may be explained either by a vasculature contamination of the kidney tissue or by the characteristic change kidney cells undergo during the culture. These possibilities need further verification. Nevertheless, these results are consistent with the hypothesis that PLN is present in vascular smooth muscle cells of the rat aorta and therefore may play a major role in cAMP-dependent Ca^{2+} accumulation in the SR of both the vascular smooth and cardiac muscle cells.

PLN from rat and rabbit are identical at the amino acid level (Fig. 3) and have 88.6% homology at the nucleotide level. When compared to canine and porcine PLN cDNAs, they are 86.2 and 86.8% homologous, respectively, with conserved substitution of a single amino acid at the second position. Glutamic

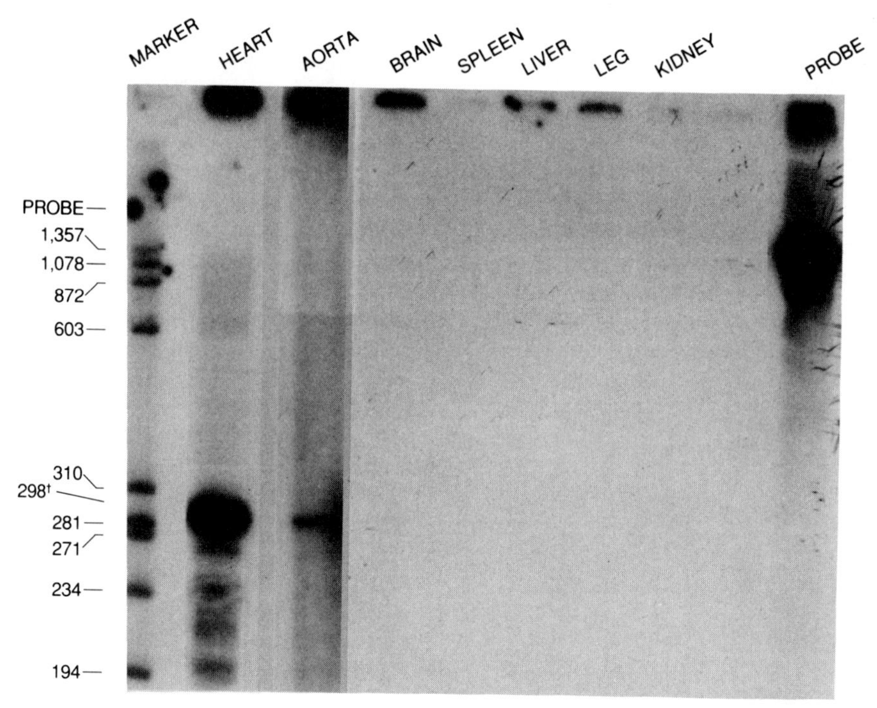

Figure 4. S1 nuclease protection assay with RNAs from various rat tissues. The probe was BspMI (number 381, also number 51 in pGEM-3Z) fragment of subclone of RC623 consisting of nucleotide number from 83 (StuI) to 392 (with all open reading frame) in HindIII site of pGEM-3Z (2,743bp). BspMI fragment (2.8kp) was 5' end labelled and hybridized with 30 μg of each RNA, and was analyzed accordingly (see Methods). The expected size of 298bp was detected in the heart (8 hr exposure) and in the aorta (24 hr).

acid (in rat and rabbit , by GAA and GAG, respectively) was replaced by aspartic acid (GAT) in canine and porcine cDNA. Since both amino acids, glutamic and arpartic acid, are negatively charged, PLN function by this substitution is most likely not significantly affected except for a possible alteration in the SR Ca^{2+}-ATPase binding site(s).

The PLN binding site of SR Ca^{2+}-ATPase (James et al., 1989) has been found in a domain just C-terminal to the aspartyl phosphate of the active site (from amino acid 367 to amino acid 342 - 360 of slow SR Ca^{2+}-ATPase). This site is unique in all isoforms of SR Ca^{2+}-ATPases; cardiac, smooth, and slow skeletal muscles, non-muscles, and even fast skeletal muscle isoforms. This site is not shared by other ATPase ion pumps which are regulated by protein phosphorylation state(s). James et al., (1989) concluded that the regulation of SR Ca^{2+}-ATPase activity depends on the differential expression of the regulatory protein, PLN, rather than depending on intrinsic structural differences, consistent with our observation.

Quantitative analysis of the tissue distribution of PLN RNAs (Fig. 4 and 5) showed that even though PLN RNAs are expressed in vascular tissue, the RNA level in the aorta is less than that in the heart. This observation of low PLN RNA levels suggests that PLN may not be coupled in a fixed stoichiometry with the SR Ca^{2+}-ATPase in all tissues. This is consistent with a report that expression of PLN and the SR Ca^{2+}-ATPase are not coordinated in hypertrophy and hyper- or hypothyroid conditions (Nagai et al., 1989). Therefore, PLN and the ATPase may be coupled, depending on the tissue, in either variable or fixed ratios. If coupling was in a fixed ratio (Louis et al., 1987), only a certain proportion of Ca^{2+}-ATPase proteins would be modulated by PLN, leaving the remaining SR Ca^{2+}-ATPases without PLN inhibition. Apparently the percentage of SR Ca^{2+}-ATPase proteins coupled to PLN appears to be less in the vascular smooth muscle than it does in cardiac muscle. This would leave a large population of SR Ca^{2+}-ATPase proteins in a loosely-regulated state by the endogenous inhibitor, PLN.

The high level of PLN expression in the heart suggests that PLN may be coupled to the majority, if not all, of cardiac SR Ca^{2+}-ATPase moieties. As a result, PLN inhibition of SR Ca^{2+}-ATPase activity may be temporally changed by its phosphorylation state, in concert with the sarcolemmal Ca^{2+} transport system(s). This would be crucial in controlling the myoplasmic [Ca^{2+}] during each heart beat. The mobilization of Ca^{2+} across the sarcolemmal membrane could therefore be modulated by the SR, which would change rapidly from uptake to release during a fraction of each excitation-contraction cycle (lasting 0.3 to 1 sec). On the other hand, "loose inhibition" of SR Ca^{2+}-ATPase by low

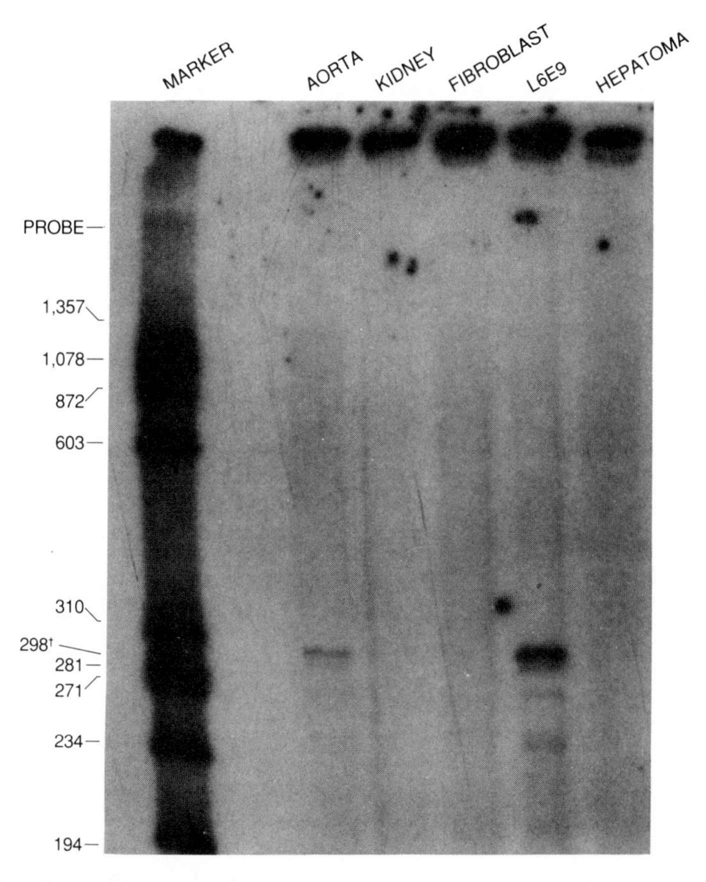

Figure 5. S1 nuclease protection assay with rat RNAs from various cultured cells. See Fig. 4 for experimental protocol. The expected size of 298bp was detected in 9 days from L_6E_9 (myogenic) and in A_7R_5 (aorta).

levels of PLN as in vascular smooth muscle may allow Ca^{2+} accumulation in the SR during depolarization-induced Ca^{2+} entry. This may explain the role of the vascular smooth muscle SR as a component of the superficial buffer barrier (Hwang and van Breemen, 1987).

Furthermore, stimulation of β-adrenergic receptor may reduce vascular smooth muscle tension via synergistic pathways involving both PLN and myosin light chain kinase phosphorylation by the cAMP-dependent protein kinase. When phosphorylated, PLN increases Ca^{2+} pumping into the SR, with a resultant decrease in $[Ca^{2+}]_i$. In addition, phosphorylation of the myosin light chain kinase reduces the Ca^{2+} sensitivity of actomyosin, which would also tend to reduce tension (Adelstein et al., 1978). The quantitative contribution of this synergistic mechanism following β-adrenergic stimulation remains to be seen.

In summary, vascular smooth muscle cells contain PLN which may modulate the SR Ca^{2+}-ATPase in a similar manner as in cardiac muscle cells. The magnitude of PLN modulation appears to be less in vascular smooth muscle cells than in cardiac, which may be very important in the processes involved in Ca^{2+} mobilization.

ACKNOWLEDGMENTS

Authors wish to thank Dr. B. Scott for his generous gift of the cardiac library, and Dr. D. Wilde for discussions. We are grateful to Mr. D. Faber for his art work, and to Ms. M. Douthat for her secretarial assistance. This work was supported by HL33730-07.

REFERENCES

Adelstein, R. S,, Conti, M. A., Hathaway, D. R., and Klee, C. B., 1978, Phosphorylation of smooth muscle myosin light chain kinase by the catalytic subunit of adenosine 3':5'-monophosphate-dependent protein kinase, *J. Biol. Chem.*, 253: 8347.

Eggermont, J. A., Wuytack, F., De Jaegere, S., Nelles, L., and Casteels, R., 1989, Evidence for two isoforms of the endoplasmic-reticulum Ca^{2+} pump in pig smooth muscle, *Biochem. J.*, 260: 757.

Ferguson, D. G., Young, E. F., Raeymaekers, L., and Kranias, E. G., 1988, Localization of phospholamban in smooth muscle using immunogold electron microscopy, *J. Cell. Biol.*, 107: 555.

Fujii, J., Lytton, J., Tada, M., and MacLennan, D. H., 1988, Rabbit cardiac and slow-twitch muscle express the same phospholamban gene, *FEBS Lett.*, 227: 51.

Fujii, J., Ueno, A., Kitano, K., Tanaka, S., Kadoma, M., and Tada, M., 1987, Complete complementary DNA-derived amino acid sequence of canine cardiac phospholamban, *J. Clin. Invest.*, 79: 301.

Hwang, K. S. and van Breemen, C., 1987, Effect of dB-c-AMP and forskolin on the ^{45}Ca influx, net Ca uptake and tension in rabbit aortic smooth muscle, *Eur. J. Pharmacol.*, 134: 155.

James, P., Inui, M., Tada, M., Chiesi, M., and Carafoli, E., 1989, Nature and site of phospholamban regulation of the Ca^{2+} pump of sarcoplasmic reticulum, *Nature*, 342: 90.

Kirchberger, M. A. and Tada, M., 1976, Effects of adenosine 3':5'-monophosphate-dependent protein kinase on sarcoplasmic reticulum isolated from cardiac and slow and fast contracting skeletal muscles, *J. Biol. Chem.*, 251: 725.

Louis, C. F., Turnquist, J., and Jarvis, B., 1987, Phospholamban stoichiometry in canine cardiac muscle sarcoplasmic reticulum, *Neurochem. Res.*, 12: 937.

Lytton, J. and MacLennan, D. H., 1988, Molecular cloning of cDNAs from human kidney coding for two alternatively spliced products of the cardiac Ca^{2+} ATPase gene, *J. Biol. Chem.*, 263: 15024.

MacLennan, D. H., Brandle, C. J., Korczak, B., and Green, N. M., 1985, Amino-acid sequence of a Ca^{2+} + Mg^{2+} -dependent ATPase from rabbit muscle sarcoplasmic reticulum, deduced from its complimentary DNA sequence, *Nature*, 316: 696.

Maniatis, T., Fritsch, E. F., and Sambrook, J., 1982, "Molecular Cloning, A laboratory manual", Cold Spring Harbor Laboratory.

Nagai, R., Zarain-Herzberg, A., Brandle, C. J., Fujii, J., Tada, M., MacLennan, D. H., Alpert, N. R., and M Periasamy, M., 1989, Regulation of myocardial Ca^{2+} ATPase and phospholamban mRNA expression in response to pressure overload and thyroid hormone, *Proc. Nat'l. Acad. Sci. U.S.A.*, 86: 2966.

Raeymaekers, L., Hofmann, F., and Casteels, R., 1988, Cyclic GMP-dependent protein kinase phosphorylates phospholamban in isolated sarcoplasmic reticulum from cardiac and smooth muscle, *Biochem. J.*, 252: 269.

Raeymaekers, L. and Jones, L., 1986, Evidence for the presence of phospholamban in the endoplasmic reticulum of smooth muscle. *Biochim. Biophys. Acta*, 882: 258.

Tada, M. and Katz, A. M., 1982, Phosphorylation of the sarcoplasmic reticulum and sarcolemma, *Ann. Rev. Physiol.*, 44: 401.

Verboomen, H., Wuytack, F., Eggermont, J. A., De Jaegere, S., Missiaen, L., Raeymaekers, L., and Casteels, R., 1989, cDNA cloning and sequencing of phospholamban from pig stomach smooth muscle, *Biochem. J.*, 262: 353.

MYOSIN HEAVY CHAIN ISOFORM PATTERNS DO NOT CORRELATE WITH FORCE-VELOCITY RELATIONSHIPS IN PULMONARY ARTERIAL COMPARED WITH SYSTEMIC ARTERIAL SMOOTH MUSCLE

C. Subah Packer, S. L. Griffith, Janet E. Roepke, Richard A. Meiss, and Rodney A. Rhoades

Department of Physiology and Biophysics
Indiana University School of Medicine
Indianapolis, IN 46202

INTRODUCTION

Velocity of shortening is dependent on the myosin heavy chain (MHC) isoform pattern in both skeletal and cardiac muscle (Pagani and Julian, 1984). Furthermore, it has been reported that a shift in MHC isoform ratio occurs with certain physiological or pathophysiological changes such as hypertrophy and/or hyperplasia of striated muscles (Litten et al., 1974). Such shifts in MHC isoform proportions accompany concomitant changes in shortening velocity and ATPase activity (Alpert et al., 1979; Alpert and Mulieri, 1980). At least two different MHC isoforms have been reported to exist in various different smooth muscles (Sparrow et al., 1987). The 200 kDa form and the 204 kDa form have been designated MHC_1 and MHC_2, respectively. The ratio of MHC_1:MHC_2 has been shown to vary dependent on smooth muscle type, animal species, stage of development, and under certain different physiological or pathophysiological conditions for the same muscle type (Sparrow et al., 1987; Mohammed and Sparrow, 1988). However, no functional correlation has yet been made between MHC isoform ratio and shortening velocity for smooth muscle. Therefore, the purpose of this study was to compare force-velocity (F-V) relationships and MHC isoform ratios from two different arterial muscles (pulmonary versus caudal) from the same species (rat).

METHODS

Rat pulmonary and caudal arterial muscle preparations for the F-V experiments were cut and oriented in the direction that produced maximum force (P_0) in response to supramaximal electrical stimulation. In the case of the caudal artery, a helical strip cut at about 20° from the transverse was the optimal preparation while for the pulmonary artery, a circumferentially-oriented preparation was found to be optimal for mechanical measurements. Each arterial preparation was mounted such that one end was attached to a servo-controlled lever system while the other end was attached to a sensitive photo-electric force transducer. The apparatus has been described in more detail previously (Meiss, 1987). Length and force were recorded simultaneously as functions of time. Limited length-tension analysis was performed on each preparation to establish the optimal resting tension (RP_0) and optimal length (l_0) for producing P_0. Then, with each preparation working at its respective l_0, a different afterload or load-clamp was applied with each consecutive contraction. Velocities of shortening were obtained by measuring the

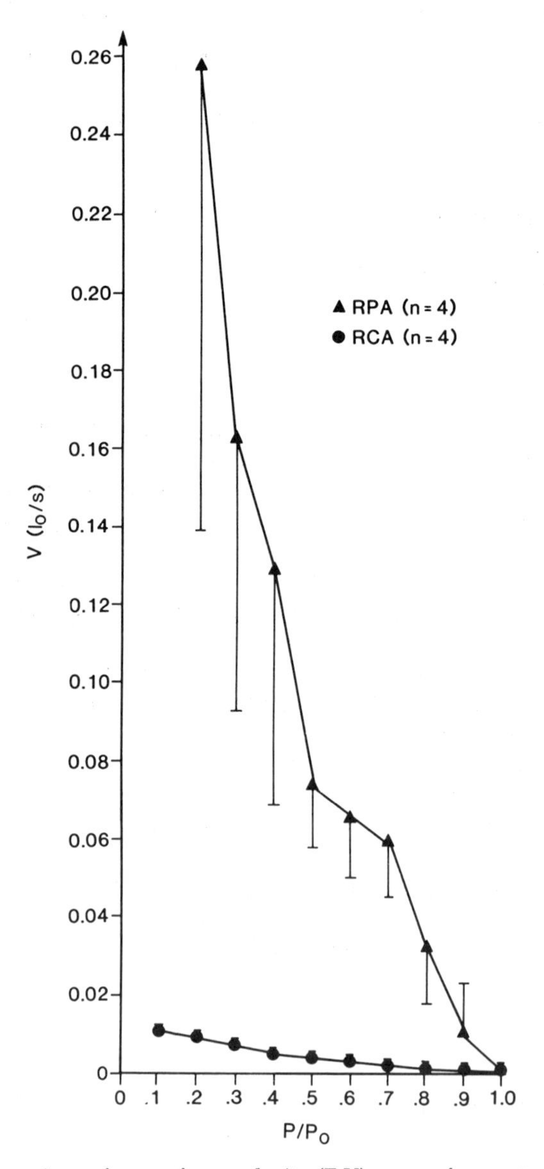

Figure 1. Comparison of mean force-velocity (F-V) curves from rat pulmonary arterial (RPA) and caudal arterial (RCA) smooth muscle. Force has been normalized as the ratio of each imposed load (P) to each respective maximum isometric load (P_0), allowing velocities of shortening to be compared for the same relative loads. The mean RPA curve is elevated above the mean RCA curve at all P/P_0 values ($p < 0.03$).

maximum slopes of the shortening traces for the various loads ranging from nearly zero load to each respective P_o. The mean pulmonary arterial and caudal arterial F-V curves were compared by one-way analysis of variance followed by Student-Newman-Keul's test. Since F-V curves for these arterial muscles are non-hyperbolic, the classical Hill equation cannot be fitted to the data. The F-V relationship for loads less than 80% P_o is exponential. Therefore, maximum velocities at zero load (V_{max}) were derived by finding mean intercept values for the exponential portions of the curves (Packer and Stephens, 1985). The mean pulmonary and caudal arterial muscle V_{max} values were compared with Student's t-test.

Arterial muscle for the biochemical experiments was frozen with liquid N_2, pulverized, acetone-dried, desiccated with a low vacuum and stored at -70° C until used. 500 µg of each sample was dissolved in 100 µl of sodium dodecyl sulfate (SDS) gel dissociation medium (200 mM Tris pH 8.0, 3% SDS, 10 mM DTT, and 0.1% bromophenol blue). The material was heated to 100° C for 30 min, sedimented (Eppendorf centrifuge), and the supernatant applied to 5% acrylamide/0.75% bis slab gels using the buffer system of Porzio and Pearson (1977). Bovine serum albumin (BSA) standards were concurrently subjected to electrophoresis, and gels were stained in Coomassie blue. For separation of the various MHC isoforms, gels were subjected to electrophoresis for about 5 hours, 4° C at 300 volts, constant voltage. Myosin content was determined by quantitative densitometric scanning, using BSA as the standard. Student's t-test was used to compare the mean pulmonary arterial and caudal arterial MHC isoform ratios.

Table 1. Comparison of Rat Caudal Arterial (RCA) and Pulmonary Arterial (RPA) Muscle Preparation Characteristics.

	RCA	RPA
P_o (in mN/cm^2)	11005 ± 948 (n = 9)	*5876 ± 459 (n = 4)
RP_o/P_o	0.40 ± 0.07 (n = 11)	0.54 ± 0.02 (n = 4)
CSA (in cm^2)	0.0009 ± 0.0001 (n=11)	*0.0080 ± 0.0008 (n=4)

*$p < 0.0001$

RESULTS

The caudal and pulmonary arterial muscle preparation characteristics are compared in Table 1. The caudal arterial strip has a smaller cross-sectional area compared with the pulmonary arterial preparation. However, the caudal arterial preparation produces significantly more stress than does the pulmonary arterial preparation.

The mean F-V curves for pulmonary and caudal arterial muscle are compared in Figures 1 and 2. In Figure 1, the data have been normalized so that velocities of shortening are compared for muscles carrying the same relative loads. In Figure 2, stress-velocity curves are compared. In either case the pulmonary arterial curve is elevated above the caudal arterial curve. In

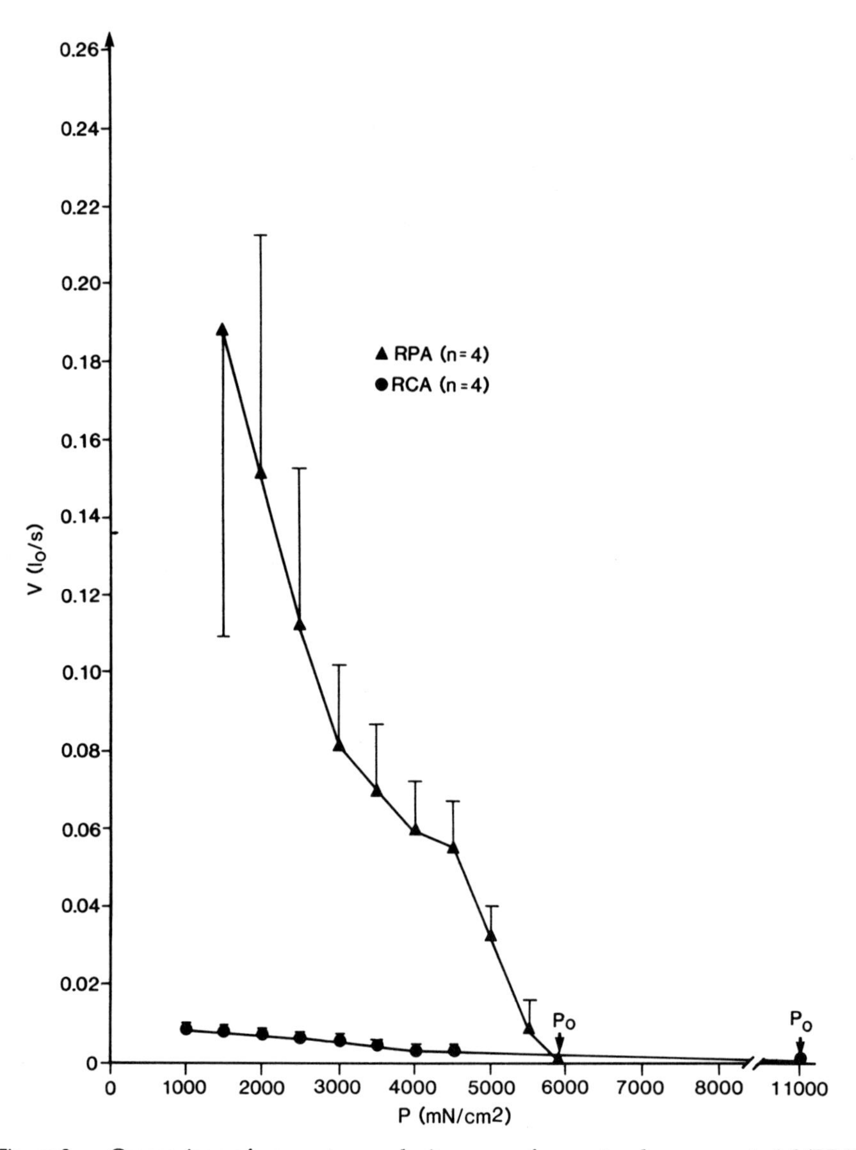

Figure 2. Comparison of mean stress-velocity curves from rat pulmonary arterial (RPA) and caudal arterial (RCA) smooth muscle. Force (P) has been normalized to muscle cross-sectional area (mN/cm^2). The RPA shortening velocity is about 20 times greater than that of the RCA for loads less than 5000 mN/cm^2 ($p < 0.02$). The RPA is unable to shorten with loads greater than about 5900 mN/cm^2 while the RCA is able to shorten with loads as great as 11000 mN/cm^2.

addition, Figure 2 shows that caudal arterial muscle can shorten against loads approximately twice as great as can pulmonary arterial muscle.

The mean extrapolated V_{max} values for caudal and pulmonary arterial muscle are compared in Table 2. The pulmonary arterial V_{max} is significantly greater than the caudal arterial V_{max}.

Finally, the mean myosin heavy chain isoform ratios are compared in Table 3. The ratio of MHC_1 to MHC_2 was the same in the pulmonary and caudal arterial muscle ($p > 0.05$). Similarly, there was no difference in proportion of non-muscle to muscle myosin heavy chains in pulmonary as compared with caudal arterial muscle ($p > 0.05$).

Table 2. Comparison of Rat Caudal Arterial (RCA) and Pulmonary Arterial (RPA) Muscle Mean Maximum Velocities of Shortening (V_{max})

	RCA	RPA
V_{max} (l_o/s)	0.017 ± 0.001 (n = 4)	*0.391 ± 0.127 (n = 6)

*$p < 0.05$

Table 3. Rat Caudal Arterial (RCA) and Pulmonary Arterial (RPA) Mean Myosin Heavy Chain (MHC) Isoform Ratios

	RCA	RPA
MHC_1/MHC_2	0.497 ± 0.016 (n = 4)	0.449 ± 0.033 (n = 4)
nmMHC / $MHC_1 + MHC_2$	0.169 ± 0.006 (n = 4)	0.165 ± 0.007 (n = 4)

Note: Each sample contained protein from 4 different rats, therefore tissue from a total of 16 individuals was used to represent each population. nmMHC = non muscle myosin isoform.

DISCUSSION

Maximum velocity of shortening is an indirect measure of the rate at which energy liberation reactions for contraction are occurring. This is significant because the rate of the formation and dissociation of actomyosin crossbridges is most likely related to the rate of ATP hydrolysis which is directly proportional to the ATPase activity of actomyosin (Bárány, 1967).

Both the maximum velocity of shortening and actomyosin ATPase activity level are known to be dependent on myosin heavy chain isoform pattern in striated muscles (Pagani and Julian, 1984). The level of actomyosin ATPase activity in smooth muscle may be due to changes in the concentration of caldesmon or calmodulin, changes in the type or amount of myosin light chain kinase, or changes in the intrinsic properties of the enzyme (i.e. a shift in myosin isoform). At least two different myosin heavy chains have been reported to exist in several different smooth muscle tissues (Sparrow et al., 1987).

In the present investigation the force-velocity relationships and myosin heavy chain isoform ratios of two different arterial smooth muscles were compared. Results of this study show that: 1) The caudal arterial preparation produces significantly more stress than the pulmonary arterial preparation. This most likely reflects a greater proportion of smooth muscle to connective tissue in the caudal arterial wall rather than greater smooth muscle contractility per se; 2) Pulmonary arterial muscle shortens faster than caudal arterial muscle when carrying the same relative loads; 3) The pulmonary arterial muscle V_{max} is greater than the caudal arterial muscle V_{max} indicating faster crossbridge cycling rate in the pulmonary arterial muscle; and 4) Pulmonary arterial and caudal arterial muscles have similar myosin heavy chain isoform ratios. Thus, there is no correlation between myosin heavy chain isoform ratio and shortening velocity in pulmonary and caudal arterial muscles. In conclusion, a cause-and-effect relationship between myosin heavy chain isoform pattern and velocity of shortening in systemic and pulmonary arterial smooth muscles does not appear to exist.

ACKNOWLEDGMENTS

This investigation was supported by The Canadian Heart Foundation, The American Heart Association, Indiana Affiliate, and NIH HL 40894. Thanks are due to Marlene King for expert typing of this paper.

REFERENCES

Alpert, N. R., Mulieri, L. A., and Litten, R. Z., 1979, Functional significance of altered myosin adenosine triphosphatase activity in enlarged hearts, *Am. J. Cardiol.*, 44: 947.

Alpert, N. R. and Mulieri, L. A., 1980, The functional significance of altered tension dependent heat in thyrotoxic myocardial hypertrophy, *Basic Res. Cardiol.*, 75: 179.

Bárány, M., 1967, ATPase activity of myosin correlated with speed of muscle shortening, *J. Gen. Physiol.*, 50: 197.

Litten, R. Z., Martin, B. J., Low, R. B., and Alpert, N. R., 1982, Altered myosin isozyme patterns from pressure-overloaded and thyrotoxic hypertrophied rabbit hearts, *Circ. Res.*, 50: 856.

Meiss, R. A., 1987, Stiffness of active smooth muscle during forced elongation, *Am. J. Physiol.*, 253: C484.

Mohammed, M. A. and Sparrow, M. P., 1988, Changes in myosin heavy chain stoichiometry in pig tracheal smooth muscle during development, *FEBS Lett.*, 228: 109.

Packer, C. S. and Stephens, N. L., 1985, Force-velocity relationships in hypertensive arterial smooth muscle, *Can. J. Physiol. Pharmacol.*, 63: 669.

Pagani, E. D. and Julian, F. J., 1984, Rabbit papillary muscle myosin isozymes and the velocity of muscle shortening, *Circ. Res.*, 54: 586.

Porzio, M. A. and Pearson, A. M., 1977, Improved resolution of myofibrillar proteins with sodium dodecyl sulfate-polyacrylamide gel electrophoresis, *Biochem. Biophys. Acta*, 490: 27.

Sparrow, M. P., Arner, A., Hellstrand, P., Morono, I., Mohammed, M. A., and Rüegg, J. C., 1987, Isoforms of myosin in smooth muscle, *in:* "Regulation and Contraction of Smooth Muscle", M. J. Siegman, A. P. Somlyo, and N. L. Stephens, eds., Alan R. Liss, New York, p. 67.

DETAILED BALANCE AND FOUR STATE MODELS OF SMOOTH MUSCLE ACTIVATION

Carlos A. Lazalde and Lloyd Barr

Department of Physiology and Biophysics
University of Illinois at Urbana
Urbana, IL 61801

INTRODUCTION

Regulation of smooth muscle contraction involves a number of chemical networks which in turn involve reactions at fixed sites. This occurs because of the organized nature of the contractile filament system. The formalisms introduced by T. L. Hill (1977) in his analyses of the hypotheses of A. F. Huxley (1957) have been useful in the study of several other motile systems and draw on the concepts of thermodynamics and statistical mechanics, and are expressed in the language of continuous time Markov chains. The definition of a state plays an important role in such analyses as does the notion of Detailed Balance. Models which do not comply with the Principle of Detailed Balance are at least inconsistent and may have no more significance than fitting a curve to a mathematical expression.

A state formalism represents a chemical reaction as a transition between two states. This requires that a represented chemical reaction be between a single reactant and a single product. To meet this requirement any reaction in the model should be considered first order, or pseudo first order. In addition, each state may contain substates. One criterion for allowing substates to be represented by a single state is that their free energies should be nearly equal, i.e. their mean lifetimes should be very brief (large rate constants) compared with the mean life time of the state; they should be close to equilibrium.

The reasons for considering the hypothesis that smooth muscle activation involves four states of myosin, are multifold. Murphy and his colleagues (Dillon et al., 1981; Hai and Murphy, 1988) first called attention to four state models by proposing a particular formulation of a four state cycle (the latch model) to simultaneously explain: 1) that during continued stimulation of swine carotid preparations force rose to a plateau while myosin phosphorylation went through a peak and then declined and also, 2) the relative economy of tonic contractions in smooth muscle. Since actin activated myosin ATPase activity is greatly enhanced by phosphorylation and phosphorylation alone is sufficient for activation of contractile activity (Walsh et al., 1982), phosphorylation is reasonably taken to be the first step in the initiation of cross bridge cycling. Now if phosphorylation is the first step it presumably triggers attachment, then it follows that a three state model cannot provide for the continuation of "active" force in the face of falling myosin phosphorylation, because the only attached state (and force generating) is phosphorylated. The existence of more than one kind of crossbridge has also been suggested on other grounds. The relative economy of a smooth muscle tonic contraction has long been an object of interest and may be associated with the phenomena of stress relaxation. On that basis, Bozler (1976) and Siegman et al. (1976) suggested that there are stretch resisting myosin-actin attachments in resting smooth muscle, a kind of non-activated crossbridge. Even earlier Rüegg (1971), while discussing "catch-like" phenomena in vertebrate smooth muscle, suggested that "this tonic state of extremely economical, tension maintenance might be due to set contractile linkages between sliding thick and thin myofilaments." By analogy with other allosteric actions (e.g. non-competitive inhibition), the fact, that myosin can bind to actin and also be phosphorylated, might be taken to indicate that myosin should exist in at least four states; M, PM, A-PM and A-M. Moreover, it seems clear that, three states are not sufficient and at least four states are needed to describe the time course of isometric force development in smooth muscle. Therefore, even if the rather specific latch model eventually fails in some fashion, these other issues require the exploration of four state cycles.

WHAT IS LEFT OUT OF FOUR STATE MODELS?

When myosin (with actin) catalyses ATP hydrolysis in systems of organized filaments it goes through a number of conformational states which result in mechanical force development between the filaments. Therefore, any four state model will of necessity lump a number of attached and unattached states together. Also, there are complexes with MLCK and MLCP which might be considered. In view of these complications, it seems reasonable to consider whether or not four states are adequate to describe smooth muscle activation. To that end, the implicit assumptions and possible kinetic behavior of four state models should be well understood.

In Figure 1 some the states considered important in the crossbridge cycle of skeletal muscle (Hibberd and Trentham, 1986) are made explicit in a four state formulation of a phosphorylation-dephosphorylation scheme for the activation of smooth muscle. Each of the four states (M, PM, A-PM, and A-M) have substates. This perhaps adds too much complexity for easy discussion but has the advantage of indicating what kind of simplification is typically assumed. Only some of the possible substates in each state are indicated. If the transitions between the substates are not fast enough (as discussed above), then the vertical transitions between attached and unattached states should include intermediate states. Another complication is that the transitions between the attached states involve the release of P_i and ADP or the uptake of ATP, which are displaced in the transition between the attached and unattached states. There is no way out of this problem using a four state formalism without introducing other confusions. For example the device shown in Figure 2 seems to incorporate some of the substate transitions in the reaction arrow, which would certainly be ambiguous unless these substates never bear a load.

Figure 3 represents an even more simplified version of reality but has the advantage that not only can it's behavior be simulated (Hai and Murphy, 1988) but the time courses of it's variables have analytic solutions, following step changes in the "rate constants". They are the sums of three exponentials

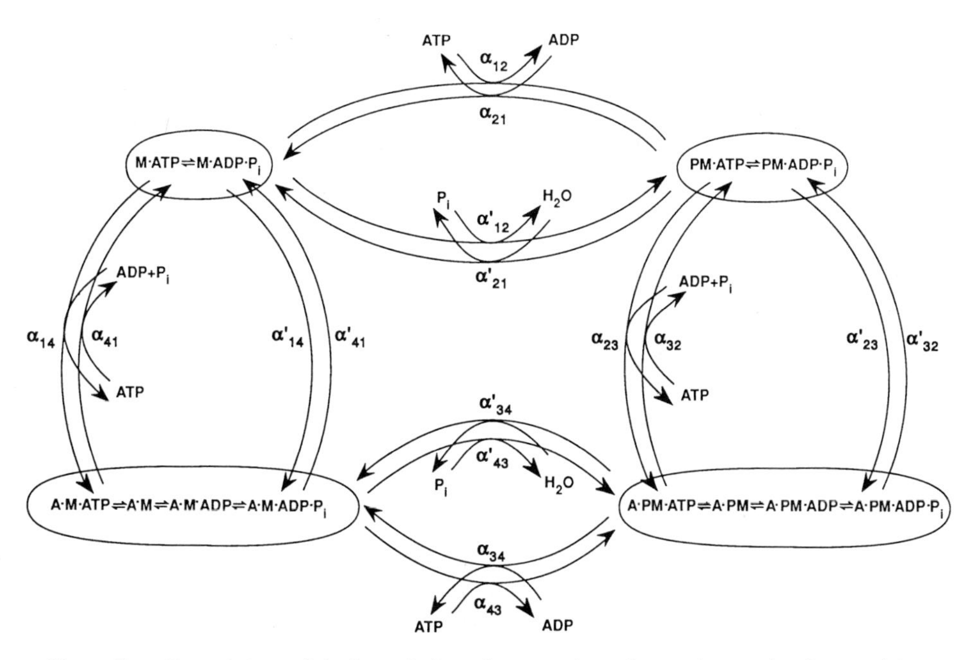

Figure 1. Four state model of regulation of contraction of smooth muscle. Some of the states considered important in contraction of skeletal muscle (Hibberd and Trentham, 1986) are shown in a four state model. Encircled state in upper left corner corresponds to an unattached dephosphorylated crossbridge. The upper right corner state corresponds to an unattached phosphorylated cross bridge. State in lower right corner corresponds to an attached phosphorylated crossbridge. State in lower left corner corresponds to an attached dephosphorylated crossbridge. Transitions with ATP and ADP at the ends of arrows represent MLCK catalysed reactions. Transitions with P_i and H_2O at the ends of arrows represent MLCP catalysed reactions. Transitions with ATP, and ADP + P_i at ends of arrows represent detachment of crossbridges. The other two transitions represent attachment of phosphorylated (α' 23 and 32) and dephosphorylated crossbridges (α' 14 and 41). Attached states are force producing states. Figure shows some of the complexity that is not shown in a simple representation of a four state model as shown in Figure 2. See text for discussion.

plus a constant. Furthermore, mathematical analysis allows the identification of the model, i.e. the finding of the rate constants of the model from experimental data and the analytic description of the models behavior (Lazalde and Barr, 1990). As will be considered below such a truncated model is still complex.

DETAILED BALANCE AND MICROSCOPIC REVERSIBILITY

The principle of detailed balance, as used in chemistry, was known to several chemists at the beginning of this century, see refs. in (Wegscheider, 1901; Fowler and Milne, 1925; Tolman, 1925), as an empirical rule that was important in the correct analysis of chemical networks. Lewis (1925) recognized that there was an important principle behind this empirical rule, and that it was applicable to any physical process at equilibrium. He called it "The Law of Entire Equilibrium". Simply stated, it requires that, when a system is at equilibrium, each elementary process occurring in the system must be at equilibrium. By equilibrium of an elementary process it is meant that the average rate in one direction, is equal to the average rate in the opposite direction. Because related ideas had been independently developed in physics, the terminology applied to these concepts varies from author to author. We use the term "Principle of Detailed Balance" as a modern name for "The Law of Entire Equilibrium". The Principle of Detailed Balance is a strong requirement. In the words of Lewis:

> "...the significance of this law may be made a little more evident by means of a crude analogy. Suppose that during a period in which there are no births or deaths the population of the several cities of the United States remains constant, the number leaving each city being balanced by the number entering it. This stationary condition would not correspond to our case of thermal equilibrium" (he is referring to his law of entire equilibrium). "We should require further, to complete the analogy, that as many people go from New York to Philadelphia as from Philadelphia to New York. If there were three railway lines between these two cities we should require that the number of passengers going by each line be equal in both directions. If some of the travel were not by railway or roads but across the country, then, if we should draw on the map two non-intersecting lines from New York to Philadelphia, we should require that the number of persons passing through any such zone in one direction be equal to the number passing through the same zone in the opposite direction. By such illustrations we may appreciate how extremely far-reaching are the consequences of the proposed law".

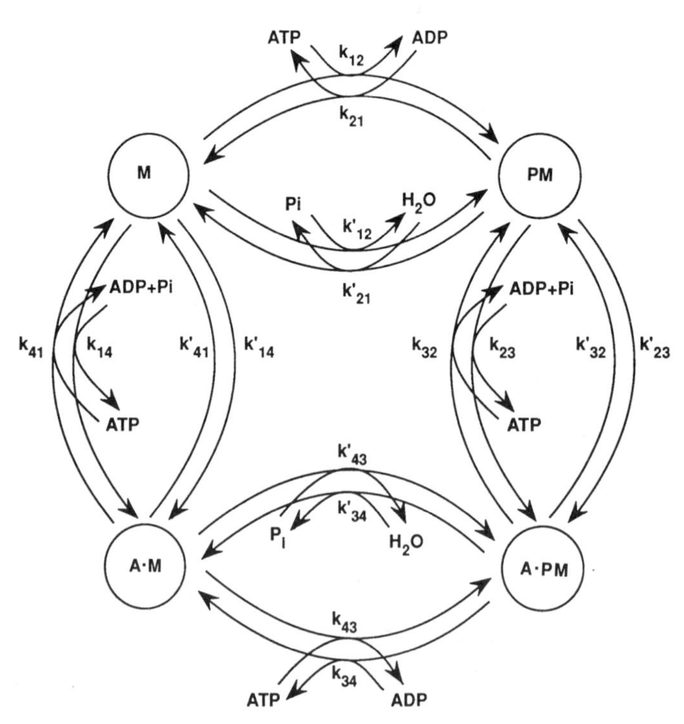

Figure 2. A simple representation of a four state model. Transition rates, α's of Figure 1, are represented by k's. States M and PM are unattached states, dephosphorylated and phosphorylated respectively. States A-M and A-PM are attached states, dephosphorylated and phosphorylated respectively. Transitions are as in Figure 1. Note that for every transition an inverse arrow shows its reverse, in accord with detailed balance.

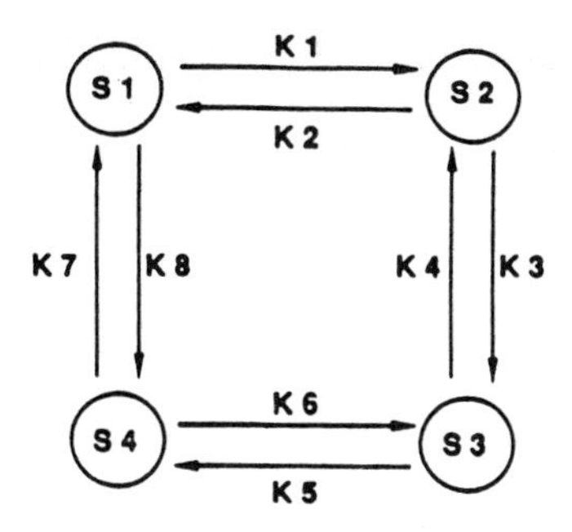

Figure 3. A simple state model. It can represent unimolecular reactions between four states, or a mathematical simplification of Figure 2. See text.

The major thrust of the Lewis paper is arguing the impossibility of maintaining equilibrium by a cyclic flux. Suppose, for example, that A goes into B, B goes into C, and C goes into A; at equilibrium, the number of molecules of A, B, and C are NA, NB, and NC. **Thermodynamics requires that, at equilibrium the chemical potentials of A, B, and C must be equal and independent of time, thus the number of molecules of A, B, and C are constant. We denote these equilibrium numbers of molecules as NA, NB, and NC respectively.** Constancy of these molecules could be obtained if x molecules of A per second go to B, x molecules of B per second go to C, and x molecules of C per second go to A. It could occur also, by going the other way around, i.e. A into C, C into B, and B into A. It can be seen that both ways require a net flux of molecules going around the cycle.

The Principle of Detailed Balance forbids the existence of such net fluxes at equilibrium, by explicitly requiring that each net flux be zero; thus, the elementary processes A $\leftrightarrow$ B, at equilibrium must have the same number of molecules of A transformed to B per unit time as there are molecules of B transformed to A per unit time; and the same would apply to the reactions between B and C, and between C and A.

Further insight into the consequences of detailed balance, can be gained by comparing more explicitly the implications of thermodynamics to those of chemical kinetics (Onsager, 1931; Denbigh, 1951; 1981; Bak, 1963; Yourgrau et al., 1982;). In the example discussed above, assume that all reactions are first order with rate constants k_{AB}, k_{BA}, k_{BC}, k_{CB}, k_{CA}, and k_{AC} where the subindeces AB, BA, and so on, refer to reactions B $\rightarrow$ A, A $\rightarrow$ B, etc. Let n_A, n_B, and n_C be the number of molecules of A, B, and C at any time, then:

$$dn_A/dt = -(k_{BA} + k_{CA})n_A + k_{AB}n_B + k_{AC}n_C \qquad (1a)$$

$$dn_B/dt = k_{BA}n_A - (k_{AB} + k_{CB})n_B + k_{BC}n_C \qquad (1b)$$

$$dn_C/dt = k_{CA}n_A + k_{CB}n_B - (k_{AC} + k_{BC})n_C \qquad (1c)$$

subject to the constraint

$$NA + NB + NC = n_A + n_B + n_C = N \qquad (1d)$$

where N is a constant equal to the total number of molecules. At equilibrium the derivatives are zero. From the solution of the equations, we get:

$$NB/NC = (k_{BC}(k_{BA}+k_{CA}) + k_{BA}k_{AC})/(k_{CB}(k_{BA}+k_{CA}) + k_{AB}k_{CA}) \tag{2}$$

and similar expressions for NC/NA and NA/NB. From thermodynamics, at equilibrium (equality of chemical potentials), assuming ideal gas mixtures, we get:

$$K1 = NB/NC, \quad K2 = NC/NA, \quad K3 = NA/NB. \tag{3}$$

where K1, K2, and K3 are the equilibrium constants of the reactions B ↔ C, C ↔ A, A ↔ B, respectively.

From equations (2) and the set (3) alone, we find that the equilibrium constants are complex functions of the rate constants. Furthermore, there is not a unique set of rate constant values for each equilibrium constant; there are only two equations, equation (2) and another from anyone of its two analogs for K2 = NC/NA and K3 = NA/NB, for six unknowns, the rate constants; this number of equations is insufficient to determine a unique set of rate constants. Some of these sets of rate constants result in a net constant flux around the cycle, one way or the other. There is an element of strangeness in the situation that kinetics and thermodynamics together allow a cyclic flux and a complex relationship between equilibrium constants and rate constants. For a single, separate reaction (A ↔ B), the situation is much simpler and clearer. At equilibrium, thermodynamics requires constant concentrations of A and B; the only way to satisfy this is having opposite rates equal. Assuming that the reaction follows first order kinetics, we obtain that the ratio of rate constants is equal to the equilibrium constant, i. e. $K_{eq} = NA/NB = k_{AB}/k_{BA}$. By analogy, one can postulate that the equilibrium constants for the three <u>simultaneous reactions</u>, are the ratio of the forward and backward rate constants:

$$K1 = k_{BC}/k_{CB}, \quad K2 = k_{CA}/k_{AC}, \quad K3 = k_{AB}/k_{BA}, \tag{4}$$

in order to examine the conditions under which this is true. Note that this postulate does not come from application of thermodynamics to the problem. The requirements of thermodynamics have been satisfied through equation (2) and its analogs. Comparing equation (2), which is an expression for K1, with K1 in the set (4), we see that the only way to reconcile both expressions is to have:

$$k_{BA}k_{AC}/k_{BC} = k_{AB}k_{CA}/k_{CB} \tag{5}$$

or,

$$k_{AC}k_{CB}k_{BA} = k_{AB}k_{BC}k_{CA} \tag{6}$$

Equation (6) is not a consequence of thermodynamics. If detailed balance is required from the reactions, we require that:

$$k_{CB}NB = k_{BC}NC, \quad k_{AC}NC = k_{CA}NA, \quad k_{BA}NA = k_{AB}NB \tag{7}$$

and we see that the set (7) combined with the set (3) gives the set (4). Equation (6) can be obtained directly from the three equations in set (7), i.e. from detailed balance. One way to describe equation 6 is to say that the product of the rate constants in one direction, is equal to the product of the rate constants in the opposite direction. Equation (6), together with equation (2) and one of its two analogs, provide three equations that the rate constants must satisfy; although there are many sets of rate constants that will satisfy these equations, none of these sets will produce a net flux around the cycle, because the rate constants will satisfy detailed balance, set (7). Importantly, set (7) establishes a relation between kinetics, k's; and thermodynamics, K's, since the equilibrium constants are related to changes in standard free energy ($K3= k_{AB}/k_{BA}= \exp(\Delta G^\circ/RT)$).

The discrepancy between expressions for equilibrium constants derived from kinetics of several reactions, and generalizations (unjustified) of a thermodynamic result for a single reaction, was first recognized by Wegscheider (1901). He found, that in a chemical network similar to the triangular one discussed above, the discrepancy was eliminated if the rate constants satisfied equation (6). But for about 25 years it was not clear why this should be so. This discrepancy was known as "Wegscheider's paradox" (Lewis, 1925; Bak, 1963), until Lewis pointed out the explanation. The term "detailed balancing " was introduced by Fowler, see (Fowler and Milne, 1925 and references therein).

The ideas involved in the Principle of Detailed Balance had been previously proposed by Tolman (1924) in the context of quantum mechanics, under the name "the principle of microscopic reversibility". Later on he used this principle as a requirement for systems at equilibrium (Tolman, 1938). While detailed balance is related to rate constants and equilibrium, i.e. to macroscopic systems, microscopic reversibility deals with molecular events. An isolated system at equilibrium looks the same at all times, macroscopically; given the immense number of particles that form a macroscopic body, it is clear that there is a very large number of possible configurations of position and velocity of the particles which is compatible with the total energy of the system. Consider the movement of a particle with velocity υ. According to statistical mechanics the probability of existence of this event depends only on the energy, and since the energy depends on the square of the velocity, the probability of the existence of a particle with velocity $-\upsilon$ is the same; but a particle with velocity $-\upsilon$ can be identified with a trajectory that is the reverse of the first one. From considerations like this, Tolman concluded that molecular processes and their reverse occur at the same average rate. Discussions of detailed balance and microscopic reversibility can be found in (Mahan, 1975; Morrissey, 1975; and references therein). They show that microscopic reversibility can also be applied to non-equilibrium situations, and derive expressions similar to detailed balance equations, that are valid out of equilibrium in some cases.

One of the most important aspects of equation (6) is that it is valid out of equilibrium! Since it involves only fixed numbers, the rate constants, equation (6) represents a constraint that is always valid even though it was obtained from considerations of equilibrium. Constancy of the concentrations etc. in living systems implies merely steady states, not equilibrium, and the only way to have steady states is to have cycles in chemical networks (Morowitz, 1965; Hill, 1977). Moreover, the most interesting cycles like those involved in the activation of smooth muscle contraction are far from equilibrium.

Consider a simple four state model. Applying detailed balance to Figure 3 we have:

$$K1\, s_1^e = K2\, s_2^e \tag{8}$$

$$K3\, s_2^e = K4\, s_3^e \tag{9}$$

$$K5\, s_3^e = K6\, s_4^e \tag{10}$$

$$K7\ s_4^e = K8\ s_1^e \qquad (11)$$

combining equations 8 - 11 we obtain"

$$K1K3K5K7 = K2K4K6K8 \qquad (12)$$

Equations (8), (9), (10), and (11) hold only at equilibrium but being a relationship between parameters (fixed numbers), equation (5) holds away from equilibrium as well, as long as the K's represent first order rate constants. We consider below cases were the rate constants are not first order. If we were asked to experimentally determine all the rate constants in Figure 3, we would need to be concerned about determination of only seven rate constants, since the remaining undetermined constant can be obtained from equation (12).

Consider Figure 2. For each conceivable transition, detailed balance requires consideration of the opposite transition. Furthermore, one has to make sure that the kinetics of the transition can be described in such a way that detailed balance at equilibrium can be applied. Suppose we say that a transition can be described by the usual irreversible Michaelis-Menten kinetics. The advantages of the previously desired relationship would be lost because it is not possible to apply detailed balance; we have to consider a reversible mechanism where rate constants are used for both directions. Application of detailed balance to the network of Figure 2 requires that all of the reactions, each one separately, be at equilibrium.

It follows from Figure 2 that:

$$M + ATP \underset{k_{21}}{\overset{k_{12}}{\leftrightarrow}} PM + ADP$$

$$M + P_i \underset{k'_{21}}{\overset{k'_{12}}{\leftrightarrow}} PM$$

.

.

.

$$A\text{-}M + ATP \underset{k_{43}}{\overset{k_{34}}{\leftrightarrow}} M + ADP + P_i$$

$$A\text{-}M \underset{k'_{43}}{\overset{k'_{34}}{\leftrightarrow}} M$$

Repeating the same tactic that yielded equation (12) from Figure 3 on the various possible loops of Figure 2, we obtain two kinds of expressions that correspond to two kinds of loops. One group of loops involves no net reaction, and the other involves a net reaction that can be either ATP hydrolysis or synthesis. An example of the first group is the following loop, going around the four states, from detailed balance we obtain:

$$k_{12}k_{23}k_{34}k_{41} = k_{21}k_{32}k_{43}k_{14} \qquad (13)$$

An example of a loop with a net reaction is the left loop of Figure 2. Applying detailed balance we obtain:

$$(k_{41}k'_{14})/(k_{14}k'_{41}) = [ADP]\,[P_i] \,/\, [ATP] = K_{ATP} \tag{14}$$

where K_{ATP} is the equilibrium constant for hydrolysis of ATP, since the concentrations of ATP, ADP, and P_i must be the equilibrium concentrations, because we are applying detailed balance <u>at equilibrium</u>. But since K_{ATP} is a fixed number, equation (14) gives a constraint between four rate constants that is always valid, even out of equilibrium; this is true because equation (14) involves just fixed numbers, for example there are no concentration terms, since they end up as an equilibrium constant.

A complete analysis of all 20 loops in Figure 2 gives 20 equations (two of the equations are equations (13) and (14)), only five being independent. This implies that experimental tests of the model in Figure 2 require determination of only eleven of the sixteen rate constants of the model.

If Figure 3 is a simplification of Figure 2 then:

$$K1 = k_{12}[ATP] + k'_{12}[P_i] \tag{15}$$

$$K2 = k_{21}[ADP] + k'_{21} \tag{16}$$

.

.

.

$$K7 = k_{34}[ATP] + k'_{34} \tag{21}$$

$$K8 = k_{43}[ADP][P_i] + k'_{43} \tag{22}$$

The network illustrated in Figure 2 can come to equilibrium and Equations (8) - (11) will hold. However, in this case K1 - K8 contain terms having equilibrium concentrations of ATP, ADP, and P_i as factors, equations (15) - (22). Equation (12) will hold at equilibrium but it will not be true out of equilibrium because the concentrations of ATP, ADP, and P_i take on arbitrary values away from equilibrium. This does not mean that detailed balance does not provide constraints valid out of equilibrium (see text after equation (14), above). It just means that equation (12) is not the right way to write the constraint. In physiological situations, it is not to be expected that ATP hydrolysis would proceed to equilibrium. However, equations (13) and (14) will hold always and have the utility indicated above. In a living cell, the components of Figure 2 may come to a steady state determined more by the relative activities of the enzymes than by the free energies of reaction which set equilibrium.

CONCLUSION

In summary, application of detailed balance to four state models of smooth muscle activation indicates that: 1) For every conceivable transition, its opposite must also be considered; 2) the nature of the transitions should be reversible, otherwise detailed balance can not be applied, and 3) all the effective rate constants in a model such as is illustrated in Figure 3 should be treated as finite and in accord with the consequences of detailed balance (equations (13),

(14), and analogs). If these points are not incorporated in a model, then experimental data should be provided to show that they are not significant. In recent models of regulation of contraction, (Driska, 1987; Hai and Murphy, 1988) points 1, 2, and 3 were ignored. As we have stressed above, detailed balance provides constraints in the rate constants that are valid out of equilibrium. Since many predictions of the models are based on the particular value of the rate constants, ignoring detailed balance disconnects these studies from standard kinetics and thermodynamics. In particular, the deletion of K8 from Figure 3, constitutes a theoretical inconsistency which, though it is part of the current "latch-bridge model" (Murphy, et al., 1990) is in conflict with the notion of latch-bridges cycling slowly and would seem to be in conflict with the complete recovery of force after a quick release (Murphy, et al., 1990) since that would imply that latch-bridges assert their share of the active force in the absence of any stretch.

Moral: "Don't burn your bridges before they are crossed."

REFERENCES

Bak, T., 1963, "Contributions to the Theory of Chemical Kinetics", Benjamin Press, New York.

Bozler, E., 1976, Mechanical properties of contractile elements of smooth muscle, *in:* "Physiology of Smooth Muscle", E. Bülbring and M. F. Shuba, eds., Raven Press, New York, p. 217.

Butler, T. M., Siegman, M. J., Mooers, S. U., and Davies, R. E., 1976, Calcium-dependent resistance to stretch and stress relaxation in resting smooth muscles, *Am. J. Physiol.*, 231: 1501.

Denbigh, K., 1951, "Thermodynamics of the Steady State", Methuen, London.

Denbigh, K., 1981, " The Principles of Chemical Equilibrium", Cambridge University Press, Cambridge.

Dillon, P. F., Askoy, M. O., Driska, S. P., and Murphy, R. A., 1981, Myosin phosphorylation and the cross-bridge cycle in arterial smooth muscle, *Science*, 211: 495.

Driska, S., 1987, High myosin light chain phosphatase activity in arterial smooth muscle: Can it explain the latch phenomenon?, *in:* "Regulation and Contraction of Smooth Muscle", M. J. Siegman, A. P. Somlyo, and N. L. Stephens, eds., Alan R. Liss, New York, p. 387.

Fowler, R. H. and Milne, E. A., 1925, A note on the principle of detailed balancing, *Proc. Nat'l. Acad. Sci. U.S.A.*, 11: 400.

Hai, C.-M. and Murphy, R. A., 1988, Cross-bridge phosphorylation and regulation of latch-bridge state in smooth muscle, *Am. J. Physiol.*, 254: C99.

Hibberd, M. G. and Trentham, D. R., 1986, Relationships between chemical and mechanical events during muscular contraction, *Ann. Rev. Biophys. Biophys. Chem.*, 15: 119.

Hill, T. L., 1977, "Free Energy Transduction in Biology", Academic Press, New York.

Huxley, A. F., 1957, Muscle structure and theories of contraction, *Prog. Biophys. Mol. Biol.*, 7: 255.

Lazalde, C. A. and Barr, L., 1990, Identification of four state models of regulation of contraction of smooth muscle, *in:* "Frontiers of Smooth Muscle Research", N. Sperelakis and J. D. Wood, eds., Wiley-Liss, New York, p. 51.

Lewis, G. N., 1925, A new principle of equilibrium, *Proc. Nat'l. Acad. Sci. U.S.A.*, 11: 179.
Mahan, B. H., 1975, Microscopic reversibility and detailed balance, *J. Chem. Ed.*, 52: 299.
Morowitz, H. J., 1966, Physical background of cycles in biological systems, *J. Theor. Biol.*, 13: 60.
Morrissey, B. W., 1975, Microscopic reversibility and detailed balance, *J. Chem. Ed.*, 52: 296.
Murphy, R. A., Rembold, C. M., and Hai, C.-M., 1990, Contraction in smooth muscle: What is Latch?, *in:* "Frontiers of Smooth Muscle Research", N. Sperelakis and J. D. Wood, eds., Wiley-Liss, New York, p. 39.
Onsager, L., 1931, Reciprocal relations in irreversible processes, *J. Phys. Rev.*, 37: 405.
Rüegg, J. C., 1971, Smooth muscle tone, *Physiol. Rev.*, 51: 201.
Tolman, R. C., 1924., Duration of molecules in quantum states, *J. Phys. Rev.*, 23: 693.
Tolman, R. C., 1925, The principle of microscopic reversibility, *Proc. Nat'l. Acad. Sci. U.S.A.*, 11: 436.
Tolman, R. C., 1938, "The principles of Statistical Mechanics", Dover reprint, 1979, Oxford University Press, Oxford.
Walsh, M. P., Bridenbaugh, R., Hartshorne, D. J., and Kerrick, W. G. L., 1982, Phosphorylation dependent activated tension in skinned gizzard muscle fibers in the absence of Ca^{++}, *J. Biol. Chem.*, 256: 5987.
Wegscheider, R., 1901, Uber simultane gleichgewichte und die Beziehungen zwischen thermodynamik und reaktionskinetik homogener systeme, *Z. Physik. Chem.*, 39: 257.
Yourgrau, W., van der Merwe, A., Raw, G., 1982, "Treatise on Irreversible and Statistical Thermophysics ", Dover, New York.

REGULATION OF THE STEP-DISTANCE IN SHORTENING MUSCLES

Avraham Oplatka

Weizmann Institute of Science
Rehovot, Israel

INTRODUCTION

It is argued that the force driving muscular shortening (ψ) differs from the force (ϕ) responsible for tension generation. ψ is associated with ATP-induced dissociation of actomyosin, whereas ϕ is due to an isomerization reaction of actomyosin, following the hydrolysis of ATP. In a shortening muscle, ATP is thus hydrolyzed after movement commences. Both forces are intimately coupled with appreciable changes in the structure of the hydration shell at the interface between the two proteins, which involves the release of stored energy. When an active muscle is allowed to shorten freely, ψ gives rise to a step- (or sliding-) distance (Δl_1) which should be a variable and its value depends on the environmental conditions. On the other hand, the step distance (Δl_2) observed upon releasing a muscle which had developed rigor tension isometrically is a constant, the value of which is related to the myosin head's length. The maximal values of the two forces (ψ_o and ϕ_o), as well as of Δl_2 are calculated on the basis of experimental data. The forces and their corresponding step distances are related through the standard free energies of the two chemical reactions responsible for them. It is claimed that the same mechanochemical mechanisms operate also in all microtube-based locomotion and force-generation systems and, furthermore, that practically the same values of ψ_o, ϕ_o, Δl_1, and Δl_2 are shared by the two types of biological energy convertors.

The mechanochemical performance of muscle fibers is characterized by two physical parameters: the maximal isometric tension (P_o) and the maximal velocity of shortening (V_o). It has generally been taken for granted that the "contractile force" developed during an isometric contraction is also the force which causes shortening. This necessarily led to the idea that, following the hydrolysis of ATP, tight actomyosin complexes undergo a force-generating conformational change which, in an isotonic contraction, will cause a change in the angle at which myosin heads bind to actin filaments, thus leading to the translocation of actin. This would limit the maximal unit displacement (the step- or sliding-distance) produced by a single ATP hydrolysis to appreciably less than twice the length (about 18 nm) of the myosin head. According to this picture (the "swinging crossbridge model", Huxley, 1990), actin is merely a passive cable: it serves as a co-factor to the enzymatically-active myosin but has no active mechanochemical role. However, several recent findings may cast doubts on this simplistic theory:

Regulation of Smooth Muscle Contraction
Edited by R.S. Moreland, Plenum Press, New York, 1991

1) Despite many years of extensive work, no evidence has been found for rotation of myosin heads (cf. Ajtai et al., 1989);

2) Work employing different experimental procedures suggests that the step distance is several times larger than anticipated (Harada et al., 1990); moreover, the value of the step distance seems not to be constant and depends strongly on the environmental conditions;

3) If, indeed, the "contractile" force observed during an isometric contraction is the force which is responsible for movement, then one should expect that changes in environmental conditions which affect the force per myosin head should lead to similar changes in the value of the step distance. As we shall see in the following, this is not always the case thus suggesting that these two molecular mechanical parameters are not directly related to each other.

4) Elastic energy seems to be stored in actin filaments in living cells due to the hydrolysis of ATP during polymerization. This energy would be available for release by actin-binding proteins such as myosin thus giving rise to mechanochemical transformations (Janmey et al., 1990). In view of the fact that the myosin:actin ratio in smooth muscle cells (and even more so in non-muscle cells) is lower than in striated muscle cells (while the concentrations of actin are comparable) one may wonder whether it could not be that actin is the major mechanochemical converter whereas myosin, in spite of its (extremely low) ATPase activity, plays a secondary role and might even be replaced by other actin-binding proteins;

5) It has been demonstrated (Tirosh et al., 1990) that isolated, free, myosin heads can enhance the translational motion of actin filaments in solution and are capable of inducing shortening and tension development in muscle preparations in which native myosin had been inactivated (Borejdo and Oplatka, 1976). These observations are difficult to interpret by the swinging crossbridge model which requires a continuous, three-dimensional, protein network not only for tension development but also for movement.

THE NATURE OF THE SLIDING DISTANCE IN ACTIVE MUSCLES

For the step distance to be a variable (possibly exceeding twice the length of a myosin head), for shortening to occur without necessitating rotation of the heads or their being part of a myosin filament, and for actin to participate actively in the relative movement of the two proteins, we have to believe that, during unloaded shortening, the myosin heads are pushed away from an actin subunit with which they had formed a complex, over a distance which is much larger than their length while they are loosely attached to many other subunits along the actin filament for each ATP molecule hydrolyzed. For such a vectorial movement to occur: a) the original tight actomyosin complex must be dissociated; b) upon dissociation, energy must be supplied in order to enable the relative movement, and possibly also work to be done if the muscle is loaded. Either or both myosin and actin must then acquire kinetic and possibly also potential energies; or c) forces opposing "sliding" must develop so as to limit the step distance

The only candidate for the dissociation of the complexes and the provision of energy to the separated proteins is ATP. This reaction involves a free energy drop which is larger than that accompanying any other stage in the enzymic actomyosin cycle and is comparable to the full free energy change of ATP hydrolysis (Kodama, 1985). It is intriguing to consider the possibility that a large part of this energy is converted into kinetic energy of the myosin heads. The acquisition of this energy might be preceded by a change in the elasticity of the actin filaments in the vicinity of the original actomyosin complexes. The movement of the heads must eventually be slowed gradually down until it comes to a standstill by: a) the mutual repulsion of the electrical double layers of the two negatively charged proteins and of their hydration shells; b) the electroendoosmotic effect due to movement; c) a chemical factor: the myosin-bound ATP must be hydrolyzed within a short time, giving rise to myosin•ADP•P_i and to its derivatives which can "chemically" bind to actin, again forming a "static" complex, this time with another actin subunit, whereas a "dynamic" complex existed during sliding with several actin monomers.

The relative movement of the two proteins during shortening can be attributed to a "pushing" force (ψ). It is clear that this force differs from the force developed in an isometrically contracting muscle by actomyosin complexes undergoing a force-generating conformational change after having hydrolyzed ATP. The movement-generating force does not require ATP splitting for its development. Actually, in this case ATP is split later on by the myosin heads while rolling along the actin filaments ("move now pay later"). Heat changes accompanying movement might, therefore, precede the liberation of the hydrolysis products, which might explain the "unexplained energy balance" during unloaded shortening (Kushmerick and Davies, 1969). The swinging crossbridge model required that "payment" by ATP hydrolysis should precede movement because it claims that "contractile force" generation is responsible for both tension generation and movement.

The step distance should, therefore, depend on the magnitude of the original free energy change (which is related to the equilibrium constant of the ATP-induced dissociation reaction), on the values of the opposing forces, and on the affinity of the myosin heads (with and without ADP and P_i) to actin. All these could be functions of temperature, ionic strength, pH, solvent composition of the medium, etc. Indeed, a) the step distance (Δl_1) assumed the value of 125 nm at 22° C whereas at 30° C its value was only 51 nm (Harada et al., 1990); b) even according to Uyeda et al. (1990) and Toyoshima et al. (1990) who claim that their data is consistent with tight binding of the myosin heads to actin during shortening (and, therefore, with tight coupling between head rotation and ATP hydrolysis), the value of the step distance may vary in the range of 7.7 - 28 nm; c) the ratio of the maximal velocity of shortening of *Tortoise iliofibularis* muscle and the rate of ATP splitting by its actomyosin *in vitro,* which is proportional to the step distance (Oplatka, 1972), decreases by a factor of 3.85 upon raising the temperature from 0° to 22° C (Bárány, 1967); d) similarly, the ratio for glycerinated psoas fibers (the ATPase activity being that of the muscle under isometric conditions) decreased from 0.3 at 6° to 0.1 at 20° C at 5 mM KCl (Yanagida, 1982); e) elevation of KCl concentration from 5 to 120 mM (Yanagida, 1982) made Δl_1 12.4 times larger and this should account for the most unexpected observation that V_o increased despite the fact that the rate of ATP hydrolysis decreased; and f) for a muscle in 20% ethylene glycol, the value of V_o was found to be only 5% of that in water, while the enzymic activity was diminished by only 40% (Maruyama et al., 1989). The presence of the organic solvent thus caused a decrease of Δl_1 by a factor of 12.

It is most interesting to note that the values of the step distance for smooth muscles are similar to those of striated muscles (Bárány, 1967; Oplatka, 1972).

The value of the step distance is thus not constant and has nothing to do with the length of the myosin head which, evidently, does not have to undergo any rotation while moving (and even if it did so, this would not necessarily mean that this is the cause for, rather than the outcome of, movement). Large values of the step distance do not "demand multiple cycles of binding and dissociation between myosin heads and actin filaments with each ATP hydrolysis", contrary to the claim made by Toyoshima et al. (1990). The model suggested above for the mechanism of movement involves real sliding, on the molecular level down to a single myosin head, whereas the sliding filament model refers to the movement past each other of myosin and actin filaments, which is merely a statement of fact and, therefore, should be considered neither as a model nor as a theory.

What is the molecular mechanism by which the dissociation of actomyosin, induced by ATP, causes the ejection of the myosin heads possessing kinetic energy? It has been reported (Rau et al., 1984) that, when DNA molecules are brought close to each other, a repulsive force develops, the value of which is independent of ionic strength. This force has been ascribed to the compression of the hydration layers of the DNA molecules. Such a force should be operating also when actin and myosin form a "static" complex in which the binding of the two proteins by hydrogen bonds and by salt linkages must overcome the repulsive force between the hydration shells, thus leading to the compression of the water. The compressive force should be transmitted to the protein backbones so that they, just like the fused hydration shells, will store energy. This potential energy should be directly converted into kinetic energy of the two proteins the moment the bonds stabilizing the complex are broken, e.g. by ATP (Oplatka, 1989). The initial values of ψ, ψ_o, are thus equal and opposite to the compressive force.

With this mechanism in mind, it is obvious that, if the myosin heads are incapable of moving freely, e.g. by forming part of a myosin filament or by being immobilized on a glass surface, the actin filaments with which they interact will undergo a translational movement past the myosin heads. The decompression induced by ATP might well lead also to the twisting and change in flexibility of the actin filaments. The spring-like forces developed may cause the ejection of the myosin heads along a pre-determined path. On *a priori* grounds, there is no reason to assume that conformational changes, rotation etc., should be limited to myosin only. In all probability, the major reasons for the neglect so far of actin as an active partner for mechanochemical activity have been: a) the consideration of actin as a biochemical co-factor which only helps the enzyme myosin (which is capable of splitting ATP by itself) in getting rid of the hydrolysis products and in exerting the force it can generate; b) the requirement by the sliding filament model of a continuous, three-dimensional, protein network for the generation and transmission of the "contractile" force; c) the fact that actin, unlike myosin, is not a fibrillar protein and is, thus, not entitled to undergo helix coil transitions such as those required by Harrington's theory (cf. Lovell et al., 1988) for the generation of the contractile force by the S2 component of myosin; d) the observation that the length of the thin filaments in a striated muscle does not change much upon activation; and e) the employment of the term "crossbridges" as the sole generators of tension and of movement.

In all probability, positive counterions such as K^+, Na^+, and Mg^{2+} are displaced from the F-actin double helix by positively-charged amino-acid residues on the myosin head surface. The cations released from in front of the moving head must be replace by others binding to F-actin behind the head. The latter can thus be considered to be sliding over the actin on an isopotential surface hence there is no thermodynamic barrier to the sliding. The mechanism of movement is thus very similar to that of a repressor molecule sliding along a DNA chain towards its target site, the operator (cf. Berg et al., 1981).

It should be most interesting to analyze the effect on shortening of a possible displacement of Ca^{2+} ions [bound to actin (in thin filament-regulated) or to myosin (in myosin-regulated) muscles]: it might give rise to a reciprocal relationship between Ca^{2+} and the mechanochemical proteins; Ca^{2+} causing movement by binding to the protein(s), while movement affecting the very binding of Ca^{2+}. Indeed, temporary dissociation of the movement-inducing Ca^{2+} ions due to movement would be in line with the Le Chatelier principle of action and reaction, according to which a system reacts to a change imposed from the outside by undergoing a process which will lead to the diminution of the perturbation. In conclusion: regulation in actomyosin systems might be a feedback process. Possible deactivation of a segment of a thin filament in the leading edge of a myosin head might, in principle, lead to a sort of stepwise or wavy shortening.

The breakdown of ATP can proceed along two routes, I and II (Eisenberg and Hill, 1985):

$$\text{(I)} \qquad \text{M-ATP} + \text{A} \rightarrow \text{M-ADP-P}_i + \text{A} \rightarrow \text{M-A+ADP+P}_i + \text{Movement } (\Delta l_1)$$

$$k_1 \uparrow$$

$$\text{M-A+ATP} \rightleftarrows \text{M-A-ATP}$$

$$k_2 \downarrow \qquad\qquad K$$

$$\text{(II)} \qquad \text{M-A-ADP-P}_i \rightarrow (\text{M-A})_1 \rightleftarrows (\text{M-A})_2 + \text{ADP+P}_i + \text{tension } (\phi_o)$$

where M and A represent, respectively, myosin head and F-actin. The possibility that reactions I and II could each serve as the basis of a <u>different mechanochemical process</u> has never been discussed. The dissociation of actomyosin complexes by ATP has been considered only as a necessary step preceding reattachment in order to generate force and movement. We may now claim that this reaction (I) leads to the development of the force ψ_o which leads to movement. In reaction II the proteins are tightly bound together, forming "static" complexes. By undergoing an isomerization reaction [$(\text{M-A})_1 \rightleftarrows (\text{M-A})_2$] (Coats et al., 1985) the so-called "contractile" force, ϕ_o, is generated and is perceived as isometric tension. The rates at which reactions I and II occur are equal to k_1[M-A-ATP] and to k_2[M-A-ATP], respectively. The observed total rate of ATP splitting is thus equal to $(k_1 + k_2)$ [M-A-ATP]. On the other hand, V_o should be proportional to k_1[M-A-ATP] only. The value of Δl_1 calculated from V_o and from the ATP turnover rate is thus proportional to $k_1/k_1{+}k_2$. It is an <u>average</u> value which must be smaller than the <u>real</u> one; the latter can be obtained only if reaction II does not occur at all. Since each of the kinetic constants could vary with temperature, ionic strength, etc. in a different fashion, the value of Δl_1 <u>cannot</u> be a constant; it is determined by the kinetic constants of the enzymic cycle, just as the capability of Ca^{2+} to regulate muscular contraction is related to kinetic constants rather than to the blocking of the interaction between myosin and actin (Payne and Rudnick, 1989). In other words: The value of Δl_1 depends on the relative importance of reactions I and II.

The ATP-induced dissociation of actomyosin is an exothermic reaction since both reactions M-A $\rightarrow$ M+A and M+ATP $\rightarrow$ M-ATP are exothermic (Woledge et al., 1985). Hence the dissociation constant and, therefore, the likelihood of reaction I, should decrease with increasing temperature. The average Δl_1 should, therefore, decrease with increasing temperature as, indeed, has been observed (Bárány, 1967; Yanagida et al., 1982., Harada et al., 1990).

Upon increasing the ionic strength, the electrical repulsive forces between the proteins should diminish and this should lead to larger Δl_1, as has been found experimentally (Yanagida et al., 1982).

The presence of an organic solvent should cause a diminution of the value of the dielectric constant (D). Since the electrical repulsive forces should be inversely proportional to D, it is to be anticipated that the resistance to motion will increase, and the value of Δl_1 will decrease, upon adding glycerol or ethylene glycol, a conclusion which has been reached above from the experimental data of Maruyama et al. (1989).

Reducing the internal pH of a muscle (either by fatigue or by lowering the pH of the medium) causes a decrease in V_o. It has been proposed that this could be due to a decrease in ATPase activity which has been observed for actomyosin (Edman and Lou, 1990). It is not impossible that part of the decrease in V_o is associated with a smaller Δl_1, due to a decrease in the number of negatively-, and an increase in the number of positively-, charged groups.

Reaction II may be considered as an internal load, similar to the way unphosphorylated myosin slows down the rate at which actin is moved by the faster cycling phosphorylated myosin heads (Warshaw et al., 1990). If extremely low concentrations of ATP are present, ADP should also act as an internal load since it forms a tight ternary complex with actomyosin

What is the molecular basis for the effect of an <u>external</u> load on the velocity of shortening? If an increasing load is applied, the velocity decreases continuously according to Hill's equation (Huxley, 1980). However, down to a velocity which is 0.5 V_o, the ATPase activity, $\dot{N}$, <u>increases</u>, which suggests that Δl_1 <u>decreases</u> (Kushmerick and Davies, 1969). Beyond that point, $\dot{N}$ decreases linearly with decreasing velocity so that, in this range also, the <u>apparent</u> value of Δl_1 decreases with decreasing velocity:

$$\Delta l_1 = \frac{v}{\dot{N}} = \frac{v}{a + bv} = \frac{1}{b + a/v}$$

where a is the value of $\dot{N}$ under isometric conditions and b is a constant. Δl_1 is thus affected by an external load similarly to the way an internal load acts. We may, therefore, assume that a load diminishes reaction I relative to reaction II or, in other words, that it increases the number of the so-called "attached cross-bridges". Indeed, evidence about the number of such bridges during rapid shortening suggests that this number is less than in an isometric contraction (Pate and Cooke, 1989).

In summary, increased temperature, decreased ionic strength, and the presence of an organic solvent or of inactive myosin heads may be considered as loads. <u>All</u> loads diminish the apparent value of Δl_1. They may affect the velocity of shortening via both Δl_1 and the rate of ATP splitting.

While the value of Δl_1 is determined by k_1/k_1+k_2, the value of P_o and, therefore, of the <u>average</u> value of ϕ_o, is proportional to the ratio of $[(M\text{-}A)_2]$ in reaction II to the total concentration of myosin heads and thus depends on the value of the equilibrium constant K in the kinetic scheme above. This has been found to increase with increasing temperature, and to decrease upon increasing the ionic strength or the hydrostatic pressure or by adding ethylene glycol, in <u>parallel</u> to the changes in P_o (Coats et al., 1985). The different way in which Δl_1 and the "contractile" force ϕ_o depend on the parameters of state is thus explained. These two fundamental molecular mechanical "constants" are not functionally related to each other, contrary to the implicit claim made by the swinging crossbridge model.

If we start from a muscle in rigor rather than from a relaxed muscle then the actomyosin complexes would have already existed, and shortening could have occurred (and work obtained) also following the addition of any substance which is capable of dissociating actomyosin, e.g. ATP under conditions at which its splitting by myosin is inhibited. Movement would have then occurred without any hydrolysis reaction, i.e., without the utilization of a so-called "high-energy" substance as a high energy source due to the splitting of a P-O bond which should actually require energy.

Since the maximal force developed by an axonemal dynein head interacting with a microtubule is strikingly similar to ϕ_o (Kamimura and Takahashi, 1981) and as the free energy liberated when a dynein-microtubule complex is dissociated by ATP does not differ much from that generating ψ_o (Porter and Johnson, 1989), it is tempting to believe that the molecular mechanism underlying generation of tension and of movement, as well as the values of ϕ_o, ψ_o, Δl_1, and Δl_2, are very similar for the actomyosin and for the dynein (kinesin, etc.) - microtubule biological mechanochemical convertors. In particular, it is probable that in the microtubule motors, as in muscle, movement is the outcome of the dissociation of the protein complex and is not due to the force coupled with the conformational change which follows or accompanies the release of the hydrolysis products of ATP, as is commonly taken for granted.

THE SLIDING DISTANCE OF MUSCLE IN RIGOR

If we allow an isometrically active muscle to consume all its ATP and creatine phosphate, rigor occurs. In this state, the muscle continuously exhibits tension, the value of which is quite close to that of the active muscle. The tension maintained by a muscle in rigor is due to the forces developed by the actomyosin complexes after each of them had broken an ATP molecule for the last time. The rigor tension is equal to the number of myosin heads interacting with actin per half sarcomere of a unit cross-sectional area times the molecular "contractile" force, ϕ_o. If we release such a muscle, in which actin and myosin form a continuous network, it will shorten slightly (White, 1970), the driving force for this shortening being ϕ_o. The change in length cannot exceed the length of a myosin head. The translocation of all the thin filaments in half a sarcomere should be equal to the translocation of a single thin filament as a result of the conformational change of a single actomyosin complex. This is nothing else but the sliding distance which has so far been considered to be the same for both actively shortening and rigor muscles. It is clear that the value of this step distance is determined by factors which differ profoundly from those governing the value of the step distance in actively shortening muscles, i.e. Δl_1 which was discussed above.

Let us calculate this step distance which we will designate Δl_2. According to White (1970), the shortening necessary to return the tension in rigor muscle fibers to zero tension once they have contracted under isometric conditions is about 1%. The value of Δl_2 is then: $\Delta l_2 \cong 1 \times 10^{-2} \times 10^3$ nm = 10 nm (10^3 nm is the length of a thin filament).

In conclusion, the value of the step distance is in accord with the swinging crossbridge model only in the case of muscle in rigor, but not for active muscle where it is a variable and could be many times larger.

Coats et al. (1985), on the basis of pressure-relaxation studies of heavy meromyosin subfragment-1 (S1) and pyrene-labelled actin in solution, have proposed the model: $M + A \rightleftarrows (M\text{-}A)_1 \rightleftarrows (M\text{-}A)_2$. They have noticed that both isometric tension generation by muscle and the transition from a weakly $(A\text{-}M)_1$ - to a strongly $(A\text{-}M)_2$ - attached state in solution are inhibited by low temperature, high pressure, high ionic strength, and the presence of ethylene glycol. This led them to suggest that tension development is associated with this reaction. One should, however, be aware of a fundamental difference between this reaction occurring in solution and the same process taking place in an isometrically "contracting" muscle: in solution the transition from state 1 to state 2 can occur freely, leading to a conformational change of the actomyosin complexes and, presumably, to a relative movement of the myosin heads with respect to the actin filaments. However, in a muscle which is held at a constant length, the isomerization process is inhibited, and this is the origin of the tension which is observed. The drop in free energy which accompanies the transition cannot then take place and the actomyosin complexes will store this energy in the form of potential chemical energy. This will be the situation also in rigor muscle fibers under tension. If such fibers are released, the equilibrium $(M\text{-}A)_1 \rightleftarrows (M\text{-}A)_2$ will be shifted in favor of state 2, as it is in solution, and shortening will take place. The stored free energy will then be liberated, thus enabling the muscle to do mechanical work if a load smaller than the original isometric force is attached. Assuming that a large part of this energy is utilized for the performance of work when the load per actomyosin complex assumes an optimal value somewhere between ϕ_o and zero, we may write, as a first approximation: $\Delta G°/N_a \cong (^1/_2\,\phi_o) \bullet \Delta l_2$; where $\Delta G°$ represents the standard free energy change of the isomerization reaction and N_a is Avogadro's number. The value of $\Delta G°$ can be calculated from the values given by Coats et al. (1985) for the equilibrium constant of the isomerization reaction and it amounts to -3.26 Kcal/mole. Substituting the value of 10 nm for Δl_2 in the last equation we then obtain: $\phi_o = 2.3$ pN.

On the other hand, employing a similar procedure for the calculation of ψ_o, we may write: $\Delta G°/N_a \cong (^1/_2\,\psi_o) \bullet \Delta l_1$; where $\Delta G°$ is the standard free energy change of the ATP-induced dissociation of actomyosin Substituting $\Delta G°$ = 7.0 Kcal/mole (Kodama, 1985) and Δl_1 = 125 nm (the value at 22° C according to Harada et al., 1990) we obtain: $\psi_o = 0.8$ pN.

The two forces are thus different not only in nature but also in value and should vary in different fashions with temperature, ionic strength etc.

REFERENCES

Ajtai, A., French, A. R., and Burghardt, T. P., 1989, Myosin cross-bridge orientation in rigor and in the presence of nucleotide studied by electron spin resonance, *Biophys. J.*, 56: 535.

Bárány, M., 1967, ATPase activity of myosin correlated with speed of muscle contraction, *J. Gen. Physiol.*, 50: 197.

Berg, D. G., Winter, R. B., and von Hippel, P. H., 1981, Diffusion-driven mechanisms of protein translocation on nucleic acids, *Biochemistry*, 20: 6929.

Borejdo, J. and Oplatka, A., 1976, Tension development in skinned glycerinated rabbit psoas fiber segments irrigated with soluble myosin fragments, *Biochim. Biophys. Acta*, 440: 241.

Coats, J. H., Criddle, A. H., and Geeves, M. A., 1985, Pressure-relaxation studies of pyrene-labelled actin and myosin subfragment 1 from rabbit skeletal muscle, *Biochem. J.*, 232: 351.

Edman, K. A. P. and Lou, L., 1990, Changes in force and stiffness induced by fatigue and intracellular acidification in frog muscle fibers, *J. Physiol.*, 424: 133.

Eisenberg, E. and Hill, T. L., 1985, Muscle contraction and free energy transduction in biological systems, *Science*, 227: 999.

Harada, Y., Sakurada, K., Aoki, T., Thomas, D. D., and Yanagida, T., 1990, Mechanochemical coupling in actomyosin energy transduction studied by *in vitro* movement assay, *J. Mol. Biol.*, 216: 49.

Huxley, A. F., 1980, "Reflections on Muscle", Liverpool Univ. Press, London.

Huxley, H. E., 1990, Sliding filaments and molecular motile systems, *J. Biol. Chem.*, 265: 8347.

Janmey, P. A., Hvidt, S., Oster, G. F., Lamb, J., Stossel, T. P., and Hartwig, J. H., 1990, Effect of ATP on actin filament stiffness, *Nature*, 347: 95.

Kamimura, S. and Takahashi, K., 1981, Direct measurement of the force of microtubule sliding in flagella, *Nature*, 293: 566.

Kodama, T., 1985, Thermodynamic analysis of muscle ATPase mechanisms, *Physiol. Rev.*, 65: 467.

Kushmerick, M. J. and Davies, R. E., 1969, The chemical energetics of muscle contraction. II. The chemistry, efficiency and power of maximally working sartorius muscles, *Proc. R. Soc. Lond. B*, 174: 315.

Lovell, S., Karr, T., and Harrington, W. F., 1988, *Proc. Nat'l. Acad. Sci. U.S.A.*, 85: 1849.

Maruyama, T., Kometani, K., and Yamada, K., 1989, Effect of ethylene glycol on contractile properties of glycerinated rabbit psoas muscle, *in:* "Muscle Energetics", R. J. Paul, G. Elzinga, and K. Yamada, eds., Alan R. Liss, New York, p. 223.

Oplatka, A., 1972, On the mechanochemistry of muscular contraction, *J. Theor. Biol.*, 34: 379.

Oplatka, A., 1989, Changes in the hydration shell of actomyosin are obligatory for tension generation and movement, *in:* "Muscle Energetics", R. J. Paul, G. Elzinga, K. Yamada, eds., Alan R. Liss, New York, p. 45.

Pate, E. and Cooke, R., 1989, A model of crossbridge action: The effects of ATP, ADP and P_i, *J. Muscle Res. Cell Motil.*, 10: 181.

Payne, M. R. and Rudnick, S. E., 1989, Regulation of vertebrate striated muscle contraction, *Trends. Biochem. Sci.*, 15: 357.

Porter, M. E. and Johnson, K. A., 1989, Dynein structure and function, *Ann. Rev. Cell Biol.*, 5: 119.

Rau, D. C., Lee, B., and Parsegian, V. A., 1984, Measurement of the repulsive force between polyelectrolyte molecules in ionic solution: Hydration forces between parallel DNA double helices, *Proc. Nat'l. Acad. Sci. U.S.A.*, 81: 2621.

Tirosh, R., Low, W. Z., and Oplatka, A., 1990, Translational motion of actin filaments in the presence of heavy meromyosin and MgATP as measured by doppler broadening of laser light scattering, *Biochim. Biophys. Acta*, 1037: 274.

Toyoshima, K. Y., Kron, S. J., and Spudich, J. A., 1990, The myosin step size: Measurement of the unit displacement per ATP hydrolyzed in an *in vitro* assay, *Proc. Nat'l. Acad. Sci. U.S.A.*, 87: 7130.

Uyeda, T. Q. P., Kron, S. J., and Spudich, J. A., 1990, Myosin step size: Estimation from slow sliding movement of actin over low densities of heavy meromyosin, *J. Mol. Biol.*, 214: 699.

Warshaw, D. M., Desrosiers, J. M., Work, S. S., and Trybus, K. M., 1990, Smooth muscle myosin crossbridge interactions modulate actin filament sliding velocity *in vitro*, *J. Cell. Biol.*, 111: 453.

White, D. C. S., 1970, Rigor contraction and the effect of various phosphate compounds on glycerinated insect flight and vertebrate muscle, *J. Physiol.*, 208: 583.

Woledge, R. C., Curtin, M. A., and Homsher, E, 1985, "Energetic aspects of muscle contraction", Academic Press, London.

Yanagida, T., Kuranaga, I., and Inoue, A., 1982, Interaction of myosin with thin-filaments during contraction and relaxation: Effect of ionic strength, *J. Biochem.*, 92: 407.

AN ANALYSIS OF LENGTH-DEPENDENT ACTIVE STIFFNESS IN SMOOTH MUSCLE STRIPS

Richard A. Meiss

Indiana University School of Medicine
Indianapolis, IN 46202

INTRODUCTION

The measured stiffness of contracting smooth muscle is strongly dependent on the level of developed force. This force-dependent stiffness is a consequence of contractile activity, and it is possible that a portion of it represents the stiffness of the population of attached crossbridges. The relationship between force and stiffness is sensitive to the particular stage of the contraction-and-relaxation cycle (Meiss, 1978), to specific external mechanical constraints imposed on the muscle (Meiss, 1987), and to the length of the muscle when the stiffness is measured (Meiss, 1978; Meiss, 1990). The character of the length-dependent stiffness relationship depends on the mechanical mode of contraction, and interpretation of these effects rests on assumptions regarding how the process of stiffness measurement interacts with changing tissue dimensions. The purpose of this paper is to characterize the difference between the length-dependent stiffness measured in isotonic and isometric contractions. Possible reasons for the differences will be considered, and a tentative model to account for the isotonic length-dependence of stiffness will be proposed.

METHODS

The strip of muscle (either mesotubarium superius or ovarian ligament from estrous rabbits) was mounted between a photoelectric force transducer and the driving arm of a servomotor (modified Cambridge Technology 300H). Square-wave electrical stimuli were delivered through the physiological saline bathing solution that was contained in a temperature-controlled bath (25° C). The sequence, duration, and type of mechanical events were determined by interaction between external control circuitry and a digital interface with a microcomputer (Zenith 158). Force and length were clamped to reference values by appropriate feedback circuitry. Output from a sine-wave generator was superimposed on the signal driving the servomotor and produced a oscillatory length perturbation whose amplitude was less than 0.5 percent of the resting muscle length. As a result of this input perturbation interacting

Regulation of Smooth Muscle Contraction
Edited by R.S. Moreland, Plenum Press, New York, 1991

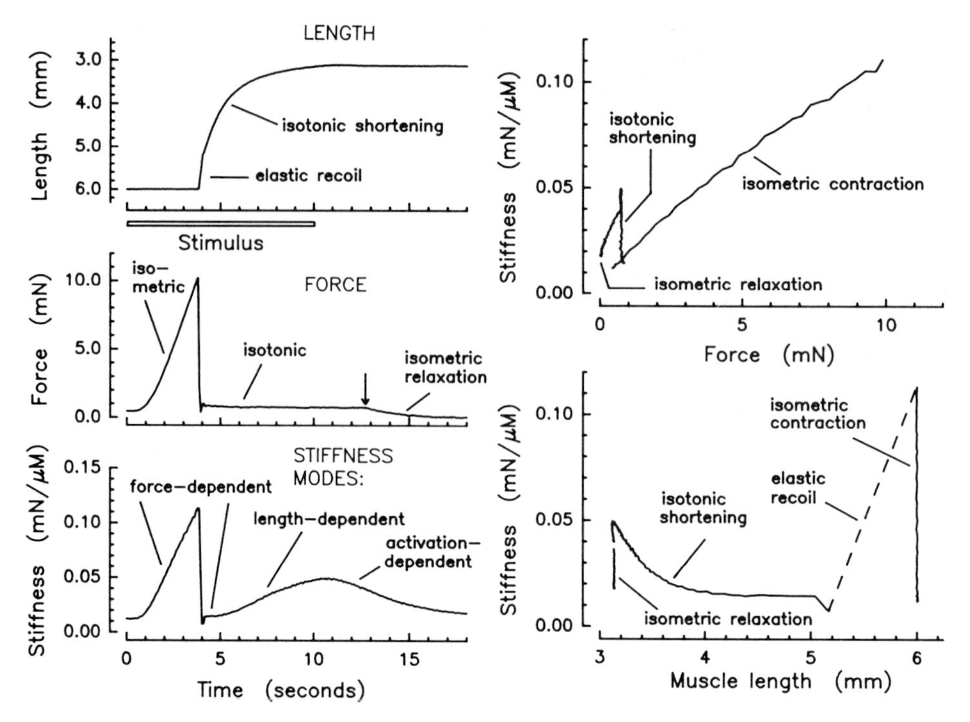

Figure 1. Typical contraction data and display of muscle stiffness.

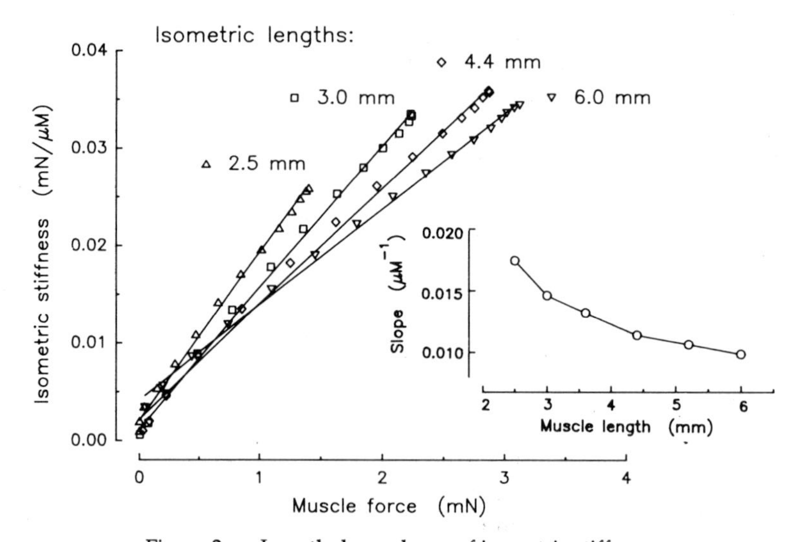

Figure 2. Length dependence of isometric stiffness.

with the stiffness of the contracting muscle, a corresponding force perturbation was produced. A digitally-controlled signal processor separated the force and length sinusoids from the primary force and length signals and produced DC signals proportional to the absolute values of the perturbations. An analog-to-digital converter (Metrabyte) digitized values of length, force, and the two perturbation amplitude signals (dF and dL) and stored the data on floppy disks

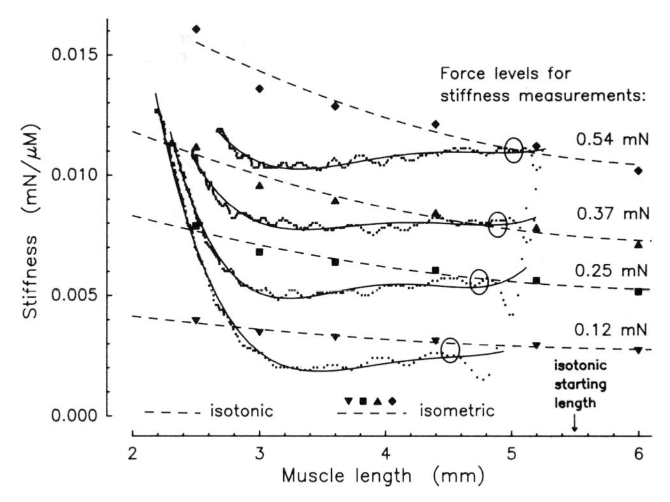

Figure 3. Isometric and isotonic length-dependencies compared.

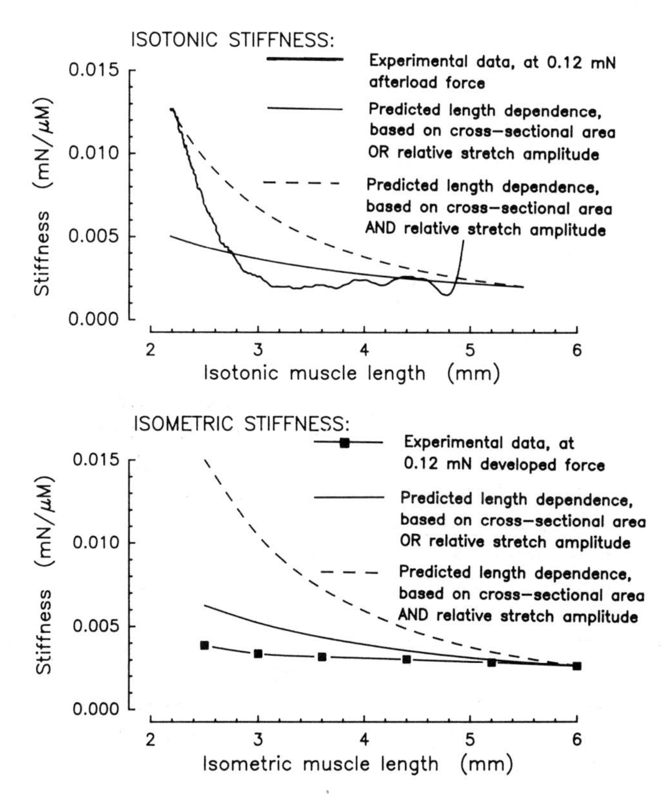

Figure 4. Muscle data compared with expected behavior.

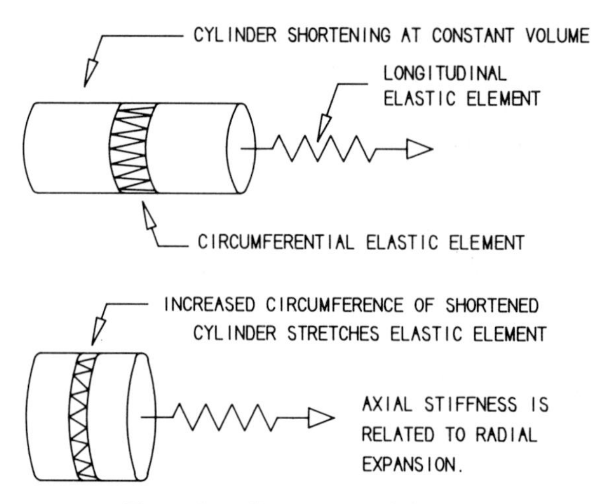

Figure 5. Components of the model.

for subsequent analysis. The stiffness of the contracting muscle was computed as the ratio dF/dL. All data traces shown in this report are reproduced from digitally-stored data.

RESULTS

At the left of Figure 1 is shown a contraction with a beginning isometric phase followed by a period of isotonic shortening at a very low afterload. The muscle stiffness (lowest panel) shows dependence on developed force during its isometric and early isotonic portions. At extreme shortening a stiffness component related to muscle length is observed. This stiffness also shows activation dependence in that it begins to fall when the stimulus ends but while both force and length are constant. At the right, the stiffness of the muscle during the whole contraction cycle is plotted. In the upper portion the stiffness is shown as a function of the muscle force. During the isometric phase the relationship is approximately linear; force-dependent stiffness is also shown in Figures 2, 3, and 4. Stiffness as a function of muscle length is shown in the lower portion. It first assumes a value proportional to the force; stiffness rises during isometric contraction, remains constant during most of the isotonic phase, and increases markedly as the muscle nears its final length. The characteristic length-dependent stiffness is also shown in Figures 3, 4, 8, and 9.

Figure 2 shows isometric contractions that were made at a series of six resting muscle lengths. During the rising phase of each contraction the stiffness was continuously measured. Data from contractions at four muscle lengths are shown. (Data from the other two were omitted for the sake of clarity.) The best-fitting straight lines were drawn through the data points and the slopes were determined. Contractions made at the shortest muscle lengths produced the steepest relationship. In the inset graph these slopes (a measure of the elastic modulus of the muscle) are plotted against the isometric muscle length. The trend to greater stiffness at shorter lengths is evident. Increased stiffness at short lengths is also shown during isotonic contraction (Figure 1, lower right panel), but the form of the relationship is quite different.

To investigate these differences, isometric contractions were made at a series of resting lengths (as above). From the force-stiffness relationships thus determined, the stiffness of the muscle at specific levels of developed force was computed. These data (solid symbols) are plotted as a function of muscle length and are fitted by an arbitrary second-degree polynomial (dashed lines). The same muscle was allowed to contract under isotonic loads (matched to the isometric developed forces) while stiffness was being measured continuously. These data are shown as dotted lines and are fitted by an arbitrary fourth-degree polynomial (solid lines). It is evident that the effect of muscle length on the stiffness differs substantially depending on the mode of length change.

DISCUSSION

In order to understand the variation of stiffness with changes in length, the behavior of several hypothetical elastic bodies was considered. The first was a "telescoping" body whose cross-sectional area would remain constant as it shortened and whose force response to a constant-amplitude length perturbation would be constant regardless of the length of the body. The middle curve shows the behavior of a body that shortens at constant cross-sectional area but with a force response to a constant-amplitude length perturbation that depends on the length of the body. Such behavior would be expected from a sliding-filament contractile mechanism with stiffness produced by a constant-sized population of attached crossbridges. An extreme alternative to this would be the length-dependence of stiffness arising from a body shortening at constant volume (and increasing cross-sectional area), with a relative force-perturbation response to a constant-amplitude length perturbation that also depended on the instantaneous length of the body. This behavior is shown as the uppermost solid line in each of the two panels of Figure 4. A compromise between these extremes could be shown by a body that shortens at increasing cross-sectional area (as in the above case) but which is not sensitive to the relative perturbation amplitude (see Fig. 4, lower solid line in each

$R = D/2;\ V = \pi R^2 L$	Radius increases as cylinder shortens because volume remains constant.
$C = 2(V\pi/L)^{1/2}$	Circumference changes as a function of cylinder length.
$F_C = \alpha e^{K_C C}$	Elastic circumferential force (tension) due to radial expansion.
$F_C = \alpha e^{2K_C(V\pi/L)^{1/2}}$	Circumferential tension in terms of cylinder length.
$F_A = F_C$	Circumferential tension is balanced against axial (compressive) force.
$S_A = \alpha K_C L^{-3/2} \pi V^{1/2} e^{2K_C(V\pi/L)^{1/2}}$	Axial stiffness as a function of circumferential stiffness, expressed in terms of cylinder length.
$S_T = K_T F_A + S_0$	Series compliance (stiffness) through which measurement is made.
$S_L = 1/[(1/S_A) + (1/S_T)]$	Resultant longitudinal stiffness of components in series.

Figure 6. Mathematical relationships among the components of the model.

panel). A similar dependence would be shown by a body that shortened at constant cross-sectional area but which was sensitive to the relative perturbation amplitude.

The curves shown in Figure 4 represent these extremes of expected behavior in comparison to experimental data from smooth muscle preparations. The upper portion shows data from isotonic shortening (cf. Figs. 1 and 3). While the total increase in apparent stiffness is ultimately as great as the expected rise in stiffness, the pathway taken is clearly different. In the lower portion (square symbols) are data from a series of isometric contractions (cf. Figs. 2 and 3). The increase in stiffness, while significant, is less than expected on the basis of either of the two cases of length-dependence considered above. It is clearly more, however, than is predicted by a length-independent constant-volume system (the first case above). These comparisons lead to (at least) two conclusions: first, that the simple models proposed are inadequate to explain the muscle behavior, and second, that the factors affecting muscle stiffness differ according to the mode of contraction (see also Fig. 3). At present there is under development a model that attempts to address the length dependence of stiffness during isotonic shortening. This model, in its current state of development, is shown in the subsequent figures. Because of the clear differences between isometric and isotonic behavior, the model will not attempt at this time to account for the isometric condition.

The essential features of the model, an idealized representation of a strip of smooth muscle tissue, are shown in Figure 5. Consider a cylinder that shortens at constant volume; its cross-sectional area will increase as it shortens, and its circumference will become greater. If an elastic element, representing connective tissue between and among cells, is arranged to lie in the circumferential direction, it will be stretched as the circumference increases because of the axial shortening. The elastic element has an exponential characteristic, becoming proportionately stiffer as its force increases due to its being stretched. This stiffness is manifested in the axial direction by consideration of the geometrical requirements of the constant-volume condition. Connections to the measuring instruments must also be made through compliant structures, and these are lumped into an exponential elastic element in series with the cylinder.

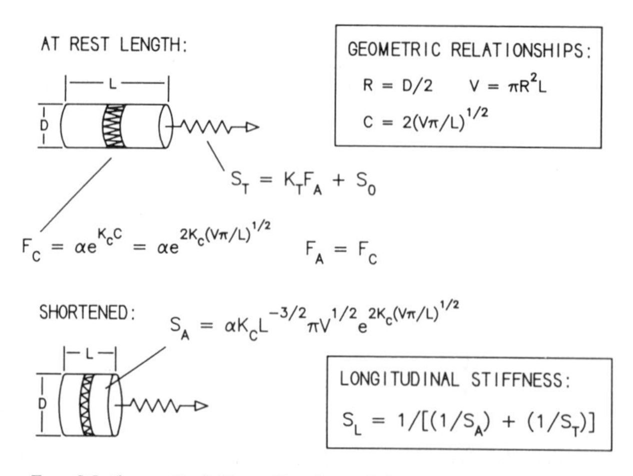

Figure 7. Mathematical identification of the components of the model.

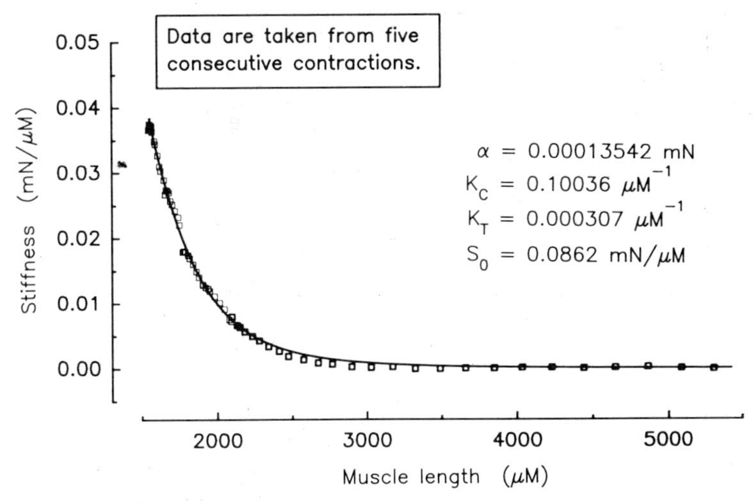

Figure 8. Fitting model equations to experimental data.

Several explicit assumptions must be made in order to construct the model:

1. The cylinder (representing muscle tissue) shortens at constant volume.
2. Circumferential expansion is opposed by an exponential elastic component (representing a radially-directed connective tissue component inherent in the muscle tissue structure).
3. Contractile force causing radial expansion is manifested as axial stiffness, since the primary locus of the activation-dependent stiffness is assumed to be attached crossbridges.
4. The axial force (compression) is balanced by circumferential force (tension).
5. Longitudinal (axial) stiffness is measured through an elastic component (in series) whose stiffness is proportional to the axial force.
6. Stiffness of the model elements in series adds as the reciprocal of the sum of the reciprocals of their individual stiffnesses (analogous to the combined value of electrical capacitors in series).

These assumptions represent first approximations to the true situation; for example, shortening may involve some changes in volume due to shifts in water or cytoplasm. The exponential formulation of the elastic elements is the simplest description that leads to reasonably realistic results. The elastic element in series may not be the complete functional equivalent of the classical series elastic element. It is additionally assumed that the model shortens at constant external force, approximating the situation for the experimental data (cf. Fig. 1).

The set of equations describing the model is shown in Figure 6. The independent variable in the set of equations is the length of the cylinder. Thus the circumference (second equation down) is formulated in terms of the cylinder length, as is the tension in its elastic element (fourth equation down). Longitudinal stiffness is thus described in terms of the first derivative of the circumferential force change with respect to cylinder length (sixth equation down). The identification of these equations with specific elements of the model is shown in Figure 7. All dimensional changes are formulated in terms of changes in the length of the cylinder. Because shortening takes place at

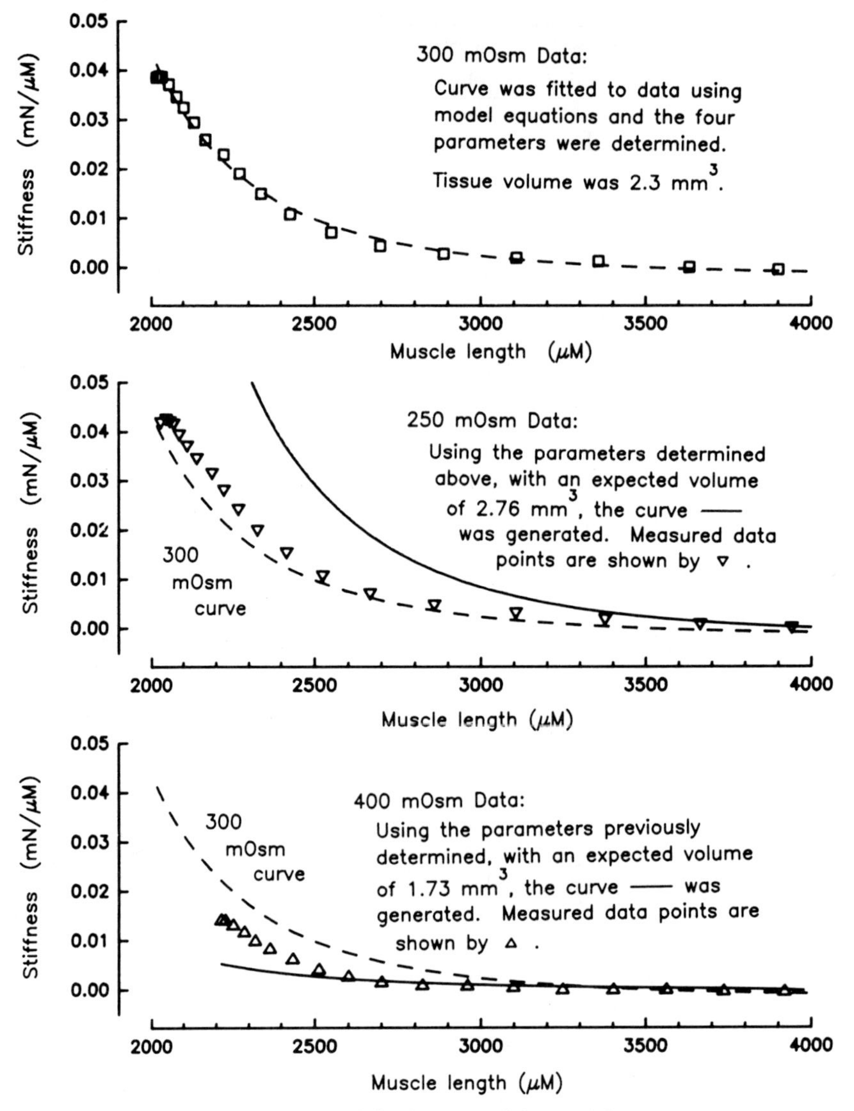

Figure 9. A further test of the model.

constant external force, the length of the longitudinal elastic element does not change.

The system of equations describing the model contains four adjustable parameters (relating to its elastic properties) as described above. To compare the model with experimental data, the computer program MINSQ (MicroMath, Salt Lake City, UT) was used. This is an iterative, non-linear, curve-fitting program that attempts to minimize the deviation between the set of equations and the experimental data by adjusting the set of parameters to their optimal values. Because independent measurements of the individual parameter values have not been made, the first goal of the modeling procedure was to attempt to reproduce the qualitative shape of the length-stiffness curves and to gain insight into the relative magnitudes of the four adjustable parameters. Limited testing of the model has been carried out as described below.

In five consecutive contractions, with varying afterloads, the length-dependent stiffness was measured (see Fig. 8). The five data sets were combined (every third data point is shown in the fig.). The best-fitting line, and the parameters that specify it, are also shown in the figure. There is apparent good agreement between the shape of the empirical curve and the line of best fit.

As a further test of the model, the following experiment was performed (see Fig. 9). Length-dependent stiffness was measured in an ovarian ligament preparation (with a weight of 2.3 mg.) in a 300 mOsm bathing solution (control conditions). The results are shown as the square data points in the upper panel. Using a value of 2.3 mm^3 for the tissue volume, the four model parameters were determined; the solid line is the resulting fit. The muscle was then placed into a 250 mOsm bathing solution; if it were a perfect osmometer, the new tissue volume would be 2.76 mm^3. The measured length-dependent stiffness is shown by the inverted triangles (lying above with 300 mOsm control curve). Using the four model parameters determined under control conditions and the predicted volume, the curve describing the length-dependent stiffness was generated. It was located well above the experimental data points. Repeating this process with a 400 mOsm bathing solution (predicted tissue volume 1.73 mm^3) produced opposite results; the data points fell below the control curve as expected, but the curve predicted on the basis of the expected volume and the control parameters fell below the data points.

CONCLUSIONS

This simple model produces a reasonable qualitative fit to the experimental data (see Fig. 8), although the precise physical meaning of the elastic constants is not clear. This makes their experimental determination rather uncertain at present. The lack of agreement between experiment and prediction in Figure 9 may be due to several factors. It is probable that the osmotically-induced changes in tissue volume are less than predicted, although the experimental data do show shifts in the expected directions. It is also possible that the parameters in the model represent factors that do change with medium osmolality (swelling of intercellular ground substances, for example). Finally, it is possible that the formulation of the model is not at all relevant to what is taking place in the shortening tissue.

A major problem with the model is its anatomical simplicity; its qualitative success does indicate that its formulation and assumptions may have some basis in reality. It should be noted that this model has some similarity with a physical model (Mullins and Gutheroth, 1965) that was proposed to account for force transmission in smooth muscle tissues; the results presented here provide some confirmation of the validity of that model. It is possible that further refinement, taking into better account the complexity of the tissue structure, will produce a model in which elastic and other tissue parameters may be determined independently of the isotonic contraction data. An anatomical basis for the postulation of intercellular connective tissue properties and arrangements has been provided by some relevant ultra-structural studies (Gabella, 1984; 1976). While the model is presently formulated in terms of extracellular elastic elements, it is also possible that intracellular cytoskeletal elements or cross-linked contractile proteins may contribute to a resistance to radial expansion; the "corkscrew" behavior of single isolated cells undergoing isotonic shortening (Warshaw et al., 1978) may also provide an element of length-dependent stiffness if such behavior is constrained by adjacent cells or connective tissue in an intact muscle. The ultimate goal of this proposed models and its future refinements is a better understanding of how tissue structure affects the mechanical function of smooth muscle, especially as it relates to the limits of isotonic shortening.

ACKNOWLEDGMENTS

This study was supported by the Dept. of OB/GYN, I. U. School of Medicine.

REFERENCES

Gabella, G., 1984, Structural apparatus for force transmission in smooth muscle, *Physiol. Rev.,* 64: 455.

Gabella, G., 1976, Quantitative morphological study of smooth muscle cells of the guinea pig taenia coli. Structural changes in smooth muscles during isotonic contraction, *Cell Tissue Res.,* 170: 161.

Meiss, R. A., 1990, The effect of tissue properties on smooth muscle mechanics, *in:* "Frontiers in Smooth Muscle Research", N. Sperelakis and J. D. Wood, eds., Wiley-Liss, New York, p. 435.

Meiss, R. A., 1978, Dynamic stiffness of rabbit mesotubarium smooth muscle: Effect of isometric length, *Am. J. Physiol.,* 3: C14.

Meiss, R. A., 1987, Stiffness of active smooth muscle during forced elongation, *Am. J. Physiol.,* 22: C484.

Mullins, G. L. and Gutheroth, W. G., 1965, A collagen net hypothesis for force transference of smooth muscle, *Nature,* 206: 592.

Warshaw, D. M., McBride, W. J., and Work, S. S., 1987, Corkscrew-like shortening in single smooth muscle cells, *Science,* 236: 1457.

EFFECTS OF MUSCLE LENGTH ON INTRACELLULAR Ca^{2+} DURING ISOMETRIC CONTRACTION OF TRACHEAL SMOOTH MUSCLE

Susan J. Gunst and Ming Fang Wu

Department of Physiology and Biophysics
Indiana University School of Medicine
Indianapolis, IN 46202

INTRODUCTION

In both smooth and striated muscles, the active force a muscle can develop is related to its length. In striated muscles, the decrease in force that occurs as muscle length is decreased below the length of maximal active tension (L_o) may result from a number of factors: length-dependent changes in the Ca^{2+}-sensitivity of contractile proteins, length-dependent changes in the amount of activator Ca^{2+} supplied to contractile proteins, and mechanical factors affecting filament overlap (Gordon et al., 1966; Taylor and Rudel, 1970; Endo, 1973; Allen and Kurihara, 1982; Stephenson and Wendt, 1984;). Studies of striated muscles, particularly cardiac muscle, suggest that as muscle length is decreased below L_o there is a decrease in the activation of the myofilaments (Jewell,1977; Allen and Kurihara,1982; Stephenson and Wendt,1984). Also, when a striated muscle is allowed to shorten during active contraction, force redevelopment immediately after the shortening is reduced. This phenomenon, termed "shortening deactivation," is dependent on Ca^{2+} and can be decreased or abolished by increasing the activation of the muscle (Edman and Kiessling, 1971; Briden and Alpert, 1972; Bodem and Sonnenblick, 1974; Ekelund and Edman, 1982).

In tracheal smooth muscle, the slope of the length-tension relationship determined by isometric contraction has also been shown to be sensitive to the degree of activation of the muscle (Kromer and Stephens, 1983; Gunst, 1986). In addition, a step decrease in muscle length imposed on tracheal muscle during active contraction decreases the rate of force redevelopment, a phenomenon that is analogous in many respects to the shortening deactivation that has been reported in striated muscles (Gunst, 1986).

In this paper, we present preliminary results of experiments in which we have explored the effects of muscle length on intracellular Ca^{2+} in canine tracheal smooth muscle during isometric contractions induced by different contractile stimuli. The Ca^{2+}-sensitive bioluminescent protein aequorin was used to monitor changes in the intracellular Ca^{2+} concentration ($[Ca^{2+}]_i$) in muscle strips. Muscle length was changed prior to activation of the muscle by electrical field stimulation (EFS), acetylcholine (ACh) or by K^+-depolarization.

Regulation of Smooth Muscle Contraction
Edited by R.S. Moreland, Plenum Press, New York, 1991

METHODS

Preparation of Tissues

Mongrel dogs weighing 20 - 25 kg were anesthetized with pentobarbital sodium and quickly exsanguinated. A 10 to 15 cm segment of extrathoracic trachea was immediately removed and immersed in 22° C physiological saline solution (PSS) of the following composition (in mM): 110 NaCl, 3.4 KCl, 2.4 $CaCl_2$, 0.8 $MgSO_4$, 25.8 $NaHCO_3$, 1.2 KH_2PO_4, and 5.6 glucose. Rectangular strips of trachealis muscle, approximately 5 mm long x 2 mm wide were dissected from the trachea after removal of the epithelium and connective tissue layer. Muscle strips were mounted vertically in 60 ml glass tissue baths containing PSS at 37° C and bubbled with 95% O_2 - 5% CO_2 to maintain a pH of 7.4. One end of each strip was fixed by a small clamp between two 7 mm platinum wire electrodes, the other end was attached by thread to an electromagnetic lever.

Muscles were equilibrated for approximately 90 min after being mounted in the tissue bath during which time each muscle was stimulated at 5 to 10 minute intervals using electrical field stimulation (EFS; 12 volts, 15 pps, 0.5 msec duration). To determine L_o, the length of maximal active force development, muscle length was increased progressively after each stimulation until the force of active contraction reached a maximum.

Muscle force and length were measured using a Cambridge Technology dual-mode servo muscle ergometer system (Model 300H). The compliance of this system was 0.030 µm/mN. The resolution of the force signal was <0.20 mN and the linearity of the position signal was 0.1%.

Introduction of Aequorin into the Tissues

The method for incorporating aequorin into the muscles was modified from that described by Morgan and Morgan (1984). The muscles were soaked in 4 successive solutions of the following composition (in mM): Solution I (at 0° C for 60 min, bubbled with O_2), 10 EGTA, 5.0 Na_2ATP, 120 KCl, 2.0 $MgCl_2$, 20 TES; Solution II (at 0° C for 120 min, bubbled with O_2), 0.1 EGTA, 5.0 Na_2ATP, 120 KCl, 2.0 $MgCl_2$, 20 TES, 0.2 mg/ml aequorin; Solution III (at 0° C for 60 min, bubbled with O_2), 0.1 EGTA, 5.0 Na_2ATP, 120 KCl, 10 $MgCl_2$, 20 TES; Solution IV (at 15° C for 60 min, bubbled with 95% O_2-5% CO_2); 3.4 KCl, 9.0 $MgCl_2$, 110 NaCl, 0.8 $MgSO_4$, 25.8 $NaHCO_3$, 1.2 KH_2PO_4, and 5.6 Dextrose. After 30 min in Solution IV, $CaCl_2$ was added gradually, to reach a final concentration of 2.4 mM. After 60 - 90 min in Solution IV, the muscle was returned to normal PSS. The muscle was then kept at 10° C overnight. The following day, the solution was gradually warmed to 37° C and the muscle was allowed to stabilize for 2 hours or more before beginning the experiment.

The aequorin used in this study was prepared in the laboratory of Dr. J. R. Blinks by methods described in Blinks, et al. (1982).

Measurement of Aequorin Light Signal

After the incorporation of aequorin into the muscle cells, muscles were lowered into a narrow cylindrical extension at the base of the glass muscle bath that protruded a short distance axially into an ellipsoidal reflector. The muscle was positioned near one focal point of the reflector; a photomultiplier tube (EMI 9635A) was mounted so that its photocathode was at the other focal point (Blinks, 1984). The tissue bath was enclosed in a light-tight apparatus while the aequorin signal was being monitored. Light (recorded in nA as photomultiplier anode current), force, and length were recorded in analog form on an oscillographic recorder (Gould, Model 2400S), and on disk in digital form using a digital oscilloscope (Nicolet model 4094B). Force and light signals were transferred to an IBM PS/2 Model 70 computer from which they were plotted using a digital plotter (Hewlett Packard model 7475A). All experiments were performed at 37° C.

Experimental Protocols

After being equilibrated at L_0 and loaded with aequorin, muscles were subjected to a series of isometric contractions at lengths alternating between L_0 and either 0.7 or 0.8 L_0. In each muscle, all contractions were induced by the same stimulus: either 10^{-4} M ACh, 60 mM KCl, or EFS. In some experiments a constant voltage (12 v), 60 Hz ac EFS was used, whereas in other experiments EFS was effected using square wave pulses at 12 V, 15 pps, and 0.5 msec duration. The latter stimulus has been shown to specifically activate intrinsic nerves (Russell, 1980). We found that its effects were blocked entirely by atropine (10^{-6} M). However, the former stimulus may have some direct activating effects on the muscle. When 60 mM KCl was used as the stimulus, 10^{-6} M atropine was included in the bathing medium to antagonize any possible effects of ACh released due to intrinsic cholinergic nerves. KCl was substituted for NaCl to prevent changes in osmolarity.

In order to obtain values for the average luminescence during a contraction (Table 1), aequorin luminescence was integrated for the first 60 sec after the onset of the stimulus and then divided by the duration (60 sec). Values for stress represent the maximum active stress obtained during that contraction. Values for dS/dt represent the peak rate of active stress development. For each experiment, values for active stress, dS/dt, and aequorin luminescence for lengths of 0.7 and 1.0 L_0 were obtained by averaging values for each parameter

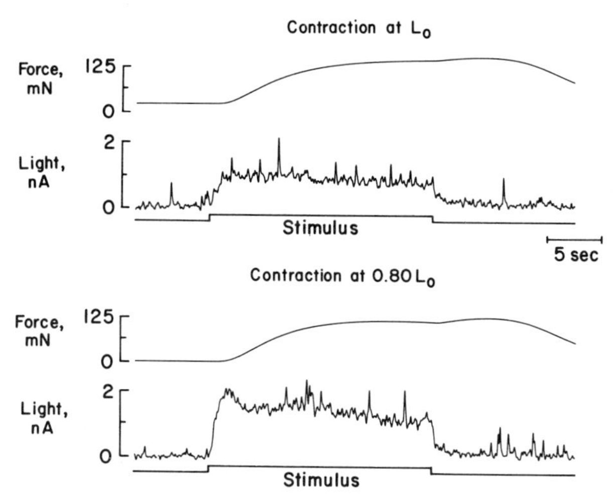

Figure 1. Aequorin luminescence and active stress during electrical field stimulation with a 12 volt, 60 Hz, ac stimulus. The aequorin light signal was higher during isometric contraction at 0.8 L_0 than during isometric contraction at L_0 during sequential contractions performed in a single tissue.

Table 1. Effect of Muscle Length on Active Stress and Aequorin Luminescence During Isometric Contractions in Canine Tracheal Smooth Muscle

	Muscle Length	Max. Active Stress (mN/mm^2)	dS/dt ($mN/mm^2/s$)	Aequorin Luminescence (nA)
Electrical Stimulation n=4	1.0 L_o	73 ± 10	5.0 ± 1.2	0.42 ± 0.07
	0.7 L_o	25 ± 2.5	1.9 ± 0.3	0.52 ± 0.05
Acetylcholine n=4	1.0 L_o	58 ± 5	2.7 ± 0.4	0.83 ± 0.07
	0.7 L_o	15.5 ± 2	0.85 ± 0.14	0.64 ± 0.07
60 mM KCl n=6	1.0 L_o	58 ± 7	2.2 ± 0.14	0.84 ± 0.20
	0.7 L_o	22 ± 3.6	1.4 ± 0.07	0.62 ± 0.78

Values represent mean ± SEM. n = number of muscles used. dS/dt represents the maximum rate of active stress development. Aequorin luminescence is computed as the average luminescence over the first 60 s of contraction.

from 2-4 contractions at each muscle length. A mean of the average values obtained from each experiment is reported. In all cases, n represents the number of muscles used to obtain the mean value.

RESULTS

Figure 1 illustrates aequorin light signals obtained during successive isometric contractions performed at L_o and at 0.8 L_o in a single muscle strip using 60 Hz ac EFS. The light signal was higher during the contraction at 0.8 L_o than during the contraction at L_o, regardless of the sequence in which the contractions were performed. The effect of muscle length on muscles contracted by stimulating only intramural nerves (EFS: 12 v square waves, 15 pps, 0.5 ms duration) was similar (Fig. 2). $[Ca^{2+}]_i$ was higher in muscles contracted at 0.7 L_o than in muscles contracted at L_o, even though active stress was much lower at 0.7 L_o than at L_o. In contrast, when muscles were contracted with 10^{-4} M ACh (Fig. 3), both $[Ca^{2+}]_i$ and active stress were higher during isometric contraction at L_o than during contraction at 0.7 L_o. This was true of both the phasic portion of the Ca^{2+} response which occurred during force development, and of the sustained portion of the Ca^{2+} response which occurred during the steady state phase of the contraction. The responses of muscles contracted with 60 mM KCl were similar to those obtained with ACh, in that both active stress and aequorin luminescence were lower at 0.7 L_o than at L_o. When a second contraction at L_o was performed with either of these stimuli immediately following the contraction at 0.7 L_o, the aequorin luminescence was again higher than that observed at 0.7 L_o. Therefore the decrease in the light signal shown in Figures 2 and 3 could not be attributed to the consumption of aequorin. When the length of an uncontracted muscle was rapidly reduced to a shorter length, in no instance was a change in the light signal observed, even when the length change was extremely large (50% L_o).

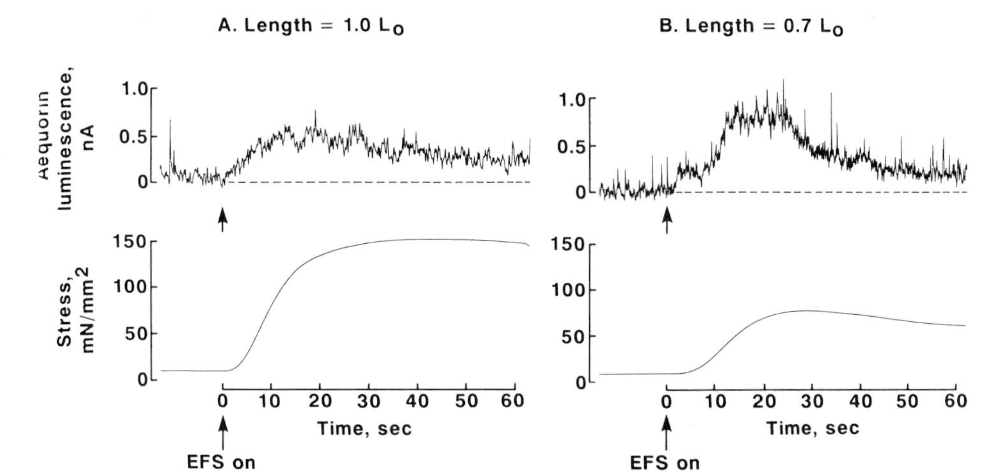

Figure 2. Aequorin luminescence and active stress during electrical field stimulation with a 12 volt, 15 pps, 0.5 msec duration square wave stimulus. The aequorin light signal was higher during isometric contraction at 0.7 L_o than during isometric contraction at L_o during sequential contractions performed in a single tissue.

The mean results of these studies are summarized in Table 1. When isometric contractions were performed at 0.7 L_o, the reduction in active stress for all three stimuli was similar; to about 25 - 35% of the active stress at L_o. When either ACh or 60 mM KCl was used as the stimulus, the average aequorin luminescence was also lower at 0.7 L_o than at L_o. However, when EFS was used as the stimulus, aequorin luminescence was higher at 0.7 L_o than at L_o. The rate of force development was also significantly different for EFS than for ACh and K^+-depolarization. When the muscles were stimulated by EFS, the peak rate of active stress development was approximately twice that obtained during stimulation with ACh or KCl.

DISCUSSION

The results of these experiments suggest that in canine tracheal smooth muscle, the effect of muscle length on aequorin luminescence during isometric contraction depends on the contractile stimulus.

When EFS was used to contract the muscles, aequorin luminescence was higher during isometric contraction at a shorter muscle length even though active stress was markedly decreased. We previously performed these experiments using a 60 Hz ac electrical stimulus, which may have had direct effects on the muscle membrane (Gunst, 1989). However, we obtained similar results in the present experiments in which we used EFS selective for neural stimulation. The latter stimulus has been shown to cause contraction in canine trachealis muscle strips entirely by activating intrinsic nerves (Russell, 1980). We found that contractions to this stimulus could be completely abolished by 10^{-6} M atropine. The fact that the results using neural stimulation were similar to those using a less selective electrical stimulus suggests that our previous results cannot be attributed to length-dependent changes in the effects of the electrical stimulus on the smooth muscle membrane.

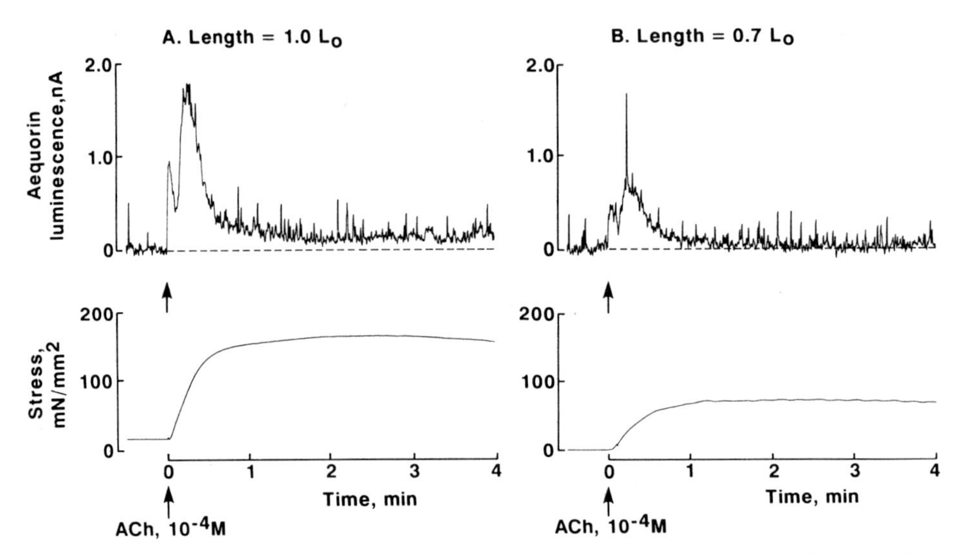

Figure 3. Aequorin luminescence and active stress during sequential isometric contractions with ACh in a single muscle strip. Both active stress and aequorin luminescence were lower when the muscle was contracted at 0.7 L_o than during contraction at L_o.

We have previously suggested that the apparent increase in $[Ca^{2+}]_i$ which we observed with decreasing muscle length and active stress during EFS might be explained by a length-dependent decrease in the Ca^{2+}-sensitivity of regulatory or contractile proteins in smooth muscle, as has been proposed for striated muscles (Allen and Blinks, 1978; Gordon and Ridgway, 1978; Allen and Kurihara, 1982; Housmans et al., 1983; Gunst, 1989). If this were true, when muscle length was reduced, less of the activator Ca^{2+} would bind to contractile or regulatory proteins thereby resulting in higher levels of free Ca^{2+} in the cytoplasm. However, this hypothesis is difficult to reconcile with the observations we made using ACh or 60 mM KCl as the contractile stimulus. When these agents were used to contract the muscles, both aequorin luminescence and active stress were lower when the length of isometric contraction was decreased.

One possible explanation for our observations is that the different results obtained with different stimuli reflect differences in the effects of muscle length on different pools of activator Ca^{2+}. However, because the excitatory innervation of the canine trachealis muscle is overwhelmingly cholinergic, both EFS and ACh act primarily by the activation of cholinergic receptors. Therefore, one would expect the sources of activator Ca^{2+} used by these two stimuli to be the same. However, the pattern of the Ca^{2+} transients evoked by the two stimuli is different (see Figs. 2 and 3), which does suggest differences in the mechanisms by which these two stimuli activate the muscle. One possibility is that EFS acts to stimulate other nerves or tissue components which are not activated by ACh, and that the activation of these components alters either the source of activator Ca^{2+} or contractile protein sensitivity to Ca^{2+}. However, we also observed that aequorin luminescence decreased during contraction at a shorter muscle length when either ACh or KCl was used as the contractile stimulus, yet these two stimuli activate the muscle by very different mechanisms.

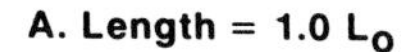

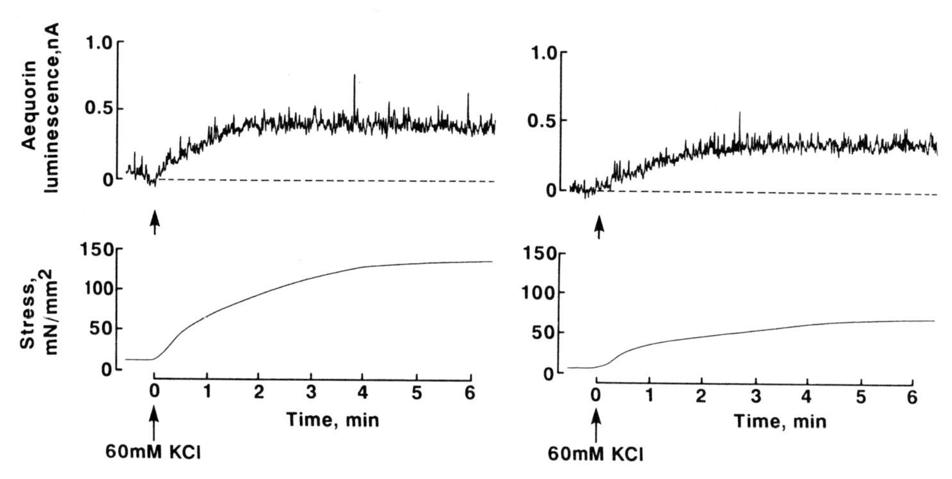

Figure 4. Aequorin luminescence and active stress during sequential isometric contractions with 60 mM KCl in a single muscle strip. Both active stress and aequorin luminescence were lower when the muscle was contracted at 0.7 L_o than during contraction at L_o.

We also considered the possibility that the increase in aequorin luminescence observed during contractions at a shorter muscle length was the result of an artifact caused by a change in the position of the muscle in the light gathering apparatus, or by the differences in the thickness of the muscle at different lengths. An increase in muscle thickness with shortening could conceivably have a quenching effect on light emission from the interior of the muscle. However, when a length change was imposed on a resting muscle, no effect on aequorin luminescence was observed, even when the magnitude of the length change was as large as 50% L_o (Gunst, 1989). In addition, that fact that a change in muscle length had opposite effects on aequorin luminescence depending on the stimulus used for contraction makes either of these explanations seem unlikely.

An alternative consideration was that changing the position of the muscle in the electric field during EFS altered the strength of the electrical stimulus reaching the tissue, perhaps resulting in an increase in the activation of the muscle at the shorter length. However, the occurrence of this type of artifact should be quite variable depending on the size and thickness of each particular muscle strip and the way in which it was positioned in the apparatus. We found the effects of muscle length on aequorin luminescence in tissues stimulated using EFS to be quite consistent among all muscles.

One striking difference between the effects of ACh and K^+-depolarization versus the effects of EFS was the rate of force development. The peak rate of active stress development (dS/dt) in muscles contracted using the exogenously added stimuli was only half that observed when EFS was the stimulus. This is probably because EFS can activate all cells in the tissue simultaneously and homogeneously, whereas the activation of cells by either ACh or K^+-depolarization depends on their respective rates of diffusion to the site of action. It is possible that the rate of activation is an important factor in determining the Ca^{2+}-sensitivity of contractile or regulatory proteins, or in determining the saturation of Ca^{2+} binding to specific proteins involved in the activation process, and that a more rapid activation of the muscle results in a higher level of intracellular free Ca^{2+} than a slower rate of activation.

ACKNOWLEDGMENTS

This work was supported by USPHS grant HL29289.

REFERENCES

Allen, D. G. and Blinks, J. R., 1978, Calcium transients in aequorin-injected frog cardiac muscle, *Nature,* 273: 509.

Allen, D. G. and Kurihara, S., 1982, The effects of muscle length on intracellular calcium transients in mammalian cardiac muscle, *J. Physiol.,* 327: 79.

Blinks, J. R., 1984, Methods for monitoring Ca^{2+} concentrations with photoproteins in living cardiac cells, *in:* "Methods for Studying Heart Membranes, Vol 2.", N. S. Dhalla, ed., CRC Press, Boca Raton, p. 237.

Blinks, J. R., Wier, W. G., Hess, P., and Prendergast, F. G., 1982, Measurement of Ca^{2+} concentrations in living cells, *Prog. Biophys. Molec. Biol.,* 40: 1.

Bodem, R. and Sonnenblick, E. H., 1974, Deactivation of contraction by quick release in the isolated papillary muscle of the cat. Effects of lever damping, caffeine, and tetanization, *Circ. Res.,* 34: 214.

Briden, K. L. and Alpert, N. R., 1972, The effect of shortening on the time-course of active state decay, *J. Gen. Physiol.,* 60: 202.

Edman, K. A. P. and Kiessling, A., 1971, The time course of the active state in relationship to sarcomere length and movement studied in single skeletal muscle fibres of the frog, *Acta Physiol. Scand.,* 81: 182.

Ekelund, M. C. and Edman, K. A. P., Shortening induced deactivation of skinned fibers of frog and mouse striated muscle, *Acta Physiol. Scand.,* 116: 189.

Endo, M., 1973, Length dependence of activation of skinned muscle fibers by calcium, *Symp. Quant. Biol.,* 37: 505.

Gordon, A. M., Huxley, A. F., and Julian, F. J., 1966, The variation in isometric tension with sarcomere length in vertebrate muscle fibers, *J. Physiol.,* 184: 170.

Gordon, A. M. and Ridgway, E. B., 1978, Calcium transients and relaxation in single muscle fibers, *Eur. J. Cardiol.,* 7: 27.

Gunst, S. J., 1986, Effect of length history on contractile behavior of canine tracheal smooth muscle, *Am. J. Physiol.,* 250: C146.

Gunst, S. J., 1989, Effects of muscle length and load on intracellular Ca^{2+} in tracheal smooth muscle, *Am. J. Physiol.,* 256: C807.

Housmans, P. R., Lee, N. K. M., Blinks, J. R., 1983, Active shortening retards the decline of the intracellular calcium transient in mammalian heart muscle, *Science,* 221: 159.

Jewell, B.R., 1977, A reexamination of the influence of muscle length on myocardial performance, *Circ. Res.,* 40: 221.

Kromer, U. and Stephens, N. L., 1983, Airway smooth muscle mechanics: reduced activation and relaxation, *J. Appl. Physiol.,* 54: 345.

Morgan, J. P. and Morgan, K. G., 1984, Alteration of cytoplasmic ionized calcium levels in smooth muscle by vasodilators in the ferret, *J. Physiol.*, 357: 539.

Russell, J. A., 1980, Innervation of airway smooth muscle in the dog, *Bull. Eur. Physiopath. Resp.*, 16: 671.

Stephenson, D. G. and Wendt, I. R., 1984, Length dependence of changes in sarcoplasmic calcium concentration and myofibrillar calcium sensitivity in striated muscle fibres, *J. Musc. Res. Cell Motil.*, 5: 243.

Taylor, S. R and Rudel, R., 1970, Striated muscle fibers: Inactivation of contraction induced by shortening, *Science*, 167: 882.

ISOTONIC SHORTENING PARAMETERS BUT NOT ISOMETRIC FORCE DEVELOPMENT ARE ALTERED IN RAGWEED POLLEN SENSITIZED CANINE BRONCHIAL SMOOTH MUSCLE

He Jiang, Kang Rao, Andrew J. Halayko, Wayne Kepron, and Newman L. Stephens

Department of Physiology
University of Manitoba
Winnipeg, Manitoba R3E 0W3 Canada

INTRODUCTION

Studies on airway smooth muscle can serve two major purposes. First, the elucidation of basic mechanisms and properties of smooth muscle contraction and its regulation can be obtained. The trachealis, for example, provides a plentiful source of relatively pure smooth muscle tissue for studies at biochemical and molecular levels. In addition, both tracheal and bronchial smooth muscles furnish us with an optimal preparation for studying mechanical properties because their respective muscle fibers are oriented in a parallel fashion. The second purpose in studying airway smooth muscle is in characterizing the pathophysiology of asthma with hopes of advancing disease management. On this tack, we have a canine model of the disease (Kepron et al., 1977), in which dogs are immunized from birth with ragweed pollen. These animals demonstrate a generalized sensitization of their smooth muscles to the pollen antigen (Antonissen et al., 1979; Antonissen et al., 1980; Kong and Stephens, 1983; Wang et al., 1990). Concomitantly, they produce high IgE anti-ragweed antibody titers and show marked increases in airflow resistance upon aerosolized, specific antigen bronchoprovocation (Becker et al., 1989).

Isotonic parameters, we believe, are the most essential indices for smooth muscle properties in hollow organs such as the airways and blood vessels. At these sites the smooth muscle component regulates the diameter of the tubes, and the resistance to flow inside, by shortening or elongating. Functionally significant alterations may be undetected if only isometric parameters are recorded. Tracheal preparations are most commonly utilized in airway smooth muscle mechanical studies, however, if the intention is to understand the bases of acute asthmatic response, as it is in our laboratory, research focus needs to be on the central bronchi, the site at which acute asthmatic response may develop (Epstein et al., 1948; Dulfano and Hewetson, 1966). One of the major problems in the study of the smooth muscle from bronchi is that it is difficult to obtain a preparation without cartilage while maintaining smooth muscle function. However, if cartilages and connective tissues are not removed valid length-tension and force-velocity measurements specifically describing smooth muscle dynamics cannot be obtained because of viscous, inertial, and elastic interferences of the cartilages. Histological analysis of the bronchi has revealed that muscle is not attached directly to cartilage in the bronchi and we have recently succeeded in removing cartilage plates from canine bronchial tissue without compromising muscle performance (Jiang and Stephens, 1990).

METHODS

As described before (Kepron at al. 1977), newborn mongrel dogs were sensitized by intraperitoneal injection of 500 μg of ragweed pollen extract in 30 mg of an aluminum hydroxide adjuvant within 24 hours of birth. Booster injections were given weekly for 8 weeks and biweekly thereafter to 16 weeks. Randomly selected littermates of these sensitized dogs received only injections of the adjuvant at the same time. Sensitization to ragweed was determined by the homologous passive cutaneous anaphylaxis test. It has been shown (Kepron et al. 1977) that all sensitized littermates with passive cutaneous anaphylaxis titers greater than 1:64 developed marked increases in specific airway resistance on bronchoprovocation with ragweed extract aerosol when compared with controls similarly challenged.

Bronchial preparations were obtained from lungs immediately after being dissected from the dogs. Non-branching 5th generation bronchi were used since the orientation of muscle fibers is mostly parallel. The connective tissues and cartilage plates were removed under a dissecting microscope (Jiang and Stephens, 1990). During the dissection, muscle strips were submersed in Krebs-Henseleit solution aerated with an 95% O_2/5% CO_2 mixture at 0° C. The solution was composed of (in mM): NaCl 115; $NaHCO_3$ 25; NaH_2PO_4 1.38; KCl 2.51; $MgSO_4$ 2.46; $CaCl_2$ 1.91; and dextrose 5.56. The muscle strip was then anchored by a clamp at the bottom of the organ bath and the upper end was tied to an electromagnetic lever system, which measures isometric force production and isotonic shortening. The muscle strips were equilibrated for approximately 2 hours in muscle baths maintained at a PO_2 of 600 Torr, a PCO_2 of 40 Torr and a pH of 7.40 at a temperature of 37° C.

Isometric forces were recorded at different muscle lengths in order to delineate length-tension relationships and to obtain maximum force production (P_o). Tensions were expressed in N/m^2 of smooth muscle tissue in the total cross-sectional area of the strips (Jiang and Stephens, 1990). The cross-sectional areas of the muscle strips were computed from images captured by a video camera system. The muscle tissue proportions were obtained by histological analysis of the sections from same strips. Comparisons were made between sensitized and control groups.

Isotonic shortening velocities were obtained by the lever system which was originally developed by Brutsaert (Brutsaert et al. 1971) for cardiac muscle and was adapted by us for use with smooth muscle (Mitchell and Stephens, 1983). The compliance of the lever system was 0.2 μm/mN, and the total equivalent moving mass was 225 mg. Because the thread connecting the muscle strip to the lever possessed compliance, the later was measured for correction of the L-T curves. The average compliance of the thread was 4.2 μm/mN. The time resolution of the lever system was 2 ms, and the sampling rate used was 200 Hz.

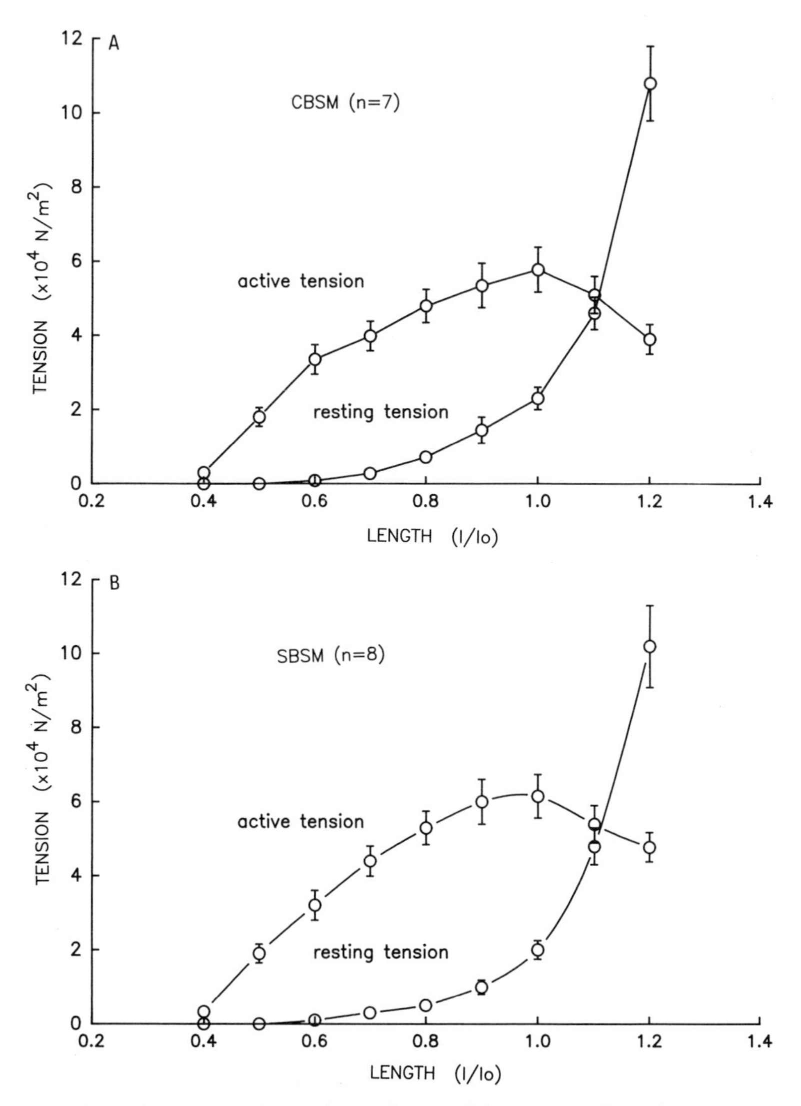

Figure 1. Length-tension relationships of control (upper panel) and sensitized (lower panel) canine bronchial smooth muscle.

The shortening velocities under given loads at given time points (2 and 8 sec) were obtained by applying quick releases from isometric contraction. From the data obtained, the F-V relationship was established and zero load shortening velocities (V_o) for control and sensitized groups were extrapolated by fitting the data with the hyperbolic Hill equation, which states that: $(P+a)(V+b) = (P_o+a)b$ (Hill, 1938). The coefficient of determination, r^2, was computed and used to determine goodness of fit.

To study the behavior of V_o as a function of time, quick releases were employed at 1 sec intervals from 2 to 8 sec in the course of a 10 sec contraction of the muscle. The measurements of velocity were made 200 msec after the onset of quick release of the isometric contraction to isotonic contraction. The maximum shortening capacities (Δl_{max}) at 2 and 8 sec of contraction were obtained directly from the isotonic shortening recordings and normalized to the optimal muscle lengths (l_o). The means of the measurements obtained

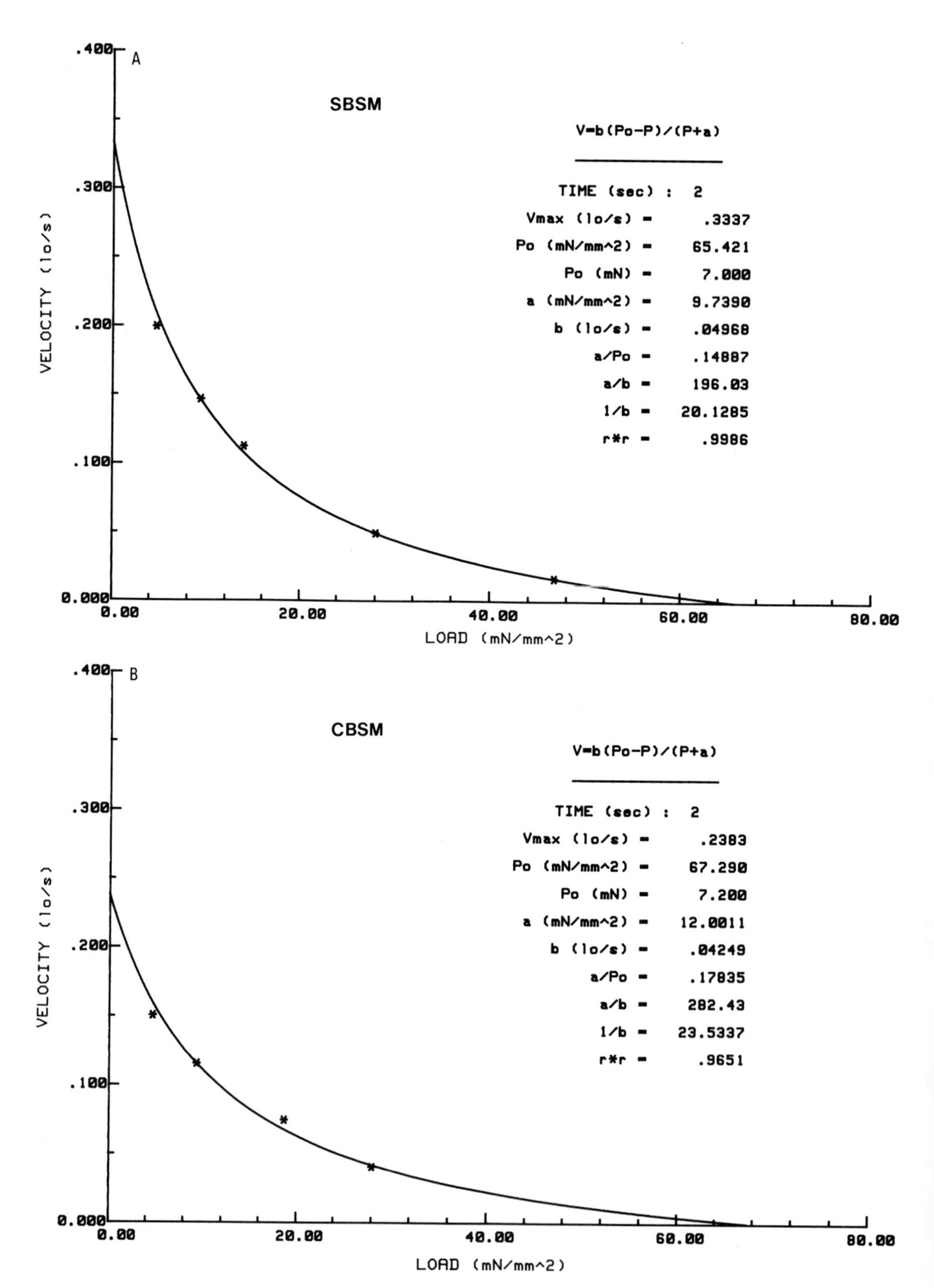

Figure 2. Force-velocity curves from sensitized and control bronchial smooth muscles.

from sensitized and control bronchial smooth muscle were compared using an unpaired, 2 tailed t-test.

RESULTS

Tensions elicited by supramaximal electrical stimulation are plotted against their lengths (Figure 1). The maximal isometric force production (P_o) for sensitized canine bronchial smooth muscle was not significantly different from that of the control ($6.15 \pm 0.58 \times 10^4$ N/m^2 for sensitized group and 5.78 ± 0.7 for control, $p > 0.05$). The slopes of the resting curves, which represent the passive stiffness of the muscle strip, were also similar indicating that passive characteristics of the two groups were not different.

The typical force-velocity curves at 2 sec from sensitized (upper panel) and control (lower panel) bronchial smooth muscles are shown in Figure 2. The mean value of V_o at 2 sec for the sensitized group was significantly higher than for controls ($p < 0.05$).

Figure 3 depicts V_o estimated from sensitized and control bronchial smooth muscles. The sensitized group had a significantly higher V_o at the 2 sec point compared to control. The estimates of V_o at other time points were equal in sensitized and control tissues.

The maximum shortening capacities are shown in Figure 4. The sensitized group shortened significantly more (0.7 ± 0.08 l/l_o) than controls (0.54 ± 0.06). The shortening of sensitized muscle within the first two seconds was also greater (0.53 ± 0.04 l/l_o) than that of the control (0.26 ± 0.03). That is to say, within the first 2 seconds, 76% of the total shortening had been completed in sensitized bronchial smooth muscle while only 48% was finished in the control group.

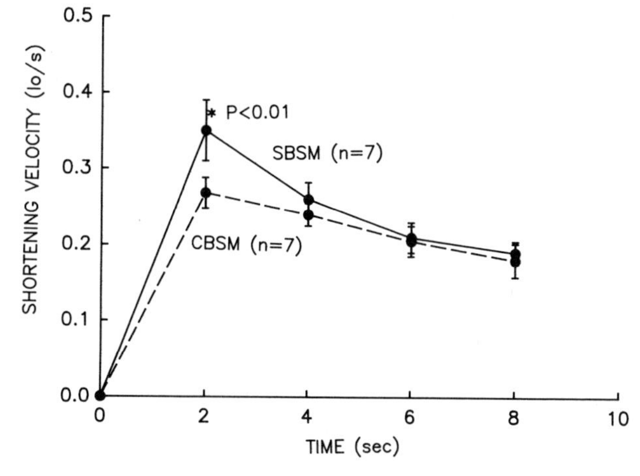

Figure 3. Maximal isotonic shortening velocities (V_o) obtained from quick releasing methods at different points in the course of contraction. Only V_o at 2 sec is significantly higher in sensitized bronchial smooth muscle (SBSM) than in control (CBSM).

DISCUSSION

It is rather surprising that most smooth muscle studies are using isometric force development as a functional index or as an index of drug effect. Isometric properties of the smooth muscle from hollow organs relate to the stiffness of the vessel wall; the role in regulating resistance is minor. The resistance to flow in a tube depends on its diameter. An increase of 29.6% in maximum shortening capacity in sensitized bronchial smooth muscle (Figure 4) is of significance with respect to bronchoconstriction because computations, using Poiseuille's equation, indicate that this increase in shortening would translate into a 129.9% increase in air flow resistance in the sensitized model (assuming all other variables held constant). Hence, flow regulation in airways or blood vessels, which are directly controlled by shortening or elongating rather than stiffening of smooth muscle, is best studied *in vitro* by evaluation of isotonic parameters such as the force-velocity relationship and shortening capacity. The justification of using isometric force production as the only functional or pharmacological index of muscle should, therefore, be questioned.

We have shown in this study that the dynamic isotonic properties of sensitized bronchial smooth muscle contraction have been altered while the force generated during isometric contraction remains unchanged. Smooth muscle contraction, unlike striated muscle, seems to be subserved by two types of crossbridges which are recruited sequentially (Siegman et al., 1977; Dillon et al., 1981). Early in a contraction, normally cycling crossbridges are activated and slowly cycling crossbridges or "latch bridges" are activated later in a contraction. We see a change in the shortening velocity of sensitized bronchial smooth muscle at 2 sec., when normally cycling crossbridges predominate, but not at 8 sec., when latch-bridges are operative, suggesting that the behavior of normally cycling crossbridges is being affected during the response to ragweed sensitization. The implication of this observation is that there are changes in some factor(s) regulating the rate of normally cycling crossbridges. Latchbridges, however, are unaltered in sensitized bronchial smooth muscle. The latter point is also illustrated by the fact that no differences exist in the force generating ability of sensitized bronchial smooth muscle compared to controls.

Actomyosin ATPase activity (i.e., the crossbridge cycling rate) can be indirectly estimated by measuring the maximum shortening velocity under zero load (V_0; Bárány, 1967). Increased V_0 in sensitized airway smooth muscle has been shown by us herewith (Figure 3) and previously (Antonissen et al., 1979; Stephens et al., 1988). The measurement of myofibrillar ATPase activity has revealed the activity to be increased in sensitized pulmonary arteries (Kong et al., 1986), and in tracheal and bronchial smooth muscles from sensitized dogs (Rao et al., this volume). The role of increased ATPase activity and increased velocity of normally cycling crossbridges in the pathophysiology of the acute asthmatic response poses an interesting scenario. During the acute asthmatic attack the central airways narrow excessively. In sensitized dogs, we have confirmed that central bronchial smooth muscle is able to shorten more than the control muscle and that this difference is manifest in the first 2 seconds of contraction. Hence, the gross steps leading to the acute increased resistance in the central bronchi of antigen challenged, sensitized dogs (Becker et al., 1989) could be: elevated myofibrillar ATPase activity => increased rate of normally cycling crossbridges => increased initial V_0 (measured herein at 2 sec) => greater degree of shortening of sensitized bronchial smooth muscle => elevated

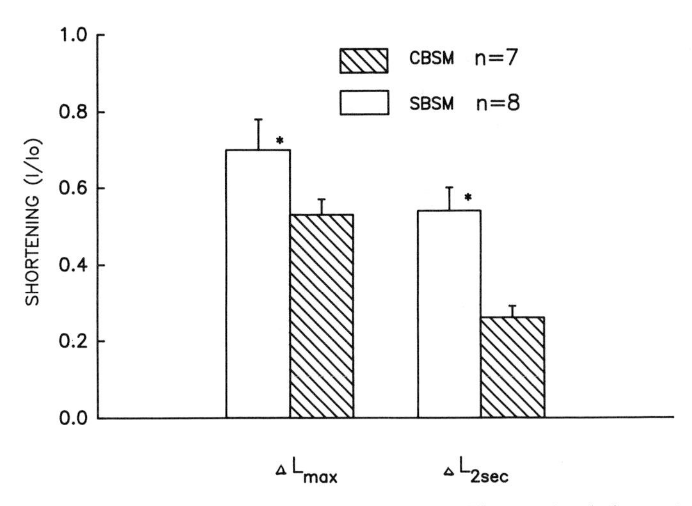

Figure 4. Isotonic shortenings under resting tensions. The maximal shortening (Δl_{max}) is significantly greater in sensitized bronchial smooth muscle (SBSM) than in control (CBSM). The shortening within the first 2 sec of contraction (Δl_{2sec}) in the sensitized group is also greater than in control. * = $p < 0.01$.

resistance to airflow in conductance airways. This sequence is, of course, merely speculative and by no means encompasses all possible factors.

An additional factor that could affect the magnitude of shortening is the compliance of the internal resistor. The evidence of it existing is forthright. If an inactivated muscle, stretched to beyond l_o, is released, it returns to l_o. If stimulation is withdrawn from a maximally shortened muscle, the muscle re-elongates to its original length. In the shortening phase, the resistor is compressed and stores potential energy. When the stimulus is removed the resistor re-expands and restores the muscle to its original length. The internal resistor (or parallel elastic component) of the sensitized tracheal smooth muscle is more compliant than that of the control (Stephens et al., 1988). This could be a factor contributing to the increased shortening of sensitized airway smooth muscle. The basic process for increased compliance of the internal resistor in sensitized muscle is unidentified. Alterations of characteristics of collagen, elastin, and other structures in the extracellular and extra-fascicular spaces are strong contenders. In the present study, however, the passive properties (resting tensions) from sensitized and control groups were similar, which could rule out the idea that changes in connective tissues are responsible. It has been reported that low doses of ionizing radiation of skinned skeletal muscle cells induced changes in two cytoskeletal proteins, nebulin and titin, resulting in decreased passive and active tensions in response in calcium (Horwits et al., 1986). Although titin and nebulin have not been found in smooth muscle, there are a large number of other smaller molecular weight cytoskeletal proteins present in smooth muscle such as filamin, desmin, vimentin, vinculin, plectin, actinin, and synemin (Rasmussen et al., 1987). All of these could contribute to the structure of the internal resistor.

CONCLUSION

Isometric force development has been shown to be insensitive in indicating changes in dynamic mechanical properties in smooth muscle. One might be able to study the properties of latchbridges by measuring isometric force production, however, only isotonic shortening can be an index of the properties of early normally cycling crossbridges in smooth muscle contraction.

The increased isotonic shortening parameters in airway smooth muscle from our ragweed sensitized dogs may account for the hyperresponsiveness which results in allergic bronchoconstriction. The excessive shortening of the airway smooth muscle to the extent observed in the present study is well enough to cause a severe airway obstruction commonly seen in an acute asthmatic attack.

ACKNOWLEDGMENTS

This project was supported by operating grants from the Council for Tobacco Research, USA and Medical Research Council of Canada. H. Jiang is the recipient of a fellowship from Medical Research Council of Canada.

REFERENCES

Antonissen, L. A., Mitchell, R. W., Kroeger, E. A., Kepron, W., Tse, K. S., and Stephens, N. L., 1979, Mechanical alterations of airway smooth muscle in a canine asthmatic model, *J. Appl. Physiol.*, 46: 681.

Antonissen, L. A., Mitchell, R. W., Kroeger, E. A., Kepron, W., Tse, K. S., Stephens, N. L., and Bergen, J., 1980, Histamine pharmacology in airway smooth muscle from a canine model of asthma, *J. Pharmacol. Exp. Ther.*, 213: 150.

Bárány, M, 1967, ATPase activity of myosin correlated with speed of muscle shortening, *J. Gen. Physiol.*, 50: 197.

Becker, A. B., Hershkovich, J., Simons, F. E. R., Simons, K. J., Lilley, M. K., and Kepron. W., 1989, Development of chronic airway hyperresponsiveness in ragweed-sensitized dogs, *J. Appl. Physiol.*, 66: 2691.

Brutsaert, D. L., Claes, V. A., and Goethals, M. A., 1971, Velocity of shortening of unloaded heart muscle and the length-tension relation, *Circ. Res.*, 29: 63.

Dillon, P. F., Aksoy, M. O., Driska, S. P., and Murphy, R. A., 1981, Myosin phosphorylation and the cross-bridge cycle in arterial smooth muscle, *Science*, 211: 495.

Dulfano, M. J. and Hewetson, J., 1966, Radiologic contributions to the nosology of obstructive lung disease entities, *Dis. Chest*, 50: 270.

Epstein, B. S., Sherman, J., and Walzer, E. E., 1948, Bronchography in asthmatic patients, with the aid of adrenalin, *Radiology*, 50: 96.

Hill, A. V., 1938, The heat of shortening and the dynamic constants of muscle, *Proc. R. Soc. Lond. B*, 126: 136.

Horwits, R., Kempner, E. S., Bishor, M. E., and Podolsky, R. J., 1986, A physiological role for titin and nebulin in skeletal muscle, *Nature*, 323: 160.

Jiang, H. and Stephens, N. L., 1990, Contractile properties of bronchial smooth muscle with and without cartilage, *J. Appl. Physiol.*, 69: 120.

Kepron, W., James, J. M., Kirk, B., Sehon, A. H., and Tse, K. S., 1977, A canine model for reaginic hypersensitivity and allergic bronchoconstriction, *J. Allergy Clin. Immunol.*, 59: 64.
Kong, S. K. and Stephens, N. L., 1983, Mechanical properties of pulmonary arteries from sensitized dogs, *J. Appl. Physiol.*, 55: 1669.
Kong, S. K., Shiu, R. P. C., and Stephens, N. L., 1986, Studies of myofibrillar ATPase in ragweed-sensitized canine pulmonary smooth muscle, *J. Appl. Physiol.*, 60: 92.
Mitchell, R. W. and Stephens, N. L., 1983, Maximum shortening velocity of smooth muscle: zero load-clamp vs afterloaded method, *J. Appl. Physiol.*, 55: 1630.
Rasmussen, H., Takuwa, Y., and Park, S., 1987, Protein kinase C in the regulation of smooth muscle contraction, *FASEB J.*, 1: 177.
Siegman, M., Butler, T. M., Mooers, S. U., and Davies, R., 1977, Mechanical and energetic correlates of isometric relaxation, *in:* "Excitation-Contraction Coupling in Smooth Muscle", R. Casteels, T. Godfraind, and J. C. Rüegg, eds., Elsevier/North Holland, New York, p. 449.
Stephens, N. L., Kong, S. K., and Seow, C. Y., 1988, Mechanisms of increased shortening of sensitized airway smooth muscle, *in:* "Mechanisms of Asthma: Pharmacology, Physiology, and Management", C. L. Armour, J. L. Black, eds., Alan R. Liss, New York, p. 231.
Wang, Z., Seow, C. Y., Kepron, W., and Stephens, N. L., 1990, Mechanical alterations in sensitized canine saphenous vein, *J. Appl. Physiol.*, 69: 171.

FREQUENCY ANALYSIS OF SKINNED INDIRECT FLIGHT MUSCLE FROM A MYOSIN LIGHT CHAIN 2 DEFICIENT MUTANT OF *DROSOPHILA MELANOGASTER* WITH A REDUCED WING BEAT FREQUENCY

Mineo Yamakawa,[1] Jeffrey Warmke,[2] Scott Falkenthal,[2]
and David Maughan[1]

[1]Department of Physiology and Biophysics
University of Vermont
Burlington, VT 05405

[2]Department of Molecular Genetics
Ohio State University
Columbus, OH 43210

INTRODUCTION

Muscle contraction is regulated by proteins associated with either the thick (myosin-containing) or thin (actin-containing) filaments. The relative importance of each set of regulatory proteins varies between muscle types (Adelstein and Eisenberg, 1980). One type of regulation, typical of smooth and non-muscle cells, involves phosphorylation of the regulatory light chain on myosin, while the other, typical of vertebrate striated muscle, involves Ca^{2+} binding to the troponin on actin.

Asynchronous flight muscle of insects has aspects of both thick and thin filament regulation (Lehman et al., 1974). Calcium binding to troponin is a prerequisite for contraction, while phosphorylation of the regulatory myosin light chain (MLC-2) enhances contraction (Persechini et al., 1985; Yamakawa et al., 1990a). It is possible that MLC-2 phosphorylation may be required to maintain oscillatory contractions which drive the wings to beat.

In *Drosophila melanogaster* a lack of phosphorylation of MLC-2 in the asynchronous indirect flight muscle (IFM) impairs flight (Takahashi et al., 1990b). Impaired flight also results from a deficiency of MLC-2 (Warmke, 1990). As a starting point for studying the role of MLC-2 in muscle contraction and regulation, we have begun to examine the mechanical properties of individual muscle cells from the MLC-2 deficient mutant *M1c2*^{38} (*E38*) (Yamakawa et al., 1990b; Warmke, et al., manuscript in preparation). We found that in *E38* the flight impairment is associated with a reduced wing beat frequency. In asynchronous flight muscle, the oscillatory contraction that drives the wing beat does not depend on continuous nervous stimulation but does depend on stretch activation, i.e., the transient tension rise associated with muscle stretch (Jewell and Rüegg, 1966; Pringle, 1978).

Stretch activation, common to many types of muscle, can be studied by frequency analysis, i.e. by using sinusoidal length perturbations (Kawai and Brandt, 1980; Thorson and White, 1983; Warmke et al., manuscript in preparation). Frequency analysis of *Drosophila* IFM enables us to correlate parameters of single fiber mechanics with the frequency of the wing beat. In this preliminary study, we have initiated frequency analysis of IFM using pseudo-random white noise length perturbations (PRWN) (Halpern and Alpert, 1971; Rossmanith, 1986; Calancie and Stein, 1987). PRWN allows us to focus in detail on that portion of the frequency response (process b: Kawai and Brandt, 1980) that relates most to the rate constant of the stretch activation. Our results indicate that the rate of process b in *E38* is slower than that in wild type, suggesting that the reduced rate of stretch activation is partly responsible for the reduced wing beat frequency.

METHODS

Single fibers of dorsal longitudinal indirect flight muscle were isolated from *Drosophila melanogaster*. Immediately after the isolation, fibers were incubated for 3 hrs (8° C) in an extraction solution which solubilizes surface membranes and organelles (50% glycerol v/v, 20 mM phosphate buffer, 1 mM sodium azide, 1 mM DTT, 2 mM $MgCl_2$, and 0.5% Triton X-100 w/v). Skinned fibers were transferred to a temperature controlled trough, and then attached to a force transducer (sensitivity, 0.6 μN/mV; natural frequency, 4 kHz) and a length controller via aluminum T-clips. Fiber length was monitored and controlled by a servo-motor (> 2.0 kHz) interfaced with a computer. Both length and force transients were recorded on video tape with a digital data recorder for later analysis. Fiber dimensions were measured using an inverted compound microscope (fiber length 160 - 350 μm; cross-sectional area 0.006 - 0.025 mm^2).

Skinned fibers were relaxed (pCa 9) and activated (pCa 4) with 18 mM MgATP, 1 mM free Mg^{2+}, 5 mM EGTA, 20 mM BES buffer (pH 7.0), 150 mM ionic strength (adjusted with K methanesulfonate). Activating solution contained 10 μM calmodulin. All experiments were carried out at ~12° C to reduce the possibility of diffusion limitation of substrate within the activated fiber. Resting fiber length was set at "0% strain" (Lund et al., 1988).

We perturbed the fiber length by applying band-width limited PRWN (0.25% fiber length peak-to-peak, <1 μm). We found the band-limited PRWN analysis to be helpful in this study since its duration is short (1.6 sec) for a given resolution of frequency steps (0.5 Hz) and range of frequency steps (0.5-1.5 kHz). The short duration of the perturbation reduces errors due to fluctuations of the tension base line and possible effects of transient changes in phosphorylation level (Moore and Stull, 1984)[1]. Another benefit of the PRWN method is the ability to calculate the coherence function (Bendat and Piersol, 1986), which is a normalized measure of the linear output dependence on input. Nonlinearity, including any noise unrelated to the input signal, reduces the coherence.

The frequency response of the fiber (length-force transfer function) was calculated by dividing the cross-spectral density function of the length and force

---

[1] In preliminary experiments, we found that calmodulin dependent increases in isometric tension occurs within 1.5 min, and that its effect is removed by relaxing the fiber. This suggests the presence of endogenous myosin light chain kinase and phosphatase activity.

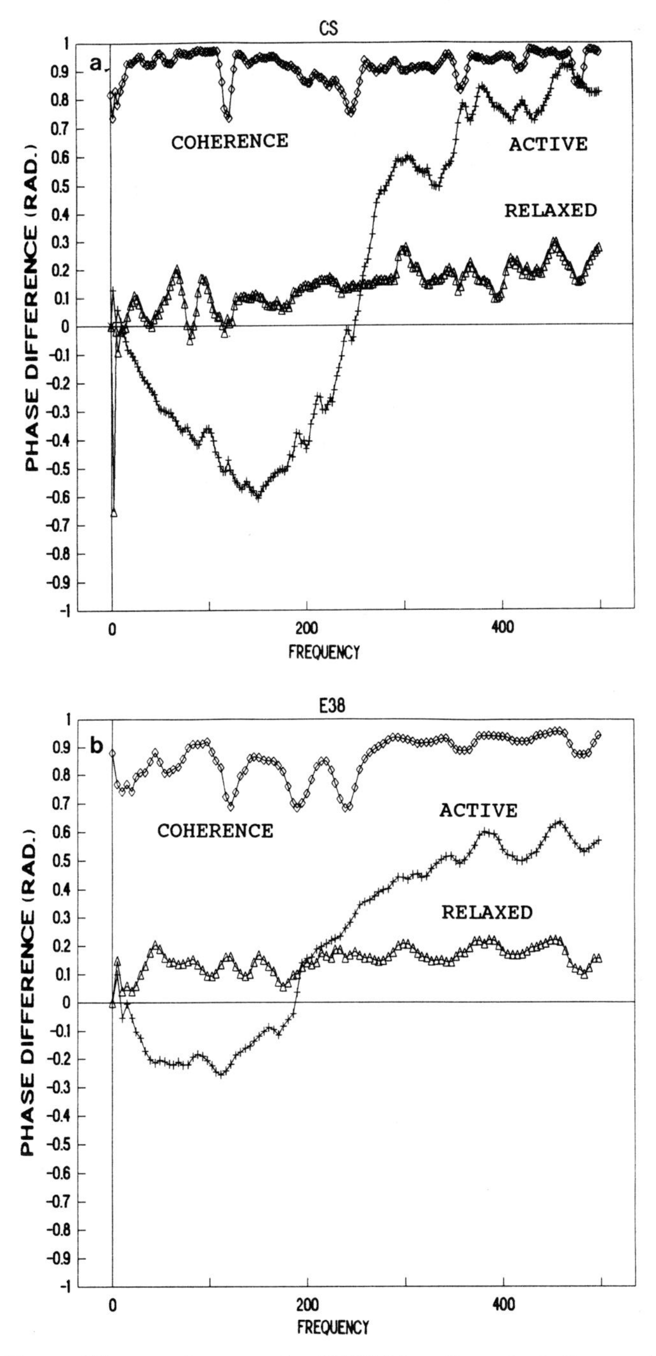

Figure 1. Phase difference between the PRWN length perturbations and the force response as a function of the frequency components. Results from a relaxed fiber (pCa 9) indicated by open triangles; from the same fiber activated (pCa 4), plus symbols. Only the frequencies between 0 and 500 Hz (the range of primary interest) are shown. The coherence plots (open diamonds) refer to the active fiber only. Note that phase difference minima are 140 - 150 Hz for *CS* and 50 - 120 Hz for *E38*.

transients by the power spectral density function of the length transient. The modulus of the length-force transfer function was used to calculate dynamic stiffness; the argument of the function gave the phase difference between force and length transients. The coherence function was calculated from the auto-spectral density and cross spectral density functions.

The wing beat of tethered flies was recorded by an optical tachometer (Unwin and Ellington, 1979) and the frequency components were extracted by a spectrum analysis routine similar to that described above.

RESULTS AND DISCUSSION

Frequency Response of Muscle Fibers from *CS* and *E38*

Figure 1 illustrates the phase difference between the length perturbation and the force response to that perturbation as a function of the frequency components of the PRWN signal. Results from both wild type, Canton-S (*CS*) (panel a) and *E38* (panel b) are given. These phase plots show the marked presence of a phase lag in active fibers from both *CS* and *E38*. There is, however, an overall increase in lower frequency components of the phase lag in *E38*, which corresponds to a slower rate of stretch activation in *E38* than in *CS*. The observed shift in minimum of phase lag from 140 - 150 Hz in *CS* to 50 - 120 Hz in *E38* and *CS*, an ~40% reduction in this case (comparison of results from three *E38* and five *CS* flies indicate a range of ~5 - 40%). This shift should be reflected in the reduction of rate of stretch activation in *E38* compared to *CS*.

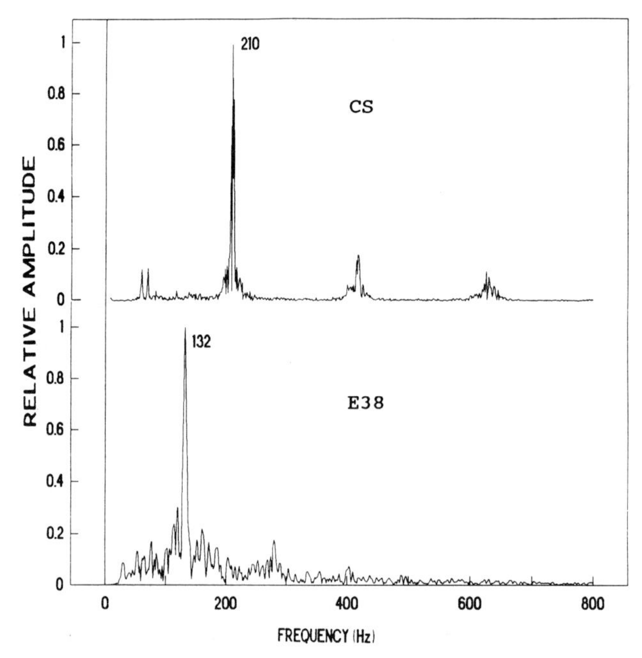

Figure 2. Wing beat frequency spectra of *CS* (upper panel) versus *E38* (lower panel). Frequencies between 7 Hz and 800 Hz are shown. Note that dominant wing beat frequencies are 210 Hz for *CS* and 132 Hz for *E38*.œ

Wing Beat Frequency of *CS* and *E38*

The wing beat frequency spectra (Fig. 2) exhibit a dominant peak and some other harmonics with smaller amplitudes. In the example shown, the dominant frequency in *CS* is 210 Hz and 132 Hz in *E38*. Comparison of results from five *CS* and four *E38* flies indicate a range in dominant frequencies of 198 - 214 Hz and 142 - 170 Hz, respectively, a reduction of ~25%. Rescued mutants (flight ability restored by P-element mediated insertion of a wild type MLC-2 gene into *E38;* Warmke et al., manuscript in preparation) show a similar range of wing beat frequencies to that of *CS* (Yamakawa and Maughan, manuscript in preparation).

Comparison of Frequency Response of the Muscle Fiber and Wing Beat Frequency

The similarity of reduction in wing beat frequency and muscle fiber frequency response suggests that the lower wing beat frequency in *E38* may be partly due to the slower rate of stretch activation in these IFM fibers. It is likely, however, that a combination of both reduced rate of stretch activation and ultrastructural disruption contributes to the flight impairment of *E38* (Yamakawa et al., 1990b).

ACKNOWLEDGMENTS

We thank Justin Molloy, Richard Schaaf and Janet Hurley for their help. This work was supported by NIH RO1 AR40234.

REFERENCES

Adelstein, R. S. and Eisenberg, E., 1980, Regulation and kinetics of the actin-myosin-ATP interaction, *Ann. Rev. Biochem.*, 49: 921.

Bendat, J. S. and Piersol, G., 1986, "Random data: analysis and measurement procedures", Wiley-Interscience, New York.

Calancie, B. and Stein, R., 1987, Measurement of rate constants for the contractile cycle of intact mammalian muscle fibers, *Biophys. J.*, 51: 149.

Halpern, W. and Alpert, N., 1971, A stochastic signal method for measuring dynamic mechanical properties of muscle, *J. Appl. Physiol.*, 31: 913.

Jewell, B. R. and Rüegg, J. C., 1966, Oscillatory contraction of insect fibrillar muscle after glycerol extraction, *Proc. R. Soc. Lond. B*, 164: 428.

Kawai M. and Brandt, P. W., 1980, Sinusoidal analysis: a high resolution method for correlating biochemical reactions with physiological processes in activated skeletal muscles of rabbit, frog, and crayfish, *J. Muscle Res. Cell Motil.*, 1: 279.

Lehman, W., Bullard, B., and Hammond, K., 1974, Calcium-dependent myosin from insect flight muscles, *J. Gen. Physiol.*, 63: 553.

Lund, J., Webb, M. R., and White, D. C. S., 1988, Changes in the ATPase activity of insect fibrillar flight muscle during sinusoidal length oscillation probed by phosphate-water oxygen exchange, *J. Biol. Chem.*, 263: 5505.

Moore, R. L. and Stull, J. T., 1984, Myosin light chain phosphorylation in fast and slow skeletal muscles in situ, *Am. J. Physiol.*, 247: C462.

Persechini, A., Stull, J. T., and Cooke, R., 1985, The effect of myosin phosphorylation on the contractile properties of skinned rabbit skeletal muscle fibers, *J. Biol. Chem.*, 260: 7951.

Pringle, J. W. S., 1978, Stretch activation of muscle: function and mechanism, *Proc. R. Soc. Lond. B*, 201: 107.

Rossmanith, G. H., 1986, Tension responses of muscle to n-step pseudorandom length reversals: a frequency domain representation, *J. Muscle Res. Cell Motil.*, 7: 299.

Stull, J. T., Bowman, B. F., Gallagher, P. J., Herring, B. P., Hsu, L., Kamm, K. E., Kubota, Y., Leachman, S. A., Sweeney, H. L., and Tansey, M. G., 1990, Myosin phosphorylation in smooth and skeletal muscles: regulation and function, *in:* "Frontiers in Smooth Muscle Research", N. Sperelakis, J. D. Wood, eds., Wiley-Liss, New York, p. 107.

Takahashi, S., Takano-Ohmuro, H., and Maruyama, K., 1990a Regulation of *Drosophila* myosin ATPase activity by phosphorylation of myosin light chains - I. wild-type fly, *Comp. Biochem. Physiol.*, 95B: 179.

Takahashi, S., Takano-Ohmuro, H., Maruyama, K., 1990b, Regulation of *Drosophila* myosin ATPase activity by phosphorylation of myosin light chains - II. Flightless mfδ_ fly, *Comp. Biochem. Physiol.*, 95B: 183.

Takano-Ohmuro H., Takahashi, S., Hirose, G. K. Maruyama, K., 1990, Phosphorylated and dephosphorylated myosin light chains of *Drosophila* fly and larva, *Comp. Biochem. Physiol.*, 95B: 171.

Thorson, J. and White, D. C. S., 1983, Role of cross-bridge distortion in the small-signal mechanical dynamics of insect and rabbit striated muscle, *J. Physiol.*, 343: 59.

Unwin, D. M. and Ellington, C. P., 1979, An optical tachometer for measurement of the wing-beat frequency of free-flying insects, *J. Exp. Biol.*, 82: 377.

Warmke, J., 1990, Genetic analysis of myosin light chain-2 function in *Drosophila melanogaster*, Ph.D. dissertation: The Ohio State University, Columbus OH.

Warmke, J. W., Kreuz, A. J., and Falkenthal, S., 1989, Co-localization of the *Drosophila melanogaster* myosin light chain-2 gene and a haplo-insufficient locus that affects flight behavior to chromosome band 99E1-3, *Genetics*, 122: 139.

Yamakawa, M., Molloy, J., Falkenthal, S., and Maughan, D, 1990a, pCa-tension curves and kinetics of stretch activation in skinned single fibers from *Drosophila melanogaster*, *Biophys. J.*, 57: 540a.

Yamakawa, M., Warmke, J., Falkenthal, S., and Maughan, D., 1990b, Properties of skinned muscle fibers from myosin light chain-2. Deficient flightless mutants of *Drosophila melanogaster*, *Biophys. J.*, 57: 411a.

CONTROL OF HCO_3-DEPENDENT EXCHANGERS BY CYCLIC NUCLEOTIDES IN VASCULAR SMOOTH MUSCLE CELLS

Robert W. Putnam, Phyllis B. Douglas, and Dianne Dewey

Department of Physiology and Biophysics
Wright State University School of Medicine
Dayton, OH 45435

INTRODUCTION

A variety of membrane transport systems responsible for the regulation of intracellular pH (pH_i) have been identified in smooth muscle (Aickin, 1986; Aalkjaer and Cragoe, 1988; Wray, 1988; Kikeri et al., 1990) and smooth muscle-like cells (Boyarsky et al.,1988a,b; Putnam,1990). These include the ubiquitous Na/H exchanger and at least two HCO_3-dependent transport systems (Fig. 1): i) a putative alkalinizing (Na + HCO_3)/Cl exchanger (although the role of Cl in this exchanger is still at issue) (Aickin and Brading, 1984; Aalkjaer and Mulvany, 1988); and ii) an acidifying Cl/HCO_3 exchanger. While these exchangers are important for determining steady state pH_i (Aalkjaer and Cragoe, 1988; Boyarsky et al., 1988a; Wray, 1988; Kikeri et al., 1990; Putnam and Grubbs, 1990), defending pH_i against acid/base disturbances (Aalkjaer and Cragoe, 1988; Boyarsky et al., 1988b; Putnam, 1990) and mediating cellular responses to external signals (Berk et al., 1987; Ganz et al., 1989), only the Na/H exchanger has been extensively studied in regard to the factors which regulate its activity. In fact, the regulation of the HCO_3-dependent transport systems is poorly studied in any cell.

In this study, we have initiated an investigation of the potential regulation of the HCO_3-dependent transport systems in smooth muscle cells by cyclic nucleotides (especially cAMP) using the adenylate cyclase activator forskolin and the phosphodiesterase inhibitor 3-isobutyl-1-methyl-xanthine (IBMX). Both rat aortic vascular smooth muscle (VSM) cells and aortic smooth muscle cells from spontaneously hypertensive rats (SHR) were used. We found that elevated cyclic nucleotide levels inhibited the (Na + HCO_3)/Cl exchanger in both types of cells, but stimulated the Na-independent Cl/HCO_3 exchanger in the VSM cells only.

MATERIALS AND METHODS

Cells were prepared using a modified version of the procedure of Harder and Sperelakis (1979). Briefly, aortae were removed from six Sprague-Dawley

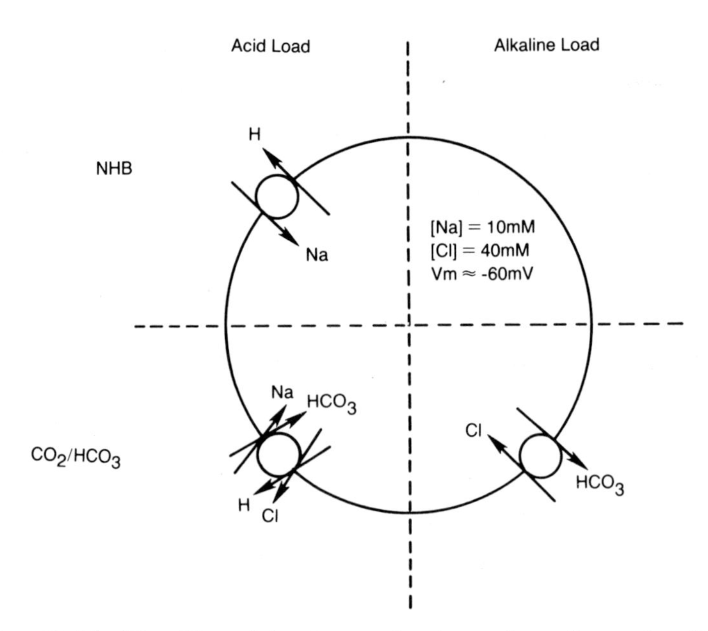

Figure 1. Model of the pH-regulating transport systems of smooth muscle cells. Values for membrane potential and intracellular ion concentrations are taken from Aickin, 1986. (Reprinted from Putnam, 1990, by permission)

or six SHR rats. The aortae were mechanically stripped of fat and connective tissue and then digested in collagenase. The tunica adventitia was peeled away as a tube; the remaining intimal and medial layers were cut into small segments and placed in a collagenase/trypsin solution. These segments were incubated in a shaking water bath with the supernatant removed into media (M199) containing bovine serum albumin (BSA) and 20% fetal bovine serum (FBS). This incubation was repeated four times and the pooled supernatants, which contained dissociated cells from the medial layer, were spun down. The resulting cell pellet was resuspended in media with BSA (serum free). Isolated smooth muscle cells were plated onto collagen-coated (rat tail type I collagen) glass chips (9 x 18 mm) at a density of 3 - 5 x 10^5 cells/ml. After 1 hour, the solution was changed to media supplemented with 2% FBS. Cells were grown for 3 - 7 days without additional media change, at which time they were 50 - 90% confluent. In addition to having a smooth muscle appearance, these cells have been shown to bind an antibody specific for smooth muscle myosin heavy chain, exhibit slow inward currents with whole cell patch recordings, and contract in response to angiotensin (R. Pun, personal communication).

Cells on the glass chips were loaded with the acetoxymethyl ester of the pH-sensitive fluorescent dye 2',7'-bis(2-carboxyethyl)-5,6-carboxyfluorescein (BCECF) by soaking the cells in a 2 μM solution for 20 min and washing in NHB (see below) for at most 1 hour before use. Intracellular pH (pH_i) was measured from the fluorescence of these dye-loaded cells as previously described (Putnam and Grubbs, 1990). The calibration curves, derived using high K solutions with nigericin (Thomas et al., 1979; Putnam and Grubbs, 1990) were $pH_i = 0.261(R_{fl}) + 5.75$ and $pH_i = 0.278(R_{fl}) + 5.57$ in VSM and SHR cells, respectively.

The solutions were as described in Putnam and Grubbs (1990). Briefly, the standard sodium N-2-hydroxyethylpiperazine-N'-2-ethanesulfonic acid-

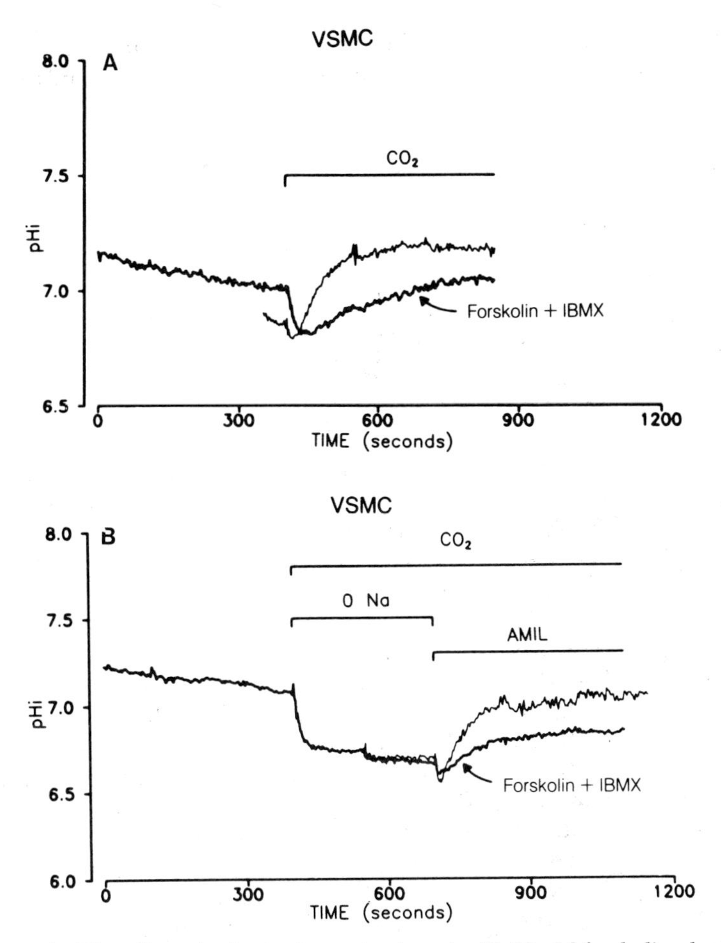

Figure 2. A. The effect of a 5 minute pre-treatment with 10 μM forskolin plus 100 μM IBMX on the rate of CO_2-induced alkalinization in VSM cells. The thin line represents the control, the heavier line a super-imposed experiment in the presence of forskolin plus IBMX. B. The effect of forskolin plus IBMX on pH recovery mediated solely by (Na + HCO_3)/Cl exchange in VSM cells. In cells acidified by CO_2 in Na-free solution, pH recovery mediated solely by (Na + HCO_3)/Cl exchange occurs upon addition of Na in the presence of amiloride. The thin line represents the control, the heavier line a super-imposed experiment in the presence of forskolin plus IBMX.

buffered solution (NHB) (in mM) contained: 127 NaCl, 4 KCl, 1 $CaCl_2$, 0.5 $MgCl_2$, 20 NaHEPES, and 5.6 glucose. The standard CO_2-HCO_3 solution contained (in mM): 127 NaCl, 4 KCl, 1 $CaCl_2$, 0.5 $MgCl_2$, 24 $NaHCO_3$, and 5.6 glucose. The pH of both solutions was 7.4 at 37° C. In Na-free solutions, all Na was replaced by N-methyl-D-glucammonium (NMDG) and in Cl-free solutions all Cl was replaced by gluconate (additional Ca and Mg gluconate were added to these solutions to maintain free Ca and Mg concentrations constant). The adenylate cyclase activator forskolin (Sigma Chem Co.) was made as a 6 mM stock solution in absolute ethanol and diluted to a final concentration of 10 μM just before use. The phosphodiesterase inhibitor IBMX (Sigma Chem Co.) was added directly to the solution to yield a final concentration of 100 μM. Amiloride, a Na/H exchange inhibitor, was a gift of Merck, Sharp, and Dohme and was used at a concentration of 1 mM.

All values are reported as the mean ± one standard error of the mean (SEM). All tests of statistical significance were two tailed Student's t-tests.

RESULTS

Steady State pH_i in VSM Cells

The initial pH_i of VSM cells in NHB was 7.04 ± 0.03 (n = 23). Upon exposure to 5% CO_2, these cells alkalinized by 0.14 pH unit ($p < 0.01$) to a new steady state pH_i of 7.18 ± 0.02 (n = 23) (Figs. 2A and 3).

The Effect of Forskolin plus IBMX on $(Na + HCO_3)/Cl$ Exchange in VSM Cells

In VSM cells, CO_2-induced alkalinization has been shown to be due in large part to $(Na + HCO_3)/Cl$ exchange (Putnam, in preparation). The rate of this CO_2-induced alkalinization is largely inhibited by a 5 minute pre-exposure to forskolin plus IBMX in NHB (Fig. 2A), suggesting that forskolin plus IBMX inhibits $(Na + HCO_3)/Cl$ exchange. Conditions under which this exchanger is solely responsible for alkalinization have been defined. If cells in NHB are exposed to Na-free CO_2 solution, they acidify to a pH_i of approximately 6.6 due to CO_2 entry, hydration, and dissociation (Fig. 2A; Roos and Boron, 1981). There is no recovery from this acidification since both alkalinizing transporters (Na/H and $(Na + HCO_3)/Cl$ exchangers) are inactive in Na-free solutions (Fig. 1; Putnam and Grubbs, 1990). Upon return of Na to the outside solution (in the presence of amiloride) only the $(Na + HCO_3)/Cl$ exchanger mediates alkalinization (Fig. 2B). The rate of alkalinization upon return of Na to the external medium was markedly inhibited in the presence of forskolin plus IBMX compared to the control rate (Fig. 2B), 0.0053 ± 0.0002 pH/s (n = 5) vs. 0.0015 ± 0.0003 pH/s (n = 7). This inhibition was significant ($p < 0.01$) and amounted to a reduction in the recovery rate of about 72% (Fig. 5A).

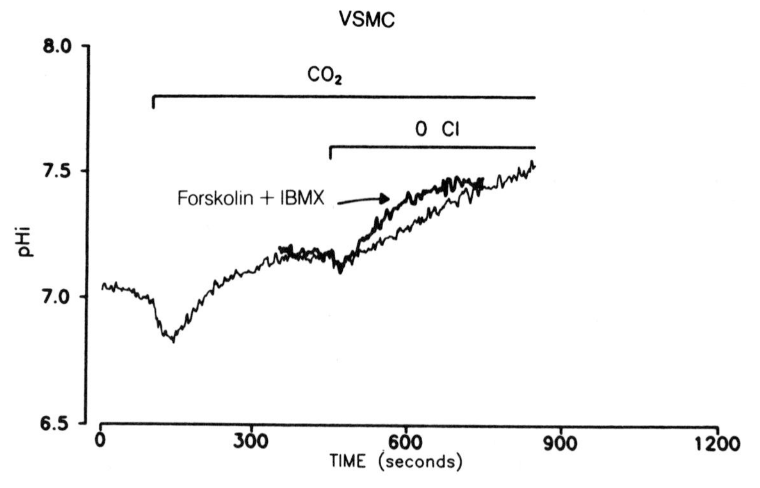

Figure 3. Effect of forskolin plus IBMX on alkalinization mediated by reversed Cl/HCO_3 exchange (upon removal of external Cl) in VSM cells. The thin line represents control, the heavier line a super-imposed experiment in the presence of forskolin plus IBMX.

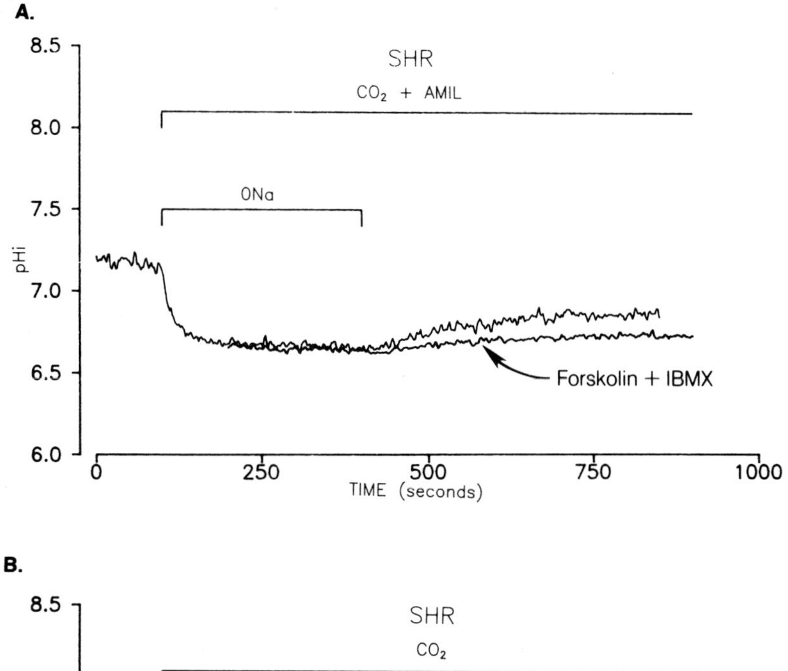

Figure 4. A. The effect of forskolin plus IBMX on pH recovery mediated solely by (Na + HCO_3)/Cl exchange in SHR cells. The thin line represents the control, the heavier line a super-imposed experiment in the presence of forskolin plus IBMX. B. The effect of forskolin plus IBMX on alkalinization mediated by reversed Cl/HCO_3 exchange (upon removal of external Cl) in SHR cells. The thin line represents control, the heavier line a super-imposed experiment in the presence of forskolin plus IBMX.

The Effect of Forskolin plus IBMX on Cl/HCO_3 Exchange in VSM Cells

In VSM cells exposed to CO_2, the removal of external Cl resulted in a rapid alkalinization (Fig. 3). The rate of alkalinization (from an initial pH_i of 7.1), presumably mediated by reversed Na-independent Cl/HCO_3 exchange upon removal of external Cl, amounted to 0.0011 ± 0.0001 pH/s (n = 5) and was stimulated by forskolin plus IBMX to a rate of 0.0017 ± 0.0002 pH/s (n = 4) (Fig. 3). This stimulation was significant ($p < 0.05$) and amounted to an increase in recovery rate of about 55% as compared to the control rate (Fig. 5A).

The Effect of Forskolin plus IBMX on pH_i in VSM Cells

In the presence of CO_2, exposure to forskolin plus IBMX reduced steady state pH_i from 7.18 to 7.09 ± 0.04 (n = 8), significantly less ($p < 0.05$) than the steady state pH_i in the absence of forskolin plus IBMX (Fig. 5B).

Steady State pH_i in SHR Cells

Experiments similar to those described above for VSM cells were performed on primary cultures of aortic smooth muscle cells from SHR. The pH_i of SHR cells in NHB was 7.24 ± 0.03 (n = 8). Thus, as previously demonstrated (Berk et al., 1989; Izzard and Heagerty, 1989), SHR cells in NHB have a significantly elevated pH_i compared to cells from normotensive controls ($p < 0.01$). Unlike VSM cells, SHR cells did not alkalinize upon exposure to 5% CO_2. In fact, steady state pH_i in CO_2-containing solutions was significantly ($p < 0.05$) more acid, 7.14 ± 0.02 (n = 8), than in NHB. Thus, in the presence of CO_2, SHR cells no longer had an alkaline pH relative to VSM cells.

The Effect of Forskolin plus IBMX on (Na + HCO_3)/Cl Exchange in SHR Cells

In SHR cells, the recovery from acidification is mediated to a lesser extent by (Na + HCO_3)/Cl exchange than in VSM cells (Fig. 4A; Putnam, unpublished observation). Regardless, there is in SHR cells a distinct Na-dependent, amiloride-insensitive alkalinization in response to CO_2-induced acidification which is characteristic of (Na + HCO_3)/Cl exchange. The initial rate of alkalinization mediated by this exchanger (0.0013 ± 0.0002 pH/s, n = 4) was inhibited by exposure to forskolin plus IBMX (0.0007 ± 0.0001 pH/s, n = 4) in SHR cells initially acidified to pH_i of approximately 6.55 (Figs. 4A). Thus, the (Na + HCO_3)/Cl exchanger is significantly ($p < 0.05$) inhibited by 46% in SHR cells (Fig. 5A).

The Effect of Forskolin plus IBMX on Cl/HCO_3 Exchange in SHR Cells

Like VSM cells, SHR cells alkalinize when, in the presence of CO_2, they are exposed to Cl-free solutions (Fig. 4B). This alkalinization occurred at a rate of 0.0014 ± 0.0002 pH/s (n = 3; from a starting pH_i of 7.06) and was unaffected ($p > 0.10$) by exposure to forskolin plus IBMX (Figs. 4B and 5A), 0.0016 ± 0.0003 pH/s (n = 5).

The Effect of Forskolin plus IBMX on pH_i in SHR Cells

In the presence of CO_2, exposure to forskolin plus IBMX did not significantly ($p > 0.10$) affect the steady state pH_i of SHR cells (Fig. 5B), 7.11 ± 0.08 (n = 5) vs. 7.14.

Intrinsic Buffering Power in VSM and SHR Cells

Comparing rates of pH change among different cells can be misleading if these cells have markedly different buffering powers (Roos and Boron, 1981). We have determined the buffering power using standard techniques (Boron,

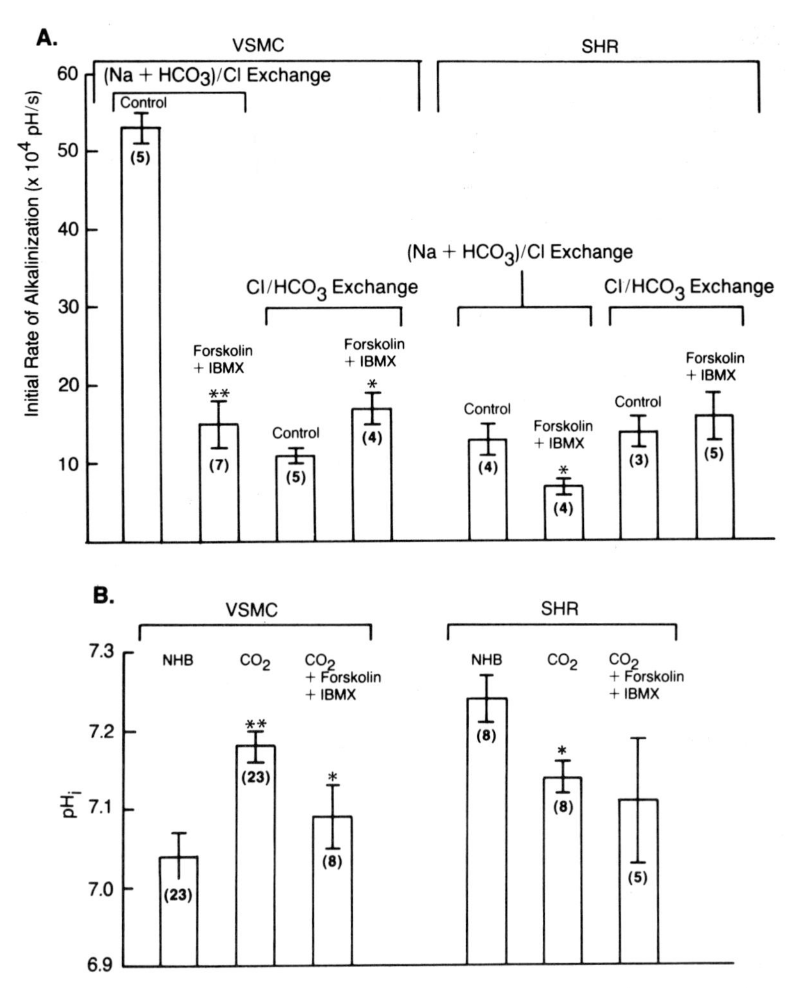

Figure 5. A. Summary of effects of forskolin plus IBMX on alkalinization rates under various conditions in VSM and SHR cells. The height of a bar indicates the mean initial rate of alkalinization ± SEM. * indicates that a bar is statistically different ($p < 0.05$) from the previous bar. ** indicates that a bar is statistically different ($p < 0.01$) from the previous bar. B. The steady state pH_i in VSM and SHR cells in NHB, 5% CO_2, or 5% CO_2 with 10 μM forskolin plus 100 μM IBMX. The height of a bar indicates the mean $pH_i \pm 1$ SEM. * indicates that a bar is statistically different ($p < 0.05$) from the previous bar. ** indicates that a bar is statistically different ($p < 0.01$) from the previous bar.

1977) involving NH_4Cl (2 - 5 mM) pulses or propionate (10 - 15 mM) pulses (data not shown). The intrinsic buffering power (β_{int}) measured with NH_4Cl pulses (at an average pH_i of 6.75 - 6.8) was 6.4 ± 0.9 (n = 3) and 5.5 ± 0.5 (n = 3) mM/pH unit in VSM and SHR cells, respectively. The β_{int} measured with propionate pulses (at pH_i between 6.8 - 7.1) was somewhat higher, 9.6 ± 0.8 (n = 5) and 9.8 ± 1.1 (n = 4) mM/pH unit in VSM and SHR cells, respectively. Thus, β_{int} is nearly identical in VSM and SHR cells and is similar to previously measured values in other smooth muscle and smooth muscle-like cells (Aickin, 1984; Boyarsky et al., 1988a; Putnam and Grubbs, 1990).

DISCUSSION

In the present study, we show that elevated cyclic nucleotide levels, induced by forskolin plus IBMX, inhibit the alkalinizing (Na + HCO_3)/Cl exchanger in both VSM and SHR cells (Figs. 2, 4A, and 5A) but activate the acidifying Cl/HCO_3 exchanger in VSM cells only (Figs. 3 and 5A). The Na-independent Cl/HCO_3 exchanger in SHR cells does not appear to be affected by elevated cyclic nucleotides (Figs. 4B and 5A). These effects of forskolin plus IBMX on the HCO_3-dependent transport systems are most likely mediated by elevated cytoplasmic cAMP. A 5 minute exposure of VSM cells to 10 μM forskolin has been shown to increase cAMP by over 25 fold while increasing cGMP less than 2 fold (Lincoln et al., 1990). However, the phosphodiesterase inhibitor IBMX could result in an elevation of both cAMP and cGMP and thus, without further experiments, it cannot be ruled out that the effects of forskolin plus IBMX are mediated by a small increase in cGMP rather than cAMP.

The effect of cAMP on pH-regulating transport systems has not been well studied but appears to be cell dependent. Elevated cAMP has been suggested to inhibit Cl/HCO_3 exchange in the A7r5 smooth muscle cell line (Vigne et al., 1988), but to weakly activate the exchanger in kidney cells (Tønnessen et al., 1990). The (Na + HCO_3)/Cl exchanger has been shown to be activated by cAMP in barnacle muscle fibers (Boron et al., 1978), but not in squid giant axons (Boron, 1985). The effects of cAMP on Na/H exchange are also cell specific. Increased cAMP has been shown to inhibit Na/H exchange from Na-transporting epithelia (Reuss and Petersen, 1985), activate the exchanger from nucleated red blood cells (Motais and Garcia-Romeu, 1988), but not affect the Na/H exchanger from frog skeletal muscle (Putnam and Roos, 1986). It is thus clear that the modulation of pH-regulating transport systems by cAMP varies depending on the cell being studied.

In smooth muscle cells, steady state pH_i is determined by the balance between acid extrusion on the alkalinizing transport systems (Na/H and (Na + HCO_3)/Cl exchangers) and acid loading on the acidifying Cl/HCO_3 exchanger (Aalkjaer and Cragoe, 1988; Ganz et al., 1989; Putnam, 1990; Putnam and Grubbs, 1990). One would thus predict that if elevated cAMP inhibits (Na + HCO_3)/Cl exchange and stimulates Cl/HCO_3 exchange that steady state pH_i should decrease. This is in fact observed in VSM cells (Fig. 5B). Under physiological conditions, forskolin plus IBMX resulted in a reduction in steady state pH_i of nearly 0.1 pH unit. In contrast, elevation of cAMP in SHR cells resulted in a small inhibition of (Na + HCO_3)/Cl exchange and no effect on Cl/HCO_3 exchange and one would thus predict very little effect of forskolin plus IBMX on steady state pH_i, as is indeed observed (Fig. 5B). This lack of effect of forskolin plus IBMX on SHR cell pH is probably due to two factors: i) the reduced effect of elevated cAMP on the HCO_3-dependent transport systems; and ii) the apparently greater role played by Na/H exchange in pH regulation in SHR cells. That Na/H exchange is more significant in the determination of steady state pH_i in SHR cells is shown by the fact that pH_i in NHB is higher in SHR than VSM cells (Fig. 5B), that the rate of (Na + HCO_3)/Cl exchange is much lower in SHR than VSM cells (Fig. 5A), and that exposure to CO_2 causes SHR cell acidification as opposed to the alkalinization seen in VSM cells (Fig. 5B). Thus, it appears that in SHR cells the Na/H exchanger is a more predominant pH-regulating transport system and the HCO_3-dependent transport systems are regulated somewhat differently than in VSM cells.

Finally, the relevance of these observations to the expression of hypertension must be addressed. Much has been made of the fact that SHR cells have a higher steady state pH_i and more active Na/H exchanger than cells from normotensive controls (Berk et al., 1989; Izzard and Heagerty, 1989). We have also observed a difference in pH_i in SHR vs. VSM cells in NHB (Fig. 5B). However, this difference in pH_i between cells from normotensive and hypertensive animals disappears when the studies are done under more normal physiological conditions, i.e. in the presence of 5% CO_2 (Fig. 5B). Thus, it is unlikely that differences in normal steady state pH_i are correlated with hypertension *in vivo*.

Several studies have suggested that changes in pH_i can affect the contractile state of smooth muscle cells (Ighoroje and Spurway, 1984; Danthuluri and Deth, 1989). It is generally observed that cellular alkalinization correlates with increased smooth muscle contraction and acidification with smooth muscle relaxation, although the relationship between pH_i and contraction can be quite complex (Danthuluri and Deth, 1989) and its significance is still unclear. Increased pH_i has also been associated with smooth muscle cell proliferation, a process which is believed to be involved in the clinical expression of hypertension (Berk et al., 1988).

In correlation with the predicted role of cellular alkalinization in smooth muscle growth and contraction, numerous studies have shown that growth-promoting agents (e.g. serum and platelet-derived growth factor) and vasoconstricting agents (e.g. endothelin, angiotensin II, and vasopressin) activate Na/H exchange and alkalinize smooth muscle cells (Berk et al., 1987; 1988). It is believed that the activation of Na/H exchange is mediated through both a protein kinase C-dependent and a Ca-dependent pathway (Huang et al., 1987). However, the significance of agonist-induced alkalinization has been questioned based on the lack of an alkalinizing effect of vasopressin on mesangial cells under physiological conditions (presence of 5% CO_2; Ganz et al., 1989). The effect of vasodilator agents (e.g. bradykinin, parathyroid hormone, and atrial natriuretic peptide) on pH_i in smooth muscle cells has not been studied, but other effects of these agents are mediated by an increase of cyclic nucleotides (Neuser and Bellemann, 1986; O'Donnell and Owen, 1986).

Based on these considerations, it could be hypothesized that vasodilators cause smooth muscle relaxation in part due to cellular acidification. This effect could be mediated by an elevation in cyclic nucleotides which might result in cellular acidification by inhibiting alkalinizing pH-regulating transport systems and activating acidifying systems. In cells from hypertensive organisms, this response of the pH-regulating transport systems to elevated cyclic nucleotides could be lost, which would prevent the acidification induced by vasodilators. Thus, a part of the pathology of hypertensive cells may be a loss of the appropriate regulatory controls of the membrane-bound pH transport systems. To test this hypothesis it will be essential to show that vasodilator agents cause acidification in normotensive cells but not in cells from hypertensive animals. It will also be important to look at the control of pH-regulating transport systems in a normotensive control that is genetically matched to the SHR strain (Izzard and Heagerty, 1989). Finally, the ability of a relatively small cytoplasmic acidification to reduce smooth muscle cell contractility must be demonstrated.

In summary, exposure of VSM cells to forskolin plus IBMX inhibited the alkalinizing (Na + HCO_3)/Cl exchanger and activated the acidifying Cl/HCO_3 exchanger, resulting in cytoplasmic acidification. In contrast, exposure of SHR cells to forskolin plus IBMX inhibited the (Na + HCO_3)/Cl exchanger (which normally has a low activity in these cells) but had no effect on the Cl/HCO_3 exchanger and thus did not alter pH_i. A model is proposed which suggests that a part of the pathology of hypertension may be due to a modification of the control of HCO_3-dependent transport systems in vascular smooth muscle cells.

ACKNOWLEDGMENTS

We gratefully acknowledge Dr. Nicholas Sperelakis for the initial gift of VSM and SHR cells and Ms. Susan Osborn for demonstrating the techniques of primary culture with VSM cells. We also acknowledge Dr. Raymond Pun for sharing unpublished data on the characterization of VSM cells. Finally, we thank Ms. Carolyn Stemmler of Merck, Sharp and Dohme for the generous gift of amiloride. This research was supported by a grant from the Ohio Affiliate of the American Heart Association and by NSF grant DIR-8812094.

REFERENCES

Aalkjaer, C. and Cragoe, Jr., E. J., 1988, Intracellular pH regulation in resting and contracting segments of rat mesenteric resistance vessels, *J. Physiol.*, 402: 391.

Aalkjaer, C. and Mulvany, M. J., 1988, Bicarbonate transport is as important for intracellular pH control in contracting rat resistance vessels as Na-H exchange, *J. Hypertension*, 6: S706.

Aickin, C. C., 1984, Direct measurement of intracellular pH and buffering power in smooth muscle cells of guinea-pig vas deferens, *J. Physiol.*, 349: 571.

Aickin, C. C., 1986, Intracellular pH regulation by vertebrate muscle, *Ann. Rev. Physiol.*, 48: 349.

Aickin, C. C. and Brading, A. F., 1984, The role of chloride-bicarbonate exchange in the regulation of intracellular chloride in guinea-pig vas deferens, *J. Physiol.*, 349: 587.

Berk, B. C., Aronow, M. S., Brock, T. A., Cragoe Jr., E. J., Gimbrone Jr., M. A., and Alexander, R. W., 1987, Angiotensin II stimulated Na^+/H^+ exchange in cultured vascular smooth muscle cells, *J. Biol. Chem.*, 262: 5057.

Berk, B. C., Canessa, M., Vallega, G., and Alexander, R. W., 1988, Agonist-mediated changes in intracellular pH: role in vascular smooth muscle cell function, *J. Cardiovasc. Pharmacol.*, 12: S104.

Berk, B. C., Vallega, G., Muslin, A. J., Gordon, H. M., Canessa, M., and Alexander, R. W., 1989, Spontaneously hypertensive rat vascular smooth muscle cells in culture exhibit increased growth and Na^+/H^+ exchange. *J. Clin. Invest.*, 83: 822.

Boron, W. F., 1977., Intracellular pH transients in giant barnacle muscle fibers, *Am. J. Physiol.*, 233: C61.

Boron, W. F., 1985, Intracellular pH-regulating mechanism of the squid axon. Relation between the external Na^+ and HCO_3^- dependences, *J. Gen. Physiol.*, 85: 325.

Boron, W. F., Russell, J., Brodwick, M., Keifer, D., and Roos, A., 1978, Influence of cyclic AMP on intracellular pH regulation and chloride fluxes in barnacle muscle fibres, *Nature*, 276: 511.

Boyarsky, G., Ganz, M. B., Sterzel, R. B., and Boron, W. F., 1988a, pH regulation in single glomerular mesangial cells. I. Acid extrusion in absence and presence of HCO_3^-, *Am. J. Physiol.*, 255: C844.

Boyarsky, G., Ganz, M. B., Sterzel, R. B., and Boron, W. F., 1988b, pH regulation in single glomerular mesangial cells. II. Na^+-dependent and -independent Cl^--HCO_3^- exchangers, *Am. J. Physiol.*, 255: C857.
Danthuluri, N. R. and Deth, R. C., 1989, Effects of intracellular alkalinization on resting and agonist-induced vascular tone, *Am. J. Physiol.*, 256: H867.
Ganz, M. B., Boyarsky G., Sterzel, R. B., and Boron, W. F., 1989, Arginine vasopressin enhances pH_i regulation in the presence of HCO_3^- by stimulating three acid-base transport systems, *Nature*, 337: 648.
Harder, D. R. and Sperelakis, N., 1979, Action potential generation in reaggregates of rat aortic smooth muscle cells in primary culture, *Blood Vessels*, 16: 186.
Huang, C.-L., Cogan, M. G., Cragoe, Jr., E. J., and Ives, H. E., 1987, Thrombin activation of the Na^+/H^+ exchanger in vascular smooth muscle cells. Evidence for a kinase C-independent pathway which is Ca^{2+}-dependent and pertussis toxin sensitive, *J. Biol. Chem.*, 262: 14134.
Ighoroje, A. D. and Spurway, N. C., 1984, How does vascular muscle in the isolated rabbit ear adapt its tone after alkaline or acid loads?, *J. Physiol.*, 357: 105P.
Izzard, A. S. and Heagerty, A. M., 1989, The measurement of internal pH in resistance arterioles: Evidence that intracellular pH is more alkaline in SHR than WKY animals, *J. Hypertension*, 7: 173.
Kikeri, D., Zeidel, M. L., Ballermann, B. J., Brenner, B. M., and Hebert., S. C., 1990, pH regulation and response to AVP in A10 cells differ markedly in the presence vs. absence of CO_2-HCO_3, *Am. J. Physiol.*, 259: C471.
Lincoln, T. M., Cornwell, T. L., and Taylor, A. E., 1990, cGMP-dependent protein kinase mediates the reduction of Ca^{2+} by cAMP in vascular smooth muscle cells, *Am. J. Physiol.*, 258: C399.
Motais, R. and Garcia-Romeu, F., 1988, Effects of catecholamines and cyclic nucleotides on Na^+/H^+ exchange, *in:* "Na/H Exchange", S. Grinstein, ed., CRC Press, Inc., Boca Raton, p. 255.
Neuser, D. and Bellemann, P., 1986, Receptor binding, cGMP stimulation and receptor desensitization by atrial natriuretic peptides in cultured A10 vascular smooth muscle cells, *FEBS Lett.*, 209: 347.
O'Donnell, M. E. and Owen, N. E., 1986, Atrial natriuretic factor stimulates Na/K/Cl cotransport in vascular smooth muscle cells, *Proc. Nat'l. Acad. Sci. U.S.A.*, 83: 6132.
Putnam, R. W., 1990, pH regulatory transport systems in a smooth muscle-like cell line, *Am. J. Physiol.*, 258: C470.
Putnam, R. W. and Grubbs, R. D., 1990, Steady-state pH_i, buffering power, and effect of CO_2 in a smooth muscle-like cell line, *Am. J. Physiol.*, 258: C461.
Putnam, R. W. and Roos, A., 1986, Na-independent pH_i recovery in frog skeletal muscle, *J. Gen. Physiol.*, 88: 47a.
Reuss, L. and Petersen, K.-U., 1985, Cyclic AMP inhibits Na^+/H^+ exchange at the apical membrane of *Necturus* gallbladder epithelium, *J. Gen. Physiol.*, 85: 409.
Roos, A. and Boron, W. F., 1981, Intracellular pH, *Physiol. Rev.*, 61: 296.
Thomas, J., Buchsbaum, R., Zimniak, A., and Racker, E., 1979, Intracellular pH measurements in Ehrlich ascites tumor cells utilizing spectroscopic probes generated *in situ*, *Biochemistry*, 18: 2210.
Tønnessen, T. I., Aas, A. T., Ludt, J., Blomhoff, H. K., and Olsnes, S., 1990, Regulation of Na^+/H^+ and Cl^-/HCO_3^- antiports in Vero cells, *J. Cell. Physiol.*, 143: 178.

Vigne, P., Breittmayer, J.-P., Frelin, C., and Lazdunski, M., 1988, Dual control of the intracellular pH in aortic smooth muscle cells by a cAMP-sensitive HCO_3^-/Cl^- antiporter and a protein kinase C-sensitive Na^+/H^+ antiporter, *J. Biol. Chem.,* 263: 18023.
Wray, S., 1988, Smooth muscle intracellular pH: Measurement, regulation and function, *Am. J. Physiol.,* 254: C213.

NOREPINEPHRINE STIMULATES INOSITOL TRISPHOSPHATE FORMATION IN RAT PULMONARY ARTERIES

Najia Jin, C. Subah Packer, Denis English,
and Rodney A. Rhoades

Departments of Physiology/Biophysics and Medicine/Pathology
Indiana University School of Medicine
Indianapolis, IN 46202

INTRODUCTION

Although the discovery of the "phosphoinositide effect" occurred over 35 years ago, its mechanisms were explored only over the last decade. It now is clear that hydrolysis of phosphoinositides generates second messengers for multiple cellular functions when receptors are activated by a wide array of hormones and agonists in a variety of cell types (for review see Rana and Hokin, 1990). Early studies focused on the role of phosphoinositides on regulation of secretion or control of release of secretory cell contents (Hokin and Hokin, 1953, 1960; Freinkel, 1957; Hokin et al., 1958, 1963; Axen et al., 1983). In recent years, studies of the physiological effects of phosphoinositide hydrolysis have extended to such cell types and tissues as cerebral cortical slices (Kendall and Nahorski, 1984), sympathetic ganglia (Bone et al., 1984), adrenal glomerulosa cells (Kojima et al., 1986), rod outer segments (Brown et al., 1987), epithelium (Anderson and Welsh, 1990), leukocytes (Bradford and Rubin, 1986), skeletal muscle (Volpe et al., 1985), and smooth muscle (Akhtar and Abdel-Latif, 1980; Bielkiewicz-vollrath et al., 1987).

While the role of phosphoinositide metabolism in secretion and other cellular functions is well established, only a few studies have provided evidence that IP_3 is, in fact, the second messenger in smooth muscle contraction. These studies have focused on only a few types of smooth muscle cells including rabbit iris (Akhtar and Abdel-Latif, 1980), guinea pig taenia caeci (Iino, 1990), bovine aorta (Lee et al., 1989), porcine coronary artery (Suematsu et al., 1984), and canine trachea (Duncan et al., 1987). Whether or not IP_3 is the universal second messenger in smooth muscle contraction is not yet known. Therefore, the purpose of this study was to quantitate the changes in inositol phosphates and the time courses of these changes in isolated rat pulmonary arterial preparations stimulated with norepinephrine (NE).

Regulation of Smooth Muscle Contraction
Edited by R.S. Moreland, Plenum Press, New York, 1991

METHODS

Male Sprague-Dawley rats (300 - 350g) were anesthetized by intraperitoneal injection of pentobarbital. Lungs and heart were removed *en bloc* and placed in ice cold Earle's balanced salt solution (EBSS; 2.4 mM $CaCl_2$, 0.8 mM $MgSO_4$, 5.4 mM KCl, 116.4 mM NaCl, 0.9 mM Na_2HPO_4, 5.5mM glucose, 26.2 mM $NaHCO_3$, and 0.03 mM phenol red). The pulmonary arteries, including main trunk and right and left branches, were isolated and slit longitudinally in order to expose more surface area. The arterial preparations were incubated with 50 μCi of ^{3}H-myo-inositol (NEN Boston, MA) for four hours in a volume of 1.5 ml gassed with 95% O_2/5%CO_2. After incubation, the tissue was washed four times with fresh EBSS and stimulated with 5 x 10^{-7} M NE (Sigma Chem. Co.). The reaction was stopped at indicated time intervals by fast-freezing of the tissue in liquid nitrogen followed by extracting in 10% cold trichloroacetic acid (TCA). The samples were homogenized in 10% TCA. Subsequently, the samples were extracted with diethyl ether to remove the TCA and the pH of the resulting aqueous phases was adjusted to approximately 8.5 by the addition of tris base (Sigma Chem. Co.). Samples were then applied to a 4.6 x 50 mm Vydac Nucleotide analysis column (Rainin Inst. Co., Woburn, MA). The analysis of inositol phosphates was accomplished by following the method described by Taylor et al. (1990). For each experiment, elution times were compared with those of guanine nucleotide standards: GMP, GDP, and GTP (Boehringer-Mannheim, Indianapolis, IN) detected with a Perkin-Elmer on-line spectrophotometer at 254 nm. Thirty fractions of each sample were collected and mixed with scintillation fluid (Fisher Scientific, Pittsburgh, PA)

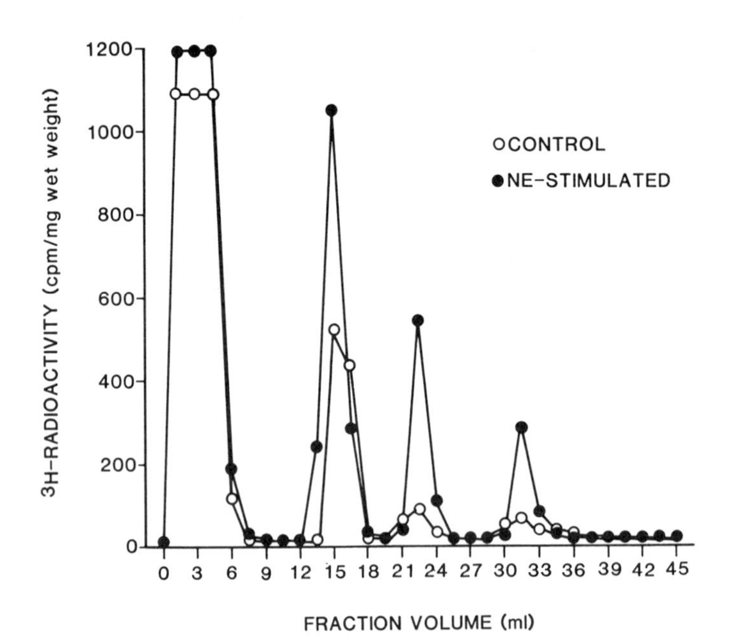

Figure 1. Analysis of inositol phosphate by HPLC. Samples (200 μl) from non-stimulated and NE-stimulated (5s) pulmonary arteries were applied to the anion-exchange column with GMP, GDP, and GTP markers. The identification of the inositol phosphates is based on comparison with the standards.

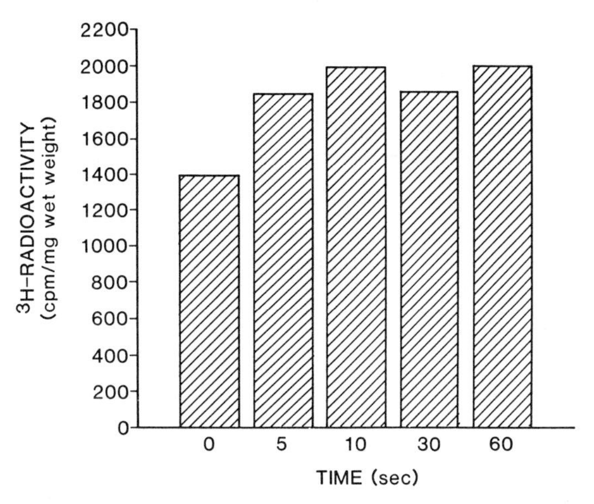

Figure 2. Time course of total inositol phosphate (i.e. $IP_1 + IP_2 + IP_3 + IP_4$) production following NE-stimulation in pulmonary arterial tissue. IP's were measured at time 0 (prior to NE-stimulation) and at 5, 10, 30, and 60 seconds following NE-stimulation.

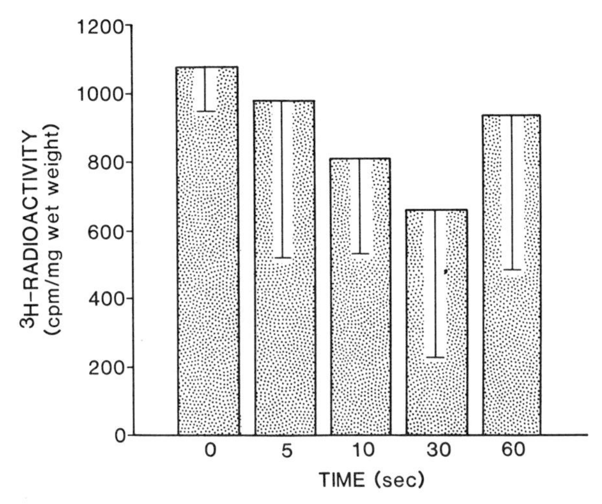

Figure 3. Time course of IP_1 levels from NE-stimulated pulmonary arterial tissue. At 5 seconds following stimulation, IP_1 was decreased compared with time 0 (prior to NE-stimulation) and appeared to continue to decrease until 30s following stimulation. IP_1 levels appeared to increase towards resting levels between 30 and 60s following NE-stimulation.

and ^3H-radioactivity was determined by liquid scintillation counting. The various inositol phosphate levels were normalized to the wet weight of individual tissue samples.

RESULTS

Figure 1 shows a typical anion-exchange high performance liquid chromatography (HPLC) resolution of inositol phosphates. Samples from non-

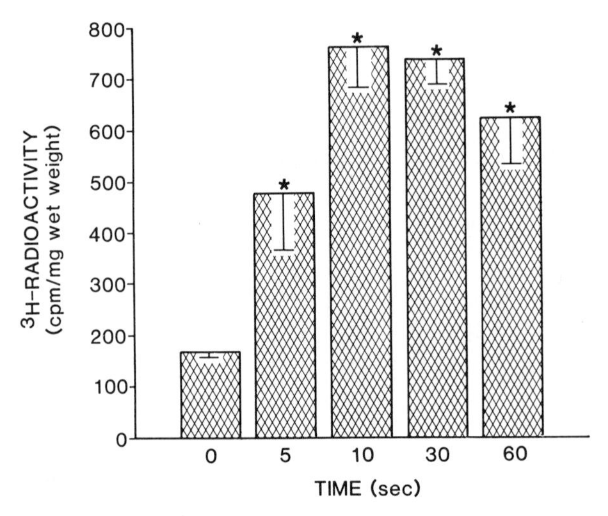

Figure 4. Time course of IP_2 levels from NE-stimulated pulmonary arterial tissue. At 5 seconds following stimulation, IP_2 was increased over resting levels ($p < 0.01$) and continued to increase until 10s following stimulation.

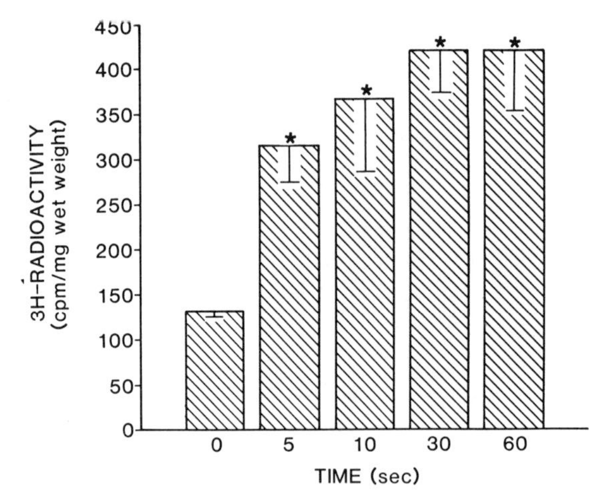

Figure 5. Time course of IP_3 levels in NE-stimulated pulmonary arterial tissue. At 5 seconds following stimulation, IP_3 was increased ($p < 0.01$) and continued to increase until 30s ($p < 0.01$).

stimulated or NE-stimulated rat pulmonary arterial tissue were applied to the column in a volume of 200 µl. The identification of inositol phosphates was based on comparison with elution time of GMP, GDP, and GTP standards. The time course of mean total inositol phosphate production following NE-stimulation is presented in Figure 2. Total inositol phosphates (IP's: $IP_1 + IP_2 + IP_3 + IP_4$) increased following NE-stimulation. However, inositol monophosphate (IP_1) appeared to decrease following stimulation as shown in Figure 3.

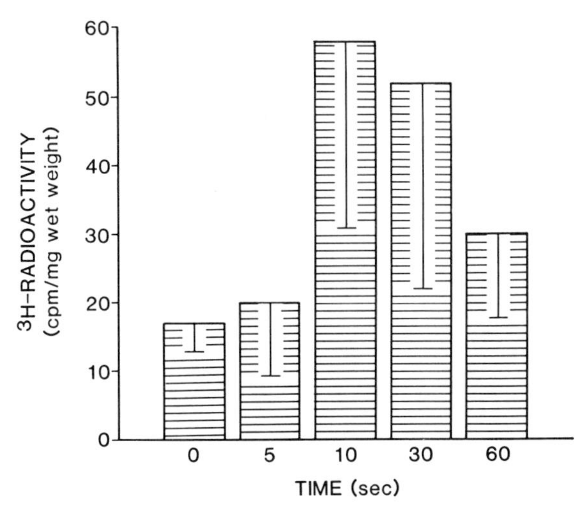

Figure 6. Time course of IP_4 levels from NE-stimulated pulmonary arterial tissue. At 10 seconds following stimulation, IP_4 was increased but appeared to decrease between 30 and 60s.

Inositol bisphosphate (IP_2) and inositol trisphosphate (IP_3; I-1,4,5-P_3 and I-1,3,4-P_3 isoforms combined) increased over resting levels at 5 seconds following NE-stimulation (see Fig. 4 and Fig. 5). IP_2 and IP_3 continued to increase until 10s and 30s, respectively, following stimulation. At 10s following stimulation, inositol tetrakisphosphate (IP_4) appeared to be increased (see Fig. 6).

DISCUSSION

The hydrolysis of polyphosphoinositides in response to stimuli (neurotransmitters or hormones) results in the formation of inositol phosphates. The breakdown of phosphoinositides, especially phosphatidylinositol 4,5-bisphosphate (PIP_2) and phosphatidylinositol 4-phosphate (PIP), is very rapid, usually observed within a few seconds of stimulation (Abdel-Latif, 1986). Results of this study show inositol phosphate levels rise rapidly following NE-stimulation in rat pulmonary arterial muscle. In particular, IP_2 and IP_3 are increased at five seconds and further increased at ten seconds following stimulation. IP_1 is decreased while IP_4 is increased at 10 seconds following stimulation.

The half time of inositol 1,4,5-trisphosphate is short, about 4 seconds, while inositol 1,3,4-trisphosphate appears to turn over much more slowly (Rana and Hokin, 1990). Therefore, the continuous increase in IP_3 10 seconds following stimulation may be due to the formation of inositol I-1,3,4-P_3. The metabolism of I-1,4,5-P_3 is carried out either by inositol trisphosphate (3)-kinase, which produces inositol tetrakisphosphate (I-1,3,4,5-P_4) or by inositol trisphosphate (5)-phosphomonoesterase which produces I-1,4-P_2 (following its action on I-1,4,5-P_3) or I-1,3,4-P_3 when I-1,3,4,5-P_4 is hydrolyzed. The fact that an increase in IP_4 was seen at a time of 10 seconds is consistent with the fact that IP_4 is generated from IP_3. IP_2 may form from either degradation of IP_3 or the phospholipase C catalyzed breakdown of PIP.

In summary, norepinephrine causes an increase in IP_3 formation in rat pulmonary arterial muscle at a time following stimulation at which second messenger formation and activity are reported to be maximal. These results are consistent with the hypothesis that IP_3 is involved as a second messenger in agonist-stimulated contraction of pulmonary arterial smooth muscle.

ACKNOWLEDGMENTS

This work was supported by NIH award HL40894. Thanks are due to Gregory Taylor for very skillful technical assistance and Marlene King for expert typing of this paper.

REFERENCES

Abdel-Latif, A. A., 1986, Calcium-mobilizing receptors, polyphosphoinositides, and the generation of second messengers, *Pharmacol. Rev.*, 38: 227.

Akhtar, R. A. and Abdel-Latif, A. A., 1980, Requirement for calcium ions in acetylcholine-stimulated phosphodiesteratic cleavage of phosphatidyl-myo-inositol 4,5-bisphosphate in rabbit iris smooth muscle, *Biochem. J.*, 142: 599.

Anderson, M. P. and Welsh, M. J., 1990, Isoproterenol, cAMP and bradykinin stimulate diacylglycerol production in airway epithelium, *Am. J. Physiol.*, 258: L294.

Axen, K. V., Schubert, U. K., Blake, A. D., and Fleischer, N., 1983, Role of Ca^{2+} in secretagogue-stimulated breakdown of phosphatidylinositol in rat pancreatic islets, *J. Clin. Invest.*, 72: 13.

Bielkiewicz-vollrath, B., Carpenter, J. R., Schulz, R., and Cook, D. A., 1987, Early production of 1,4,5-inositol trisphosphate and 1,3,4,5-inositol tetrakisphosphate by histamine and carbachol in ileal smooth muscle, *Mol. Pharmacol.*, 31: 513.

Bone, E. A., Fretten, P., Palmer, S., Kirk, C. J., and Mitchell, R. H., 1984, Rapid accumulation of inositol phosphates in isolated rat superior cervical sympathetic ganglia exposed to V_1-vasopressin and muscarinic cholinergic stimuli, *Biochem. J.*, 221: 803.

Bradford, P. G. and Rubin, R. P., 1986, Quantitative changes in inositol 1,4,5-trisphosphate in chemoattractant-stimulated neutrophils, *J. Biol. Chem.*, 261: 15644.

Brown, J. E., Blazynski, C., and Cohen, A. I., 1987, Light induces a rapid and transient increase in inositol-trisphophate in toad rod outer segments, *Biochem. Biophys. Res. Commun.*, 146: 1392.

Duncan, R. A., Krzanowski Jr., J. J., Davis, J. S., Polson, J. B., Coffey, R. G., Shimoda, T., and Szentivanyi, A., 1987, Polyphosphoinositide metabolism in canine tracheal smooth muscle (CTSM) in response to a cholinergic stimulus, *Biochem. Pharmacol.*, 36: 307

Freinkel, N., 1957, Pathways of thyroidal phosphate metabolism: The effect of pituitary thyrotropin upon the phospholipids of the sheep thyroid gland. *Endocrinology*, 61: 448.

Hokin, L. E. and Hokin, M. R., 1960, Studies on the carrier function of phosphatidic acid in sodium transport. I. The turnover of phosphatidic acid and phosphoinositide in avian salt gland on stimulation of secretion, *J. Gen. Physiol.*, 44: 61.

Hokin, L. E., Hokin, M. R., and Lobeck, C. C., 1963, Effects of acetylcholine on the incorporation of ^{32}P into the phospholipids in slices of skin from children with and without cystic fibrosis of the pancreas, *J. Clin. Invest.*, 42: 1232.

Hokin, M. R. and Hokin, L. E., 1953, Enzyme secretion and the incorporation of ^{32}P into the phospholipids of pancreas slices, *J. Biol. Chem.*, 203: 967.

Hokin, M. R., Hokin, L. E., Saffran, M., Schally, A. V., and Zimmermann, B. U., 1958, Phospholipid and the secretion of adrenocorticotropin and of corticosteriods, *J. Biol. Chem.*, 233: 811.

Iino, M., 1990, Biphasic Ca^{2+} dependence of inositol 1,4,5-trisphosphate-induced Ca release in smooth muscle cells of the guinea pig taenia caeci, *J. Gen. Physiol.*, 95: 1103.

Kendall, D. A. and Nahorski, S. R., 1984, Inositol phospholipid hydrolysis in rat cerebral cortical slices. II. Calcium requirement., *J. Neurochem.*, 42: 1388.

Kojima, I., Shibata, H., and Ogata, E., 1986, Pertussis toxin blocks angiotensin II-induced calcium influx but not inositol trisphosphate production in adrenal glomerulosa cells, *FEBS Lett.*, 204: 347.

Lee, T. S., Chao, T., Hu, K. Q., and King, G. L., 1989, Endothelin stimulates a sustained 1,2-diacylglycerol increase and protein kinase C activation in bovine aortic smooth muscle cells, *Biochem. Biophys. Res. Commun.*, 162: 381.

Rana, R. S. and Hokin, L. E., 1990, Role of phosphoinositides in transmembrane signaling, *Physiol. Rev.*, 70: 115.

Suematsu, E., Harita, M., Hashimoto, T., and Kuriyama, H., 1984, Inositol 1,4,5-trisphosphate releases Ca^{2+} from intracellular store sites in skinned single cells of porcine coronary artery, *Biochem. Biophys. Res. Commun.*, 120: 481.

Taylor, G. S., Carcia, J. G. N., Dukes, R., and English, D., 1990, High-performance liquid chromatographic analysis of radiolabeled inositol phosphates, *Anal. Biochem.*, 188: 118.

Volpe, P., Salviati, G., Divigilio, F., and Pozzan, T., 1985 Inositol 1,4,5-trisphophate induced calcium release from the sarcoplasmic reticulum of skeletal muscle, *Nature*, 316: 347.

TIME-DEPENDENT DECREASE IN Ca^{2+}-SENSITIVITY IN "PHASIC SMOOTH MUSCLE"

Hiroshi Ozaki, William T. Gerthoffer,[1] Nelson G. Publicover, and Kenton M. Sanders

Departments of Physiology and [1]Pharmacology
University of Nevada School of Medicine
Reno, NV 89557

INTRODUCTION

Recent studies using the Ca^{2+}-indicators, aequorin and fura-2, support the concept that increased intracellular Ca^{2+} ($[Ca^{2+}]_i$) leads to force development in smooth muscle (for reviews, see Karaki, 1989; Somlyo and Himpens, 1989). For example, in guinea-pig taenia caecum a close correlation exists between $[Ca^{2+}]_i$ and muscle tension (Ozaki et al., 1988; Mitsui and Karaki, 1990). However, in some smooth muscles, the relationship between $[Ca^{2+}]_i$ and force development appears to depend upon the method of stimulation. For a given increase in $[Ca^{2+}]_i$, agonists such as norepinephrine, histamine, prostaglandins, and endothelin in vascular smooth muscle (Morgan and Morgan, 1984; DeFeo and Morgan, 1985; Sato et al., 1988; Sakata et al., 1989; Mori et al., 1990; Ozaki et al., 1990a), and carbachol in trachea (Gerthoffer et al., 1990; Ozaki et al., 1990b) induce greater contractions than simple depolarization with elevated external K^+. These findings suggest that the Ca^{2+}-sensitivity of the contractile elements may be increased by certain agonists. Although the mechanism of Ca^{2+}-sensitization has not been clarified, the agonist-induced activation of protein kinase C and subsequent phosphorylation of specific protein(s) may be involved.

Other mechanisms may reduce the Ca^{2+}-sensitivity of the contractile elements. For example, cAMP, cGMP, and compounds that cause the synthesis of these second messengers, such as isoproterenol, forskolin, and nitroprusside, reduce tonic contraction of vascular or tracheal muscle with little or no change in $[Ca^{2+}]_i$ (for review, see Karaki, 1989). During the contraction-relaxation cycle of isolated toad stomach muscle cells, Yagi et al. (1981) have shown that the relationship between $[Ca^{2+}]_i$ and force is represented by a clockwise hysteresis loop indicating that Ca^{2+}-sensitivity decreases with time. Himpens and Casteels (1990) have also reported that Ca^{2+}-sensitivity decreased during contractions of the guinea-pig ileum elicited by elevated external K^+ solutions. Sakata et al. (1989) suggested that a decrease in Ca^{2+}-sensitivity may occur during the early phase of contractions induced by endothelin in rat aorta.

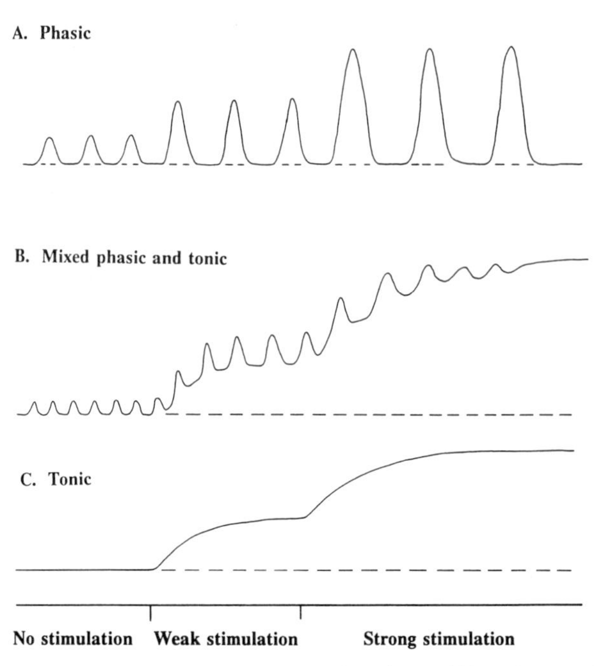

Figure 1. Representative contractile patterns for phasic (A) and tonic (C) muscles as described by Golenhofen (1976, 1981) (see Text). There are also examples of muscles that show a "mixed" pattern of contraction (panel B) where force development is characterized by phasic contractions superimposed upon a tonic rise in baseline tension.

We have studied whether changes in Ca^{2+}-sensitivity play a role in regulating the rhythmic contractions of gastrointestinal muscles such as the gastric antrum. Contractions in the distal stomach are organized into peristaltic waves that spread toward the pyloric sphincter. *In vitro,* stimulation by moderately elevated external K^+, acetylcholine (ACh), or pentagastrin increases the amplitude of phasic contractions, with little or no elevation in tone (for review, see Szurszewski, 1987). Golenhofen (1976, 1981) classified this type of muscle as "phasic muscle", and contrasted its activity with "tonic muscles", such as many vascular and airway smooth muscles which maintain force over time without visible rhythmic fluctuations (i.e. generate tone) (Fig. 1). Some smooth muscles are not purely phasic or tonic, but instead generate an intermediate pattern of contraction that is a mixture of phasic contractions superimposed upon tonic contraction. Muscles of the gastric fundus, guinea-pig taenia caecum, longitudinal layer of the ileum, and some veins are examples of the latter type.

RESULTS

Antral circular muscle is electrically and mechanically active in the absence of extrinsic stimuli (for review, see Sanders and Publicover, 1989). The occurrence of electrical slow wave activity leads to the development of tension, and the amplitude and duration of slow waves is an important determinant in the strength of contractions elicited (Morgan and Szurszewski, 1980). This is

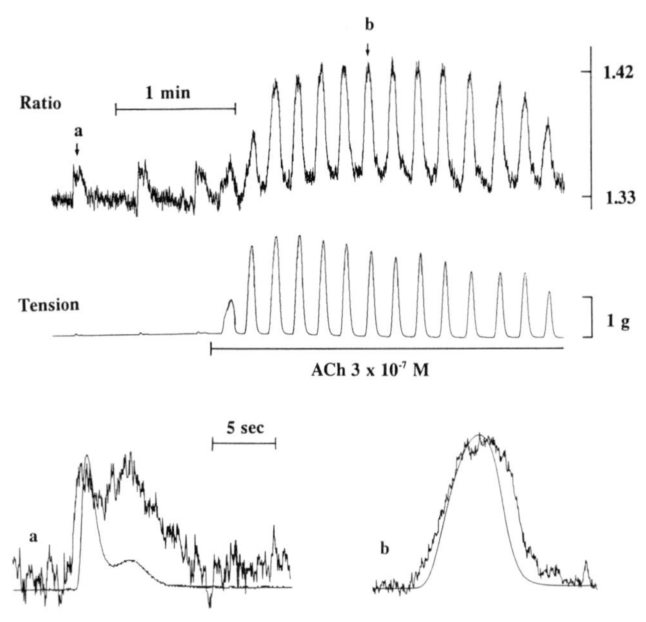

Figure 2. $[Ca^{2+}]_i$-force relationship of circular muscle from the canine antrum in response to ACh. Before adding ACh (3 x 10^{-7} M), small Ca^{2+} transients (top trace) and contractions (bottom trace) were observed. Addition of ACh increased the amplitude and frequency of Ca^{2+} transients, and increased resting $[Ca^{2+}]_i$. The increase in the amplitude of the Ca^{2+} transients caused an increase in the amplitude of phasic contractions. However, the increase in baseline $[Ca^{2+}]_i$ had no effect on resting tension. The lower traces show Ca^{2+} transients and contractions at an expanded scale before and during ACh (scales were normalized to 100% of maximum response). $[Ca^{2+}]_i$ was monitored by monitoring fluorescence ratio (405/500 nm fluorescence ratio excited at 340 nm) in indo-1/AM treated antral tissue while simultaneously measuring muscle tension.

because the plateau phase of slow waves is associated with sustained activation of the dihydropyridine-sensitive Ca^{2+} current (Vogalis et al., 1991), and the Ca^{2+} that enters cells during slow waves results in a rise in $[Ca^{2+}]_i$ above the threshold for contraction (H. Ozaki et al., 1991). Figure 2 shows an example of the changes in $[Ca^{2+}]_i$ and muscle tension that occur in circular muscle of the canine antrum. Force development was initiated when $[Ca^{2+}]_i$ reached approximately 10 - 30% of the peak of the Ca^{2+} transient, and this occurred approximately 0.3 sec after the first resolvable increase in $[Ca^{2+}]_i$. Under these recording conditions, force reached a maximum within 0.2 sec after the maximum level of $[Ca^{2+}]_i$ was reached. Ca^{2+} transients and contractions consisted of two phases. The first phase of the Ca^{2+} transient and contraction corresponded to the upstroke of depolarization, and the second phase corresponded to the plateau of depolarization (H. Ozaki et al., 1991). Although the magnitude of the second phase of the Ca^{2+} transient was equal to (Fig. 2A) or greater than (Fig. 2B) the first phase, the second contraction was, in these cases, smaller than the first contraction. Furthermore, in both cases, muscle tension decreased more rapidly than the decrease in $[Ca^{2+}]_i$. These results suggest that Ca^{2+}-sensitivity of the contractile element may decrease as a

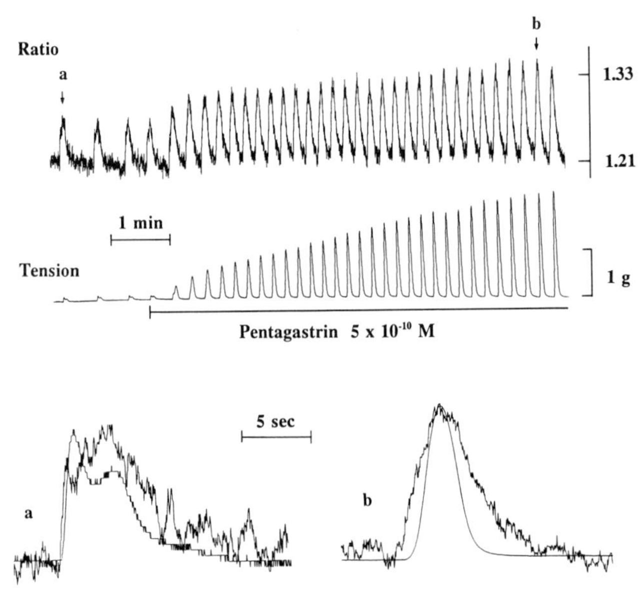

Figure 3. $[Ca^{2+}]_i$-force relationship of circular muscle from the canine antrum in response to pentagastrin. Before adding pentagastrin (5 x 10^{-10} M), small Ca^{2+} transients (top trace) and contractions (bottom trace) were observed. Addition of pentagastrin increased the amplitude and frequency of Ca^{2+} transients, and increased resting $[Ca^{2+}]_i$. The increase in the amplitude of the Ca^{2+} transients caused an increase in the amplitude of phasic contractions. However, the increase in baseline $[Ca^{2+}]_i$ had no effect on resting tension. The lower traces show Ca^{2+} transients and contractions at an expanded scale before and during pentagastrin (scales were normalized to 100% of maximum response). $[Ca^{2+}]_i$ and tension were measured as in Fig. 2.

function of time during spontaneous phasic contractions. This hypothesis is further supported by studies of agonist induced changes in $[Ca^{2+}]_i$.

ACh (3 x 10^{-7} M; see Fig. 2) and pentagastrin (5 x 10^{-10} M; see Fig. 3) raised baseline or "resting" $[Ca^{2+}]_i$, increased the amplitude of Ca^{2+} transients, and increased the force of phasic contractions. Despite the increase in the level of resting $[Ca^{2+}]_i$, ACh and pentagastrin did not increase the level of tension between phasic contractions. In the presence of high K^+ (40.4 mM), the increase in baseline $[Ca^{2+}]_i$ averaged 180% of the peak increase in $[Ca^{2+}]_i$ observed during spontaneous Ca^{2+} transients (referred to as 100% response). This magnitude change in $[Ca^{2+}]_i$ during spontaneous activity, however, induced only an 80% increase in resting tension of the reference transient contraction (data not shown). The phasic contractions induced by ACh and pentagastrin also decreased more rapidly than $[Ca^{2+}]_i$. These results suggest that contractile force is not simply proportional to $[Ca^{2+}]_i$ and that the Ca^{2+}-sensitivity of the contractile regulatory mechanisms may change during continuous elevation of $[Ca^{2+}]_i$. One explanation for these observations might be that an inactivation mechanism decreases the sensitivity of the contractile element to Ca^{2+}.

DISCUSSION

Smooth muscle contraction is thought to be initiated by myosin light chain (MLC) phosphorylation via the activation of Ca^{2+}/calmodulin-dependent MLC kinase (for reviews see, Kamm and Stull, 1985; Hartshorne, 1987). According to this model, relaxation is thought to be initiated by the reduction of $[Ca^{2+}]_i$ and resultant decrease in MLC kinase activity. When MLC kinase activity decreases, myosin is dephosphorylated by MLC phosphatase. At present the regulation of dephosphorylation has not been clarified, and most investigators have assumed that dephosphorylation occurs by an unregulated process (for review, see Hartshorne, 1987).

From studies of the guinea-pig ileum, Somlyo and Somlyo (1990) have suggested that dephosphorylation of myosin may be enhanced by activation of phosphatases, because the fall in tension was greater than would be predicted by

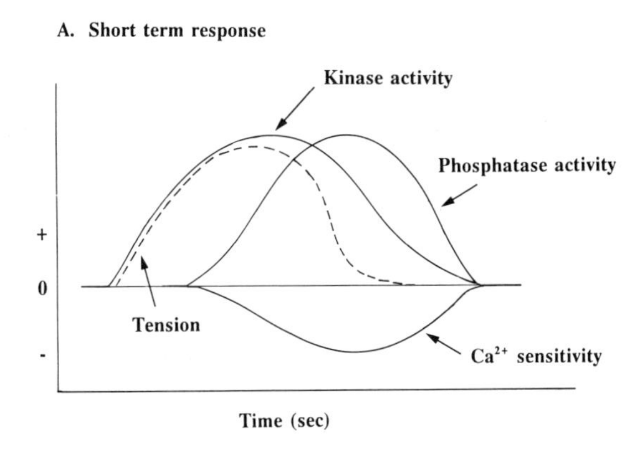

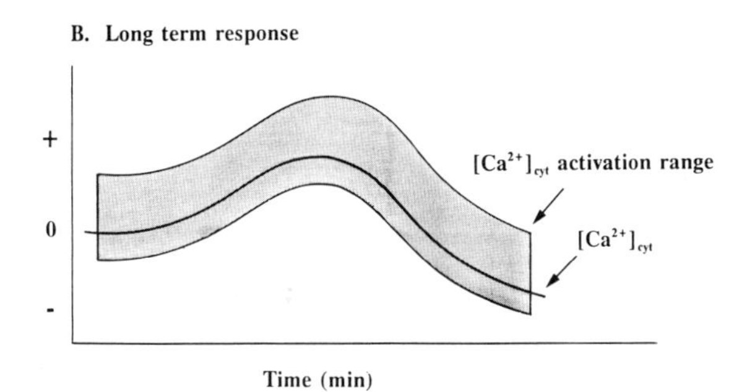

Figure 4. Changes in the Ca^{2+}-sensitivity in response to $[Ca^{2+}]_i$. Panel A shows a hypothetical model relating the change in Ca^{2+}-sensitivity to proposed changes in the activity of MLC kinase and phosphatase enzymes. Initially, the entry of Ca^{2+} activates the kinase and tension (dotted line) develops. With time, the phosphatase is also activated and this reduces the degree of phosphorylation of myosin and reduces tension. This can occur despite a maintained level of $[Ca^{2+}]_i$ and therefore constitutes a decrease in the Ca^{2+} sensitivity of the contractile element. Panel B shows the proposed long term effects of enhanced or depressed $[Ca^{2+}]_i$. As $[Ca^{2+}]_i$ increases, the threshold and range of $[Ca^{2+}]_i$ (shaded area) that would result in contraction is shifted (i.e. decrease in Ca^{2+} sensitivity). This figure also proposes that a long-term decrease in $[Ca^{2+}]_i$ may result in an increase in Ca^{2+}-sensitivity.

the decline in $[Ca^{2+}]_i$. To further study the mechanism of the Ca^{2+}-desensitization in canine antrum, we have measured MLC phosphorylation during phasic contractions induced by ACh (H. Ozaki and W. T. Gerthoffer, unpublished observations). In these contractions the change in MLC phosphorylation during the relaxation phase closely correlated with the change in muscle tension but not with $[Ca^{2+}]_i$. The rates of relaxation and MLC dephosphorylation were significantly greater than the decrease in $[Ca^{2+}]_i$.

At present it is unclear what mediates the decrease in Ca^{2+}-sensitivity in gastric muscles, but it is possible that MLC phosphorylation (activation) and dephosphorylation (inactivation) may both be regulated by elevation of $[Ca^{2+}]_i$. When $[Ca^{2+}]_i$ increases, the initial response may be MLC phosphorylation and contraction. With time, elevated $[Ca^{2+}]_i$ might lead to an increase in the activity of the phosphatase(s) that dephosphorylate myosin. The relatively rapid increase in phosphorylation and delay in the acceleration of dephosphorylation would tend to produce a time-dependent decrease in the sensitivity of the contractile element to Ca^{2+} as the "activation-inactivation balance" shifts from an initially high level of MLC phosphorylation to an increasingly higher rate of dephosphorylation (see Fig. 4A). This shift in the "activation-inactivation balance" may explain why muscle tension decreases more rapidly than $[Ca^{2+}]_i$. Differences in the time-dependence and Ca^{2+}-dependence of the activation and inactivation processes may determine the amplitude of contractions in phasic muscles.

A time-dependent change in the "activation-inactivation balance" might also explain why the slow increase in baseline $[Ca^{2+}]_i$ in response to ACh, pentagastrin, and high K^+ did not induce tonic contraction. Since the reduction in Ca^{2+}-sensitivity occurs within a few seconds (i.e. within the time-course of phasic contractions), a slow increase in $[Ca^{2+}]_i$ may not be fast enough to initiate the activation process (MLC phosphorylation) before the enhancement of the inactivation process (MLC dephosphorylation). Although the increase in $[Ca^{2+}]_i$ may increase the rate of MLC phosphorylation, an increase in the rate of dephosphorylation may create a condition where there is no net increase in MLC phosphorylation. Thus, a slow increase in $[Ca^{2+}]_i$ might result in a shift in the threshold level of Ca^{2+} necessary to elicit the development of force (Fig. 4B). In guinea-pig ileum (Himpens et al., 1989), swine carotid artery (Rembold, 1989), and rat aorta (H. Karaki, personal communication), readdition of Ca^{2+} (lower concentration) in the continuous presence of high K^+ induced sustained increases in $[Ca^{2+}]_i$ without substantial mechanical responses. In these experiments, the increase in $[Ca^{2+}]_i$ by readdition of less-than-physiological concentrations of Ca^{2+} was slower than when high K^+ is applied in the presence of normal external Ca^{2+}. With a slow rise in $[Ca^{2+}]_i$, the inactivation process we have described may have shifted the Ca^{2+} threshold for contraction to higher levels and inhibited the development of force.

Relaxation, resulting from dephosphorylation of MLC, may be mediated by more than a single type of phosphatase. In order to evaluate this possibility, we have studied the effects of phosphatase inhibitor, calyculin-A (H. Ozaki, unpublished observations). In gizzard smooth muscle a Type-1 phosphatase mainly contributes to the dephosphorylation of MLC (Ishihara et al., 1989). This phosphatase is very sensitive to calyculin-A (IC_{50} for phosphorylated MLC, 1.6×10^{-9} M; Ishihara et al., 1989). In canine antral muscles calyculin-A induced a slow contraction and a significant increase in MLC phosphorylation without a significant increase in $[Ca^{2+}]_i$. In the presence of calyculin-A, Ca^{2+}

transients and phasic contractions persisted and were superimposed upon the tonic contraction, but calyculin-A changed the characteristics of the relaxation of phasic contractions. In calyculin-A, relaxation was biphasic, consisting of an initial, rapid phase of relaxation that was not significantly affected by calyculin-A, and a second phase, the rate of which was greatly reduced by calyculin-A. The two phases of relaxation may be due to the action of two phosphatases, and only one of these may be sensitive to calyculin-A. Calyculin-A is far less effective as an inhibitor of Type-2B phosphatase (IC_{50} for phosphorylated MLC, 7.2×10^{-7} M; Ishihara et al., unpublished observation). The Type-2B phosphatase is Ca^{2+}/calmodulin-dependent and dephosphorylates skeletal muscle MLC *in vitro* (Ingebritsen and Cohen, 1983; Stewart et al., 1983). At the dose of calyculin-A used in the present study, it is possible that the initial dephosphorylation (rapid phase of relaxation) is due to Type-2B phosphatase, and the slower relaxation may have been due to dephosphorylation by residual Type-1 phosphatase or an unknown calyculin-A-sensitive phosphatase.

In summary, in "phasic muscle" of canine antrum, the Ca^{2+}-sensitivity of the contractile element changes in a Ca^{2+}- and time-dependent manner during the time-course of phasic contractions. It is conceivable that a Ca^{2+}-dependent phosphatase may mediate this mechanism. Ca^{2+}-desensitization may facilitate the rhythmic, peristaltic contractions of the distal stomach and protect this region of the stomach from summation of contractions or tonic contraction.

ACKNOWLEDGMENTS

This work was supported by NIH grants DK 32176 to NGP and KMS; and DK 41315 to WTG and KMS.

REFERENCES

DeFeo, T. T. and Morgan, K. G., 1985, Calcium-force relationship as detected by aequorin in two different vascular smooth muscle of the ferret, *J. Physiol.*, 369: 269.

Gerthoffer, W. T., Murphey, K. A., and Gunst, S. J., 1989, Aequorin luminescence, myosin phosphorylation, and active stress in tracheal smooth muscle, *Am. J. Physiol.*, 257: C1062.

Golenhofen, K., 1976, Theory of P and T systems for calcium activation in smooth muscle, *in:* "Physiology of Smooth Muscle", E. Bülbring and M. F. Shuba, eds., Raven Press, New York, p. 197.

Golenhofen, K., 1981, Differentiation of calcium activation processes in smooth muscle using selective antagonists, *in:* "Smooth Muscle: An Assessment of Current Knowledge", E. Bülbring, A. F. Brading, A. W. Jones, and T. Tomita, eds., University of Texas Press, Austin, p. 157.

Hartshorne D. J., 1987, Biochemistry of the contractile process in smooth muscle, *in:* "Physiology of Gastrointestinal Tract", L. R. Johnson, ed., Raven Press, New York, p. 423.

Himpens, B., Matthjis, G., and Somlyo, A. P., 1989, Desensitization to cytosolic Ca^{2+} and Ca^{2+} sensitivity in guinea-pig ileum and rabbit pulmonary artery, *J. Physiol.*, 413: 489.

Himpens, B. and Casteels, R., 1990, Different effects of depolarization and muscarinic stimulation on the Ca^{2+}/force relationship during the contraction-relaxation cycle in the guinea-pig ileum, *Pflügers Arch.*, 416: 28.

Ingebritsen, T. S. and Cohen, P., 1983, The protein phosphatases involved in cellular regulation. 1. Classification and substrate specificities, *Eur. J. Biochem.*, 132: 255.

Ishihara, H., Martin, B. L., Brautigan, D. L., Karaki, H., Ozaki, H., Kato, Y., Fusetani, N., Watabe, S., Hashimoto, K., Uemura, D., and Hartshorne, D. J., 1989, Calyculin A and okadaic acid: Inhibitors of protein phosphatase activity, *Biochem. Biophys. Res. Commun.*, 159: 871.

Kamm, K. E. and Stull, J. T., 1985, The function of myosin and myosin light chain kinase phosphorylation in smooth muscle, *Ann. Rev. Pharmacol. Toxicol.*, 25: 593.

Karaki, H., 1989, Ca^{2+} localization and sensitivity in vascular smooth muscle, *Trends Pharmacol. Sci.*, 10: 320.

Mitsui, M. and Karaki, H., 1990, Dual effects of carbachol on cytosolic Ca^{2+} and contraction in the intestinal smooth muscle, *Am. J. Physiol.*, 258: C787.

Morgan, J. P. and Morgan, K. G., 1984, Stimulus-specific patterns of intracellular calcium levels in smooth muscle of ferret portal vein, *J. Physiol.*, 351: 312.

Morgan, K. G. and Szurszewski, J. H., 1980, Mechanism of phasic and tonic actions of pentagastrin on canine gastric smooth muscle, *J. Physiol.*, 301: 229.

Mori, T., Yanagisawa, T., and Taira, N., 1990, Histamine increases vascular tone and intracellular calcium level using both intracellular and extracellular calcium in porcine coronary arteries, *Jpn. J. Pharmacol.*, 52: 263.

Ozaki, H., Satoh, T., Karaki, H., and Ishida, Y., 1988, Regulation of metabolism and contraction by cytoplasmic calcium in the intestinal smooth muscle, *J. Biol. Chem.*, 263: 14074.

Ozaki, H., Ohyama, T., Sato, K., and Karaki, H., 1990a, Ca^{2+} dependent and independent mechanism of sustained contraction in vascular smooth muscle of rat aorta, *Jpn. J. Pharmacol.*, 52: 509.

Ozaki, H., Kwon, S.-C., Tajimi, M., and Karaki, H., 1990b, Changes in cytosolic Ca^{2+} and contraction induced by various stimulants and relaxants in canine tracheal smooth muscle, *Pflügers Arch.*, 416: 351.

Ozaki, H., Stevens, R. J., Blondfield, D. P., Publicover, N. G., and Sanders, K. M., 1991, Simultaneous measurement of membrane potential, cytosolic Ca^{2+} and tension in intact smooth muscle, *Am. J. Physiol.*, 260: C917.

Rembold, C. M., 1989, Desensitization of swine arterial smooth muscle to transplasmalemmal Ca^{2+} influx, *J. Physiol.*, 416: 273.

Sakata, K., Ozaki, H., Kwon, S.-C., and Karaki, H., 1989, Effects of endothelin on the mechanical activity and cytosolic calcium levels of various types of smooth muscle, *Br. J. Pharmacol.*, 98: 483.

Sanders, K. M. and Publicover, N. G., 1989, Electrophysiology of the gastric musculature, *in:* "Handbook of Physiology, The Gastrointestinal System, Vol I.", S. G. Schultz and J. D. Wood., eds., The American Physiological Society, Bethesda, p. 187.

Sato, K., Ozaki, H., and Karaki, H., 1988, Changes in cytosolic calcium level in vascular smooth muscle strip measured simultaneously with contraction using fluorescent calcium indicator fura-2, *J. Pharmacol. Exp. Ther.*, 246: 294.

Somlyo, A. P. and Himpens, B., 1989, Cell calcium and its regulation in smooth muscle, *FASEB J.*, 3: 2266.
Somlyo, A. P. and Somlyo, A. V., 1990, Flash photolysis studies of excitation-contraction coupling, regulation and contraction in smooth muscle, *Ann. Rev. Physiol.*, 52: 857.
Stewart, A. A., Ingebritsen, T. S., and Cohen, P., 1983, The protein phosphatases involved in cellular regulation. 5. Purification of and properties of a Ca^{2+}- and calmodulin-dependent protein phosphatase (2B) from rabbit skeletal muscle, *Eur. J. Biochem.*, 132: 289.
Szurszewski, J. H., 1987, Electrical basis for gastrointestinal motility, *in:* "Physiology of Gastrointestinal Tract", L. R., Johnson, ed., Raven Press, New York, p. 383.
Vogalis, F., Publicover, N. G., Hume, J., and Sanders, K. M., 1991, Relationship between calcium current and cytosolic calcium concentration in canine gastric smooth muscle cells, *Am. J. Physiol.*, 260: C1012.
Yagi, S., Becker P. L., and Fay, F. S., 1988, Relationship between force and Ca^{2+} concentration in smooth muscle as revealed by measurement on single cells, *Proc. Nat'l. Acad. Sci. U.S.A.*, 85: 4109.

CALCIUM-DEPENDENT AND INDEPENDENT MECHANISMS OF CONTRACTION IN CANINE LINGUAL ARTERY TO U-46619

Stan S. Greenberg, Ye Wang, Jianming Xie, Freidrich P. J. Diecke, Fred A. Curro, Lisa Smartz, and Louis Rammazzatto

Department of Physiology
UMDNJ-New Jersey Medical School
Newark, NJ 07103

Department of Dental Research
St. Josephs Hospital
Paterson, NJ 07586

INTRODUCTION

The thromboxane A_2/prostaglandin H_2 (TXA_2/PGH_2) mimetic, U-46619, is a potent platelet aggregating agent and constrictor of vascular smooth muscle. Aggregation and contraction to U-46619, and other TXA_2/PGH_2 mimetics, such as U-44069 and carbocyclic TXA_2, are associated with both an influx of extracellular Ca^{2+} and a release of Ca^{2+} from binding and sequestration sites within the platelet and arterial smooth muscle (Greenberg, 1981; Loutzenhiser and van Breemen, 1981; Dorn et al., 1987; Santoian et al., 1987). These mechanisms were shown by the ability of both calcium channel blocking agents and intracellular calcium antagonists, such as dantrolene sodium and TMB-8, to inhibit contraction of vascular smooth muscle to TXA_2/PGH_2 receptor mimetics and by the ability of these mimetics to increase ^{45}Ca efflux without affecting ^{45}Ca uptake (Ally et al., 1978; Greenberg, 1981; Loutzenhiser and Van Breemen, 1981; Angerio et al., 1982; Dorn et al., 1987; Santoian et al., 1987). The pool of Ca^{2+} utilized for the physiologic and pharmacologic actions of U-46619 appears to depend on the site and species from which the arteries and platelets are obtained (Greenberg, 1981; Loutzenhiser and Van Breemen, 1981; Angerio et al., 1982; Dorn et al., 1987; Santoian et al., 1987; Verheyen et al., 1989).

The contraction of vascular smooth muscle to U-46619 appears mediated by two binding sites linked to distinct effector systems: a common binding site shared with primary prostaglandins and a secondary, definitive site which can be used to discriminate between the binding of a TXA_2/PGH_2 receptor agonist from its antagonist (Halushka et al., 1989; Hanasaki and Arita, 1989). One receptor to which U-46619 appears to bind may be linked to influx of extracellular Ca^{2+}, whereas the second receptor may be linked to activation of phospholipase C and the resultant formation of IP_3 and DAG and mobilization

Regulation of Smooth Muscle Contraction
Edited by R.S. Moreland, Plenum Press, New York, 1991

of intracellular Ca^{2+} (Rasmussen et al., 1987; Mene et al., 1988). However, recent studies suggested that the canine lingual artery (LA) behaves differently to U-46619 than porcine and canine systemic vessels in that it appears to contract by a calcium-independent mechanism (Greenberg et al., 1991; Wang et al., 1991). This study evaluates the mechanism by which U-46619 contracts the canine LA.

MATERIALS AND METHODS

Dogs of either sex (10 - 23 kg) were anesthetized with sodium pentobarbital (35 mg/kg, i.v.). Paired rings of normal and endothelium-rubbed LA and mesenteric artery (MA) were prepared for isometric recording of force development with methods previously described in detail (Wang et al., 1991).

The blood vessels were incubated for 3 hr in normal physiological salt solution (PSS), PSS containing either EGTA (2 mM), caffeine (5 mM), or ryanodine (20 μM), to deplete calcium from the SR, or PSS containing both EGTA and caffeine or ryanodine. This was followed by complete cumulative concentration-response curves for U-46619. An additional series of studies were performed on four paired rings of LA in which one artery served as the control tissue for sequential, complete concentration response curves for KCl, NE, and U-46619 and the remaining three paired LA were pretreated for 30 min before and throughout the experiment with increasing concentrations of vehicle, nitrendipine (0.05, 0.5, or 5 μM), or dichlorobenzamil (0.1, 1, or 10 μM). When steady state responses to U-46619 (1 μM) were evident, the muscle chambers were washed rigorously with the appropriate PSS and the tension of the maintained contraction to U-46619 followed for 150 min. This procedure was repeated on LA incubated overnight in oxygenated PSS (pH 7.4, 4° C) containing vehicle, staurosporine (0.1 μM and 1 μM), amphotericin-B (10 μM), or PMA (0.5 μM).

Triton X-100 Skinned Lingual Artery

Strips of LA were exposed for 2 hr to a skinning solution containing 1% Triton X-100, followed by relaxing solution, and then by test solutions. The test solutions contained variable amounts of free Ca^{2+} or U-46619. All solutions contained the protease inhibitors leupeptin and chymostatin (2 μg/ml). Complete, cumulative concentration response curves to free Ca^{2+} were obtained. The muscle chambers were washed with imidazole-PSS containing 5 mM EGTA and the cumulative concentration response curve for free Ca^{2+} evaluated before and 20 min after addition of buffer or 0.1 or 10 μM U-46619. In a separate series of experiments, cumulative concentration response curves were obtained to U-46619.

RESULTS AND DISCUSSION

Responses of LA and MA to U-46619

U-46619 contracted the LA and MA. Under the conditions of these experiment, the contractions of LA and MA to U-46619 consisted of three phases: a rapid contraction, a sustained contraction and, after washout of drug, a maintained contraction. The absolute magnitude of the steady state

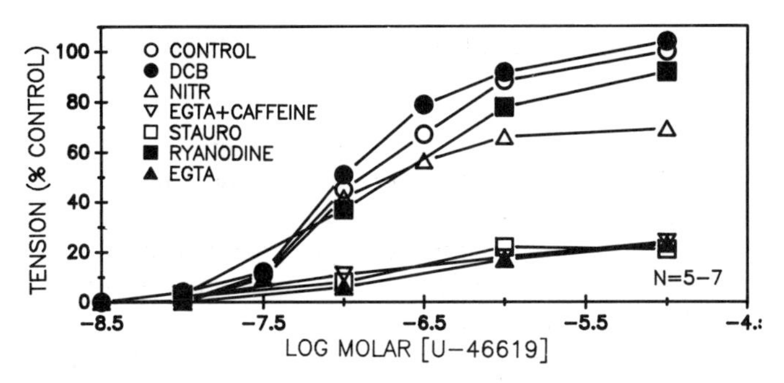

Figure 1. Effect of 2 mM EGTA (pCa = 6.8), nitrendipine (NITR, 0.5 μM), staurosporine (STAURO, 0.1 μM), ryanodine (20 μM), caffeine (5 mM) + 2 mM EGTA, and dichlorobenzamil (DCB, 50 μM) on the contraction of canine LA to U-46619. The ordinate is the response to U-46619 expressed as a percent of control maximum.

contraction was variable among LA obtained from different dogs. When U-46619 was washed out of the muscle chamber the MA slowly relaxed but the LA remained contracted for at least 2 - 3 hr. The contraction remaining after washout of U-46619 is hereafter referred to as "maintained contraction". Endothelium-rubbing inhibited acetylcholine-induced relaxation of the LA and MA but did not affect any contractions to U-46619 (Greenberg et al., 1991).

In LA the initial magnitude of the rapid and sustained contraction to U-46619 appears to be mediated primarily by influx of extracellular Ca^{2+} through receptor operated calcium channels and by indirect activation of protein kinase C. Incubation of these arteries with 2 mM EGTA or staurosporine, a potent, relatively selective inhibitor of PKC for 30 min, equally inhibited the contractions to U-46619. However, nitrendipine had little effect on the contractions to U-46619 (Fig. 1). DAG and PKC, rather than IP_3, appear to play a major role in U-46619 induced contraction of the LA since both EGTA and staurosporine inhibited 80% of the contraction of the LA to U-46619 while ryanodine and caffeine have little effect (Fig. 1).

Agonists such as phenylephrine produce a rapid initial contraction of rabbit aorta followed by a sustained contraction. In contrast, activators of PKC produce only a slow tonic contractile response in rabbit aorta (Rasmussen et al., 1987; Khalil and van Breemen, 1988). The sustained contractions to both agonists are reduced, but not completely abolished in Ca^{2+}-free solution containing 2 mM EGTA. The authors concluded that these findings emphasized the importance of Ca^{2+} influx and suggested only a minor role for PKC in maintaining tonic contractions of vascular smooth muscle. Similar conclusions were reached by Rembold and Weaver (1990) in porcine carotid artery, using histamine as an agonist. In contrast, tension development in the LA appears to be mediated by DAG-induced stimulation of PKC as well as calcium-calmodulin dependent activation of myosin light chain kinase. DAG may also act independently of PKC. 1,2-*sn*-Dioctanoylglycerol, an analog of DAG, increases cytosolic Ca^{2+} and acidification of T lymphocytes through a PKC-independent process (Ebanks et al., 1989). Moreover, clearly evident in Figure 1 is that despite the efficacy of the inhibitors as antagonists of U-46619 induced contraction, approximately 13 - 20% of the contraction persists even in LA pretreated with a combination of EGTA and caffeine or staurosporine.

Modulation of the Maintained Contraction of MA and LA after Washout of U-46619

The contractions of the LA to U-46619 fail to demonstrate a relaxation phase when U-46619 is washed out of the muscle chamber. The LA remains contracted for at least 2 - 3 hr after rigorous washout of U-46619 (Fig. 2). Addition of 2 mM EGTA to PSS does not affect the maintained contraction of the LA to U-46619. In contrast, when the MA is used for comparison, EGTA produces almost complete relaxation of the maintained contraction of this blood vessel (Figure 2). Moreover, neither staurosporine (0.1 μM), ryanodine (20 μM), or dichlorobenzamil (50 μM) relax LA when added during the maintained contraction phase of the response to U-46619 (Fig. 2). Similarly, as shown in Table 1, pretreatment of LA with these pharmacologic interventions reduces the absolute tension reached during the sustained and maintained contraction to U-46619, but does not affect the duration of the maintained contraction to U-46619. Thus, the duration of the maintained contraction of the LA to U-46619 appears to be independent of Ca^{2+} and PKC.

Table 1. Effect of Pretreatment With Inhibitors On Contractions to 10 μM U-46619

Pretreatment	Sustained	Maintained	% Sustained*
	g ± SEM		% ± SEM
Control	2.8 ± 0.1	2.3 ± 0.4	87 ± 4
EGTA, 2 mM	0.43 ± 0.05	0.39 ± 0.08	92 ± 5
Nitrendipine, 0.5 μM	2.4 ± 0.5	2.1 ± 0.4	85 ± 6
Ryanodine, 20 μM	2.1 ± 0.6	1.8 ± 0.7	83 ± 7
Caffeine, 5 mM	3.1 ± 0.7	2.7 ± 0.8	84 ± 8
Caffeine + EGTA	0.41 ± 0.04	0.35 ± 0.06	81 ± 8
Dichlorobenzamil, 50 μM	2.3 ± 0.4	2.0 ± 0.5	86 ± 5
Staurosporine, 0.1 μM	0.77 ± 0.38	0.69 ± 0.43	84 ± 8
Amphotericin B, 10 μM	1.53 ± 0.37	1.38 ± 0.46	88 ± 9

*percent of sustained tension remaining after 60 minutes of washout.

Triton X-100 Skinned Lingual Artery

U-46619 contracts the Triton X-100 skinned LA in the presence of 0.15 μM of Ca^{2+} but does not enhance the sensitivity to Ca^{2+} (Fig. 3), whereas PKC-mediated contractions are postulated to sensitize the myofilaments to Ca^{2+} (Rasmussen et al., 1987). These findings support the conclusion that the maintained contraction of the LA to U-46619 cannot result from direct or indirect stimulation PKC or sensitization of the contractile filaments to Ca^{2+}. These findings emphasize the direct contractile filament activation in both intact and skinned LA in response to U-46619-mediated contractions and for maintained tonic contractions following washout of U-46619.

It may be argued that the maintained tension of both the intact and Triton X-100 skinned LA following washout of U-46619 from the muscle chamber simply results from a slow dissociation of U-46619 from its receptor. This is a simplistic and attractive explanation because of the lipophilicity of U-46619. However, the concept that the maintained contraction of the LA must be dependent on an agonist-membrane receptor interaction is not supported by the data because U-46619 contracts the Triton X-100 skinned artery which is

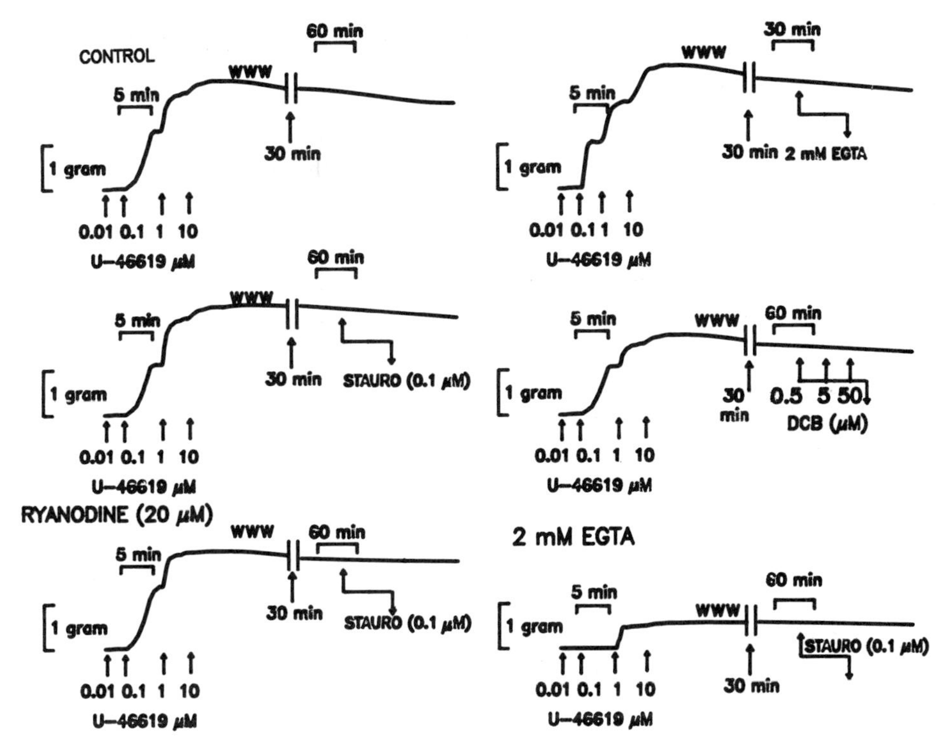

Figure 2. A series of representative tracings showing the phasic, sustained, and maintained contractions of intact LA to U-46619 (given at arrows), and the effect of addition of 2 mM EGTA, 2 mM EGTA + 5 mM caffeine, staurosporine, ryanodine, and dichlorobenzamil on the maintained contraction to U-4661. W=wash of muscle. Drugs were added and removed at the up and down arrows.

devoid of a functional plasmamembrane. Slow dissociation of U-46619 from its receptor binding sites on the arterial smooth muscle membrane may account for part of the long duration of the maintained contraction in both preparations. However, in the skinned artery, the receptor must be on the contractile proteins. Moreover, if the maintained contraction of LA to U-46619 is dependent on calcium influx, release of calcium from SR, or conventional activation of PKC, then the binding of U-46619 to its receptor site on the LA should not result in maintained contraction when these pathways are inactivated or inhibited. Yet small contractions can be elicited after depletion of extracellular and intracellular Ca^{2+}, inhibition of PKC, and after disruption of the cell membrane. Thus, it is unlikely that simple binding of U-46619 to its receptor solely accounts for the maintained contraction in either intact or Triton X-100 skinned LA. Therefore, U-46619 apparently contracts the canine LA by a membrane-receptor mediated mechanism and by a direct effect on the contractile filaments.

Our findings demonstrate that the contraction of the LA to U-46619 consists of: 1) a calcium and PKC-dependent component which accounts for approximately 80% of the maximum response; 2) a calcium and PKC-independent component which accounts for the residual 20% of the contraction; and 3) a calcium and PKC-independent maintained contractions which persists for hours after washout of U-46619 from the muscle chamber.

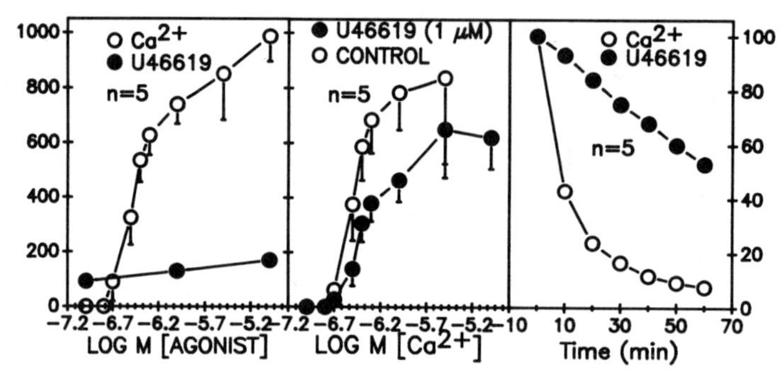

Figure 3. Cumulative-concentration response curves for Ca^{2+} and U-46619 in Triton X-100 skinned LA (Left) and for Ca^{2+} in the absence and presence of 1 μM U-46619 (Center). Right panel: Time for tension to return to control values after addition of relaxing solution to Ca^{2+} or U-46619 contracted skinned LA.

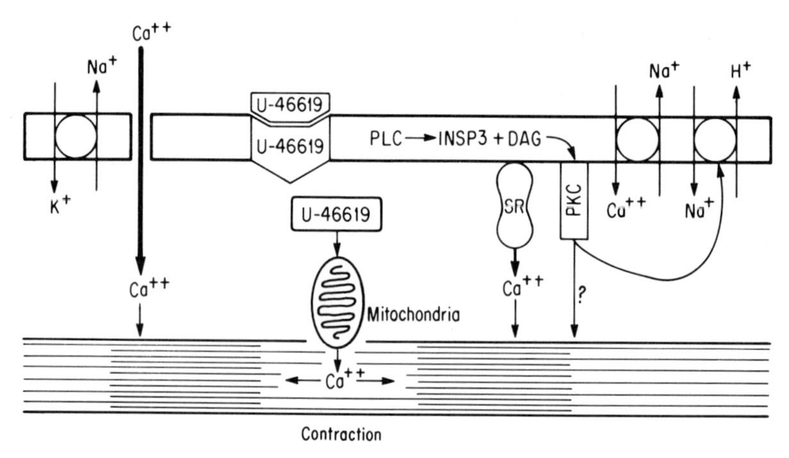

Figure 4. Postulated scheme for the mechanism of action of U-46619 induced contraction in canine lingual artery.

The absolute magnitude of both the calcium and PKC-dependent sustained contraction and the maintained contraction is initially determined by the availability of extracellular Ca^{2+} and the integrity of PKC. However, the maintained contraction disencumbers itself from this dependency and becomes independent of PKC and the calcium pools affected by ryanodine, caffeine, and EGTA. Because U-46619 contracts the Triton X-100 skinned LA but does not enhance its sensitivity to Ca^{2+} we conclude that the calcium and PKC-independent and maintained contractions of the LA to U-46619 result in part from direct stimulation of the contractile filaments by U-46619 (Figure 4). Further studies are now ongoing to evaluate these possibilities.

ACKNOWLEDGMENTS

The investigators would like to thank Dr. Alexander Fabiato for providing us with a working copy of his program for the calculation of free ions in a solution containing multiple metals and ligands. We also thank

Berlex Laboratories, Inc. Cedar Knolls, NJ for providing many of the dog tongues used in this study. This research was supported in part by grant from the New Jersey Affiliate of the American Heart Association NJHA-89-002. Dr. Xie and Dr. Wang are recipients of post-doctoral fellowships of the New Jersey Affiliate of the American Heart Association.

REFERENCES

Ally, A. I., Horrobin, D. F., Manku, M. S., Morgan, R. O., Karmazyn, M, Karmali, R. A., and Cunnane, S. C., 1978, Dantrolene blocks intracellular calcium release in smooth muscle: Competitive antagonism of thromboxane A2, *Can. J. Physiol. Pharmacol.*, 56: 520.

Angerio, A. D., Fitzpatrick, T. M., Kot, P. A., Ramwell, P. W., Rose, J. C., and Santoian, E. C., 1982, Effect of TMB-8 on the pulmonary vasoconstrictor action of prostaglandin F_2 alpha and the thromboxane mimic, U 46619, *Br. J. Pharmacol.*, 77: 55.

Dorn, G. W., Sens, D., Chaikhouni, A., Mais, D., and Halushka, P. V., 1987, Cultured human vascular smooth muscle cells with functional thromboxane A_2 receptors: Measurement of U46619-induced ^{45}calcium efflux, *Circ. Res.*, 60: 952.

Ebanks, R., Roifman, C., Mellors, A., and Mills, G. B., 1989, The diacylglycerol analogue, 1,2-sn-dioctanoylglycerol, induces an increase in cytosolic free Ca^{2+} and cytosolic acidification of T lymphocytes through a protein kinase C-independent process, *Biochem. J.*, 258: 689.

Greenberg, S., 1981, Effect of prostacyclin and 9α, 11α-epoxymethanoprostaglandin H_2 on calcium and magnesium fluxes and tension development in canine intralobar pulmonary arteries, *J. Pharmacol. Exp. Ther.*, 219: 326.

Greenberg, S., Wang, Y., Xie, J., Diecke, F. P. J., Smartz, L., Rammazzatto, L., and Curro, F. A., 1991, Thromboxane released during trauma and inflammation may contribute to tongue and periodontium dysfu nction, *J. Dental Res.*, in press.

Halushka, P. V., Mais, D. E., Mayeux, P. R., and Morinelli, T. A., 1989, Thromboxane, prostaglandin and leukotriene receptors, *Ann. Rev. Pharmacol. Toxicol.*, 29: 213.

Hanasaki, K. and Arita, H. A., 1989, A common binding site for primary prostanoids in vascular smooth muscles: A definitive discrimination of the binding for thromboxane A_2/prostaglandin H_2 receptor agonist from its antagonist, *Biochim. Biophys. Acta,* 1013: 28.

Khalil, R. A. and van Breemen, C., 1988, Sustained contraction of vascular smooth muscle: Calcium influx or C-kinase activation?, *J. Pharmacol. Exp. Ther.*, 244: 537.

Loutzenhiser, R. and van Breemen, C., 1981, Mechanism of activation of isolated rabbit aorta by PGH_2 analogue U-44069, *Am. J. Physiol.*, 241: C243.

Mene, P., Dubyak, G. R., Abboud, H. E., Scarpa, A., and Dunn, M. J., 1988, Phospholipase C activation by prostaglandins and thromboxane A_2 in cultured mesangial cells. *Am. J. Physiol.*, 255: F1059.

Rasmussen, H., Takuwa, Y., and Park, S., 1987, Protein kinase C in the regulation of smooth muscle contraction, *FASEB J.*, 1: 177.

Rembold, C. M. and Weaver, B. A., 1990, [Ca^{2+}], not diacylglycerol, is the primary regulator of sustained swine arterial smooth muscle contraction, *Hypertension,* 15: 692.

Santoian, E. C., Angerio, A. D., Fitzpatrick, T. M., Rose, J. C., Ramwell, P. W., and Kot, P. A., 1987, Effects of intracellular and extracellular calcium blockers on the pulmonary vascular responses to PGF_2 alpha and U46619, *Angiology,* 38: 51.

Verheyen, A., Lauwers, F., Vlaminckx, E., and De Clerck, F., 1989, Differential vasoreactivity to the thromboxane A_2 mimic U-46619 of collateral and normal peripheral blood vessels in in situ perfused rat hindquarters, *Blood Vessels,* 26: 165.

Wang Y., Xie, J., Greenberg, S., Diecke, F.P.J., Smartz, L., Rammazzatto, L., and Curro, F.A., 1991, Evidence for Ca^{2+} and protein kinase C dependent and independent contraction of lingual artery, *J. Pharmacol. Exp. Ther.,* in press.

MODE OF ACTION OF NEUROTRANSMITTERS ON DEPOLARIZED SMOOTH MUSCLE

Gertrude Falk

Department of Physiology
University College
London, WC1E 6BT England

Department of Pharmacology
University of Washington
Seattle, WA 98195

INTRODUCTION

That neurotransmitters could evoke tension changes of depolarized, electrically inexcitable smooth muscle was observed by Evans, Schild, and Thesleff (1958), an example of what was subsequently termed pharmacomechanical coupling by Somlyo and Somlyo (1968).

The aim of this study was to examine systematically the relation of membrane potential to contraction in a variety of intestinal and uterine smooth muscles and the way in which this relationship is modified by neurotransmitters.

A brief report of this work was published earlier (Falk and Landa, 1960).

METHODS

Isolated preparations of taenia coli of guinea pig and rabbit, uteri of the non-pregnant rat, guinea pig, and rabbit, and of the pregnant rat and mouse, ileum of guinea pig, rabbit, and rat were mounted in a chamber so that they were under a small resting tension. Usually an equilibration period of one hour was allowed before the start of recording.

Isometric tension was measured by means of a strain gauge, D.C. amplifier, and direct writing recorder. All results are reported as increases in tension above resting tension. Recording of membrane potentials with capillary microelectrodes was conventional. The electrodes were flexibly mounted by means of a 25 μm tungsten wire.

Normal Tyrode's solution consisted of (in mM): 137 NaCl, 13 $NaHCO_3$, 2.7 $CaCl_2$, 2.7 KCl, 0.5 $MgCl_2$, 3.5 NaH_2PO_4, and 5.6 glucose. K_2SO_4-Tyrode's, saturated with $CaSO_4$, contained (in mM): 126.5 K_2SO_4, 3.6 $KHCO_3$, 1.1 $CaCl_2$, 1.0 $MgCl_2$, and 5.6 glucose. KCl-Tyrode's had a similar composition to the latter with the exception that K_2SO_4 was replaced by its isosmotic equivalent, 168 mM

Regulation of Smooth Muscle Contraction
Edited by R.S. Moreland, Plenum Press, New York, 1991

KCl, and $CaSO_4$ was omitted. Unless otherwise noted in the text, solutions containing other concentrations of K^+ were prepared by diluting the standard isotonic K_2SO_4 or KCl-Tyrode's with sucrose Tyrode's in which the potassium salt was replaced by an osmotic equivalent of sucrose. A gas mixture of 95% O_2 - 5% CO_2 was passed through the perfusing solutions. All experiments were carried out at room temperature (23° - 28° C).

The following drugs were used: acetylcholine bromide, histamine phosphate, epinephrine hydrochloride, and papaverine hydrochloride.

RESULTS

Relation of External K^+ to Membrane Potential

Since the membrane potential was controlled by varying the external K^+ concentration it was necessary to define quantitatively the relationship of K^+ concentration to the membrane potential. Membrane potentials were determined in guinea pig taenia coli, pregnant and non-pregnant rat uteri, depolarized by varying K^+. The potential was linearly related to the log K^+ concentration above about 20 mM K^+ with a slope of 30 - 40 mV per tenfold change of K^+.

Relationship of Tension to Membrane Potential

It was somewhat difficult to determine the threshold membrane potential for smooth muscles since many were spontaneously active or become so following small depolarization. When normal Tyrode's solution was

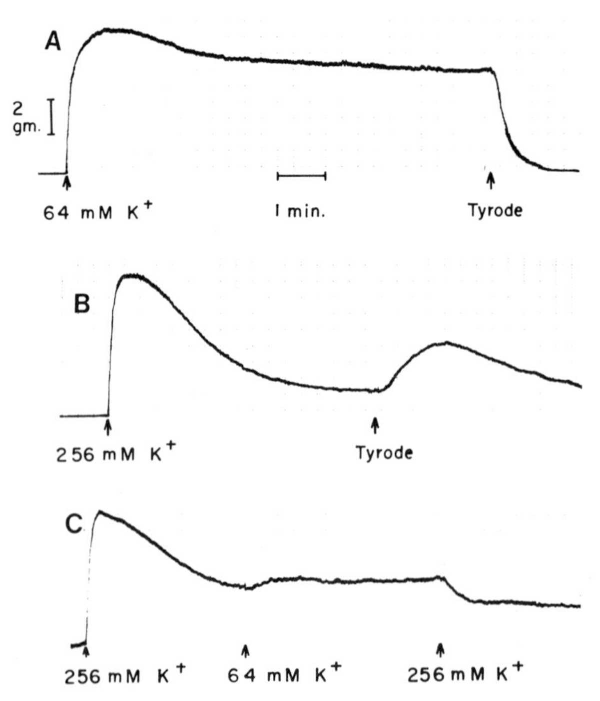

Figure 1. Effects of changes in membrane potentials on tension (anestrus rat uterus). The muscle was not spontaneously active.

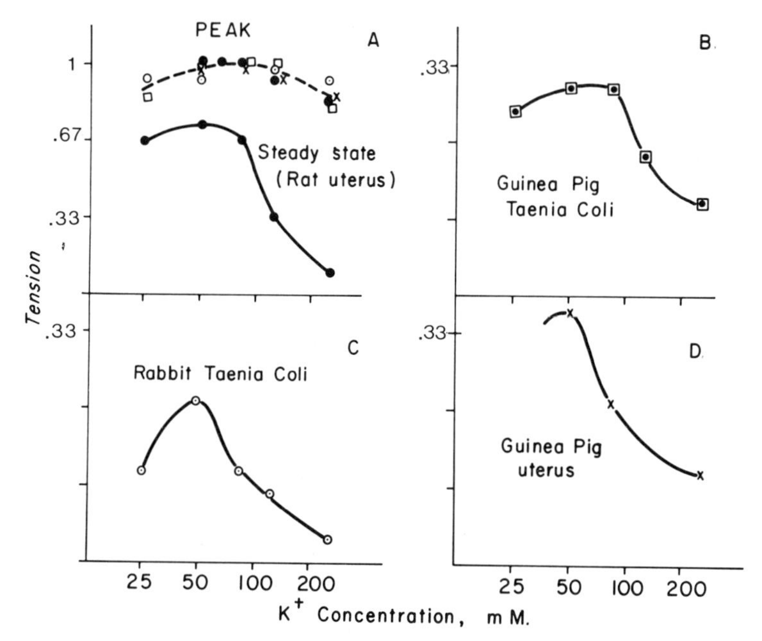

Figure 2. Variation of peak (transient) and steady state tension with log K^+ (as sulfate). Mean values for 3 - 5 preparation of each muscle type, normalized to maximum tension. (- - - - peak transient, —— steady state)

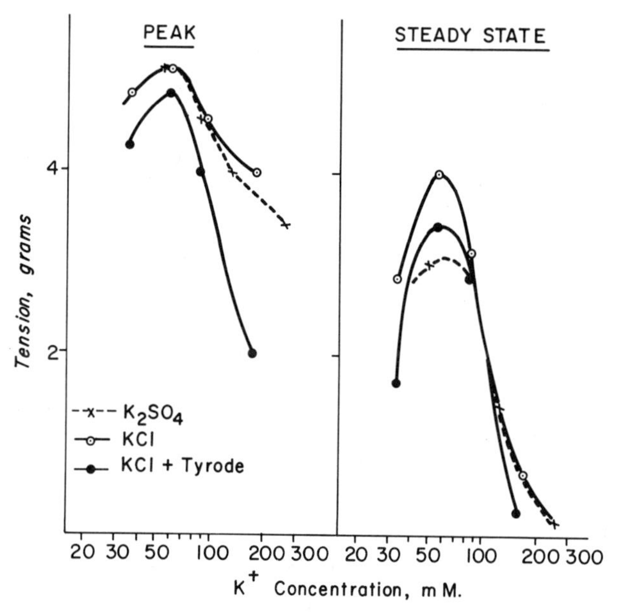

Figure 3. Comparison of peak transient and maintained tension resulting from elevated K as sulfate or as chloride. Non-pregnant rat uterus.

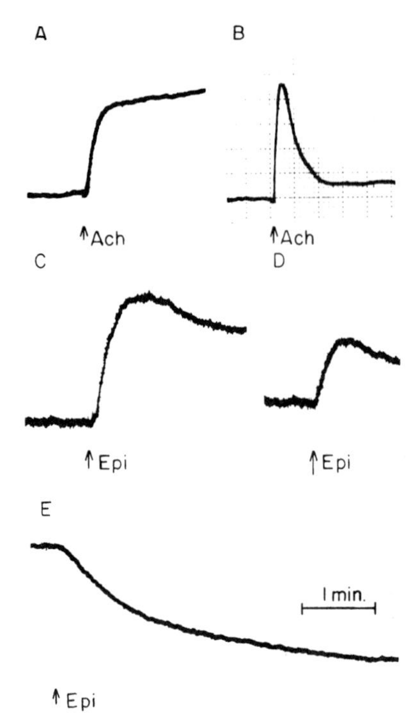

Figure 4. Typical responses of depolarized and normally polarized muscles to acetylcholine and epinephrine. A and B: non-pregnant rat uterus and guinea pig ileum, respectively, equilibrated in isotonic KCl. Contraction initiated by 5 x 10^{-5} M acetylcholine. C and D: Non-pregnant rabbit uterus equilibrated in normal Tyrode's or isotonic K_2SO_4, respectively. Contraction initiated with 5 x 10^{-5} M epinephrine. E: Non-pregnant rat uterus equilibrated in 51 mM K^+ which results in maximum sustained tension. Epinephrine concentration was 10^{-7} M.

replaced by a solution containing 20 - 25 mM K^+, there was initially an increased frequency of spike activity and oscillatory contractions became superimposed on a baseline of increased tension. After some 3 - 5 minutes the oscillations in the tension recorded disappeared as did electrical activity but tension remained above the resting level. Spikes persisted at lower K^+ concentration. Twenty mM K^+ changed the membrane potential to about -30 mV which was taken to be mechanical threshold.

On depolarization by K^+ in excess of 25 mM, tension rose rapidly to a peak from which it gradually fell to a lower level which was maintained as long as the muscle was kept depolarized. Figure 1 shows the tension produced by a rat uterus depolarized by 68 mM K^+ to a membrane potential of -18 mV or by 256 mM K^+ which changed the membrane potential to -3 mV. Although the peak tension was the same, relaxation was more rapid with the larger depolarization. Moreover, steady state tension was maintained at a higher level with the lower concentration of K^+.

The transient and steady state response to depolarization of a variety of smooth muscles were studied over a range of K^+ concentrations. The results are summarized in Figures 2 and 3. Transient peak tension was a steep function of depolarization when threshold was exceeded. The relation between steady state tension and K^+ concentration was qualitatively similar among

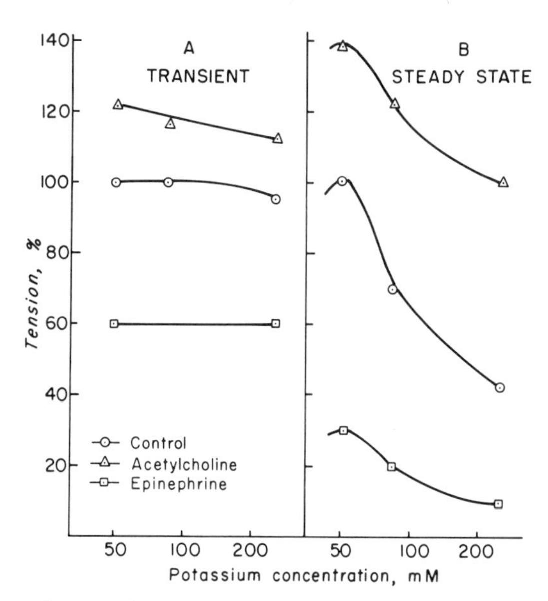

Figure 5. Effects of acetylcholine and epinephrine on transient peak tension and on steady state tension of depolarized pregnant rat uterus. Curves normalized to maximum control values of transient or steady state tension, respectively. Acetylcholine concentration = 5 x 10^{-6} M; epinephrine concentration = 5 x 10^{-7} M.

different types of mammalian smooth muscles. Steady state tension was a maximum with depolarizations resulting from about 50 - 60 mM K^+ and declined sharply with concentrations in excess of 100 mM. The question arises whether the apparent optimum depolarization for steady state tension results from some form of regenerative electrical activity. However, no membrane potential oscillations or spike activity were observed (guinea pig taenia coli or rat uterus). Steady state tension was maximum at a membrane potential of -20 mV. The qualitative relation of membrane potential to tension was similar whether K_2SO_4 or KCl was used, which would rule out the possibility that the fall in tension with larger depolarizations might be due to a change in the level of ionized Ca in the external medium.

If there is an optimum depolarization for maintained steady state tension, one might expect that there would be a range of membrane potentials over which an increase in internal negativity (hyperpolarization) would result in a rise in tension. That this is indeed the case is illustrated in Figure 1C where it can be seen that the muscle which had partially relaxed when depolarized by 256 mM K^+ re-developed tension when the membrane potential was shifted by 68 mM K^+. The converse, relaxation when 68 mM K^+ was replaced by 256 mM K^+ is also illustrated at the end of record C.

The existence of an optimum depolarization for maintained tension would also explain an otherwise paradoxical effect seen when replacing high K^+ solutions by normal Tyrode's (Fig. 1). Not unexpectedly tension fell on replacing 68 mM K^+ by normal Tyrode's (Fig. 1, record A). However, tension rose transiently when 256 mM K^+ was replaced by normal Tyrode's (Fig. 1, record B). The rise in tension was not due to spiking activity which was absent during this period and for a subsequent period of several minutes. It seems reasonable to suppose that this transient increase in tension on replacement of high K^+ solutions by normal Tyrode's solution was a reflection of the steady

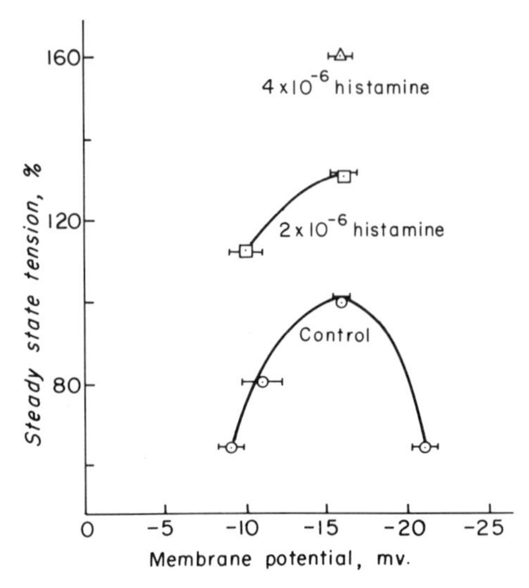

Figure 6. Effect of histamine on steady state tension of depolarized guinea pig taenia coli.

relation between tension and depolarization. Owing to the time for diffusion of K^+ out of the extracellular space, the return of the membrane potential to its normal polarized state would not be instantaneous so that, as the membrane began to repolarize, tension would rise until further repolarization led to relaxation.

Effect of Neurotransmitters on Tension

Figure 4 illustrates typical changes of tension evoked by acetylcholine and epinephrine of a variety of smooth muscles, both depolarized and non-depolarized. Interestingly, epinephrine which causes contraction of normally polarized rabbit uterus also contracts depolarized muscle. The way the relation between membrane potential and tension is altered by neurotransmitters is illustrated in Figure 5 and 6. Transient peak tension was increased by acetylcholine and decreased by epinephrine at all suprathreshold depolarizations of the pregnant rat uterus near term (Fig. 5). There was a nearly uniform increment or decrement of steady state tension in the presence of acetylcholine or epinephrine, respectively, for all suprathreshold depolarizations. Most significantly, there was no shift in the optimum depolarization for steady state tension. As may be seen in Figure 6, histamine increased steady state tension of depolarized guinea pig taenia coli in a dose-dependent manner.

The question arises whether drugs which produce large changes of tension of depolarized smooth muscle alter the relation between K^+ concentration and membrane potential. No change of membrane potential at any given K^+ level was observed, as K^+ was varied between 50 and 250 mM K^+ (either as the chloride or sulfate salt). The observations were made on pregnant rat and mice and the guinea pig taenia coli,, and drugs tested were acetylcholine epinephrine, histamine, and papaverine.

DISCUSSION

The regulation of contraction by membrane potential in a wide variety of non-vascular mammalian smooth muscles has the following common properties: (1) Transient contraction on depolarization followed by relaxation to a lower maintained level of tension; (2) More rapid rates of relaxation with increasing depolarization; and (3) A non-monotonic relation between depolarization and steady state tension with maximum force developed when the membrane potential is set near -20 mV. Similar observations have been made by Gabella (1978) and Lydrup and Hellstrand (1990). Neurotransmitters such as acetylcholine and epinephrine produce a nearly constant change of force at all suprathreshold depolarizations without otherwise altering the membrane potential for maximum sustained force.

These results raise important questions about the regulation of contraction of smooth muscle. Since force development upon depolarization depends on external Ca, one might suppose that the transient contraction followed by relaxation results from a voltage and time dependent inactivation of Ca channels (and of release from internal stores). But is one to assume that the existence of an optimum depolarization for maintained force mirrors changes in internal free Ca levels? Clearly experimental data are required over a wide range of membrane potentials in the presence and absence of neurotransmitters. It seems problematic that there is no displacement of the optimal depolarization for maintained tension produced by for example acetylcholine, if it increases $[Ca^{2+}]_i$ (by Ca entry through receptor-activated channels and release from intracellular stores via second messengers) and also shifts the pCa_i-force relationship towards lower Ca levels (Kitazawa, et al., 1989).

REFERENCES

Evans, D. H. L., Schild, H. O., and Thesleff, S., 1958, Effect of drugs on depolarized plain muscle, *J. Physiol.*, 143: 474.

Falk, G. and Landa, J. F., 1960, Mode of action of drugs on depolarized smooth muscle, *Pharmacologist*, 2: 68.

Gabella, G., 1978, Effect of potassium on the mechanical activity of taenia coli, uterus and portal vein of the guinea-pig, *Quart. J. Exp. Physiol.*, 63: 125.

Kitazawa, T., Kobayashi, S., Horiuti, K., Somlyo, A. V., and Somlyo, A. P, 1989, Receptor coupled, permeabilized smooth muscle: Role of the phosphatidylinositol cascade, G proteins and modulation of the contractile response to Ca^{2+}, *J. Biol. Chem.*, 264: 5339.

Lydrup, M.-L. and Hellstrand, P., 1990, Effects of extracellular K^+ and Ca^{2+} on membrane potential, contraction and $^{86}Rb^+$ efflux in guinea-pig mesotubarium, *Pflügers Arch.*, 415: 664.

Somlyo, A. V. and Somlyo, A. P., 1968, Electromechanical and pharmacomechanical coupling in vascular smooth muscle, *J. Pharmacol. Exp. Ther.*, 159: 129.

IN VITRO SYNAPTIC TRANSMISSION IN SYMPATHETIC NEURON-VASCULAR SMOOTH MUSCLE CO-CULTURES

Donald G. Ferguson, Stephanie A. Lewis,
and Raymund Y. K. Pun

Department of Physiology and Biophysics
University of Cincinnati College of Medicine
Cincinnati, OH 45267

INTRODUCTION

It is well established that nerves can have a direct and dramatic influence over the phenotypic expression and behavior of target tissues. This has been most thoroughly documented in the skeletal muscles, where denervation results in a reduction of resting membrane potential, appearance of tetrodotoxin-insensitive Na^+ channels, and development of supersensitivity associated with the spread of acetylcholine (ACh) -receptors to extrajunctional membranes (Patrick et al., 1978; Fambrough, 1979). Furthermore, the type of innervation and pattern of stimulation on the skeletal muscle fibers can change the phenotype of the muscle, i.e. from a fast to slow type (Eisenberg et al., 1984).

In contrast to skeletal muscle, in which all muscle cells are individually innervated by branches of a motor nerve, the relationship between the autonomic nerves and vascular smooth muscle (VSM) cells may vary. In small arterioles with only one or a few layers of muscle cells forming the vascular wall, most VSM cells receive more or less direct innervation. However, in the large distributing arteries with multilayered smooth muscle cells, the density of innervation is considerably less, and presumably is largely confined to the outermost cells bordering the adventitia (Burnstock, 1975; Hirst and Edwards, 1989). As with skeletal muscle, the pattern of innervation has a pronounced effect on the VSM chemosensitivity. Directly innervated fine arterioles are relatively insensitive to bath applied norepinephrine (NE), responding only to relatively high concentrations, presumably similar to that attained at or near junctional sites. In contrast, the large distributing arteries with considerably lower density of innervation are some 100-fold more sensitive to exogenously applied NE, suggesting either higher density and/or higher affinity of α-adrenoreceptors (Bolton and Large, 1986; Hirst and Edwards, 1989).

In addition to the implied effects of innervation on smooth muscle membrane chemosensitivity, there are also indications that nerves influence electrical membrane properties and expression of voltage-gated channels in

Regulation of Smooth Muscle Contraction
Edited by R.S. Moreland, Plenum Press, New York, 1991

VSM cells. Development of supersensitivity following denervation and decentralization is associated with a decrease in membrane potential (Fleming et al., 1973; Aprigliano and Hermsmeyer, 1977; Fleming, 1978). Spike threshold did not alter so that a lesser degree of depolarization would now induce contraction (Fleming, 1978). Cross-transplantation experiments revealed that transplanted VSMs from 2-weeks old normotensive animals developed membrane potentials which are similar to those of the hypertensive hosts, whereas transplanted VSMs from 12-16-weeks old normotensive animals did not alter their membrane potentials (Campbell et al., 1981). This alteration in membrane potential appeared to be related to the re-innervation of the transplanted VSMs by the host sympathetic nerves (Abel and Hermsmeyer, 1981).

There is some evidence to suggest that ionic channels are also affected by innervation. Edwards and Hirst (1988) reported that densely innervated segments of cerebral arterioles exhibit stable potentials of about -70 mV, but less densely innervated segments have a low and unstable resting membrane potential of about -40 mV. These effects seem to be related to the presence of inward rectifier K^+ channels in the innervated segments.

Under experimental conditions, electrical stimulation of the perivascular nerves in most systemic arteries and arterioles evokes depolarizing, excitatory junctional potentials (e.j.p.s) which consist of fast and slow components. The fast component has a relatively short latency (50 -70 msec), peaks at bout 100 msec, and decays with a time constant of about 200 - 700 msec. The slow depolarization, which often requires repetitive nerve stimulation, has a longer latency (several hundred msec) and may last for several seconds (see Bolton and Large, 1986; Hirst and Edwards, 1989).

All of the above studies have examined the interactions between nerve and VSM at the tissue level, however, considerably less is known about the influence of nerve on VSM at the cellular and molecular level. It is technically difficult to perform voltage-clamp experiments in VSM cells *in vivo*. Not only are VSM cells small, which limits their accessibility to electrical recording, but VSM form a syncytium and are electrically coupled, which makes adequate and proper control of membrane potential extremely difficult. It is possible to perform voltage-clamp experiments in single cells, either acutely isolated, or in cultures to characterize their voltage-dependent ionic conductances. Unfortunately, the vigorous dissociation procedure invariably severs the connecting nerve fibers, which makes activation of the synapse and the study of synaptic interactions between neurons and VSM cells impossible. Co-culture model systems, which have been previously used to study noradrenergic synaptic transmission of central synapses (Pun et al., 1985), can circumvent this limitation. In co-cultures, it should be possible to re-introduce a source of innervation, and an identified synapse will be available for study. To this end, we have developed a sympathetic nerve-VSM co-culture preparation.

METHODS

Mesenteric Smooth Muscle Cultures

Following the isolation of the mesenteric blood vessels (both arteries and veins from newborn, and only arteries from adults) from the animals, the tissue was treated with a mixture of collagenase and trypsin for 5 min at 37° C.

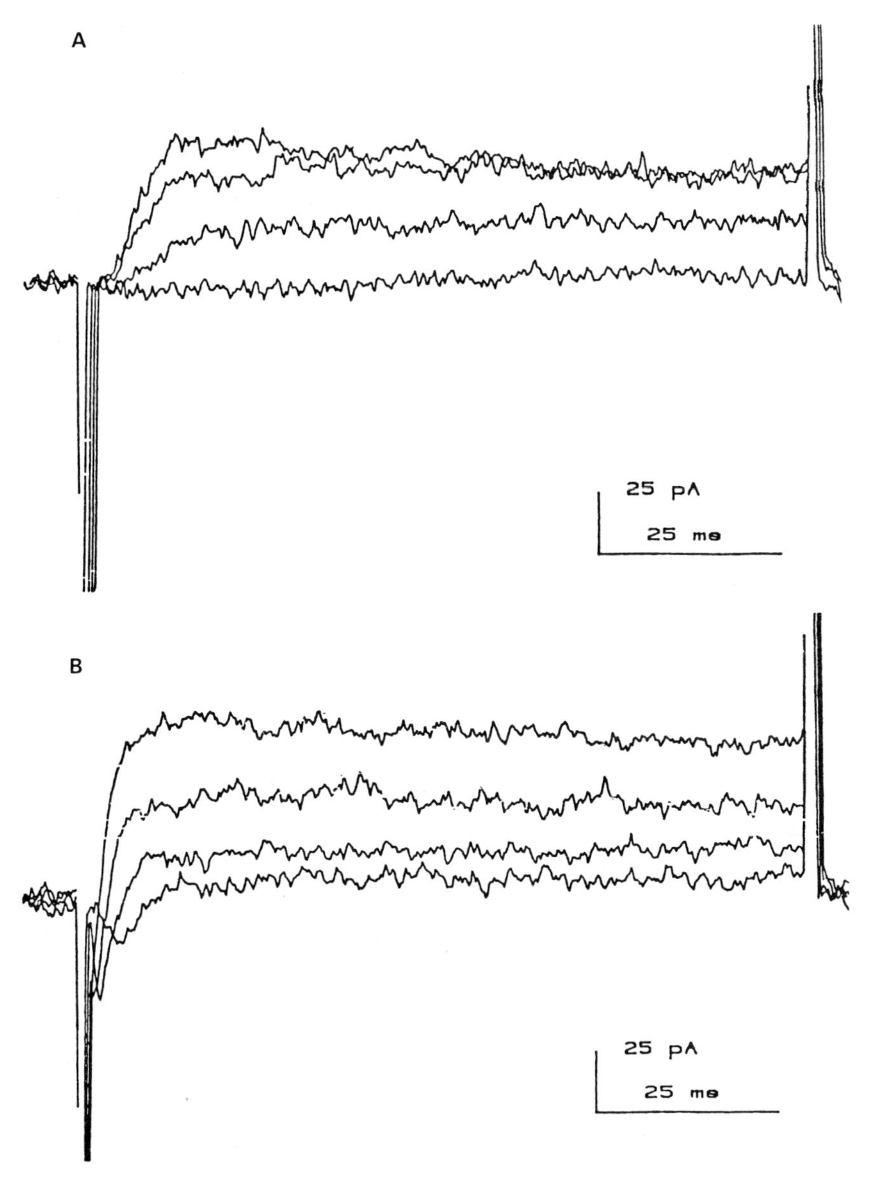

Figure 1. Currents recorded from mesenteric smooth muscle cells after 1 day (A) and 4 days (B) in culture. Cells were voltage-clamped at a holding potential of -80 mV. Depolarizing voltage steps from -20 to +20 mV evoked mainly outward currents in day - 1 cultures. In day - 4 cultures, small inward currents which peaked at +10 mV could be recorded as well.

In the adult, the fat surrounding the arteries was removed before enzyme treatment. The use of neonate versus adult tissues allows for the comparison between innervation and re-innervation since the neonate is not innervated by the sympathetic nerves. The preparation was then suspended in 5 ml of growth medium [Dulbecco's Modified Eagle Medium (DMEM)] supplemented with 10% fetal bovine serum (FBS) and mechanically dissociated by trituration through a pasteur pipette. The cloudy supernatant was collected and put into cold growth medium. The precipitate was resuspended in the enzyme mixture and the procedure repeated. The entire collection of vessels was usually dissociated after 2 - 3 digestions. The supernatants thus collected were centrifuged for 4 min at 1000 rpm. Cells were then resuspended, counted, and plated onto collagen coated dishes at a low density. Tissues from adults usually took a longer period of incubation (about 15 - 20 min per digestion and about 4 digestions).

Neuronal-Smooth Muscle Co-Cultures

Following isolation of the ganglia, the encapsulating sheath was carefully removed. Ganglia were placed in a Ca^{2+}/Mg^{2+} free solution, minced and treated with 0.25% trypsin for 15 min at 37° C. The enzyme-treated ganglia were then suspended in DMEM medium supplemented with 10% FBS, and dissociated mechanically. Nerve growth factor (50 ng/ml) was added to the cells. The suspension was then plated onto the smooth muscle cells at a concentration of approximately one ganglion per plate. Cells were maintained in a humidified atmosphere of 95% air: 5% CO_2 at 37° C and used for experiments after 1 - 4 days in culture.

Electrophysiological Experiments

Detailed descriptions for the methodology and procedure for intracellular recordings have been published (Pun, 1988; Pixley and Pun, 1990; Pun and Behbehani, 1990). Briefly, the bathing solution was changed from the growth medium into a N-2-hydroxyethyl piperazine-N'-2-ethanesulfonic acid (HEPES)-buffered high Na^+ solution, containing Ca^{2+} (2 mM) and Mg^{2+} (1 mM). VSM cells were recorded with blunt-tipped pipettes (patch-pipettes), and filled with a high K^+ aspartate or gluconate-ethylene glycol-bis(b-amino-ethylether) N,N,N',N'-tetra-acetic acid (EGTA) buffered solution. Sympathetic neurons were recorded with conventional, sharp-tipped pipettes containing either 3 M KAc or horse radish peroxidase (HRP) solution (4% HRP in 150 mM KCl). The neuron was stimulated with depolarizing current pulses, either single or repetitive, to evoke single or multiple action potentials. In some experiments, HRP was injected into the neuron at the end of the recordings. The cells were then fixed with glutaraldehye, reacted with diaminobenzidine (DAB), and later used for electron microscopy for identification of synapses (see Pun et al., 1986 for experimental details). Simultaneous recordings of sympathetic neuron and VSM were performed using an Axoclamp-2A Amplifier. Membrane potentials were digitized, taped with a video casette recorder (Unitrade), and later played back for analysis. Voltage-clamp experiments with VSM cells were obtained using a List-EPC7 Amplifier. Current records were filtered with a Bessel filter, and recorded by a computer (DEC 11-23). All experiments were performed at room temperature (22 - 24° C).

Electron Microscopy

Electron microscopy was performed on the co-cultures to verify that synapses between the axons and VSM cells can be visualized ultrastructurally.

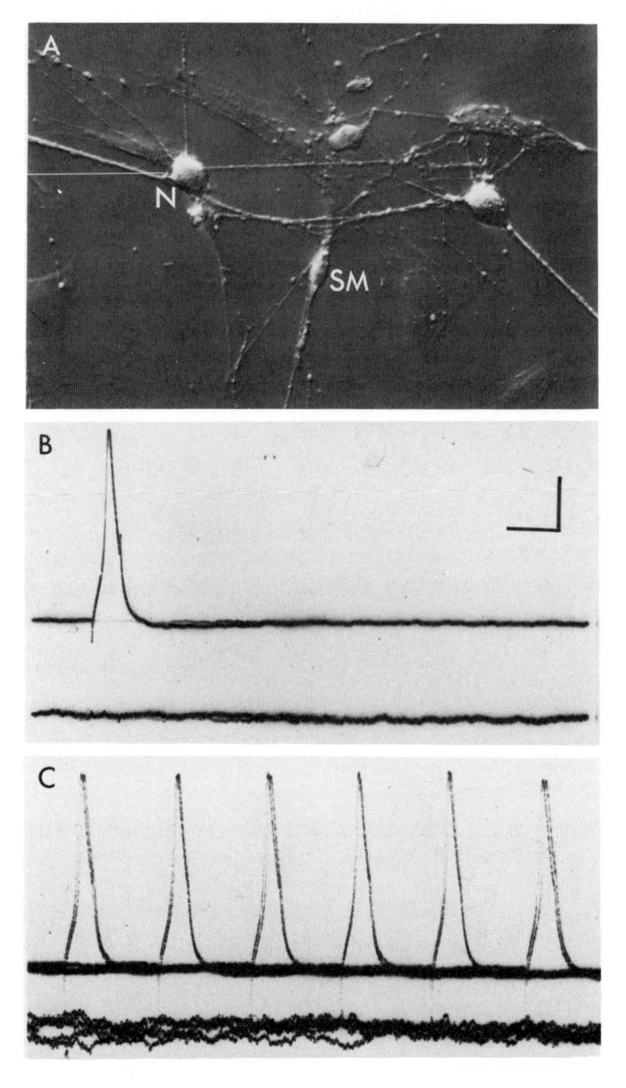

Figure 2. Non-functional interaction between sympathetic neuron and mesenteric smooth muscle cell. A) Recordings were obtained from the sympathetic neuron (N) and the smooth muscle cell (SM) shown in photomicrograph. Oscilloscope tracings of continuous voltage recordings demonstrate that neither a single stimulation (B) nor repetitive stimulation (C) elicits any e.j.p.s. The membrane potential of the smooth muscle cell was -54 mV. Calibration: Presynaptic, 20 mV; Postjunctional, 5 mV; 20 msec.

Cells were fixed in cacodylate-buffered 2.5% glutaraldehyde (pH 7.4), washed in buffer, postfixed in buffered 1% osmium tetroxide, washed again in buffer, dehydrated in graded series of alcohols, and embedded in Epon epoxy resin. After polymerization of the plastic, the cultured cells were split off the culture plates by freezing in liquid nitrogen and areas selected by phase microscopic examination were sectioned using a Porter-Blum MT2B ultramicrotome. Thin sections were stained with lead citrate and uranyl acetate and examined in a Zeiss EM10 transmission electron microscope.

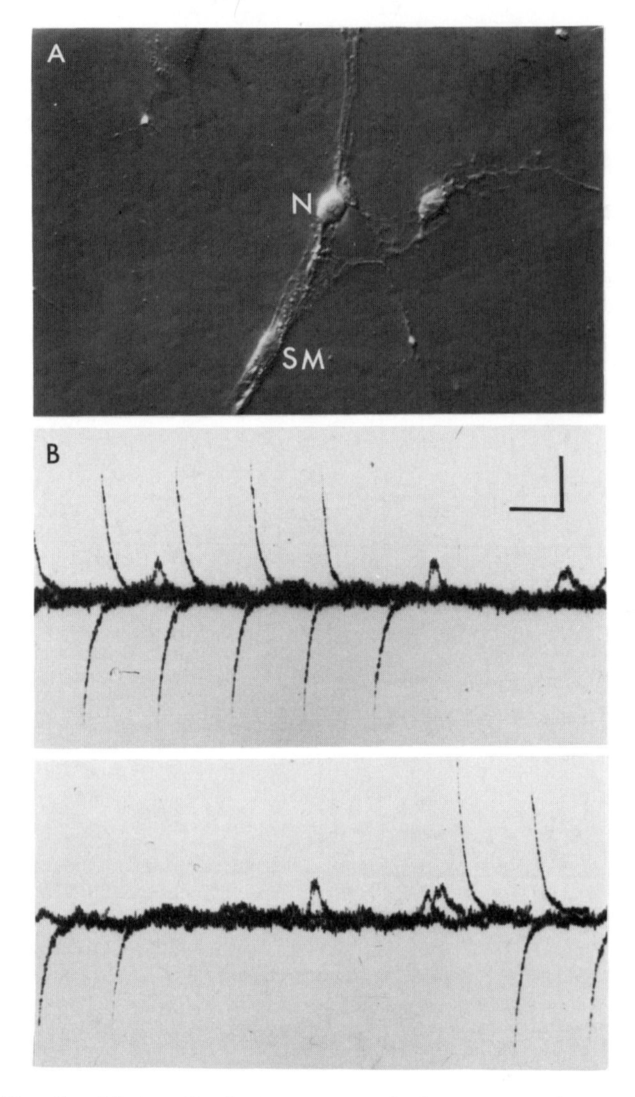

Figure 3. Functional interaction between sympathetic neuron and mesenteric smooth muscle cell. A) Evoked e.j.p.s between the single sympathetic neuron (N) and the single mesenteric smooth muscle cell (SM) shown in the photomicrograph. B) The neuron was stimulated by a hyperpolarizing pulse to evoke an off-spike (not shown). Repetitive stimulation of the neuron elicited e.j.p.s of approximately 1 mV. Membrane potential of the SM was between -50 and -60 mV. Calibration: 2 mV, 50 msec.

RESULTS AND DISCUSSION

Membrane electrical properties of mesenteric VSM cells were estimated from whole-cell recordings under both current- and voltage-clamp conditions. VSM cells have a resting membrane potential of -49.4 ± 3.1 mV (mean ± SEM, n = 8). The input resistance of the cell, measured from the current deflection to a 5 mV pulse, was 0.21 ± 0.04 GΩ (n = 10). The capacitance, obtained from the capacitative transient to a 5 mV pulse, was 20.24 ± 2.39 pF (n = 10). After four days in culture, VSM cells exhibited outward and inward currents upon depolarization of the membrane potential under voltage-clamp conditions

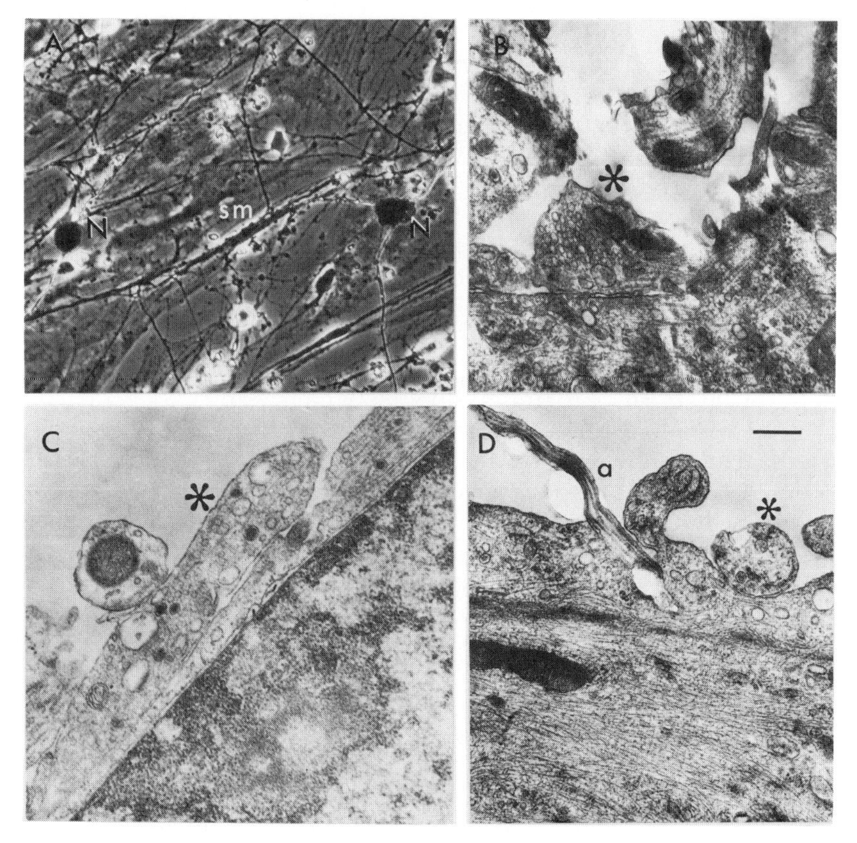

Figure 4. Morphology of sympathetic neuron-VSM co-cultures. A) Light micrograph of a co-culture in which the neurons (N) have been treated with HRP/DAB; sm = VSM cell. B) Electron micrograph of the VSM cell from A, in which a synapse (*) is clearly recognizable. C and D) Electron micrographs of VSM from the same culture dish as the cells depicted in Figures 2 and 3. Synaptic terminals of various sizes are present (*) and an axon (a) can be observed in proximity to the underlying VSM . Bar = 425 nm (B and D); 330 nm (C).

(Fig. 1), though in the first 2 - 3 days, only outward currents were evident. Since we were more interested in evaluating whether functional synapses are formed between single sympathetic neurons and single VSM cells, we concentrated on determining if excitatory junctional potentials (e.j.p.s) can be recorded in the muscle cells following stimulation of the neuron.

In co-cultures, under light microscopic examination, neuronal processes of about 1 μm can be discerned coursing over and around muscle cells (Figs. 2A and 3A). In about 70% of the successful simultaneous recordings made (stable recordings from a neuron and a VSM cell), elicitation of action potentials in the neuron failed to evoke any detectable change in the membrane potential of the VSM cell. This lack of response was observed with either a single pulse or by repetitive stimulation (Figs. 2B and 2C). In the remaining cell pairs studied, a small e.j.p. of amplitude about 1 - 5 mV could be recorded. An example of this recording is shown in Figure 3B. On some occasions, facilitation of the e.j.p. amplitudes was observed as has been described for guinea pig vas deferens preparation (data not shown). In one cell pair, stimulation of the neuron led to generation of action potentials in the VSM cell (data not shown). Our data indicate that functional synapses are formed between neurons and VSM cells in co-culture.

In order to confirm the presence of synapses we examined the co-cultures using electron microscopy. In some experiments the neurons were injected through the stimulating electrode with HRP and during preparation for EM the cells were treated with DAB to develop dark black staining (Fig. 4A). This served to pinpoint the cells of interest for ultrastructural study and synapses were observed to be present (Fig. 4B). Ultrastructurally recognizable synaptic terminals were also observed (Figs. 4C and 4D) in the co-cultures. The terminals appeared to contain organelles which had the ultrastructural characteristics of cholinergic, adrenergic, and even peptidergic synaptic vesicles.

The co-cultures which we describe here provide a system in which a detailed examination of synaptic transmission may be profitably approached at junctions formed between a single presynaptic sympathetic neuron and a single postjunctional VSM cell. For example, measurement of synaptic currents in voltage clamped postjunctional muscle cells should permit accurate determinations of reversal potentials, time course (latency and decay) of synaptic currents, and their pharmacology. Furthermore, responses to applied ATP and NE and those elicited by nerve stimulation can be directly compared in the same cell. Additional studies will allow the unequivocal establishment of the effects of sympathetic nerves on the expression of voltage-dependent ionic channels, the expression of neurotransmitter receptors, and expression of proteins in VSM cells. Finally, the identity of the neurotransmitter(s) responsible for the generation of e.j.p.s in VSM can also be established. This information should provide us with a better understanding of the role of sympathetic nerves in the regulation of VSM function and present an integrated view of neural regulation of VSM biology and electrophysiology.

ACKNOWLEDGMENTS

The authors wish to thank Susan Osborne, Nancy Rosen, and Lisa Wukusick for expert technical assistance with the cell culture procedures. These studies were funded in part by NIH HL34779 (DGF) and NSF DCB 8812562 (RYKP).

REFERENCES

Abel, P. W. and Hermsmeyer, K., 1981, Sympathetic cross-innervation of SHR and genetic controls suggests a trophic influence on vascular muscle membranes, *Circ. Res.*, 49: 1311.

Aprigliano, O. and Hermsmeyer, K., 1977, Trophic influence on the sympathetic nervous system on the rat portal vein, *Circ. Res.*, 41: 198.

Bolton, T. B. and Large, W. A., Are junctional potentials essential: Dual mechanism of smooth muscle cell activation by transmitter released from autonomic nerves, *Quart. J. Exp. Physiol.*, 71: 1.

Burnstock, G., 1975, Innervation of vascular smooth muscle: histochemistry and electron microscopy, *Clin. Exp. Pharmacol. Suppl.*, 2: 7.

Campbell, R. C., Chamley-Campbell, J., Short, N., Robinson, R. B., and Hermsmeyer, K., 1981, Effect of cross-transplantation on normotensive and spontaneously hypertensive rat arterial muscle membrane, *Hypertension*, 3: 534.

Edwards, F. R. and Hirst, G. D. S., 1988, Inward rectification on rat rat cerebral arterioles: Involvement of potassium ions on autoregulation, *J. Physiol.*, 404: 455.

Eisenberg, B. R., Brown, J. M. C., and Salmons, S., 1984, Restoration of fast muscle characteristics following cessation of chronic stimulation: The ultrastructural of slow-to-fast transformation, *Cell Tissue Res.*, 238: 221.

Fambrough, D. H., 1979, Control of acetylcholine receptors in skeletal muscle, *Physiol. Rev.*, 59: 165.

Fleming, W. W., 1978, The trophic influence of autonomic receptors on electrical properties of the cell membrane in smooth muscle, *Life Sci.*, 22: 1223.

Fleming, W. W., McPhillips, J. J., and Westfall, D., 1973, Postjunctional supersensitivity and subsensitivity of excitable tissues to drugs, *Erg. der Physiol.*, 68: 55.

Hirst, G. D. S. and Edwards F. R., 1989, Sympathetic neuroeffector transmission in arteries and arterioles, *Physiol. Rev.*, 60: 546.

Patrick, J., Heinemann, S., and Schubert, D., 1978, Biology of cultured nerve and muscle, *Ann. Rev. Neurosci.*, 1: 417.

Pixley, S. K. and Pun, R. Y. K., 1990, Cultured rat olfactory neurons are excitable and respond to odors, *Dev. Brain Res.*, 53: 125.

Pun, R. Y. K., 1988, Voltage clamping with single microelectrodes: Comparison of the discontinuous mode and continuous mode using the Axoclamp 2A amplifier, *Mol. Cell. Biochem.*, 80: 109.

Pun, R. Y. K. and Behbehani, M. M., 1990, A rapidly inactivating Ca^{2+}-dependent K^{+} current in pheochromocytoma cells (PC12) of the rat, *Pflügers Arch.*, 415: 425.

Pun, R. Y. K., Marshall, K. C., Hendelman, W. J., Guthrie, P. B., and Nelson, P. G., 1985, Noradrenergic responses of spinal neurons in locus coeruleus-spinal cord co-cultures, *J. Neurosci.*, 5: 181.

PLATELET ACTIVATING FACTOR CAUSES RELAXATION OF ISOLATED PULMONARY ARTERY AND AORTA

Najia Jin, C. Subah Packer, and Rodney A. Rhoades

Department of Physiology and Biophysics
Indiana University School of Medicine
Indianapolis, IN 46202

INTRODUCTION

Platelet activating factor (PAF) is a lipid chemical mediator produced by a variety of activated cells, including basophils, eosinophils, platelets, monocytes, endothelial cells, macrophages, and neutrophils (Tamura, 1987; Vercellotti, 1988). A number of pathophysiological responses, such as inflammatory and allergic reactions, bronchoconstriction, platelet aggregation, pulmonary hypertension, anaphylactic shock, endotoxic shock, and systemic hypotension are mediated by PAF (Benveniste and Vargaftig, 1983; Heffner, 1983a; 1983b; Terashita, 1985; 1987; Hanahan, 1986; Barnes, 1988). These pathological conditions are linked to changes in smooth muscle tone. For example, bronchoconstriction is due to the increase in tone of airway smooth muscle and hypotension is due to the decrease in tone of vascular smooth muscle.

Injection of PAF via femoral veins or oral administration of PAF causes a systemic hypotensive response in intact animals (Page, 1984; Terashita, 1987). Blood flow distribution studies also indicate a decrease in vascular resistance after administration of PAF (Goldstein, 1984). Several antagonists of PAF have been shown to reverse the PAF-induced hypotension in rat (Terashita, 1987) and guinea pig (Feuerstein, 1984). The fact that the hypotensive effect of PAF cannot be blocked by specific cholinergic, histaminergic, or adrenergic receptor antagonists (Lai, 1983), suggests that this hypotensive effect is mainly due to an endothelium-dependent vasodilation (Page, 1984). Direct evidence of endothelium-mediated PAF-induced hypotension has not yet been reported. In contrast to systemic hypotension, pulmonary hypertension was observed upon PAF administration (Heffner 1983a; 1983b; Hamasaki, 1984). However, the mechanisms for the PAF-induced pulmonary hypertension at the smooth muscle level is not known. The purpose of this study was to investigate the direct effects of PAF on isolated pulmonary arterial and aortic muscles and to determine whether these effects are mediated by endothelium.

Regulation of Smooth Muscle Contraction
Edited by R.S. Moreland, Plenum Press, New York, 1991

METHODS

Male Sprague-Dawley rats (250-275g) were anesthetized by intraperitoneal injection of pentobarbitol. Lungs and heart were removed *en bloc* and placed in ice cold Earle's balanced salt solution (EBSS; 2.4 mM $CaCl_2$, 0.8 mM $MgSO_4$, 5.4 mM KCl, 116.4 mM NaCl, 0.9 mM Na_2HPO_4, 5.5 mM glucose, 26.2 mM $NaHCO_3$, and 0.03 mM phenol red). The right and left pulmonary arterial branches and the thoracic aorta were isolated, cleaned of connective and fatty tissue under the microscope, and cut into rings 2.5 - 3.0 mm in length. In some rings the endothelium was rubbed by threading the rings onto a lightly sanded steel rod and gently rotating the rings several times. The rings were immersed in tissue baths filled with 10 ml EBSS and equilibrated for 1 hour. The presence of an intact functional endothelium was tested by determining the effect of 5×10^{-6} M acetylcholine (ACh) on a 5×10^{-7} M norepinephrine-induced contraction. ACh is known to cause relaxation of NE-precontracted vascular smooth muscle only in preparations that have a functional endothelium. Then the rings were contracted with 80 mM KCl. The maximum force produced in response to 80 mM KCl was designated P_o and all other responses were normalized as a percent of P_o for each preparation.

PAF (1-O-alkyl-2-acetyl-snglyceryl-3-phosphorylcholine; Sigma Chem. Co.) stocked in chloroform, was evaporated with nitrogen, and resuspended in 5% ethanol/95% water. In some experiments PAF (10^{-5} M) was added to each bath while pulmonary arterial or aortic rings were at resting tension and then the arterial rings were subsequently contracted with high K^+ or NE. In other experiments, PAF was added while the muscle was at the peak of a high K^+-induced contraction.

RESULTS

PAF (10^{-8} M to 10^{-5} M; much higher doses than that inducing platelet aggregation) caused neither contraction nor relaxation of either pulmonary arterial or aortic muscle when the muscle was initially under resting conditions (i.e. at resting tone).

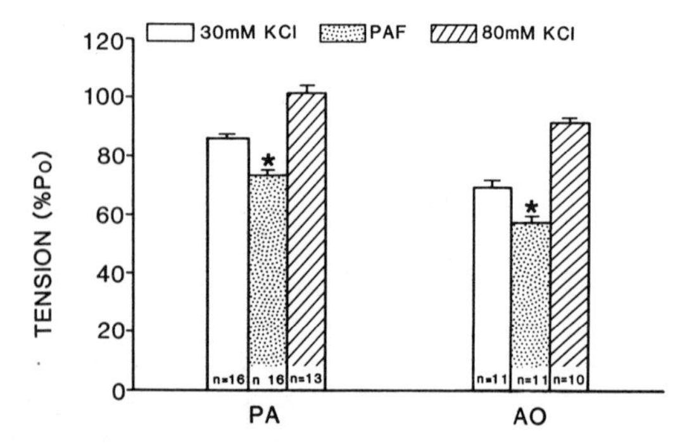

Figure 1. Effect of PAF on precontracted pulmonary arterial (PA) and aortic (AO) rings. The rings were precontracted with 30 mM KCl and then 10^{-5} M PAF was added at the peak of the KCl response. PAF caused partial relaxation ($p < 0.01$) in both pulmonary arterial and aortic rings.

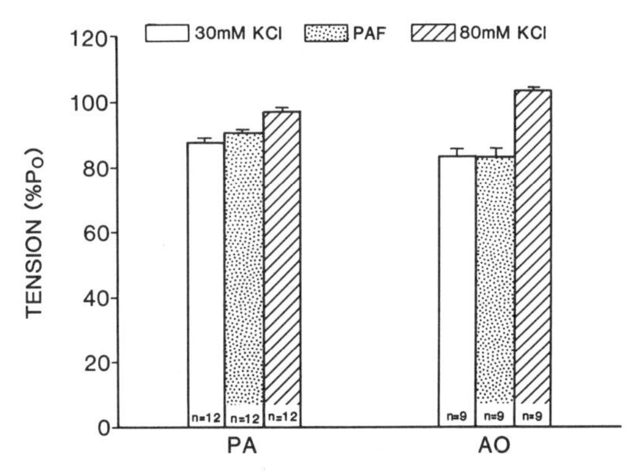

Figure 2. Effect of PAF on KCl-precontracted rings without endothelium. Removal of endothelium prevented the PAF-induced partial relaxation ($p < 0.001$) in both pulmonary arterial and aortic rings.

The effect of PAF on submaximal KCl precontracted pulmonary arterial and aortic rings is shown in Figure 1. When the precontractile force reached a plateau, PAF was added. 10^{-5} M PAF caused partial relaxation in both pulmonary arterial and aortic rings. Removal of endothelium prevented the PAF-induced relaxation as shown in Figure 2.

The effect of PAF exposure on the responses of pulmonary arteries to high K^+ or NE is shown in Figure 3. The pulmonary arterial rings were contracted with 30 mM KCl or 5×10^{-5} M NE. The peaks of these responses were considered to be 100% of force development for the particular doses of each respective agonist. Following KCl or NE wash-out and complete relaxation, the pulmonary arterial rings were treated with 10^{-5} M PAF. Five minutes later, rings were contracted with 30 mM KCl or 5×10^{-5} M NE again.

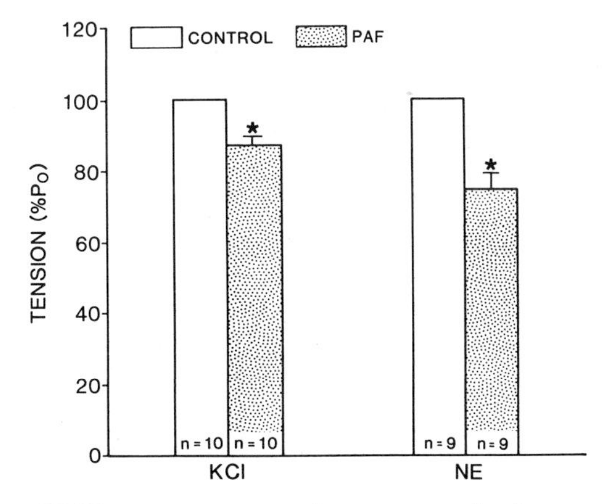

Figure 3. Effects of PAF pretreatment on the response of pulmonary arteries to KCl or NE. 5 min after 10^{-5} M PAF was added to the bath, rings were contracted with 30 mM KCl or 5×10^{-7} M NE. PAF attenuated the force production of pulmonary arteries in response to either KCl or NE ($p < 0.001$).

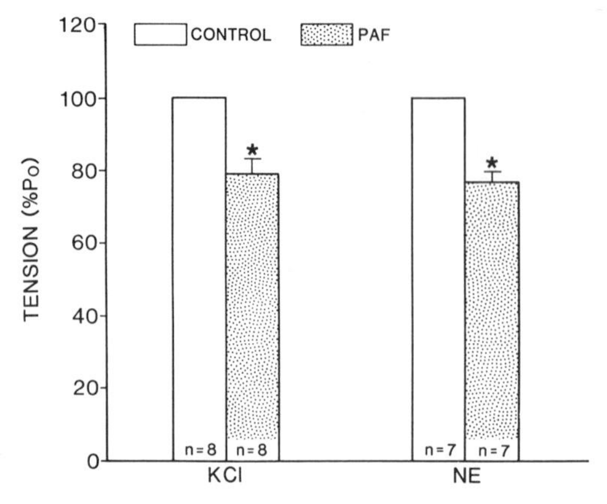

Figure 4. Effects of PAF pretreatment on the response of aorta to KCl or NE. 5 min after 10^{-5} M PAF was added to the baths, aortic rings were contracted with either 30 mM KCl or 5×10^{-7} M NE. PAF attenuated the force production of aorta in response to KCl or NE ($p < 0.01$).

The force production of pulmonary arterial muscle in response to KCl or NE was attenuated by the PAF pretreatment.

Similarly, the responses of aortic muscle to KCl or NE were also attenuated by PAF pretreatment as shown in Figure 4. Finally, Figure 5 shows that removal of the endothelium prevented the PAF-induced attenuation of force production in response to NE in both pulmonary arterial and aortic rings.

DISCUSSION

In the current study, PAF caused relaxation of isolated pulmonary arterial and aortic rings, suggesting that the hypotensive effect of PAF is due, at least in part, to its direct effect on vessel calibre. Results of this study also

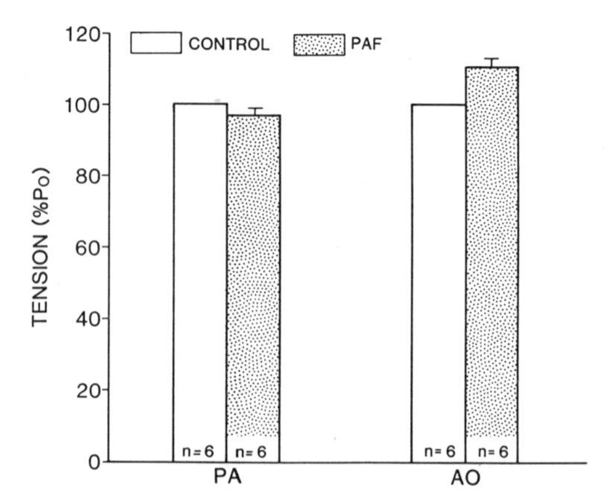

Figure 5. Effects of PAF on the response of arterial rings without endothelium to NE. Removal of endothelium prevented PAF-induced attenuation of force production in response to NE in both pulmonary arterial and aortic rings ($p < 0.001$).

demonstrated that PAF-induced relaxation of arterial rings is mediated by the endothelium. In addition exposure to PAF diminishes the reactivity of arterial muscle to various agonists. Specifically, PAF pretreatment suppressed the force production of pulmonary arterial and aortic rings in response to KCl or NE. Since contractile responses induced by either membrane depolarization or receptor activation are both inhibited by PAF pretreatment, PAF may be acting at the level of the contractile or regulatory apparatus of smooth muscle.

In contrast to some previous studies which reported that PAF caused pulmonary hypertension, our experiments showed that PAF caused relaxation of pulmonary arteries as it did of aorta. Heffner (1983a; 1983b) showed that PAF caused acute bronchoconstriction, pulmonary hypertension, and edema mediated by the formation of leukotrienes and thromboxane. These studies were carried out on isolated rat, rabbit, cat, and guinea pig lung tissue (Heffner, 1983b; Chen, 1990) or in whole animal (Ohar, 1990). In these preparations, PAF has a vast range of target cells, such as neutrophils, platelets, basophils, and alveolar II cells. The fact that PAF caused pulmonary hypertension in isolated lung or in whole animals likely reflects the combined effects of interactions between PAF and a variety of those cell types. In the current investigation, PAF-induced relaxation reflects the effects of PAF on pulmonary arterial smooth muscle and endothelial cells only.

In summary, PAF has no effect on resting tension of pulmonary arterial and aortic muscles, but causes relaxation of precontracted isolated arterial muscles. The PAF-induced relaxation is endothelium dependent. PAF pretreatment attenuates the responses of pulmonary arterial and aortic muscle to high K^+ or NE. These effects of PAF may conceivably play a role in endotoxic and/or anaphylactic shock and PAF-induced hypotension but do not contribute to PAF-induced pulmonary hypertension.

ACKNOWLEDGMENTS

This work was supported by NIH award HL36745. Thanks are due to Marlene King for expert typing of this paper.

REFERENCES

Barnes, D. J., Chung, K. F., and Page, C. P., 1988, Platelet-activating factor as a mediator of allergic disease, *J. Allerg. Clin. Immunol.*, 81: 919.

Benveniste, J. and Vargaftig, B. B., 1983, Platelet-activating factor: An ether lipid with biological activity, *in:* "Ether Lipids- Biochemical and Biomedical Aspects", H. K. Mangold, F. Paltauf, eds., Academic Press, New York, p. 355.

Chen, C. R., Voelkel, N. F., and Chang, S. W., 1990, PAF potentiates protamine-induced lung edema: Role of pulmonary venoconstriction, *J. Appl. Physiol.*, 68: 1059.

Feuerstein, G., Lux, W. E., Snyder, F., Ezra, D., and Faden, A. I., 1984, Hypotension produced by platelet-activating factor is reversed by thyrotropine-releasing hormone, *Circ. Shock*, 13: 255.

Goldstein, B. M., Gabel, R. A., Huggins, F. J., Cervoni, P., and Crandall, D.J., 1984, Effect of platelet-activating factor (PAF) on blood flow distribution in the spontaneously hypertensive rat, *Life Sci.*, 35: 1373.

Hamasaki, Y., Mojarad, M., Saga, T., Tai, H. H., and Said, S. I., 1984, Platelet-activating factor raises airway and vascular pressure and induces edema in lungs perfused with platelet free solution, *Am. Rev. Resp. Dis.*, 129: 742.
Hanahan, D. J., 1986, Platelet activating factor: A biologically active phosphoglyceride, *Ann. Rev. Biochem.*, 55: 483.
Heffner, J. E., Shoemaker, S. A., Canham E. M., Patel, M., McMurtry, I. F., Morris, H. G., and Repine, J. E., 1983a, Platelet-induced pulmonary hypertension and edema, *Chest*, 83: 78s.
Heffner, J. E., Shoemaker, S. A., Canham, E. M., Patel, M., McMurtry, I. F., Morris, H. G., and Repine, J. E., 1983b, Acetyl glyceryl ether phosphoryl choline-stimulated human platelets cause pulmonary hypertension and edema in isolated rabbit lungs, *J. Clin. Invest.*, 71: 351.
Lai, F. M., Shepherd, C. A., Cervoni, P., and Wissner, A., 1983, Hypotensive and vasodilatory activity of (+)1-O-octadecyl-2-acetyl glyceryl-3-phosphorylcholine in the normotensive rat, *Life Sci.*, 32: 1159.
Ohar, J. A., Pyle, J. A., Waller, K. S., Hyers, T. M., Webster, R. O., and Lagunoff, D., 1990, A rabbit model of pulmonary hypertension induced by the synthetic platelet-activating factor acetylglyceryl ether phosphorylcholine, *Am. Rev. Resp. Dis.*, 141: 104.
Page, C. P., Archer, C. B., Paul, W., and Morley, J., 1984, Paf-acether: a mediator of inflammation and asthma, *Trends Pharmacol. Sci.*, 5: 239.
Tamura, N., Agrawal, D. K., Suliaman, F. A., and Townley, R. G., 1987, Effects of platelet activating factor on the chemotaxis of normodense eosinophils from normal subjects, *Biochem. Biophys. Res. Commun.*, 142: 638.
Terashita, Z., Imura, Y., Nishikowa, K., and Sumida, S., 1985, Beneficial effects of (Rs)-2-methoxy-3-(Octadecylcarbamoyloxy)propyl 2-(3-thiazolio) ethyl phosphate, a specific PAF antagonist, in endotoxic and anaphylactic shock, *Adv. Prostaglan. Throm. Leuk. Res.*, 15: 715.
Terashita, Z., Imura, Y., Takatani, M., Tsushima, S., and Nishikowa, K., 1987, CV-6209, a highly potent antagonist of platelet activating factor in vitro and in vivo, *J. Pharmacol. Exp. Ther.*, 242: 263.
Vercellotti, G. M., Yin, H. Q., Gustafson, K. S., R. D.,and Jacob H. S., 1988, Platelet-activating factor primes neutrophil responses to agonists: role in promoting neutrophil-mediated endothelial damage, *Blood*, 71: 1100.

THE EFFECT OF CALCIUM ANTAGONIST ON NOREPINEPHRINE-INDUCED ^{86}Rb EFFLUX IN OBESE ZUCKER RATS

Jacquelyn M. Smith, Dennis J. Paulson, and Sandra T. Labak

Department of Physiology
Chicago College of Osteopathic Medicine
Downers Grove, IL 60515

INTRODUCTION

The obese Zucker rat is a model for obesity which exhibits an insulin resistance characterized by hyperinsulinemia and mild hyperglycemia (Kurtz, et al., 1989). There is also an increased incidence of hypertension in the obese Zucker rat which is a common observation in human obesity. Although a variety of factors have been proposed to account for the elevated blood pressure (Dornfeld et al., 1987; Levy et al., 1989), the observation that total peripheral resistance is elevated in the hypertensive obese rats (Shehin et al., 1989) suggests that vasoconstriction of vascular smooth muscle may be a contributing factor. Vascular smooth muscle from obese Zucker rats exhibits an increased sensitivity to a variety of agonists including phenylephrine, serotonin, and potassium (Zemel et al., 1990). This non-specific supersensitivity suggests that the enhanced vascular reactivity is due to a functional alteration in vascular smooth muscle itself rather than via selective receptor-mediated events. Insulin has both direct and indirect effects on calcium metabolism in a variety of tissues (Levy et al., 1989). Since both insulin resistance and hyperinsulinemia occur in the obese Zucker rat, it has been suggested that altered calcium metabolism in vascular smooth muscle may underlie the increased vascular reactivity. The recent report that both the CaATPase activity and ^{45}Ca efflux from aorta of obese Zucker rat are decreased compared to normotensive lean controls is consistent with this proposal. Recently it has been reported that insulin also modulates potential sensitive calcium channels (POC) in some tissues (Zierler and Fong, 1988). Calcium entry through this pathway plays a dominant role in both depolarization-induced as well as receptor-mediated stimulation of vascular smooth muscle (Cauvin et al., 1983). Therefore, it was of interest to compare the effect of the calcium antagonist diltiazem, which blocks calcium entry through the POC, on norepinephrine stimulation of aorta from obese Zucker and age-matched lean control rats.

METHODS

40 week old female obese Zucker and age-matched lean control rats had been maintained on normal rat chow and given tap water to drink. Systolic blood pressure was determined by the tail-cuff method. On the morning of the experiment, the animals were anesthetized with ketamine (80 mg/kg) and xylazine (8 mg/kg), the aorta removed, placed in dissection solution, and cleaned of fat and connective tissue. After two rings (3.5 mm wide) were removed for contractile experiments, the two aortic sections were slit lengthwise, the endothelium removed and each piece mounted on a stainless steel holder. The tissues were incubated for 3 hrs in physiological salt solution (PSS) to which ^{86}Rb (10 μCi/ml, New England Nuclear, Boston MA) had been added. Although ^{86}Rb quantitatively underestimates the movement of K^+ through basal and calcium-activated K^+ channels, qualitatively the responses are similar and Rb substitution does not alter estimation of NE sensitivity (Smith et al., 1986).

Solutions

The PSS had the following composition (in mM): Na^+ 146.2, K^+ 5.0, Mg^{2+} 1.2, Ca^{2+} 1.5, Cl^- 143.9, HCO_3^- 13.5, H_2PO_4 1.2, and glucose 11.4. Dissection solution was K^+-free and the Ca^{2+} was reduced to 0.25 mM. Solutions were gassed with a 97% O_2 - 3% CO_2 mixture to obtain a pH of 7.4. Propranolol (3 μM), EDTA (0.1 mM), and ascorbic acid (1 mM) were added to norepinephrine (NE) solutions. NE and diltiazem (DZ) were dissolved in PSS and water, respectively. All drugs were obtained from Sigma Chemical Company (St. Louis, MO.).

^{86}Rb Efflux

At the end of the 3 hr equilibration period, each aortic section was passed through tubes containing non-radioactive PSS ± DZ (10 μM) for 40 min to obtain the ^{86}Rb rate constant (k) during steady state. Each aortic section was then passed through PSS ± DZ containing NE for 10 min followed by a 20 min wash before exposure to the next higher NE concentration. A gamma spectrophotometer (Packard Inst. Co., Downers Grove, IL) was used to count the ^{86}Rb activity and washout curves were computed (IBM) as described previously (Smith et al., 1987). The basal rate constant, k, for ^{86}Rb, (or fraction of Rb exchanged/min^{-1}) was the average of the rate constants obtained during the 30 and 40 min period of the efflux. The response to NE (Δk) was determined by subtracting the rate constant obtained during the last 5 min of exposure to each NE concentration from the rate constant obtained during the preceding wash period. Each Δk was normalized in terms of the maximal response to NE (Δk_{max}) and the median effective concentration (EC_{50}) for NE obtained by interpolation between the log dose just below and just above the 50% response.

Contraction

The two aortic rings were cleaned, the endothelium removed, and the rings stretched 33% of their resting diameter by means of a micrometer mounted on a force transducer. The rings were equilibrated in PSS at 37° C for 90 min. 10 μM DZ was added to the PSS of one ring for 40 min and the effect of

DZ on basal tension was determined at this time. NE was then added to the bath of each aortic ring in a cumulative fashion and the contractile response to NE was measured once the response had stabilized (usually in 10 min). A NE concentration-response curve was constructed and the EC_{50} calculated as described above.

Statistics

Results are expressed as mean ± SEM. Log values were used to make statistical comparisons of the EC_{50} values and the geometric means are presented. Significance at the level of $p < 0.05$ was determined using the Student's t-test.

Table 1. Effect of diltiazem (DZ, 10 μM) on basal and norepinephrine (NE) - induced ^{86}Rb efflux in aorta from obese Zucker and lean control rats.

	OBESE		LEAN	
	- DZ	DZ	- DZ	DZ
basal k for ^{86}Rb efflux, min^{-1}	0.0116 ± 0.0003	0.0098 ± 0.0002*	0.0126 ± 0.0004	0.0114 ± 0.00058
§NE EC_{50}, nM	6.30 ± 0.54*	16.6 ± 8.1	19.0 ± 5.0	27.2 ± 8.9
NE Δk_{max}	0.0071 ± 0.0004	0.0035 ± 0.0003	0.0075 ± 0.0006	0.0036 ± 0.0005

§Obese (n = 8) and Lean (n=6) as some EC_{50} values could not be accurately calculated. * denotes significant difference between Obese and Lean groups evaluated in the presence or absence of DZ, $p < 0.05$.

RESULTS

The 40 week old female obese (533 ± 21 gms) rats had a higher (Fig. 1) systolic blood pressure (140 ± 5 mmHg) than the age-matched lean (98 ± 3 gms) controls (98 ± 3 mmHg, $p < 0.001$). The summary of ^{86}Rb efflux in obese and lean rats is presented in Table 1. The basal ^{86}Rb efflux from aorta of obese and lean rats was similar (p = NS). The EC_{50} for the NE-induced stimulation of ^{86}Rb efflux was shifted to the left in aorta from obese rats (6.3 ± 0.54 nM) compared to the NE EC_{50} for ^{86}Rb efflux in lean rats (19.0 ± 5.0 nM; $p < 0.01$). However, when the response of the tissue to NE was evaluated in the presence of the calcium antagonist diltiazem (10 μM), this difference was eliminated and the NE EC_{50} for ^{86}Rb efflux (16.6 ± 8.1 nM and 27.2 ± 8.9 nM for obese and lean respectively, p = NS) was similar (Fig. 2). DZ also had a differential effect on the basal ^{86}Rb efflux in aorta from obese rat: the basal ^{86}Rb efflux from aorta of obese rats was lower compared to that from the lean rats ($p < 0.02$) in the presence of DZ (Table 1).

The results obtained for NE-stimulated tension were similar. The NE EC_{50} was shifted to the left in aorta from the obese rat (2.49 ± 0.63 nM) compared to the NE EC_{50} obtained for aorta from the lean rat (4.69 ± 0.59 nM, p

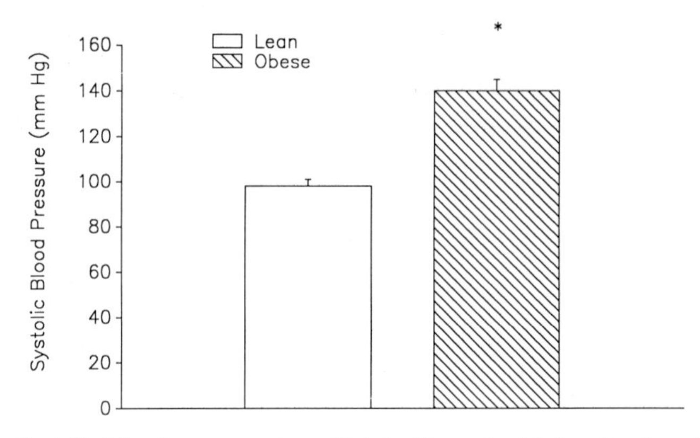

Figure 1. Systolic blood pressure (mm Hg) in 40 week old female obese Zucker rats (hatched bar, n = 9) and age-matched lean control rats (open bar, n = 9). * represents significant difference between the two groups at $p < 0.001$.

< 0.05) and this difference was eliminated in the presence of DZ (31.6 ± 6.9 vs 26.9 ± 5.0 nM for obese and lean rats, respectively; p = NS). The calcium antagonist also produced a greater decrease (Δ grams, 0.56 ± 0.3) in the basal (no agonists present) contractile recording from obese than lean rat (Δ grams, 0.11 ± 0.03, $p < 0.05$).

DISCUSSION

The female obese Zucker rats evaluated in this study exhibited an elevated systolic blood pressure (Fig. 1) compared to their age-matched lean controls. The degree of elevation is similar to that reported for both male (Shehin et al., 1989; Zemel et al., 1990) and female (Kurtz et al, 1989) Zucker rats when blood pressure was measured directly. It is not clear what factors underlie the development of this moderate hypertension although hyperphagia (Kurtz et al., 1989), obesity itself (Michaelis et al., 1984), and sodium retention (Kurtz et al., 1989) are unlikely candidates. Since an increased diastolic pressure contributes to the systolic hypertension in the obese animals (Shehin et al., 1989), it has been proposed that vasoconstriction of vascular smooth muscle may be an important factor. The results obtained from the present study are consistent with this possibility. The EC_{50} for NE stimulation of ^{86}Rb efflux (Table 1) or contraction is lower in the obese than lean rat reflecting an increased sensitivity to this catecholamine. Although the origin of the NE supersensitivity may be receptor-mediated, the fact that the increased vascular reactivity also extends to a variety of agents including KCl, serotonin and phenylephrine (Zemel et al., 1990) suggests that there is a functional alteration in the vascular smooth muscle from obese Zucker rat which produces a non-specific supersensitivity. The obese Zucker rat is characterized by an insulin resistance and hyperinsulinemia. Since insulin modulates active calcium transport in many tissues, it is feasible that altered calcium metabolism in vascular smooth muscle results in an enhanced vascular reactivity to agonist stimulation. Insulin has been shown to directly stimulate the Ca^{2+}-ATPase which actively transports calcium out of the tissue.

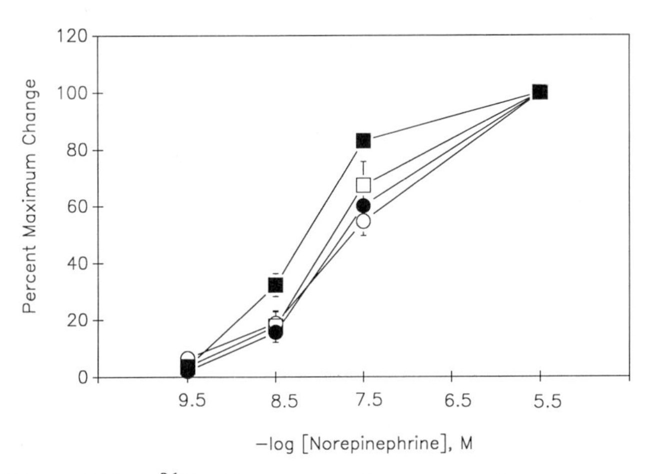

Figure 2. Effect of NE on ^{86}Rb efflux, expressed as percent of maximum response to NE, in the absence (closed symbols) and presence (open symbols) of 10 μM diltiazem, in aorta from obese Zucker (■, ❑) and age-matched lean control (●, ○) rats. Symbols represent mean data plus SEM. Data used to construct this curve obtained only from those experiments (O = 8; L = 6) for which EC_{50} could be accurately calculated.

Insulin stimulation of Ca^{2+}-ATPase is depressed in the obese rat (Levy et al., 1989) and both the efflux of ^{45}Ca as well as Ca^{2+}-ATPase activity are diminished in aorta from Zucker rats (Shehin et al., 1989) suggesting that calcium extrusion from vascular smooth muscle may be impaired. The results from the present study also indicate that there may be increased calcium entry into this tissue during norepinephrine stimulation: the calcium antagonist diltiazem, an agent which blocks calcium entry through depolarization-sensitive calcium channels (Cauvin et al., 1983), eliminates the enhanced sensitivity to NE in aorta from obese Zucker rats. In the presence of the calcium antagonist, the EC_{50}, for NE stimulated ^{86}Rb efflux and contraction is similar in obese and lean rats (Table 1). Similar findings have been reported in experimental models of hypertension in which calcium antagonists inhibit the response of hypertensive vessels to NE stimulation to a much greater extent than in control vessels (Jones and Smith, 1986; Smith and Myers, 1990). The magnitude of depolarization during NE stimulation is determined by the coupling between those conductances that depolarize (Ca^{2+}, Na^+, Cl^-) and those that hyperpolarize (K^+) (Smith and Jones, 1985). Disruption of this coupling may allow greater depolarization of the tissue and thus increased calcium entry through these channels. There has been a report that the NE- stimulated increase in K^+ efflux is diminished in aged obese rats (Cox and Kikta, 1989) although in the present study, the maximal increase in NE-stimulated ^{86}Rb efflux is similar in both groups (Table 1).

In many (Smith et al., 1987; Smith and Jones, 1990; Smith and Myers, 1990) but not all (Smith and Jones, 1983) models of hypertension, the efflux of K^+ is markedly elevated and a basal tension develops in the absence of agonist stimulation. This basal tension is eliminated and the K^+ efflux returned to control values in the presence of calcium antagonist (Smith and Jones, 1990). Since calcium-activated K^+ channels and tension are activated when intracellular calcium is elevated, these results imply that under "resting" conditions, there is enhanced calcium entry into the tissue through potential-operated calcium channels which elevates intracellular calcium and stimulates calcium-dependent processes. The results from the present study suggest that a similar alteration may occur in the vascular smooth muscle from obese Zucker rat. Although the steady-state ^{86}Rb efflux is similar, the aorta from obese Zucker rat exhibits an increased sensitivity to the calcium antagonist diltiazem under basal conditions: diltiazem decreased basal tension and ^{86}Rb efflux significantly more in aorta from obese rat that the lean control animals (Table 1). Thus, in the obese animals, an increased calcium entry as well as decreased calcium extrusion may contribute to the functional alterations observed in the vascular smooth muscle. Diltiazem also decreased basal ^{86}Rb efflux and tension, although to a lesser extent, in aorta from lean rats. This finding suggests that 40 week old lean rats exhibit some abnormalities in vascular smooth muscle that parallel those in the obese animals. Consistent with this observation is the report that both calcium efflux and Ca^{2+}-ATPase activity is decreased in old (60 weeks) lean rats which coincides with a mild elevation in systolic blood pressure (Zemel et al., 1989) and these changes have been attributed to latent expression of the Fa gene in the lean rat.

In conclusion, there is an increase in basal tension, calcium-dependent ^{86}Rb efflux, and sensitivity to norepinephrine in aorta from obese rat which were eliminated in the presence of a calcium antagonist. These findings suggest that calcium entry through potential-operated calcium channels is an important factor in the functional alterations that develops in vascular smooth muscle from obese Zucker rat.

ACKNOWLEDGMENTS

This study was supported by Biomedical Research Support Fund (NIH 2 S07 RR05972-02) and Chicago College of Osteopathic Medicine.

REFERENCES

Cauvin, C., Loutzenhizer, R., van Breemen, C., 1983, Mechanisms of calcium antagonist-induced vasodilation, *Ann. Rev Pharmacol Toxicol.*, 23: 373.

Cox, R. H. and Kikta, D. C., 1989, Comparison of norepinephrine (NE) activated ^{86}Rb efflux in arteries from genetically obese Zucker rats, *FASEB J.*, 3: A1009.

Dornfeld, L. P., Maxwell, M. H., Waks, A., and Tuck, M., 1987, Mechanisms of hypertension in obesity, *Kidney Internat'l.*, 32: S254.

Jones, A. W. and Smith, J. S., 1986, Altered Ca-dependent fluxes of ^{42}K in rat aorta during aldosterone-salt hypertension, *in:* "Recent Advances in Arterial Diseases: Atherosclerosis, Hypertension, and Vasospasm", T. N. Tulenko and R. H. Cox, eds., Alan R. Liss, New York, p. 265.

Kurtz, T. Z., Morris, R. C., and Pershadsingh, H. A., 1989, The Zucker fatty rat as a genetic model of obesity and hypertension, *Hypertension*, 13: 896.

Levy, J., Zemel, M. B., and Sowers, J. R., 1989, Role of cellular calcium metabolism in abnormal glucose metabolism and diabetic hypertension, *Am. J. Med.*, 87(Suppl 6A): 7S.

Michaelis, O. E., Ellwood, K. C., Judge, J. M., Schoene, N. W., and Hansen, C. T., 1984, Effect of dietary sucrose on the SHR/N corpulent rat: A new model of insulin-dependent diabetes, *Am. J. Clin. Nutr.*, 39: 612.

Shehin, S. E., Sowers, J. R., and Zemel, M. B., 1989, Impaired vascular smooth muscle ^{45}Ca efflux and hypertension in Zucker obese rats, *J. Vasc. Med. Biol.*, 1: 278.

Smith, J. M., and Jones, A. W., 1983, Potassium turnover and norepinephrine sensitivity in the thoracic aorta of the Dahl rat, *Proc. Soc. Exp. Biol. Med.*, 174: 291.

Smith, J. M. and Jones, A. W., 1985, Calcium-dependent fluxes of potassium-42 and chloride-36 during norepinephrine activation of rat aorta, *Circ. Res.*, 56: 507.

Smith, J. M. and Jones, A. W., 1990, Calcium antagonists inhibit elevated potassium efflux from aorta of aldosterone-salt hypertensive rats, *Hypertension*, 15: 78.

Smith, J. M., Jones, S. B., Bylund, D. B., and Jones, A. W., 1987, Characterization of the α-1 adrenergic receptors in the thoracic aorta of control and aldosterone hypertensive rats: Correlation of radioligand binding with potassium efflux and contraction, *J. Pharmacol. Exp. Ther.*, 241: 882.

Smith, J. M. and Myers, J. H., 1990, Elevated calcium-dependent K^+ efflux in aorta from stroke prone-spontaneously hypertensive rat, *in:* "Frontiers in Smooth Muscle Research", N. Sperelakis and J. D. Wood, eds., Wiley-Liss, New York, p. 539.

Smith, J. M., Sanchez, A. A., and Jones, A. W., 1986, Comparison of rubidium-86 and potassium-42 fluxes in rat aorta, *Blood Vessels*, 23: 297.

Zemel, M. B., Reddy, S., Shehin, S. E., Lockette, W., and Sowers, J. R., 1990, Vascular reactivity in Zucker obese rats: Role of insulin resistance, *J Vasc. Med. Biol.*, 2: 81.

Zemel, M. B, Reddy, S., and Sowers, J. R., 1990, Insulin attenuation of vasoconstrictor responses to phenylephrine in Zucker lean and obese rats, *Am. J. Hypertension*, 3: 19A.

Zemel, M. B., Shehin, S. E., Chiou, S.-Y., and Sowers, J. R., 1989, Blood pressure regulation and ^{45}Ca flux in aging Zucker rats, *FASEB J.*, 2: A702.

Zemel, M. B., Sowers, J. R., Shehin, S., Walsh, M. F., and Levy, J., 1990, Impaired calcium metabolism associated with hypertension in Zucker obese rats, *Metabolism*, 39: 1.

Zierler, K. and Fong, S.-W., 1988, Insulin acts on Na, K and Ca currents, *Trans. Assoc. Am. Physiol.*, 101: 320.

EFFECTS OF PINACIDIL, CROMAKALIM, AND NICORANDIL ON POTASSIUM CURRENTS OF RAT BASILAR ARTERY SMOOTH MUSCLE

He Zhang, Norman Stockbridge, and Bryce Weir

Department of Surgery
University of Alberta
Edmonton, Alberta, Canada

INTRODUCTION

Pinacidil, nicorandil, and cromakalim are antihypertensive agents, which relax peripheral vascular smooth muscle by hyperpolarizing smooth muscle cells towards the potassium equilibrium potential (Weir and Weston, 1986; Southerton et al., 1988). These drugs increase $^{86}Rb^{+}$ or $^{42}K^{+}$ efflux and reverse or attenuate the effects of a wide variety of vasoconstrictors (Hamilton et al., 1989). Patch-clamping studies show that these drugs act on several types of potassium channels. In cardiac myocytes, nicorandil (Hiraoka and Fan, 1989), pinacidil (Escande et al., 1989), and cromakalim (Sanguinetti et al., 1988) usually activate a background, time- and voltage-independent potassium current which is blocked by physiological levels of ATP. In arterial and venous smooth muscles, these agents activate either large-conductance calcium-dependent potassium channels (Gelband et al., 1989) or ATP-dependent potassium channels (Standen et al., 1989). Nicorandil has a dual action, a stimulation of guanylate cyclase and an increase in potassium conductance (Newgreen et al., 1988).

Some previous studies have been directed at the actions of these potassium channel openers on intact cerebral vessels. Pinacidil has been shown to be less effective at relaxing canine cerebral arteries than coronary, renal, and mesenteric arteries in an isometric tension study (Toda et al., 1985), nicorandil was found to hyperpolarize canine cerebral artery and partially reverse canine cerebral vasospasm (Harder et al., 1987), and cromakalim improved rabbit brain blood flow (Hof et al., 1988).

In this study, we used whole-cell and cell-free patch-clamp techniques to assess the potential effects of these drugs in cerebral vasculatures and to determine what types of potassium channels were activated by these agents in enzymatically isolated smooth muscle cells from rat basilar artery. The potential use of these drugs in the treatment of cerebral disorders, especially cerebral vasospasm after subarachnoid hemorrhage, is also discussed.

Regulation of Smooth Muscle Contraction
Edited by R.S. Moreland, Plenum Press, New York, 1991

MATERIALS AND METHODS

Isolation of Rat Basilar Artery Vascular Smooth Muscle Cells

Sprague-Dawley rats were anesthetized with halothane and decapitated. The basilar arteries were removed to a medium consisting of (in mM): NaCl 130, KCl 5, $CaCl_2$ 0.8, $MgCl_2$ 1.3, glucose 5, HEPES 10, penicillin (100 U/ml), and streptomycin (0.1 g/l). Arteries were then cleaned of connective tissue and small side branches. The basilar artery was moved to a medium in which the $CaCl_2$ was reduced to 0.2 mM and to which was added collagenase (Type II, 0.5 g/l), elastase (0.5 g/l), hyaluronidase (Type IV-S, 0.5 g/l), and deoxyribonuclease I (0.1 g/l). The arteries were cut into 0.2 mm rings and incubated for 1h at room temperature. The rings were transferred to fresh solution containing $CaCl_2$ (0.2 mM), trypsin inhibitor (0.5 g/l), and deoxyribonuclease I (0.1 g/l) and then triturated gently. Cells were plated on glass cover slips and stored at 4° C in saline containing $CaCl_2$ (0.8 mM) and essentially fatty-acid-free bovine serum albumin (2 g/l).

Whole-Cell Recording Techniques

Whole cell currents were recorded using standard techniques (Hamill et al., 1981) and an Axopatch 1C patch clamp amplifier (Axon Instruments). The patch pipettes had tip resistances of 1 - 4 MΩ and seal resistances of over 1 GΩ. Series resistance compensation was not employed. Under whole cell patch clamp conditions, the cells had an input resistance of 5 - 10 GΩ and a capacitance of 20 - 30 pF. The membrane potentials and current signals were stored on a laboratory computer. Leakage conductance was determined by applying small, hyperpolarizing voltage steps. None of the records shown were leakage corrected. The normal bath solution for the whole cell recordings was (in mM): NaCl 130, KCl 5.4, $MgCl_2$ 1.2, $CaCl_2$ 1.8 (I_{KCa}) or 0 (I_K), HEPES 10, glucose 5.2, and the pH was adjusted to 7.4 with NaOH. Pipettes were filled with (in mM): KCl 139, $MgCl_2$ 0.5, $CaCl_2$ 0.1 (I_{KCa}) or 0 (I_K), EGTA 0.09 (I_{KCa}) or EDTA 10 (I_K), HEPES 10, glucose 10, and the pH was adjusted to 7.4 with KOH. Bath solution changes were made with a syringe pump (Sage model 351), modified for simultaneous withdrawal and injection. All experiments were done at room temperature (19 - 22° C).

Single Channel Recording Techniques

Single channel current recordings were conducted at room temperature in the inside-out configuration (Hamill et al., 1981). The bath (intracellular) solution had the following composition (in mM): KCl 140, $CaCl_2$ 0.008, $MgCl_2$ 0.5, HEPES 10, EGTA 0.09, glucose 10, and the pH was adjusted to 7.4 with KOH. The free calcium concentration in this solution was calculated (Stockbridge, 1987) to be 1.5 nM. The pipette medium (extracellular solution) contained (in mM): KCl 140, $CaCl_2$ 0.1, $MgCl_2$ 0.5, HEPES 10, glucose 10, pH 7.4 with KOH. Single channel recordings were made with the Axopatch 1C amplifier and stored digitally on video tape (Neurocorder model 384 digitizer and Sony SL-700 VCR). On playback, data were redigitized at 10 KHz with a DT-2841-F (Data Translation) analog-digital converter on an AT-compatible computer. Amplitude histograms were made from contiguous segments of data about 5 min long. Channel openings were detected as crossings of 50% or 150% thresholds between histogram peaks.

Drugs

Drugs were pinacidil (Leo), cromakalim (Beecham), and nicorandil (Upjohn). Other agents were purchased from Sigma Chem. Co. Stock solutions of cromakalim and pinacidil (10 mM) were made up in 95% ethanol. Nicorandil (10 mM) was dissolved in HCl at pH 3 or 95% ethanol. Control experiments established that the vehicle was not responsible for changes in potassium currents that were observed.

RESULTS

Whole-Cell Recording

A small inwardly rectifying potassium current and spontaneous transient outward currents were visible in some records. Because of the variable nature of their appearance, the effect of potassium channel openers on these currents was not studied.

Delayed rectifier. With no added calcium in the pipette and bath solutions, a potassium current (Fig. 1) was obtained. This current activated at about -30 mV and showed outward rectification above +50 mV. This current showed slow sigmoidal activation and no inactivation over the 150 msec time course of the command potential. The amplitude of this current was substantially reduced by exposure to procaine and was almost completely eliminated by strychnine. These characteristics were similar to those of the I_K observed in many other preparations (Hille, 1984) and it will be subsequently referred to by that name. The maximum magnitude of this current was a few hundred picoamperes.

The effects of nicorandil, pinacidil, and cromakalim were determined on I_K. In 18 separate experiments, nicorandil (n = 7), pinacidil (n = 7), and

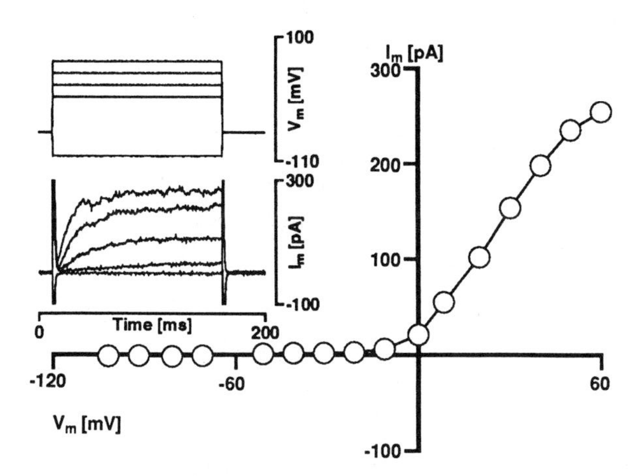

Figure 1. I_K from whole-cell patch-clamp experiments of enzymatically isolated rat basilar artery smooth muscle cells. The inset shows voltage commands (upper traces) and current records (bottom traces). The steady-state current-voltage relationship shows a voltage- and time-dependent potassium current (O). No inward current or inward rectifier current is evident in these records.

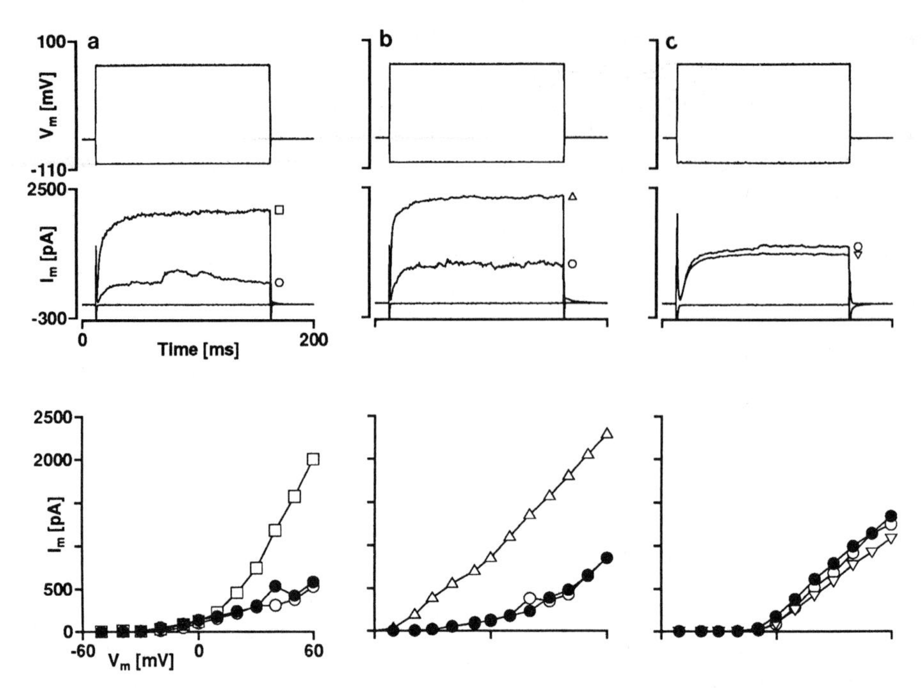

Figure 2. Effects of pinacidil, cromakalim, and nicorandil on I_{KCa} from whole-cell recordings. The upper part of the figure show voltage steps (upper traces) and current records (bottom traces). Current-voltage plots show I_{KCa} measured in the normal bath solution (O), in the presence of: a: 200 μM pinacidil (□), b: 100 μM cromakalim (Δ), or c: 400 μM nicorandil (∇), and after wash out (●). Each figure shows results from one cell. Pinacidil and cromakalim reversibly increased I_{KCa}. Nicorandil had no effect.

cromakalim (n = 4) exerted no effect on the I_K. They modified neither the amplitude nor the steady-state activation curve of this current. The concentrations of nicorandil were increased cumulatively from 100 - 900 μM; cromakalim and pinacidil were tried from 50 - 150 μM and 100 - 250 μM, respectively.

Calcium-dependent potassium current. With the high calcium pipette and bath solutions, the potassium current was much larger, usually several nanoamperes (Fig. 2, control). The current activated beginning at about -50 to -30 mV. At any given potential, this current activated more quickly than did the I_K. Like the I_K, this current showed no inactivation in 150 msec. The current was not blocked by exposure to 1 mM procaine, but was blocked by TEA (1 mM). This current, which includes some contribution from the much smaller I_K, is henceforth referred to as I_{KCa}.

Application of cromakalim and pinacidil to the external solutions increased the magnitude of the outward potassium currents in the high calcium solutions, as shown in Figures 2a and b. At 1 - 10 μM, pinacidil and cromakalim had no detectable effect on potassium currents in four cells. Pinacidil at 200 μM produced an approximately four-fold increase in the current elicited by a voltage step to +60 mV. In five such experiments, all cells showed increased outward potassium currents upon addition of pinacidil to the bath. In these five cells, a concentration of 100 - 200 μM pinacidil produced

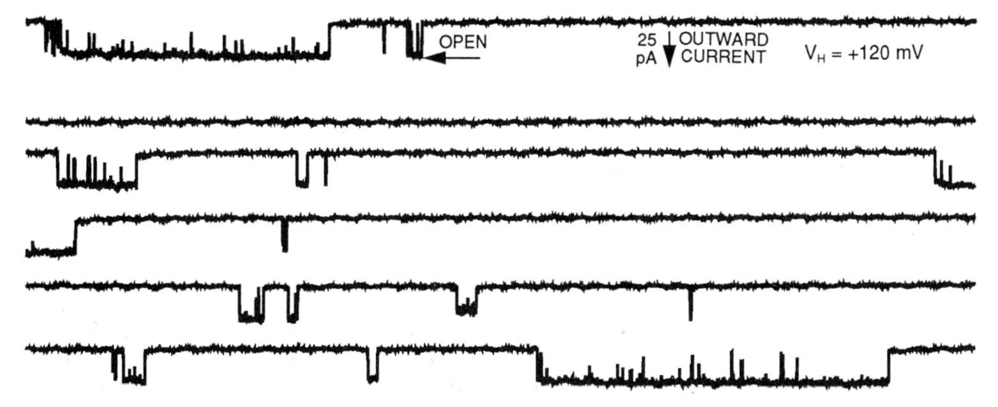

Figure 3. Large conductance calcium-dependent potassium channels from cell-free inside-out membrane patches in symmetrical 140 mM potassium. Bath free calcium concentration was estimated to be 1.5 nM. Downward deflections are channel openings. The membrane was voltage-clamped to +120 mV. The channel had a conductance of about 220 pS. Some openings were less than the full height; these probably reflect subconductance levels of the calcium-dependent potassium channel.

increases in peak potassium current of 50 - 400% (+60 mV). In each of these experiments, after washout with the normal bath solution, currents decreased to near control levels. Four cells exposed to cromakalim experienced increases of the potassium current. The mean concentration required to double the potassium current was about 100 µM. A complete reversal of the current amplitude to control levels was obtained in two of four cells upon washout of the drug. In one other experiment, the cell became leaky during the washout and in the fourth experiment, the current did not decrease during washout.

We did not observe significant increases in potassium current produced by nicorandil in 7 experiments (Figure 2c). Outward currents remained the same as control in six cells at 1 - 900 µM. Only one cell at 100 µM showed a small increase, but the increase was not reversible with washout.

Single Channel Recording

In inside-out membrane patches, calcium-dependent potassium channels were easily identified by their large conductance (220 pS), high sensitivity to the cytoplasmic calcium concentration, and voltage-dependence (Stuenkel, 1989). Figure 3 shows a short segment of a current recording obtained with a holding potential of +120 mV. At potentials positive to 0 mV, these channels were recorded as downward deflections. Some other conductances less than 200 pS were also observed, but were not subjected to study.

All such patches contained multiple potassium channels. The probability of channel openings was reduced by using a low free calcium concentration (estimated to be 1.5 nM) and by holding the patch at about +60 mV. The average number of channel openings was expressed as P_X; a value of 1% P_X indicates that an average of 0.01 channels were opened throughout a certain period. In all such recordings, there were still more than one channel per patch. Consequently, without knowing how many channels, these opening percentages could not be converted to single channel opening probability.

However, since each patch was used as its own control, we were still able to show that the drugs increased the channel opening probability.

After 5 min control recordings, the potassium channel openers were added to the bath (intracellular) medium. At each given concentration, channel activities were recorded for 5 min. In seven inside-out patches (Figure 4a and b), pinacidil (n = 4) and cromakalim (n = 3) all increased the channel open-state probability in a dose-dependent manner. These drugs did not alter channel conductance nor its reversal potential. The increase produced by 200 μM pinacidil was from 2-fold to 5-fold. The largest increases of P_x were produced with cromakalim at 250 - 300 μM: 1.7 to 7.9-fold. At lower concentrations (50 - 200 μM), cromakalim exerted no significant effects. Results obtained with nicorandil (Figure 4c; n = 8) were more variable. At 50 - 300 μM, nicorandil produced a dose-dependent increase of 1.4- to 2.7-fold in 4 patches. Four other patches demonstrated increases of P_x at 100 μM (in the first 5 min). Then, with increases in the nicorandil concentration, two patches exhibited no further changes in channel activities and the other two showed fewer openings than in the control solution.

DISCUSSION

In these experiments, the I_K and the I_{KCa} were identified from the whole cell recording of enzymatically isolated rat basilar artery smooth muscle cells. I_K and I_{KCa} were separated from one another pharmacologically and by controlling the intracellular calcium concentration. The I_K current, recorded in a bath containing no added calcium and with a pipette containing EDTA and no added calcium, was typically a few hundred picoamperes in maximum amplitude. Procaine and strychnine were effective blockers of this current. With 1.8 mM calcium in the bath and 90 μM EGTA + 100 μM calcium in the pipette, I_{KCa} of up to several nanoamperes could be elicited. This current was insensitive to procaine and strychnine, but was blocked by TEA in the bath. All

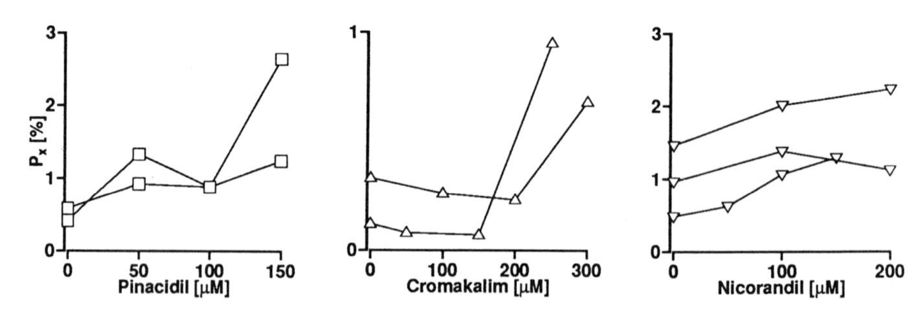

Figure 4. Dose-dependent effects of pinacidil, cromakalim, and nicorandil on calcium-dependent potassium channel opening percentages from cell-free inside-out membrane patches in symmetrical 140 mM potassium. Bath free calcium concentration was estimated to be 1.5 nM. The membrane was voltage-clamped to +60 mV. The concentration of pinacidil, cromakalim, and nicorandil is shown on the abscissa; the channel opening percentage P_x is shown on the ordinate. Each curve represents data from one patch. Cromakalim and pinacidil produced dose-dependent increases in the channel opening percentages. Effects obtained with nicorandil were variable.

recordings obtained with cell-free inside-out membrane patches contained multiple potassium channels with a conductance of about 220 pS in symmetrical potassium. These typical maxi-K channels (Stuenkel, 1989) probably underlie most of the I_{KCa} detected in whole cell recordings.

Table 1. Action of potassium channel agonists related to specific potassium channels.

Tissue	Drug	[μM]	ATP	Ca^{2+}	Other	Tech	Ref
Cardiac myocytes							
Guinea-pig ventricles	C	100			+	WC	Osterrieder, 1988
	C	30-300	+			WC,SC	Escande et al., 1988
	P	10	+			WC	Arena & Kass, 1989
	P	10-500	+			WC,SC	Escande et al., 1989
Rabbit ventricle	N	10-300	+			WC,SC	Hiraoka & Fan, 1989
G.-pig papillary m.	C	10	+			WC	Sanguinetti et al., 1988
Aorta smooth muscle							
Rabbit, Bovine	C	0.1-1		+		WC,SC	Kusano et al., 1989
Rabbit	C	0.05-10		+		SC	Gelband et al., 1989,1990
	C	?		+		WC	Economos et al., 1990
	P	0.1-10		+		SC	Gelband et al., 1990
	P	0.01-1		+		WC	Economos et al., 1990
Arterial smooth muscle							
Human mesenteric	C	1		+		SC	Klöckner et al., 1989
Rabbit, rat mesenteric	C	1	+			SC	Standen et al., 1989
Rat basilar	C	100-300		+		WC,SC	this chapter
	P	100-200		+		WC,SC	this chapter
	N	50-300		+?		SC	this chapter
Venous smooth muscle							
Rabbit portal	C	10			+	WC	Beech & Bolton, 1989
	C	1		+		WC,SC	Hu et al., 1990
Rat azygous	P	10		+		WC,SC	Hermsmeyer, 1988

The drugs were C = cromakalim, P = pinacidil, and N = nicorandil. The potassium currents associated with the drugs were ATP = time- and voltage-independent potassium current inhibited by ATP, Ca^{2+} = large conductance calcium-dependent potassium current, and other = unidentified potassium currents. A '+' indicates that the drug acted as an agonist. The techniques were WC = whole-cell patch-clamp and SC = single channel.

Table 1 summarizes voltage-clamp data which relate the action of potassium channel agonists to specific potassium channels. In cardiac muscles, cromakalim (Escande et al., 1988; Osterrieder, 1988; Saguinetti et al., 1988), nicorandil (Hiraoka and Fan, 1989), and pinacidil (Arena and Kass, 1989) all activate a potassium conductance which is non-inactivating and voltage-independent, except for some outward rectification with large depolarizations. This current is now known to be mediated by channels which are blocked at physiological levels of intracellular ATP or other nucleotide triphosphates (Kakei et al., 1985). At 10 μM to a few hundred μM, cromakalim (Escande et al., 1988), nicorandil (Hiraoka and Fan, 1989), and pinacidil (Arena and Kass, 1989; Escande et al., 1989) increase this current by increasing the single channel open probability. One study (Kakei et al., 1986) demonstrated an effect of nicorandil (100-2000 μM) on a time- and voltage-independent potassium channel other than the ATP-dependent potassium channel in cardiac myocytes.

In vascular smooth muscle, these drugs, at concentrations from 0.01 to 10 μM, exert their effects on either the calcium-dependent potassium channels or the ATP-dependent potassium channels. In the rabbit aorta, cromakalim (Kusano et al., 1987; Gelband et al., 1989) and pinacidil (Economos et al., 1990; Gelband et al., 1990) increase the macroscopic potassium conductance or the opening probability of single large conductance calcium-dependent potassium channels. Similar results have been obtained with cromakalim on vascular smooth muscle cells obtained from the human mesenteric artery (Klöckner et al., 1989) and rabbit portal vein (Hu et al., 1990). Pinacidil has similar effects on smooth muscle cells from the rat azygous vein (Hermsmeyer, 1988). There are other reports of cromakalim and pinacidil activation of another potassium current in vascular smooth muscle. In the rabbit portal vein (Beech and Bolton, 1989), cromakalim activated a potassium current for which the identity is unclear. In another study of single channel recordings from the rabbit mesenteric artery (Standen et al., 1989), cromakalim clearly activated ATP-dependent potassium channels.

In our experiments, neither the pipette solution used in whole cell recordings nor the bath solution used in recordings from inside-out membrane patches contained nucleotide triphosphates. Since time- and voltage-independent current obtained under these conditions was small, we did not try to identify whether there is an ATP-dependent potassium channel by changing the concentration of cytoplasmic ATP or using any specific antagonists of ATP-dependent potassium channels. There is evidence for such a conductance in some vascular smooth muscles (Standen et al., 1989), but some of this evidence depends upon the use of glybenclamide. Gelband (1990) has shown that glybenclamide is not a specific blocker of ATP-dependent potassium channels, but also blocks calcium-dependent potassium channels in rabbit aorta smooth muscle.

We showed that these potassium channel openers have no effects on the I_K from enzymatically isolated smooth muscle cells of rat basilar artery. At concentrations above those required to produce increases in the I_{KCa}, no effect on the amplitude, time course, or voltage-sensitivity of the I_K was found. The results from this study of whole-cell and single channel recordings suggested that large conductance calcium-dependent potassium channels were activated by a high concentration of pinacidil and cromakalim in rat basilar artery. In whole-cell studies, outward currents decreased to control levels after the drugs were washed out, showing that the effects of pinacidil and cromakalim were reversible. Recordings from cell-free membrane patches confirmed that the effects of these drugs on the calcium-dependent potassium channels were direct and did not require changes in the intracellular calcium concentration. Pinacidil and cromakalim at higher concentrations showed increases of channel opening percentages, which is consistent with the above mentioned reports using lower concentrations in other species or tissues (Kusano et al., 1987; Gelband et al., 1989, 1990; Economos et al., 1990).

The situation with nicorandil is evidently more complex. We know of no study which has concluded that nicorandil activated a calcium-dependent potassium channel. Hiraoka and Fan (1989) suggested that nicorandil exerted its actions on ATP-dependent potassium channels in rabbit ventricular myocytes and ruled out the possible involvement of I_K and I_{KCa}. They found that nicorandil mildly depressed the I_{KCa} by an average of 16%. They also found that nicorandil's effects did not last long and that spontaneous recovery to a level about 20 to 40% of the peak level occurred after 2 to 5 min, despite the continued presence of nicorandil. In our whole cell recordings, nicorandil, at concentrations from 1 to 900 μM, did not significantly increase the outward currents. In our single channel recordings, nicorandil increased the open probability only during the first 5 minutes. At higher concentrations (and later times), channel activities varied considerably. This may represent the time-dependent effects of nicorandil noted by Hiraoka and Fan (1989). A lack of effect on whole-cell currents is consistent with a previous report that nicorandil (100 μM) had no effect on the membrane potential or input resistance of smooth muscle cells studied in the intact guinea-pig basilar artery (Fujiwara and Kuriyama, 1983).

There is evidence that cerebral vasospasm following subarachnoid hemorrhage is accompanied by depolarization of the arterial smooth muscle cells (Harder et al., 1987). This depolarization would be expected to open voltage-dependent calcium channels and thereby raise the intracellular calcium concentration. An intracellular calcium antagonist, HA 1077, blocks development of vasospasm in a canine subarachnoid hemorrhage model (Shibuya et al., 1988). It is reasonable to expect that drugs which prevent the development of smooth muscle cell depolarization (Ahnfelt-Rønne, 1988) would also be effective in the treatment of vasospasm. Potassium channel agonists were suggested as prime candidates for such drugs (Harder et al., 1987). Our results from rat basilar artery smooth muscle cells suggest that pinacidil and cromakalim only at high concentrations activated calcium-dependent potassium channels, which may result in cell membrane hyperpolarization. Since cromakalim was reported to be more or equally effective in the basilar artery than in the thoracic aorta of the rabbit (Grant and O'Hara, 1989), and since much higher concentrations of these drugs were needed to open calcium-dependent potassium channels in the rat basilar artery than in peripheral arteries (Table 1), the search for additional mechanisms deserves further investigation.

ACKNOWLEDGMENTS

We wish to express our thanks to Dr. G. Maljkovic for valuable help. We also thank Beecham Pharmaceuticals (cromakalim), Leo Laboratories (pinacidil), and Upjohn Co. (nicorandil) for providing potassium channel agonists. This work was supported by a grant from the Canadian Heart and Stroke Foundation (N.S.) and by NIH grant R01 NS25957-01 (B.W.). N.S. is an Alberta Heritage Foundation Scholar.

REFERENCES

Ahnfelt-Rønne, I., 1988, Pinacidil, Preclinical investigations, *Drugs*, 36: 4.

Arena, J. P. and Kass, R. S., 1989, Activation of ATP-sensitive K channels in heart cells by pinacidil: dependence on ATP, *Am. J. Physiol.*, 257: H2092.

Beech, D. J. and Bolton, T. B., 1989, Properties of the cromakalim-induced potassium conductance in smooth muscle cells isolated from the rabbit portal vein, *Br. J. Pharmacol.*, 98: 851.

Economos, D., Peyrow, M., Escande, D., and Bkaily, G., 1990, Effects of K^+ openers, pinacidil and BRL 34915 on K^+ current of aortic single cells, *Biophys. J.*, 57: 508a.

Escande, D., Thuringer, D., Leguern, S., and Cavero, I., 1988, The potassium channel opener cromakalim (BRL 34915) activates ATP-dependent K^+ channels in isolated cardiac myocytes, *Biochem. Biophys. Res. Commun.*, 154: 620.

Escande, D., Thuringer, D., Le Guern, S., Courteix, J., Laville, M., and Cavero, I., 1989, Potassium channel openers act through activation of ATP-sensitive K^+ channels in guinea-pig cardiac myocytes, *Pflügers Arch.*, 414: 669.

Fujiwara, S. and Kuriyama, H., 1983, Effects of agents that modulate potassium permeability in smooth muscle cells of the guinea-pig basilar artery, *Br. J. Pharmacol.*, 79: 23.

Gelband, C. H., Lodge, N. J., and van Breemen, C., 1989, A Ca^{2+}-activated K^+ channel from rabbit aorta: Modulation by cromakalim, *Eur. J. Pharmacol.*, 167: 201.

Gelband, C. H., McCollough, J. R., and van Breemen, C., 1990, Modulation of vascular Ca^{2+}-activated K^+ channels by cromakalim, pinacidil, and glyburide, *Biophys. J.*, 57: 509a.

Grant, T. L. and O'Hara, K., 1989, Comparison of the effects of cromakalim (BRL 34915) on rabbit isolated basilar artery and thoracic aorta, *Br. J. Pharmacol.*, 98: 721p.

Hamill, O. P., Marty, A., Neher, E., Sakmann, B., and Sigworth, F. J., 1981, Improved patch-clamp techniques for high-resolution current recording from cells and cell-free membrane patches, *Pflügers Arch.*, 391: 85.

Hamilton, T. C. and Weston, A. H., 1989, Cromakalim, nicorandil and pinacidil: novel drugs which open potassium channels in smooth muscle, *Gen. Pharmacol.*, 20: 1.

Harder, D. R., Dernbach, P., and Waters, A., 1987, Possible cellular mechanism for cerebral vasospasm after experimental subarachnoid hemorrhage in the dog, *J. Clin. Invest.*, 80: 875.

Hermsmeyer, R. K.. 1988, Pinacidil actions on ion channels in vascular muscle, *J. Cardiovasc. Pharmacol.*, 12(Suppl. 2): S17.

Hille, B., 1984, "Ionic Channels of Excitable Membranes", Sinauer Assoc. Inc., Sunderland, MA.

Hiraoka, M. and Fan, Z., 1989, Activation of ATP-sensitive outward K^+ current by nicorandil (2-nicotinamidoethyl nitrate) in isolated ventricular myocytes, *J. Pharmacol. Exp. Ther.*, 250: 278.

Hof, R. P., Quast, U., Cook, N. S., and Blarer, S., 1988, Mechanism of action and systemic and regional hemodynamics of the potassium channel activator BRL 34915 and its enantiomers, *Circ. Res.*, 62: 679.

Hu, S., Kim, H. S., and Weiss, G. B., 1990, Effects of BRL 34915 and P 1060 on the Ca^{++}-activated K^+ channel in rabbit portal vein cells, *Biophys. J.*, 57: 307a.

Kakei, M., Noma, A., and Shibasaki, T., 1985, Properties of adenosine triphosphate-regulated potassium channels in guinea-pig ventricular cells, *J. Physiol.*, 363: 441.

Kakei, M., Yoshinaga, M., Saito, K., and Tanaka, H., 1986, The potassium current activated by 2-nicotinamidoethyl nitrate (nicorandil) in single ventricular cells of guinea-pig, *Proc. R. Soc. Lond. B.*, 229: 331.

Klöckner, U., Trieschmann, U., and Isenberg, G., 1989, Pharmacological modulation of calcium and potassium channels in isolated vascular smooth muscle cells, *Drug Res.*, 39: 120.

Kusano, K., Barros, F., Katz, G., Garcia, M., Kaczorowski, G., and Reuben, J. P., 1987, Modulation of K channel activity in aortic smooth muscle by BRL 34915 and a scorpion toxin, *Biophys. J.*, 51: 55a.

Newgreen, D. T., Bray, K. M., Southerton, J. S., and Weston, A. H., 1988, The relationship between K-channel opening and cGMP concentration in rat aorta, *Pflügers Arch.*, 411: 198.

Osterrieder, W., 1988, Modification of K^+ conductance of heart cell membrane by BRL 34915, *Naunyn-Schmied. Arch. Pharmacol.*, 337: 93.

Sanguinetti, M. C., Scott, A. L., Zingaro, G. J., and Siegl, P. K. S., 1988, BRL 34915 (cromakalim) activates ATP-sensitive K^+ current in cardiac muscle, *Proc. Nat'l. Acad. Sci. U.S.A.*, 85: 8360.

Shibuya, M., Sukuki, Y., Takayasu, M., Asano, T., Harada, T., Ikegaki, I., Satoh, S., and Hidaka, H., 1988, The effects of intracellular calcium antagonist HA 1077 on delayed cerebral vasospasm in dogs, *Acta Neurochir.*, 90: 53.

Southerton, J. S., Weston, A. H., Bray, K. M., Newgreen, D. T., and Taylor, S. G., 1988, The potassium channel opening action of pinacidil; studies using biochemical, ion flux and electrophysiological techniques, *Naunyn-Schmied. Arch. Pharmacol.*, 338: 310.

Standen, N. B., Quayle, J. M., Davies, N. W., Brayden, J. E., Huang, Y., and Nelson, M. T., 1989, Hyperpolarizing vasodilators activate ATP-sensitive K^+ channels in arterial smooth muscle, *Science*, 245: 177.

Stockbridge, N., 1987, EGTA, *Comput. Biol. Med.*, 17: 299.

Stuenkel, E. L., 1989, Single potassium channels recorded from vascular smooth muscle cells, *Am. J. Physiol.*, 257: H760.

Toda, N., Nakajima, S., Miyazaki, M., and Ueda, M., 1985, Vasodilatation induced by pinacidil in dogs: comparison with hydralazine and nifedipine, *J. Cardiovasc. Pharmacol.*, 7: 1118.

Weir, S. W. and Weston, A. H., 1986, The effects of BRL 34915 and nicorandil on electrical and mechanical activity and on $^{86}Rb^+$ efflux in rat blood vessels, *Br. J. Pharmacol.*, 88: 121.

Participants

Mark O. Aksoy
Department of Medicine
Temple University School of Medicine
3420 N. Broad Street
Philadelphia, PA 19140

Mohammed Al-Hassini
Department of Physiology and Biophysics
Indiana University Medical School
635 Barnhill Drive
Indianapolis, IN 46202

Julianne Bacsik
Harvard Medical School
10B Gardner Road
Cambridge, MA 02139

*Carl Baron
Department of Physiology
University of Pennsylvania
School of Medicine
Philadelphia, PA 19104

*Lloyd Barr
Department of Physiology and Biophysics
University of Illinois
Urbana, IL 61801

Robert J. Barsotti
Bockus Research Institute
Graduate Hospital
415 S. 19th Street
Philadelphia, PA 19146

*Christopher Benham
Department of Pharmacology
SmithKline and French Research Ltd.
The Frythe, Welwyn, Herts
AL6 9AR England

Joshua Berlin
Bockus Research Institute
Graduate Hospital
415 S. 19th Street
Philadelphia, PA 19146

Bruce Blyth
Division of Urology
Children's Hospital of Philadelphia
34th Street and Civic Center Boulevard
Philadelphia, PA 19104

Donald. C. Bode
Sterling Research Group
81 Columbia Turnpike
Rensselaer, NY 12144

*David F. Bohr
Department of Physiology
Univ of Michigan School of Medicine
7710 Medical Science Building II
Ann Arbor, MI 48109

John P. Boyle
University of Pennsylvania
School of Veterinary Medicine
3800 Spruce Street
Philadelphia, PA 19104

Verena Briner
Univ of Colorado Health Science Ctr
Renal Division, 4200 East 9th Ave
Denver, CO 80262

Frank V. Brozovich
Bockus Research Institute
Graduate Hospital
415 S, 19th Street
Philadelphia, PA 19146

*Thomas M. Butler
Department of Physiology
Jefferson Medical College
1020 Locust Street
Philadelphia, PA 19107

Shuang Cai
Univ of Cincinnati College of Med
Dept of Physiology and Biophysics
231 Bethesda Avenue
Cincinnati, Ohio 45267-0576

Graham LeM. Campbell
Ctr Excellence Cardiovascular Studies
Graduate Hospital
415 S. 19th Street
Philadelphia, PA 19146

Mary E. Carsten
Dept of Obstetrics and Gynecology
UCLA, A5-38 Rehabilitation Center
1000 Veteran Avenue
Los Angeles, CA 90024

Sushanta K. Chakder
Department of Medicine
Jefferson Medical College
1025 Walnut Street
Philadelphia, PA 19107

Meng Chen
Dept of Physiology and Biochemistry
Medical College of Pennsylvania
3300 Henry Avenue
Philadelphia, PA 19129

Hsien C. Cheng
Pharmacological Sciences
Merrell Dow Research Institute
Cincinnati, OH 45215

Linda Cheng
Lab of Cardiovascular Science
Gerontology Research Center
NIH, NIA
4940 Eastern Avenue
Baltimore, MD 21224

Pasquale Chitano
Department of Physiology
University of Manitoba
770 Bannatyne Ave
Winnipeg, Manitoba R3E OW3, Canada

*Jacqueline Cilea
Bockus Research Institute
Graduate Hospital
415 S. 19th Street
Philadelphia, PA 19146

*Ronald F. Coburn
Department of Physiology
University of Pennsylvania
School of Medicine
Philadelphia, PA 19104

David Cook
Department of Pharmacology
University of Alberta
Edmonton, Alberta
Canada T6G 2H7

*Robert H. Cox
Graduate Hospital
Bockus Research Institute
415 S. 19th Street
Philadelphia, PA 19146

John Cummings
Department of Physiology
Temple University School of Medicine
3420 N. Broad Street
Philadelphia, PA 19140

Gerald D'Angelo
Dept of Physiology and Biophysics
University of Vermont School of Medicine
Given Building
Burlington, VT 05405

Amy Davidoff
Bockus Research Institute
Graduate Hospital
415 S. 19th Street
Philadelphia, PA 19146

*Anna Dominiczak
Department of Physiology
Univ of Michigan School of Medicine
7710 Medical Science Building II
Ann Arbor, MI 48109

Steven P. Driska
Department of Physiology
Temple University
School of Medicine
3420 N. Broad Street
Philadelphia, PA 19140

David Ewalt
Division of Urology
Children's Hospital of Philadelphia
34th and Civic Center Boulevard
Philadelphia, PA 19104

*Gertrude Falk
Department of Physiology
University College London
Gower Street
London WCIE 6BT, England

*Donald G. Ferguson
Dept of Physiology and Biophysics
University of Cincinnati
231 Bethesda Avenue
Cincinnati, OH 45220

John T. Fisher
Depts of Physiol and Anesthesiology
Queen's University
Kingston, Ontario, Canada

John Fulginiti, III
Department of Surgery
One Graduate Plaza
Graduate Hospital
Philadelphia, PA 19146

*Louise Garone
Maryland Biotechnology Institute
Department of Biochemistry
Univ. of Maryland School of Med
Baltimore, MD

*William T. Gerthoffer
Department of Pharmacology
Univ. of Nevada School of Medicine
Reno, NV 89557-0046

*Ming Cui Gong
Department of Physiology
Univ of Virginia School of Medicine
Box 449, Jordan Hall
Charlottesville, VA 22908

*Steven L. Griffith
Department of Physiology and Biophysics
Indiana Univ. School of Medicine
635 Barnhill Drive
Indianapolis, IN 46202

Hong Gu
Bockus Research Institute
Graduate Hospital
415 S. 19th Street
Philadelphia, PA 19146

*Susan Gunst
Department of Physiology and Biophysics
Indiana University School of Medicine
635 Barnhill Ave
Indianapolis, IN 46223

Joe R. Haeberle
Department of Physiology and Biophysics
University of Vermont College of Medicine
Given Building
Burlington, VT 05405

David Harris
Department of Physiology and Biophysics
University of Vermont School of Medicine
Given Building
Burlington, VT 05405

*David J. Hartshorne
Department of Nutrition and Food Science
University of Arizona
309 Agricultural Science Building
Tucson, AZ 85721

*Per Hellstrand
Department of Physiology and Biophysics
University of Lund
Sölvegatan 19, S-223 62
Lund, Sweden

Kelly Hester
Department of Medical Pharmacology
Texas A&M Univ College of Medicine
College Station, TX 77843-1114

John Hexam
Department of Anesthesia
Univ of Pennsylvania School of Medicine
University Hospital
Philadelphia, PA 19103

Robert B. Hill
Department of Zoology
University of Rhode Island
Kingston, RI 02881

*Kwang Hwang
Univ of Michigan School of Medicine
R4038 Kresge II
Ann Arbor, MI 48109

*Mitsuo Ikebe
Department of Physiology
Case Western Reserve University
School of Medicine
Cleveland, OH 44106

*He Jiang
Department of Physiology
University of Manitoba
770 Bannatyne Ave
Winnipeg, Manitoba R3E OW3
Canada

*Najia Jin
Dept of Physiology and Biophysics
Indiana Univ School of Medicine
635 Barnhill Drive
Indianapolis, IN 46202

Börje E. G. Johansson
Hässle Research Laboratories
S-431 83
Mölndal
Sweden

Mark D. Johnson
Department of Pharmacology
Medical College of Pennsylvania
3200 Henry Avenue
Philadelphia, PA 19129

*Kristine E. Kamm
Department of Physiology
Univ Texas Southwestern Med Center
5323 Harry Hines Boulevard
Dallas, TX 75235-9040

C. Y. Kao
Department of Pharmacology
SUNY Downstate Medical Center
450 Clarkson Avenue
Brooklyn, NY 11203

W. Glenn L. Kerrick
Dept of Physiology and Biophysics
Univ of Miami School of Medicine
1600 NW 10th Avenue
PO Box 016430
Miami, FL 33101

*Toshio Kitazawa
Department of Physiology
Univ of Virginia School of Medicine
Box 449
Charlottesville, VA 22908

Eio Koh
Laboratory of Cardiovascular Science
Gerontology Research Center
NIH, NIA
4940 Eastern Avenue
Baltimore, MD 21224

Michael Kotlikoff
University of Pennsylvania
School of Veterinary Medicine
3800 Spruce Street
Philadelphia, PA 19146

Ingrid Krampetz
Department of Physiology and Biophysics
Indiana University School of Medicine
635 Barnhill Drive
Indianapolis, IN 46202

Jacob Krier
Department of Physiology
111 Giltner Hall
Michigan State University
East Lansing, MI 48824

Umberto Kucich
Department of Medicine Research
Graduate Hospital
415 S. 19th Street
Philadelphia, PA 19146

*Hirosi Kuriyama
Department of Pharmacology
Faculty of Medicine
Kyushu University
Fukuoka 812, Japan

Edward F. LaBelle
Bockus Research Institute
Graduate Hospital
415 S. 19th St
Philadelphia, PA 19146

Jane Lalli
Physics Department
Drexel University
Philadelphia, PA 19104

David R. Larach
Department of Anesthesia
Pennsylvania State University College of Medicine
PO Box 850
Hershey, PA 17033

*Carlos Lazalde
Department of Physiology and Biophysics
University of Illinois
Urbana, IL 61801

Robert M. Levin
Division of Urology
University of PA Hospital
3400 Spruce St
Philadelphia, PA 19104

*Stephanie Lewis
Dept of Physiology and Biophysics
Univ of Cincinnati College of Med
231 Bethesda Avenue
Cincinnati, OH 45267

Victor K. Lin
Division of Urology
Univ Texas Southwestern Med Center
5323 Harry Hines Boulevard
Dallas, TX 75235-9031

Nicholas J. Lodge
Department of Pharmacology
Wyeth-Ayerst Research
CN 8000
Princeton, NJ 08543

Mỹ G. Mahoney
Department of Biochemistry
University of Massachusetts
Amherst, MA 01002

G.M. Makhlouf
Department of Medicine
Medical College of Virginia
Box 711 MCV Station
Richmond, VA 23298

*Anne F. Martin
Dept of Physiology and Biophysics
Univ. of Cincinnati College of Med
231 Bethesda Avenue
Cincinnati, OH 45267-0575

Hunter Martin
Bockus Research Institute
Graduate Hospital
415 S. 19th Street
Philadelphia, PA 19146

Mary McLane
Department of Physiology
Temple University School of Medicine
Broad and Ontario Streets
Philadelphia, PA 19140

*Anthony R. Means
Department of Cell Biology
Baylor College of Medicine
One Baylor Plaza
Houston, TX 77030

*Richard A. Meiss
Department of Obstetrics and Gynecology
Indiana University School of Medicine
1001 Walnut St
Indianapolis, IN 46223

*Gerhard Meissner
Department of Biochemistry
Univ of North Carolina at Chapel Hill
CB-7260
Chapel Hill, NC 27599

Linda Merkel
Rorer Central Research
620 Allendale Road
King of Prussia, PA 19406

Penny Monical
Connective Tissue Research Institute
University City Science Center
3624 Market Street
Philadelphia, PA 19104

*Susan Mooers
Department of Physiology
Jefferson Medical College
1020 Locust Street
Philadelphia, PA 19107

*Edwin D. W. Moore
Department of Physiology
University of Massachusetts Medical Center
55 Lake Avenue North
Worcester, MA 01605

*Robert S. Moreland
Bockus Research Institute
Graduate Hospital
415 S. 19th Street
Philadelphia, PA 19146

*Suzanne Moreland
Department of Pharmacology
Bristol-Myers Squibb Pharmaceutical Res Inst
P. O. Box 4000
Princeton, NJ 08543

*Kathleen G. Morgan
Dept of Cell and Molec Physiology in Medicine
Beth Israel Hospital
Harvard Medical School
330 Brookline Avenue
Boston, MA 02215

*Richard A. Murphy
Department of Physiology
Univ of Virginia School of Medicine
Box 449
Charlottesville, VA 22908

*Srinivasa Narayan
Department of Physiology
Jefferson Medical College
1020 Locust Street
Philadelphia, PA 19107

Paul Nemir, Jr
Department of Surgery
Graduate Hospital
One Graduate Plaza
Philadelphia, PA 19146

Eiichiro Nishie
Department of Physiology
Univ of Virginia School of Medicine
Jordan Hall
Charlottesville, VA 22908

*Junji Nishimura
Dept of Molec and Cell Pharmacology
University of Miami School of Med
P. O. Box 016189
Miami, FL 33101

*Avraham Oplatka
Department of Polymer Research
Weizmann Institute of Science
Rehovot 76100
Israel

*Hiroshi Ozaki
Department of Physiology
Univ of Nevada School of Medicine
Reno, NV 89557

*C. Subah Packer
Dept of Physiology and Biophysics
Indiana Univ. School of Medicine
635 Barnhill Drive
Indianapolis, IN 46202

Philip Palade
Dept of Physiology and Biophysics
University of Texas Medical Branch
Galveston, TX 77550

*Richard J. Paul
Dept of Physiology and Biophysics
Univ of Cincinnati College of Med
231 Bethesda Avenue
Cincinnati, OH 45267

Rebecca R. Pauly
Laboratory of Cardiovascular Science
Gerontology Research Ctr, NIH, NIA
4940 Eastern Avenue
Baltimore, MD 21224

John Pawlowski
Cardiovascular Division
Beth Israel Hospital
330 Brookline Avenue
Boston, MA 02215

Robert D. Phair
Department of Biomedical Engineering
Johns Hopkins School of Medicine
720 Rutland Ave
Baltimore, MD 21205

*Robert W. Putnam
Department of Physiology and Biophysics
Wright State University School of Medicine
Dayton, OH 45435

*Kang Rao
Department of Physiology
University of Manitoba
770 Bannatyne Ave
Winnipeg, Manitoba R3E OW3, Canada

Satish Rattan
Department of Medicine
Jefferson Medical College
1025 Walnut Street
Philadelphia, PA 19107

*Rodney A. Rhoades
Department of Physiology and Biophysics
Indiana Univ. School of Medicine
635 Barnhill Drive
Indianapolis, IN 46202

Daniel Rock
Dept of Physiology and Biochemistry
Medical College of Pennsylvania
3300 Henry Avenue
Philadelphia, PA 19129

James P. Ryan
Department of Physiology
Temple University School of Medicine
3420 N. Broad Street
Philadelphia, PA 19140

*Charles M. Schworer
Weis Center for Research
Geisinger Clinic
North Academy Avenue
Danville, PA 17822

Roberta Secrest
Department of Pharmacology
Merrell Dow Research Institute
9550 N. Zionsville Road
PO Box 68470
Indianapolis, IN 46268

*Charles L. Seidel
Department of Medicine
Section of Cardiovascular Science
Baylor College of Medicine
One Baylor Plaza
Houston, TX 77030

*Marion J. Siegman
Department of Physiology
Jefferson Medical College
1020 Locust Street
Philadelphia, PA 19107

*Harold A. Singer
Weis Center for Research
Geisinger Clinic
North Academy Avenue
Danville, PA 17822

Linda L. Slakey
Department of Biochemistry
University of Massachusetts
Amherst, MA 01003

*Jacquelyn M. Smith
Department of Physiology
Chicago College of Osteopathic Med
555 31st Street
Downers Grove, IL 60515

James P. Smith
Bockus Research Institute
Graduate Hospital
415 S. 19th Street
Philadelphia, PA 19146

*Andrew P. Somlyo
Department of Physiology
Univ of Virginia School of Medicine
Box 449, Jordan Hall
Charlottesville, Virginia 22908

*Avril V. Somlyo
Department of Physiology
Univ of Virginia School of Medicine
Box 449, Jordan Hall
Charlottesville, VA 22908

David Stepp
Dept of Physiology and Biochemistry
Medical College of Pennsylvania
3300 Henry Avenue
Philadelphia, PA 19129

*Newman L. Stephens
Department of Physiology
Univ of Manitoba Facility of Med
770 Bannatyne Avenue
Winnipeg, Manitoba R3E 0W3
Canada

*James T. Stull
Department of Physiology
Univ Texas Southwestern Medical Center
5323 Harry Hines Boulevard
Dallas, Texas 75235-9040

Eiichi Suematsu
Department of Medicine
Beth Israel Hospital
330 Brookline Ave
Boston, MA 02215

H. Lee Sweeney
Department of Physiology
Univ of Pennsylvania School of Medicine
37th Street and Hamilton Walk
Philadelphia, PA 19104

Mary Swider
Department of Physiology
Temple University School of Medicine
3420 N. Broad Street
Philadelphia, PA 19140

*Pawel Szymanski
Univ of Cincinnati College of Medicine
Department of Physiology and Biophysics
231 Bethesda Avenue
Cincinnati, OH 45267-0576

*Malú G. Tansey
Department of Physiology
Univ Texas Southwestern Medical Center
5323 Harry Hines Boulevard
Dallas, TX 75235-9040

Michael Tomasic
University of Pennsylvania
School of Veterinary Medicine
3800 Spruce Street
Philadelphia, PA 19104

Snezana Trajkovic
Bockus Research Institute
Graduate Hospital
415 S. 19th Street
Philadelphia, PA 19146

Jeffrey Travis
Rorer Central Research
640 Allendale Road
King of Prussia, PA 19406

Laura A. Trinkle
Bockus Research Institute
Graduate Hospital
415 S. 19th Street
Philadelphia, PA 19146

Betty M. Twarog
Bockus Research Institute
Graduate Hospital
415 S. 19th Street
Philadelphia, PA 19146

*Cornelis van Breemen
Dept of Cell and Molec Pharmacology
Univ of Miami School of Medicine
P. O. Box 016189
Miami, FL 33101

*Tapan Vyas
Department of Physiology
Jefferson Medical College
1020 Locust Street
Philadelphia, PA 19107

*Michael P. Walsh
Department of Medical Biochemistry
University of Calgary
3330 Hospital Drive NW
Calgary, Alberta T2N 1N4, Canada

Michael Wang
Department of Physiology
Temple University School of Medicine
3420 N. Broad Street
Philadelphia, PA 19140

Robert L. Wardle
The Ohio State University
1900 Coffey Road
Columbus, OH 43210

*David M. Warshaw
Department of Physiology
Univ of Vermont School of Medicine
Given Medical Building
Burlington, VT 05405

Robert J. Washabau
Department of Clinical Studies
University of Pennsylvania
School of Veterinary Medicine
Philadelphia, PA 19104

George Weinbaum
Department of Medicine Research
Graduate Hospital
415 S. 19th Street
Philadelphia, PA 19146

Kristene Whitmore
Division of Urology
Graduate Hospital
One Graduate Plaza
Philadelphia, PA 19146

*Ann Word
Dept. Green Center
Univ Texas Southwestern Medical Center
5323 Harry Hines Boulevard
Dallas, TX 75235-9051

*Jianming Xie
Department of Physiology
UMDNJ-New Jersey Medical School
185 S. Orange St - H607 MSB
Newark, NJ

*Mineo Yamakawa
Department of Physiology and Biophysics
University of Vermont School of Medicine
Given Building
Burlington, VT 05405

Stephen Zderic
Division of Urology
Children's Hospital of Philadelphia
34th Street and Civic Center Boulevard
Philadelphia, PA 19104

Chong Zhang
Univ of Cincinnati College of Med
Dept of Physiology and Biophysics
231 Bethesda Avenue
Cincinnati, OH 45267-0576

*He Zhang
Department of Surgery
University of Alberta
Edmonton, Alberta T6G 2B7
Canada

Andreas Ziegler
Department of Physiology
Univ of Virginia School of Medicine
Box 449, Jordan Hall
Charlottesville, VA 22908

*Contributor

INDEX